Advances in Experimental Medicine and Biology

For further volumes:
http://www.springer.com/series/5584

Lucio Annunziato
Editor

Sodium Calcium Exchange: A Growing Spectrum of Pathophysiological Implications

Proceedings of the 6th International Conference on Sodium Calcium Exchange

Springer

Editor
Lucio Annunziato
Division of Pharmacology
Department of Neuroscience
School of Medicine
"Federico II" University of Naples
Naples, Italy

ISBN 978-1-4614-4755-9 ISBN 978-1-4614-4756-6 (eBook)
DOI 10.1007/978-1-4614-4756-6
Springer New York Heidelberg Dordrecht London

Library of Congress Control Number: 2012951339

Printed on acid-free paper

Springer is part of Springer Science+Business Media (www.springer.com)

This book is dedicated to Mordecai Blaunstein and Kenneth Philipson as a tribute to their outstanding scientific carreers: most of the current knowledge on NCX is due to their passionate work over the last 45 years and this conference would not have been possible without their memorable contribution.

Preface

From October 1 to October 5, 2011 more than 100 scientists from 21 countries of all the continents conveyed in the nice, little village of Lacco Ameno in the Island of Ischia in the Gulf of Naples, Italy, to take part to the 6th International Conference on Sodium Calcium Exchange. This book reports the text of the lectures of the Conference and the state-of-art of the topic discussed and it intends to represent a long lasting memory of the magic atmosphere of the meeting where science and friendship perfectly matched in a lovely natural setting.

The 6th International Conference on Sodium Calcium Exchange is the last of a series that begun with the meeting held in Stowe 1987, England and, thereafter, every 4–5 years each time in a different location. Nowadays, 35 years after this first meeting, the International Conference on Sodium Calcium Exchange is an highly expected appointment for the NCX community where new data are presented and fruitful discussions take place. The last 30 years experienced big changes in the field of NCX that, at the very beginning was a partially characterized protein object of the interest of few highly focused scientists and nowadays is a protein characterized even in its three dimensional structure with an ever growing spectrum of pathophysiological implications. Two lectures, given at the 6th International Conference on Sodium Calcium Exchange and reported in the present volume, gave an enthusiastic overview of the tremendous progress in the knowledge on the exchanger with the words of two leaders in the field, Mordecai Blaunstein and Kenneth Philipson to whom we intend to dedicate this book as a tribute to their outstanding scientific careers.

Looking back to the volumes of the Proceedings of the previous conferences it can be easily realized that all the milestones in the field have been documented in these books which, therefore, represent an important part of the literature on the exchanger. We are confident that the same will also happen for the present book that uncovers the most striking new findings on NCX that emerged since the previous Conference on Sodium Calcium Exchange, such as the structural dissection of the molecular determinants of Ca^{2+} sensitivity of the exchanger, the epigenetic regulation of *ncx1* gene, the molecular identification of the mitochondrial Sodium Exchanger, and the discovery of NCX in unexpected anatomical locations such as the female reproductive tract.

As a final remark we would like to emphasize that many of the participants to the Conference were young scientists under 35 years: it looks like that besides the outstanding past of the fathers of NCX we should also celebrate the birth of a new generation of NCX proselytes that we hope will keep high the tradition of the NCX Conferences also in the future.

We would like to express our gratitude for their precious help in the organization of the meeting to all the members of the International (Luis Beaugé, Lorella MT Canzoniero, Ernesto Carafoli, Gianfranco Di Renzo, David Eisner, André Herchuelz, Takahiro Iwamoto, Lung-Sen Kao, Daniel Khananshvili, Jonathan Lytton, Kenneth D Philipson, Hannah Rahamimoff, John Reeves, Paul Schnetkamp, Karin R Sipido, Dandan Sun, Bruno Trimarco, Jin Zhang) and Local (Lorella MT Canzoniero, Mauro Cataldi, Gianfranco Di Renzo, Pasquale Molinaro, Anna Pannaccione, Giuseppe Pignataro, Antonella Scorziello, Agnese Secondo, Maurizio Taglialatela) Organizing Committees. Special thanks also to Mauro Cataldi for helping in the revision of the chapters and in the organization of the book, to Springer for taking care of the publication of the book, and to Studio Grafico Ciotola, Naples, Italy, for preparing all the illustrations of the book.

Lucio Annunziato, M.D.
Chairman of the Conference

Contents

Contributors

Andrey Y. Abramov Department of Molecular Neuroscience, UCL Institute of Neurology, Queen Square, London, WC1N 3BG, UK

Károly Acsai Division of Cardiovascular Pharmacology, Hungarian Academy of Sciences, Szeged, Hungary

Haider F. Altimimi Department of Physiology and Pharmacology, Hotchkiss Brain Institute, University of Calgary, Calgary, Alberta, Canada

Beum-Soo An Laboratory of Veterinary Biochemistry and Molecular Biology, College of Veterinary Medicine, Chungbuk National University, Cheongju, Chungbuk, Republic of Korea

Lucio Annunziato Division of Pharmacology, Department of Neuroscience, School of Medicine, "Federico II" University of Naples, Naples, Italy

Gudrun Antoons Laboratory of Experimental Cardiology, Department of Cardiovascular Diseases, University of Leuven, Leuven, Belgium

Department of Cardiology, Medical University of Graz, Graz, Austria

Sergey G. Baryshnikov Department of Physiology, University of Maryland, Medical School, Baltimore, MD, USA

Luis Beaugé Laboratorio de Biofísica, Instituto de Investigación Médica "Mercedes y Martín Ferreyra" (INIMEC-CONICET), Córdoba, Argentina

Jillian C. Belrose Department of Anatomy and Cell Biology, University of Western Ontario, London, ON, Canada

Robarts Research Institute, Molecular Brain Research Group, University of Western Ontario, London, ON, Canada

Gustavo Benaim Laboratorio de Señalización Celular, Instituto de Estudios Avanzados (IDEA), Caracas, Venezuela

Graciela Berberián Laboratorio de Biofísica, Instituto de Investigación Médica "Mercedes y Martín Ferreyra" (INIMEC-CONICET), Córdoba, Argentina

Virginie Bito Laboratory of Experimental Cardiology, Department of Cardiovascular Diseases, University of Leuven, Leuven, Belgium

Mordecai P. Blaustein Department of Physiology and Medicine, University of Maryland School of Medicine, Baltimore, MD, USA

Mariana Bollo Laboratorio de Biofísica, Instituto de Investigación Médica "Mercedes y Martín Ferreyra" (INIMEC-CONICET), Córdoba, Argentina

Francesca Boscia Division of Pharmacology, Department of Neuroscience, School of Medicine, "Federico II" University of Naples, Naples, Italy

Marie Caldwell Department of Neurology, University of Pittsburgh, Pittsburgh, PA, USA

Carlo Caputo Laboratorio de Fisiología Celular, Centro de Biofísica, Instituto Venezolano de Investigaciones Científicas (IVIC), Caracas, Venezuela

Annalisa Carlucci Department of Molecular and Cellular Biology and Pathology "L. Califano", Federico II University of Naples, Naples, Italy

Antonella Casamassa Division of Pharmacology, Department of Neuroscience, School of Medicine, "Federico II" University of Naples, Naples, Italy

Mauro Cataldi Division of Pharmacology, Department of Neuroscience, School of Medicine, "Federico II" University of Naples, Naples, Italy

Tung O. Chan Center of Translational Medicine, Temple University School of Medicine, Philadelphia, PA, USA

Olga Chernysh Gazes Cardiac Research Institute, Division of Cardiology, Department of Medicine, Medical University of South Carolina, SC, USA

Joseph Y. Cheung Center of Translational Medicine, Temple University School of Medicine, Philadelphia, PA, USA

Kyung-Chul Choi Laboratory of Veterinary Biochemistry and Molecular Biology, College of Veterinary Medicine, Chungbuk National University, Chungbuk, Republic of Korea

Xiang-Ping Chu Department of Basic Medical Science, School of Medicine, University of Missouri-Kansas City, Kansas City, MO, USA

Claudia Colina Laboratorio de Fisiología Celular, Centro de Biofísica, Instituto Venezolano de Investigaciones Científicas (IVIC), Caracas, Venezuela

Ellis Cooper Department of Physiology, McGill University, Montreal, QC, Canada

Alexandra Cousido Department of Structural Biology and Genomics, IGBMC, CNRS, INSERM, Université de Strasbourg, Illkirch, France

Ornella Cuomo Division of Pharmacology, Department of Neuroscience, School of Medicine, "Federico II" University of Naples, Naples, Italy

Carla D'Avanzo Division of Pharmacology, Department of Neuroscience, School of Medicine, "Federico II" University of Naples, Naples, Italy

Tony Dadd Unilever R&D, Colworth Science Park, Sharnbrook, Bedfordshire, UK

Emma Deas Department of Molecular Neuroscience, UCL Institute of Neurology, Queen Square, London, UK

Gianfranco Di Renzo Division of Pharmacology, Department of Neuroscience, School of Medicine, "Federico II" University of Naples, Naples, Italy

Reinaldo DiPolo Laboratorio de Fisiología Celular, Centro de Biofísica, Instituto Venezolano de Investigaciones Científicas (IVIC), Caracas, Venezuela

Benayahu Elbaz Department of Biochemistry and Molecular Biology, Hebrew University-Hadassah Medical School, Jerusalem, Israel

Daniela Elmaz Department of Biochemistry and Molecular Biology, Hebrew University-Hadassah Medical School, Jerusalem, Israel

Ruth M. Empson Department of Physiology, Brain Health Research Centre, University of Otago, Dunedin, New Zealand

Marina Eskin-Schwartz Department of Biochemistry and Molecular Biology, Hebrew University-Hadassah Medical School, Jerusalem, Israel

Alba Esposito Division of Pharmacology, Department of Neuroscience, School of Medicine, "Federico II" University of Naples, Naples, Italy

Elga Esposito Division of Pharmacology, Department of Neuroscience, School of Medicine, "Federico II" University of Naples, Naples, Italy

Arthur M. Feldman Center of Translational Medicine, Temple University School of Medicine, Philadelphia, PA, USA

Antonio Feliciello Department of Molecular and Cellular Biology and Pathology "L. Califano", Federico II University of Naples, Naples, Italy

Patrizia Ferrari Prassis-sigma tau Research Institute, Settimo Milanese, Milan, Italy

Michael Fine Department of Physiology, University of Texas Southwestern Medical Center, TX, USA

Luigi Formisano Division of Pharmacology, Department of Neuroscience, "Federico II" University of Naples, Naples, Italy

Gary Frazier Department of Physiology, University of Texas Southwestern Medical Center, TX, USA

Erhe Gao Center of Translational Medicine, Temple University School of Medicine, Philadelphia, PA, USA

Moshe Giladi Department of Physiology and Pharmacology, Tel-Aviv University, Ramat-Aviv, Tel-Aviv, Israel

Rebecca S. Ginger Unilever R&D, Colworth Science Park, Sharnbrook, Bedfordshire, UK

Joshua I. Goldhaber Cedars-Sinai Heart Institute, Los Angeles, CA, USA

Vera A. Golovina Department of Physiology, University of Maryland, Medical School, Baltimore, MD, USA

Francescopaolo Granata Division of Clinical Immunology and Allergy, School of Medicine, University of Naples Federico II, Naples, Italy

Martin R. Green Unilever R&D, Colworth Science Park, Sharnbrook, Bedfordshire, UK

Natascia Guida Division of Pharmacology, Department of Neuroscience, School of Medicine, "Federico II" University of Naples, Naples, Italy

David Gunn Unilever R&D, Colworth Science Park, Sharnbrook, Bedfordshire, UK

John M. Hamlyn Department of Physiology, University of Maryland Medical School, Baltimore, MD, USA

André Herchuelz Laboratoire de Pharmacodynamie et de Thérapeutique, Université Libre de Bruxelles (ULB), Faculté de Médicine, Brussels, Belgium

Masaki Hikida Center for Innovation in Immunoregulative Technology and Therapeutics, Graduate School of Medicine, Kyoto University, Yoshida-konoe, Sakyo-ku, Kyoto, Japan

Mark Hilge Center for Cellular Imaging and Nano Analytics (C-CINA), Biozentrum, University Basel, Mattenstrasse, Basel, Switzerland

Donald W. Hilgemann Department of Physiology, University of Texas Southwestern Medical Center, TX, USA

Sang-Hwan Hyun Laboratory of Veterinary Biochemistry and Molecular Biology, College of Veterinary Medicine, Chungbuk National University, Cheongju, Chungbuk, Republic of Korea

Masataka Ifuku Depertment of Integrative Physiology, Graduate School of Medical Sciences, Kyushu University, Fukuoka, Japan

Luca Lignitto Department of Molecular and Cellular Biology and Pathology "L. Califano", Federico II University of Naples, Naples, Italy

Alina Ilie Department of Physiology, McGill University, Montreal, QC, Canada

Michael F. Jackson Robarts Research Institute, Molecular Brain Research Group, University of Western Ontario, London, ON, Canada

Department of Physiology and Pharmacology, University of Western Ontario, London, ON, Canada

Erica Jaffe Laboratorio de Fisiología Celular, Centro de Biofísica, Instituto Venezolano de Investigaciones Científicas (IVIC), Caracas, Venezuela

Eui-Bae Jeung Laboratory of Veterinary Biochemistry and Molecular Biology, College of Veterinary Medicine, Chungbuk National University, Cheongju, Chungbuk, Republic of Korea

Lin Jiang Laboratoire de Pharmacodynamie et de Thérapeutique, Université Libre de Bruxelles (ULB), Faculté de Médicine, Brussels, Belgium

Tushare Jinadasa Department of Physiology, McGill University, Montreal, QC, Canada

Eui-Man Jung Laboratory of Veterinary Biochemistry and Molecular Biology, College of Veterinary Medicine, Chungbuk National University, Cheongju, Chungbuk, Republic of Korea

Lung-Sen Kao Department of Life Sciences and Institute of Genome Sciences, National Yang-Ming University, Taipei, Taiwan, Republic of China

Eiji Karashima Department of Physiology, University of Maryland Medical School, Baltimore, MD, USA

Harinath Kasiganesan Division of Cardiology, Department of Medicine, Gazes Cardiac Research Institute, Medical University of South Carolina, Charleston, SC, USA

Daniel Khananshvili Department of Physiology and Pharmacology, Tel-Aviv University, Ramat-Aviv, Tel-Aviv, Israel

Mahdi Khatib Department of Biochemistry and Molecular Biology, Hebrew University-Hadassah Medical School, Jerusalem, Israel

Bongju Kim Center for Innovation in Immunoregulative Technology and Therapeutics, Graduate School of Medicine, Kyoto University, Yoshida-konoe, Sakyo-ku, Kyoto, Japan

Dong Kim Department of Neurology, University of Pittsburgh, Pittsburgh, PA, USA

Denise Kimbrough Gazes Cardiac Research Institute, Division of Cardiology, Department of Medicine, Medical University of South Carolina, 114 Doughty Street, Charleston, SC, USA

Sergei Kirischuk Institute of Physiology and Pathophysiology, Universal Medical Center of the Johannes Gutenberg University Mainz, Mainz, Germany

Walter J. Koch Center of Translational Medicine, Temple University School of Medicine, Philadelphia, PA, USA

Orie Koga Center for Innovation in Immunoregulative Technology and Therapeutics, Graduate School of Medicine, Kyoto University, Yoshida-konoe, Sakyo-ku, Kyoto, Japan

Dmitri O. Levitsky Unité de Fonctionnalité et Ingénierie des Protéines, FRE-CNRS, Faculté des Sciences et des Techniques, Université de Nantes, Nantes Cedex, France

Mona S. Li Division of Cardiology, Department of Medicine, Gazes Cardiac Research Institute, Medical University of South Carolina, Charleston, SC, USA

Fei-Ling Lim Unilever R&D, Colworth Science Park, Sharnbrook, Bedfordshire, UK

Mei-Jung Lin Department of Physiology, University of Texas Southwestern Medical Center, TX, USA

Cristina I. Linde Department of Physiology, University of Maryland Medical School, Baltimore, MD, USA

Stefania Loffredo Division of Clinical Immunology and Allergy, School of Medicine, University of Naples Federico II, Naples, Italy

Viktoria Lukashova Department of Physiology, McGill University, Montreal, QC, Canada

John F. MacDonald Department of Anatomy and Cell Biology, University of Western Ontario, London, ON, Canada

Robarts Research Institute, Molecular Brain Research Group, University of Western Ontario, London, ON, Canada

Department of Physiology and Pharmacology, University of Western Ontario, London, ON, Canada

Niall Macquaide Laboratory of Experimental Cardiology, Department of Cardiovascular Diseases, University of Leuven, Leuven, Belgium

Santhosh K. Mani Division of Cardiology, Department of Medicine, Gazes Cardiac Research Institute, Medical University of South Carolina, Charleston, SC, USA

Gianni Marone Division of Clinical Immunology and Allergy, School of Medicine, University of Naples Federico II, Naples, Italy

Center for Basic and Clinical Immunology Research (CISI), Naples, Italy

Carmela Matrone Division of Pharmacology, Department of Neuroscience, School of Medicine, "Federico II" University of Naples, Naples, Italy

Satoshi Matsuoka Center for Innovation in Immunoregulative Technology and Therapeutics, Graduate School of Medicine, Kyoto University, Yoshida-konoe, Sakyo-ku, Kyoto, Japan

Donald R. Menick Division of Cardiology, Department of Medicine, Gazes Cardiac Research Institute, Medical University of South Carolina, Charleston, SC, USA

Pasquale Molinaro Division of Pharmacology, Department of Neuroscience, School of Medicine, "Federico II" University of Naples, Naples, Italy

Yuki Mori Laboratory of Pathophysiology, Graduate School of Pharmaceutical Sciences, Kyushu University, Fukuoka, Japan

Nathalie Pachera Faculté de Médicine, Laboratoire de Pharmacodynamie et de Thérapeutique, Université Libre de Bruxelles (ULB), Brussels, Belgium

Evrard Nguidjoe Faculté de Médicine, Laboratoire de Pharmacodynamie et de Thérapeutique, Université Libre de Bruxelles (ULB), Brussels, Belgium

Debora A. Nicoll Department of Physiology and the Cardiovascular Research Laboratory, David Geffen School of Medicine at UCLA, Los Angeles, CA, USA

Mami Noda Laboratory of Pathophysiology, Graduate School of Pharmaceutical Sciences, Kyushu University, Fukuoka, Japan

John Orlowski Department of Physiology, McGill University, Montreal, QC, H3G 1Y6, Canada

Michela Ottolia Department of Physiology and the Cardiovascular Research Laboratory, David Geffen School of Medicine at UCLA, Los Angeles, CA, USA

Anna Pannaccione Division of Pharmacology, Department of Neuroscience, School of Medicine, "Federico II" University of Naples, Naples, Italy

Vladimir Parpura IKERBASQUE, Basque Foundation for Science, Bilbao, Spain

Department of Neurosciences, University of the Basque Country UPV/EHU, Leioa, Spain

Department of Neurobiology, Center for Glial Biology in Medicine, Civitan International Research Center, Atomic Force Microscopy & Nanotechnology Laboratories, and Evelyn F. McKnight Brain Institute, University of Alabama, Birmingham, USA

Department of Biotechnology, University of Rijeka, Rijeka, Croatia

Angelica Petraroli Division of Clinical Immunology and Allergy, School of Medicine, University of Naples Federico II, Naples, Italy

Kenneth D. Philipson Department of Physiology and the Cardiovascular Research Laboratory, David Geffen School of Medicine at UCLA, Los Angeles, CA, USA

Giuseppe Pignataro Division of Pharmacology, Department of Neuroscience, School of Medicine, "Federico II" University of Naples, Naples, Italy

Alberto Podjarny Department of Structural Biology and Genomics, IGBMC, CNRS, INSERM, Université de Strasbourg, Illkirch, France

Maria V. Pulina Department of Physiology, University of Maryland School of Medicine, Baltimore, MD, USA

Joseph E. Rabinowitz Center of Translational Medicine, Temple University School of Medicine, Philadelphia, PA, USA

Hannah Rahamimoff Department of Biochemistry and Molecular Biology, Hebrew University-Hadassah Medical School, Jerusalem, Israel

Magaly Ramos Laboratorio de Fisiología Celular, Centro de Biofísica, Instituto Venezolano de Investigaciones Científicas (IVIC), Caracas, Venezuela

Ludivine Renaud Division of Cardiology, Department of Medicine, Gazes Cardiac Research Institute, Medical University of South Carolina, Charleston, SC, USA

Héctor Rojas Laboratorio de Fisiología Celular, Centro de Biofísica, Instituto Venezolano de Investigaciones Científicas (IVIC), Caracas, Venezuela

Chris J. Roome Department of Physiology, Brain Health Research Centre, University of Otago, Dunedin, New Zealand

Melanie Sandel Unilever R&D, Colworth Science Park, Sharnbrook, Bedfordshire, UK

Claudia Savoia Division of Pharmacology, Department of Neuroscience, School of Medicine, "Federico II" University of Naples, Naples, Italy

Magdalena Sawicka Unilever R&D, Colworth Science Park, Sharnbrook, Bedfordshire, UK

Paul P.M. Schnetkamp Department of Physiology and Pharmacology, Hotchkiss Brain Institute, University of Calgary, Calgary, AB, Canada

Antonella Scorziello Division of Pharmacology, Department of Neuroscience, School of Medicine, "Federico II" University of Naples, Naples, Italy

Agnese Secondo Division of Pharmacology, Department of Neuroscience, School of Medicine, "Federico II" University of Naples, Naples, Italy

Yejie Shi Department of Neurology, University of Pittsburgh, Pittsburgh, PA, USA

Karin R. Sipido Laboratory of Experimental Cardiology, Department of Cardiovascular Diseases, University of Leuven, Leuven, Belgium

Rossana Sirabella Fondazione IRCCS SDN, Naples, Italy

Maria Josè Sisalli Division of Pharmacology, Department of Neuroscience, School of Medicine, "Federico II" University of Naples, Naples, Italy

Sophie Sokolow Laboratory of Pharmacology and Therapeutics, Universitè Libre de Bruxelles, Brussels, Gosselies, Belgium

Jianliang Song Center of Translational Medicine, Temple University School of Medicine, Philadelphia, PA, USA

Rosaria I. Staiano Division of Clinical Immunology and Allergy, School of Medicine, University of Naples Federico II, Naples, Italy

Dandan Sun Department of Neurology, University of Pittsburgh, Pittsburgh, PA, USA

Robert T. Szerencsei Department of Physiology and Pharmacology, Hotchkiss Brain Institute, University of Calgary, Calgary, Alberta, Canada

Masayuki Takahashi Unité de Fonctionnalité et Ingénierie des Protéines, Faculté des Sciences et des Techniques, Université de Nantes, Nantes, France

Ayako Takeuchi Department of Physiology and Biophysics, Graduate School of Medicine, Kyoto University, Yoshida-konoe, Sakyo-ku, Kyoto, Japan

Shuilong Tong Center for Membrane Biology, Department of Biochemistry and Molecular Biology, University of Texas Houston Medical School, Houston, TX, USA

Massimo Triggiani Division of Clinical Immunology and Allergy, School of Medicine, University of Naples Federico II, Naples, Italy

Center for Basic and Clinical Immunology Research (CISI), Naples, Italy

Michael Valitsky Department of Biochemistry and Molecular Biology, Hebrew University-Hadassah Medical School, Jerusalem, Israel

Valeria Valsecchi Division of Pharmacology, Department of Neuroscience, School of Medicine, "Federico II" University of Naples, Naples, Italy

David Verbich Department of Physiology, McGill University, Montreal, QC, Canada

Alexei Verkhratsky Faculty of Life Sciences, The University of Manchester, Manchester, UK

IKERBASQUE, Basque Foundation for Science, Bilbao, Spain

Department of Neurosciences, University of the Basque Country UPV/EHU, Leioa, Spain

Davide Viggiano Division of Pharmacology, Department of Neuroscience, School of Medicine, "Federico II" University of Naples, Naples, Italy

Antonio Vinciguerra Division of Pharmacology, Department of Neuroscience, School of Medicine, "Federico II" University of Naples, Naples, Italy

Hao-Ran Wang Department of Physiology, University of Texas Southwestern Medical Center, TX, USA

JuFang Wang Center of Translational Medicine, Temple University School of Medicine, Philadelphia, PA, USA

Stephen Wilson Unilever R&D, Colworth Science Park, Sharnbrook, Bedfordshire, UK

Nicholas W. Wood Department of Molecular Neuroscience, UCL Institute of Neurology, Queen Square, London, UK

Alison Wood-Kaczmar Department of Molecular Neuroscience, UCL Institute of Neurology, Queen Square, London, UK

Mousheng Wu Center for Membrane Biology, Department of Biochemistry and Molecular Biology, the University of Texas Houston Medical School, Houston, TX, USA

Yu-Feng Xie Robarts Research Institute, Molecular Brain Research Group, University of Western Ontario, ON, Canada

Zhi-Gang Xiong Department of Neurobiology, Morehouse School of Medicine, Atlanta, GA, USA

Hyun Yang Laboratory of Veterinary Biochemistry and Molecular Biology, College of Veterinary Medicine, Chungbuk National University, Cheongju, Chungbuk, Republic of Korea

Ya-Chi Yang Institute of Biochemistry and Molecular Biology, National Yang-Ming University, Taipei, Taiwan, Republic of China

Nicola Zambrano Department of Biochemistry and Medical Biotechnology, "Federico II" University of Naples, Naples, Italy

Jin Zhang Department of Physiology, University of Maryland School of Medicine, Baltimore, MD, USA

Xue-Qian Zhang Center of Translational Medicine, Temple University School of Medicine, Philadelphia, PA, USA

Lei Zheng Center for Membrane Biology, Department of Biochemistry and Molecular Biology, The University of Texas Houston Medical School, Houston, TX, USA

A. Zulian Department of Physiology, University of Maryland Medical School, Baltimore, MD, USA

Participants to the 6th Conference on Sodium Calcium Exchange, Lacco Ameno, Ischia, Naples, Italy, October 1–5, 2011

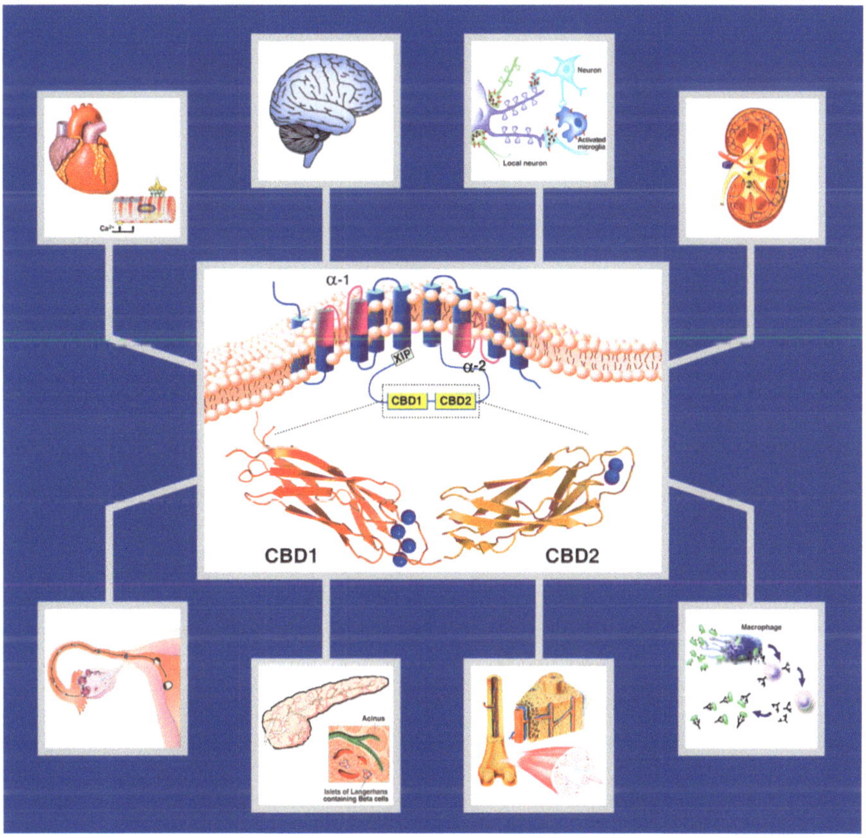

Part I

Historical Perspective

Livin' with NCX and Lovin' It: A 45 Year Romance

1

Mordecai P. Blaustein

Sit down before fact as a little child, be prepared to give up every conceived notion, follow humbly wherever, whatever abysses nature leads, or you will learn nothing.

Thomas H. Huxley,
letter to C. Kingsley, September 20, 1863

Abstract

This conference commemorates, almost to the day, the 45th anniversary of the discovery of the Na^+/Ca^{2+} exchanger (NCX). The discovery was serendipitous, as is so often the case with scientific breakthroughs. Indeed, that is what is so fascinating and romantic about scientific research. I will describe the discovery of NCX, but will begin by explaining how I got there, and will then discuss how the discovery influenced my career path.

Keywords

Cardiotonic steroids • Hypertension • PLasmERosomes • Synaptosomes • Vascular smooth muscle

1.1 For the Love of Physiology

I was introduced to cell physiology by Howard Schneiderman, a distinguished insect physiologist and developmental biologist, during my undergraduate days at Cornell University. I was interested in neurophysiology and the mind-brain problem but came under the spell of Daniel Tosteson when I was a medical student at Washington University in St. Louis. Dan convinced me to work on the Na^+,K^+-ATPase ("sodium pump"), which had just been discovered (Skou 1957).

I spent a year and a half in Dan's lab studying the red blood cell cardiotonic steroid-sensitive Na^+ pump. In 1963, after completing medical school and an internship at Boston City Hospital, I was offered a naval commission to work at the US Naval Medical Research Institute in Bethesda, MD (much better than a tour in Vietnam!).

Thus, I returned to neurophysiology, and under David Goldman (of the Goldman-Hodgkin-Katz

In memory of Peter F. Baker, David E. Goldman, Alan L. Hodgkin, Howard A. Schneiderman, Daniel C. Tosteson, and Mani Matter.

M.P. Blaustein (✉)
Departments of Physiology and Medicine, University of Maryland School of Medicine, 655 West Baltimore Street, Baltimore, MD 210201, USA
e-mail: mblaustein@som.umaryland.edu

L. Annunziato (ed.), *Sodium Calcium Exchange: A Growing Spectrum of Pathophysiological Implications*, Advances in Experimental Medicine and Biology 961, DOI 10.1007/978-1-4614-4756-6_1,

equation), I studied the effects of divalent cations and anesthetics on lobster nerve conduction (Blaustein and Goldman 1966; Blaustein 1968). I also was fortunate to spend a few weeks at Woods Hole with John Moore and Toshio Narahashi working on tetrodotoxin's action on squid axons (Moore et al. 1967).

I was planning to continue my career in cellular neurophysiology and arranged for a position in Alan Hodgkin's laboratory in Cambridge, England, with a special fellowship from the NIH. My family and I arrived in Cambridge in late August of 1966. After a family trip to Vienna for the International Congress of Biophysics, I left my wife Ellen and our two children (ages 3 and 5) in Cambridge and headed off to the Laboratory of the Marine Biological Association in Plymouth, England, for the fall squid season.

1.2 All Hands to the Pump

My expectation was to study squid axon electrophysiology, but Peter Baker, Alan's junior associate, a lecturer at Emmanuel College, Cambridge, took a mini-sabbatical that fall, and he wanted all Plymouth squid researchers to work on the Na^+ pump. I was paired up with Rick Steinhardt, Richard Keynes' postdoctoral fellow, and we were tasked with studying the activation of the Na^+ pump by external cations. Richard came to Plymouth at the end of September; he showed us how to dissect squid axons (not knowing of my prior experience) and how to measure $^{22}Na^+$ efflux after injecting the giant axons (0.8–1.2-mm diameter) with a microsyringe that he and Alan designed (Hodgkin and Keynes 1956). Richard then went off to Homburg (Saar), Germany, to teach in a course on membrane biophysics organized by Hermann Passow and Robert Stampfli (more about this later).

Rick and I began our $^{22}Na^+$ efflux experiments on squid axons, and we rapidly identified a component that depended on external K^+ and was blocked by ouabain, i.e., the Na^+ pump component. When we removed external Na^+ and K^+ (Na_o and K_o, respectively), preparatory to adding back one monovalent cation at a time, we exposed a very large $^{22}Na^+$ efflux that did not depend upon K_o and was not blocked by ouabain (Fig. 1.1).

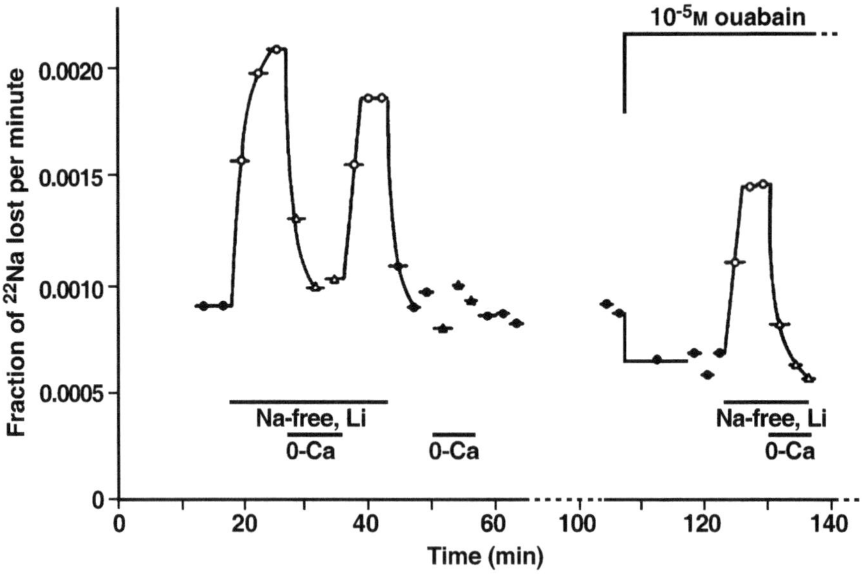

Fig. 1.1 *Reduction of $[Na^+]_o$ activates a large Ca_o-dependent, ouabain-resistant $^{22}Na^+$ efflux in squid axons with high $[Na^+]_i$.* Replacement of 460 mM NaCl in the artificial seawater (ASW) by LiCl greatly increased $^{22}Na^+$ efflux, measured as the fraction of $^{22}Na^+$ lost per minute. The increment was abolished by removal of the 11 mM $CaCl_2$ in the ASW ($MgCl_2$ was increased from 55 to 66 mM). This is evidence of "Na_o-Ca_o antagonism" and indirect evidence for Na^+/Ca^{2+} exchange. Ouabain (10^{-5} M; added at 107 min) reduced the $^{22}Na^+$ efflux in control Na ASW (i.e., it inhibited the Na^+ pump), but it had no effect on the large Ca_o-dependent $^{22}Na^+$ efflux in Li ASW) (Reprinted from Baker et al. (1969), with permission)

This efflux persisted when the external NaCl and KCl were replaced by sucrose, LiCl, or choline Cl^-. We consulted with Peter Baker, who agreed that internal Na^+ could be exiting with an anion or exchanging for a cation. It was easiest to remove the other external cations, Ca^{2+} and Mg^{2+}, so we first removed the Ca^{2+} – et voila! The large Na^+ efflux was reversibly abolished, i.e., the Na^+ efflux was external Ca^{2+}-dependent (Fig. 1.1). Removal of external Mg^{2+} had negligible effect on the Na^+ efflux, so we had our answer: Na/Ca exchange! This was just the first month of my fellowship!

Peter had read Ralph Niedergerke's articles on Na^+-Ca^{2+} interactions in frog cardiac muscle, and he suggested that they might be of interest. There was no time to read, however, we were working 14–16-h days during October because I feared that a gale would interrupt our daily squid supply; the large squid did not survive in the relatively small holding tanks. In fact, the squid usually were injured in the nets and would die before the collection boat docked. Therefore, as soon as squid were caught, the fishermen removed the head and internal organs and placed the mantle (containing the giant axons) in a thermos of iced seawater. With the Na^+ pumps thus turned off, the axons slowly gained Na^+; the axons did not last overnight. As we shall see, this rise in the intracellular Na^+ concentration, $[Na^+]_i$, was fortuitous for the Ca^{2+} influx experiments.

In the meantime, we ordered some $^{45}Ca^{2+}$ to test the Na^+/Ca^{2+} exchange idea directly. The $^{45}Ca^{2+}$ arrived the first week in November, but just before that, we had a gale. Finally, I had a chance to catch my breath. It was a miserable, stormy afternoon, and the laboratory building was deserted; I completed my data analysis for the last experiments and sauntered down the hall to the library. As soon as I started to read the description (Luttgau and Niedergerke 1958) of extracellular Na^+-Ca^{2+} ($[Na^+]_o$-$[Ca^{2+}]_o$) antagonism and its influence on frog cardiac contraction (reduced $[Na^+]_o$ induces cardiac contraction), I got very excited. I immediately recognized that NCX must be widely distributed in both tissues and species, including vertebrate heart. Therefore, since NCX apparently functions in the heart, it is the missing link to the puzzle that had stumped me ever since my first studies on the Na^+ pump and, as an intern, my use of digitalis to treat patients with heart failure: How does Na^+ pump inhibition by cardiotonic steroids increase the force of contraction of the heart? Because of both my clinical and research experiences, I frequently thought about this enigma.

Here was the answer: raising $[Na^+]_i$ promotes net Ca^{2+} gain by NCX and thereby enhances cardiac contraction. That "Eureka! moment" was even more thrilling than the discovery of NCX itself. I was, for a brief time, the only one in the world who understood how cardiotonic steroids enhance cardiac contraction! I was so exhilarated that I went off, alone, to the nearby Green Lantern restaurant, for a fine celebratory dinner with a bottle of claret. Then, slightly inebriated, I returned to the lab to reread Luttgau-Niedergerke, to be sure I was not delusional. It was a great day!

1.3 $^{45}Ca^{2+}$ Flux Studies: Verification of Na^+/Ca^{2+} Exchange

The following Monday afternoon, Alan Hodgkin came down to Plymouth to see how I was getting on. After we dissected a few axons for the evening's $^{22}Na^+$ efflux experiments, he and I went to dinner with Trevor Shaw, another Plymouth squidder. We talked about the NCX, including my explanation of how Na^+ pump inhibitors exert their cardiotonic effect. Alan asked a few questions but was, otherwise, impassive. I was crestfallen. How could he fail to be enthused by the story? Two days later, however, Alan asked me if I would mind if he remained in Plymouth to perform the $^{45}Ca^{2+}$ flux experiments with me. Would I mind? I was ecstatic! We had won him over.

Alan and I performed the first influx experiments the next Monday. Axons were incubated for 1 h in artificial seawater (ASW, the external fluid) containing either NaCl or LiCl as the predominant salt and labeled with $^{45}Ca^{2+}$. The axons were then washed in tracer-free solution, and the axoplasm

Table 1.1 Effects of external cations and of ouabain on $^{45}Ca^{2+}$ influx in squid axons. Reduction of $[Na^+]o$ increased $^{45}Ca^{2+}$ influx. This is evidence of "Nao-Cao antagonism." The increase in Ca^{2+} influx was much greater in the axons from refrigerated squid mantles (with a high $[Na^+]i$) than in axons from live squid (with a lower $[Na^+]i$). This is direct evidence of Na^+/Ca^{2+} exchange (Data from Baker et al. (1969))

	Ca influx (p-mole/cm^2 s)		
External solution	Mean ± S.E.	Range	*n*
A. Axons from refrigerated mantles			
Collected results[a]			
Na seawater, 0 and 10-K, ± ouabain	0.15 ± 0.02	0.04–0.57	30
Li seawater, 0 and 10-K, ± ouabain	4.33 ± 0.43	0.90–9.50	29
Dextrose seawater, 0 and 10-K, ± ouabain	2.46 ± 0.29	0.80–5.30	15
B. Axons from live squid			
10-K, Na seawater	0.23	–	1
10-K, Li seawater	0.66 ± 0.48	0.16–1.60	3

n is the number of axons in each group

[a]Li seawater contained 460 mM LiCl in place of NaCl; dextrose seawater contained 720 mM dextrose in place of 460 mM NaCl. In K-free ("0 K") seawaters, the 10 mM KCl was replaced by 10 mM NaCl or LiCl or 20 mM dextrose. Because neither replacement of external K^+ nor addition of 10^{-5} M ouabain affected the Ca^{2+} influx, the 0 K and ouabain treatment data were not separated

was extruded and weighed, and ^{45}Ca activity was measured in an old Panax counter with a Nixie tube display. After the first sample from NaCl ASW, with a low count, axoplasm from a LiCl ASW axon was counted: the Nixie tubes lit up with a very high count. Because $[Na^+]_i$ was high in these axons from refrigerated mantles, the differences between the Ca^{2+} influx from Na^+ ASW and Li^+ ASW were large and easy to detect. Alan's eyes twinkled as he pulled his pipe from his mouth and broke into a broad grin. Reduction of $[Na^+]_o$ increased Ca^{2+} influx (Table 1.1). Na^+/Ca^{2+} exchange was confirmed! At 2 am, after measuring several replicate samples, we celebrated with a little "medicinal" Scotch before heading off to bed. Another great day!

When I wrote up the results for publication, Alan was reluctant to include the proposed explanation for the cardiotonic action of cardiac glycosides. Therefore, while Alan was on a lecture tour in Eastern Europe, Peter Baker and I contrived to submit the manuscript, including the cardiotonic steroid hypothesis (Baker et al. 1969). [Note: *J Physiol* authorship was alphabetical in those days; there was no jockeying for "first" or "senior" authorship.] Our explanation for the cardiotonic steroid effect was later verified with Ca^{2+} measurements (Wier and Hess 1984; Altamirano et al. 2006), NCX-knockout mice (Reuter et al. 2002), and NCX blockers (Tanaka et al. 2007).

1.4 NCX on the Continent

At the very same time that we were performing these experiments, the fall of 1966, Harald Reuter, from Mainz, Germany, was attending the Membrane Biophysics Lab Techniques course in Homburg. There, he learned about Na^+/Na^+ exchange, using tracer $^{22}Na^+$, from Peter Caldwell and Richard Keynes (who did not yet know of the results in Plymouth). After completing the course, Harald, a cardiac pharmacologist who was familiar with the Luttgau-Niedergerke articles, immediately set out to look for NCX in cardiac muscle. And, of course, he found it (Reuter and Seitz 1968). He knew what he was looking for; we were simply lucky, albeit prepared to recognize it when we saw it! As Louis Pasteur put it, "*le hazard ne favorise que les esprits prepares.*"

Harald and I first met at the International Physiology Congress in Washington, DC, in August 1968. I had just returned from England and was then heading to St. Louis to take a faculty position in the Department of Physiology and Biophysics at Washington University. It was a very amicable meeting that soon led to extremely beneficial consequences. Before describing those consequences, however, I should mention some other fallout from our Cambridge years.

1.5 Cambridge Collateral: Ellen's Contributions

While I was off squidding in Plymouth, my "squidow" (the local name for squidder spouses), Ellen, when she was not antiquing or rubbing church brasses, met several Cambridge sabbatical spouses. Upon my return to Cambridge in mid-December, after the 1966 squid season, Ellen invited Marcella and Len Ross for dinner. Len, an anatomist from Philadelphia, told me of his work with Victor Whittaker on the structure of isolated nerve endings, "synaptosomes."

I was fascinated and hypothesized (to myself) that the terminals might reseal and become functionally competent; I planned to test this idea when I returned to the States. In fact, my colleagues and I subsequently showed that synaptosomes do reseal and that they function like intact terminals. They have functional ion transport mechanisms, including Na^+ pumps and NCX, and they generate ion gradients and membrane potentials (Blaustein and Wiesmann 1970; Blaustein and Goldring 1975; Fontana et al. 1995). Synaptosomes can be triggered to release neurotransmitters and to recycle synaptic vesicle membranes (Fried and Blaustein 1976; Drapeau and Blaustein 1983), and they sequester Ca^{2+} in the endoplasmic reticulum (Kendrick et al. 1977). We discovered voltage-gated Ca^{2+} channels in synaptosomes that were resistant to dihydropyridines (Nachshen and Blaustein 1979); this was the first evidence for N-, P-, and Q-type Ca^{2+} channels (Catterall 1998; Cao and Tsien 2010) that play such an important role in brain physiology. We also used synaptosomes as an assay for the identification of a number of novel Na^+ and K^+ channel inhibitors (Krueger et al. 1980; Blaustein et al. 1991). In all, we published more than 65 articles on synaptosomes, including 6 in *Nature* or *PNAS*. In an extension of the synaptosome studies, we are currently testing the NCX-mediated effects of nanomolar ouabain on Ca^{2+} signaling in cultured neurons and astrocytes. Thus, Ellen's friendship with Marcella Ross paid off in a line of research that has continued to the present day.

1.6 Smooth Muscle Is Not Such Smooth Sailing

During our second year in Cambridge, again thanks to Ellen's initial contact while I was in Plymouth, she and I became very friendly with our next-door neighbors, Joy and Mani Matter, from Bern, Switzerland. Mani was the Bern town counsel and was spending a year studying international law in Cambridge; Joy was a teacher who later became Minister of Education for the Canton of Bern. We maintained contact when we returned to our respective home countries, and I visited the Matters in 1970, after presenting my first paper on NCX in synaptosomes at a meeting in Sweden. Ellen, too, was anxious to see the Matters again, so I arranged for a NATO Fellowship-funded mini-sabbatical with Harald Reuter, who had just become chairman of the Department of Pharmacology in Bern (what a fortunate coincidence!). And off we went to Bern, from May to September of 1971. On Harald's suggestion, I agreed to look for NCX in vascular smooth muscle. The experiments were affirmative from the very beginning.

We demonstrated antagonism between Na_o and Ca_o (reduced $[Na^+]_o$ promoted Ca^{2+} entry and arterial contraction) (Fig. 1.2a) as well as Na_o-dependent $^{45}Ca^{2+}$ extrusion from guinea pig aortic smooth muscle (Fig. 1.2b).

In a manuscript describing these results, we suggested that NCX helps to regulate vascular tone. We also postulated a close approximation between the plasma membrane (PM) NCX and the sarcoplasmic reticulum; this idea foreshadowed the "buffer-barrier" hypothesis (van Breemen et al. 1986, 1995) and the "PLasmERosome" model (Blaustein et al. 1998; Arnon et al. 2000b) of local Ca^{2+} control. The

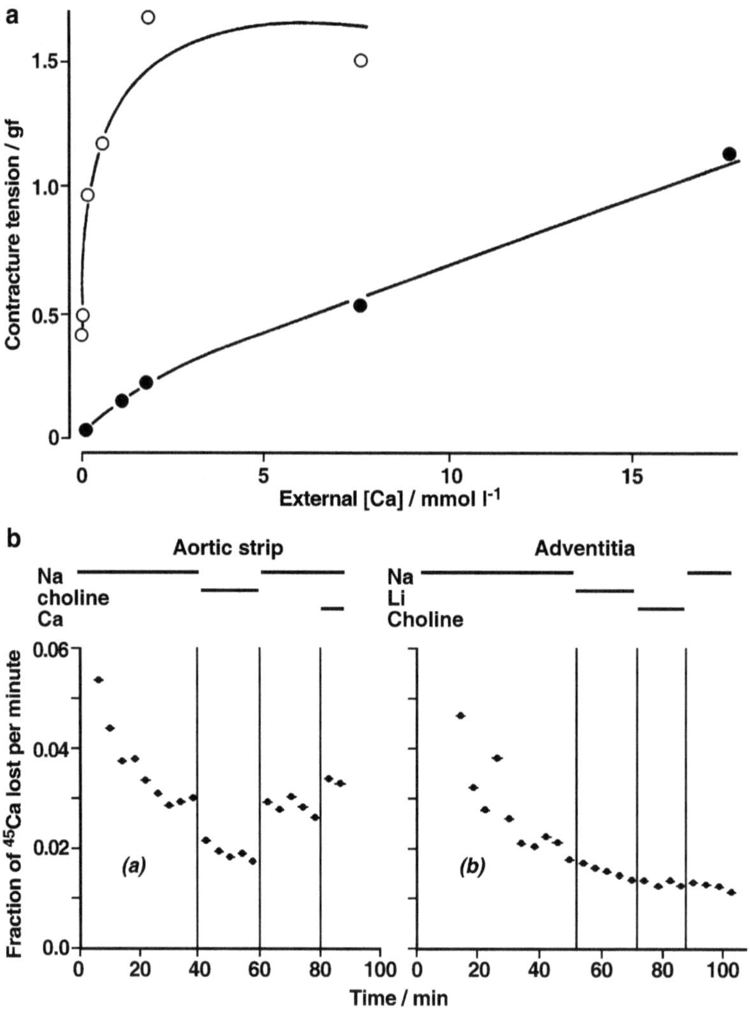

Fig. 1.2 (**a**) *Effect of the external Ca^{2+} concentration, $[Ca^{2+}]_o$, on contraction of rabbit aorta.* Contractile tension (gf) is increased by reduction of $[Na^+]_o$. Either 80% (filled circles) or 100% (open circles) of the NaCl in the standard medium (137 mM) was replaced by equiosmotic sucrose. The greater effectiveness of low $[Ca^{2+}]_o$ in inducing contraction when $[Na^+]_o$ was lowered is evidence of "Na_o-Ca_o antagonism." (**b**) *Reduction of $[Na^+]_o$ decreases $^{45}Ca^{2+}$ efflux (fraction of $^{45}Ca^{2+}$ lost per min) in a rabbit aortic strip (a), but not in the isolated adventitia (b).* The NaCl in the standard medium (137 mM) was replaced by equimolar LiCl or choline-Cl during the periods indicated by the bars at the *top*. To avoid influx of Ca^{2+} and dilution of intracellular tracer $^{45}Ca^{2+}$, the efflux (extracellular) solutions were Ca-free and contained 0.5 mM EGTA, except during the period indicated by the "Ca" bar at the end of the experiment on the *left*. The Na_o-dependent $^{45}Ca^{2+}$ efflux in intact aortic strip is evidence of Na^+/Ca^{2+} exchange (Reprinted from Reuter et al. (1973), with permission)

manuscript (Blaustein and Reuter) was rejected by *Nature*, *Science*, and *Experientia*. It was published in a non-refereed journal, *Phil Trans Roy Soc Lond*, because Harald was invited to present the results at a Royal Society conference on smooth muscle (Reuter et al. 1973); the article has been cited ~300 times.

This novel idea that NCX is functionally important in vascular smooth muscle was, at first, ignored or even disparaged (Somlyo et al. 1986; Murphy 1988). Views about this concept slowly began to change more than two decades after its discovery (Murphy 1993; Somlyo and Somlyo 1994), once the NCX was cloned (Nicoll et al. 1990) and could be readily identified by immunocytochemistry and immunoblotting (Vemuri et al. 1990; Blaustein et al. 1992; Juhaszova et al. 1994).

Tragically, Mani Matter was killed in an auto accident in 1972. Therefore, we were most fortunate to have spent time in Bern in 1971. Mani had become a Swiss folk hero because, as an avocation, he founded the Bernese Troubadours, who wrote and sang humorous and satirical songs in Swiss-German dialect that are still popular today.

1.7 PLasmERosomes: A Structural Basis for Functional Coupling of Transport Proteins

The NCX, like many other Na^+-coupled transporters, is driven by the Na^+ electrochemical gradient across the PM (Blaustein and Lederer 1999). Nevertheless, we did not anticipate that the NCX would be colocalized with certain other transporters in PM microdomains adjacent to "junctional" sarcoplasmic or endoplasmic reticulum (jS/ER) in a variety of cell types including neurons, glia, and arterial myocytes (Juhaszova and Blaustein 1997b). In fact, we were somewhat astonished (but should not have been) that the NCX was confined to the PM microdomains that also contained the Na^+ pumps with a high ouabain affinity catalytic subunit ($\alpha 2$ or $\alpha 3$, depending upon cell type) (Juhaszova and Blaustein 1997a). Moreover, both immunocytochemical and functional studies indicated that the much more prevalent Na^+ pumps with an $\alpha 1$ catalytic subunit that, in rodents, has low ouabain affinity, were apparently excluded from these microdomains (Lee et al. 2006; Song et al. 2006). Another mechanism that extrudes Ca^{2+} from most types of cells, the PM Ca^{2+}-ATPase (Ca^{2+} pump), may be excluded from these PM microdomains (Lencesova et al. 2004).

Some other Na^+ and Ca^{2+} transporters also colocalize to these microdomains, notably TRPC proteins that are components of receptor- and store-operated channels (ROCs and SOCs, respectively) (Golovina 2005; Lee et al. 2006; Zulian et al. 2010a). Many ROCs and SOCs are relatively nonselective cation channels, permeable to both Na^+ and Ca^{2+} (Arnon et al. 2000a; Owsianik et al. 2006).

These findings provided a structural basis for the observations that the high ouabain affinity Na^+ pumps, NCX, ROCs and SOCs, and the adjacent jS/ER function cooperatively to regulate Ca^{2+} signaling in a variety of cell types (Arnon et al. 2000a, b; Eder et al. 2005; Poburko et al. 2007). I called this PM microdomain-jS/ER complex, the "PLasmERosome" (Fig. 1.3). PLasmERosomes appear to be widely distributed fundamental units that play a key role in cell Ca^{2+} signaling. They can modulate Ca^{2+} signaling differently in different types of cells, depending upon (1) the specific complement of TRPC proteins and perhaps other channel proteins (e.g., voltage-gated Ca^{2+} channels, TRPM proteins) and (2) various regulatory molecules that modulate the NCX and other transporters in the PLasmERosomes such as protein kinases, PIP_2 (phosphatidylinositol bisphosphate), and calmodulin.

Not surprisingly, then, these mechanisms may also play prominent roles in pathophysiological processes (Blaustein 1977), as discussed in the example below.

1.8 Linkage to Hypertension and Other Heretical Concepts

I learned from Alan Hodgkin that if you perform careful, controlled experiments, you should be very confident about your own data. So, ignoring the naysayers, I continued to ponder the role of NCX in vascular smooth muscle and even began to think that it might play a role in hypertension (Blaustein 1974). This led me to propose a naïve but useful hypothesis about the roles of a hypothetical endogenous cardiotonic steroid, its Na^+ pump receptor, and NCX, in the pathogenesis of salt-dependent hypertension (Blaustein 1977). That was before NCX and the Na^+ pump were cloned and their isoforms discovered, and of course, the endogenous

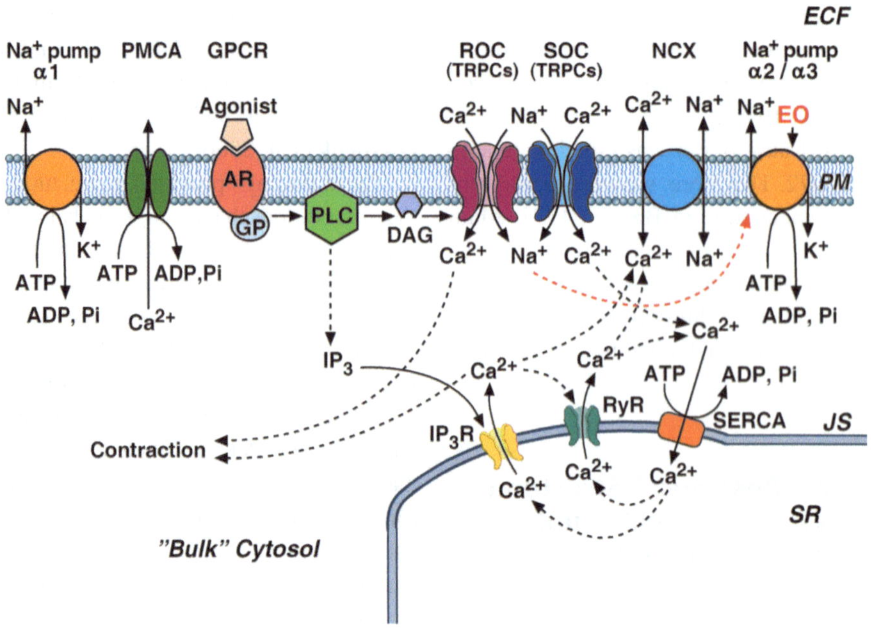

Fig. 1.3 *Diagram of ouabain-/EO-regulated α2 Na⁺ pump-modulated Ca²⁺ signaling at the plasma membrane-sarco/endoplasmic reticulum junction (PLasmERosome) in arterial smooth muscle.* Endogenous ouabain (EO) reduces Na^+ extrusion by the α2 Na^+ pump, thereby increasing local $[Na^+]$ in the junctional space (JS) and reducing Ca^{2+} extrusion by the Na/Ca exchanger (NCX). This enhances Ca^{2+} signaling and, in arterial smooth muscle, contraction. *ADP* adenosine diphosphate, *AR* agonist receptor, *ATP* adenosine triphosphate, *DAG* diacylglycerol, *ECF* extracellular fluid, *GP* G-protein, *GPCR* G-protein-coupled receptor, IP_3 inositol trisphosphate, IP_3R IP_3 receptor, *Pi* inorganic phosphate, *PLC* phospholipase C, *PMCA* plasma membrane Ca^{2+} pump, *SERCA* sarco-/endoplasmic reticulum Ca^{2+} pump, *SR* sarcoplasmic reticulum, *SOC* store-operated channel, *ROC* receptor-operated channel (both composed of TRPC proteins) (Reprinted from Blaustein et al. (2011), with permission)

cardiotonic steroid was then just a figment of our imagination (the 'holy spirit').

Despite the previously mentioned skepticism of vascular biologists, however, this hypertension hypothesis aroused great interest in the idea of an endogenous ligand for the Na^+ pump ouabain-binding site. In particular, it excited one young man, John Hamlyn, who was a preteen-ager in Plymouth, England, when we discovered the NCX there. John came to my laboratory in 1980, as a postdoctoral fellow, with the idea of trying to purify this hypothetical agent.

Ignoring the enormous difficulties, the fierce competition, and the repeated funding rejections from NIH, our team (John and our colleagues at the Upjohn Company) plunged ahead. In 1991, we purified from two tons of human plasma, and identified, by mass spectroscopy, an endogenous compound that, astonishingly, is indistinguishable from plant ouabain, i.e., "endogenous ouabain" or EO (Hamlyn et al. 1991). That paper, too, was rejected by *Nature*: one reviewer had a "*gut feeling*" that our result was an artifact, although another reviewer offered to write a "*News and Views*" article about the discovery. The *Nature* editor, John Maddox, had just taken the heat for publishing articles on "cold fusion" and "infinite dilution"; he apparently did not want to take a chance on being burned once more. Fortunately, our friend, Joseph Hoffman, a Na^+ pump expert, agreed to communicate the manuscript to *PNAS*.

Again, many colleagues were skeptical (and some still are), even though the identification of EO was replicated in bovine adrenal and hypothalamic samples by three distinguished groups (Tamura et al. 1994; Schneider et al. 1998; Kawamura et al. 1999).

Nevertheless, our Upjohn coworkers, who performed the original mass spectrometry, convinced us of the correctness of the result. Although the EO biosynthetic pathway has not yet been elucidated (funding bodies are not interested in such mundane studies), John Hamlyn and colleagues demonstrated net synthesis of EO by primary cultured bovine adrenocortical cells (Laredo et al. 1994). Furthermore, plasma EO levels are very low in adrenalectomized rodents (Hamlyn et al. 1991) and in humans with adrenocortical insufficiency (Sophocleous et al. 2003).

Ideas about the involvement of EO in the pathogenesis of some forms of hypertension have come a long way since my first, rather simplistic (but perhaps not so far-fetched) thoughts on the subject (Blaustein 1977). Recent work is summarized to two reviews (Blaustein and Hamlyn 2010; Blaustein et al. 2011).

1.9 NCX: A Key Player in the Pathogenesis of Hypertension?

I have already alluded to the central role of NCX in Ca^{2+} signaling that results from its colocalization with certain (ouabain-sensitive) Na^{+} pumps and ROCs and SOCs in PM microdomains at PM-S/ER junctions (PLasmERosomes) (Fig. 1.3). This fit with the simplistic notion that if circulating EO levels are elevated in patients with hypertension (Rossi et al. 1995), we would expect the PLasmERosome mechanisms to amplify the Ca^{2+} signals. In fact, this is precisely what we observed in isolated, pressurized rodent small arteries: Ca^{2+} signals and myogenic constriction were augmented by ouabain with an EC_{50} of ~1 nM (Zhang et al. 2005). Could this account for the increased peripheral vascular resistance and elevation of blood pressure, as I initially proposed (Blaustein 1977)?

Alas, life is not so simple! (A good thing, because that is what keeps us scientists in business). As my colleague Vera Golovina discovered, the situation in hypertension is much more complex. She and her associates were studying arterial smooth muscle acutely isolated from rats in which hypertension was induced by prolonged subcutaneous infusion of ouabain ("ouabain hypertension"). They found that NCX1 (the vascular isoform) and TRPC6 protein (a component of ROCs) were greatly overexpressed in the myocytes and that this was reflected in augmented Ca^{2+} signaling (Pulina et al. 2010). The effects were not due to the elevation of blood pressure: both the protein upregulation and the augmented Ca^{2+} signaling were also induced in primary cultured normal rat arterial myocytes incubated with 100 nM ouabain for 72–96 h. [In an important twist, digoxin did not upregulate NCX1 or TRPC6 expression (Zulian et al. 2010b), and unlike ouabain, it does not induce hypertension in rats (Manunta et al. 2000) – but that is another, as yet incomplete, story.]

Vera and her colleagues also found that NCX1 and TRPC6 were both markedly upregulated in arterial myocytes from Dahl salt-sensitive rats fed a high salt diet to induce hypertension (Table 1.2) and in myocytes from Milan hypertensive rats, but not the normotensive strain (Zulian et al. 2010a). In fact, arterial myocyte NCX1 and either TRPC6 or TRPC3 (another component of ROCs) are upregulated in all models of hypertension that have been tested to date (Table 1.2).

An editorial commentary (Giachini and Tostes 2010) on the articles from Vera's laboratory stated that "*Given the... interrelationship among TRPC6, Na^{+} influx, NCX, and Ca^{2+} influx, it seems that this is... the key to better understand the role of Na^{+} in hypertension, vascular reactivity, and blood pressure regulation.*" Vindication! This is the diametric opposite of the view that "*The Na/Ca exchanger does not play an essential role in the regulation of cytoplasmic Ca^{2+} in smooth muscle*" (Somlyo et al. 1986). The upregulation of NCX1 and TRPC6 expression may, in fact, be a key mechanism underlying the phenomenon of "whole-body autoregulation" (Guyton 1989) that sustains the increased peripheral vascular resistance and elevated BP in hypertension.

This, of course, is not the end of the story but just the beginning of the next chapter. The fun in

Table 1.2 Expression of Na/Ca exchanger-1 (NCX1) and some TRPC protein components of receptor-operated channels (ROCs) is increased in several hypertensive animal models and in human primary pulmonary hypertension (Modified from Blaustein et al. (2011))

		Artery smooth muscle			
			ROC		
	Hypertension	NCX1	TRPC3	TRPC6	Reference
1	Ouabain* (vs. vehicle and digoxin)	↑		↑	Pulina et al. (2010), Zulian et al. (2010b)
2	DOCA-salt*	ND**		↑	Bae et al. (2007)
3	Milan hypertension* (vs. Milan NT)***	↑		↑	Zulian et al. (2010a)
4	SHR (vs. WKY)***	↑	↑		Taniguchi et al. (2004), Liu et al. (2009)
5	Dahl salt-sensitive/high (vs. low) salt*	↑		↑	Golovina (unpublished)
6	$NCX1^{SM.Tg/Tg}$***	↑		↑	Iwamoto et al. (2004), Zhang and Blaustein (unpublished)
7	Angiotensin II	↑		ND**	Chen et al. (unpublished)
8	Human primary pulmonary hypertension	↑	↑	↑	Yu et al. (2004), Zhang et al. (2007a, b)

* hypertension associated with high lasma ouabain/Eo levels.
** not determined.
*** Abbreviations: NT, normotensive control; SHR, spontaneously hypertensive rat; WKY, Wistar-Kyoto normotensive control for SHR; $NCX1^{SMTg/Tg}$, smooth muscle-specific NCX1 overexpressor mouse

research is that every new discovery leads to new questions. In this case, we must now ask: "What is the mechanism by which ouabain upregulates NCX1 and TRPC6 expression?" The upregulation appears to be mediated by a novel activity of Na^+ pumps first elucidated by Zi-Jian Xie and Amir Askari, namely, ouabain-induced activation of protein kinase signaling cascades (Xie and Askari 2002; Liu et al. 2007, 2011; Liu and Xie 2010). [Zi-Jian's new idea, that Na^+ pumps could do something other than pump Na^+ and K^+, was greeted with profound skepticism – and I was originally one of the skeptics – but the data speak for themselves!] It is the anticipation of learning something new at any moment that keeps me young and keeps me going. That is the lesson I most enjoy conveying to students; it is the essence (and the excitement) of scientific research. The fact that such moments are not everyday occurrences makes them so special.

1.10 In Closing

I have been most fortunate to have had the opportunity to engage in this exciting adventure. I was well prepared by my teachers, but I could not have had such success without wonderful students, colleagues, and collaborators. My wife, Ellen, in addition to providing extraordinary support, also played a critical part (mentioned above) that neither she nor I could have imagined a priori. And, of course, none of this would have been possible without the many years of financial support from the National Institutes of Health and the American Heart Association.

> After all, in spite of opinion, prejudice, or error, Time will fix the real value upon this discovery, and determine whether I have imposed upon myself and others, or contributed to the benefit of science and mankind.
>
> William Withering, *"An Account of the Foxglove and Some of its Medical Uses"*
>
> Birmingham, England, July 1, 1785.

Acknowledgments I thank Suzanne Ventura for assistance with the manuscript. Supported by NIH/NHLB, NCX. Supported by NIH/NHLBI grants HL-045215, HL-078870, and HL-107555 and NIH/NINDS grant NS-16106.

References

J. Altamirano, Y. Li, J. DeSantiago, V. Piacentino 3rd, S.R. Houser, D.M. Bers, The inotropic effect of cardioactive glycosides in ventricular myocytes requires Na^+-Ca^{2+} exchanger function. J. Physiol. **575**, 845–854 (2006)

A. Arnon, J.M. Hamlyn, M.P. Blaustein, Na^+ entry via store-operated channels modulates Ca^{2+} signaling in arterial myocytes. Am. J. Physiol. Cell Physiol. **278**, C163–C173 (2000a)

A. Arnon, J.M. Hamlyn, M.P. Blaustein, Ouabain augments Ca^{2+} transients in arterial smooth muscle without raising cytosolic Na^+. Am. J. Physiol. Heart Circ. Physiol. **279**, H679–H691 (2000b)

Y.M. Bae, A. Kim, Y.J. Lee, W. Lim, Y.H. Noh, E.J. Kim, J. Kim, T.K. Kim, S.W. Park, B. Kim, S.I. Cho, D.K. Kim, W.K. Ho, Enhancement of receptor-operated cation current and TRPC6 expression in arterial smooth muscle cells of deoxycorticosterone acetate-salt hypertensive rats. J. Hypertens. **25**, 809–817 (2007)

P.F. Baker, M.P. Blaustein, A.L. Hodgkin, R.A. Steinhardt, The influence of calcium on sodium efflux in squid axons. J. Physiol. **200**, 431–458 (1969)

M.P. Blaustein, Barbiturates block sodium and potassium conductance increases in voltage- clamped lobster axons. J. Gen. Physiol. **51**, 293–307 (1968)

M.P. Blaustein, The interrelationship between sodium and calcium fluxes across cell membranes. Rev. Physiol. Biochem. Pharmacol. **70**, 33–82 (1974)

M.P. Blaustein, Sodium ions, calcium ions, blood pressure regulation, and hypertension: a reassessment and a hypothesis. Am. J. Physiol. **232**, C165–C173 (1977)

M.P. Blaustein, D.E. Goldman, Action of anionic and cationic nerve-blocking agents: experiment and interpretation. Science **153**, 429–432 (1966)

M.P. Blaustein, J.M. Goldring, Membrane potentials in pinched-off presynaptic nerve terminals monitored with a fluorescent probe: evidence that synaptosomes have potassium diffusion potentials. J. Physiol. **247**, 589–615 (1975)

M.P. Blaustein, J.M. Hamlyn, Signaling mechanisms that link salt retention to hypertension: endogenous ouabain, the Na^+ pump, the Na^+/Ca^{2+} exchanger and TRPC proteins. Biochim. Biophys. Acta **1802**, 1219–1229 (2010)

M.P. Blaustein, W.J. Lederer, Sodium/calcium exchange: its physiological implications. Physiol. Rev. **79**, 763–854 (1999)

M.P. Blaustein, W.P. Wiesmann, Effect of sodium ions on calcium movements in isolated synaptic terminals. Proc. Natl. Acad. Sci. U. S. A. **66**, 664–671 (1970)

M.P. Blaustein, R.S. Rogowski, M.J. Schneider, B.K. Krueger, Polypeptide toxins from the venoms of Old World and New World scorpions preferentially block different potassium channels. Mol. Pharmacol. **40**, 932–942 (1991)

M.P. Blaustein, A. Ambesi, R.J. Bloch, W.F. Goldman, M. Juhaszova, G.E. Lindenmayer, D.N. Weiss, Regulation of vascular smooth muscle contractility: roles of the sarcoplasmic reticulum (SR) and the sodium/calcium exchanger. Jpn. J. Pharmacol. **58**(Suppl 2), 107P–114P (1992)

M.P. Blaustein, M. Juhaszova, V.A. Golovina, The cellular mechanism of action of cardiotonic steroids: a new hypothesis. Clin. Exp. Hypertens. **20**, 691–703 (1998)

M.P. Blaustein, F.H. Leenen, L. Chen, V.A. Golovina, J.M. Hamlyn, T.L. Pallone, J.W. Van Huysse, J. Zhang, W.G. Wier, How NaCl raises blood pressure: a new paradigm for the pathogenesis of salt-dependent hypertension. Am. J. Physiol. Heart. Circ. Physiol. **302**, H3031-H1049 (2012)

Y.Q. Cao, R.W. Tsien, Different relationship of N- and P/Q-type Ca^{2+} channels to channel- interacting slots in controlling neurotransmission at cultured hippocampal synapses. J. Neurosci. **30**, 4536–4546 (2010)

W.A. Catterall, Structure and function of neuronal Ca^{2+} channels and their role in neurotransmitter release. Cell. Calcium. **24**, 307–323 (1998)

P. Drapeau, M.P. Blaustein, Initial release of [3 H]dopamine from rat striatal synaptosomes: correlation with calcium entry. J. Neurosci. **3**, 703–713 (1983)

P. Eder, M. Poteser, C. Romanin, K. Groschner, Na^+ entry and modulation of Na^+/Ca^{2+} exchange as a key mechanism of TRPC signaling. Pflugers Arch. **451**, 99–104 (2005)

G. Fontana, R.S. Rogowski, M.P. Blaustein, Kinetic properties of the sodium-calcium exchanger in rat brain synaptosomes. J. Physiol. **485**, 349–364 (1995)

R.C. Fried, M.P. Blaustein, Synaptic vesicle recycling in synaptosomes in vitro. Nature **261**, 255–256 (1976)

F.R. Giachini, R.C. Tostes, Does Na^+ really play a role in Ca^{2+} homeostasis in hypertension? Am. J. Physiol. Heart Circ. Physiol. **299**, H602–H604 (2010)

V.A. Golovina, Visualization of localized store-operated calcium entry in mouse astrocytes. Close proximity to the endoplasmic reticulum. J. Physiol. **564**, 737–749 (2005)

A.C. Guyton, Dominant role of the kidneys and accessory role of whole-body autoregulation in the pathogenesis of hypertension. Am. J. Hypertens. **2**, 575–585 (1989)

J.M. Hamlyn, M.P. Blaustein, S. Bova, D.W. DuCharme, D.W. Harris, F. Mandel, W.R. Mathews, J.H. Ludens, Identification and characterization of a ouabain-like compound from human plasma. Proc. Natl. Acad. Sci. U. S. A. **88**, 6259–6263 (1991)

A.L. Hodgkin, R.D. Keynes, Experiments on the injection of substances into squid giant axons by means of a microsyringe. J. Physiol. **131**, 592–616 (1956)

T. Iwamoto, S. Kita, J. Zhang, M.P. Blaustein, Y. Arai, S. Yoshida, K. Wakimoto, I. Komuro, T. Katsuragi, Salt-sensitive hypertension is triggered by Ca^{2+} entry via Na^+/Ca^{2+} exchanger type-1 in vascular smooth muscle. Nature Med. **10**, 1193–1199 (2004)

M. Juhaszova, M.P. Blaustein, Distinct distribution of different Na^+ pump alpha subunit isoforms in plasmalemma. Physiological implications. Ann. N. Y. Acad. Sci. **834**, 524–536 (1997a)

M. Juhaszova, M.P. Blaustein, Na^+ pump low and high ouabain affinity alpha subunit isoforms are differently distributed in cells. Proc. Natl. Acad. Sci. U. S. A. **94**, 1800–1805 (1997b)

M. Juhaszova, A. Ambesi, G.E. Lindenmayer, R.J. Bloch, M.P. Blaustein, Na^+-Ca^{2+} exchanger in arteries: identification by immunoblotting and immunofluorescence microscopy. Am. J. Physiol. **266**, C234–C242 (1994)

A. Kawamura, J. Guo, Y. Itagaki, C. Bell, Y. Wang, G.T. Haupert Jr., S. Magil, R.T. Gallagher, N. Berova, K.

Nakanishi, On the structure of endogenous ouabain. Proc. Natl. Acad. Sci. U. S. A. **96**, 6654–6659 (1999)
N.C. Kendrick, M.P. Blaustein, R.C. Fried, R.W. Ratzlaff, ATP-dependent calcium storage in presynaptic nerve terminals. Nature **265**, 246–248 (1977)
B.K. Krueger, M.P. Blaustein, R.W. Ratzlaff, Sodium channels in presynaptic nerve terminals. Regulation by neurotoxins. J. Gen. Physiol. **76**, 287–313 (1980)
J. Laredo, B.P. Hamilton, J.M. Hamlyn, Ouabain is secreted by bovine adrenocortical cells. Endocrinology **135**, 794–797 (1994)
M.Y. Lee, H. Song, J. Nakai, M. Ohkura, M.I. Kotlikoff, S.P. Kinsey, V.A. Golovina, M.P. Blaustein, Local subplasma membrane Ca^{2+} signals detected by a tethered Ca^{2+} sensor. Proc. Natl. Acad. Sci. U. S. A. **103**, 13232–13237 (2006)
L. Lencesova, A. O'Neill, W.G. Resneck, R.J. Bloch, M.P. Blaustein, Plasma membrane- cytoskeleton-endoplasmic reticulum complexes in neurons and astrocytes. J. Biol. Chem. **279**, 2885–2893 (2004)
J. Liu, Z.J. Xie, The sodium pump and cardiotonic steroids-induced signal transduction protein kinases and calcium-signaling microdomain in regulation of transporter trafficking. Biochim. Biophys. Acta **1802**, 1237–1245 (2010)
L. Liu, X. Zhao, S.V. Pierre, A. Askari, Association of PI3K-Akt signaling pathway with digitalis- induced hypertrophy of cardiac myocytes. Am. J. Physiol. Cell Physiol. **293**, C1489–C1497 (2007)
D. Liu, D. Yang, H. He, X. Chen, T. Cao, X. Feng, L. Ma, Z. Luo, L. Wang, Z. Yan, Z. Zhu, M. Tepel, Increased transient receptor potential canonical type 3 channels in vasculature from hypertensive rats. Hypertension **53**, 70–76 (2009)
L. Liu, A.V. Ivanov, M.E. Gable, F. Jolivel, G.A. Morrill, A. Askari, Comparative properties of caveolar and noncaveolar preparations of kidney Na^+/K^+-ATPase. Biochemistry **50**, 8664–8673 (2011)
H.C. Luttgau, R. Niedergerke, The antagonism between Ca and Na ions on the frog's heart. J. Physiol. **143**, 486–505 (1958)
P. Manunta, J. Hamilton, A.C. Rogowski, B.P. Hamilton, J.M. Hamlyn, Chronic hypertension induced by ouabain but not digoxin in the rat: antihypertensive effect of digoxin and digitoxin. Hypertens. Res. **23**(Suppl), S77–S85 (2000)
J.W. Moore, M.P. Blaustein, N.C. Anderson, T. Narahashi, Basis of tetrodotoxin's selectivity in blockage of squid axons. J. Gen. Physiol. **50**, 1401–1411 (1967)
R.A. Murphy, Contraction of muscle cells, in *Physiology*, ed. by R.M. Berne, M.N. Levy, 2nd edn. (Mosby, Washington, DC, 1988), pp. 315–342
R.A. Murphy, Smooth muscle, in *Physiology*, ed. by R.M. Berne, M.N. Levy, 3rd edn. (Mosby Year Book, St. Louis, 1993), pp. 309–324
D.A. Nachshen, M.P. Blaustein, The effects of some organic "calcium antagonists" on calcium influx in presynaptic nerve terminals. Mol. Pharmacol. **16**, 576–586 (1979)
D.A. Nicoll, S. Longoni, K.D. Philipson, Molecular cloning and functional expression of the cardiac sarcolemmal Na^+-Ca^{2+} exchanger. Science **250**, 562–565 (1990)
G. Owsianik, K. Talavera, T. Voets, B. Nilius, Permeation and selectivity of TRP channels. Annu. Rev. Physiol. **68**, 685–717 (2006)
D. Poburko, C.H. Liao, V.S. Lemos, E. Lin, Y. Maruyama, W.C. Cole, C. van Breemen, Transient receptor potential channel 6-mediated, localized cytosolic $[Na^+]$ transients drive Na^+/Ca^{2+} exchanger-mediated Ca^{2+} entry in purinergically stimulated aorta smooth muscle cells. Circ. Res. **101**, 1030–1038 (2007)
M.V. Pulina, A. Zulian, R. Berra-Romani, O. Beskina, A. Mazzocco-Spezzia, S.G. Baryshnikov, I. Papparella, J.M. Hamlyn, M.P. Blaustein, V.A. Golovina, Upregulation of Na^+ and Ca^{2+} transporters in arterial smooth muscle from ouabain-induced hypertensive rats. Am. J. Physiol. Heart Circ. Physiol. **298**, H263–H274 (2010)
H. Reuter, N. Seitz, The dependence of calcium efflux from cardiac muscle on temperature and external ion composition. J. Physiol. **195**, 451–470 (1968)
H. Reuter, M.P. Blaustein, G. Haeusler, Na-Ca exchange and tension development in arterial smooth muscle. Philos. Trans. R. Soc. Lond. B Biol. Sci. **265**, 87–94 (1973)
H. Reuter, S.A. Henderson, T. Han, R.S. Ross, J.I. Goldhaber, K.D. Philipson, The Na^+-Ca^{2+} exchanger is essential for the action of cardiac glycosides. Circ. Res. **90**, 305–308 (2002)
G. Rossi, P. Manunta, J.M. Hamlyn, E. Pavan, R. De Toni, A. Semplicini, A.C. Pessina, Immunoreactive endogenous ouabain in primary aldosteronism and essential hypertension: relationship with plasma renin, aldosterone and blood pressure levels. J. Hypertens. **13**, 1181–1191 (1995)
R. Schneider, V. Wray, M. Nimtz, W.D. Lehmann, U. Kirch, R. Antolovic, W. Schoner, Bovine adrenals contain, in addition to ouabain, a second inhibitor of the sodium pump. J. Biol. Chem. **273**, 784–792 (1998)
J.C. Skou, The influence of some cations on an adenosine triphosphatase from peripheral nerves. Biochim. Biophys. Acta **23**, 394–401 (1957)
A.P. Somlyo, A.V. Somlyo, Signal transduction and regulation in smooth muscle. Nature **372**, 231–236 (1994)
A.P. Somlyo, R. Broderick, A.V. Somlyo, Calcium and sodium in vascular smooth muscle. Ann. N. Y. Acad. Sci. **488**, 228–239 (1986)
H. Song, M.Y. Lee, S.P. Kinsey, D.J. Weber, M.P. Blaustein, An N-terminal sequence targets and tethers Na^+ pump $\alpha 2$ subunits to specialized plasma membrane microdomains. J. Biol. Chem. **281**, 12929–12940 (2006)
A. Sophocleous, I. Elmatzoglou, A. Souvatzoglou, Circulating endogenous digitalis-like factor(s) (EDLF) in man is derived from the adrenals and its secretion is ACTH-dependent. J. Endocrinol. Invest. **26**, 668–674 (2003)

M. Tamura, F. Konishi, M. Sakakibara, T. Inagami, Large scale purification of an endogenous Na^+/K^+-pump inhibitor from bovine adrenal glands, in *The Sodium Pump: Structure, Mechanism, Hormonal Control and Its Role in Disease*, ed. by E. Bamberg, W. Schoner (Steinkopff, Darmstadt, 1994), pp. 763–766

H. Tanaka, H. Shimada, I. Namekata, T. Kawanishi, N. Iida-Tanaka, K. Shigenobu, Involvement of the Na^+/Ca^{2+} exchanger in ouabain-induced inotropy and arrhythmogenesis in guinea-pig myocardium as revealed by SEA0400. J. Pharmacol. Sci. **103**, 241–246 (2007)

S. Taniguchi, K. Furukawa, S. Sasamura, Y. Ohizumi, K. Seya, S. Motomura, Gene expression and functional activity of sodium/calcium exchanger enhanced in vascular smooth muscle cells of spontaneously hypertensive rats. J. Cardiovasc. Pharmacol. **43**, 629–637 (2004)

C. van Breemen, S. Lukeman, P. Leijten, H. Yamamoto, R. Loutzenhiser, The role of superficial SR in modulating force development induced by Ca entry into arterial smooth muscle. J. Cardiovasc. Pharmacol. **8**(Suppl 8), S111–S116 (1986)

C. van Breemen, Q. Chen, I. Laher, Superficial buffer barrier function of smooth muscle sarcoplasmic reticulum. Trends Pharmacol. Sci. **16**, 98–105 (1995)

R. Vemuri, M.E. Haberland, D. Fong, K.D. Philipson, Identification of the cardiac sarcolemmal Na^+-Ca^{2+} exchanger using monoclonal antibodies. J. Membr. Biol. **118**, 279–283 (1990)

W.G. Wier, P. Hess, Excitation-contraction coupling in cardiac Purkinje fibers. Effects of cardiotonic steroids on the intracellular $[Ca^{2+}]$ transient, membrane potential, and contraction. J. Gen. Physiol. **83**, 395–415 (1984)

Z. Xie, A. Askari, Na^+/K^+-ATPase as a signal transducer. Eur. J. Biochem. **269**, 2434–2439 (2002)

Y. Yu, I. Fantozzi, C.V. Remillard, J.W. Landsberg, N. Kunichika, O. Platoshyn, D.D. Tigno, P.A. Thistlethwaite, L.J. Rubin, J.X. Yuan, Enhanced expression of transient receptor potential channels in idiopathic pulmonary arterial hypertension. Proc. Natl. Acad. Sci. U. S. A. **101**, 13861–13866 (2004)

Z. Zhang, C. Cao, W. Lee-Kwon, T.L. Pallone, Descending vasa recta pericytes express voltage operated Na^+ conductance in the rat. J. Physiol. **567**, 445–457 (2005)

S. Zhang, H. Dong, L.J. Rubin, J.X. Yuan, Upregulation of Na^+/Ca^{2+} exchanger contributes to the enhanced Ca^{2+} entry in pulmonary artery smooth muscle cells from patients with idiopathic pulmonary arterial hypertension. Am. J. Physiol. Cell Physiol. **292**, C2297–C2305 (2007a)

S. Zhang, H.H. Patel, F. Murray, C.V. Remillard, C. Schach, P.A. Thistlethwaite, P.A. Insel, J.X. Yuan, Pulmonary artery smooth muscle cells from normal subjects and IPAH patients show divergent cAMP-mediated effects on TRPC expression and capacitative Ca^{2+} entry. Am. J. Physiol. Lung Cell. Mol. Physiol. **292**, L1202–L1210 (2007b)

A. Zulian, S.G. Baryshnikov, C.I. Linde, J.M. Hamlyn, P. Ferrari, V.A. Golovina, Upregulation of Na^+/Ca^{2+} exchanger and TRPC6 contributes to abnormal Ca^{2+} homeostasis in arterial smooth muscle cells from Milan hypertensive rats. Am. J. Physiol. Heart Circ. Physiol. **299**, H624–H633 (2010a)

A. Zulian, M.V. Pulina, C.I. Linde, S. Baryshnikov, J.M. Hamlyn, M.P. Blaustein, V.A. Golovina, Digoxin blocks ouabain-induced hypertension by preventing up-regulation of the Na/Ca exchanger-1/TRPC6 Ca^{2+} signaling pathway. Hypertension **56**, e57 (2010b)

20 Years from NCX Purification and Cloning: Milestones

2

Debora A. Nicoll, Michela Ottolia,
Joshua I. Goldhaber, and Kenneth D. Philipson

Abstract

The Na^+/Ca^{2+} exchanger protein was first isolated from cardiac sarcolemma in 1988 and cloned in 1990. This allowed study of Na^+/Ca^{2+} exchange at the molecular level to begin. I will review the story leading to the cloning of NCX and the research that resulted from this event. This will include structure-function studies such as determination of the numbers of transmembrane segments and topological arrangement. Information on ion transport sites has been gathered from site-directed mutagenesis. The regions involved in Ca^{2+} regulation have been identified, analyzed, and crystallized.

We have also generated genetically altered mice to study the role of NCX in the myocardium. Of special interest are mice with atrial- or ventricular-specific KO of NCX that reveal new information on the role of NCX in excitation-contraction coupling and in cardiac pacemaker activity.

Keywords

Na^+/Ca^{2+} exchange • Ca^{2+} transport • Excitation-contraction coupling • Ca^{2+} regulation • Na^+ transport • Cardiac pacemaking

2.1 Introduction

This chapter will provide a brief overview of the progress in the development and use of molecular tools to study the Na^+/Ca^{2+} exchanger, NCX1. Focus will be on advances made in the laboratory of the author. This is not to mean that major breakthroughs did not occur in other laboratories. For example, Na^+/Ca^{2+} exchange research was initiated by seminal measurements in cardiac tissue (Reuter and Seitz 1968) and in the squid axon

D.A. Nicoll • M. Ottolia • K.D. Philipson (✉)
Departments of Physiology and Medicine and the Cardiovascular Research Laboratories MRL 3-645, David Geffen School of Medicine at UCLA, Los Angeles, CA 90095 1760, USA
e-mail: kphilipson@mednet.ucla.edu

J.I. Goldhaber
Cedars-Sinai Heart Institute, 8700 Beverly Boulevard Davis Research Building 2065, Los Angeles, CA 90048, USA

L. Annunziato (ed.), *Sodium Calcium Exchange: A Growing Spectrum of Pathophysiological Implications*, Advances in Experimental Medicine and Biology 961, DOI 10.1007/978-1-4614-4756-6_2,

(Baker et al. 1969). Study of regulation of the exchanger was largely initiated and developed by a large body of work from Reinaldo DiPolo and Luis Beauge (2006). Reeves and Sutko (1979) stimulated the application of biochemical techniques to study Na^+/Ca^{2+} exchange with the development of an assay using cardiac sarcolemmal vesicles. Na^+/Ca^{2+} exchange research received a strong boost by the development of the giant excised patch technique by Don Hilgemann (1990). All of these advances were essential to stimulate the research described below.

The exchanger that has received the most attention at the molecular level is that present in cardiac sarcolemma, now known as NCX1, which will be the focus of this chapter. The application of molecular tools to cardiac physiology has also been a productive area of research and some examples will be presented.

2.2 Isolation of the Na^+/Ca^{2+} Exchange Protein

Isolation of the Na^+/Ca^{2+} exchange protein was a nontrivial task. The exchanger is in low abundance and membrane proteins are almost always difficult with which to work. The protein must first be solubilized with detergent prior to any fractionation. In the case of the exchanger, there was no way to identify the protein except by function. So, every fraction from the separation procedures needed to be reconstituted into liposomes to be assayed for activity using a $^{45}Ca^{2+}$ uptake assay.

The successful isolation procedure (Philipson et al. 1988) identified a 120 kDa protein as the Na^+/Ca^{2+} exchanger. The protein migrated as a 160 kDa band under nonreducing conditions due to disulfide bond formation. Proteolysis reduced the size to 70 kDa though functional exchange activity was retained. However, the quantity of protein that could be isolated was only several micrograms – not enough for most biochemical approaches. The one useful purpose for the isolated exchanger protein was to produce anti-exchanger polyclonal antibodies.

2.3 Cloning of the Na^+/Ca^{2+} Exchanger

We made use of our polyclonal antibody to clone the Na^+/Ca^{2+} exchanger (Nicoll et al. 1990). We screened a λ phage expression library to find a partial clone that expressed a portion of a protein that reacted with the antibodies. After much effort, we found a full-length clone by screening an unamplified homemade library. At this point, we were unsure if the protein encoded by the clone was truly the Na^+/Ca^{2+} exchanger. This was because the identity of the clone was dependent on the quality of our antibodies and the antigen used to produce the antibodies had not been completely pure. cRNA was synthesized from the clone and injected into oocytes. The injected oocytes displayed Na^+ gradient-dependent uptake of $^{45}Ca^{2+}$ confirming that the first exchanger, NCX1, had been cloned (Nicoll et al. 1990). Subsequently, we cloned NCX2 (Li et al. 1994) and NCX3 (Nicoll et al. 1996b). We and others have noticed the presence of several splice variants of the NCX clones.

DNA sequencing revealed an open-reading frame of 970 amino acids with multiple hydrophobic segments. The initial 32 amino acids represent a cleaved leader peptide and are not present in the mature protein. One site (Asn9) is glycosylated. Hydropathy analysis suggested the presence of 12 transmembrane segments (TMSs) though experimental analysis indicates 9 TMSs (Nicoll et al. 1999; Iwamoto et al. 2000). A large intracellular hydrophilic loop was present between the N- and C-terminal groups of TMSs (Fig. 2.1).

We noted a region at the beginning of the large intracellular loop with characteristics similar to a calmodulin-binding region. Subsequently, we found that a 20 amino acid peptide with this sequence was a potent inhibitor of NCX (Li et al. 1991). The peptide is known as XIP (exchanger inhibitory peptide). It appears that the endogenous XIP region of NCX is involved in autoregulation though the exact role of the XIP region is still not fully defined. Mutations of the XIP region have major effects on a process known as Na^+-dependent inactivation (Matsuoka et al. 1997).

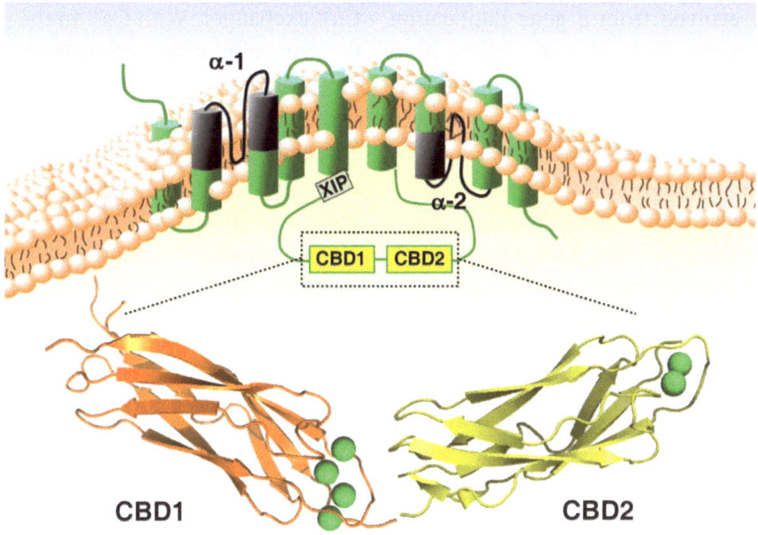

Fig. 2.1 *Two-dimensional model of the Na⁺/Ca²⁺ exchanger (NCX1)*. The extracellular surface is on *top* and shown are 9 TMSs. The α repeat regions (α-1 and α-2), critical for transport, are in *black*. The XIP region is involved in autoregulation and binds PIP_2. The two Ca^{2+}-binding domains (CBD1 and CBD2) regulate exchange activity. The structures of CBD1 and CBD2 are known and are shown as ribbon diagrams as determined by X-ray crystallography. *Green* balls represent bound Ca^{2+}

Another feature of the large intracellular loop is a region of extensive alternative splicing. Different combinations of six small exons denoted A through F are expressed in a tissue specific manner. Exons A and B are mutually exclusive and one or the other is always present. The presence of a specific splice variant bestows distinctive regulatory properties to an exchanger though the effects are often subtle and of unclear physiological significance.

2.4 Structure and Function

A current model of the exchanger NCX1 with 9 TMSs and a large intracellular loop is shown in Fig. 2.1. The model is based on cysteine accessibility experiments and epitope mapping (Nicoll et al. 1999; Iwamoto et al. 2000). Interestingly, a bacterial homologue of NCX has been determined to have 10 TMSs (Saaf et al. 2001). It remains to be confirmed whether the mammalian and bacterial homologues truly have different topologies.

2.5 Helix Packing

The topology studies give us an idea of the two dimensional arrangement of NCX. We have used a crosslinking approach to begin to obtain information on the three dimensional organization of the TMSs. The approach determines the proximity of an individual TMS to other TMSs. A series of studies provides a model (Ren et al. 2010) in which the N-terminal group of TMSs is segregated from the C-terminal group of TMSs.

2.6 α Repeats and Mutational Analysis

The α repeats (α-1 and α-2) of NCX are two regions of intramolecular homology (Schwartz and Benzer 1997). As shown in Fig. 2.1, α 1 spans portions of TMSs 2 and 3 while α-2 encompasses a portion of TMS7 and a portion of a reentrant loop between TMSs 7 and 8. The intramolecular homology strongly suggests that the exchanger evolved from a gene duplication event. The fact that only the α repeats have retained intramolecular homology through evolution suggests that these regions have key roles in the ion translocation process. Mutational analysis confirms this assertion. Mutation of several residues in the α repeats has strong effects on transport properties (Nicoll et al. 1996a; Ottolia et al. 2005; also see chapter by M. Ottolia and K.D. Philipson in this volume). We have characterized mutations of the α repeats that eliminate exchange activity or alter Na^+ and Ca^{2+} affinities.

It is also striking that homologous α repeats are present in all classes of the cation/Ca^{2+} superfamily (Cai and Lytton 2004). Thus, the α repeats display both intra- and intermolecular homology. The α repeats are present, for example, in the NCXs, NCKXs, and in prokaryotic exchangers.

2.7 Regulation

The most prominent regulatory mechanisms of NCX1 are induced by Na^+ and Ca^{2+}. That is, in addition to being the substrates for transport, Na^+ and Ca^{2+} have separate regulatory roles. Although regulation of Na^+/Ca^{2+} exchange activity by Ca^{2+} has been studied for quite some time (DiPolo 1979), the development of the giant excised patch technique by Hilgemann (1990) greatly facilitated study of regulation at the molecular level. The use of the giant patch to measure exchanger currents allowed Na^+ regulation (Na^+-dependent inactivation) of the exchanger to be described for the first time. Although experimentally Na^+ regulation can be quite prominent, the physiological role of this process is unknown. Na^+ regulation may be involved in the effects of PIP_2 on exchange activity (Hilgemann and Ball 1996). It was fortuitous that the cloning of NCX and the development of the giant patch were both described in 1990. The giant patch approach was well suited for analysis of mutant exchangers expressed in *Xenopus* oocytes.

Ca^{2+} regulation is also most readily studied in giant excised patches. In a typical experiment, a giant patch of plasma membrane is excised from an oocyte expressing a wild type or mutant Na^+/Ca^{2+} exchanger. With Ca^{2+} in the patch pipette at the extracellular surface, an exchange current can be elicited by rapid application of Na^+ into the bath at the intracellular surface. However, having Na^+ and Ca^{2+} on opposite sides of the membrane is not sufficient by itself to initiate exchange. A low concentration of Ca^{2+} must be added to the Na^+ to bind to a regulatory site to activate exchange activity (Hilgemann 1990). With an exchanger clone in hand, we set out to locate the Ca^{2+} regulatory site.

We analyzed an exchanger from which we had deleted a large portion of the intracellular loop (Matsuoka et al. 1993). Loop deletion eliminated Ca^{2+} regulation and we speculated that a Ca^{2+} regulatory site was present on the large intracellular loop. This was the first published study using a mutated Na^+/Ca^{2+} exchanger. Subsequently, we employed the $^{45}Ca^{2+}$ overlay technique to localize a Ca^{2+}-binding region within the large intracellular loop (Levitski et al. 1994). We found a region of about 140 amino acids that bound Ca^{2+} with high affinity. This region has excellent overlap with what is now known as the first Ca^{2+}-binding domain (CBD1). Mutations of specific acidic amino acid residues eliminated Ca^{2+} binding.

Although we had found a Ca^{2+}-binding region, we initially had no evidence that this region was involved in Ca^{2+} regulation. To address this question, we combined site-directed mutagenesis with electrophysiological measurements using the excised patch technique. We found an excellent correlation between mutations that eliminated the binding of $^{45}Ca^{2+}$ and mutations that disrupted Ca^{2+} regulation (Matsuoka et al. 1995). Thus, we were confident that we had identified a region of the Na^+/Ca^{2+} exchanger that was involved in Ca^{2+} regulation.

An advance in the understanding of NCX Ca^{2+} regulation was the discovery by Hilge et al. (2006) that there was a second Ca^{2+} binding region adjoining the region that we had initially identified. The two Ca^{2+}-binding regions are now known as Ca^{2+}-binding domains 1 and 2 (CBD1 and CBD2). These two domains are the same as the ß repeats noted by Schwartz and Benzer (1997) from sequence analysis. Hilge et al. (2006) also determined structures of CBD1 and CBD2 by NMR. The overall structures of CBD1 and CBD2 are similar comprised of a ß sandwich fold with the Ca^{2+} bound by connecting loops. Shortly thereafter, we determined crystal structures of both CBDs that detailed features of the Ca^{2+}-binding sites (Nicoll et al. 2006; Besserer et al. 2007). Knowing the nature of the Ca^{2+}-binding sites allowed detailed mutational studies to probe the relative roles of CBD1 and CBD2 (Ottolia et al. 2009).

CBD1 is the more dominant player in Ca^{2+} regulation and is primary in determining the apparent Ca^{2+} affinity for regulation. We find that an intact CBD2 is also essential in regulation though the exact role of CBD2 Ca^{2+} binding is unclear. Overall, our current understanding of NCX Ca^{2+} regulation is incomplete. Essentially nothing is known of how the binding of regulatory Ca^{2+} is transduced to the transmembrane segments.

We have been successful in using fluorescent resonance energy transfer (FRET) techniques to study Ca^{2+}-induced conformational changes of the exchanger. We have used two approaches. First, we have coexpressed in *Xenopus* oocytes full-length exchangers in which either CFP or YFP has been inserted into the large intracellular loop (John et al. 2011). We then monitor fluorescence generated at the plasma membrane using a novel preparation of sheets of plasma membrane (Ottolia et al. 2007). We detect FRET indicating that exchangers are in close proximity to one another. Analysis indicates the presence of NCX dimers. A striking finding was that the interaction between exchangers was very sensitive to the level of regulatory Ca^{2+}. That is, conformational changes induced by the binding of Ca^{2+} changed the proximity of at least a portion of the intracellular loops of adjacent exchangers. Further research will resolve the significance of this finding.

Second, we have applied FRET to the study of individual CBDs (John et al. 2011). We expressed CBD1, CBD2, or CBD12 (CBD1 and CBD2 linked as occurs in the intact exchanger) with CFP and YFP attached to both the N- and C-termini of the protein. Each construct was targeted to the plasma membrane of the oocyte with an appropriate tag. This allowed us to ascertain the Ca^{2+} dependence of Ca^{2+} binding by monitoring conformational changes by FRET. As expected from previous studies, CBD1 bound Ca^{2+} with a higher affinity than CBD2. Unexpectedly, CBD12 bound Ca^{2+} with a substantially higher affinity than either CBD1 or CBD2. The data were most consistent with a model in which the interaction of CBD2 with CBD1 increased the binding affinity of CBD1 for Ca^{2+}. The higher binding affinity of CBD12 reflects the Ca^{2+}-binding properties of the intact Na^+/Ca^{2+} exchanger.

2.8 Cardiac-Specific Knockout of the Na^+/Ca^{2+} Exchanger

In addition to structure/function studies, we also do physiological studies using mice with genetically altered levels of the Na^+/Ca^{2+} exchanger. Some of our more interesting experiments have used mice with a cardiac-specific knockout of the exchanger. Three studies are of special interest:

1. We created a ventricular-specific knockout of the Na^+/Ca^{2+} exchanger (Henderson et al. 2004). Ca^{2+} enters cardiac myocytes with each excitation through the voltage-dependent Ca^{2+} channel. To maintain Ca^{2+} homeostasis, an equal amount of Ca^{2+} must be extruded from the cell and the exchanger is the primary Ca^{2+} efflux mechanism. Surprisingly, the mouse survives under these conditions but the myocardium undergoes remarkable adaptations to allow survival. One striking adaptation is that Ca^{2+} influx is markedly decreased by 80% by downregulation of the L type Ca^{2+} current and by shortening of the action potential (Pott et al. 2005). With a diminution of Ca^{2+} influx, a powerful Ca^{2+} efflux mechanism is no longer needed. Under these conditions, the PMCA is an adequate Ca^{2+} efflux mechanism. The mechanism of downregulation of the Ca^{2+} channel is of interest: In the absence of the exchanger, Ca^{2+} appears to accumulate in the diadic cleft and directly inactivate Ca^{2+} channels. Thus, we have uncovered an important autoregulatory mechanism. When Ca^{2+} efflux is limited, there is feedback to limit Ca^{2+} influx (Pott et al. 2007).

2. There has been a controversy in the field of cardiac excitation-contraction coupling regarding the role of the Na^+/Ca^{2+} exchanger in Ca^{2+} influx. More specifically, it has been hypothesized that Na^+ influx during the action potential causes a local increase in Na^+ that transiently reverses the exchanger and the resultant Ca^{2+} influx contributes to the Ca^{2+}-induced Ca^{2+} release process (LeBlanc and Hume 1990). To test this controversial hypothesis, we examined the effects of blockade of the Na^+ channel in myocytes isolated from wild type mice and from mice with a cardiac-specific KO of the Na^+/Ca^{2+} exchanger. Blockade of the Na^+ channel had a substantial effect on the action potential-induced Ca^{2+} transient in wild type myocytes but had no effect on the KO myocytes (Larbig et al. 2010; also see chapter by J. Goldhaber et al. in this volume). The results strongly support a role for Ca^{2+} influx mediated by the Na^+/Ca^{2+} exchanger as an important contributor to cardiac contraction.
3. The KO experiments described above make use of a ventricular-specific KO of the exchanger. We have now also generated a mouse with an atrial-specific KO of the Na^+/Ca^{2+} exchanger including the sino-atrial (SA) node. This is of particular interest as there has been much recent, and controversial, work indicating a key role for the exchanger in generating the pacemaker current of the SA node (Lakatta et al. 2010). Consistent with this hypothesis, we find the SA node of the KO mice to be electrically silent and no P wave (due to atrial depolarization) is evident on electrocardiograms. These results need to be confirmed but preliminarily our results are consistent with the exchanger being an essential component of cardiac pacemaking.

2.9 Final Comment

The cloning of NCX1 in 1990 opened the door to molecular studies of the Na^+/Ca^{2+} exchanger. It will be exciting to follow futures advances in our understanding of this intriguing molecule.

Acknowledgements We have had many talented colleagues over the years. Some of those who participated in the experiments presented here include Drs. Xiaoyan Ren, Satoshi Matsuoka, Larry Hryshko, Dmitri Levitski, Don Hilgemann, Jeff Abramson, Christian Pott, Robert Larbig, and Sabine Groenke.

References

P.F. Baker, M.P. Blaustein, A.L. Hodgkin, R.A. Steinhardt, The influence of calcium on sodium efflux in squid axons. J. Physiol. **200**, 431–458 (1969)

G.M. Besserer, M. Ottolia, D.A. Nicoll, V. Chaptal, D. Cascio, K.D. Philipson, J. Abramson, The second Ca^{2+}-binding domain of the Na^+-Ca^{2+} exchanger is essential for regulation: Crystal structures and mutational analysis. Proc. Natl. Acad. Sci. U. S. A. **104**, 18467–18472 (2007)

X. Cai, J. Lytton, The cation/Ca^{2+} exchanger superfamily: phylogenetic analysis and structural implications. Mol. Biol. Evol. **21**, 1692–1703 (2004)

R. DiPolo, Calcium influx in internally dialyzed squid giant axons. J. Gen. Physiol. **73**, 91–113 (1979)

R. DiPolo, L. Beauge, Sodium/calcium exchanger: influence of metabolic regulation on ion carrier interactions. Physiol. Rev. **86**, 155–203 (2006)

S.A. Henderson, J.I. Goldhaber, J.M. So, T. Han, C. Motter, A. Ngo, C. Chantawansri, M.R. Ritter, M. Friedlander, D.A. Nicoll, J.S. Frank, M.C. Jordan, K.P. Roos, R.S. Ross, K.D. Philipson, Functional adult myocardium in the absence of Na^+-Ca^{2+} exchange: cardiac-specific knockout of NCX1. Circ. Res. **95**, 604–611 (2004)

M. Hilge, J. Aelen, G.W. Vuister, Ca^{2+} regulation in the Na^+/Ca^{2+} exchanger involves two markedly different Ca^{2+} sensors. Mol. Cell **22**, 15–25 (2006)

D.W. Hilgemann, Regulation and deregulation of cardiac Na^+-Ca^{2+} exchange in giant excised sarcolemmal membrane patches. Nature **344**, 242–245 (1990)

D.W. Hilgemann, R. Ball, Regulation of cardiac Na^+, Ca^{2+} exchange and KATP potassium channels by PIP2. Science **273**, 956–959 (1996)

T. Iwamoto, A. Uehara, I. Imanaga, M. Shigekawa, The Na^+/Ca^{2+} exchanger NCX1 has oppositely oriented reentrant loop domains that contain conserved aspartic acids whose mutation alters its apparent Ca^{2+} affinity. J. Biol. Chem. **275**, 38571–38580 (2000)

S.A. John, B. Ribalet, J.N. Weiss, K.D. Philipson, M. Ottolia, Ca^{2+}-dependent structural rearrangements within Na^+-Ca^{2+} exchanger dimers. Proc. Natl. Acad. Sci. U. S. A. **108**, 1699–1704 (2011)

E.G. Lakatta, V.A. Maltsev, T.M. Vinogradova, A coupled SYSTEM of intracellular Ca^{2+} clocks and surface membrane voltage clocks controls the timekeeping mechanism of the heart's pacemaker. Circ. Res. **106**, 659–673 (2010)

R. Larbig, N. Torres, J.H. Bridge, J.I. Goldhaber, K.D. Philipson, Anno activation of reverse Na^+-Ca^{2+} exchange by the Na^+ current augments the cardiac Ca^{2+} transient: evidence from NCX knockout mice. J. Physiol. **588**, 3267–3276 (2010)

N. Leblanc, J.R. Hume, Sodium current-induced release of calcium from cardiac sarcoplasmic reticulum. Science **248**, 372–376 (1990)

D.O. Levitsky, D.A. Nicoll, K.D. Philipson, Identification of the high affinity Ca^{2+}-binding domain of the cardiac Na^+-Ca^{2+} exchanger. J. Biol. Chem. **269**, 22847–22852 (1994)

Z. Li, D.A. Nicoll, A. Collins, D.W. Hilgemann, A.G. Filoteo, J.T. Penniston, J.N. Weiss, J.M. Tomich, K.D. Philipson, Identification of a peptide inhibitor of the cardiac sarcolemmal Na^+-Ca^{2+} exchanger. J. Biol. Chem. **266**, 1014–1020 (1991)

Z. Li, S. Matsuoka, L.V. Hryshko, D.A. Nicoll, M.M. Bersohn, E.P. Burke, R.P. Lifton, K.D. Philipson, Cloning of the NCX2 isoform of the plasma membrane Na^+-Ca^{2+} exchanger. J. Biol. Chem. **269**, 17434–17439 (1994)

S. Matsuoka, D.A. Nicoll, R.F. Reilly, D.W. Hilgemann, K.D. Philipson, Initial localization of regulatory regions of the cardiac sarcolemmal Na^+-Ca^{2+} exchanger. Proc. Natl. Acad. Sci. U. S. A. **90**, 3870–3874 (1993)

S. Matsuoka, D.A. Nicoll, L.V. Hryshko, D.O. Levitsky, J.N. Weiss, K.D. Philipson, Regulation of the cardiac Na^+-Ca^{2+} exchanger by Ca^{2+}. Mutational analysis of the Ca^{2+}-binding domain. J. Gen. Physiol. **105**, 403–420 (1995)

S. Matsuoka, D.A. Nicoll, Z. He, K.D. Philipson, Regulation of cardiac Na^+-Ca^{2+} exchanger by the endogenous XIP region. J. Gen. Physiol. **109**, 273–286 (1997)

D.A. Nicoll, S. Longoni, K.D. Philipson, Molecular cloning and functional expression of the cardiac sarcolemmal Na^+-Ca^{2+} exchanger. Science **250**, 562–565 (1990)

D.A. Nicoll, L.V. Hryshko, S. Matsuoka, J.S. Frank, K.D. Philipson, Mutation of amino acid residues in the putative transmembrane segments of the cardiac sarcolemmal Na^+-Ca^{2+} exchanger. J. Biol. Chem. **271**, 13385–13391 (1996a)

D.A. Nicoll, B.D. Quednau, Z. Qui, Y.R. Xia, A.J. Lusis, K.D. Philipson, Cloning of a third mammalian Na^+-Ca^{2+} exchanger, NCX3. J. Biol. Chem. **271**, 24914–24921 (1996b)

D.A. Nicoll, M. Ottolia, L. Lu, Y. Lu, K.D. Philipson, A new topological model of the cardiac sarcolemmal Na^+-Ca^{2+} exchanger. J. Biol. Chem. **274**, 910–917 (1999)

D.A. Nicoll, M. Sawaya, S. Kwon, D. Cascio, K.D. Philipson, J. Abramson, The crystal structure of the primary Ca^{2+} sensor of the Na^+/Ca^{2+} exchanger reveals a novel Ca^{2+} binding motif. J. Biol. Chem. **281**, 21577–21581 (2006)

M. Ottolia, D.A. Nicoll, K.D. Philipson, Mutational analysis of the alpha-1 repeat of the cardiac Na^+-Ca^{2+} exchanger. J. Biol. Chem. **280**, 1061–1069 (2005)

M. Ottolia, K.D. Philipson, S. John, Xenopus oocyte plasma membrane sheets for FRET analysis. Am. J. Physiol. Cell Physiol. **292**, C1519–C1522 (2007)

M. Ottolia, D.A. Nicoll, K.D. Philipson, Roles of two Ca^{2+}-binding domains in regulation of the cardiac Na^+-Ca^{2+} exchanger. J. Biol. Chem. **284**, 32735–32741 (2009)

K.D. Philipson, S. Longoni, R. Ward, Purification of the cardiac Na^+-Ca^{2+} exchange protein. Biochim. Biophys. Acta **945**, 298–306 (1988)

C. Pott, K.D. Philipson, J.I. Goldhaber, Excitation-contraction coupling in Na^+-Ca^{2+} exchanger knockout mice: reduced transsarcolemmal Ca^{2+} flux. Circ. Res. **97**, 1288–1295 (2005)

C. Pott, M. Yip, J.I. Goldhaber, K.D. Philipson, Regulation of cardiac L-type Ca^{2+} current in Na^+-Ca^{2+} exchanger knockout mice: functional coupling of the Ca^{2+} channel and the Na^+-Ca^{2+} exchanger. Biophys. J. **92**, 1431–1437 (2007)

J.P. Reeves, J.L. Sutko, Sodium-calcium ion exchange in cardiac membrane vesicles. Proc. Natl. Acad. Sci. U. S. A. **76**, 590–594 (1979)

X. Ren, D.A. Nicoll, L. Xu, Z. Qu, K.D. Philipson, Transmembrane segment packing of the Na^+/Ca^{2+} exchanger investigated with chemical cross-linkers. Biochemistry **49**, 8585–8591 (2010)

H. Reuter, N. Seitz, The dependence of calcium efflux from cardiac muscle on temperature and external ion composition. J. Physiol. **195**, 451–470 (1968)

A. Saaf, L. Baars, G. von Heijne, The internal repeats in the Na^+/Ca^{2+} exchanger-related *Escherichia coli* protein YrbG have opposite membrane topologies. J. Biol. Chem. **276**, 18905–18907 (2001)

E.M. Schwarz, S. Benzer, Calx, a Na-Ca exchanger gene of *Drosophila* melanogaster. Proc. Natl. Acad. Sci. U. S. A. **94**, 10249–10254 (1997)

Part II

Structural and Functional Aspects of NCX

Ca^{2+} Regulation in the Na^+/Ca^{2+} Exchanger Features a Dual Electrostatic Switch Mechanism

3

Mark Hilge

Abstract

Ion transport performed by the Na^+/Ca^{2+} exchanger (NCX) is regulated via its cytosolic Ca^{2+}-binding domains, CBD1 and CBD2, which act as sensors for intracellular Ca^{2+}. Striking differences in the electrostatic potential of the Ca^{2+}-bound and Ca^{2+}-free forms turn the CBD1 and CBD2 Ca^{2+}-binding sites into electrostatic switches similar to those of C_2 domains. Binding of Ca^{2+} with high affinity to CBD1 induces a conformational change that is relayed to the transmembrane domain and thereby initiates Na^+/Ca^{2+} exchange. The Ca^{2+} concentration at which this conformational change occurs is determined by the Ca^{2+} affinities of the strictly conserved CBD1 Ca^{2+}-binding sites that are modulated by an adjacent, variable region of CBD2. In contrast, the Ca^{2+}-binding properties of CBD2 depend on the isoform and the type of residues in the Ca^{2+}-binding sites, encoded by a mutually exclusive exon. This second electrostatic switch, formed by CBD2, appears to be required for sustained Na^+/Ca^{2+} exchange and may allow tailored, tissue-specific exchange activities.

Keywords

Na^+/Ca^{2+} exchanger • Ca^{2+} binding domain • Ca^{2+} sensor • Ca^{2+} regulation • Electrostatic switch • Alternative splicing

3.1 General Overview

Over the last 5 years, high-resolution NMR (Hilge et al., 2006, 2009) and X-ray (Besserer et al., 2007; Nicoll et al., 2006; Wu et al., 2009, 2010, 2011) structures of a large part of the regulatory loop in the Na^+/Ca^{2+} exchanger (NCX) have added a new dimension to our understanding of the regulation of this Ca^{2+} transporter. Structurally, NCX consists of an α-helical transmembrane domain and a large, approximately 500-residue-long, cytosolic loop. There are two Ca^{2+}-binding domains in this loop, in the following referred to as CBD1 and CBD2 (Hilge et al., 2006). These domains regulate ion transport

M. Hilge (✉)
Center for Cellular Imaging and Nano Analytics (C-CINA), Biozentrum, University Basel, Mattenstrasse 26, Basel CH-4058, Switzerland
e-mail: mark.hilge@unibas.ch

L. Annunziato (ed.), *Sodium Calcium Exchange: A Growing Spectrum of Pathophysiological Implications*, Advances in Experimental Medicine and Biology 961, DOI 10.1007/978-1-4614-4756-6_3,

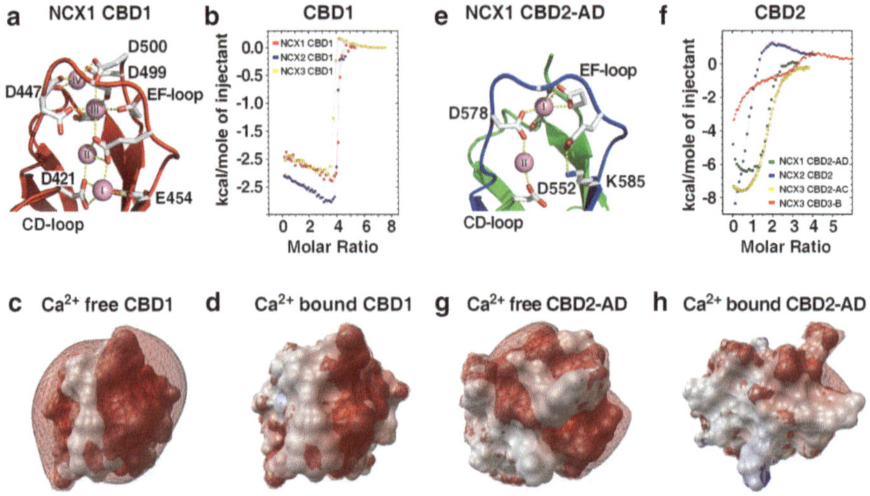

Fig. 3.1 (**a**) and (**e**) display the CBD1 (PDB code: 2DPK) and CBD2 (PDB code: 2QVM) Ca^{2+}-binding sites in the Ca^{2+}-bound form. (**b**) and (**e**) display the binding isotherms of the various forms of CBD1 and CBD2. (**c**), (**d**), (**g**) and (**h**) show the electrostatic potentials of the CBD1 and CBD2 Ca^{2+}-binding sites in the absence and presence of Ca^{2+}. The red and blue meshes represent isocontours of the electrostatic potential at −5 and +5 kT/e (Figure modified from Hilge (2009))

across the plasma membrane. Their structure has been determined in both the Ca^{2+}-bound (Figs. 3.1a, d, e, h) and the Ca^{2+}-free forms (Figs. 3.1c, g) revealing a β-sandwich architecture. Subsequent X-ray structures have provided a more detailed view of the Ca^{2+}-binding sites and in the case of CBD1, revealed the presence of four Ca^{2+} ions (Fig. 3.1a) bound in an arrangement reminiscent of C_2 domains (Nicoll et al., 2006). In contrast, due to Lys585 and the absence of acidic counterparts of Asp499 and Asp500 of CBD1, the Ca^{2+}-binding sites of CBD2-AD (Fig. 3.1e), where AD denotes the presence of residues encoded by exons A and D, show a reduced density of negative charges. Hence, CBD2-AD coordinates only two Ca^{2+} ions (Besserer et al., 2007; Hilge et al., 2006). In the NCX1 CBD2-AD structure (Fig. 3.1e), residues encoded by the mutually exclusive exon A or B contribute half of the ion-coordinating atoms in the Ca^{2+}-binding sites. Residues encoded by the cassette exons C–F on the other hand are located at the opposite end of the domain facing the CBD1 Ca^{2+}-binding sites. Furthermore, analysis of the Ca^{2+}-binding sites by [^{1}H, ^{15}N]-HSQC NMR spectra suggested that there is a substantial loss of structural integrity in the absence of Ca^{2+} for CBD1 but only a modest degree of disorder in the Ca^{2+}-binding sites of CBD2-AD (Hilge et al., 2006). These preliminary indications were confirmed by an X-ray structure of the Ca^{2+}-free form of CBD1 (PDB code: 3E9T, chains C and D) (Wu et al., 2010), where no electron density was visible for residues corresponding to Asp447-Glu451 and Asp498-Asp500, respectively. In case of CBD2-AD, the NMR structure of the apo-form (PDB code: 2KLS) (Hilge et al., 2009) showed that Lys585 crucially points into Ca^{2+}-binding site II where it forms a salt bridge with Asp552 and thereby stabilizes the Ca^{2+}-binding sites. Intriguingly, this NMR structure determined at neutral pH does not reflect a state as rigid and locked as the one suggested by a recent X-ray structure of CBD2-AD (PDB code: 2QVK) (Besserer et al., 2007) obtained from crystals grown at pH 4.9. Since the largest functional differences among all described forms of NCX occur between NCX1 splice variants encoded by the mutually exclusive exons A and B, it was tempting to also determine the structure of CBD2-BD. Strikingly, the Ca^{2+}-binding sites of this splice variant were completely disordered, even in the presence of 10 mM $CaCl_2$ (Hilge et al., 2009). The most recent structural contribution describes the interface between CBD1 and CBD2 in the *Drosophila* exchangers CalX1.1 and CalX1.2 in the Ca^{2+}-bound form (Wu et al., 2011). Remarkably, CalX is the only exchanger that is inhibited rather than activated upon Ca^{2+} binding to the CBDs.

Using the CBDs as constraints resulted in a model for the intact exchanger in which the Ca^{2+}-binding sites of CBD1 are approximately 90 Å away from the transport Ca^{2+}-binding site in the transmembrane domain. In contrast, the Ca^{2+}-binding sites of CBD2 are close to a predicted third domain that shows homology with α-catenin.

3.2 Ca^{2+}-Binding Determinants of CBD1 and CBD2

The thermodynamic properties of the Ca^{2+}-binding sites of CBD1 and CBD2 have been determined by an extensive analysis using isothermal titration calorimetry (ITC). Initial sequence comparison of all available Na^+/Ca^{2+} exchanger sequences revealed strict conservation of the Ca^{2+} coordinating residues in CBD1. The expected similarity in their Ca^{2+}-binding properties among the three isoforms was confirmed by the titration of CBD1 of canine NCX1 and mouse NCX2 and NCX3. Indeed, the binding isotherms (Fig. 3.1b) were very similar and consistent with the coordination of four Ca^{2+} ions as in the Ca^{2+}-bound CBD1 X-ray structure of NCX1 (Nicoll et al. 2006). However, due to the steepness of the binding isotherms caused by three high-affinity Ca^{2+}-binding sites, it was not possible to confidently determine the individual macroscopic binding constants, but only an approximate range of 100–600 nM. These affinities make CBD1 the primary Ca^{2+} sensor and thus responsible for Ca^{2+} activation in NCX (Hilgemann et al., 1992b).

In strong contrast to CBD1, sequence analysis of CBD2 has revealed that residue types at positions 552, 578 and 585 located around Ca^{2+}-binding site II (Fig. 3.1e and Table 3.1) can vary depending on the isoform and splice variant. In particular, in NCX1 CBD2-AD, Asp578 crucially coordinates Ca^{2+} ions in sites I and II, while Lys585 prevents binding of a second Ca^{2+} ion on the side of the CD-loop as in CBD1 (Fig. 3.1a). Decisively, NCX1 CBD2-BD possesses an arginine and a cysteine residue at positions 578 and 585, respectively, that are both disordered in the CBD2-BD structure. In NCX2, for which no splice variants exist, replacement of Asp552 by a histidine residue probably removes one of the two sole Ca^{2+}-coordinating atoms and thereby eliminates site II in this isoform (Fig. 3.1e and Table 3.1). Three splice variants have been reported for NCX3, namely, NCX3 CBD2-AC, CBD2-B and CBD2-BC. NCX3 CBD2-AC has the same Ca^{2+}-coordinating residue types as NCX1 CBD2-AD and also appears to bind two Ca^{2+} ions. NCX3 CBD2-B on the other hand has largely adopted the Ca^{2+} coordination scheme of CBD1 between the CD- and EF-loops (Fig. 3.1a and Table 3.1) and therefore probably binds three Ca^{2+} ions. The impact of the different residue types has been experimentally characterized by ITC analyses of NCX1 CBD2-AD and CBD2-BD, NCX2 CBD2, NCX3 CBD2-AC and NCX3 CBD2-B suggesting the binding of 0–3 Ca^{2+} ions (Fig. 3.1f and Table 3.1). In agreement with the NMR structure, no heat pulses could be detected for NCX1 CBD2-BD in response to $CaCl_2$ additions as high as 100 mM, thus reflecting the inability of exon B-containing NCX1 splice variants to specifically bind Ca^{2+}. Hence, exon B-containing variants of NCX1 and NCX3 are fundamentally different from each other, while their exon A-containing variants apparently coordinate two Ca^{2+} ions in a very similar manner. Control experiments with mutants, designed to restore the Ca^{2+} coordination scheme of NCX1 CBD2-AD, revealed similar macroscopic Ca^{2+}-binding constants and indirectly confirmed the assigned functions of the isoform and splice variant-specific residues (Hilge et al., 2009).

3.3 Electrostatic Switches in CBD1 and CBD2

Changes in electrostatic potential are regarded as switches that can promote structural rearrangements (Shao et al., 1997). For instance, Ca^{2+}-binding increases the electrostatic potential of

Table 3.1 ITC-derived stoichiometries and binding constants

	552	578	585	# Ca^{2+}	K_d [μM]
NCX1 CBD2-AD	Asp	Asp	Lys	2	1, 9
NCX1 CBD2-BD	Asp	Arg	Cys	0	–
NCX2 CBD2	His	Asp	Asp	1	13
NCX3 CBD2-AC	Asp	Asp	Lys	2	0.25, 16
NCX3 CBD2-B	Asp	Glu	Glu	3	4–20
	421	447	454		
NCX1 CBD1	Asp	Asp	Glu	4	0.1–0.6

Ca^{2+}-binding sites and thereby favours attraction to negatively charged areas that in the absence of Ca^{2+} may cause repulsion. The similarities of CBD1 and CBD2 with C_2 domains and different numbers of Ca^{2+}-binding sites prompted the analysis of their electrostatic potentials in the absence and presence of Ca^{2+} (Fig. 3.1c, d, g, h). In the absence of Ca^{2+}, the Ca^{2+}-binding sites of CBD1 form a strongly negative electrostatic potential (Fig. 3.1c), caused by the extensive cluster of aspartic and glutamic acid residues (Asp421, Asp446, Asp447, Asp448, Asp498, Asp499, Asp500, Glu385, Glu451 and Glu454) and the lack of any basic residues (Fig. 3.1a). The negative potential is drastically reduced when four Ca^{2+} ions bind (Fig. 3.1d). In comparison, the CBD2-AD electrostatic potential of the Ca^{2+}-free form (Fig. 3.1g) is considerably less negative due to the lower number of acidic residues and the presence of basic residues, Arg547, Lys583 and Lys585, at the Ca^{2+}-binding site II. In the presence of Ca^{2+}, the electrostatic potential at the CBD2-AD Ca^{2+}-binding sites is approximately neutral (Fig. 3.1h). Strikingly, in exon B-encoded variants of NCX1, the electrostatic potential of CBD2 remains unchanged due to its inability to bind Ca^{2+}, thus rendering these splice variants unable to form an electrostatic switch. While Ca^{2+}-binding causes a drastic change in the potential at the CBD1 Ca^{2+}-binding sites, the electrostatic potential in CBD2 strongly depends on the isoform and splice variant as different numbers and affinities of the Ca^{2+}-binding sites determine the strength of the electrostatic switch. Predicted by SAXS analysis (Hilge et al., 2009) and supported by homology modelling of CBD12-AD using the CBD12 structures from *Drosophila* (Wu et al., 2011), the CBD1 Ca^{2+}-binding sites are likely to be close to a highly acidic region consisting of four consecutive glutamate residues (residues 622–625) in the FG-loop of CBD2. In the absence of Ca^{2+}, the high density of negative charges of the CBD1 Ca^{2+}-binding sites and these four glutamate residues may well cause repulsion, which might be alleviated upon Ca^{2+} binding. Strikingly, in the CalX exchanger, three of the four glutamate residues are absent while they are strictly conserved among mammalian NCX. As otherwise all interactions at the interface of CBD1 and CBD2 appear to be conserved between mammalian and CalX exchangers, it is tempting to speculate that residues 622–625 may be the reason for the opposite response of the exchangers to Ca^{2+} binding.

3.4 Function of Residues Encoded by the Cassette Exons

Besides the Ca^{2+}-binding sites of CBD1 and CBD2, residues encoded by the cassette exons may contribute to ion transport regulation in NCX. To investigate this possibility CBD12-AD and CBD12-ACDEF, two-domain constructs containing CBD1 and CBD2, representing the brain and heart splice variants, were titrated with $CaCl_2$. Based on the Ca^{2+}-bound structures of the individual CBDs, binding of six Ca^{2+} ions is expected for both constructs. While this was indeed the case for CBD12-AD, strikingly the ACDEF form appeared to entirely lack the high-affinity component of CBD1 and displayed only medium-affinity Ca^{2+} binding (Fig. 3.2a). This implied that residues encoded by the additional

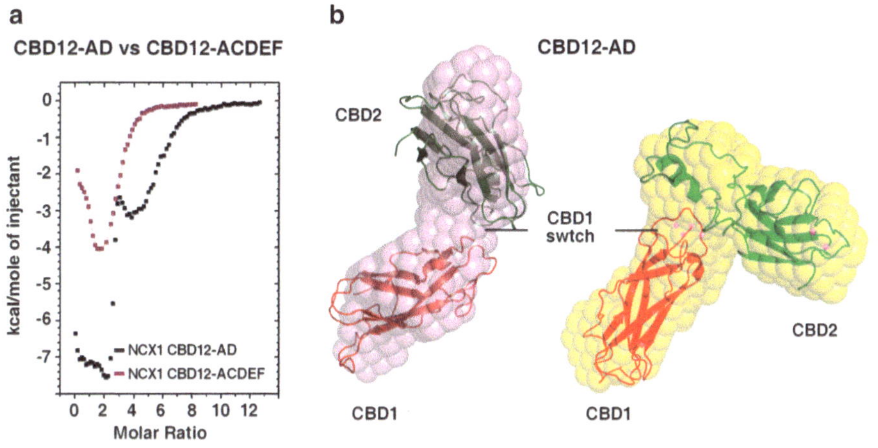

Fig. 3.2 (**a**) ITC analyses of brain CBD12-AD and heart CBD12-ACDEF. (**b**) Ab initio SAXS shape reconstructions of CBD12-AD in the absence (pink beads) and presence of Ca^{2+} (yellow beads) revealing a conformational change induced by Ca^{2+} binding to CBD1. Independently calculated rigid body models are superimposed on the beads (Figure modified from Hilge (2009))

cassette exons C, E and F, located in proximity to the CBD1 Ca^{2+}-binding sites, either destroy the high-affinity sites in CBD1 or increase their affinities substantially. In the latter case, the CBD1 Ca^{2+}-binding sites are probably already occupied prior to the start of titration. Applying microparticle-induced X-ray emission (microPIXE) using the method of Garman and Grime (Garman and Grime 2005) on CBD12-AD and CBD12-ACDEF constructs used for the ITC experiments revealed the presence of zero and four Ca^{2+} ions, respectively. Hence, Ca^{2+} is apparently trapped at the CBD1 Ca^{2+}-binding sites in CBD12-ACDEF, thus hinting at substantially higher Ca^{2+} affinities compared to those of CBD12-AD. This is in line with the fact that the ACDEF splice variant is predominantly found in the heart where high Ca^{2+} fluxes are required. Therefore, this exchanger probably enters the activated state at considerably lower intracellular Ca^{2+} concentrations or is even permanently activated.

3.5 Ca^{2+} Induces a Twist in the Hinge Between CBD1 and CBD2

To explore the effects of Ca^{2+} binding more globally, small-angle X-ray scattering (SAXS) was employed on CBD12-AD using ab initio reconstruction and rigid body modelling approaches. Independently, both approaches revealed a substantial conformational change in the form of a twist in orientation between CBD1 and CBD2 accompanied by a compaction of the domains upon Ca^{2+} binding (Fig. 3.2b). Although this conformational change may be constrained by the N- and C-terminal linker regions to the catenin-like domain (CLD) in the intact exchanger, the SAXS models likely reflect effects of Ca^{2+} binding and release at the CBD1 Ca^{2+}-binding sites.

3.6 Link to Electrophysiological Data

In pioneering work, Hilgemann and co-workers dissected the regulatory properties of cardiac NCX into Ca^{2+}-dependent activation and Na^+-dependent inactivation using the giant patch method (Hilgemann et al., 1992a, b). Ion transport of NCX was shown to be activated upon binding of Ca^{2+} ions at submicromolar Ca^{2+} concentrations, while Na^+ concentrations rising above 15 mM can lead to Na^+-dependent inactivation. However, NCX1 splice variants encoded by exon A can alleviate Na^+-dependent inactivation at elevated intracellular Ca^{2+} concentrations, while those variants encoded by the mutually

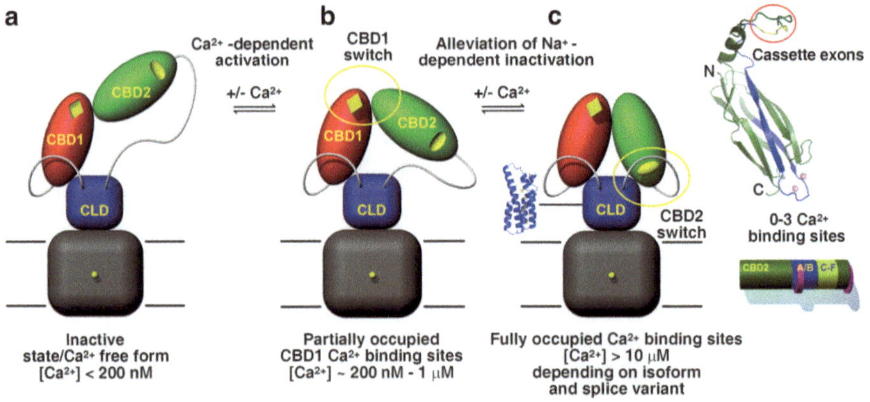

Fig. 3.3 Hypothetical dual electrostatic switch mechanism of Ca^{2+} regulation in NCX using a model consisting of four domains: Transmembrane domain (grey; residues 1–216 and 706–903), CLD (blue; residues 217–370 and 651–705), CBD1 (red; residues 371–500) and CBD2 (green; residues 501–650) with the numbering based on the canine NCX1 AD-splice variant (NCX1.4). (**a**) Inactive, Ca^{2+}-free NCX in extended conformation. (**b**) Submicromolar Ca^{2+} concentrations induce a conformational change via the electrostatic switch in CBD1 that results in a compaction of the Ca^{2+}-binding domains and probably reduces tension on the linker regions to the CLD. (**c**) Binding of Ca^{2+} to CBD2 allows sustained Na^{+}/Ca^{2+} exchange and removes counteracting Na^{+}-dependent inactivation. The CBD2 ribbon diagram (right) depicts the location of the regions encoded by the mutually exclusive exon A or B (blue) and the cassette exons (yellow). Whether Ca^{2+} binding to CBD2 induces yet another conformational change or just promotes an interaction with the CLD or the adjacent, highly conserved linker region remains elusive (Figure modified from Hilge (2009))

exclusive exon B cannot (Dyck et al., 1999; Hurtado et al., 2006). Intriguingly, exchanger variants encoded by exon A seem to be almost exclusively expressed in excitable cells, where high Ca^{2+} fluxes are needed, while exon B-encoded variants are generally restricted to non-excitable cells (Quednau et al., 1997).

3.7 Conclusions

The combination of available structural and biophysical data with results from functional studies resulted in the proposal of a dual electrostatic switch mechanism for Ca^{2+} regulation in NCX (Fig. 3.3a, b, c). Upon rise of intracellular Ca^{2+}, Ca^{2+} ions initially bind to the CBD1 Ca^{2+}-binding sites, thereby substantially increasing their electrostatic potential. This leads to a conformational change accompanied by a compaction of the CBDs as visualized by SAXS analysis (Fig. 3.2b). As a consequence, the tension on the N- and/or C-terminal linker regions to the CLD that relays Ca^{2+} binding and releases events to the transmembrane domain is most likely reduced. This initial activation step (Fig. 3.3b) is referred to as Ca^{2+}-dependent activation and initiates Na^{+}/Ca^{2+} exchange. The Ca^{2+} concentration at which NCX becomes activated probably depends on the residues encoded by the cassette exons that modulate the affinities of the CBD1 Ca^{2+}-binding sites. To sustain Na^{+}/Ca^{2+} exchange, Ca^{2+} ions must also bind to CBD2 (Fig. 3.3c), a process that is isoform and splice variant-specific. Functional and thermodynamic data for exchangers that bind 0–3 Ca^{2+} ions at the CBD2 Ca^{2+}-binding sites as well as distinct macroscopic Ca^{2+}-binding constants ranging from 250 nM to 20 μM (Table 3.1) strongly indicate that the number and affinity of Ca^{2+}-binding sites in CBD2 determine its capability to overcome Na^{+}-dependent inactivation. Hence, this isoform- and splice variant-specific Ca^{2+} relief of Na^{+}-dependent inactivation may allow the optimal adaptation of Na^{+}/Ca^{2+} exchange to meet special needs within the cell or a specific tissue. Finally, a drop in Ca^{2+} concentration below the CBD2 and CBD1 Ca^{2+} affinities reverses the previous steps and arrests NCX in an inactive state (Fig. 3.3a).

Acknowledgements I thank Jan Aelen, Drs. Anastassis Perrakis and Geerten W. Vuister for their help in the lab and NMR measurements. I would like to thank Dr. Shirley A. Müller for critically reading the manuscript. This work was supported by grants of the Netherlands Organisation for Scientific Research (NWO; grants # 700.55.443 and # 700.57.101), the Swiss National Science Foundation (NCCRs Struct. Biol. and TransCure) and the Swiss Initiative for Systems Biology (SystemsX.ch).

References

G.M. Besserer, M. Ottolia, D.A. Nicoll, V. Chaptal, D. Cascio, K.D. Philipson, J. Abramson, The second Ca^{2+}-binding domain of the Na^+ Ca^{2+} exchanger is essential for regulation: crystal structures and mutational analysis. Proc. Natl. Acad. Sci. U. S. A. **104**, 18467–18472 (2007)

C. Dyck, A. Omelchenko, C.L. Elias, B.D. Quednau, K.D. Philipson, M. Hnatowich, L.V. Hryshko, Ionic regulatory properties of brain and kidney splice variants of the NCX1 Na^+-Ca^{2+} exchanger. J. Gen. Physiol. **114**, 701–711 (1999)

E.F. Garman, G.W. Grime, Elemental analysis of proteins by microPIXE. Prog. Biophys. Mol. Biol. **89**, 173–205 (2005)

M. Hilge, J. Aelen, G.W. Vuister, Ca^{2+} regulation in the Na^+/Ca^{2+} exchanger involves two markedly different Ca^{2+} sensors. Mol. Cell **22**, 15–25 (2006)

M. Hilge, J. Aelen, A. Foarce, A. Perrakis, G.W. Vuister, Ca^{2+} regulation in the Na^+/Ca^{2+} exchanger features a dual electrostatic switch mechanism. Proc. Natl. Acad. Sci. U. S. A. **106**, 14333–14338 (2009)

D.W. Hilgemann, A. Collins, S. Matsuoka, Steady-state and dynamic properties of cardiac sodium-calcium exchange. Secondary modulation by cytoplasmic calcium and ATP. J. Gen. Physiol. **100**, 933–961 (1992a)

D.W. Hilgemann, S. Matsuoka, G.A. Nagel, A. Collins, Steady-state and dynamic properties of cardiac sodium-calcium exchange. Sodium-dependent inactivation. J. Gen. Physiol. **100**, 905–932 (1992b)

C. Hurtado, M. Prociuk, T.G. Maddaford, E. Dibrov, N. Mesaeli, L.V. Hryshko, G.N. Pierce, Cells expressing unique Na^+/Ca^{2+} exchange (NCX1) splice variants exhibit different susceptibilities to Ca^{2+} overload. Am. J. Physiol. Heart Circ. Physiol. **290**, H2155–H2162 (2006)

D.A. Nicoll, M.R. Sawaya, S. Kwon, D. Cascio, K.D. Philipson, J. Abramson, The crystal structure of the primary Ca^{2+} sensor of the Na^+/Ca^{2+} exchanger reveals a novel Ca^{2+} binding motif. J. Biol. Chem. **281**, 21577–21581 (2006)

B.D. Quednau, D.A. Nicoll, K.D. Philipson, Tissue specificity and alternative splicing of the Na^+/Ca^{2+} exchanger isoforms NCX1, NCX2, and NCX3 in rat. Am. J. Physiol. **272**, C1250–C1261 (1997)

X. Shao, C. Li, I. Fernandez, X. Zhang, T.C. Sudhof, J. Rizo, Synaptotagmin-syntaxin interaction: the C2 domain as a Ca^{2+}-dependent electrostatic switch. Neuron **18**, 133–142 (1997)

M. Wu, M. Wang, J. Nix, L.V. Hryshko, L. Zheng, Crystal structure of CBD2 from the Drosophila Na^+/Ca^{2+} exchanger: diversity of Ca^{2+} regulation and its alternative splicing modification. J. Mol. Biol. **387**, 104–112 (2009)

M. Wu, H.D. Le, M. Wang, V. Yurkov, A. Omelchenko, M. Hnatowich, J. Nix, L.V. Hryshko, L. Zheng, Crystal structures of progressive Ca^{2+} binding states of the Ca^{2+} sensor Ca^{2+} binding domain 1 (CBD1) from the CALX Na^+/Ca^{2+} exchanger reveal incremental conformational transitions. J. Biol. Chem. **285**, 2554–2561 (2010)

M. Wu, S. Tong, J. Gonzalez, V. Jayaraman, J.L. Spudich, L. Zheng, Structural basis of the Ca^{2+} inhibitory mechanism of Drosophila Na^+/Ca^{2+} exchanger CALX and its modification by alternative splicing. Structure **19**, 1509–1517 (2011)

4 Molecular Determinants of Allosteric Regulation in NCX Proteins

Moshe Giladi and Daniel Khananshvili

Abstract

Allosteric activation of NCX involves the binding of cytosolic Ca^{2+} to regulatory domains CBD1 and CBD2. Previous studies with isolated CBD12 and full-size NCX identified synergistic interactions between the two CBD domains that modify the affinity and kinetic properties of Ca^{2+} sensing, although it remains unclear how the Ca^{2+}-binding signal is decoded and propagates to transmembrane domains. Biophysical analyses (X-ray, SAXS, and stopped-flow techniques) of isolated preparations of CBD1, CBD2, and CBD12 have shown that Ca^{2+} binding to Ca3-Ca4 sites of CBD1 results in interdomain tethering of CBDs through specific amino acids on CBD1 (Asp499 and Asp500) and CBD2 (Arg532 and Asp565). Mutant analyses of isolated CBDs suggest that the two-domain interface governs Ca^{2+}-driven conformational alignment of CBDs, resulting in slow dissociation of Ca^{2+} from CBD12, and thus, it mediates Ca^{2+}-induced conformational transitions associated with allosteric signal transmission. Specifically, occupation of Ca3-Ca4 sites by Ca^{2+} induces disorder-to-order transition owing to charge neutralization and coordination, thereby constraining CBD conformational freedom, rigidifying the NCX1 f-loop, and triggering allosteric signal transmission to the membrane domain. The newly found interdomain switch is highly conserved among NCX isoform/splice variants, although some additional structural motifs may shape the regulatory specificity of NCX variants.

Keywords

NCX • Allosteric regulation • Interdomain Ca^{2+} switch • Disorder-to-order transition

M. Giladi • D. Khananshvili (✉)
Department of Physiology and Pharmacology,
Tel-Aviv University, Ramat-Aviv, Tel-Aviv 69978, Israel
e-mail: dhanan@post.tau.ac.il

L. Annunziato (ed.), *Sodium Calcium Exchange: A Growing Spectrum of Pathophysiological Implications*, Advances in Experimental Medicine and Biology 961, DOI 10.1007/978-1-4614-4756-6_4,

4.1 Introduction

The plasma membrane Ca^{2+}-ATPase and Na^+/Ca^{2+} exchanger systems are two major systems that extrude Ca^{2+} from the cell, although their partial contribution to Ca^{2+} homeostasis differs in distinct cell types (Carafoli 1987; Berridge et al. 2003). In general, the Ca^{2+}-ATPase is a "housekeeping" system, which creates and maintains a primary Ca^{2+} gradient across the cell membrane, whereas the Na^+-Ca^{2+} system responds to transient changes in $[Ca^{2+}]_i$ under ever-varying regulatory conditions (Carafoli 1987; Khananshvili 1998; Blaustein and Lederer 1999). To fulfill the physiological demands of various cell types, the Na^+/Ca^{2+} exchanger isoforms (NCX1–3) and their splice variants are expressed in a tissue-specific manner (Philipson and Nicoll 2000; Lytton 2007; Khananshvili 2012). NCX proteins are regulated by cytosolic Ca^{2+} and Na^+ ions at sites that are not directly involved in ion translocation (Hilgemann et al. 1992a, b; Matsuoka et al. 1993, 1995). A rise in cytosolic $[Na^+]$ rapidly stimulates and then inactivates the exchanger, whereas a rise in cytosolic $[Ca^{2+}]$ activates NCX and relieves the Na^+-dependent inactivation (Hilgemann et al. 1992a, b). Moreover, NCX is extremely sensitive to cytosolic acidification (DiPolo and Beauge 1982, 2006; Doering and Lederer 1994; Doering et al. 1996).

NCX proteins consist of nine transmembrane segments (TMSs) with a large cytoplasmic f-loop between TMSs 5 and 6 composed of two adjacent Ca^{2+}-binding domains, CBD1 and CBD2 (Philipson and Nicoll 2000; Levitsky et al. 1994; Hilge et al. 2006). Two CBD domains are connected through a short linker to form a "head-to-tail"-oriented two-domain tandem, CBD12 (Hilge et al. 2006, 2009). Allosteric regulation of NCX requires Ca^{2+} to interact with the CBD domains (Weber et al. 2001; Nicoll et al. 2007; Reeves and Condrescu 2008; Boyman et al. 2011).

Allosteric interaction of cytosolic Ca^{2+} with the CBD1 and CBD2 domains is a major mode of NCX regulation (Hilgemann et al. 1992a, b; Weber et al. 2001; Ginsburg et al. 2002). Different splice variants of NCX exhibit significant differences in their dynamic properties of $[Ca^{2+}]$-dependent allosteric activation (Hryshko, et al. 1996; Dyck et al. 1999; Dunn et al. 2002), although the underlying mechanisms are poorly understood. Owing to large and rapid changes in cytosolic $[Ca^{2+}]$ during the action potential, the $[Ca^{2+}]$-dependent allosteric activation of NCX is especially important in excitable tissues (Hilgemann et al. 1992a, b; Weber et al. 2001; Boyman et al. 2008). For example, in cardiomyocytes, only ~5% of the maximal NCX current is detected at resting $[Ca^{2+}]_i$ levels, whereas the rise of $[Ca^{2+}]_i$ to 1–2 μM recruits nearly 100% of NCX-mediated current (Boyman et al. 2011).

High-resolution crystal structures of isolated CBD1 and CBD2 domains reveal four Ca^{2+}-binding sites (Ca1-Ca4) on CBD1 and two Ca^{2+} sites (CaI-CaII) on CBD2 (Nicoll et al. 2006; Besserer et al. 2007). The CBDs share a similar structure with seven antiparallel β-strands arranged as an immunoglobulin-like β-sandwich (Hilge et al. 2006; Nicoll et al. 2006; Besserer et al. 2007) and C_2-type folding (Rizo and Südhof 1998). The splicing segment is located on CBD2, meaning that CBD2 may control the regulatory specificity of NCX variants (Hilge et al. 2006). Thus, splice variants containing exon A (in excitable tissues) can relieve Na^+-dependent inactivation at high $[Ca^{2+}]_i$, whereas those variants comprising the exon B (in non-excitable tissues) lack this ability (Hilge et al. 2009; Hryshko 2008).

FRET studies demonstrated that the CBD domains of NCX can undergo significant conformational changes within EC coupling in the living cell (Ottolia et al. 2004; Chaptal et al. 2007), although the underlying molecular mechanisms remain unclear. Only three of the six Ca^{2+}-binding sites of CBD12 (Ca3 and Ca4 on CBD1 and CaI on CBD2) actually contribute to $[Ca^{2+}]$-dependent regulation of the intact NCX1 protein (Nicoll et al. 2006; Besserer et al. 2007; Ottolia et al. 2009). Two high-affinity sites of CBD1 (Ca3 and Ca4) play a critical role in determining the affinity for $[Ca^{2+}]$-dependent regulation, whereas the primary Ca^{2+} sensor of CBD2 (CaI) can control Ca^{2+}-dependent alleviation of Na^+-dependent inactivation of full-size NCX (Nicoll et al. 2006; Besserer et al. 2007; Ottolia et al. 2009, 2010).

Interestingly, CBD domains undergo Ca^{2+}-induced reorientation of two CBD domains without significant changes in a core structure of CBDs (Hilge et al. 2009).

4.2 The Two-Domain CBD Tandem Is a Wide-Range Dynamic Sensor

4.2.1 Equilibrium and Kinetic Properties of Six Ca^{2+} Sites in Isolated CBD1 and CBD2

By using protein ultrafiltration techniques, we found that isolated CBD1 contains two high-affinity sites (Ca3-Ca4) with K_d=0.05–0.2 μM and two low-affinity (Ca1-Ca2) sites with K_d>20 μM (Boyman et al. 2009, 2011; Giladi et al. 2010). The primary (CaI) and secondary (CaII) sites of CBD2 exhibit moderate (K_d=~5 μM) and low (K_d>20 μM) affinities, respectively (Boyman et al. 2009, 2011). Isolated CBD1 and CBD2 have a capacity for cooperative binding of Ca^{2+}, manifested by a Hill coefficient of n_H=~2. The Ca^{2+}-binding parameters of the CBD1, CBD2, and CBD12 domains are summarized in Table 4.1.

Table 4.1 Kinetic and equilibrium properties of isolated CBD1, CBD2, and CBD12 domains

	CBD1				*CBD2*	
Site	Ca1	Ca2	Ca3	Ca4	CaI	CaII
K_d (μM)	7	21	0.1	0.03	10	45
k_{off} (s^{-1})	> 300	> 300	18	18	125	> 300
	CBD12					
Site	Ca1	Ca2	Ca3	Ca4	CaI	CaII
K_d (μM)	0.3	180	0.06	0.02	10	45
k_{off} (s^{-1})	> 300	> 300	5	0.5	280	> 300

Purified preparations of CBD1, CBD2-AD, and CBD12-AD were obtained as previously described (Boyman et al. 2009, 2011; Giladi et al. 2010). The K_d values were derived from $^{45}Ca^{2+}$ titration curves, and the k_{off} values were obtained from stopped-flow assays. Owing to technical limitations, k_{off} values larger than 300 s^{-1} cannot be measured

Early studies with full-size NCX in a cellular system (Wei et al. 2002; Ottolia et al. 2004) and the f-loop fragments (Levitsky et al. 1994; Levitsky et al. 1996) have shown that Mg^{2+} ions compete with Ca^{2+} ions in binding to high-affinity regulatory sites of NCX. Recent studies have demonstrated that Mg^{2+} decreases the affinity of Ca3-Ca4 sites at CBD1, whereas it increases the affinity at the CaI site of CBD2 owing to occupation of the CaII site by Mg^{2+} (Boyman et al. 2009; Breukels et al. 2011). Therefore, in the absence of Mg^{2+}, the primary Ca^{2+} sensor of CBD1 (Ca3-Ca4 sites) has 20–25 times higher affinity than the primary sensor of CBD2 (CaI site), whereas in the presence of 2–5 mM Mg^{2+}, the difference in the affinity is "only" 3–7-fold (Boyman et al. 2009). These Mg^{2+}-dependent effects may have physiological relevance because accumulating data indicate that the CaII site of CBD2 is a "genuine" Mg^{2+} site (Boyman et al. 2009; Breukels et al. 2011).

Electrophysiological data suggest that both CBDs can contribute to [Ca^{2+}]-dependent regulation of NCX, whereas the Ca3-Ca4 sites of CBD1 largely govern allosteric activation, with $K_{0.5}$~0.3 μM. Moreover, the CaI site of CBD2 alleviates Na^+-dependent inactivation, with $K_{0.5}$~5μM (Chaptal et al. 2007; Ottolia et al. 2010). Notably, 100 mM Na^+ had no effect on Ca^{2+}-binding affinity or stoichiometry to isolated CBD1, CBD2, or CBD12 (Boyman et al. 2009). This is consistent with the claim that Na^+-dependent inactivation is due to Na^+ binding to transport sites. The possibility remains, however, that the putative "regulatory" Na^+ site is located on the N-terminal of the f-loop connecting CBD2 with TMS6.

The stopped-flow measurements demonstrated that the primary sensors at CBD1 (Ca3-Ca4 sites) and CBD2 (CaI) have strikingly diverse Ca^{2+} off kinetics (Boyman et al. 2009, 2011; Giladi et al. 2010). The monophasic kinetics of Ca^{2+} dissociation was observed either in isolated CBD1 or CBD2, whereas the dissociation of two Ca^{2+} ions from the Ca3-Ca4 sites of CBD1 is much slower (k_{obs} = 15–20 s^{-1}) than the dissociation of one Ca^{2+} ion from the CaI site of CBD2 (k_{obs} = 150–300 s^{-1}) (Boyman et al. 2009). Notably, the Ca^{2+} off-rates of three low-affinity sites (Ca1 and Ca2 on CBD1 and CaII on CBD2) of CBD12 are so fast that they cannot be measured by stopped-flow techniques. The K_d and k_{off} values of isolated CBD preparations are summarized in Table 4.1.

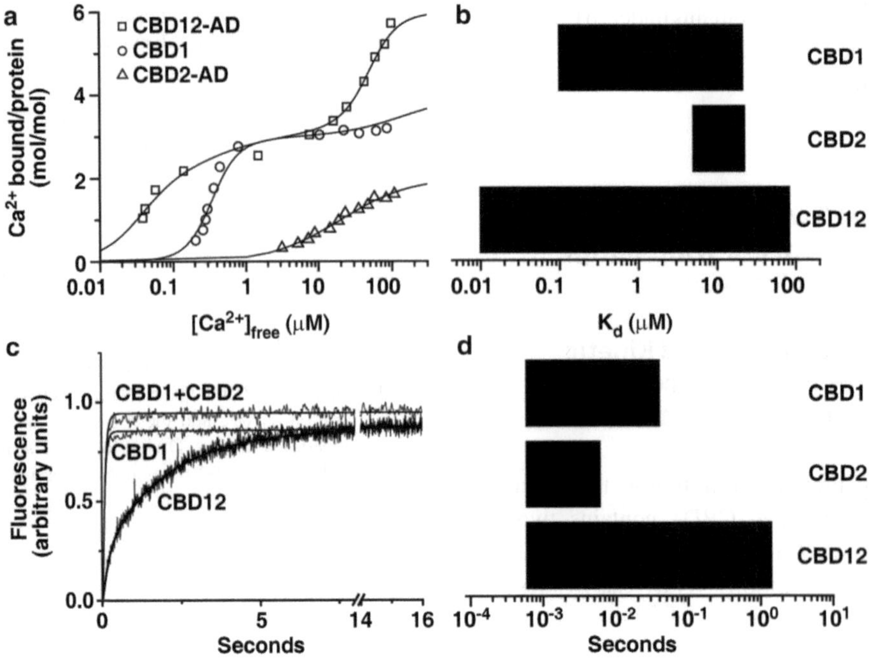

Fig. 4.1 *Ca^{2+}-binding properties of isolated CBD domains*. (**a**) $^{45}Ca^{2+}$ titration curves of CBD1, CBD2-AD, and CBD12-AD. The data were fitted using Adair equations with the appropriate number of sites. (**b**) The range of K_d values of CBD1, CBD2, or CBD12. (**c**) Stopped-flow measurements of Ca^{2+} dissociation kinetics from CBD1, CBD12, and isolated CBD1 + isolated CBD2. The representative traces of CBD1 or CBD1 + CBD2 were fitted to a single exponential curve with $k_f = 14.8 \pm 0.07$ s^{-1} and $k_f = 13.1 \pm 0.07$ s^{-1}, respectively. The representative trace of CBD12 was fitted to a double exponential curve with $k_f = 5.2 \pm 0.08$ s^{-1} and $k_s = 0.52 \pm 0.001$ s^{-1}. (**d**) The range of $t_{1/2}$ values for Ca^{2+} dissociation from CBD1, CBD2, or CBD12

4.2.2 Slow Dissociation of Occluded Ca^{2+} Occurs in CBD12 but not in CBD1 or CBD2

The fundamental finding is that Ca^{2+} off-rates of Ca3-Ca4 sites are strikingly different in isolated CBD1 and CBD12. Namely, the dissociation of two Ca^{2+} ions from the Ca3-Ca4 sites of CBD1 is fast and mono-exponential ($k_f = 15$–20 s^{-1}), whereas the sequential (biphasic) dissociation of two Ca^{2+} ions from the Ca3-Ca4 sites of CBD12 comprises the fast ($k_f = 5$ s^{-1}) and slow ($k_s \sim 0.5$ s^{-1}) phases (Giladi et al. 2010; Boyman et al. 2011). Therefore, the dissociation of one Ca^{2+} ion from the Ca3-Ca4 sites of CBD12 is ~50 times slower than the Ca^{2+} dissociation from the same sites in isolated CBD1 (Fig. 4.1). The slow dissociation (designated as "occluded" Ca^{2+}) is completely abolished in the CBD12-7Ala mutant (with a longer linker), thereby suggesting that the linker-dependent interdomain interactions extend the dynamic range of the CBD12 primary sensor (Giladi et al. 2010). Moreover, a recent study demonstrated that the Ca^{2+} affinity of the Ca3-Ca4 sites is ~10 times higher in isolated CBD12 than in isolated CBD1 (Boyman et al. 2011). This finding is consistent with FRET studies showing that CBD12 has a higher affinity for Ca^{2+} than does CBD1 (John et al. 2011). Therefore, CBD2 interacts with CBD1 to modify the properties of the Ca3-Ca4 sites on CBD1, whereas the other sites largely retain their dynamic properties. In conclusion, the dynamic range of three regulatory sites of CBD12 are specifically modified, conserved, diversified, and integrated in the context of the two-domain tandem, thus producing a wide-range dynamic sensor that can fulfill diverse physiological needs.

4.2.3 Integration of Three Regulatory Sites of CBD12 in a Wide-Range Dynamic Sensor

Figure 4.1 summarizes the affinity and kinetic boundaries of three functional Ca^{2+} sites of CBD12 (Ca3 and Ca4 on CBD1 and CaI on CBD2). Interestingly, the kinetic boundaries of three regulatory sites are much broader (k_{off}=0.3 s^{-1}–300 s^{-1}) than the affinity boundaries (K_d=0.2–5 μM) of the same sites (Boyman et al. 2009, 2011; Giladi et al. 2010). Since the K_d and k_{off} values were measured, the second-order rate constants for Ca^{2+} binding to these sites were calculated and were found to be in the range of $k_{on}=10^6$–10^8 $M^{-1}s^{-1}$. These kinetic parameters are nearly ideal for dynamic sensing of varying $[Ca^{2+}]_i$ values ranging from milliseconds to seconds in accordance with the physiological demands of specific cell types. This may have biological relevance because the Ca^{2+}-dependent regulation of NCX might require different dynamic features in excitable and non-excitable cells. It is possible that during the depolarization of excitable cells, the CaI site of CBD2 rapidly responds to $[Ca^{2+}]_i$ rise under the cell membrane to overcome $[Na^+]$-dependent inactivation of NCX.

Different lines of evidence suggest that NCX is activated allosterically by Ca^{2+} when the $[Ca^{2+}]_i$ level rises in a cellular system (Weber et al. 2001; Pogwizd et al. 2001; Ginsburg et al. 2002; Sobie et al. 2002), although the details of this allosteric process are still not fully understood (Cannell and Soeller 2004). It is established, however, that the higher the I_{NCX} value at any given $[Ca^{2+}]_i$, the more rapid the Ca^{2+} extrusion and the greater the arrhythmic potential of any perturbation of $[Ca^{2+}]_i$ (Pogwizd et al. 2001). Thus, the dynamics of NCX modulation by allosteric activation is critical for normal Ca^{2+} signaling and stability in the heart. We speculate that the slow dissociation of "occluded" Ca^{2+} from the Ca3-Ca4 sites of CBD12 (and hence in the intact NCX protein) may underlie the slow removal of the allosteric effect of Ca^{2+} on I_{NCX} (otherwise known as I_2 inactivation).

Our findings revealed that the short linker between the two CBDs not only specifically diversifies and preserves the dynamic properties of specific Ca^{2+} sites in the context of CBD12; it also effectively integrates the dynamic properties of all three regulatory sites (Ca3, Ca4, and CaI). The rationale behind this is to provide a broad-range dynamic sensor capable of dynamically regulating NCX proteins within a timescale spanning over four orders of magnitude, from milliseconds to seconds (Fig. 4.1). The kinetically diverse Ca^{2+} sensors may represent the physical basis for differential Ca^{2+} sensing in various cellular compartments. For example, Ca^{2+} interaction with a "rapid" CaI site of CBD2 may respond to rapid transient swings of $[Ca^{2+}]$ that occur within <10 msec, when the $[Ca^{2+}]$ levels can transiently reach 300–600-μM levels within a few milliseconds in dyadic cleft (Sobie et al. 2002). The dynamic properties of the CaI site are in good agreement with the rapid alleviation of Na^+-dependent inactivation of NCX, which may occur at the initial stages of action potential. On the other hand, the Ca3-Ca4 sites can effectively sense the bulk cytosolic $[Ca^{2+}]$ changes within the action potential duration (10–300 msec).

4.3 Synergistic Interactions Between CBDs in Isolated CBD12 and NCX

4.3.1 CBD2 Interacts with CBD1 to Modulate a High-Affinity Sensor (Ca3-Ca4 Sites)

Different lines of evidence suggest that interdomain interactions in isolated CBD12 and intact NCX modify the equilibrium binding and kinetic properties of the primary Ca^{2+} sensor (Ca3-Ca4) on CBD1. A slow dissociation of "occluded" Ca^{2+} (0.5 s^{-1}) was observed in CBD12, but not in isolated CBD1 or CBD2 (Fig 4.1), i.e., the Ca^{2+} off-rates of CBD1 and CBD2 are ~50 and ~500 times faster, respectively, than in CBD12 (Boyman et al. 2009; Giladi et al. 2010). Moreover, the Ca3-Ca4 sites have ~10 times higher affinity in CBD12 than in CBD1 (John et al. 2011; Boyman et al. 2011), meaning that CBD2 somehow modifies the intrinsic properties of the Ca3-Ca4 sites on CBD1. Slow conformational transitions were observed by FRET in isolated CBD12 or NCX upon Ca^{2+} removal, suggesting that slow conformational transitions may associate with slow Ca^{2+} dissociation (John et al. 2011). SAXS analysis of isolated CBD12 revealed Ca^{2+}-dependent reorientation of CBDs upon Ca^{2+} binding to CBD1 (Hilge et al. 2009). These findings raise an intriguing question: How does Ca^{2+} binding to the primary sensor promote conformational transitions, and in turn, how does this contribute to allosteric regulation?

4.3.2 The Interdomain Linker Controls Synergistic Interactions Between CBDs

For resolving the role of the interdomain CBD1-CBD2 linker, we systematically analyzed the Ca^{2+} titration curves and Ca^{2+} off-rates in isolated preparations of CBD1, CBD2, and CBD12 (Boyman et al. 2009, 2011; Giladi et al. 2012). We found that the linker-controlled interactions in the two-domain CBD tandem selectively retain, modify, and integrate dynamic assets of three regulatory sites of CBD12 to allow NCX regulation within a broad-range time interval (Giladi et al. 2010; Boyman et al. 2011). More specifically, the linker-dependent constraints decelerate 30–70 times the Ca^{2+} off-rates of one Ca^{2+} ion (Giladi et al. 2010) while increasing ~10 times the affinity at specific sites of CBD1, most probably at the Ca4 site. Notably, the intrinsic properties of the remaining two Ca^{2+} sites located on CBD1 (Ca3) and CBD2 (CaI) largely retain their intrinsic properties.

The current findings are consistent with the notion that the linker defines the sequential dissociation of two Ca^{2+} ions from the Ca3-Ca4 sites, whereas the dissociation of the first ("fast") Ca^{2+} ion is followed by dissociation of the second ("occluded") Ca^{2+} ion, displaying characteristically slow kinetics. The underlying biological relevance of these findings is that the linker-dependent modification of Ca^{2+} on/off kinetics at a specific regulatory site can diversify and widen the dynamic range of Ca^{2+} sensing of CBD12 by nearly two orders of magnitude (while retaining a submicromolar range required for regulation). Thus, the linker-dependent interactions extend the dynamic range of regulation for up to 10–20 s.

The structure-functional relationships of the CBD1-CBD2 linker are of special interest in light of the interdomain interactions found in the two-domain tandems of other regulatory proteins (Stahelin et al. 2005). For example, in analogy with CBD12 (Salinas et al. 2011), the Ca^{2+} binding to a two-domain construct of cadherin restricts the linker motions substantially (Häussinger et al. 2002). Nevertheless, the Ca^{2+}-binding modes of cadherins and CBD12 are dissimilar. Namely, the Ca^{2+} binding to the two-domain cadherin construct involves direct interactions with residues in the linker region, whereas the binding of Ca^{2+} to sites Ca3 and Ca4 in CBD1 involves ligation with residues 498–500 that directly precede linker residues 501–503 (see below). Analogously to CBD12, in rabphilin-3A the C2A-C2B linker results in ~10-fold increase in Ca^{2+} affinity through interdomain interactions. However, in contrast to CBD12, the linker of rabphilin-3A itself contributes to the ligation sphere (Montaville et al. 2007).

4.3.3 Isolated CBD12 Features Ca^{2+}-Sensing Properties of Full-Size NCX

The principal question is as follows: How do we know that the interdomain interactions found on isolated preparations of CBD12 are relevant for full-size NCX in a cellular system? Recent findings in several laboratories suggest that the CBD1-CBD2 linker governs synergistic interactions between two CBD domains either in isolated CBD12 (Giladi et al. 2010; John et al. 2011; Salinas et al. 2011) or full-size NCX (Ottolia et al. 2009, 2010). The significance of these findings is that the isolated CBD12 and intact NCX might share common molecular mechanisms for Ca^{2+} sensing and translation of allosteric massages upon Ca^{2+} binding to the primary sensor on CBD1. Therefore, isolated CBD12 is a good candidate for convoluting the molecular mechanisms associated with allosteric regulation

(Giladi et al. 2010; John et al. 2011; Salinas et al. 2011). Electrophysiological studies showed that inserting seven alanines between H501 and A502 in the linker (7Ala) decreases the affinity of [Ca^{2+}]-dependent activation of full-size NCX in a cellular system (Ottolia et al. 2010). We found that the 7Ala insertion in the isolated CBD12 construct results in a complete loss of slow dissociation of occluded Ca^{2+} (Giladi et al. 2010). A recent NMR study demonstrated that Ca^{2+} binding to Ca3-Ca4 sites of isolated CBD12 results in decreased conformational flexibility of the linker, thereby restricting interdomain motions of CBDs (Salinas et al. 2011).

4.4 Regulatory Switch at the Interface of the Two CBD Domains

4.4.1 The CBD12-E454 Mutant Resembles the Properties of CBD12-WT and NCX

As we previously summarized, accumulating data from several laboratories provided strong evidence that the synergistic interactions between the two CBDs strongly modify the dynamic properties of Ca^{2+} sensing either in isolated CBD12 or in full-size NCX. Despite this progress, the following question remains: How do two CBDs communicate with each other to decode information upon Ca^{2+} binding to the primary sensor (Ca3-Ca4 sites) on CBD1? We posit that the two-domain interface encodes the primary module for decoding the allosteric signal. Although the X-ray structures of isolated CBD1 and CBD2 are available (Nicoll et al. 2006; Besserer et al. 2007), this information cannot resolve molecular determinants of interdomain coupling.

To evaluate the functional properties of CBD12-E454K, we tested its capacity for Ca^{2+}-binding affinity and dissociation kinetics. This mutant binds a total of ~4 Ca^{2+} ions, of which two are bound with high affinity. This stoichiometry is in line with our previous studies that demonstrated a binding of ~2 Ca^{2+} ions to both CBD1-E454K and CBD2-WT (Giladi et al. 2010, 2012) and with the fact that Ca^{2+} is bound with very low affinity (Boyman et al. 2009). Ca^{2+} dissociation kinetics revealed that CBD12-E454K maintains a slow phase of Ca^{2+} dissociation, whereas the fast phase is 10–20 times accelerated, in agreement with the ~10-fold higher K_d value. These data suggest that the interdomain interactions underlying Ca^{2+} occlusion in CBD12 are intact in CBD12-E454K, in agreement with the electrophysiological properties of full-size NCX-E454K in a cellular system (Chaptal et al. 2009).

For crystallization of the two-domain tandem, we employed the E454K mutant because the E454K substitution replaces the Ca^{2+} ion at the Ca1 site of CBD1, thus, stabilizing protein structure through charge compensation (Chaptal et al. 2009). Moreover, the full-size NCX1-E454 exhibits WT [Ca^{2+}]-dependent regulation (Chaptal et al. 2009), whereas the Ca3-Ca4 sites exhibit WT affinity for Ca^{2+} in an isolated CBD1-E454K (Giladi et al. 2010).

4.4.2 The Crystal Structure Reveals an Extended Orientation of CBDs

We crystallized the CBD12-E454K mutant in the presence of 10 mM $CaCl_2$ and resolved its crystal structure at 2.7 Å resolution (Giladi et al. 2012). As shown in Fig. 4.3, the overall structure of CBD12-E454K comprises an extended orientation of the two-domain tandem, exhibiting a buried surface area of ~360 Å^2 at the interface. Similar features were found for CALX1.1 and CALX1.2 (Wu et al. 2011). Such a small contact area is consistent with our previous notion, suggesting that in the absence of the linker, the two CBD domains cannot form a stable complex in solution (Giladi et al. 2010). In agreement with the crystal structure of CBD1-E454K (Chaptal et al. 2009), our structure reveals three Ca^{2+} ions in CBD1 that correspond to Ca2, Ca3, and Ca4 in CBD1-WT. The low-solvent accessibility indicates the structural role of Ca^{2+} binding for CBD1, which loses its structural integrity in the apo form. Notably, there is no indication of bound Ca^{2+} ions in CBD2 in our structure.

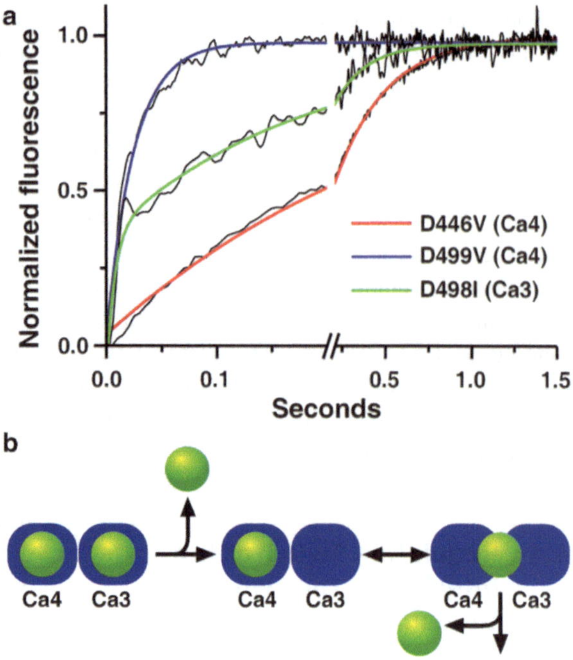

Fig. 4.2 *Sequential dissociation of two Ca^{2+} ions from high-affinity sites of CBD12*. (**a**) Ca^{2+} dissociation kinetics from Ca3 site mutant D498I and from Ca4 site mutants D446V and D499V. Ca^{2+} dissociation is accelerated in all mutants, but is monophasic only in Ca4 site mutants. (**b**) A model for sequential dissociation of two Ca^{2+} ions from the Ca3-Ca4 sites of CBD12

Similarly to previously published crystal structures of isolated CBD domains (Nicoll et al. 2006; Besserer et al. 2007), the FG loops of both domains (residues 469–481 and 600–618) are not fully visualized due to the presumed high flexibility of these areas. It has been shown by NMR that the CBD2 FG-loop contains an α-helix (Hilge et al. 2006) that spans residues 625 through 630 (Hilge et al. 2009). Our structure reveals a helix spanning residues 620–632, the side chains of which contribute to the interdomain interface. Our kinetic analyses of Ca3 and Ca4 site mutants suggest that sequential dissociation of two Ca^{2+} ions from the high-affinity sensor of CBD12 involves the dissociation of one Ca^{2+} ion from the Ca3 site first, whereas the second Ca^{2+} ion becomes "occluded" (Fig. 4.2), probably due to increased negative charge density at Ca3-C4 sites attracting the second Ca^{2+} ion.

4.4.3 Identification of the Interdomain Ca^{2+} Switch and Its Structure-Functional Properties

The most striking property of the two-domain crystal structure is that two CBD domains are tethered through a network of interdomain salt bridges involving two amino acids from each domain, D499 and D500 (on CBD1) and R532 and D565 (on CBD2). Moreover, the arrangement of this interdomain module is entirely coupled to the specific coordination of two Ca^{2+} ions at the Ca3-Ca4 sites on CBD1 (Fig. 4.3). Therefore, the most important feature of this newly found interdomain Ca^{2+} switch is that CBDs are tethered in a Ca^{2+}-dependent manner through specific amino acids located on CBD1 (D499 and D500) and CBD2 (R532 and D565) (Fig. 4.3b). The buried R532 is a central residue in the network, tethering D565 on CBD2 and D499 and D500 in CBD1. Most importantly, this bifurcated salt bridge supports Ca^{2+} coordination with D499 and D500, part of the Ca3-Ca4 sites. Therefore, Ca^{2+} binding to Ca3-Ca4 couples CBDs through the interface to restrict interdomain flexibility.

Although the crystal structures of apo-CBD1 or apo-CBD12 are unavailable, NMR data suggest that the D499 and D500 side chains are highly disordered in the apo form (Hilge et al. 2006, 2009; Johnson et al. 2008; Salinas et al. 2011). Recent NMR studies on an isolated CBD12 have shown that Ca^{2+} binding to CBD1 drastically constrains the linker's flexibility and

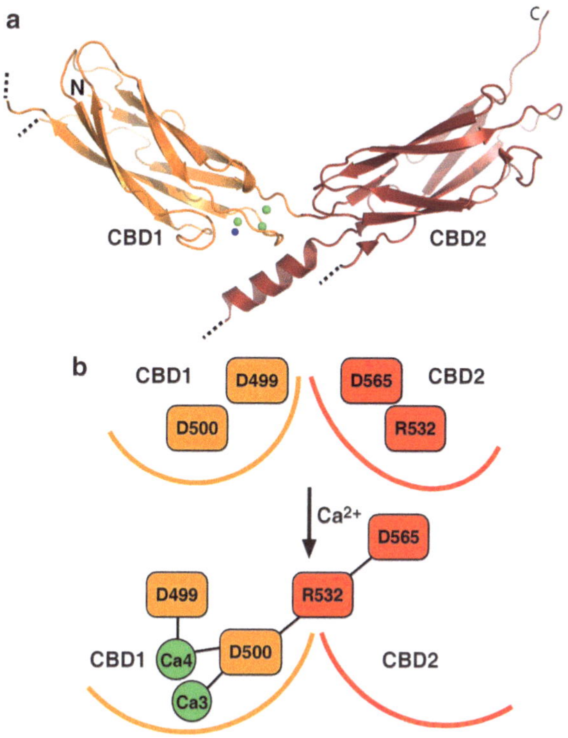

Fig. 4.3 *Interdomain interactions in CBD12.* (**a**) Overall structure of the CBD12-E454K mutant. *Dashed lines* represent segments in the structure that were not observable. (**b**) Scheme of Ca^{2+}-dependent interactions between the CBD domains

thus stabilizes a more rigid conformation of the two-domain tandem (Salinas et al. 2011). Altogether, occupation of Ca3-Ca4 sites by Ca^{2+} orients side chains D499 and D500, which pair with R532, which in turn, restricts linker flexibility and interdomain motions.

4.4.4 The Interface Between the Two CBD Domains Governs Ca^{2+}-Induced Transitions

As observed in the crystal structure, two charged residues from CBD2, R532 in the B-C loop and D565 in the D-E loop, participate in an electrostatic network involving D499 and D500, which coordinate Ca3-Ca4 on CBD1 (Giladi et al. 2012). To evaluate the functional role of the proposed network, we tested the effect of the R532 and D565 mutations on Ca^{2+}-binding affinity and off-rates. CBD12-R532A exhibits ~5 times lower affinity for Ca^{2+} at the Ca3-Ca4 sites, as compared with CBD12-WT (Fig. 4.3). The Ca^{2+} slow dissociation is abolished in CBD12-R532A, whereas the Ca^{2+}-binding curve of this mutant is similar to the Ca^{2+}-binding curve observed for CBD1+CBD2 (Giladi et al. 2010). Thus, in CBD12, R532 governs Ca^{2+} occlusion at the Ca3-Ca4 sites of CBD1 through specific interdomain interactions. Although R532 is not directly involved in Ca^{2+} coordination, CBD12-D565A exhibits four high-affinity sites instead of the two observed in CBD12-WT.

To investigate the structural consequences of mutations that abolish Ca^{2+} occlusion, we utilized the SAXS technique (Hilge et al. 2009) to detect Ca^{2+}-dependent reorientation of CBDs. Indeed, CBD12-7Ala lacks not only Ca^{2+} occlusion but also Ca^{2+}-dependent reorientation of the CBDs (Giladi et al. 2012). Like CBD12-7Ala, CBD12-R532A does not display Ca^{2+}-dependent reorientation of the CBDs. Both of these mutants adopt only the extended Ca^{2+}-free conformations, correlating with their lack of Ca^{2+} occlusion. Electrophysiological experiments have shown that NCX1-E454K is regulated by Ca^{2+} as NCX1-WT (Chaptal et al. 2009). Our SAXS analysis, however, revealed no Ca^{2+} dependency for CBD orientation in this mutant. In contrast to CBD12-7Ala and CBD12-R532A (that adopt only the Ca^{2+}-free conformation), CBD12-E454K features only the Ca^{2+}-bound conformation. Therefore, the E454K substitution replaces the Ca^{2+} ion at the Ca1 site of CBD1, thus stabilizing the "Ca^{2+}-bound" conformation of protein through a charge compensation mechanism (Chaptal et al. 2009). Therefore, the amino acids involved in interdomain tethering of CBDs are also important for Ca^{2+} occlusion and Ca^{2+}-induced reorientation of CBD domains, meaning that the interdomain Ca^{2+} switch is responsible for Ca^{2+}-dependent conformational transitions in NCX. If so, the Ca^{2+}-driven interdomain switch can effectively restrict CBD motions upon Ca^{2+} binding to CBD1 while generating a more rigid conformation of CBD12 to propagate the allosteric signal (Salinas et al. 2011; Breukels and Vuister 2010).

4.5 Involvement of CBDs in Ca^{2+} Signaling and EC Coupling

4.5.1 Ca^{2+} Interactions with CBDs are Highly Sensitive to Cytosolic Protons

The cardiac Na^+/Ca^{2+} exchange is extremely sensitive to cytosolic pH (Philipson et al. 1982; Doering and Lederer 1994; Doering et al. 1996). A critical question is as follows: Can protons interact with CBDs to affect Ca^{2+}-dependent allosteric regulation of NCX under physiologically related conditions? To address this question, we examined proton actions on NCX function by applying state-of-the-art electrophysiological, imaging, and biochemical approaches on intact cardiomyocytes and isolated preparations of CBD1, CBD2, and CBD12 proteins (Boyman et al. 2011). We found that protons compete with Ca^{2+} for binding to CBD1 and CBD2 and that this competition does not involve Na^+ (Boyman et al. 2009, 2011). The apparent pK_a values for proton interaction with CBD1 and CBD2 are within the physiological range of cytoplasmic pH 7.2, meaning that CBDs may serve as a dual Ca^{2+}/H^+ sensor. Therefore, protons can effectively regulate NCX when cytosolic pH drops to 6.9 under acidosis/ischemia conditions.

Our surprising finding is the very large Hill coefficient for Ca^{2+}-dependent activation of I_{NCX} exhibiting $n_H = 8$ at both pH 7.2 and pH 6.9 in intact cardiomyocytes (Boyman et al. 2011). We are unable to fully account for this high degree of cooperativity. In fact, we cannot explain the observed high values of n_H even if we were to assume the involvement of all six Ca^{2+}-binding sites of CBD12 (which apparently is not the case). The extreme cooperativity of NCX does, however, have important physiological implications; it indicates high sensitivity of the Ca^{2+} extrusion to $[Ca^{2+}]_i$. This is consistent with the remarkable role of NCX in regulating $[Ca^{2+}]_i$. An interesting possibility is that Ca^{2+}-induced disorder-to-order conformational transitions in the f-loop may cause a dimerization of NCX (John et al. 2011) which in turn contributes to the observed high levels of cooperativity for allosteric activation.

Although a mild cytosolic acidification (from pH 7.2 to 6.9) decreases several fold the affinity of regulatory Ca^{2+}, the relevant pH changes do not affect the degree of cooperativity (n_H) or the maximal capacity of either I_{NCX} or Ca^{2+}-binding stoichiometry in isolated CBDs (Boyman et al. 2011). Interestingly, the Ca^{2+}-binding sites of CBDs bind multiple Ca^{2+} ions in proximity (four Ca^{2+} sites of CBD1 are separated by 3–4 Å) and exhibit positive cooperativity for Ca^{2+} binding despite the strong charge repulsion between the adjacent divalent cations (Nicoll et al. 2006; Besserer et al. 2007; Wu et al. 2009, 2010). The close adjacency of Ca^{2+} sites in CBDs is consistent with the sharp dependence of Ca^{2+} binding on pH. The binding of the first Ca^{2+} ion may partially (or fully) deprotonate the coordinating residue(s), thereby enabling the next Ca^{2+} ion to bind to the remaining site(s). A similar mechanism was proposed for the C2 domain of phospholipase A2, in which two Ca^{2+} sites are separated by 4.1 Å (Malmberg et al. 2004).

The kinetics of allosteric activation of NCX is important because the higher the I_{NCX} at any given $[Ca^{2+}]_i$, the more rapid is the Ca^{2+} extrusion and the greater is the arrhythmic potential of any perturbation of $[Ca^{2+}]_i$ that may occur (Pogwizd et al. 2001). When NCX activation by the allosteric effect of Ca^{2+} occurs, the activation kinetics are very fast, <50 msec (Weber et al. 2001; Ginsburg et al. 2002). The longevity of the effect (i.e., the kinetics of the deactivation of the allosteric effects of Ca^{2+} on NCX) is uncertain and could be slow compared with the period of the heartbeat. For example, Hilgemann and colleagues suggested that the deactivation process may be as long as 10 seconds (Hilgemann et al. 1992a, b). Nevertheless, as ions that compete with Ca^{2+} to bind at CBD1 and CBD2, protons may regulate NCX, and thus, protons may serve as an endogenous antiarrhythmic agent by reducing I_{NCX} at any given $[Ca^{2+}]_i$ value without having any action on NCX thermodynamics.

4.5.2 Involvement of Two CBD Ca^{2+} Sensors in Localized Ca^{2+} Signaling

The two-domain Ca^{2+} sensors of NCX may enable the transporter to respond over a wide range of $[Ca^{2+}]_i$ values and longevity (see above). If so, this may provide NCX with the dynamic flexibility it needs to respond to both local $[Ca^{2+}]_i$ and cell-wide $[Ca^{2+}]_i$ signals (the timescales of which are very different). The spatial organization of the Ca^{2+}-signaling proteins may result in the exposure of NCX involved in EC coupling to a wide range of local $[Ca^{2+}]_i$. Because the L-type Ca^{2+} channels (LTCC) are largely located facing the junctional SR (jSR) across a nanoscopic subspace (15 nm), it is estimated that when an LTCC opens, the "subspace" calcium ($[Ca^{2+}]_{SS}$) increases briefly from a diastolic level of 100 nM to about 10 μM (Weber et al. 2001; Sobie et al. 2002; Cannell and Soeller 2004). The NCX proteins near the jSR will be exposed to this high $[Ca^{2+}]_{SS}$. If a Ca^{2+} spark is triggered by the LTCC opening, $[Ca^{2+}]_{SS}$ is estimated to increase 10-fold to about 100 μM. If the NCX protein transport function is modulated by $[Ca^{2+}]_{SS}$, then NCX needs "sensors" that can respond to $[Ca^{2+}]$ over a range of about 100 nM to 100 μM – 1,000-fold change in concentration. By having two high-affinity sites (Ca3-Ca4) on CBD1, one moderate site on CBD2 (CaI site) and three low-affinity sites (Ca1 and Ca2 sites on CBD1 and CaII on CBD2), the range of sensitivity of NCX to Ca^{2+} in and around jSR is fully covered. Although the function of the three low-affinity sites is unknown, the remaining three sites of CBD12 provide NCX with sensors over an extremely wide range of $[Ca^{2+}]_i$ values and may enable NCX to respond to $[Ca^{2+}]_i$ signals during EC coupling. The inhibition of Ca^{2+} binding to CBD12 by protons may extend the effects of acidification to the full range of cellular and subcellular Ca^{2+} signaling.

In light of the present findings, it is tempting to speculate that the Na^+/Ca^{2+} exchanger is allosterically "switched off" when $[Ca^{2+}]_i$ falls sufficiently, since at sub-physiological $[Ca^{2+}]_i$ values, NCX may enter a state in which no net Ca^{2+} transport occurs (Hilgemann et al. 1992b). The level of $[Ca^{2+}]_i$ at which NCX enters its "off" mode, based on the experiments we conducted recently (Boyman et al. 2011), may be modulated by $[H^+]_i$ since protons appear to act as competitive inhibitors of Ca^{2+} interaction at CBD1 and CBD2 sites. If true, myocardial ischemia, which can dramatically increase $[H^+]_i$ values, may appreciably affect Ca^{2+} transport and Ca^{2+}-dependent arrhythmogenesis. Note, however, that the time course of such effects may be critical since during acidosis, cellular Ca^{2+} loading is likely to occur by NCX blocking, but this same NCX blocking will inhibit I_{NCX} and its potential Ca^{2+} activation (Harrison et al. 1992; Choi et al. 2000). Following the acidosis, however, the large increase in cellular Ca^{2+} will better activate I_{NCX} and thus enhance Ca^{2+}-activated arrhythmogenic currents.

4.6 Conclusions

1. We developed new experimental approaches for measuring the equilibrium and kinetic parameters governing Ca^{2+} interaction with isolated proteins of CBD1, CBD2, and CBD12 and established the K_d and k_{off} values of three functionally important sites, Ca3-Ca4 (on CBD1) and CaI (on CBD2).

2. We found that the interdomain CBD1-CBD2 linker governs synergistic interactions between CBD2 and CBD1, which results in modifying the affinity and kinetic properties of a primary Ca^{2+} sensor (Ca3-Ca4) in the context of a two-domain CBD12 tandem. Moreover, the CBD1-CBD2 linker specifically alters, retains, and integrates the dynamic properties of three regulatory sites on CBD12, thus providing a wide-range Ca^{2+} sensor that can respond to large changes in cytosolic $[Ca^{2+}]=0.1–10$ μM within a time window of 10^{-3}–10 s.
3. We solved the crystal structure of a two-domain tandem mutant of the NCX1 brain splice variant and found that CBD domains communicate through a complex network of electrostatic interactions at the interface of two CBD domains. There is an excellent correlation between the in vitro tests of the dynamic assets of CBD12 (slow Ca^{2+} dissociation and Ca^{2+}-dependent alignment of CBDs) and structural data revealing Ca^{2+}-driven communication between CBDs. Most importantly, occupation of Ca3-Ca4 sites by Ca^{2+} enables CBD tethering and thus, constrains their conformational freedom, consequently triggering allosteric signal transmission. Additional structural motifs are likely to be involved in transmitting the regulatory signal in NCX variants.
4. We found that despite the fact the primary Ca^{2+} sensors of CBD1 (Ca3-Ca4 sites) and CBD2 (CaI site) have a distinct affinity for Ca^{2+} binding, these differences become nearly equal ($K_d=0.8–3$μM) in the presence of physiological $[Mg^{2+}]$ concentrations owing to the decreased affinity of Ca3-Ca4 sites at CBD1 and the increased affinity of CaI at CBD2 (through the occupation of the CaII site). We concluded that the CaII site of CBD2 is the Mg^{2+}-binding site under physiologically relevant conditions.
5. We found that in intact cardiomyocytes the $[Ca^{2+}]$-dependent allosteric activation of I_{NCX} is very sensitive to mild cytosolic acidification, whereas protons decrease the affinity of Ca^{2+}-dependent activation, but not the maximal levels of I_{NCX} current. In intact cardiomyocytes, the $[Ca^{2+}]$-dependent allosteric activation of I_{NCX} has an unusually high degree of cooperativity, displaying a Hill coefficient of $n_H=\sim8$. Parallel experiments with isolated CBD1, CBD2, and CBD12 preparations revealed that protons compete with Ca^{2+} for binding at CBD regulatory sites. These findings raise the possibility that CBD12 is a dual pH/Ca^{2+}. In contrast to protons, the Na^+ ion does not directly interact with CBDs.

Acknowledgments This work was partially funded by the Israeli Ministry of Health Grant # 2010-3-6266, the USA-Israeli Binational Research Grant # 2009-334, and the Israel Science Foundation Grant # 23/10. Financial support from the Bernstein Foundation is highly appreciated.

References

M.J. Berridge, M.D. Bootman, H.L. Roderick, Calcium signalling: dynamics, homeostasis and remodelling. Nature Rev. Mol. Cell. Biol. **4**, 517–529 (2003)

G.M. Besserer, M. Ottolia, D.A. Nicoll, V. Chaptal, D. Cascio, K.D. Philipson, J. Abramson, The second Ca^{2+}-binding domain of the Na^+/Ca^{2+} exchanger is essential for regulation: crystal structures and mutational analysis. Proc. Natl. Acad. Sci. U. S. A. **104**, 18467–18472 (2007)

M.P. Blaustein, W.J. Lederer, Sodium/calcium exchange: its physiological implications. Physiol. Rev. **79**, 763–854 (1999)

L. Boyman, R. Hiller, W.J. Lederer, D. Khananshvili, Direct loading of the purified endogenous inhibitor into the cytoplasm of patched cardiomyocytes blocks the ion currents and calcium transport through the NCX1 protein. Biochemistry **47**, 6602–6611 (2008)

L. Boyman, H. Mikhasenko, R. Hiller, D. Khananshvili, Kinetic and equilibrium properties of regulatory calcium sensors of NCX1 protein. J. Biol. Chem. **284**, 6185–6193 (2009)

L. Boyman, B.M. Hagen, M. Giladi, R. Hiller, W.J. Lederer, D. Khananshvili, Proton sensing Ca^{2+} binding domains regulate the cardiac Na^+/Ca^{2+} exchanger. J. Biol. Chem. **286**, 28811–28820 (2011)

V. Breukels, G.W. Vuister, Binding of calcium is sensed structurally and dynamically throughout the second calcium-binding domain of the sodium/calcium exchanger. Proteins **78**, 1813–1824 (2010)

V. Breukels et al., The second Ca^{2+}-binding domain of NCX1 binds Mg^{2+} with high affinity. Biochemistry **50**, 8804–8812 (2011)

M.B. Cannell, C. Soeller, Analysing cardiac excitation-contraction coupling with mathematical models of local control. Prog. Biophys. Mol. Biol. **85**, 141–162 (2004)

E. Carafoli, Intracellular calcium homeostasis. Ann. Rev. Biochem. **56**, 395–433 (1987)

V. Chaptal, G. Mercado-Besserer, M. Ottolia, D.A. Nicoll, D. Cascio, K.D. Philipson, J. Abramson, How does regulatory Ca^{2+} regulate the Na^+-Ca^{2+} exchanger? Channels **1**, 397–399 (2007)

V. Chaptal, M. Ottolia, G. Mercado-Besserer, D.A. Nicoll, K.D. Philipson, J. Abramson, Structure and functional analysis of a Ca^{2+} sensor mutant of the Na^+/Ca^{2+} exchanger. J. Biol. Chem. **284**, 14688–14692 (2009)

H.S. Choi, A.W. Trafford, C.H. Orchard, D.A. Eisner, The effect of acidosis on systolic Ca^{2+} and sarcoplasmic reticulum calcium content in isolated rat ventricular myocytes. J. Physiol. (Lond.) **529**, 661–668 (2000)

R. DiPolo, L. Beauge, The effects of pH on Ca^{2+} extrusion mechanisms in dialyzed squid axons. Biochim. Biophys. Acta **688**, 237–245 (1982)

R. DiPolo, L. Beauge, Sodium/calcium exchanger: influence of metabolic regulation on ion carrier interactions. Physiol. Rev. **86**, 155–203 (2006)

A.E. Doering, W.J. Lederer, The action of Na^+ as a cofactor in the inhibition by cytoplasmic protons of the cardiac Na^+-Ca^{2+} exchanger in the guinea-pig. J. Physiol. (Lond.) **480**, 9–20 (1994)

A.E. Doering, D.A. Eisner, W.J. Lederer, Cardiac Na-Ca exchange and pH. Ann. N. Y. Acad. Sci. **779**, 182–198 (1996)

J. Dunn, C.L. Elias, H.D. Le, A. Omelchenko, L.V. Hryshko, J. Lytton, The molecular determinants of ionic regulatory differences between brain and kidney Na^+/Ca^{2+} exchanger (NCX1) isoforms. J. Biol. Chem. **277**, 33957–33962 (2002)

C. Dyck, A. Omelchenko, C.L. Elias, B.D. Quednau, K.D. Philipson, M. Hnatowich, L.V. Hryshko, Ionic regulatory properties of brain and kidney splice variants of the NCX1 Na^+-Ca^{2+} exchanger. J. Gen. Physiol. **114**, 701–711 (1999)

M. Giladi, L. Boyman, H. Mikhasenko, R. Hiller, D. Khananshvili, Essential role of the CBD1-CBD2 linker in slow dissociation of Ca^{2+} from the regulatory two-domain tandem of NCX1. J. Biol. Chem. **285**, 28117–28125 (2010)

M. Giladi, Y. Sasson, X. Fang, R. Hiller, T. Buki, Y.X. Wang, J.A. Hirsh, D. Khananshvili, A common Ca^{2+}-driven interdomain module governs eukariotic NCX regulation. PloS One **7**(6): e39985 (2012)

K.S. Ginsburg et al., Simultaneous measurement of [Na]i, [Ca]i, and INCX in intact cardiac myocytes. Ann. N. Y. Acad. Sci. **976**, 157–158 (2002)

S.M. Harrison, J.E. Frampton, E. McCall, M.R. Boyett, C.H. Orchard, Contraction and intracellular Ca^{2+}, Na^+, and H^+ during acidosis in rat ventricular myocytes. Am. J. Physiol. **262**, C348–C357 (1992)

D. Häussinger, T. Ahrens, H.J. Sass, O. Pertz, J. Engel, S. Grzesiek, Calcium-dependent homoassociation of E-cadherin by NMR spectroscopy: changes in mobility, conformation and mapping of contact regions. J. Mol. Biol. **324**, 823–839 (2002)

M. Hilge, J. Aelen, G.W. Vuister, Ca^{2+} regulation in the Na^+/Ca^{2+} exchanger involves two markedly different Ca^{2+} sensors. Mol. Cell **22**, 15–25 (2006)

M. Hilge, J. Aelen, A. Foarce, A. Perrakis, G.W. Vuister, Ca^{2+} regulation in the Na^+/Ca^{2+} exchanger features a dual electrostatic switch mechanism. Proc. Natl. Acad. Sci. U. S. A. **106**, 14333–14338 (2009)

D.W. Hilgemann, S. Matsuoka, G.A. Nagel, A. Collins, Steady-state and dynamic properties of cardiac sodium-calcium exchange. Sodium-dependent inactivation. J. Gen. Physiol. **100**, 905–932 (1992a)

D.W. Hilgemann, A. Collins, S. Matsuoka, Steady-state and dynamic properties of cardiac sodium-calcium exchange. Secondary modulation by cytoplasmic calcium and ATP. J. Gen. Physiol. **100**, 933–961 (1992b)

L. Hryshko, What regulates Na^+-Ca^{2+} exchanger? Focus on "sodium-dependent inactivation of sodium-calcium exchange in transfected Chinese hamster ovary cells". Am. J. Physiol. Cell Physiol. **295**, C869–C871 (2008)

L.V. Hryshko, S. Matsuoka, D.A. Nicoll, J.N. Weiss, E.M. Schwarz, S. Benzer, K.D. Philipson, Anomalous regulation of the *Drosophila* Na^+-Ca^{2+} exchanger by Ca^{2+}. J. Gen. Physiol. **108**, 67–74 (1996)

S.A. John, B. Ribalet, J.N. Weiss, K.D. Philipson, M. Ottolia, Ca^{2+}-dependent structural rearrangements within Na^+-Ca^{2+} exchanger dimers. Proc. Natl. Acad. Sci. U. S. A. **108**, 1699–1704 (2011)

E. Johnson, L. Brüschweiler-Li, S.A. Showalter, G.W. Vuister, F. Zhang, R. Brüschweiler, Structure and dynamics of Ca^{2+}-binding domain 1 of the Na^+/Ca^{2+} exchanger in the presence and in the absence of Ca^{2+}. J. Mol. Biol. **377**, 945–955 (2008)

D. Khananshvili, Structure, mechanism and regulation of the cardiac sarcolemma Na^+-Ca^{2+} exchanger. Mol. Cell. Biol. **23B**, 309–356 (1998)

D. Khananshvili, The SLC8 gene family of sodium-calcium exchangers (NCX) – structure, function, and regulation in health and disease. Mol. Asp. Med. (2012) in press

D.O. Levitsky, D.A. Nicoll, K.D. Philipson, Identification of the high affinity Ca^{2+}-binding domain of the cardiac Na^+-Ca^{2+} exchanger. J. Biol. Chem. **269**, 22847–22852 (1994)

D.O. Levitsky et al., Cooperative interaction between Ca^{2+} binding sites in the hydrophilic loop of the Na^+-Ca^{2+} exchanger. Mol. Cell. Biochem. **160–161**, 27–32 (1996)

J. Lytton, Na^+/Ca^{2+} exchangers: three mammalian gene families control Ca^{2+} transport. Biochem. J. **406**, 365–382 (2007)

N.J. Malmberg, S. Varma, E. Jakobsson, J.J. Falke, Ca^{2+} activation of the cPLA2 C2 domain: ordered binding of two Ca^{2+} ions with positive cooperativity. Biochemistry **43**, 16320–16328 (2004)

S. Matsuoka, D.A. Nicoll, R.F. Reilly, D.W. Hilgemann, K.D. Philipson, Initial localization of regulatory regions of the cardiac sarcolemmal Na^+-Ca^{2+} exchanger. Proc. Natl. Acad. Sci. U. S. A. **90**, 3870–3874 (1993)

S. Matsuoka, D.A. Nicoll, L.V. Hryshko, D.O. Levitsky, J.N. Weiss, K.D. Philipson, Mutant analysis of the Ca^{2+} binding domain. J. Gen. Physiol. **105**, 403–420 (1995)

P. Montaville, C. Schlicker, A. Leonov, M. Zweckstetter, G.M. Sheldrick, S. Becker, The C2A-C2B linker defines the high affinity Ca^{2+} binding mode of rabphilin-3A. J. Biol. Chem. **282**, 5015–5025 (2007)

D.A. Nicoll, M.R. Sawaya, S. Kwon, D. Cascio, K.D. Philipson, J. Abramson, The crystal structure of the primary Ca^{2+} sensor of the Na^+/Ca^{2+} exchanger reveals a novel Ca^{2+} binding motif. J. Biol. Chem. **281**, 21577–21581 (2006)

D.A. Nicoll, X. Ren, M. Ottolia, M. Phillips, A.R. Parades, J. Abramson, K.D. Philipson, What we know about the structure of NCX1 and how it relates to its function. Ann. NY Acad. Sci. **1099**, 1–6 (2007)

M. Ottolia, K.D. Philipson, S. John, Conformational changes of the Ca^{2+} regulatory site of the Na^+-Ca^{2+} exchanger detected by FRET. Biophys. J. **87**, 899–906 (2004)

M. Ottolia, D.A. Nicoll, K.D. Philipson, Roles of two Ca^{2+}-binding domains in regulation of the cardiac Na^+-Ca^{2+} exchanger. J. Biol. Chem. **284**, 32735–32741 (2009)

M. Ottolia, D.A. Nicoll, S. John, K.D. Philipson, Interactions between Ca^{2+} binding domains of the Na^+-Ca^{2+} exchanger and secondary regulation. Channels **4**, 1–4 (2010)

K.D. Philipson, D.A. Nicoll, Sodium-calcium exchange: a molecular perspective. Annu. Rev. Physiol. **62**, 111–133 (2000)

K.D. Philipson, M.M. Bersohn, A.Y. Nishimoto, Effects of pH on Na^+-Ca^{2+} exchange in canine cardiac sarcolemmal vesicles. Circ. Res. **50**, 287–293 (1982)

S.M. Pogwizd, K. Schlotthauer, L. Li, W. Yuan, D.M. Bers, Arrhythmogenesis and contractile dysfunction in heart failure: roles of sodium-calcium exchange, inward rectifier potassium current, and residual beta-adrenergic responsiveness. Circ. Res. **88**, 1159–1167 (2001)

J. Reeves, M. Condrescu, Ionic regulation of the cardiac sodium-calcium exchanger. Channels **2**, 322–328 (2008)

J. Rizo, T.C. Südhof, C2-domains, structure and function of a universal Ca^{2+}-binding domain. J. Biol. Chem. **273**, 15879–15882 (1998)

P.K. Salinas, L. Brüschweiler-Li, E. Johnson, R. Brüschweiler, Ca^{2+} binding alters the inter-domain flexibility between the cytoplasmic calcium-binding domains in the Na^+/Ca^{2+} exchanger. J. Biol. Chem. **286**, 32123–32131 (2011)

E.A. Sobie, K.W. Dilly, J. dos Santos Cruz, W.J. Lederer, M.S. Jafri, Termination of cardiac Ca^{2+} sparks: an investigative mathematical model of calcium-induced calcium release. Biophys. J. **83**, 59–78 (2002)

R.V. Stahelin, J. Wang, N.R. Blatner, J.D. Rafter, D. Murray, W. Cho, The origin of C1A-C2 interdomain interactions in protein kinase C. J. Biol. Chem. **280**, 36452–36463 (2005)

C.R. Weber, K.S. Ginsburg, K.D. Philipson et al., Allosteric regulation of Na/Ca exchange current by cytosolic Ca in intact cardiomyocytes. J. Gen. Physiol. **117**, 119–131 (2001)

S. Wei et al., Cytosolic free magnesium modulates Na/Ca exchange currents in pig myocytes. Cardiovasc. Res. **53**, 334–340 (2002)

M. Wu, M. Wang, J. Nix, L.V. Hryshko, L. Zheng, Crystal structure of CBD2 from the *Drosophila* Na^+/Ca^{2+} exchanger: diversity of Ca^{2+} regulation and its alternative splicing modification. J. Mol. Biol. **387**, 104–112 (2009)

M. Wu et al., Crystal structures of progressive Ca^{2+} binding states of the Ca^{2+} sensor Ca^{2+} binding domain 1 (CBD1) from the CALX Na^+/Ca^{2+} exchanger reveal incremental conformational transitions. J. Biol. Chem. **285**, 2554–2561 (2010)

M. Wu et al., Structural basis of the Ca^{2+} inhibitory mechanism of *Drosophila* Na/Ca exchanger CALX and its modification by alternative splicing. Structure **19**, 1509–1517 (2011)

NCX1: Mechanism of Transport

5

Michela Ottolia and Kenneth D. Philipson

Abstract

The plasma membrane Na^+/Ca^{2+} exchanger (NCX) plays a critical role in the maintenance of Ca^{2+} homeostasis in a variety of tissues. NCX accomplishes this task by either lowering or increasing the intracellular Ca^{2+} concentration, a process which depends on electrochemical gradients. During each cycle, three Na^+ are transported in the opposite direction to one Ca^{2+}, resulting in an electrogenic transport that can be measured as an ionic current.

The residues involved in ion translocation are unknown. A residue thought to be important for Na^+ and/or Ca^{2+} transport, Ser110, was replaced with a cysteine, and the properties of the resulting exchanger mutant were analyzed using the giant patch technique. Data indicate that this residue, located in transmembrane segment 2 (part of the α-1 repeat), is important for both Na^+ and Ca^{2+} translocations. Using cysteine susceptibility analysis, we demonstrated that Ser110 is exposed to the cytoplasm when the exchanger is in the inward state configuration.

Keywords

Sodium-calcium exchange • α repeats • Ion translocation

5.1 Introduction

There are three mammalian isoforms of the Na^+/Ca^{2+} exchanger: NCX1, NCX2, and NCX3, which are found in many tissues where they play a major role in Ca^{2+} extrusion. The cardiac isoform NCX1 (NCX throughout this chapter) has been studied in the most molecular detail, and its fundamental role in contractile events has been extensively investigated (Philipson et al. 2002; Pogwizd and Bers 2002).

M. Ottolia (✉) • K.D. Philipson
Department of Physiology and the Cardiovascular Research Laboratory, David Geffen School of Medicine at UCLA, Cardiovascular Research Laboratory MRL 3-645, Los Angeles, CA 90095-1760, USA
e-mail: ottolia@mednet.ucla.edu

L. Annunziato (ed.), *Sodium Calcium Exchange: A Growing Spectrum of Pathophysiological Implications*, Advances in Experimental Medicine and Biology 961, DOI 10.1007/978-1-4614-4756-6_5,

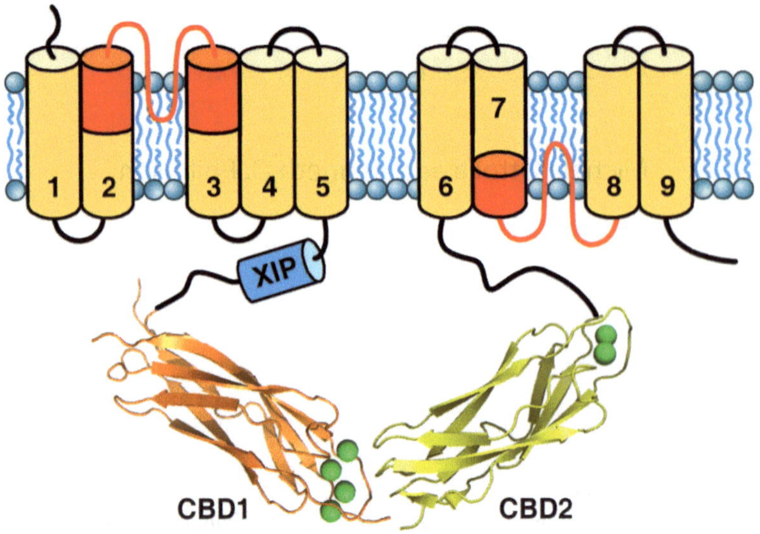

Fig. 5.1 *Secondary structure of the cardiac* Na^+/Ca^{2+} *exchanger*. Important regions are highlighted. The α repeats, involved in ion translocation, are shown in *red*. In the large cytoplasmic loop are regulatory domains such as the XIP region (*dark blue*), involved in Na^+-dependent inactivation, and the Ca^{2+}-binding domains (CBD1 and CBD2). Binding of Ca^{2+} to these two domains activates the exchanger and removes the Na^+-dependent inactivation. The crystal structures of the first and second regulatory domains are shown in an arbitrary orientation

Cloning of the cardiac Na^+/Ca^{2+} exchanger in 1990 has led to our current understanding of the exchanger secondary structure (Fig. 5.1). NCX is organized into three domains. The first N-terminal domain consists of transmembrane segments (TMSs) 1–5. The large cytoplasmic loop connecting TMS5 to TMS6 is the second domain, while the remaining four TMSs in the carboxyl terminus constitute the third component of NCX structure. This topology was determined mainly from cysteine accessibility mutagenesis and epitope mapping studies (Iwamoto et al. 1999; Nicoll et al. 1999; Iwamoto et al. 2000). Using membrane impermeable reagents, both techniques allow determination of whether strategically positioned cysteine residues or other tags are extracellular or intracellular. NCX is a secondary transporter that utilizes the Na^+ gradient to move Ca^{2+} against its concentration gradient. During each reaction cycle, one Ca^{2+} is moved out of the cell, while three Na^+ are transported into the cytoplasm. This is considered the normal or forward reaction mode of NCX. Depending on the net electrochemical driving force, NCX may also transport Ca^{2+} into the cell, so-called the "reverse" mode. The regions of NCX involved in ion translocation are not well defined. Evidence suggests that two highly conserved signature sequences within the TMSs (indicated as "α repeats"; Fig. 5.1) are central to ion translocation, since mutations at these sites affect transport (Nicoll et al. 1996; Ottolia et al. 2005). Additional studies are required to determine which residues coordinate the transported ions.

In this chapter, we will summarize recent developments toward the understanding of NCX modes of operation with special emphasis on the mechanism of ion transport. Additionally, we will present new data indicating that residue Ser^{110} of TMS2 of the cardiac Na^+/Ca^{2+} exchanger is important for ion translocation.

5.2 Results

A feature of NCX, and the Ca^{2+}-cation exchanger family in general, is the presence of two highly conserved motifs identified as alpha repeats (α repeats) (Schwarz and Benzer 1997, 1999). Alpha-1 repeat (α-1 repeat) consists of a portion of TMSs 2–3 and its connecting loop. Alpha-2 repeat (α-2 repeat) is made up from TMS 7 and

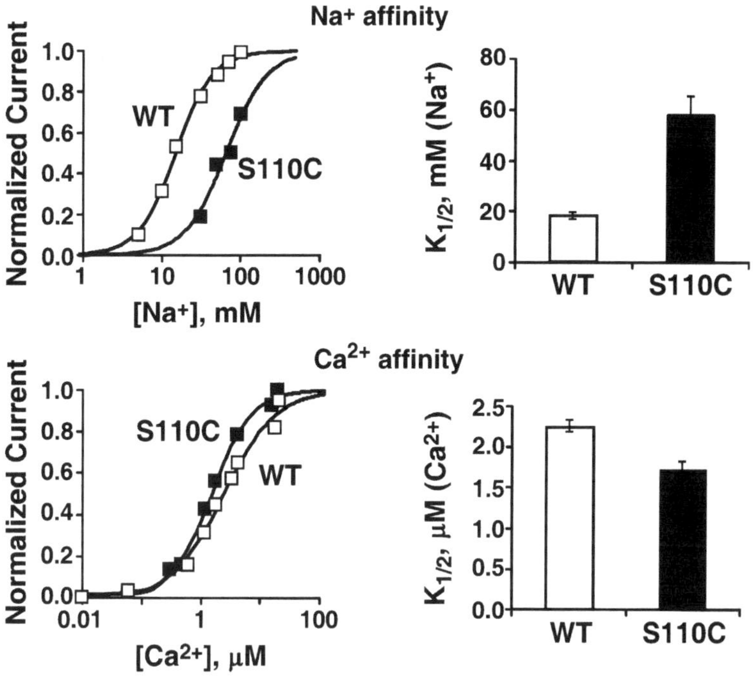

Fig. 5.2 *Mutation of Ser[110] alters both Na^+ and Ca^{2+} transport*. The WT and S110C exchangers were expressed in *Xenopus laevis* oocytes. Outward currents were generated by perfusing the intracellular surface of the patch with different Na^+ concentrations in the presence of 8 mM Ca^{2+} within the pipette (external surface). To assess the intracellular dependency of NCX to Ca^{2+}, 100 mM Na^+ was present in the pipette, while Ca^{2+} was varied in cytoplasm. Before recording, the patches were treated with chymotrypsin to remove ionic regulation. Shown are representative Na^+ (*top right*) and Ca^{2+} (*bottom right*) dependency curves. Average values from 2 to 5 experiments are shown on the *left*. The modest change in the Ca^{2+} affinity was significant. *Black* and *white circles* indicate wild-type and mutant exchangers, respectively. Experiments were conducted at $V_H = 0$ mV at 35 °C

the following intracellular loop (Fig. 5.1) (Philipson et al. 2002). The alpha repeats have opposite orientation (Iwamoto et al. 1999; Nicoll et al. 1999) similar to that observed in other transporters and have been associated with ion translocation (Compton et al. 2010). The connecting loops of the α repeats are modeled to invaginate into the membrane to form reentrant loops. Similar structures are found lining the ion translocation pathway in ion channels supporting the hypothesis that the α repeats may form an ion translocation pathway within NCX. Electrophysiological studies have corroborated the role of the second reentrant loop of the α-2 repeat in ion translocation since mutagenesis in this region alters both Na^+ and Ca^{2+} affinities of NCX (Shigekawa et al. 2002). In contrast, only residues within the transmembrane portion of TMS2 of the α-1 repeat have been found important for NCX activity (Nicoll et al. 1996), as assessed by $^{45}Ca^{2+}$ uptake, while the loop is not essential for transport (Ottolia et al. 2005).

Despite these investigations, the residues that coordinate the transported Na^+ and Ca^{2+} remain unidentified. By combining mutagenesis and electrophysiology, we are currently investigating which of the residues lining TMSs 2 and 7 of the cardiac Na^+/Ca^{2+} exchanger are important for ion translocation. As an example of the mutagenesis studies that we are conducting, Fig. 5.2 shows the altered transport properties of an exchanger with residue Ser[110] mutated to cysteine. The mutant exchanger S110C was expressed in *Xenopus laevis* oocytes, and ion currents were measured using the giant patch technique in the inside-out configuration (Ottolia et al. 2005). Prior to recording, the cytoplasmic side of the patch was briefly exposed to chymotrypsin, thereby cleaving the

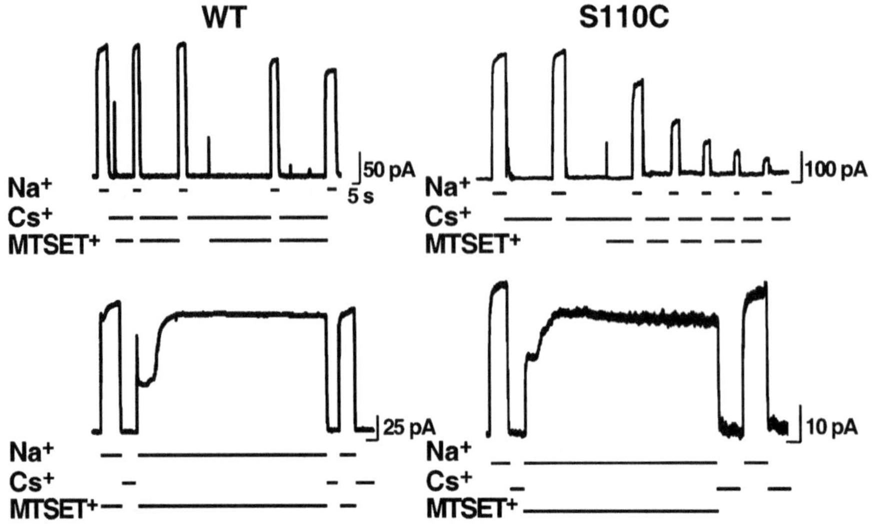

Fig. 5.3 *State-dependent accessibility of S110C*. Shown are outward currents recorded from oocytes expressing either the WT (*left*) or S110C (*right*) mutant exchangers. Intracellular $MTSET^+$ (5 mM) was applied in either the presence (*bottom*) or absence of Na^+ (*top*, replaced with equal amounts of Cs^+). Note that $MTSET^+$ does not affect ionic currents recorded from the WT, indicating that the endogenous cysteines are either inaccessible or functionally unimportant. In contrast, ionic currents recorded from S110C were blocked by $MTSET^+$ when perfused in presence of Cs^+ (*top right*). In this condition, the exchanger is held in the inward configuration state and does not transport. $MTSET^+$ block was not observed during cycling (*bottom right*) showing that Na^+ alters S^{110} accessibility

large intracellular loop eliminating both Na^+ and Ca^{2+} secondary regulations, without altering transport properties (Hilgemann et al. 1991a). This allows ion transport to be studied independently from the effects of Na^+ and Ca^{2+} ions on secondary regulation. As shown, mutation at S110C drastically decreases the affinity of NCX for cytoplasmic Na^+ while increasing the affinity for Ca^{2+}. This new evidence supports the involvement of residues Ser^{110} of TMS2 in ion transport.

To corroborate the role of Ser^{110} in ion translocation, we determined its accessibility to the membrane impermeant sulfhydryl reagent $MTSET^+$ applied from the cytoplasmic side during transport or when NCX is restricted to the inward state configuration. If this residue contributes to ion binding, its accessibility may change in either or both conditions.

Outward currents were recorded from oocytes expressing either WT or S110C exchangers. Inside-out excised patches were then exposed to cytoplasmic $MTSET^+$ (5 mM) in the absence or presence of Na^+. In the absence of Na^+, the transported binding sites are exposed to the cytoplasm waiting for the substrate (inward state configuration). Under these conditions, S110C currents decreased overtime upon perfusion of $MTSET^+$ due to its interaction with the introduced cysteine (Fig. 5.3), while the WT exchanger was insensitive to $MTSET^+$. Alternatively, $MTSET^+$ was applied simultaneously with Na^+, conditions that allow NCX to cycle through the different states of the reaction cycle. Under these conditions, $MTSET^+$ application prevented inhibition of the S110C current, indicating that the accessibility of Ser^{110} is state dependent. Similar analyses of the remaining residues of TMS2 and TMS7 are currently being investigated.

5.3 Conclusion

With the cloning of the Na^+/Ca^{2+} exchanger (Nicoll et al. 1990) and the development of the giant patch technique, detailed molecular studies of the protein and the understanding of its modes of regulation and transport have been made possible.

It is generally accepted that to exchange ions, NCX follows a consecutive mechanism where the occupancy of the transport site(s) by Na^+

(or Ca^{2+}) triggers conformational changes, causing the binding sites to shuttle between opposite sides of the membrane (Hilgemann et al. 1991b; Matsuoka and Hilgemann 1992). The regions lining the ion passageways are unknown as are the residues that coordinate transported Na^+ and Ca^{2+}. Evidence suggests that two highly conserved signature sequences within the TMSs (indicated as "α repeats"; Fig. 5.1) are central to ion translocation, since mutations at these locations affect activity as determined by $^{45}Ca^{2+}$ uptake (Nicoll et al. 1996). Cross-linking experiments show that the α repeats are in close proximity and may form an ion translocation pathway (Qiu et al. 2001).

In this work, we analyzed the role of residue Ser^{110} in ion transport. Mutation of this residue to cysteine drastically decreased the affinity of NCX for cytoplasmic Na^+ while significantly increasing the sensitivity to intracellular Ca^{2+}. These data indicate that Ser^{110} is crucial for activity and suggests residue Ser^{110} as an important component of the Na^+ and or Ca^{2+}-binding sites.

Two other observations strengthen the likely importance of this residue in ion translocation. First, Ser^{110} is highly conserved among the exchanger family, and second, structural studies of other unrelated secondary Na^+-coupled transporters reveal that contiguous Ser and/or Thr residues contribute carboxyl oxygens for Na^+ coordination (Krishnamurthy et al. 2009). Also, recent studies show that the corresponding residue in Na^+/Ca^{2+}-K^+ exchanger alters both the Na^+ and Ca^{2+}/K^+ affinities (Altimimi et al. 2010).

We sought further evidence to support our hypothesis that Ser^{110} is involved in ion binding by determining its accessibility during the Na^+/Ca^{2+} exchange cycle. If this portion of NCX binds either Na^+ or Ca^{2+}, it is expected to undergo conformational changes. Accordingly, analysis of S110C indicates that this residue faces the cytoplasm when the exchanger binding sites are forced toward the intracellular environment (inward state configuration), but not during ion exchange. This residue is modeled to be deep within the plasma membrane, and its accessibility to intracellular $MTSET^+$ suggests the presence of a cytoplasmic ion pathway or cavity used by Na^+ and/or Ca^{2+} to reach S110C.

Together these results demonstrate the important role of Ser^{110} for activity. We are currently investigating the other regions of NCX to determine which residues are critical for activity and to define those forming the ion translocation pathway of NCX.

References

H.F. Altimimi, E.H. Fung, R.J. Winkfein, P.P. Schnetkamp, Residues contributing to the Na+-binding pocket of the SLC24 Na^+/Ca^{2+}-K^+ Exchanger NCKX2. J. Biol. Chem. **285**, 15245–15255 (2010)

E.L. Compton, E.M. Taylor, J.A. Mindell, The 3-4 loop of an archaeal glutamate transporter homolog experiences ligand-induced structural changes and is essential for transport. Proc. Natl. Acad. Sci. U. S. A. **107**, 12840–12845 (2010)

D.W. Hilgemann, A. Collins, D.P. Cash, G.A. Nagel, Cardiac Na^+-Ca^{2+} exchange system in giant membrane patches. Ann. N. Y. Acad. Sci. **639**, 126–139 (1991a)

D.W. Hilgemann, D.A. Nicoll, K.D. Philipson, Charge movement during Na^+ translocation by native and cloned cardiac Na^+/Ca^{2+} exchanger. Nature **35**, 715–718 (1991b)

T. Iwamoto, T.Y. Nakamura, Y. Pan, A. Uehara, I. Imanaga, M. Shigekawa, Unique topology of the internal repeats in the cardiac Na^+/Ca^{2+} exchanger. FEBS Lett. **446**, 264–268 (1999)

T. Iwamoto, A. Uehara, I. Imanaga, M. Shigekawa, The Na^+/Ca^{2+} exchanger NCX1 has oppositely oriented reentrant loop domains that contain conserved aspartic acids whose mutation alters its apparent Ca^{2+} affinity. J. Biol. Chem. **275**, 38571–38580 (2000)

H. Krishnamurthy, C.L. Piscitelli, E. Gouaux, Unlocking the molecular secrets of sodium-coupled transporters. Nature **459**, 347–355 (2009)

S. Matsuoka, D.W. Hilgemann, Steady-state and dynamic properties of cardiac sodium-calcium exchange. Ion and voltage dependencies of the transport cycle. J. Gen. Physiol. **100**, 963–1001 (1992)

D.A. Nicoll, S. Longoni, K.D. Philipson, Molecular cloning and functional expression of the cardiac sarcolemmal Na^+-Ca^{2+} exchanger. Science **250**, 562–565 (1990)

D.A. Nicoll, L.V. Hryshko, S. Matsuoka, J.S. Frank, K.D. Philipson, Mutation of amino acid residues in the putative transmembrane segments of the cardiac sarcolemmal Na^+-Ca^{2+} exchanger. J. Biol. Chem. **271**, 13385–13391 (1996)

D.A. Nicoll, M. Ottolia, L. Lu, Y. Lu, K.D. Philipson, A new topological model of the cardiac sarcolemmal Na^+-Ca^{2+} exchanger. J. Biol. Chem. **274**, 910–917 (1999)

M. Ottolia, D.A. Nicoll, K.D. Philipson, Mutational analysis of the alpha-1 repeat of the cardiac Na^+-Ca^{2+} exchanger. J. Biol. Chem. **280**, 1061–1069 (2005)

K.D. Philipson, D.A. Nicoll, M. Ottolia, B.D. Quednau, H. Reuter, S. John, Z. Qiu, The Na^+/Ca^{2+} exchange molecule: an overview. Ann. N. Y. Acad. Sci. **976**, 1–10 (2002)

S.M. Pogwizd, D.M. Bers, Na/Ca exchange in heart failure: contractile dysfunction and arrhythmogenesis. Ann. N. Y. Acad. Sci. **976**, 454–465 (2002)

Z. Qiu, D.A. Nicoll, K.D. Philipson, Helix packing of functionally important regions of the cardiac Na^+-Ca^{2+} exchanger. J. Biol. Chem. **276**, 194–199 (2001)

E.M. Schwarz, S. Benzer, Calx, a Na^+-Ca^{2+} exchanger gene of *Drosophila* melanogaster. Proc. Natl. Acad. Sci. U. S. A. **94**, 10249–10254 (1997)

E. Schwarz, S. Benzer, The recently reported NIbeta domain is already known as the Calx-beta motif. Trends Biochem. Sci. **24**, 260 (1999)

M. Shigekawa, T. Iwamoto, A. Uehara, S. Kita, Probing ion binding sites in the Na^+/Ca^{2+} exchanger. Ann. N. Y. Acad. Sci. **976**, 19–30 (2002)

Structural Studies of the Ca^{2+} Regulatory Domain of *Drosophila* Na^{+}/Ca^{2+} Exchanger CALX

6

Lei Zheng, Mousheng Wu, and Shuilong Tong

Abstract

CALX, the NCX homolog in Drosophila, involves in light-mediated Ca^{2+} homeostasis in sensory neuronal cells. CALX exhibits a unique negative Ca^{2+} regulatory property mediated by Ca^{2+} binding at its intracellular regulatory domain. Our structural studies of individual CBD1 or CBD2 domain reveal that CBD1 is the only Ca^{2+} binding domain in CALX. Crystal structures of the entire Ca^{2+} regulatory domain CBD12 from two alternative splicing isoforms, CALX1.1 and CALX1.2, demonstrate that CBD1 and CBD2 form an open V-shaped conformation with four Ca^{2+} ions bound on the CBD domain interface. The structures together with Ca^{2+} binding analyses strongly argue that the Ca^{2+} inhibition of CALX is achieved by interdomain conformational change induced by Ca^{2+} binding at CBD1. The conformational difference between the two isoforms also raises a hypothesis that alternative splicing residues adjust the interdomain orientation angle between CBD1 and CBD2 to modify the Ca^{2+} regulatory property of the exchanger. These studies not only establish structural basis to understand the inhibitory Ca^{2+} regulation and the alternative splicing modification of CALX, but also shed light on the general Ca^{2+} regulatory mechanism of other mammalian NCX proteins.

Keywords

Drosophila • CALX • Sodium-calcium exchange

L. Zheng (✉) • M. Wu • S. Tong
Center for Membrane Biology,
Department of Biochemistry and Molecular Biology,
The University of Texas Houston Medical School,
Houston, TX 77030, USA
e-mail: lei.zheng@uth.tmc.edu

L. Annunziato (ed.), *Sodium Calcium Exchange: A Growing Spectrum of Pathophysiological Implications*, Advances in Experimental Medicine and Biology 961, DOI 10.1007/978-1-4614-4756-6_6,

6.1 Introduction

6.1.1 Na^+/Ca^{2+} Exchanger CALX

Sodium-calcium exchangers (NCXs) constitute a large Ca^{2+} transporter protein family existing in many eukaryotic cells (Cai and Lytton 2004). They are energized by a counter-Na^+ electrochemical gradient to perform Ca^{2+} flux across the plasma membrane. The importance of NCXs in maintaining Ca^{2+} homeostasis and in regulating Ca^{2+}-mediated signaling has been well demonstrated in many tissues including cardiomyocytes and neurons (Bers 2001; Blaustein et al. 1996).

CALX is the *Drosophila* NCX homolog protein mainly expressed in the photoreceptor cells (Schwarz and Benzer 1997; Hryshko et al. 1996; Wang et al. 2005). It plays a pivotal role in *Drosophila* visual signaling transduction (Wang and Montell 2007). In fly rhabdomeres, the intracellular Ca^{2+} concentration is maintained at a low level of ~160 nM at resting state. Light stimulation mediated via rhodopsin triggers activation of photoreception cation channels, primarily TRP channels. As a result, the intracellular Ca^{2+} level may be instantly elevated to 200 μM by more than 10^3 fold (Ranganathan et al. 1991). To adapt the rapid photo-stimulated response, an efficient Ca^{2+} efflux system is required to reset the intracellular Ca^{2+} concentration back to the resting state level. CALX colocalized with TRP channels on the photoreceptor cell membrane provides a primary mechanism to fulfill Ca^{2+} efflux out of the cells and to achieve dynamic Ca^{2+} homeostatic turnover. *CalX* inactivation impairs Ca^{2+} efflux, sustaining abnormally high Ca^{2+} concentration induced by light stimuli in the cytosol. This impairment is associated with profound reduction of light stimulating amplification, hyper-adaptation and retinal degeneration, probably caused by Ca^{2+} overload (Wang et al. 2005).

CALX shares high sequence homology (52%) and also a conserved structural motif with other mammalian NCX proteins. Ten transmembrane helices (TMs) are predicted to form a membrane domain for Na^+/Ca^{2+} exchange and a large intracellular regulatory region of about 500 amino acid residues splits TM 5 and 6 (Schwarz and Benzer 1997). Electrophysiological characterization in Oocytes suggested that CALX shares a similar mechanism to that of other mammalian NCXs to perform Na^+/Ca^{2+} exchange on its membrane domain (Hryshko et al. 1996).

6.1.2 Ca^{2+} Regulatory Feature of the NCX Proteins

Na^+/Ca^{2+} exchange activity of NCX is regulated by many factors including its substrates Ca^{2+} and Na^+. Ca^{2+} regulation remains one of the most interesting features (Philipson and Nicoll 2000) since it controls the exchanger activity by sensing any change of the intracellular Ca^{2+} condition to fulfil instant request for cellular Ca^{2+} extrusion. Ca^{2+} regulation is exclusively achieved through the intracellular regulatory region of the exchanger (Matsuoka et al. 1993). Two specific Ca^{2+} binding domains, CBD1 and CBD2, have been characterized within the intracellular region (Hilge et al. 2006; Nicoll et al. 2006; Besserer et al. 2007). In spite of the high sequence homology of the CBD domains, CALX exhibits a unique negative Ca^{2+} regulatory property in contrast to the positive effect of other characterized NCXs (Hryshko et al. 1996), making CALX an ideal model to study the regulatory diversity of the NCX proteins.

The fundamental hypothesis for the Ca^{2+} regulatory mechanism is that Ca^{2+} binding on the intracellular CBD domains induces a large protein conformational change, which is tranduced to its membrane domain to control Na^+/Ca^{2+} exchange. This hypothesis is supported by a FRET study with fluorescent protein inserted in the full length NCX1 protein. A large conformational change within the intracellular domain was observed by regulatory Ca^{2+} induction (Ottolia et al. 2004). However, a controversial opinion has been raised with another study using a different approach. i.e. an NMR study with isolated CBD1 of NCX1 only detected subtle protein motions which is limited to the local Ca^{2+} binding site (Johnson et al. 2008). This argument suggests a novel mechanism governing the reversible protein conformational transition mediated by Ca^{2+} binding/unbinding within the CBD domains.

6.1.3 Alternative Splicing Modification

The Ca^{2+} regulatory property of NCXs is significantly modified by alternative splicing process. Two splicing variants, CALX1.1 and CALX1.2, have identified in *Drosophila*, which only differ by five residues (Omelchenko et al. 1998). Although the functional relevance of the alternative splicing in *Drosophila* visual Ca^{2+} signaling is unknown, electrophysiological study in oocytes has shown markedly distinct Ca^{2+} regulatory properties between these two variants. In CALX1.1, both peak and steady currents exhibit high sensitivity to regulatory Ca^{2+}. In contrast, CALX1.2 displays higher affinity for Ca^{2+} but much smaller contents of the inhibitory effect, which is also limited to the peak current (Omelchenko et al. 1998). Mammalian NCXs appear to undergo greater splicing processing, resulting in more splicing variants than CALX (Dyck et al. 1999). Regardless of the different Ca^{2+} regulatory features between CALX and mammalian NCXs, all of the alternative splicing occurs within the 2F–2G loop of CBD2 (Wu et al. 2009). How those alternative residues on CBD2 affect the Ca^{2+} regulatory property of the exchanger remains unknown. Studying this feature will provide insight into how alternative splicing modulates the Ca^{2+} regulatory feature of NCX to adapt Ca^{2+} homeostasis in different tissues.

6.2 Structure of CBD2 Domain of CALX

6.2.1 Overall Conformation of CBD2 Structure

Both CBD1 and CBD2 domains of CALX exhibit high sequence similarity in comparison with their counterparts from canine NCX1 (Wu et al. 2009). Sequence alignments suggest that CALX has an abnormal Ca^{2+} binding site on CBD2 in contrast to a conserved Ca^{2+} binding site on CBD1. To confirm any structural difference of CBD2 between CALX and NCX1, we determined crystal structure of the CBD2 domain of CALX1.1 (Wu et al. 2009).

Similar to the CBD2 structure of canine NCX1, the CALX-CBD2 structure displays an immunoglobulin-like conformation. It contains two antiparallel β sheets formed by β strands 2A, 2B, and 2G, and β strands 2C, 2D, 2E and 2F, respectively (Fig. 6.1a). The two CBD2 structures are well superimposed, as indicated by a marginal root mean square deviation (*rmsd*) of 0.89 Å by comparing their 97 C_α atoms. Including a downstream fragment (amino acid residues 709–730) in the CBD2 construct is important for the crystallization. This short α-helix is expected to connect directly with the adjacent TM6 of the membrane domain. The high similarity of these

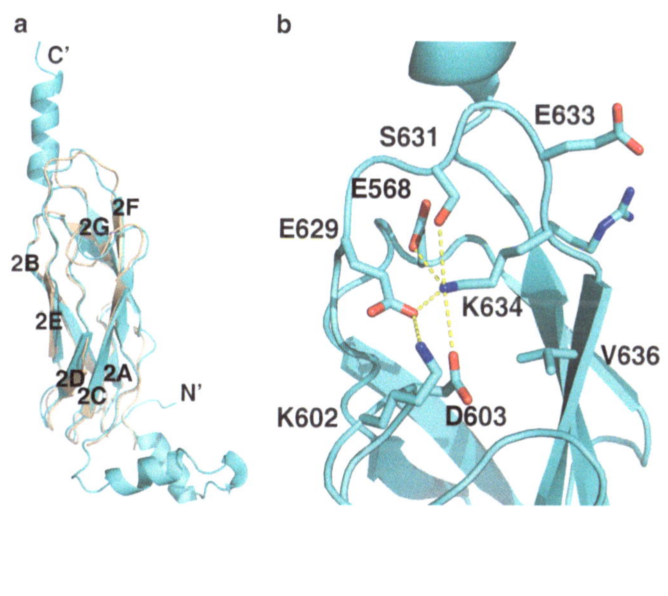

Fig. 6.1 *Structure of the CBD2 domain from CALX1.1.* (**a**) Superimposition of CBD2 structures from CALX1.1 (*cyan*) and NCX1.1 (*gray*) depicted as cartoons. (**b**) The conformation of the inactive Ca^{2+} binding site of CALX1.1 CBD2. The residues drawn as sticks interact each other via H-bonds shown as *yellow dash lines*

two CBD2 structures from mammal or insect suggests structural conservation of the CBD2 domain in the NCX protein family.

6.2.2 Conformation of the Inactive Ca^{2+} Binding Site in CALX-CBD2

However, significant conformational difference was observed in the putative Ca^{2+} binding site of CBD2. In NCX1-CBD2 structure, two Ca^{2+} ions were observed in its distal loop region (Hilge et al. 2006; Besserer et al. 2007). However, no Ca^{2+} was found in the analogous region of CALX1.1-CBD2 in presence of 2 mM Ca^{2+} in the crystallization buffer. In contrast to that of NCX1, the predicted Ca^{2+} binding region of CALX-CBD2 undergoes a dramatic structural rearrangement mainly owing to flipping-over of the 2E–2F loop residues across the predicted Ca^{2+} binding site. Resulting in disruption of the Ca^{2+} binding site (Fig. 6.1b) (Wu et al. 2009).

This structural rearrangement might be initialized by a charge compensation effect due to substitution of a lysine at position 636 by a valine residue in CALX. Three carboxylate residues (E603, E568, and E633) that are involved in Ca^{2+} binding in NCX1-CBD2 point toward the center of the predicted Ca^{2+} binding site. Instead of coordinating Ca^{2+}, these carboxylate residues are stabilized by two lysine residues (K602 and K634) via an extensive electrostatic interaction network, consequently occluding any Ca^{2+} access to the Ca^{2+} binding site (Fig. 6.1b). This abolished Ca^{2+} binding conformation is unlikely caused by crystallization constraint, but reflects the native conformation of CBD2 of CALX since no Ca^{2+} binding was detected with the purified CALX-CBD2 protein in the isothermal titration calorimetric (ITC) Ca^{2+} binding assay (Wu et al. 2009). It is still unclear that this abnormal CBD2 conformation is causal for the negative Ca^{2+} regulatory feature of CALX. These evidences provide evidence on the structural diversity of the CBD2 conformation in the NCX proteins. The inactivation of the CBD2 domain is also indicative of a critical role of the CBD1 in the Ca^{2+} regulatory mechanism of CALX since it is the only intracellular region interacting with regulatory Ca^{2+} in this *Drosophila* exchanger protein.

6.3 Structure of the CBD1 Domain of CALX

6.3.1 Overall Conformation of CBD1 Structure

We first determined the structure of the Ca^{2+} bound form of CALX-CBD1 (Wu et al. 2010). The CBD1 domain exhibits a high structural similarity with CBD2, suggesting a gene duplication occurred within NCX molecule. The structure shows that CALX-CBD1 has an immunoglobulin-like conformation formed by two antiparallel β sheets consisting of β-strands 1A, 1B and 1G and β strands 1C, 1D, 1E and 1F, respectively (Fig. 6.2a). CALX-CBD1 shares 60% sequence identity with that of canine NCX1-CBD1 (Wu et al. 2009). As expected, these two CBD1 structures can be superimposed with a *rmsd* value of 0.8 Å for 113 aligned C_α atoms. The major difference between these two CBD1 structures is at their 1F–1G loops. CALX has a rather short 1F–1G loop containing only 9 residues, in contrast to its longer counterpart from NCX1 containing 28 residues.

6.3.2 Conformation of the Ca^{2+} Binding Site at CBD1

In addition to the overall structure similarity, the fly CALX-CBD1 domain exhibit nearly identical conformation at the Ca^{2+} binding site to that from mammalian NCX-CBD1 (Wu et al. 2010). Four Ca^{2+} ions are found in the distal loops of the β sandwich, forming a linear orientation with ~4 Å distance spacing between them (Fig. 6.2b). At the binding site, nine carboxylate residues assemble together to form an extensive electronegative moiety to coordinate Ca^{2+}. The 1E–1F loop is the major structural component in the Ca^{2+} binding site with five residues involved in Ca^{2+} coordination, including D515, D516, E520 and the carbonyl oxygen atoms of D517 and V518. Compared to the

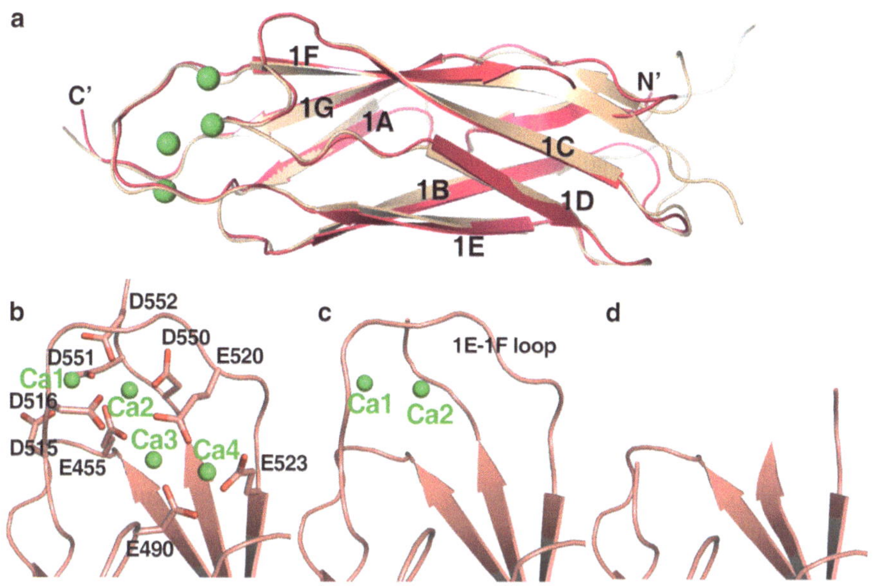

Fig. 6.2 *Structure of the CBD1 domain from CALX1.1.* (**a**) Superimposition of CBD1 domain structures from CALX1.1 (*red*) and NCX1.1 (*gold*) shown as cartoons. Four Ca^{2+} (Ca 1–4) are displayed as *green* spheres. (**b**) The conformation of the Ca^{2+} binding site of CALX CBD1. Nine carboxylate residues for Ca^{2+} coordination are drawn as sticks and labeled. (**c**) The two-Ca binding state of CBD1 structure. (**d**) The apo form (0-Ca) of CBD1 structure

penta- or hexa-coordinated Ca-1 or Ca-2, Ca-4 is only tri-coordinated by D490, E523, and E520. The highest thermal B factor suggests that Ca-4 is the most mobile species in the Ca^{2+} binding conformation (Wu et al. 2010). E455 appears to be the critical residue in the binding site. It is located at a central position of the binding site to coordinate three Ca^{2+} ions (Ca 1-3) simultaneously.

6.3.3 Progressive Ca^{2+} Binding Conformations of CBD1

Protein conformational change induced by regulatory Ca^{2+} binding at the intracellular domain is important in maintaining the Ca^{2+} regulatory properties of both CALX and NCX1. Given that the membrane domain structure is unavailable, it would be important to demonstrate the protein conformational change mediated by Ca^{2+} binding/unbinding to gain insight into how this regulatory motion is triggered on the intracellular side of the molecule. Therefore, we crystallized the CBD1 protein pretreated with EDTA and then determined the apo form structure of CBD1 (Wu et al. 2010).

In the crystal lattice of CALX-CBD1, four CBD1 molecules assemble in antiparallel in an asymmetric unit. While these four CBD1 molecules show no overall conformational change, their Ca^{2+} binding sites exhibit three distinct Ca^{2+} binding conformations: (a) the four-Ca binding state with its Ca^{2+} binding conformation identical to that seen in the Ca^{2+} bound form; (b) the two-Ca binding state containing only Ca-1 and Ca-2 (the primary Ca^{2+} pair) (Fig. 6.2c); and (c) the apo form, in which no Ca^{2+} were found. The most interesting conformation is the two-Ca state since the absence of Ca-3 and Ca-4 (the secondary Ca^{2+} pair) does not lead to any significant structural change compared to the four-Ca binding state (Fig. 6.2d). Most residues in the Ca^{2+} binding site except E520 are clearly resolved, suggesting the secondary Ca pair is less prominent in maintaining the Ca^{2+} binding site (Wu et al. 2010). In sharp contrast, significant conformational changes was observed in the apo form. Extraction of the primary Ca^{2+} pair results in disruption of nearly all residues at the Ca^{2+} binding site including the entire 1E–1F loop (D516–E522). These three Ca^{2+} binding states not only vividly demonstrate a progressive Ca^{2+} binding picture at CBD1 but also strongly suggest the predominating role of the primary Ca^{2+} pair in stabilizing the entire Ca^{2+} binding region. To be noted, our structural observations of the local conformational change are in line with the protein secondary structure analysis using CD spectroscopy (Wu et al. 2010).

To evaluate the functional roles of the two Ca^{2+} pair and to confirm these different Ca^{2+} binding states existing in the Ca^{2+} regulatory mechanism, we mutated E455 and that E520 to either glutamine or alanine, given that E455 appears to be critical for the primary Ca^{2+} binding and that E520 is the only residue effected by the dissociation of the secondary Ca^{2+} pair in the two-Ca binding state. With collaboration of Dr. Larry Hryshko, these mutants were measured using giant patch-clamping technique in *Xenopus laevis* oocytes to examine their effects on the regulatory Ca^{2+} feature of CALX (Wu et al. 2010). The results clearly reveal unequal roles of these two glutamate residues in the Ca^{2+}-dependent regulatory mechanism. E520 exhibits far less importance in mediating the Ca^{2+} regulatory response than E455. The E520A mutant only showed a slight reduction (two- to threefold) in the inhibitory potency of regulatory Ca^{2+} for peak and steady-state currents. In sharp contrast, the E455A mutation causes a dramatic interfere of the Ca^{2+} regulatory response, i.e. the apparent regulatory Ca^{2+} affinity of E455A for the peak currents is reduced by ~20-fold. A further reduction of 35-fold was observed in the steady-state current. These results consistent with our structural observation clearly illustrate the significance of the residue E455 in the Ca^{2+} regulatory mechanism of CALX and lend support to the hypothesis of the progressive Ca^{2+} binding on CBD1.

6.4 Structures of CBD12 Domain of CALX

6.4.1 Overall Conformation of CBD12

Above structural and functional analyses is consistent with the NMR study that Ca^{2+} binding induces local conformational change of CBD1 (Johnson et al. 2008), raising significant interest in how the local protein motion is transduced to the adjacent membrane domain. To perform the Ca^{2+} regulatory role of CBD1, CBD2 may be its critical partner. Given its proximity with CBD1, particularly its Ca^{2+} binding site. To explore any interaction between these two CBD domains in the intracellular regulatory region, we determined crystal structure of the Ca^{2+}-bound CBD12 (naturally linked CBD1 and CBD2) from CALX1.1 (Wu et al. 2011).

In the CBD12 structure, CBD1 and CBD2 exhibit a V-shaped "soaring eagle" open conformation. The two CBD "wings" are orientated in a 115° angle around the linker residue H553 (Fig. 6.3a). This open conformation results in a large separation of 70 Å between the N- and C-termini. The 2F–2G loop of CBD2, where alternative splicing occurs, forms two helices (H1 and H2) nearly perpendicular to the plane of the two "wings" and lies under the domain hinge region as the "eagle" body. This open V-shaped conformation has been further confirmed in solution, using a luminescence resonance energy transfer (LRET)-based distance measurements (Wu et al. 2011), suggesting that it may represent the native conformation of CBD12 in CALX.

6.4.2 Conformation of the Ca^{2+} Binding Site in Domain Interface

In the CBD12 structure, four Ca^{2+} ions were found at the Ca^{2+} binding site of CBD1 in a conformation nearly identical to that previously observed in the individual CBD1 structure (Figs. 6.2a and 6.3a). CBD1 and CBD2 are directly interacted due to no linker in between. Such a tight orientation inevitably sandwiches the Ca^{2+} binding site of CBD1 on the domain interface. The domain interface is primarily contributed by the 1E–1F loop, the major player in the Ca^{2+}-mediated protein conformational change of CBD1 (Fig. 6.2).

Owing to the high resolution of the CBD12 structure, the domain interface was clearly resolved (Fig. 6.3b). On one side, the 1E–1F loop directly coordinates the four Ca^{2+} ions. On another side, it is structurally backed by the perpendicular

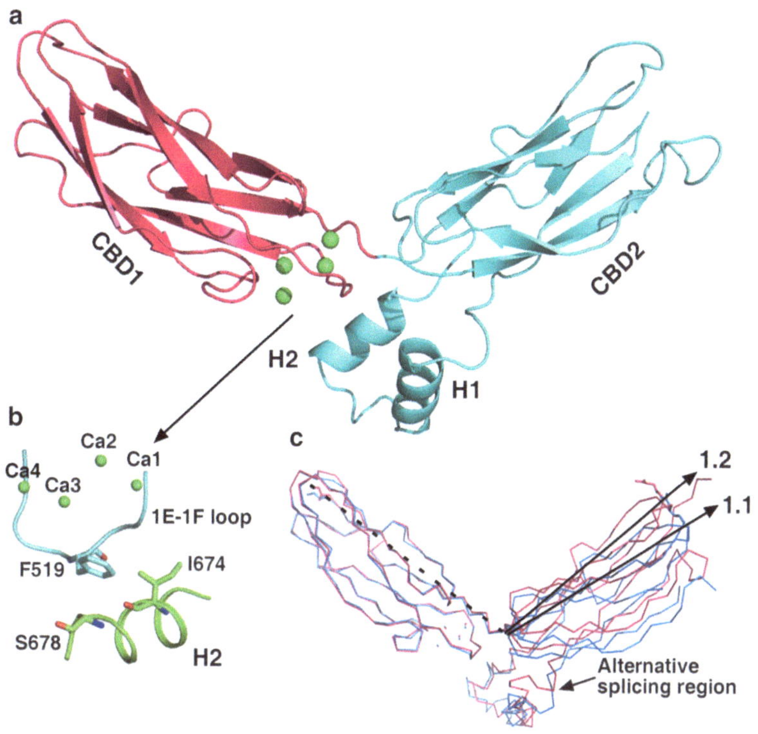

Fig. 6.3 *Structures of CBD12 domain from CALX1.1 and CALX1.2*. (**a**) Overall conformation of CBD12 structure from CALX1.1 with CBD1 (*cyan*) and CBD2 (*red*) naturally linked. Four Ca^{2+} (Ca 1–4) are displayed as *green* spheres. (**b**) The domain interface formed by the 1E–1F loop and H2 helix. (**c**) Superimposition of CBD12 structures shown as ribbons from two CALX alternative splicing variants 1.1 (*blue*) and 1.2 (*red*). Two CBD1 domains are superimposed. Their individual CBD2 domain orientation angles are indicated as *arrows*

H2 helix. F519 on the outer arc of the 1E–1F loop directly interacts with I674 and S678 of the H2 helix. It would be expected that a rigid body conformational change of the two CBD "wings" will be induced if the hinge conformation was destabilized by Ca^{2+} extraction. Mutations of each individual residues on the interface strongly interfere the Ca^{2+} binding behavior of the CBD12 domain protein (Wu et al. 2011). The unequal roles of the four Ca^{2+} ions at the binding site can also be rationalized by their positions at the domain interface. The primary Ca^{2+} pair is close to the hinge region to anchor the 1E–1F loop with the physical linker on the interface, whereas the secondary Ca pair is located at a position closed to the surface. Given the importance of the primary Ca^{2+} pair (Wu et al. 2010), it is anticipated that stabilizing the hinge region by the primary Ca^{2+} pair is critical for the CBD12 domain conformational change.

6.5 Ca^{2+} Regulatory Mechanism of CALX

Our structural and functional analyses support a hypothesis for Ca^{2+} regulatory mechanism of CALX (Wu et al. 2011): (1) CALX remains active in absence of Ca^{2+} at CBD12. (2) Upon increase of intracellular Ca^{2+} concentration, Ca^{2+} accesses the primary Ca^{2+} pair position to stabilize the 1E–1F loop, which will consequently reinforce the domain interface between CBD1 and CBD2. The reposition of CBD1 or CBD2 domain conformation will deliver a protein motion to trigger closure of the membrane domain. (3) Further increasing intracellular Ca^{2+} concentration will lead to occupancy of the secondary Ca^{2+} pair, which will progressively enhance the CBD12 domain conformation to turn off its Na^+/Ca^{2+} translocation pathway on the membrane domain.

6.6 Conformational Difference Between Two Alternative Splicing Variants

Alternative splicing occurs within the region between the H1 helix and the β strand 2F of CBD2. In the CBD12 structure of CALX1.1, this splicing region has no direct contact with the Ca^{2+} binding site of CBD1 (Fig. 6.3a). The CBD12 proteins from the two variants, 1.1 and 1.2, also exhibit similar Ca^{2+} binding activity (Wu et al. 2011). To demonstrate any structural modification of the CBD12 conformation by alternative splicing, we determined crystal structure of the CBD12 domain protein from another splicing variant CALX1.2 (Wu et al. 2011).

The CBD12 structure of CALX1.2 shows a similar V-shaped conformation and a similar Ca^{2+} binding conformation as seen in the CALX1.1 structure (Wu et al. 2011). However, the angle between the CBD1 and CBD2 "wings" undergoes a considerable change, i.e., the CBD2 of CALX1.2 rotates along the hinge axis of H553 further upward by 9° and results in a movement of the C-terminal R693 by 6.6 Å (Fig. 6.3c). This movement is clearly caused by the alternative splicing region. In CALX1.1, the five alternative residues constitute a 3_{10} end of the helix 1 (H1). In contrast, the alternative residues extend the H1 helix by one additional turn toward the 2B–2C loop of CBD2 in CALX1.2. The interface between the H2 and the 1E–1F loop of CBD1 remains unaffected. Thus, the conformational changes occurring on the loop between the H1 helix and the β-strand body of CBD2 eventually lead the rigid CBD2 domain to undergo a closure rotation around the hinge. This pair of CBD12 structures provides direct evidence for the hypothesis that alternative splicing modifies the CBD domain orientational angle to alter the regulatory Ca^{2+} signal transduction to other structural components within the exchanger molecule.

6.7 Conclusions

CALX is the major Ca^{2+} efflux system in *Drosophila* sensory neurons. CALX exhibits a novel negative Ca^{2+} regulatory feature different from other characterized mammalian NCXs. Our structural studies reveal that CBD1 is the only Ca^{2+} binding site in its intracellular regulatory region and the Ca^{2+} binding site of its CBD2 was inactivated by local structural rearrangement. The multiple Ca^{2+} binding states of CBD1 illucidate a progressive Ca^{2+} binding mechanism at CBD1, in which the primary Ca^{2+} pair plays an essential role in triggering protein conformational change. The location of the conformational changes on CBD1 reflects a chorus of multiple domains in the Ca^{2+} regulatory mechanism, as demonstrated by the CBD12 structures. The individual CBD1 and the naturally linked CBD12 structures jointly support the critical role of the 1E–1F loop of CBD1 in triggering protein conformational change by Ca^{2+} binding. Its central position on the interdomain interface enables communication between the Ca^{2+} binding site of CBD1 with CBD2 and further other coupled regions. We also illustrate, with strong structural evidence, that alternative splicing may induce rigid body movement between CBD1 and CBD2 to alter the Ca^{2+} regulatory behavior. This observation provides a new insight to understand the interplay of the diverse regulatory features of the exchanger. Our structural and functional analyses did not fully answer the question about the negative Ca^{2+} regulatory feature of CALX. However, given the conserved sequence and structural features of the CBD domains between CALX and NCXs, our structural studies may offer a general model to study the Ca^{2+} regulatory mechanism of other Na^+/Ca^{2+} exchanger proteins.

References

D.M. Bers, *Excitation–Contraction Coupling and Cardiac Contractile Force* (Kluwer, Boston, 2001), pp. 71–92

G.M. Besserer, M. Ottolia, D.A. Nicoll, V. Chaptal, D. Cascio, K.D. Philipson, J. Abramson, The second Ca^{2+}-binding domain of the Na^+-Ca^{2+} exchanger is essential for regulation: crystal structures and mutational analysis. Proc. Natl. Acad. Sci. U. S. A. **104**, 18467–18472 (2007)

M.P. Blaustein, G. Fontana, R.S. Rogowski, The Na^+-Ca^{2+} exchanger in rat brain synaptosomes. Kinetics and regulation. Ann. N. Y. Acad. Sci. **779**, 300–317 (1996)

X. Cai, J. Lytton, The Cation/Ca Exchanger Superfamily: Phylogenetic Analysis and Structural Implications. Mol. Biol. Evol. **21**, 1692–1703 (2004)

C. Dyck, A. Omelchenko, C.L. Elias, B.D. Quednau, K.D. Philipson, M. Hnatowich, L.V. Hryshko, Ionic regulatory properties of brain and kidney splice variants of the NCX1 Na^+-Ca^{2+} exchanger. J. Gen. Physiol. **114**, 701–711 (1999)

M. Hilge, J. Aelen, G.W. Vuister, Ca^{2+} regulation in the Na^+/Ca^{2+} exchanger involves two markedly different Ca^{2+} sensors. Mol. Cell **22**, 15–25 (2006)

L.V. Hryshko, S. Matsuoka, D.A. Nicoll, J.N. Weiss, E.M. Schwarz, S. Benzer, K.D. Philipson, Anomalous regulation of the Drosophila Na^+-Ca^{2+} exchanger by Ca^{2+}. J. Gen. Physiol. **108**, 67–74 (1996)

E. Johnson, L. Bruschweiler-Li, S.A. Showalter, G.W. Vuister, F. Zhang, R. Brüschweiler, Structure and dynamics of Ca^{2+}-binding domain 1 of the Na^+/Ca^{2+} exchanger in the presence and in the absence of Ca^{2+}. J. Mol. Biol. **377**, 945–955 (2008)

S. Matsuoka, D.A. Nicoll, R.F. Reilly, D.W. Hilgemann, K.D. Philipson, Initial localization of regulatory regions of the cardiac sarcolemmal Na^+-Ca^{2+} exchanger. Proc. Natl. Acad. Sci. U. S. A. **90**, 3870–3874 (1993)

D.A. Nicoll, M.R. Sawaya, S. Kwon, K.D. Cascio, D.A. Philipson, J. Abramson, The crystal structure of the primary Ca^{2+} sensor of the Na^+/Ca^{2+} exchanger reveals a novel Ca^{2+} binding motif. J. Biol. Chem. **281**, 21577–21581 (2006)

A. Omelchenko, C. Dyck, M. Hnatowich, J. Buchko, D.A. Nicoll, K.D. Philipson, L.V. Hryshko, Functional differences in ionic regulation between alternatively spliced isoforms of the Na^+-Ca^{2+} exchanger from *Drosophila melanogaster*. J. Gen. Physiol. **111**, 691–702 (1998)

M. Ottolia, K.D. Philipson, S. John, Conformational changes of the Ca^{2+} regulatory site of the Na^+-Ca^{2+} exchanger detected by FRET. Biophys. J. **87**, 899–906 (2004)

K.D. Philipson, D.A. Nicoll, Sodium-calcium exchange: a molecular perspective. Annu rev. physiol. **62**, 111–133 (2000)

R. Ranganathan, G.L. Harris, C.F. Stevens, C.S. Zuker, A Drosophila mutant defective in extracellular calcium-dependent photoreceptor deactivation and desensitization. Nature **354**, 230–232 (1991)

E.M. Schwarz, S. Benzer, Calx, a Na-Ca exchanger gene of *Drosophila melanogaster*. Proc. Natl. Acad. Sci. U. S. A. **94**, 10249–10254 (1997)

T. Wang, C. Montell, Phototransduction and retinal degeneration in *Drosophila*. Pflugers Arch-Eur. J. Physiol. **454**, 821–847 (2007)

T. Wang, H. Xu, J. Oberwinkler, Y. Gu, R.C. Hardie, C. Montell, Light activation, adaptation, and cell survival functions of the Na^+/Ca^{2+} exchanger CalX. Neuron **45**, 367–378 (2005)

M. Wu, M. Wang, J. Nix, L.V. Hryshko, L. Zheng, Crystal structure of CBD2 from the *Drosophila* Na^+/Ca^{2+} exchanger: diversity of Ca^{2+} regulation and its alternative splicing modification. J. Mol. Biol. **387**, 104–112 (2009)

M. Wu, H.D. Le, M. Wang, V. Yurkov, A. Omelchenko, M. Hnatowich, J. Nix, L.V. Hryshko, L. Zheng, Crystal structures of progressive Ca^{2+} binding states of the Ca^{2+} sensor CBD1 from the CALX Na^+/Ca^{2+} exchanger reveal incremental conformational transitions. J. Biol. Chem. **285**, 2554–2561 (2010)

M. Wu, S. Tong, J. Gonzalez, V. Jayaraman, J.L. Spudich, L. Zheng, Structural basis of the Ca^{2+} inhibitory mechanism of *Drosophila* Na^+/Ca^{2+} exchanger CALX and its modification by alternative splicing. Structure **19**, 1509–1517 (2011)

Interplay of Ca^{2+} and Mg^{2+} in Sodium-Calcium Exchanger and in Other Ca^{2+}-Binding Proteins: Magnesium, Watchdog That Blocks Each Turn if Able

7

Dmitri O. Levitsky and Masayuki Takahashi

Abstract

Sodium-calcium exchange across plasma membrane is regulated by intracellular calcium ions. The sodium-calcium exchanger (NCX1) is activated by successive saturation of numerous Ca^{2+}-binding sites located in the intracellular loop of the protein. The progressive saturation of the binding domain CBD12 by Ca^{2+} results in a series of conformational changes of CBD12 as well as of entire NCX1 molecule. Like other soluble and membrane Ca^{2+}-binding proteins, NCX1 can also be regulated by Mg^{2+} that antagonises Ca^{2+} at the level of divalent cation-binding sites. This chapter summarises data on Mg^{2+} impacts in the cells. Regulatory action of Mg^{2+} on intracellular Ca^{2+}-dependent processes can be achieved due to changes of its cytoplasmic level, which take place in the range of $[Mg^{2+}]_i$ from 0.5 to 3 mM. Under normal conditions, these changes are ensured by activation of plasmalemmal Mg^{2+} transport systems and by variations in ATP level in cytoplasm. In heart and in brain, some pathological conditions, such as hypoxia, ischemia and ischemia followed by reperfusion, are associated with an important increase in intracellular Ca^{2+}. The tissue damage due to Ca^{2+} overload may be prevented by Mg^{2+}. The protective actions of Mg^{2+} can be achieved due to its ability to compete with Ca^{2+} for the binding sites in a number of proteins responsible for the rise in intracellular free Ca^{2+}, including NCX1, in case when the reverse mode of Na^+/Ca^{2+} exchange becomes predominant. Saturation of CBD12 by Mg^{2+} results in important changes of NCX1 conformation. Modulating actions of Mg^{2+} on the conformation of NCX1 were detected at a narrow range of Mg^{2+} concentration, from 0.5 to 1 mM. These data support an idea that variations of intracellular Mg^{2+} could modify transmembrane Ca^{2+} movements ensured by NCX1.

D.O. Levitsky (✉) • M. Takahashi
Unité de Fonctionnalité et Ingénierie des Protéines,
FRE-CNRS 3478, Faculté des Sciences
et des Techniques, Université de Nantes,
Nantes Cedex 03 44037, France
e-mail: dmitri.levitsky@univ-nantes.fr

L. Annunziato (ed.), *Sodium Calcium Exchange: A Growing Spectrum of Pathophysiological Implications*,
Advances in Experimental Medicine and Biology 961, DOI 10.1007/978-1-4614-4756-6_7,

Keywords

Calcium • Magnesium • Na^+/Ca^{2+} exchanger • NCX1 • Calcium-/magnesium-binding proteins • Ca^{2+}-binding sites

7.1 Introduction

The Na^+/Ca^{2+} exchanger (NCX) represents a major transporter assuring Ca^{2+} efflux from mammalian cells (Blaustein and Lederer, 1999). Like other membrane proteins involved in intracellular Ca^{2+} homeostasis, NCX possesses numerous cation-binding sites of high affinity which can be saturated in the range of free Ca^{2+} concentrations below 1 μM providing regulation of the protein activity. In two neighbouring Ca^{2+}-binding domains (CBD1 and CBD2) located in a large intracellular loop of NCX1, six Ca^{2+}-binding sites were identified (Besserer et al. 2007; Hilge et al. 2006; Nicoll et al. 2006). In most proteins regulated by changes of intracellular Ca^{2+}, the cation-binding sites can also interact with Mg^{2+}, though it occurs at three to four orders higher concentrations. A priori, these sites under some circumstances may be at least partially occupied by Mg^{2+}. Thus, a logical question arises: Could the high-affinity Ca^{2+}-binding sites be rather considered as Ca^{2+}-/Mg^{2+}-binding sites (or Ca^{2+}-binding, Mg^{2+}-modulated sites)?

The position of Mg^{2+} in cell hierarchy is quite different from that of Ca^{2+}. The second most abundant intracellular cation, Mg^{2+}, serves as a cofactor in numerous enzyme reactions and is essential in reactions requiring ATP. A major function of Ca^{2+}, in addition to its fundamental role in EC coupling and exocytosis, is to mediate in a concentration-dependent manner activity of prokaryotic and eukaryotic cell, by acting as a second messenger. While mechanisms of Ca^{2+} entry and extrusion are well characterised in many cell types, no reliable information is available for magnesium outward transport systems. It is generally accepted that these systems, even if they exist in mammalian tissues, are poorly developed. Furthermore, Mg^{2+} that crosses the digitonin-permeabilised membrane and enters the cell remains exclusively in cytosolic compartment, while Ca^{2+} is rapidly accumulated in cell organelles (Grubbs et al. 1984, 1991).

Mg^{2+} is generally accepted to antagonise Ca^{2+} in the cells, competing, in particular, at the level of soluble Ca^{2+}-binding proteins. With few exceptions, amino acid residues, involved in coordination of Ca^{2+}, participate also in Mg^{2+} binding. Mg^{2+} promotes a structural rearrangement of the binding domain. Classical examples are competitive binding of these cations to EF-hand-type proteins: calmodulin, troponin C and parvalbumin. As summarised by Allouche et al. (1999), effects of displacing Ca^{2+} by Mg^{2+} include decreasing of coordination number from seven to six oxygen atoms and the modification of the binding site from pentagonal bipyramidal configuration to octahedral configuration. Since most of the data concerning interaction of Ca^{2+} and Mg^{2+} with the binding sites have been obtained using purified preparations of EF-hand proteins, the physiological input of Mg^{2+} as a modulator of Ca^{2+}-binding sites in the living cell is still to determine. First of all, for any Ca^{2+}-binding proteins, reported affinity constants for Mg^{2+} vary considerably. Secondly, in many cases, competitive actions of Mg^{2+} have been revealed at five to ten millimolar or even larger concentrations. On the other hand, if the binding affinity Mg^{2+} of a site was too high, at physiological Ca^{2+} concentrations, this site would always exists in a Mg^{2+}-occupied form. In both cases, any speculation concerning possible modulatory role of Mg^{2+} should be lifted.

Investigation of potential Ca^{2+}-/Mg^{2+}-binding sites in membrane proteins using traditional biochemical and biophysical techniques is limited to soluble fragments corresponding to intracellular loops of the protein. On the other hand, electrophysiological approaches may bring valuable information on Ca^{2+}/Mg^{2+} competitive interactions in conditions close to the situation existing in the living cell. Na^+/Ca^{2+} exchanger is one of the plasmalemmal proteins possessing high-affinity calcium-binding sites. Many studies indicate that this protein may also be modulated by Mg^{2+}.

The objective of this chapter is to review actions of Mg^{2+} on conformational states of some membrane proteins, in particular on NCX. An accent was made on possible roles of Ca^{2+}/Mg^{2+} competition for common binding site in normal tissue as well as in pathophysiological conditions.

7.2 Regulatory Ca^{2+}-Binding Domains of NCX (Historical Preview)

Sodium-calcium exchange is a ubiquitous mechanism providing the exchange of $3Na^+/1Ca^{2+}$ between cytoplasm and extracellular medium. In most tissues, it operates in a "forward" way corresponding to inward current and thus to calcium exit from the cell (Blaustein and Lederer 1999). Under some conditions, however, a reverse mode of Na^+/Ca^{2+} exchange can be activated. In cardiac muscle, one of the isoforms of the sodium-calcium exchanger, NCX1, represents a major system of extrusion of Ca^{2+} that enters the cell by L-type Ca^{2+} channels (Cav1.2). Like other Ca^{2+} transport systems which are involved in excitation-contraction coupling (Cav1.2 and Ca^{2+} release channels of SR), NCX1 is regulated by intracellular Ca^{2+}. This is achieved by Ca^{2+} binding to numerous sites located in the middle portion of a large intracellular loop of NCX1.

The existence of a Ca^{2+}-binding domain has been postulated from the data obtained in electrophysiological studies performed on voltage-clamped cardiomyocytes whose cytoplasmic ionic composition could be changed by using an intracellular perfusion technique (Kimura et al. 1986). The major finding of this study was that the outward current could be completely blocked in the absence of cytoplasmic Ca^{2+}. Further studies indicated that a potential Ca^{2+}-binding domain is located in the NCX1 intracellular loop, since it is the Ca^{2+}-dependent regulation but not the ionic current was abolished after chymotrypsin treatment (Matsuoka et al. 1993).

A priori, this hypothetical Ca^{2+}-binding domain should possess high-affinity binding sites. Indeed, in inside-out oocytes expressing full-length NCX1, the outward exchange current could be activated in the presence of one μM Ca^{2+}. Concerning the localisation of this domain, experiments on deletion mutants allowed narrowing the region of interest to a stretch of about 300 amino acids (Matsuoka et al. 1993).

The latter highly informative work of K. Philipson and D. Hilgemann groups presented in short paper was a starting point in the search for a high-affinity Ca^{2+}-binding domain. Different techniques and approaches were employed: Expression of fusion proteins and 6xHis-Tag purified mutants, $^{45}Ca^{2+}$ overlays of nitrocellulose blots, Ca/EGTA-induced shifts of protein mobility in SDS-polyacrylamide gels, and equilibrium dialysis at different free Ca^{2+} concentrations. The calcium-binding domain was localised in a middle portion of NCX1 loop and positioned between 240 and 685 amino acids (Levitsky et al. 1994). This region is characterised by the presence of several acidic sequences which could, potentially, represent Ca^{2+}-binding sites. However, none of them corresponded to the consensual EF-hand motif present in Ca^{2+} high-affinity proteins. $^{45}Ca^{2+}$ overlay of the blots with fusion proteins truncated from COOH terminus and NH_2 terminus allowed to narrow this region to a segment of 151 amino acids (371–508). Deletions inside this segment or its shortening from COOH or NH_2 terminus abolished or significantly reduced $^{45}Ca^{2+}$ binding. This segment contains two unusual acidic sequences each possessing three consecutive aspartates: DDDIFEEDENFL and DDDHAGIFTFEE, separated by a four dozen residues.

More recent NMR and crystallographic studies (Besserer et al. 2007; Hilge et al. 2006; Nicoll et al. 2006) indicate that NCX1 regulatory Ca^{2+}-binding domain (CBD) represents two structurally associated domains: a primary high-affinity CBD1 (with four binding sites) and secondary, regulatory CBD2 (two binding sites).

7.3 Membrane Proteins Implicated in Excitation-Contraction Coupling (ECC)

The major membrane proteins assuring ECC in striated muscles, L-type Ca^{2+} channels, Ca^{2+}-ATPase and ryanodine receptor of SR, and NCX, are modulated by physiological Mg^{2+} concentrations. In most cases, saturation of binding sites in the regulatory domains of these proteins by micromolar Ca^{2+} activates Ca^{2+} transport. Generally, these high-affinity Ca^{2+}-binding sites are of high affinity with respect to Mg^{2+}, and these divalent cations compete for the occupancy of the same sites.

7.3.1 L-type Ca^{2+} Channels

An essential element in EC coupling in heart, a pore-forming alfa1C subunit of Cav1.2 channel, possesses in its N-terminal portion a consensual EF-hand motif: E-helix-**DPEAKGRIKHLD**-F-helix (XYZ-Y-X-Z residues ensuring the coordination are highlighted in bold). This sequence is not very typical for high-affinity calcium-binding sites. Indeed, the site is saturated at relatively high Ca^{2+} concentrations, tens of μmoles/L. In spite of that (or due to), this region is a good candidate for a modulating Mg^{2+}-binding site.

One of the characteristics of the calcium current governed by the Cav1.2 channel is that it inactivates in a calcium- and a voltage-dependent manner (Lee et al. 1985).

Several reports indicate that an increase in intracellular magnesium inhibits the inward Ca^{2+} current in isolated ventricular myocytes (Wang et al. 2004; White and Hartzell, 1988). As shown in a recent study, Mg^{2+} affects both V-dependent current inactivation and the current density. Upon changing Mg_i^{2+} from 0.8 to 2.4 mM, a 1.5-fold enhancement of VDI and threefold decrease in the current density was observed (Brunet et al. 2009). These modulating effects of Mg^{2+} would reduce Ca^{2+} overload and a priori should protect ischemic heart cells.

According to the model proposed by Brunet et al. 2009, two types of current inactivation are due to conformational changes induced by Ca^{2+} calmodulin (via a distant C-terminal site IQ motif; Ca^{2+}-dependent inactivation) and EF-hand motif (at the level of the proximal C-terminal region; VDI).

7.3.2 Ryanodine Receptor (RYR)

Another key element in ECC is Ca^{2+}-release channel located in the terminal region of sarcoplasmic reticulum (SR), just in proximity to DHPR (skeletal muscle) and L-type channel (cardiac muscle). It has been called "ryanodine receptor" due to its ability to fix and to be activated or inhibited, depending on the experimental conditions, by a plant alkaloid ryanodine (Meissner, 1986). In cardiac muscle, RYR assures a rapid rise of intracellular Ca^{2+} by the mechanism of calcium-induced calcium release from the SR and the major contractile response.

The modulation of RYR2 (cardiac) activity is achieved by numerous mechanisms, in particular by Ca^{2+}, ATP and Mg^{2+}. A bell-shaped curve of Ca^{2+} dependence of the channel's open probability (Bezprozvanny et al. 1991) indicates existence of two types of Ca^{2+}-binding sites in RYR: of high affinity (activating) and of low affinity (inhibiting). Magnesium ions, at millimolar concentrations, produce a double inhibiting action, competing with Ca^{2+} for high-affinity Ca^{2+} activation sites (KmMg 0.25 mM, cooperative interaction with n_H 2) and interacting with a low-affinity inhibiting site (KmMg 0.76 mM) (Zahradnikova et al. 2004). Due to relatively high affinities for both types of the sites, Mg^{2+} seems to be able to modulate RYR activity at physiological concentrations. Indeed, it decreases the open probability of RYR channel (Po) of the RYR channel in a quite narrow concentrational range: Po = 0.57 at 0.6 mM, 0.37 at 0.94 mM, and 0.25 at 1.33 mM.

7.3.3 "Big Potassium" Channel (BK, Sto1)

This "maxiK" channel is expressed in most tissues. Its regulatory role in calcium signalling was shown for brain and smooth muscle cells

(Blatz and Magleby, 1987). As for the heart and skeletal muscle, the impacts of variations in BK expression on ECC have not been yet investigated. The operation of this high-conductance potassium channel is also under control of both voltage and Ca^{2+}, but unlike L-type Ca^{2+} channels, the BK channels, instead of being inactivated, could be activated by depolarisation and by increase in intracellular Ca^{2+}. The mechanism of Ca^{2+} stimulation includes saturation of numerous (as much as four) Ca^{2+}-specific high-affinity sites, presumably located in a long COOH-terminal intracellular portion of the protein. In the absence of Ca^{2+}, this "tail region" interacts with the "core" BK region, and the current cannot be activated at physiologically reasonable levels of depolarisation. Saturation by Ca^{2+} of the high-affinity sites relieves this inhibition (Schreiber et al., 1999; Zhang et al., 2001). In addition, non-selective divalent ion-binding sites in the "tail region" can be saturated at millimolar Ca^{2+} concentrations, assuring a further shift of activating depolarising voltage to more negative potentials (Zhang et al., 2001). Millimolar Mg^{2+} modulates voltage dependence by interacting with the two sets of Ca^{2+}-binding domains: It weakly competes with Ca^{2+} at high-affinity Ca^{2+}-binding sites and promotes current activation by interacting with a low-affinity (relatively nonselective) divalent cation-binding domain. In the latter case, instead of competing with Ca^{2+}, Mg^{2+} potentiates Ca^{2+} activating effects, acting as allosteric regulator (Golowasch et al., 1986), increasing the Hill coefficient for Ca^{2+}-dependent activation from 1.7 to 3.2 (at 5 mM) and 4.2 (at 10 mM).

The presence of numerous binding sites of different affinities for Ca^{2+} and Mg^{2+} does not allow formulating a simple mechanism of BK activation. A model proposed by Zhang et al. (2001) postulates existing of 8 independent binding sites and as much as 250 open and closed channel states. So far, only one of the sequences located in the "tail" of Slo1 was identified as a Ca^{2+}-binding site. This highly acidic region, named "calcium bowl", contains in particular 5 successive aspartates: VQFLDQDD*DD*DPDTELY (Schreiber et al., 1999). The latter sequence could represent one of the Ca^{2+}-binding sites. Indeed, substitution of any asterisk-marked aspartates decreased $^{45}Ca^{2+}$ binding and disrupted Ca^{2+} sensing (Bao et al., 2004).

To summarise this section, effects of increasing Mg^{2+} on gating behaviour of BK channel would promote membrane hyperpolarisation, a process that in most excitable tissues corresponds to moderation/inhibition major cell functions.

7.4 Mg^{2+} Transport Systems

Prokaryotes express quite a few Mg^{2+} transport systems. In salmonella, for example, three Mg^{2+} transporters have been identified, and two of them, MgtA and MgtB, are also expressed in *E. coli*. Surprisingly, expression of MgtA and MgtB in salmonella seems to be regulated by extracellular Mg^{2+} at transcriptional level. Uptake of Mg^{2+} increased when the cells were grown in a low-magnesium medium. To ensure the adaptation to a low-Mg^{2+} environment, an 800-fold increase in MgtB transcription was shown at a very low (1μM) Mg^{2+} extracellular concentration (Snavely et al., 1991). Thus, the divalent cation-binding sites of these transporters should be considered as Mg-Ca rather than Ca-Mg-binding sites.

Attempts were made to find Mg^{2+} transport proteins in mammalian cells. So far, two dozen of genes encoding potential Mg^{2+} transporters were identified (for review, see Quamme (2010)), and seven of them are expressed in plasma membrane. These plasmalemmal proteins seem to function as divalent cation importers. Two of them, MagT1 and NIPA2, were found to be highly selective for Mg^{2+}. Their role in magnesium homeostasis is not clarified. Other transporters, TRPM6 and TRPM7, have a unique feature, possessing both channel and kinase activity (chanzyme). Ubiquitously expressed TRPM7 does not discriminate Ca^{2+} and Mg^{2+}. It may be partially responsible for cell damage due to Ca^{2+} overload representing, on the other hand, a mechanism allowing Mg^{2+} entry into the cell.

If we take into consideration approximate levels of extra- and intracellular Mg^{2+} (about 1 mM and 0.5 mM) as well as average resting membrane potential (−60 mV), we inevitably arrive to the conclusion that Mg^{2+} equilibrium potential, 10 mV, largely favours movement of this ion from extracellular space to the cytoplasm. Since under normal physiological conditions the ratio extra-/intracellular Mg^{2+} is kept more or less constant, from 2 to 3, one has to postulate existence of systems providing active Mg^{2+} extrusion from the cell. They could correspond to mechanisms of primary or secondary active transport, for example, to exchangers Na^{+}/Mg^{2+} or Ca^{2+}/Mg^{2+}. A nonelectrogenic Na^{+}/Mg^{2+} exchanger can be involved in Mg^{2+} extrusion from red blood cells and hepatocytes (Cefaratti and Romani, 2011; Ebel et al., 2004). Though there is no strong evidence of expression of the cation/Mg^{2+} exchangers in other mammalian cells, several experiments are in favour of sodium/magnesium- and calcium/magnesium-modulated Mg^{2+} redistribution between extra- and intracellular compartments. Thus, in heart, β-adrenergic stimulation provokes an important loss of intracellular Mg^{2+} (Romani et al., 1993). The Mg^{2+} efflux from cardiomyocytes was found to decrease at lowering extracellular Na^{+} and Ca^{2+} concentration. Existence of Na^{+}/Mg^{2+} exchange and possibly an ATP-dependent Mg^{2+} efflux system was also postulated for colon smooth muscle cells (Nakayama and Tomita, 1991).

7.5 Magnesium Intracellular Levels Under Normal and Pathological Conditions

It has been known for 70 years (Eichelberger et al., 1942) that intracellular level of total Mg^{2+} largely exceeds that present in extracellular fluids (about 10 mmol/kg, skeletal muscle vs. 1 mmol/L, serum). It is accepted that in cells at rest, the total Mg^{2+} level (10 mM) is quite constant (Gasser et al., 2005). However, major fraction of cell Mg^{2+} is sequestered in cell organelles, while cytoplasmic Mg^{2+} is bound to adenine nucleotides. It is to mention that 85–90% of cytoplasmic ATP exist in a form of MgATP (Grubbs et al., 1991). All available data, with rare exceptions, indicate that ionised cytoplasmic Mg^{2+} does not exceed 1 mM: **0.37** mM, hepatocytes (Corkey et al., 1986); **0.5** mM, kidney, distal convoluted tubule (Dai et al., 2001); **0.6** mM, vascular SMC, primary culture (Touyz and Schiffrin, 1996); **0.73** mM, papillary muscle (Gasser et al., 2005); and **1.5** mM, sensory neurones (Henrich and Buckler, 2008).

In pathological situations, changes of intracellular magnesium have been reported in a number of organs, including heart and brain.

7.5.1 Heart

Depending on experimental model, cytosolic-free Mg^{2+} concentration increases (ischemic conditions) or decreases (heart failure).

The most accurate estimations of intracellular free Mg^{2+} in cardiomyocytes under normal and ischemic conditions were obtained using ^{31}P-NMR (Headrick and Willis, 1991) and ^{19}F-NMR (Headrick and Willis, 1991) techniques. The fluctuations of intracellular Mg^{2+} are closely related to changes in ATP level. Thus, a transient hypoxia of rat heart induces a decrease of intracellular ATP level and a 2.5-fold increase in free cytoplasmic Mg^{2+}, from 0.7 to about 1.75 mM (Headrick and Willis, 1991). Similar results were obtained in experiments on isolated rat heart under conditions of ischemia followed by reperfusion (Murphy et al., 1989). ATP intracellular level decreased by 50% after 15 min of global ischemia, and it was not restored by reperfusion. Fluctuations in Mg^{2+} level were much more pronounced. The concentration of free Mg_i^{2+} gradually increased in ischemic heart from 0.6 to 2.1 mM and was partially restored by reflow (1.5 mM). This partial normalisation of intracellular Mg^{2+} indicates that some mechanisms, other simple changes in ATP level, contribute under those conditions to variations of free Mg^{2+}. They may correspond to its buffering in intracellular organelles and/or to Mg^{2+} extrusion from the cells.

At congestive non-ischemic heart failure, expression of major H^{+}, Na^{+} and Ca^{2+} transport systems is drastically changed and the entire ionic homeostasis is perturbed. The decrease in the level of SERCA is partially compensated by an elevation of the level of NCX1 (Pogwizd et al., 1999; Reinecke et al., 1996). It was suggested that an increase in NCX1 expression in the failing heart would increase a risk of arrhythmias (Pogwizd et al., 1999).

After induction of heart failure in dogs, pronounced decrease of the level of cytoplasmic Mg^{2+} in ventricular cardiomyocytes takes place, from 1 mM in normal dogs to 0.5 mM in those with heart failure (Haigney et al., 1998). Experiments on current-clamped myocytes dialysed with 0.5 mM and 1.0 mM Mg^{2+} indicated shortening of repolarisation phase at higher Mg_i^{2+}. It was hypothesised that a decreased Mg_i^{2+} would contribute to prolongation of the action potential in the failing myocardium (Haigney et al., 1998).

7.5.2 Brain

Even short periods of brain ischemia may produce irreversible damage of neurons. The cerebral ischemic damage is partially due to an activation of Ca^{2+} entry pathways, in particular through glutamate NMDA receptors (Paschen, 1996), that leads to cell Ca^{2+} overload and activation of Ca^{2+}-dependent proteases and phospholipases. Under hypoxic/ischemic conditions, damaging effects of elevated Ca_i^{2+} are likely to be attenuated by Mg^{2+}. Indeed, its level increases by 15–20% after a brief anoxia due to a fall in ATP_i (Henrich and Buckler, 2008). Another source of Mg^{2+} in hypoxic and ischemic neurons is extracellular. One-hour oxygen-glucose deprivation of hippocampal neuron results in 1.5-fold increase of intracellular Mg^{2+} and a decrease of extracellular Mg^{2+} (Zhang et al., 2011). After blocking ubiquitous divalent transporter TRPM7 or its silencing with shRNA, the increase of Mg_i^{2+} induced by oxygen-glucose deprivation or by ischemia was attenuated. Numerous studies on animal models as well as clinical trials indicate that elevated extracellular Mg^{2+} protects neurons from ischemic damage (for review, see Muir (2002)).

7.5.3 Magnesium and Apoptosis

Numerous studies indicate a key role of rising in cytoplasmic Ca^{2+} level in apoptosis. During Fas-initiated induced apoptosis of B cells, both Ca^{2+}- and Mg^{2+}-free concentrations increase simultaneously. Two Mg^{2+} pools were distinguished: from 0.1 to 0.3 mM and from 0.5 to 0.7 mM, respectively (Chien et al., 1999). It is hypothesised that elevated free intracellular Mg^{2+} is not a consequence of mitochondrial membrane damage and a collapse of membrane potential but a factor essential for apoptosis. If this suggestion is proved, the results would indicate that relatively small changes in free Mg^{2+} produce important effects on the cell viability. This is one of rare examples of synergetic actions of Ca^{2+} and Mg^{2+} in the cell.

7.6 Interplay of Ca^{2+} and Mg^{2+} During Conformational Transitions of NCX1

7.6.1 Magnesium as a Modulator of Calcium Binding

For the first time, the competition between Ca^{2+} and Mg^{2+} was shown in $^{45}Ca^{2+}$ overlay experiments (Levitsky et al. 1994). Nitrocellulose strips containing the exchanger CBD1 domain (region 240 to 532 a/a) were preincubated at different ratios $^{45}Ca^{2+}$/EGTA in the presence and the absence of Mg^{2+}. The variations in $^{45}Ca^{2+}$ binding to the fusion protein were determined by scintillation counting. The approach used for estimating Ca^{2+} sensitivity of the binding sites looks at least strange. Indeed, one has to postulate the protein conserves its binding properties after (1) its prior denaturation in SDS solution, (2) migration in PAAG, and (3) electrical field transfer from the gel and its fixation on a solid porous support. Surprisingly, the curves of the dependence-bound $^{45}Ca^{2+}$/free Ca^{2+} obtained using this technique corresponded to those obtained later in "more native" conditions after equilibrium dialysis of 6xHis-tag purified CBD1 (Levitsky et al., 1996). Both techniques revealed 50% saturation of the binding sites at 0.2 μM free Ca^{2+}.

The Ca^{2+} dependence of binding to the fusion protein immobilised on the nitrocellulose shifted rightwards upon increasing Mg^{2+} concentration which resulted in gradual elevation of Kd values (pH 7.2): 0.3 μM (at 0.2 mM Mg^{2+}), 0.6 μM (at 1 mM Mg^{2+}), 2 μM (3 mM Mg^{2+}) and 2.5 μM (5 mM Mg^{2+}). Thus, this inhibition occurred at physiological range of Mg^{2+} concentrations (a calculated value Kd/Ki was found to be 0.3 mM). An interesting passing remark about this study concerns influence of Mg^{2+} on the Hill coefficient of Ca^{2+} binding. In fact, Mg^{2+} apparently increases n_H from 1.7 (at 0.2 mM Mg^{2+}) to 2.7 (at 5.0 mM). No precise explanation of this effect was suggested. Nevertheless, two modes of Mg^{2+} action on the Ca^{2+}-binding domain (competitive inhibition and increase in cooperativity of Ca^{2+} binding) would indicate existence of at least two sets of Mg^{2+}-binding sites. Their occupancy by one or two Mg^{2+} should a priori modify conformation of the entire binding domain. As to effects on the exchanger activity, these data did not allow to go further a speculation that successive saturation of the regulatory sites by Mg^{2+} would increase (potentiate) or block the exchange current.

These findings fall into oblivion, presumably forever. The Mg^{2+} effects were not mentioned until appearance of the publication of Wei et al. (2002) who showed direct inhibiting effects of Mg^{2+} on Na^+/Ca^{2+} exchanger activity in isolated cardiomyocytes. It is to indicate that these effects were not due to variations in intracellular levels of ATP or MgATP, they were not sensitive to variations of Ca_i^{2+}. A crucial point was that by increasing Mg_i^{2+} (from 0.13 mM to 1.25 mM) inhibited both exchange currents: by 30% (outward current, equivalent to Ca^{2+} entry) and by 60% (inward current). It was thus concluded that effects of Mg^{2+} are due to its competition with Ca^{2+} at the regulatory Ca^{2+}-binding domain of NCX1.

7.6.2 Calcium-/Magnesium-Dependent Conformational Transitions in CBD12 Revealed by Variations of Tyrosine Fluorescence

Fluorescence measurements of intrinsic tryptophan and tyrosine residues are widely used to follow subtle variations in the conformation of peptides and purified proteins. Exact conformational states of protein cannot be determined by fluorescence changes, which are related to aromatic amino acids exposure to solvent and intermolecular energy transfers. Nevertheless, if we take as example calcium-binding proteins, they would indicate at which concentrations of free Ca^{2+} conformational transitions take place.

The construct CBD12, corresponding to a splice variant AD (279 amino acids) of NCX, lacks tryptophans but has four tyrosines. One of them is located in proximity to the first aspartate triplet. The variations in fluoresce intensity of this particular tyrosine could reflect Ca^{2+}-induced changes of CBD12 conformation (though impact of three other tyrosines is not excluded). As seen from Fig. 7.1a, emission spectra of CBD12 after its excitation at 270 nm have a maximum at 303 nm, which is typical for tyrosine molecules. The intensity of fluorescence decrease in the presence of 35 μM Ca^{2+} when the CBD12 is in Ca^{2+}-saturated form, and it increases when free Ca^{2+} concentration is fixed at level of about 1 μM.

Figure 7.1b shows changes in tyrosine fluorescence upon a wide range of free Ca^{2+} concentrations. At 2 to 8 μM Ca^{2+}, the fluorescence varies only slightly, but at lower concentrations of free Ca^{2+}, it becomes quite unstable with successive tyrosine(s) "unmasking and hiding", indicating transitions between numerous conformational states of the protein.

Effects of Mg^{2+} on tyrosine exposure (Fig. 7.1c) were tested at a free Ca^{2+} concentration corresponding to the beginning of the mentioned interval of fluorescence instability (1.4 μM, marked by asterisk in Fig. 7.1b). At this Ca^{2+} concentration, increasing of Mg^{2+} up to 0.7 mM does not modify fluorescence intensity. At higher Mg^{2+} concentrations (from 0.7 to 4.5 mM), a progressive increase of fluorescence intensity takes place.

7.6.3 CBD12 Electrophoretic Mobility of CBD12 in Ca^{2+}-Saturated and Desaturated Forms

As seen from Fig. 7.2a, the construct CBD12 in a Ca^{2+}-saturated form (well 1) is detected in the gel as a major sharp band. Less intense band of

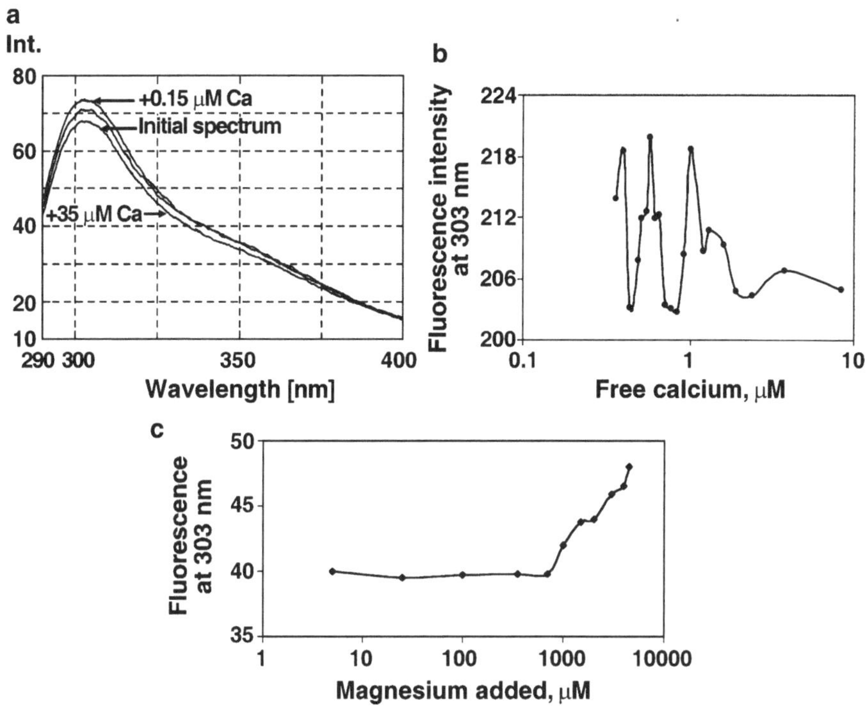

Fig. 7.1 *Fluorescence of tyrosine(s) in CBD12.* (**a**) Effects of Ca^{2+} and Ca^{2+}/EGTA on fluorescence emission spectra of CBD12. The spectrum in the middle corresponds to a measurement in the presence of 50 mM Tris-HCL (pH 7.2), 0.1 M NaCl and 1 mM azide, without additions of Ca^{2+} and EGTA. (**b**) Changes of fluorescence intensity CBD12 at 303 nm upon gradual decrease of free Ca^{2+} from 0.36 to 8.4 μM. Small aliquots of EGTA solutions were added to the buffer containing 0.5 mM Ca^{2+}. Dilutions of the medium were taken into account. (**c**) Effects of increasing magnesium on the tyrosine fluorescence in the presence of 1.4 μM free Ca^{2+}. In a submicromolar range of free Ca^{2+}, fluorescence becomes quite unstable: tyrosine(s) unmasking is followed by its hiding and new unmasking. This may reflect successive modifications of CBD12 conformation

lower mobility corresponds to the protein dimer (compare its mobility with that of BSA, last well). The bands of distinct electrophoretic mobility are detected in the gel after preincubation of the protein in the presence of Ca^{2+}/EGTA. The increase in the ratio EGTA/Ca^{2+} induces a gradual disappearance of a Ca^{2+}-saturated form and emerging of a new band of lower mobility. This indicates an important conformational reorganisation of the protein during saturation/desaturation of the binding sites. Two distinct protein conformations could be clearly distinguished. One of them corresponds to a fast-moving, presumably "compact" Ca^{2+}-saturated form and another one to low-mobility forms of the Ca^{2+}-desaturated polypeptide. This type of experiment does not allow us to make conclusions concerning the degree of saturation of CBD12 by Ca^{2+} at each Ca/EGTA ratio. The protein band of lower mobility may represent a calcium-free form, partially saturated forms or mixture of both.

7.6.4 Effects of Mg^{2+} on Mobility of CBD12 in the SDS-Polyacrylamide Gels

As indicated above, preincubation of CBD12 in the presence of submicromolar concentrations of free Ca^{2+} results in emerging of a low-electrophoretic mobility protein band corresponding to a less compact form of the protein. Figure 7.2b shows

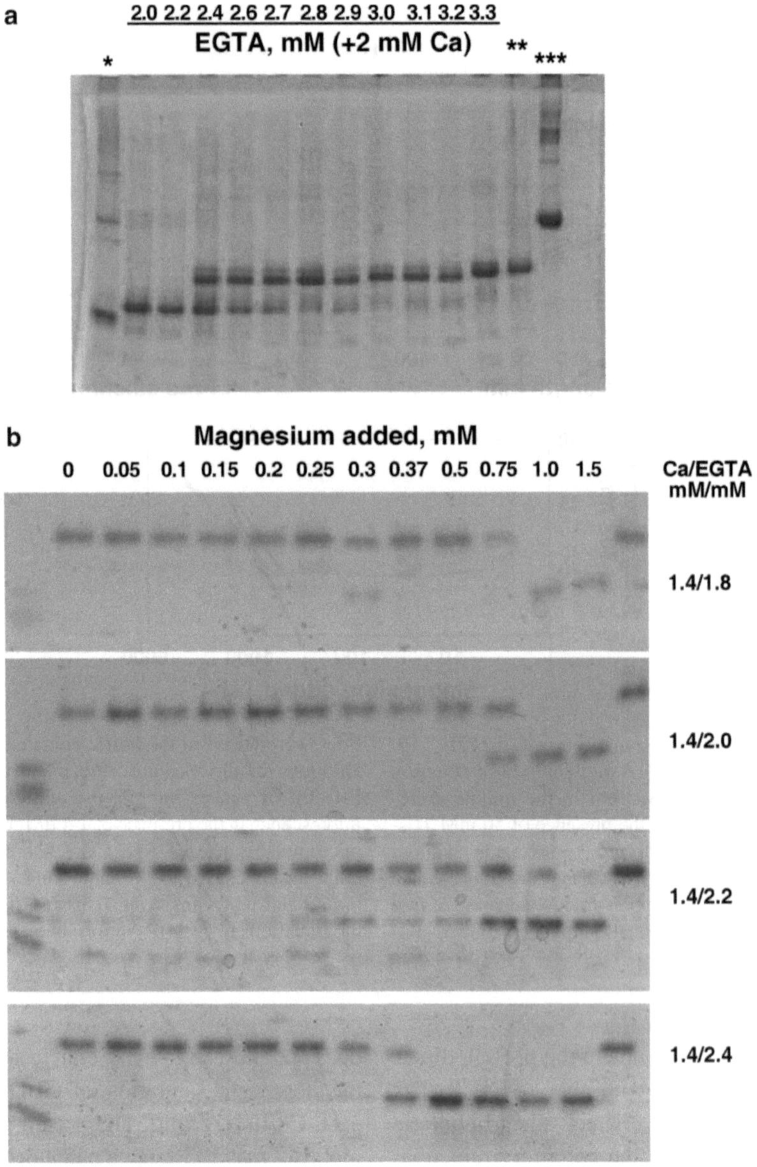

Fig. 7.2 *Ca^{2+}/Mg^{2+}-induced shifts of electrophoretic mobility of CBD12*. (**a**) *Ca^{2+}/EGTA titration.* Prior to electrophoresis, the protein was preincubated in the presence of 1.4 mM Ca^{2+} and different EGTA concentrations. The first lane: no additions (in the presence of contaminant Ca^{2+} only). A weak upper band corresponds to CBD12 dimer (cf. last lane: ***BSA). The lane marked by *double asterisk*: the protein was preincubated in the presence of 2 mM EGTA. (**b**) Effects of increasing magnesium concentration on electrophoretic mobility of CBD12 at 4 ratios Ca^{2+}/EGTA. The first lane: without Ca^{2+} or EGTA in the sample loading buffer. The last lane: 2 mM EGTA was added. Remark: Mg^{2+} does not change the protein mobility in the absence of calcium ions. The gels were stained with Coomassie Brilliant Blue

that progressive increase in magnesium in the sample buffer results in appearance of a new band of higher mobility. The latter becomes predominant after protein preincubation with 1 mM Mg^{2+} (at 1.4 Ca/1.8 EGTA ratio) or with 0.37 mM Mg^{2+} (at decreased free Ca^{2+}, the ratio 1.4/2.4). This data are in line with the model of direct competition between Ca^{2+} and Mg^{2+} for their binding sites.

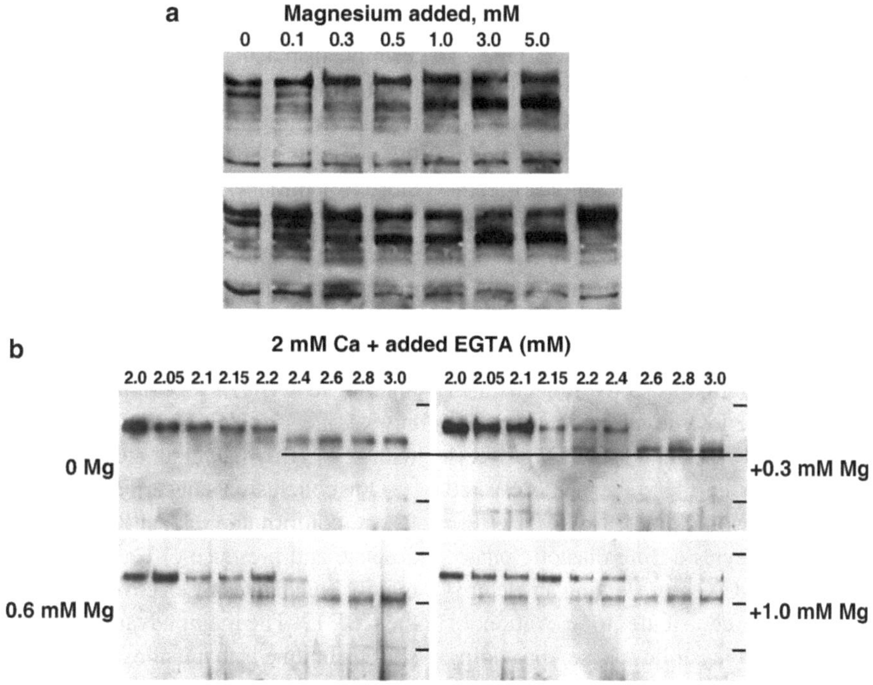

Fig. 7.3 *Western blotting of rabbit heart homogenates using anti-NCX1 mAb R3F1.* (**a**) Effects of increasing Mg^{2+} on the electrophoretic mobility of entire NCX molecule. Prior to SDS-PAGE, the samples were preincubated in the presence of different Mg^{2+} concentrations at two ratios Ca^{2+}/EGTA. (**b**) Effects of decreasing free Ca^{2+} in the absence of Mg^{2+} and at three different Mg^{2+} concentrations. The position of a "Ca^{2+}/Mg^{2+} band" is indicated. Concentration, a new protein band appears, becoming predominant at 1–5 mM Mg^{2+}. (**b**) Desaturation of Ca^{2+}-binding site(s) leads to appearance of new NCX1 bands, including those revealed only in the presence of Mg^{2+} (as indicated by a *parallel line*). It is to note that in the case of "cardiac" isoform of NCX1 (splicing variant ACDEF), Ca^{2+}-free forms possess higher electrophoretic mobilities. Remark: Mg^{2+} does not change the protein mobility in the absence of calcium ions (not shown)

7.6.5 Effects of Mg^{2+} on the Conformational State of NCX1

Figure 7.3a indicates that Ca^{2+}/Mg^{2+} modifications of the conformation of CBD12 have an influence on electrophoretic mobility of the whole NCX molecule (cardiac exchanger with exons ACDEF). At low free calcium concentrations, a band of higher mobility becomes predominant at after preincubation of the sample with 0.5 to 1.0 mM Mg^{2+}, and the transition of the protein from a partially Ca^{2+}-saturated form to Ca^{2+}/Mg^{2+} forms occurs even at lower Mg^{2+} concentrations.

The results of an inverse titration experiment (decreasing free calcium at fixed Mg^{2+} concentrations) also indicate that conformational changes of NCX1 take place submillimolar concentrations of Mg^{2+}. The range of the effective concentrations of Mg^{2+} fairly corresponds to that existing in the cell. It is not too high (otherwise the possibility of Mg^{2+} influence on Na^+/Ca^{2+} exchange has to be excluded), and it is not too low (otherwise certain binding sites would be permanently occupied by Mg^{2+}, and the latter could not assure modulation of Ca^{2+} binding).

7.7 Physiological Implications

As indicated above, ionic homeostasis in the hypoxic, ischemic and failing heart is greatly perturbed. The changes include rise in intracellular Ca^{2+}, H^+ and Na^+, breakdown of ATP, and changes in Mg^{2+} levels. These events are provoked or followed by variations in expression of major ion transport systems, such as SERCA (decrease), Na^+/H^+ antiporter and NCX (increase). One of the major consequences of these perturbations is a Ca^{2+}-induced cell damage. Sodium-dependent extrusion of Ca^{2+} from the cardiomyocytes under these conditions would be an only efficient mechanism to protect the heart from Ca^{2+} overload. What would be an impact of the rise in intracellular Mg^{2+} on the redistribution of Ca^{2+}? Two scenarios could be envisaged. Increased NCX1 expression in the pathological heart may be considered as cardioprotective mechanism activated in response to a downregulation of SR Ca^{2+}-ATPase. In this case, an elevated Mg^{2+} level would certainly worsen the state of the cardiac muscle since this would decrease Ca^{2+} efflux from the cardiomyocytes during diastolic phase. On the other hand, NCX, representing an electrogenic system, may contribute to elevation of intracellular Ca^{2+} at systolic phase, providing outward ionic current and importing Ca^{2+} through reverse mode of Na^+/Ca^{2+} exchange. In ischemic heart and under hypoxic conditions, when intracellular acidosis activates mechanism of Na^+ entry into the cardiomyocytes through Na^+/H^+ exchanger (NHE1), the elevated Na_i^+ would favour operation of NCX in the reverse mode of Na^+/Ca^{2+} exchange. In this case, high Mg^{2+} would attenuate Ca^{2+} overload of the myocardium. The protective effect of elevated Mg^{2+} may be especially pronounced under conditions of heart ischemia followed by reperfusion when reactive oxygen species further aggravate the situation inducing large diffusion of Ca^{2+} into the myoplasm. The latter Ca^{2+} influx seems to be mediated by NCX, functioning in a reverse mode.

7.8 Discussion

Quite similar to other Ca^{2+}-binding proteins, NCX can be modulated by Mg^{2+} at concentrations close to those existing in the animal cell. The exact mechanism underlying conformational transitions of CBD12 induced by Mg^{2+} is to be solved. We are evidently too far from stating that Mg^{2+}-bound configuration of CBD12 has a particular impact on the conformation of entire NCX1 molecule. If Mg^{2+} effects are limited only to a simple displacing of Ca^{2+} from CBD12, a partial desaturation regulatory Ca^{2+}-binding sites should result in inhibition of Na^+/Ca^{2+} exchange. Indeed, this inhibitory effect of Mg^{2+} (at a concentration close to 1 mM) was clearly demonstrated on isolated cardiomyocytes (Wei et al., 2002). However, activating actions of Mg^{2+} at a low concentration range should not be excluded, in particular if one takes into account existence of numerous Mg^{2+}-binding sites in CBD12. Indeed, results on $^{45}Ca^{2+}$ binding to CBD12 at equilibrium indicate that any of the six Ca^{2+}-binding sites may interact with Mg^{2+} (Boyman et al., 2009).

Our conclusion concerning existence of different conformational states of NCX1 in the absence and presence of Ca^{2+} and Mg^{2+} is based mainly on changes of electrophoretic mobility of the CBD12. The major advantage of gel mobility shift technique is that it allows easy identification of calcium-binding proteins in polyacrylamide gels after running purified protein samples. This analysis can be also performed on crude tissue homogenates using Western blotting method. A prerequisite is that the protein upon Ca^{2+} saturation/desaturation of the binding sites changes its conformation taking more compact (closed, "globular") or more open configuration that would influence its mobility in the gel. The major application of this approach is the comparison of two extreme states: Ca^{2+}-free sites and saturated sites. Once entered into the gel, the protein rests "frozen" in one of the conformations and further moves as a single narrow band. SDS that is present in the sample loading buffer seems to have minor effects on the conformational transitions which could be partially explained by repulsive actions of highly acidic Ca^{2+}-binding sites on the negatively charged detergent. Using different ratios Ca^{2+}/EGTA, this technique also allows estimate changes in affinity of the Ca^{2+}-binding sites in different ionic environment, in our case in the presence of Mg^{2+}. One of the limitations of gel mobility shift technique is that exact values of Kd_{Ca} of the sites could not be determined. This is due, in particular, to uncertainty concerning pH values in the gel loading solution (which represents in fact a mixture of alkaline electrophoresis buffer and neutral sample buffer) just before entering the protein into the concentrating gel. Indeed, the stability of Ca/EGTA complex is highly pH sensitive, and the calculated free Ca^{2+} values at a ratio 2 mM EGTA/1.3 mM Ca^{2+} (see Fig. 7.1) would differ considerably in a quite narrow pH range, being 0.6 μM (at pH 7.0), 0.38 (pH 7.1), 0.24 (pH 7.2) and 0.15 (pH 7.3).

7.9 Conclusions

Numerous physiological, epidemiological and clinical studies indicate that increasing concentrations of Mg^{2+} are beneficial for organism under normal and pathological conditions. It is generally accepted that, owing to functioning of plasmalemmal Mg^{2+} channels and transporters, changes in intracellular Mg^{2+} level follow variations of its concentration in the blood. Under pathological conditions, leading to Ca^{2+} overload, increased intracellular Mg^{2+} may have protective effects on the cell. The data accumulated so far indicate that the protective action of Mg^{2+} is realised, in particular, due to the competitive displacing of Ca^{2+} from its high-affinity binding sites. The competitive Ca^{2+}/Mg^{2+} interaction seems to be realised at the level of divalent cation-binding domains in any protein that is responsible for regulation of intracellular free Ca^{2+} concentration.

Acknowledgements We thank M. Hilge for providing NCX1 plasmids. We are thankful to our students E. Foucault, Ou. Louahdi, A. Menou, L. Diakite and M. Mekideche for their participation in a number of experiments.

References

D. Allouche, J. Parello, Y.H. Sanejouand, Ca^{2+}/Mg^{2+} exchange in parvalbumin and other EF-hand proteins. A theoretical study. J. Mol. Biol. **285**, 857–873 (1999)

L. Bao, C. Kaldany, E.C. Holmstrand, D.H. Cox, Mapping the BKCa channel's "Ca^{2+} bowl": side-chains essential for Ca^{2+} sensing. J. Gen. Physiol. **123**, 475–489 (2004)

G.M. Besserer, M. Ottolia, D.A. Nicoll, V. Chaptal, D. Cascio, K.D. Philipson, J. Abramson, The second Ca^{2+}-binding domain of the Na^+ Ca^{2+} exchanger is essential for regulation: crystal structures and mutational analysis. Proc. Natl. Acad. Sci. U. S. A. **104**, 18467–18472 (2007)

I. Bezprozvanny, J. Watras, B.E. Ehrlich, Bell-shaped calcium-response curves of Ins(1,4,5)P3- and calcium-gated channels from endoplasmic reticulum of cerebellum. Nature **351**, 751–754 (1991)

A.L. Blatz, K.L. Magleby, Calcium-activated potassium channels. Trends Neurosci. **10**, 463–467 (1987)

M.P. Blaustein, W.J. Lederer, Sodium/calcium exchange: its physiological implications. Physiol. Rev. **79**, 763–854 (1999)

L. Boyman, H. Mikhasenko, R. Hiller, D. Khananshvili, Kinetic and equilibrium properties of regulatory calcium sensors of NCX1 protein. J. Biol. Chem. **284**, 6185–6193 (2009)

S. Brunet, T. Scheuer, W.A. Catterall, Cooperative regulation of Ca(v)1.2 channels by intracellular Mg^{2+}, the proximal C-terminal EF-hand, and the distal C-terminal domain. J. Gen. Physiol. **134**, 81–94 (2009)

C. Cefaratti, A. Romani, Modulation of Na^+/Mg^{2+} exchanger stoichiometry ratio by Cl ions in basolateral rat liver plasma membrane vesicles. Mol. Cell. Biochem. **351**, 133–142 (2011)

M.M. Chien, K.E. Zahradka, M.K. Newell, J.H. Freed, Fas-induced B cell apoptosis requires an increase in free cytosolic magnesium as an early event. J. Biol. Chem. **274**, 7059–7066 (1999)

B.E. Corkey, J. Duszynski, T.L. Rich, B. Matschinsky, J.R. Williamson, Regulation of free and bound magnesium in rat hepatocytes and isolated mitochondria. J. Biol. Chem. **261**, 2567–2574 (1986)

L.J. Dai, G. Ritchie, D. Kerstan, H.S. Kang, D.E. Cole, G.A. Quamme, Magnesium transport in the renal distal convoluted tubule. Physiol. Rev. **81**, 51–84 (2001)

H. Ebel, R. Kreis, T. Gunther, Regulation of Na^+/Mg^{2+} antiport in rat erythrocytes. Biochim. Biophys. Acta **1664**, 150–160 (2004)

L. Eichelberger, F.C. McLean, W.A. Catterall, The distribution of calcium and magnesium between the cells and the extracellular fluids of skeletal muscle and liver in dogs. J. Biol. Chem. **142**, 467–476 (1942)

S. Gasser, N. Bareza, E. Scheer, D. Pruthi, R. Gasser, U.E. Spichiger Keller, E. Toferer, Free intracellular magnesium remains uninfluenced by changes of extracellular magnesium in cardiac guinea pig papillary muscle. J. Clin. Basic Cardiol. **8**, 29–32 (2005)

J. Golowasch, A. Kirkwood, C. Miller, Allosteric effects of Mg^{2+} on the gating of Ca^{2+}-activated K^+ channels from mammalian skeletal muscle. J. Exp. Biol. **124**, 5–13 (1986)

R.D. Grubbs, S.D. Collins, M.E. Maguire, Differential compartmentation of magnesium and calcium in murine S49 lymphoma cells. J. Biol. Chem. **259**, 12184–12192 (1984)

R.D. Grubbs, P.A. Beltz, K.L. Koss, Practical considerations for using mag-fura-2 to measure cytosolic free magnesium. Magnes. Trace Elem. **10**, 142–150 (1991)

M.C. Haigney, S. Wei, S. Kaab, E. Griffiths, R. Berger, R. Tunin, D. Kass, W.G. Fisher, B. Silver, H. Silverman, Loss of cardiac magnesium in experimental heart failure prolongs and destabilizes repolarization in dogs. J. Am. Coll. Cardiol. **31**, 701–706 (1998)

J.P. Headrick, R.J. Willis, Cytosolic free magnesium in stimulated, hypoxic, and underperfused rat heart. J. Mol. Cell. Cardiol. **23**, 991–999 (1991)

M. Henrich, K.J. Buckler, Effects of anoxia, aglycemia, and acidosis on cytosolic Mg^{2+}, ATP, and pH in rat sensory neurons. Am. J. Physiol. Cell Physiol. **294**, C280–C294 (2008)

M. Hilge, J. Aelen, G.W. Vuister, Ca^{2+} regulation in the Na^+/Ca^{2+} exchanger involves two markedly different Ca^{2+} sensors. Mol. Cell **22**, 15–25 (2006)

J. Kimura, A. Noma, H. Irisawa, Na-Ca exchange current in mammalian heart cells. Nature **319**, 596–597 (1986)

C.O. Lee, P. Abete, M. Pecker, J.K. Sonn, M. Vassalle, Strophanthidin inotropy: role of intracellular sodium ion activity and sodium-calcium exchange. J. Mol. Cell. Cardiol. **17**, 1043–1053 (1985)

D.O. Levitsky, D.A. Nicoll, K.D. Philipson, Identification of the high affinity Ca^{2+}-binding domain of the cardiac Na^+-Ca^{2+} exchanger. J. Biol. Chem. **269**, 22847–22852 (1994)

D.O. Levitsky, B. Fraysse, C. Leoty, D.A. Nicoll, K.D. Philipson, Cooperative interaction between Ca^{2+} binding sites in the hydrophilic loop of the Na^+-Ca^{2+} exchanger. Mol. Cell. Biochem. **160–161**, 27–32 (1996)

S. Matsuoka, D.A. Nicoll, R.F. Reilly, D.W. Hilgemann, K.D. Philipson, Initial localization of regulatory regions of the cardiac sarcolemmal Na^+-Ca^{2+} exchanger. Proc. Natl. Acad. Sci. U. S. A. **90**, 3870–3874 (1993)

G. Meissner, Ryanodine activation and inhibition of the Ca^{2+} release channel of sarcoplasmic reticulum. J. Biol. Chem. **261**, 6300–6306 (1986)

K.W. Muir, Magnesium in stroke treatment. Postgrad. Med. J. **78**, 641–645 (2002)

E. Murphy, C. Steenbergen, L.A. Levy, B. Raju, R.E. London, Cytosolic free magnesium levels in ischemic rat heart. J. Biol. Chem. **264**, 5622–5627 (1989)

S. Nakayama, T. Tomita, Regulation of intracellular free magnesium concentration in the taenia of guinea-pig caecum. J. Physiol. **435**, 559–572 (1991)

D.A. Nicoll, M.R. Sawaya, S. Kwon, D. Cascio, K.D. Philipson, J. Abramson, The crystal structure of the primary Ca^{2+} sensor of the Na^+/Ca^{2+} exchanger reveals a novel Ca^{2+} binding motif. J. Biol. Chem. **281**, 21577–21581 (2006)

W. Paschen, Glutamate excitotoxicity in transient global cerebral ischemia. Acta Neurobiol. Exp. (Wars) **56**, 313–322 (1996)

S.M. Pogwizd, M. Qi, W. Yuan, A.M. Samarel, D.M. Bers, Upregulation of Na^+/Ca^{2+} exchanger expression and function in an arrhythmogenic rabbit model of heart failure. Circ. Res. **85**, 1009–1019 (1999)

G.A. Quamme, Molecular identification of ancient and modern mammalian magnesium transporters. Am. J. Physiol. Cell Physiol. **298**, C407–C429 (2010)

H. Reinecke, R. Studer, R. Vetter, J. Holtz, H. Drexler, Cardiac Na^+/Ca^{2+} exchange activity in patients with end-stage heart failure. Cardiovasc. Res. **31**, 48–54 (1996)

A. Romani, C. Marfella, A. Scarpa, Regulation of magnesium uptake and release in the heart and in isolated ventricular myocytes. Circ. Res. **72**, 1139–1148 (1993)

M. Schreiber, A. Yuan, L. Salkoff, Transplantable sites confer calcium sensitivity to BK channels. Nat. Neurosci. **2**, 416–421 (1999)

M.D. Snavely, S.A. Gravina, T.T. Cheung, C.G. Miller, M.E. Maguire, Magnesium transport in *Salmonella typhimurium*. Regulation of mgtA and mgtB expression. J. Biol. Chem. **266**, 824–829 (1991)

R.M. Touyz, E.L. Schiffrin, Angiotensin II and vasopressin modulate intracellular free magnesium in vascular smooth muscle cells through Na^+-dependent protein kinase C pathways. J. Biol. Chem. **271**, 24353–24358 (1996)

M. Wang, M. Tashiro, J.R. Berlin, Regulation of L-type calcium current by intracellular magnesium in rat cardiac myocytes. J. Physiol. **555**, 383–396 (2004)

S.K. Wei, J.F. Quigley, S.U. Hanlon, B. O'Rourke, M.C. Haigney, Cytosolic free magnesium modulates Na/Ca exchange currents in pig myocytes. Cardiovasc. Res. **53**, 334–340 (2002)

R.E. White, H.C. Hartzell, Effects of intracellular free magnesium on calcium current in isolated cardiac myocytes. Science **239**, 778–780 (1988)

A. Zahradnikova, Z. Kubalova, J. Pavelkova, S. Gyorke, I. Zahradnik, Activation of calcium release assessed by calcium release-induced inactivation of calcium current in rat cardiac myocytes. Am. J. Physiol. Cell Physiol. **286**, C330–C341 (2004)

X. Zhang, C.R. Solaro, C.J. Lingle, Allosteric regulation of BK channel gating by Ca^{2+} and Mg^{2+} through a nonselective, low affinity divalent cation site. J. Gen. Physiol. **118**, 607–636 (2001)

J. Zhang, F. Zhao, Y. Zhao, J. Wang, L. Pei, N. Sun, J. Shi, Hypoxia induces an increase in intracellular magnesium via transient receptor potential melastatin 7 (TRPM7) channels in rat hippocampal neurons in vitro. J. Biol. Chem. **286**, 20194–20207 (2011)

Part III

Structural and Functional Aspects of NCKX

Functional and Structural Properties of the NCKX2 Na^+-Ca^{2+}/K^+ Exchanger: A Comparison with the NCX1 Na^+/Ca^{2+} Exchanger

8

Haider F. Altimimi, Robert T. Szerencsei, and Paul P.M. Schnetkamp

Abstract

Na^+/Ca^{2+}-K^+ exchangers (NCKX), alongside the more widely known Na^+/Ca^{2+} exchangers (NCX), are important players in the cellular Ca^{2+} toolkit. But, unlike NCX, much less is known about the physiological roles of NCKX, while emergent evidence indicates that NCKX has highly specialized functions in cells and tissues where it is expressed. As their name implies, there are functional similarities in the properties of the two Ca^{2+} exchanger families, but there are specific differences as well. Here, we compare and contrast their key functional properties of ionic dependence and affinities, as well as report on the effects of KB-R7943 – a compound that is widely used to differentiate the two exchangers. We also review structural similarities and differences between the two exchangers. The aim is to draw attention to key differences that will aid in differentiating the two exchangers in physiological contexts where both exist but perhaps play distinct roles.

Keywords

Na^+/Ca^{2+} exchanger • Na^+/Ca^{2+}-K^+ exchanger • NCX • NCKX • KB-R7943 • SLC24

8.1 Introduction

Calcium plays a ubiquitous role in eukaryotic intracellular signalling. To maintain a high signal-to-noise ratio, cells maintain their resting Ca^{2+} at very low levels relative to the extracellular environment through the concerted action of buffering cytosolic Ca^{2+}, sequestration in organelles and extrusion – the latter is most essential for long-term homeostasis. Two mechanisms exist to extrude Ca^{2+}, the

H.F. Altimimi
Department of Physiology and Pharmacology, Hotchkiss Brain Institute, University of Calgary, Calgary, Alberta, Canada

Centre for Research in Neuroscience, McGill University, Montreal, Quebec, Canada

R.T. Szerencsei • P.P.M. Schnetkamp (✉)
Department of Physiology and Pharmacology, Hotchkiss Brain Institute, University of Calgary, Calgary, Alberta, Canada
e-mail: pschnetk@ucalgary.ca

L. Annunziato (ed.), *Sodium Calcium Exchange: A Growing Spectrum of Pathophysiological Implications*, Advances in Experimental Medicine and Biology 961, DOI 10.1007/978-1-4614-4756-6_8,

ATP-driven plasma membrane Ca^{2+} ATPase (PMCA) in humans represented by four genes *ATP2B1-4* (Strehler and Zacharias 2001) and Na^+-driven Na^+/Ca^{2+} exchangers (NCX) in humans represented by three genes *SLC8A1-3* (Quednau et al. 2004) and the later discovered (see below) Na^+/Ca^{2+}-K^+ exchangers (NCKX) in humans represented by five genes *SLC24A1-5* (Schnetkamp 2004).

PMCA has a high affinity for Ca_i^{2+}, less than 0.5 μM, while Na^+/Ca^{2+} exchangers have a lower affinity for Ca_i^{2+} (in the range of 1–5 μM); however, Na^+/Ca^{2+} exchangers have a higher turnover rate than PMCA (Blaustein and Lederer 1999). Hence, what is thought to differentiate these two mechanisms physiologically is that PMCA regulates resting Ca_i^{2+}, while Na^+/Ca^{2+} exchangers handle the larger, more dynamic fluxes of Ca^{2+} which are most prominent in excitable tissue. But while both plasma membrane Ca^{2+} pump and Na^+/Ca^{2+} exchange were known to the scientific community since the 1960s, it was a relatively later discovery which revealed that retinal rod outer segments express a unique Na^+/Ca^{2+} exchanger which co-transports K^+ with Ca^{2+} (Schnetkamp et al. 1989; Cervetto et al. 1989). The subsequent cloning of multiple members of this family – NCKX – which have a tissue distribution that is often overlapping with NCX (Li and Lytton 2002; Papa et al. 2003; Minelli et al. 2007) has raised the question of whether these two distinct proteins play redundant roles in Ca^{2+} regulation and cellular physiology. It is also clear that in many tissues functional expression of both NCX and NCKX was observed within the same cell (Roberts and Bose 2011; Yang et al. 2011; Kiedrowski et al. 2004; Li et al. 2006; Pan et al. 2008).

The purpose of this chapter is to compare functional as well as structural features of NCKX against NCX, to highlight similarities in their function and address differences, with the hope of drawing more attention to the distinction between the two, so that future studies can help us to further understand the important roles that these two families of plasma membrane Ca^{2+} transporters play in normal cell function and in pathophysiology.

8.1.1 Roles of NCX and NCKX in Cell Physiology

While numerous studies have investigated the physiological roles of NCX, especially in myocardial cells (for review, see Blaustein and Lederer (1999)), relatively little detail is known of the in situ contributions of NCKX outside of retinal photoreceptors (Schnetkamp 1995); in recent years, however, novel specific roles for NCKX proteins have been emerging in pigmentation in epidermal melanocytes and the retinal pigment epithelium (Lamason et al. 2005; Vogel et al. 2008), motor learning and memory (Li et al. 2006) and olfaction (Stephan et al. 2011). In the brain, where both NCX and NCKX are abundantly expressed, the investigation of the role of NCX2 (*SLC8A2*) by genetic ablation revealed a deficit in neuronal Ca_i^{2+} clearance, associated with a lowering of the threshold for induction of long-term potentiation (LTP) in the hippocampus – a process that is thought to be critical for the formation of new spatial memories – hence, animals deficient in NCX2 displayed enhanced learning and memory performance relative to controls (Jeon et al. 2003). On the other hand, knockout of NCKX2 (SLC24A2) – which is also highly expressed in the hippocampus – also perturbed Ca^{2+} fluxes in neurons, but led to elimination of LTP expression and an associated deficiency in spatial working memory (Li et al. 2006). These studies indicate that in fact NCX and NCKX could serve distinct roles, potentially through mechanistic differences in their modes of operation. However, to begin to address differences in the modes of operation and physiological functions of NCX and NCKX in tissues or cells where both are expressed, it would be desirable to have specific pharmacological tools which can be used to antagonize the function of one set of proteins while leaving the other functional and importantly that the elimination of function can be implemented on a timescale relevant to the operation of NCX and NCKX in Ca_i^{2+} clearance.

8.1.2 Small Compound Inhibitors of Na$^+$/Ca$^+$ Exchangers

One of the first small compound inhibitors developed to target NCX was KB-R7943, an isothiurea derivative (Iwamato et al. 1996). Because this initial study reported that KB-R7943 preferentially inhibits the Ca^{2+} influx mode of the bidirectional NCX, numerous studies have since used the compound to demonstrate beneficial effects of application of the drug under posited pathological conditions (typically under perturbed ionic conditions) in which NCX is expected to predominantly mediate unabated Ca^{2+} influx into the cytosol, leading to toxic Ca_i^{2+} accumulation (reviewed in Amran et al. (2003)). However, it is clear by now that KB-R7943 is far from a specific NCX inhibitor; many ion channels and transporters have been shown to be inhibited by the compound at concentrations equivalent to, or lower than, those used to inhibit the Ca^{2+} influx mode of NCX (e.g. Barrientos et al. 2009 and references therein). Moreover, the extent to which KB-R7943 inhibits NCX is variable under different experimental conditions (Iwamoto et al. 1996; Linck et al. 1998; Elias et al. 2001). Additionally, questions have been raised as to the apparent paradox of the compound acting more potently on the Ca^{2+} import mode of transport over the Ca^{2+} efflux mode (Iwamoto et al. 1996; Kimura et al. 1999); further insight has been gained on the mechanisms of action of KB-R7943 since these early experiments, and it is now evident that the compound acts on specific kinetic states of NCX (Bouchard et al. 2004; Lee et al. 2004), discussed further below. Other, more potent, compounds have been developed to target NCX since the development of KB-R7943; SEA0400, for example, is effective at much lower concentrations (Matsuda et al. 2001) and appears to be substantially more selective (Tanaka et al. 2002), but nonetheless has been shown to have non-specific actions (Reuter et al. 2002).

Thus far, no commercially available small compound has been found to appreciably inhibit the function of NCKX proteins, except for 3′, 5′ dichlorobenzamil (Nicol et al. 1987), tetracaine and L-cis diltiazem (Schnetkamp et al. 1989), all of which are non-specific. Since it was reported that KB-R7943 acts selectively to inhibit NCX over NCKX (Iwamoto et al. 2001), other investigators have used sensitivity to the compound as a diagnostic to attribute Na$^+$/Ca^{2+} exchange activity in a given preparation to NCX over NCKX (Czyz and Kiedrowski 2002; Kiedrowski et al. 2004; Wu et al. 2008). On the other hand, studies have reported on KB-R7943-sensitive Ca^{2+} transport in invertebrate spermatozoa (Su and Vacquier 2002; Islam et al. 2006), as well as mammalian platelets (Takano et al. 2001), and that the putative target of KB-R7943 in those cells is of the NCKX type. This has prompted us to re-examine the effect of KB-R79473 on NCKX; herein, we test the compound on NCKX2 as well as NCKX1, since, previously, only NCKX2 had been directly tested for sensitivity to KB-R7943 (Iwamoto et al. 2001). We also present data demonstrating directly the key functional difference between NCX and NCKX in K$^+$ dependence and transport, as well as affinity for Na$^+$ and Ca^{2+}. This is followed by a comparative review of structural features of these two exchangers, with NCKX2 serving as the model NCKX as it is the best studied member of the family in terms of structure-function relationships and NCX1 as the best studied member of the NCX family.

8.2 Functional Comparison of NCX1 and NCKX2

To clearly demonstrate functional differences between NCX and NCKX, we have undertaken measurements of K$^+$ dependence and Na$^+$ and Ca^{2+} affinity under defined conditions of heterologous expression in HEK293 cells; for NCKX2, we used the human clone (Prinsen et al. 2000), while for NCX1 we used the canine clone (Nicoll et al. 1990). Details of our assays have been previously published elsewhere (Szerencsei et al. 2001; Kang et al. 2005a; Altimimi and Schnetkamp 2007; Altimimi et al. 2010).

8.2.1 K$^+$ Dependence and Transport

Figure 8.1 illustrates the key difference between NCX and NCKX – the absolute requirement for K$^+$ as a co-transported substrate in the latter. As in

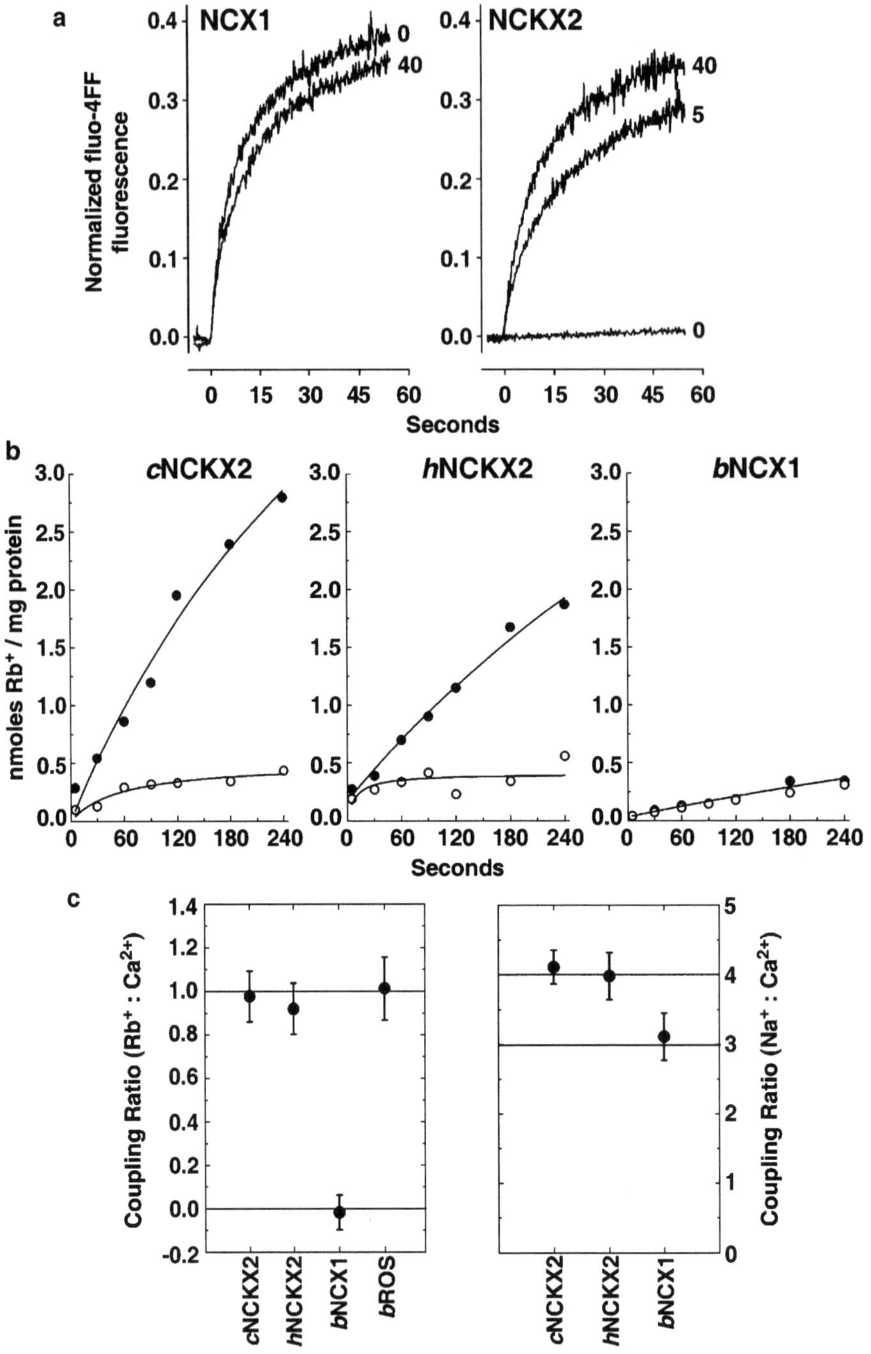

Fig. 8.1 *K^+ dependency and transport stoichiometry of NCX1 and NCKX2 expressed in HEK 293 cells*: (**a**) Ca^{2+} influx through reverse Na^+/Ca^{2+} exchange was initiated at time zero by addition of 250 μM $[Ca^{2+}]_o$ in the presence of the indicated $[K^+]_o$ (mM). Experimental conditions and protocols are as described in the text and in more detail in Kang et al. (2005a). (**b**) Ca^{2+}-dependent ^{86}Rb uptake in High Five cells expressing chicken NCKX2, human NCKX2 and bovine NCX1. ^{86}Rb uptake was initiated at time zero by addition of ^{86}Rb, 0.4 mM RbCl in the presence of 0.4 mM EDTA (*open circles*) or 0.4 mM $CaCl_2$ (*filled circles*) in a medium containing 150 mM choline chloride and 20 mM HEPES (pH 7.4). (**c**) Rb/Ca and Na/Ca coupling ratios were obtained in High Five cells expressing chicken NCKX2, human NCKX2, bovine NCX1 or bovine rod outer segments (ROS which express NCKX1) as described in detail in Szerencsei et al. (2002). Average values (± standard deviation) are illustrated representing 8–15 separate experiments (panels (**b**) and (**c**) were taken with slight modifications from Szerencsei et al. (2001), with permission from the publisher)

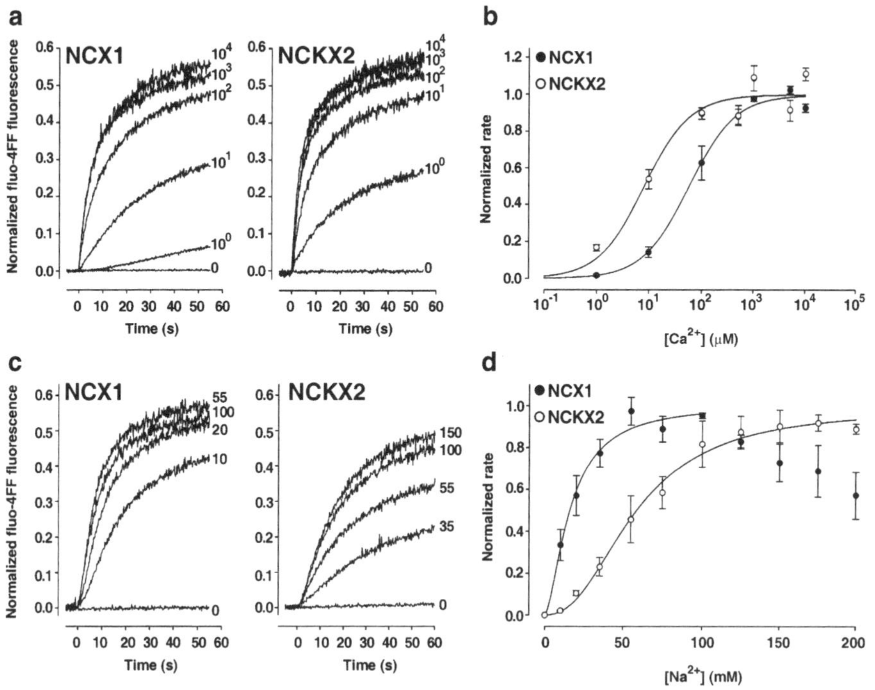

Fig. 8.2 *Na^+ and Ca^{2+} dependencies of NCX1 and NCKX2 expressed in HEK293 cells.* (**a**) Ca^{2+} influx through reverse Na^+/Ca^{2+} exchange was initiated at time zero by addition of the indicated $[Ca^{2+}]_o$ (μM). Experimental conditions and protocols are as described in the text and in more detail in Kang et al. (2005a). (**b**) Average initial rate of Ca^{2+} influx (± standard error of the mean for all subsequent error bars) is plotted versus $[Ca^{2+}]_o$ for three experiments. (**c**) Ca^{2+} influx through reverse Na^+/Ca^{2+} exchange was initiated at time zero by addition of the indicated $[Na^+]_i$ (mM). Experimental conditions and protocols are as described in the text and in more detail in Altimimi et al. (2010). (**d**) Average initial rate of Ca^{2+} influx is plotted versus $[Na^+]_i$ for three experiments

most functional studies on NCX and NCKX, we measured Na_i^+-dependent Ca^{2+} influx or reverse Na^+/Ca^{2+} exchange. Ca^{2+} influx mode was initiated in NCX1-transfected, fluo-4FF-loaded HEK293 cells by the addition of Ca_o^{2+} in the presence or absence of K_o^+ in a buffered medium where Li^+ is the major constituent cation; there was no difference in initial rate of Ca^{2+} influx between the presence or absence of K^+ (Fig. 8.1a). NCKX2-transfected HEK293 cells in the same Li^+ medium, on the other hand, did not show Ca^{2+} influx on addition of Ca_o^{2+} until the addition of K_o^+ at time zero (Fig. 8.1a). While the free $[Ca^{2+}]_i$ measurements shown in Fig. 8.1a illustrate K^+-dependence, they do not signify K^+-cotransport, which is the hallmark of NCKX; Fig. 8.1b shows Ca^{2+}-activated Rb^+ (a substitute for K^+) co-transport in two different NCKX clones, chicken and human, compared against bovine NCX, where it is clear that only NCKX mediates Ca^{2+}-activated transmembrane Rb^+ transport (from Szerencsei et al. (2001)). Figure 8.1b also demonstrates a direct determination of the stoichiometry of transport in NCKX with 1 K^+ ion coupled to the transport of 1 Ca^{2+} ion, in exchange for 4 Na^+ ions, while NCX transports 1 Ca^{2+} ion in exchange for 3 Na^+ ions (from Szerencsei et al. (2001)).

8.2.2 Ca^{2+} Affinity

To compare Ca^{2+} affinity of the two exchangers, Ca^{2+} influx mode was initiated in NCX1 or NCKX2-transfected, fluo-4FF-loaded HEK293 cells by the addition of different $[Ca^{2+}]_o$ in the presence of K_o^+ in a buffered medium where Li^+ is the major constituent cation (Kang et al. 2005a; Altimimi and Schnetkamp 2007) and changes in free $[Ca^{2+}]_i$ were monitored (Fig. 8.2a). Note that at 1 μM $[Ca^{2+}]_o$ NCX1-mediated Ca^{2+} influx was comparatively very low and commenced after a lag of a few seconds. This is indicative of the catalytic requirement of NCX1 for $[Ca^{2+}]_i$, which binds to domains within its large intracellular loop to activate the exchanger (Hilgemann et al. 1992). NCKX2 on the other hand does not appear to require $[Ca^{2+}]_i$ to activate the exchanger, as seen in Fig. 8.2a; addition of 1 μM Ca_o^{2+} results, without any lag, in rapid influx at a rate 15–20% of Vmax. The K_m values for $[Ca^{2+}]_o$ derived under these conditions were 58±8 μM and 8±3 μM for NCX1 and NCKX2, respectively (Fig. 8.2b).

8.2.3 Na+ Affinity

Na^+ affinity of the exchangers was measured using an assay based on the use of gramicidin as a means of controlling $[Na^+]_i$ (Altimimi et al. 2010); the assays illustrated in Fig. 8.2c, d were carried out in the presence of 0.5 mM Ca_o^{2+} in a buffered medium where K^+ was the major constituent cation and Ca^{2+} influx was initiated by the addition of Na^+. From the exemplar traces illustrated in Fig. 8.2c, it is also apparent that Ca^{2+} influx in NCX1-transfected HEK293 cells follows a sigmoidal trajectory, indicative of the lag required for catalytic Ca_i^{2+} to fully activate NCX1 by binding to its intracellular loop; this does not appear for NCKX2-mediated Ca^{2+} influx. From these measurements, we find that NCX1 under these conditions has a higher affinity for Na_i^+ than NCKX2; the K_m values are 16±2 mM and 58±3 mM, respectively (Fig. 8.2d).

These properties may dictate differences in the modes of operation of NCX and NCKX in their native environments and may be at play in tissues and cells where both are expressed in the same compartments. The catalytic requirement of NCX for Ca_i^{2+} to fully activate the exchanger, which factors into our measurements of Ca^{2+} and Na^+ affinity herein, may limit its contribution under basal conditions when $[Ca^{2+}]_i$ is at low levels – near resting $[Ca^{2+}]_i$. Given that NCKX does not require Ca_i^{2+} for its activation, NCKX may be positioned as the "intermediate" Ca^{2+} extrusion mechanism in between PMCA which is posited to be fully operational at resting $[Ca^{2+}]_i$ and NCX which is fully operational when $[Ca^{2+}]_i$ reaches higher levels required to occupy its Ca^{2+}-binding domains. In this context, it is pertinent to highlight the study of Kim et al. (2005) who dissected the contributions of plasma membrane Ca^{2+} transporters to Ca^{2+} clearance in neuronal axon terminals at the calyx of Held when $[Ca^{2+}]_i$ loading was modest, in the range of<2 μM; NCKX made the biggest contribution to Ca_i^{2+} clearance at 42%, followed by NCX and PMCA at 26% and 23%, respectively.

8.2.4 Comparison of the Effect of KB-R7943 on NCX and NCKX

The commonly used drug KB-R7943 was tested on NCX1 and NCKX2, as well as dolphin NCKX1 (Cooper et al. 1999). Transfected HEK293 cells were placed in a medium of Li^+ and treated with various concentrations of KB-R7943 10s prior to addition of Ca^{2+} in the case of NCX, or addition of K^+ (in the presence of Ca_o^{2+}) for NCKX1 and NCKX2. Figure 8.3a illustrates exemplar Ca^{2+} influx traces for all three exchangers. We noted a marked difference of the effect of KB-R7943 on the initial rates of Ca^{2+} influx compared with steady state free $[Ca^{2+}]_i$ levels in the case of NCX1, where even at the highest concentration of the drug tested, the rate of Ca^{2+} influx appeared unchanged from control (Fig. 8.3b). However, there was an abrupt plateau phase in the free $[Ca^{2+}]_i$ signal in NCX1-transfected cells, appearing only at the higher concentrations tested – 30 μM (not shown) and 50 μM (Fig. 8.3a). These results are quantified both in terms of effect of KB-R7943 on initial rates (Fig. 8.3b) as well as on the steady-state level of free Ca_i^{2+} achieved by the three exchangers (Fig. 8.3c).

From these results, it is clear that under our assay conditions, KB-R7943 produces only a modest effect on NCX1 and is only effective at relatively high concentrations. While our assay here used ensemble fluo-4FF measurements of free $[Ca^{2+}]_i$ in HEK293 cells, an equally low efficacy of KB-R7943 on NCX1 has been

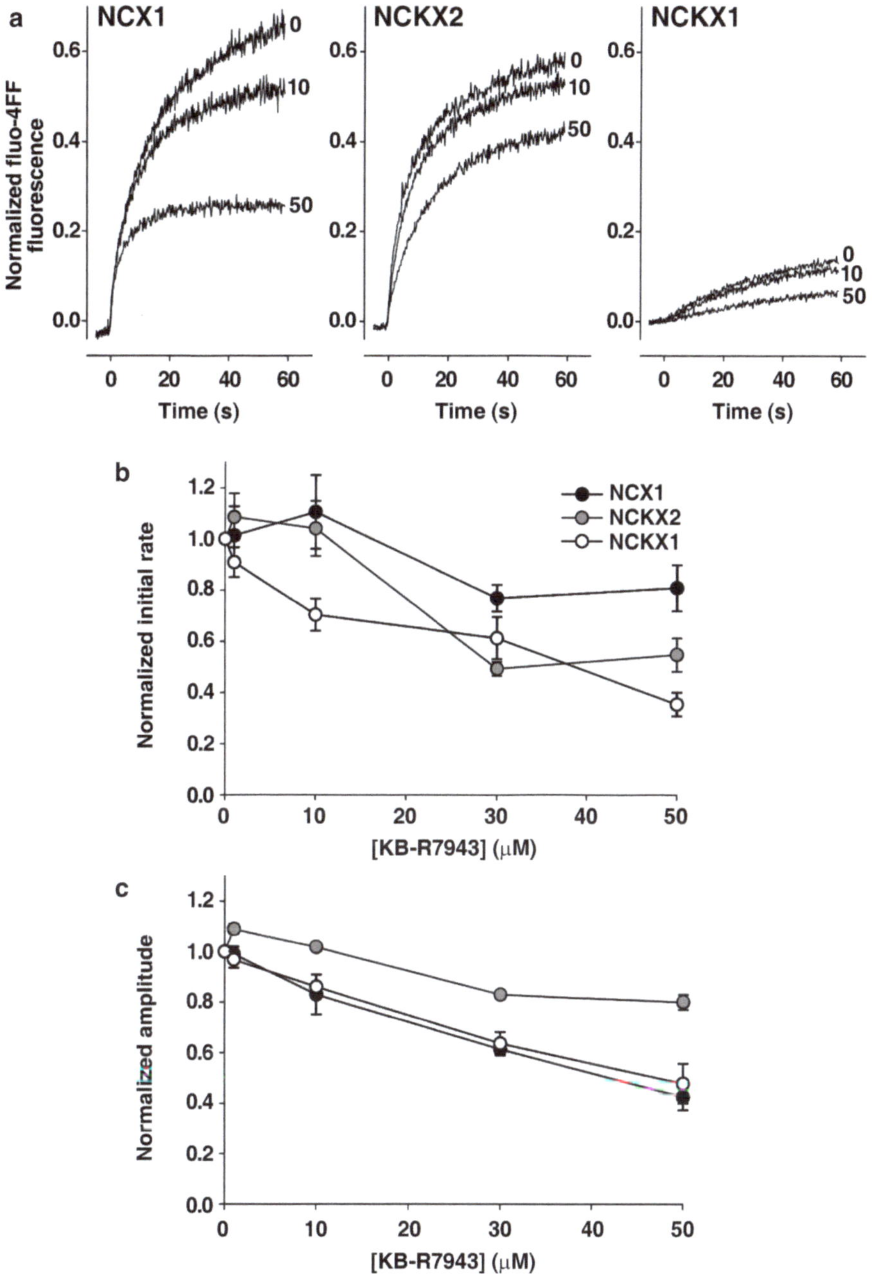

Fig. 8.3 *Effect of KB R7943 on NCKX1, NCKX2 or NCX1 expressed in HEK293 cells.* (**a**) Ca^{2+} influx through reverse Na^+/Ca^{2+} exchange was initiated at time zero by addition of 250 μM $[Ca^{2+}]_o$ in the presence of the indicated concentrations (in μM) of KB-R7943. (**b**) Initial rate of Ca^{2+} influx as a function of KB-R7943 concentration. (**c**) Steady-state level of increase in $[Ca^{2+}]_i$ as a function of KB-R7943 concentration

reported previously in a study that employed radioactive $^{45}Ca^{2+}$ flux in both NCX1-transfected BHK cells (20% inhibition by KB-R7943 at 30 μM) and membrane vesicles isolated from those cells (30% inhibition by KB-R7943 at 10 μM) (Linck et al. 1998). One possible explanation for the relatively low potency of KB-R7943 inhibition observed in our measurements is that the time of

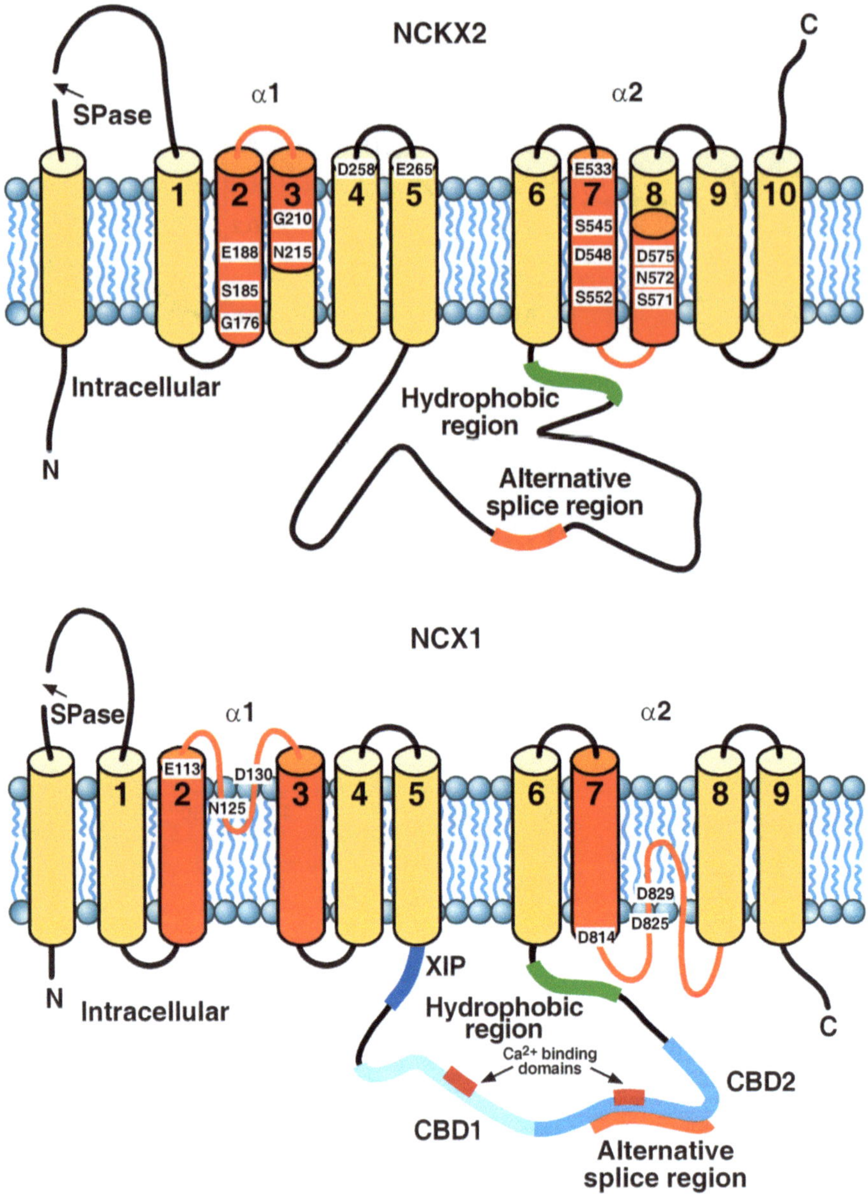

Fig. 8.4 Current topological models of NCKX2 (*top*) and NCX1 (*bottom*)

KB-R7943 pre-incubation before commencing Ca^{2+} influx through the exchanger was brief (10s in the data illustrated in Fig. 8.3); however, we tested longer pre-incubations – as long as 5 min – and found no appreciable difference in efficacy of inhibition (data not shown). Also arguing against time being the factor are the results of Linck et al. (1998), where cells were incubated with KB-R7943 for 20 min before assaying for Ca^{2+} transport.

From various studies on the effect of KB-R7943 on NCX, mechanistic insight has been gained on the structural elements important for binding (Iwamoto et al. 2001), as well as the likely kinetic state which KB-R7943 impacts (Iwamoto et al. 2004). While it was found that mutations in the α-2 repeat of the exchanger (see Fig. 8.4) abrogated sensitivity of NCX1 to KB-R7943 (Iwamoto et al. 2001), analysis of a chimeric exchanger containing the α-1 repeat of NCX1, but the α-2 repeat of NCKX2, showed equivalent inhibition by KB-R7943 to that of wild-type NCX1; wild-type NCKX2 was not sensitive to KB-R7943 in that same study (Iwamoto et al. 2004). This suggested that earlier findings on the site of KB-R7943 interaction with NCX1 may be a result of an allosteric effect of the drug, rather than a direct effect (Iwamoto et al. 2001). The same authors also investigated the effects of KB-R7943, as well as SEA0400, on mutant NCX1 exchangers with altered regulatory properties. A prominent kinetic feature of NCX1 is Na^+-dependent inactivation, also termed I_1 inactivation, which is most evident in patch-clamp electrophysiological measurements as a decay of Na_i^+-dependent outward current to a steady state that is 10–70% of peak outward current (Hilgemann and Collins 1992). A mutant of NCX with enhanced I_1 inactivation showed greater sensitivity to KB-R7943 (as well as SEA0400), while another mutant with no apparent I_1 inactivation displayed insensitivity to KB-R7943 (Iwamoto et al. 2004). Similar findings were obtained in other detailed studies on the mechanism of SEA0400-mediated inhibition of NCX1 (Bouchard et al. 2004; Lee et al. 2004).

When examining the effect of KB-R7943 on NCKX2, we noted that the drug had only a small effect on the steady-state level of free $[Ca^{2+}]_i$ attained at ~1 min from initiation of Ca^{2+} influx. However, analysis of the initial rate of change in free $[Ca^{2+}]_i$ revealed a more pronounced inhibition produced by KB-R7943 at 30 and 50 μM. NCKX1, on the other hand, appeared to be more sensitive to KB-R7943; there was equivalent inhibition to that seen in NCKX2 when examining initial rates of change in free $[Ca^{2+}]_i$, and the steady-state level of $[Ca^{2+}]_i$ was also decreased to the same extent as that seen in NCX1-transfected HEK293 cells (Fig. 8.3b).

In light of the mechanistic insight gained for the effects of KB-R7943 and SEA0400 on NCX1, it is plausible that the inhibition we observed with NCKX1 (or on the rate of change in $[Ca^{2+}]_i$ mediated by NCKX2) is due to some interaction of KB-R7943 with specific kinetic states of NCKX. In HEK293 cells, we found that NCKX2 displayed kinetic features reminiscent of Na^+-dependent inactivation described for NCX1, but under the conditions of our assay, inactivation was only evident for the Ca^{2+} extrusion mode. Like I_1 inactivation in NCX1, NCKX2 inactivated with exposure to high $[Na^+]_i$ was relieved by decreasing $[Ca^{2+}]_o$ (which favours outward-facing conformation exchangers, thereby decreasing high Na_i^+ exposure), and mutants with increased Na_i^+ affinity displayed inactivation at lower $[Na^+]$ than wild-type NCKX2 (Altimimi and Schnetkamp 2007).

In summary, we found that KB-R7943 is a poor agent for unambiguously differentiating the contributions of NCX from NCKX for several reasons: (a) its efficacy in inhibiting NCX is variable under different assay conditions; (b) it may in fact impact the function of NCKX, depending on isoform present; while we have tested the compound here on NCKX1 and NCKX2, it is not yet known what the effects of the compound are on NCKX3-5, and (c) as found by many other investigators, KB-R7943 can interfere with many other ion channels and transporters, many of which transport Ca^{2+}, further complicating the assignment of perturbations in Ca^{2+} signalling caused by the compound as the result of inhibition of NCX.

8.3 Comparing NCX and NCKX Sequences

Both NCX and NCKX are intrinsic membrane proteins. The predominant structural motifs of the membrane-spanning domains of such proteins are invariably alpha-helical segments of ~20 residues that traverse the membrane (transmembrane segments or TMS). Hydrophobicity analysis of all NCX and NCKX sequences reveals the presence of twelve hydrophobic segments that could constitute TMS. The first is located at the N-terminus and thought to be a cleavable signal peptide. By placing tags before and after the putative cleavage site, we determined that the putative signal peptide of NCKX2 was only partially cleaved, resulting in two populations of NCKX2 protein: one full-length and one with the signal peptide cleaved as evidenced by a characteristic two-band pattern seen in Western blots (Kang and Schnetkamp 2003). The situation with

NCKX1 was more complex, as partial signal peptide cleavage was seen for dolphin NCKX1 but not for chicken NCKX1. Additionally, deletion of the signal peptide in both NCKX2 and dolphin NCKX1 prevented trafficking to the plasma membrane (Kang and Schnetkamp 2003). The consequence of partial cleavage of the signal peptide is that after expression of various NCKX1 and NCKX2 cDNAs in cell lines, a significant fraction of expressed NCKX protein is localized within the cell rather than in the plasma membrane. It remains to be established whether this is an artefact of overexpression of NCKX cDNA in cell lines or if NCKX proteins may function in intracellular organelles as well. This intriguing possibility is made more likely by the observation that NCKX5, critical for pigmentation in epidermal melanocytes and the retinal pigment epithelium, is not found in the plasma membrane but localized exclusively within the cell, most likely in the trans-Golgi network, although its precise function remains to be elucidated (Lamason et al. 2005; Ginger et al. 2008).

Little sequence conservation is observed for the signal peptides of the five NCKX isoforms, whereas the remaining eleven hydrophobic segments are the only sequence elements that show significant sequence conservation among all NCKX isoforms. These eleven hydrophobic segments are grouped in two sets of five and six putative TMS, respectively, and separated by a large hydrophilic loop thought to be located in the cytoplasm. This large hydrophilic loop is not or is poorly conserved among NCKX isoforms and ranges from more than 400 residues in mammalian NCKX1 to approximately 100 residues in NCKX5. Very little has been elucidated about the role of the large cytoplasmic loop in NCKX function except that it is not directly involved in either cation transport or cation selectivity, that is, the TMS domains are both necessary and sufficient for Na^+/Ca^{2+}-K^+ exchange transport (Szerencsei et al. 2000). In the same study, we also showed that the transport properties observed for mammalian NCKX1 were very similar to those observed for a distantly related NCKX cloned from *C. elegans*. The highest degree of sequence conservation was observed in four of the eleven hydrophobic segments which make up the two so-called alpha repeats which are thought to have arisen from an ancient gene duplication event (Schwarz and Benzer 1997). This suggests that the alpha repeats contain most of the residues important for cation binding and cation transport. Much of the work in our laboratory over the past ten years has focused on elucidating the role of the TMS in NCKX ion transport function and determining a topological model for their arrangement.

8.3.1 Topological Models of NCX and NCKX

We have proposed a topological model for human NCKX2 (Fig. 8.4) based on a combination of results from two methods: (1) determining the accessibility of substituted cysteine residues to small externally applied hydrophilic cysteine-modifying reagents (e.g. MTSET) and (2) inserting glycosylation sites in the short loops connecting putative TMS (Kinjo et al. 2003). This model places the alpha repeats in an inverted configuration while the short C-terminal loop faces the extracellular space. Although actual sequence similarity between NCX and NCKX is extremely limited to two short stretches of ~35 residues that make up the core of each of the two alpha repeats, the hydropathy analysis of NCX1 reveals a very similar pattern of eleven hydrophobic segments. However, the current topological model of NCX1 (Nicoll et al. 1999; Iwamoto et al. 2000) differs considerably from that of NCKX2 (Fig. 8.4) due to the presence of two re-entrant loops, one in each of the two alpha repeats. The first re-entrant loop is located in the region linking TMS2 to TMS3, while the second re-entrant loop replaces TMS8, inverting the orientation of the two TMS closest to the C-terminal and thus placing the C-terminus of NCX1 in the cytoplasm.

Unlike the case for NCKX proteins, many studies have addressed regulatory features imposed on Na^+/Ca^{2+} exchange transport by distinct sequences contained in the large cytosolic loop of NCX1, for example, the binding domain of the XIP peptide and sequences responsible for Na^+-dependent inactivation and secondary activation by cytosolic Ca^{2+}. Both NMR and X-ray crystal structures have been obtained for these domains (see other chapters in this volume). We have described Na^+-dependent inactivation for NCKX2 that shares some characteristics with Na^+-dependent inactivation seen in NCX1 (i.e. occupancy of the cation transport sites by Na^+), but it remains to be established if sequence elements in the cytosolic loop of NCKX2 participate in this process (Altimimi and Schnetkamp 2007).

8.3.2 Residues Important for NCX- and NCKX-Mediated Cation Transport

We carried out scanning mutagenesis of the two alpha repeats of NCKX2 to identify residues important for Na^+/Ca^{2+}-K^+ exchange transport. We also examined all aspartate and glutamate residues found in the TMS as such acidic residues are commonly found to be critical for Ca^{2+} and Na^+ transport (Kang et al. 2005a, b; Winkfein et al. 2003). All of these residues were examined for changes in total activity (V_{max}) and changes in K_m for Na^+, while a subset was examined for changes in the K_m for Ca^{2+} and K^+ (a full scan is currently in progress). Some of the most important residues are highlighted in Fig. 8.4. It is probably no surprise that these residues are mostly negatively charged residues or polar residues that could provide cation-coordinating oxygen atoms, but also include two glycine residues. Moreover, all the residues involved are conserved in most if not all NCKX sequences currently in the database. Furthermore, all the residues depicted here as located in the membrane interior are conserved between NCX and NCKX sequences with one notable exception. An aspartate residue is found in all NCKX sequences at the position equivalent to D575 in human NCKX2, whereas an asparagine is found in all NCX sequences at this position. We showed that the D575N (or C) substitution in human NCKX2 renders the mutant NCKX2 protein independent of K^+ (as is the case with NCX proteins), and we suggest this is an essential residue for K^+ binding to NCKX (Kang et al. 2005b). For most of the residues shown here that are conserved between NCX1 and NCKX2, the activity of mutant NCX1 in which any of these residues were replaced was very low in comparison to wild type and did not permit an analysis of shifts in K_m's for Ca^{2+} or Na^+. We proposed that E188 and D548 in NCKX2 (and the equivalent E113 and D814 in NCX1) are the two main Ca^{2+}-coordinating residues based on three observations: (1) these two residues are conserved in all NCX and NCKX sequences, (2) they are the only two acidic residues in NCKX2 for which removal of the charge led to a complete abolition of transport (<0.2%) and (3) the charge-conservative E188D and D548E substitutions resulted in mutant NCKX2 proteins that displayed the largest shifts in Ca^{2+} K_m (Kang et al. 2005a). The more peripherally located acidic residues (e.g. D258, E265, E533) are conserved in most NCKX sequences, while NCX has different conserved acidic residues in more peripheral locations that can affect Ca^{2+} K_m values (Iwamoto et al. 2000). Such residues may not be directly involved in the Ca^{2+} binding site of NCX or NCKX but may influence K_m values by increasing the local [Ca^{2+}] due to electrostatic attraction. The two glycine residues shown in Fig. 8.4a (G176 and G210) are very sensitive to substitution as even the very conservative Gly to Ala substitution results in a greater than 90% inhibition of the V_{max} (Winkfein et al. 2003; Altimimi et al. 2010). Critical glycine residues often are in positions that either require hingelike movement of two helical segments or indicate helix-helix contacts.

8.4 Conclusions

Na^+/Ca^{2+} exchangers, NCX, and Na^+/Ca^{2+}-K^+ exchangers, NCKX, both play important roles in physiology as part of the cellular Ca^{2+} toolkit. While it may be convenient to lump the two mechanisms as one that mostly mediates Ca^{2+} extrusion via Na^+/Ca^{2+} exchange, the data and review of the literature we presented here hopefully will have convinced the reader that these two Ca^{2+} transporters are in fact quite distinct, both structurally and functionally. Especially when considering cells in which both NCX and NCKX are expressed, we believe that the differences between NCX and NCKX could serve mechanistically distinct functions. With this in mind, we hope that future studies will continue to shed light on the specific physiological roles that NCX and NCKX play in cellular physiology.

Acknowledgements The work presented here was supported by an operating grant (MOP 81327) from the Canadian Institutes for Health Research (to PPMS).

References

H.F. Altimimi, P.P.M. Schnetkamp, Na^+-dependent inactivation of the retinal cone/brain Na^+/Ca^{2+}-K^+ exchanger NCKX2. J. Biol. Chem. **282**, 3720–3729 (2007)

H.F. Altimimi, E.H. Fung, R.J. Winkfein, P.P. Schnetkamp, Residues contributing to the Na^+-binding pocket of the SLC24 Na^+/Ca^{2+}-K^+ Exchanger NCKX2. J. Biol. Chem. **285**, 15245–15255 (2010)

M.S. Amran, N. Homma, K. Hashimoto, Pharmacology of KB-R7943: a Na^+-Ca^{2+} exchange inhibitor. Card. Drug Rev. **21**, 255–276 (2003)

G. Barrientos, D.D. Bose, W. Feng, I. Padilla, I.N. Pessah, The Na^+/Ca^{2+} exchange inhibitor 2-(2-(4-(4-nitrobenzyloxy)phenyl)ethyl) isothiourea methanesulfonate (KB-R7943) also blocks ryanodine receptors type 1 (RyR1) and type 2 (RyR2) channels. Mol. Pharmacol. **76**, 560–568 (2009)

M.P. Blaustein, W.J. Lederer, Sodium/calcium exchange: its physiological implications. Physiol. Rev. **79**, 763–854 (1999)

R. Bouchard, A. Omelchenko, H.D. Le, P. Choptiany, T. Matsuda, A. Baba, K. Takahashi, D.A. Nicoll, K.D. Philipson, M. Hnatowich, L.V. Hryshko, Effects of SEA0400 on mutant NCX1.1 Na^+-Ca^{2+} exchangers with altered ionic regulation. Mol. Pharmacol. **65**, 802–810 (2004)

L. Cervetto, L. Lagnado, R.J. Perry, D.W. Robinson, P.A. McNaughton, Extrusion of calcium from rod outer segments is driven by both sodium and potassium gradients. Nature **337**, 740–743 (1989)

C.B. Cooper, R.J. Winkfein, R.T. Szerencsei, P.P.M. Schnetkamp, cDNA-cloning and functional expression of the dolphin retinal rod Na-Ca+K exchanger NCKX1: comparison with the functionally silent bovine NCKX1. Biochemistry **38**, 6276–6283 (1999)

A. Czyz, L. Kiedrowski, In depolarized and glucose-deprived neurons, Na^+ influx reverses plasmalemmal K^+-independent and K^+-independent Na^+/Ca^{2+} exchangers and contributes to NMDA Excitotoxicity. J. Neurochem. **83**, 1321–1328 (2002)

C.L. Elias, A. Lukas, S. Shurraw, J. Scott, A. Omelchenko, G.J. Gross, M. Hnatowich, L.V. Hryshko, Inhibition of Na^+/Ca^{2+} exchange by KB-R7943: transport mode selectivity and antiarrhythmic consequences. Am. J. Physiol. Heart Circ. Physiol. **281**, H1334–H1345 (2001)

R.S. Ginger, S.E. Askew, R.M. Ogborne, S. Wilson, D. Ferdinando, T. Dadd, A.M. Smith, S. Kazi, R.T. Szerencsei, R.J. Winkfein, P.P. Schnetkamp, M.R. Green, SLC24A5 encodes a trans-Golgi network protein with potassium-dependent sodium-calcium exchange activity that regulates human epidermal melanogenesis. J. Biol. Chem. **283**, 5486–5495 (2008)

D.W. Hilgemann, A. Collins, Mechanism of cardiac Na^+-Ca^{2+} exchange current stimulation by MgATP: possible involvement of aminophospholipid translocase. J. Physiol. **454**, 59–82 (1992)

D.W. Hilgemann, S. Matsuoka, G.A. Nagel, A. Collins, Steady state and dynamic properties of cardiac sodium-calcium exchange: sodium-dependent inactivation. J. Gen. Physiol. **100**, 905–932 (1992)

M.S. Islam, O. Kawase, S. Hase, H. Minakata, M. Hoshi, M. Matsumoto, Na^+/Ca^{2+} exchanger contributes to asterosap-induced elevation of intracellular Ca^{2+} concentration in starfish spermatozoa. Zygote **14**, 133–141 (2006)

T. Iwamoto, T. Watano, M. Shigekawa, A novel isothiourea derivative selectively inhibits the reverse mode of Na^+/Ca^{2+} exchange in cells expressing NCX1. J. Biol. Chem. **271**, 22391–22397 (1996)

T. Iwamoto, A. Uehara, I. Imanaga, M. Shigekawa, The Na^+/Ca^{2+} exchanger NCX1 has oppositely oriented reentrant loop domains that contain conserved aspartic acids whose mutation alters its apparent Ca^{2+} affinity. J. Biol. Chem. **275**, 38571–38580 (2000)

T. Iwamoto, S. Kita, A. Uehara, Y. Inoue, Y. Taniguchi, I. Imanaga, M. Shigekawa, Structural domains influencing sensitivity to isothiourea derivative inhibitor KB-R7943 in cardiac Na^+/Ca^{2+} exchanger. Mol. Pharmacol. **59**, 524–531 (2001)

T. Iwamoto, S. Kita, A. Uehara, I. Imanaga, T. Matsuda, A. Baba, T. Katsuragi, Molecular determinants of Na^+/Ca^{2+} exchange (NCX1) inhibition by SEA0400. J. Biol. Chem. **279**, 7544–7553 (2004)

D. Jeon, Y.M. Yang, M.J. Jeong, K.D. Philipson, H. Rhim, H.S. Shin, Enhanced learning and memory in mice lacking Na^+/Ca^{2+} exchanger 2. Neuron **38**, 965–976 (2003)

K.-J. Kang, P.P.M. Schnetkamp, Signal sequence cleavage and plasma membrane targeting of the rod NCKX1 and cone NCKX2 Na^+/Ca^{2+}-K^+ exchangers. Biochemistry **42**, 9438–9445 (2003)

K.-J. Kang, T.G. Kinjo, R.T. Szerencsei, P.P.M. Schnetkamp, Residues contributing to the Ca^{2+} and K^+ binding pocket of the NCKX2 Na^+/Ca^{2+}-K^+ exchanger. J. Biol. Chem. **280**, 6823–6833 (2005a)

K.-J. Kang, Y. Shibukawa, R.T. Szerencsei, P.P.M. Schnetkamp, Substitution of a single residue, Asp575, renders the NCKX2 K^+-dependent Na^+/Ca^{2+} exchanger independent of K^+. J. Biol. Chem. **280**, 6834–6839 (2005b)

L. Kiedrowski, A. Czyz, G. Baranauskas, X.F. Li, J. Lytton, Differential contribution of plasmalemmal Na/Ca exchange isoforms to sodium-dependent calcium influx and NMDA excitotoxicity in depolarized neurons. J. Neurochem. **90**, 117–128 (2004)

M.H. Kim, N. Korogod, R. Schneggenburger, W.K. Ho, S.-H. Lee, Interplay between Na^+/Ca^{2+} exchangers and mitochondria in Ca^{2+} clearance at the calyx of Held. J. Neurosci. **25**, 6057–6065 (2005)

J. Kimura, E.M. Jeanclos, R.J. Donnelly, J. Lytton, J.P. Reeves, A. Aviv, Physiological and molecular characterization of the Na^+/Ca^{2+} exchanger in human platelets. Am. J. Physiol. Heart Circ. Physiol. **277**, H911–H917 (1999)

T.G. Kinjo, R.T. Szerencsei, R.J. Winkfein, K.-J. Kang, P.P.M. Schnetkamp, Topology of the retinal cone NCKX2 Na/Ca-K exchanger. Biochemistry **42**, 2485–2491 (2003)

R.L. Lamason, M.A. Mohideen, J.R. Mest, A.C. Wong, H.L. Norton, M.C. Aros, M.J. Jurynec, X. Mao, V.R. Humphreville, J.E. Humbert, S. Sinha, J.L. Moore, P. Jagadeeswaran, W. Zhao, G. Ning, I. Makalowska, P.M. McKeigue, D. O'donnell, R. Kittles, E.J. Parra, N.J. Mangini, D.J. Grunwald, M.D. Shriver, V.A. Canfield, K.C. Cheng, SLC24A5, a putative cation exchanger, affects pigmentation in zebrafish and humans. Science **310**, 1782–1786 (2005)

C. Lee, N.S. Visen, N.S. Dhalla, H.D. Le, M. Isaac, P. Choptiany, G. Gross, A. Omelchenko, T. Matsuda, A. Baba, K. Takahashi, M. Hnatowich, L.V. Hryshko, InhibitoryprofileofSEA0400[2-[4-[(2,5-difluoropheny) methoxy]phenoxy]-5-ethoxyaniline] assessed on the cardiac Na^+-Ca^{2+} exchanger, NCX1.1. J. Pharmacol. Exp. Ther. **311**, 748–757 (2004)

X.F. Li, J. Lytton, Differential expression of Na/Ca exchanger and Na/Ca+K exchanger transcripts in rat brain. Ann. N. Y. Acad. Sci. **976**, 64–66 (2002)

X.F. Li, L. Kiedrowski, F. Tremblay, F.R. Fernandez, M. Perizzolo, R.J. Winkfein, R.W. Turner, J.S. Bains, D.E. Rancourt, J. Lytton, Importance of K^+-dependent Na^+/Ca^{2+} exchanger 2, NCKX2, in motor learning and memory. J. Biol. Chem. **281**, 6273–8262 (2006)

B. Linck, Z. Qiu, Z. He, Q. Tong, D.W. Hilgemann, K.D. Philipson, Functional comparison of the three isoforms of the Na^+/Ca^{2+} exchanger (NCX1, NCX2, NCX3). Am. J. Physiol. **274**, C415–C423 (1998)

T. Matsuda, N. Arakawa, K. Takuma, Y. Kishida, Y. Kawasaki, M. Sakaue, K. Takahashi, T. Takahashi, T. Suzuki, T. Ota, A. Hamano-Takahashi, M. Onishi, Y. Tanaka, K. Kameo, A. Baba, SEA0400, a novel and selective inhibitor of the Na^+-Ca^{2+} exchanger, attenuates reperfusion injury in the in vitro and in vivo cerebral ischemic models. J. Pharmacol. Exp. Ther. **298**, 249–256 (2001)

A. Minelli, P. Castaldo, P. Gobbi, S. Salucci, S. Magi, S. Amoroso, Cellular and subcellular localization of Na^+-Ca^{2+} exchanger protein isoforms, NCX1, NCX2, and NCX3 in cerebral cortex and hippocampus of adult rats. Cell Calcium **41**, 221–234 (2007)

G.D. Nicol, P.P.M. Schnetkamp, Y. Saimi, E.J. Cragoe Jr., M.D. Bownds, A derivative of amiloride blocks both the light- and cyclic GMP-regulated conductances in rod photoreceptors. J. Gen. Physiol. **90**, 651–669 (1987)

D.A. Nicoll, S. Longoni, K.D. Philipson, Molecular cloning and functional expression of the cardiac sarcolemmal Na^+-Ca^{2+} exchanger. Science **250**, 562–565 (1990)

D.A. Nicoll, M. Ottolia, L. Lu, Y. Lu, K.D. Philipson, A new topological model of the cardiac sarcolemmal Na^+-Ca^{2+} exchanger. J. Biol. Chem. **274**, 910–917 (1999)

C.Y. Pan, L.L. Tsai, J.H. Jiang, L.W. Chen, L.S. Kao, The co-presence of Na^+/Ca^{2+}-K^+ exchanger and Na^+/Ca^{2+} exchanger in bovine adrenal chromaffin cells. J. Neurochem. **107**, 658–667 (2008)

M. Papa, A. Canitano, F. Boscia, P. Castaldo, S. Sellitti, H. Porzig, M. Taglialatela, L. Annunziato, Differential expression of the Na^+-Ca^{2+} exchanger transcripts and proteins in rat brain regions. J. Comp. Neurol. **461**, 31–48 (2003)

C.F.M. Prinsen, R.T. Szerencsei, P.P.M. Schnetkamp, Molecular cloning and functional expression the potassium-dependent sodium-calcium exchanger from human and chicken retinal cone photoreceptors. J. Neurosci. **20**, 1424–1434 (2000)

B.D. Quednau, D.A. Nicoll, K.D. Philipson, The sodium/calcium exchanger family-SLC8. Eur. J. Physiol. **447**, 543–548 (2004)

H. Reuter, S.A. Henderson, T. Han, T. Matsuda, A. Baba, R.S. Ross, J.I. Goldhaber, K.D. Philipson, Knockout mice for pharmacological screening: testing the specificity of Na^+-Ca^{2+} exchange inhibitors. Circ. Res. **91**, 90–92 (2002)

D.E. Roberts, R. Bose, Molecular and functional characterization of the human platelet Na^+/Ca^{2+} exchangers. Br. J. Pharmacol. **165**, 922–936 (2011)

P.P.M. Schnetkamp, Calcium homeostasis in vertebrate retinal rod outer segments. Cell Calcium **18**, 322–330 (1995)

P.P.M. Schnetkamp, The SLC24 Na^+/Ca^{2+}-K^+ exchanger family: vision and beyond. Eur. J. Physiol. **447**, 683–688 (2004)

P.P.M. Schnetkamp, D.K. Basu, R.T. Szerencsei, Na-Ca exchange in the outer segments of bovine rod photoreceptors requires and transports potassium. Am. J. Physiol. Cell Physiol. **257**, C153–C157 (1989)

E.M. Schwarz, S. Benzer, Calx, a Na-Ca exchanger gene of *Drosophila* melanogaster. Proc. Natl. Acad. Sci. U. S. A. **94**, 10249–10254 (1997)

A.B. Stephan, S. Tobochnik, M. Dibattista, C.M. Wall, J. Reisert, H. Zhao, The Na^+/Ca^{2+} exchanger NCKX4 governs termination and adaptation of the mammalian olfactory response. Nat. Neurosci. **15**, 131–137 (2011)

E.E. Strehler, D.A. Zacharias, Role of alternative splicing in generating isoform diversity among plasma membrane calcium pumps. Phys. Rev. **81**, 21–50 (2001)

Y.H. Su, V.D. Vacquier, A flagellar K^+-dependent Na^+/Ca^{2+} exchanger keeps Ca^{2+}low in sea urchin spermatozoa. Proc. Natl. Acad. Sci. U. S. A. **99**, 6743–6748 (2002)

R.T. Szerencsei, J.E. Tucker, C.B. Cooper, R.J. Winkfein, P.J. Farrell, K. Iatrou, P.P.M. Schnetkamp, Minimal domain requirement for cation transport by the potassium-dependent Na/Ca-K exchanger: comparison with an NCKX paralog from *Caenorhabditis elegans*. J. Biol. Chem. **275**, 669–676 (2000)

R.T. Szerencsei, C.F.M. Prinsen, P.P.M. Schnetkamp, The stoichiometry of the retinal cone Na/Ca-K exchanger heterologously expressed in insect cells: comparison with the bovine heart Na/Ca exchanger. Biochemistry **40**, 6009–6015 (2001)

R.T. Szerencsei, R.J. Winkfein, C.B. Cooper, C. Prinsen, T.G. Kinjo, K. Kang, P.P. Schnetkamp, The Na/Ca-K exchanger gene family. Ann. N. Y. Acad. Sci. **976**, 41–52 (2002)

S. Takano, J. Kimura, T. Ono, Inhibition of aggregation of rabbit and human platelets induced by adrenaline and 5-hydroxytryptamine by KB-R7943, a Na^+/Ca^{2+} exchange inhibitor. Br. J. Pharmacol. **132**, 1383–1388 (2001)

H. Tanaka, K. Nishimaru, T. Aikawa, W. Hirayama, Y. Tanaka, K. Shigenobu, Effect of SEA0400, a novel inhibitor of sodium-calcium exchanger, on myocardial ionic currents. Br. J. Pharmacol. **135**, 1096–1100 (2002)

P. Vogel, R.W. Read, R.B. Vance, K.A. Platt, K. Troughton, D.S. Rice, Ocular albinism and hypopigmentation defects in Slc24a5-/- mice. Vet. Pathol. **45**, 264–279 (2008)

R.J. Winkfein, R.T. Szerencsei, T.G. Kinjo, K.-J. Kang, M. Perizzolo, L. Eisner, P.P.M. Schnetkamp, Scanning mutagenesis of the alpha repeats and of the transmembrane acidic residues of the human retinal cone Na/Ca-K exchanger. Biochemistry **42**, 543–552 (2003)

M.P. Wu, L.S. Kao, H.T. Liao, C.Y. Pan, Reverse mode Na^+/Ca^{2+} exchangers trigger the release of Ca^{2+} from intracellular Ca^{2+} stores in cultured rat embryonic cortical neurons. Brain Res. **1201**, 41–51 (2008)

H. Yang, T.H. Kim, H.H. Lee, K.C. Choi, E.B. Jeung, Distinct expression of the calcium exchangers, NCKX3 and NCX1, and their regulation by steroid in the human endometrium during the menstrual cycle. Reprod. Sci. **18**, 577–585 (2011)

NCKX5, a Natural Regulator of Human Skin Colour Variation, Regulates the Expression of Key Pigment Genes MC1R and Alpha-MSH and Alters Cholesterol Homeostasis in Normal Human Melanocytes

9

Stephen Wilson*, Rebecca S. Ginger*, Tony Dadd, David Gunn, Fei-Ling Lim, Magdalena Sawicka, Melanie Sandel, Paul P.M. Schnetkamp, and Martin R. Green

Abstract

Natural human skin colour is determined both by environmental exposure to ultraviolet light and through inherited genetic variation in a very limited number of genes. Variation of a non-synonymous single-nucleotide polymorphism (nsSNP; rs1426654) in the gene (SLC24A5) encoding the NCKX5 protein is associated with differences in constitutive skin colour in South Asians. The nsSNP encodes the substitution of alanine for threonine at residue 111 (A111T) near a transmembrane region required for exchanger activity, a region which is highly conserved across different species and between NCKX family members. We have shown that NCKX5 is located at the *trans*-Golgi network of melanocytes and functions as a potassium-dependent sodium-calcium exchanger. When heterologously expressed, the 111T variant of NCKX5 shows significantly lower exchanger activity than the A111 variant. We have postulated that lower

*These authors contributed equally to the work.

S. Wilson • R.S. Ginger (✉) • T. Dadd • D. Gunn
F.-L. Lim • M. Sawicka • M. Sandel • M.R. Green
Unilever R&D, Colworth Science Park,
Sharnbrook, Bedfordshire, MK44 1LQ, UK
e-mail: rebecca.ginger@unilever.com

P.P.M. Schnetkamp
Department of Physiology and Pharmacology,
Hotchkiss Brain Institute, University of Calgary,
Calgary, Alberta, Canada

L. Annunziato (ed.), *Sodium Calcium Exchange: A Growing Spectrum of Pathophysiological Implications*, Advances in Experimental Medicine and Biology 961, DOI 10.1007/978-1-4614-4756-6_9,

exchanger activity causes the reduced melanogenesis and lighter skin in Thr111-positive individuals. We used gene expression microarrays with qPCR replication and validation to assess the impact of siRNA-mediated knockdown of SLC24A5 on the transcriptome of cultured normal human melanocytes (NHM). Very few genes associated with melanogenesis were altered at the transcript level except for MC1R, suggesting that SLC24A5 interacts with at least one well-characterized melanogenic signalling pathway. More surprisingly, the expression of a number of cholesterol homeostatic genes was altered after SLC24A5 knockdown, and the total cholesterol content of NHM was increased. Cholesterol has previously been identified as a potential melanogenic regulator, and our data imply that NCKX5 exchanger function influences natural variation in skin pigmentation via a novel, unknown mechanism affecting cellular sterol levels.

Keywords
Pigmentation • Cholesterol • Melanocyte • *trans*-Golgi

9.1 Introduction

Cutaneous pigmentation has been the object of intense research by biologists and anthropologists for many years, but the complex interplay between the numerous genes involved in pigment production and distribution is still poorly understood (Yamaguchi and Hearing 2009). A more recent addition to the extensive list of genes involved in pigmentation is SLC24A5, a gene which gives rise to the protein product NCKX5 (Lamason et al. 2005; Stokowski et al. 2007). Using a high-density whole-genome array platform, we identified a non-synonymous single-nucleotide polymorphism (nsSNP) within SLC24A5 (rs1426654) as having a major impact on skin colour variation in subjects of South Asian ancestry (Stokowski et al. 2007). The nsSNP encodes an amino acid substitution from alanine (the ancestral allele) to threonine at amino acid 111 of the protein (pA111T). Independently, a mutation within the zebrafish orthologue of SLC24A5 resulting in a truncated, presumed inactive gene product was shown to be responsible for the phenotype of the *golden* zebrafish, characterized by hypopigmentation with a diminished number, size and density of melanosomes (Lamason et al. 2005). Using data from the human HapMap project (International HapMap Consortium 2005), this paper also reported that the alternate alleles of the nsSNP in human SLC24A5 were present at very different frequencies in populations of African and European ancestry and that in an admixed African-American population, the threonine-encoding allele was associated with lighter skin.

Homology predictions have placed SLC24A5 within a family of five members, SLC24A1-5 (Schnetkamp 2004), encoding ion exchangers whose general function is to maintain intracellular calcium homeostasis. The ion exchangers NCKX1–4 utilize ion gradients across the plasma membrane to exchange sodium for calcium plus potassium with a stoichiometry of 4:1:1 (Altimimi and Schnetkamp 2007a). Although the ion exchanger stoichiometry of NCKX5 has not yet been confirmed, its ion dependence has been established (Ginger et al. 2008) as has a reduced activity conferred by the threonine nsSNP allele. The same study also unexpectedly defined the protein localization of NCKX5 to be within or in close proximity to the *trans*-Golgi network (TGN). Although it remains possible that NCKX5 is present at very low abundance in melanosomes, as has been suggested (Chi et al. 2006), this alternate localization has not yet been established and is not supported by our antibody-based studies.

We also found no evidence for NCKX5 localization in the plasma membrane based on protein or activity assays (Wilson and Schnetkamp, unpublished observations). In animals, SLC24A5 gene knockout is characterized phenotypically by a reduction in the abundance and density of melanosomes, implying that NCKX5 is involved in the process of melanosome biogenesis (Lamason et al. 2005; Vogel et al. 2008).

Our studies demonstrated reduced protein abundance of several key melanosome-localized proteins in response to SLC24A5 knockdown, including tyrosinase (TYR), tyrosinase-related protein 1 (TYRP1) and PMEL17 (Fig. 7, Ginger et al. 2008), while, interestingly, the protein abundance of lysosome-associated membrane protein 1 (LAMP1) was increased. The intracellular localization of NCKX5 suggests that its role might be distinct from other family members. A central function of the TGN is the modification and sorting of proteins for compartmentalization and/or secretion (Klemm et al. 2009). As such, it is possible that NCKX5 is involved in protein sorting and/or trafficking. NCKX5 has also been proposed to have a role in melanosomal calcium and/or pH maintenance (Lamason et al. 2005), a hypothesis which is not incompatible with NCKX5 acting within the TGN. Higher resolution localization of NCKX5 might illuminate a more precise role in these processes.

MC1R is a primary regulator of many melanocyte functions including melanogenesis and the tanning response. Its regulation of these functions is via both transcriptional and post-translational mechanisms (Hida et al. 2009). Polymorphisms within the MC1R gene have been linked to variation in pigmentation phenotypes including red hair, fair skin and a poor tanning response (Beaumont et al. 2008). MC1R-mediated signal transduction is initiated by binding of α-melanocyte-stimulating hormone (αMSH) to MC1R, leading to the activation of adenylate cyclase and cyclic 3′, 5′-adenosine monophosphate (cAMP) synthesis (Scott et al. 2002). Agouti signal protein (ASIP) is a natural antagonist of MC1R opposing the effects of αMSH, inhibiting its capacity to initiate the signal transduction cascade and consequent eumelanogenesis (Abdel-Malek et al. 2001). A functional link has recently been suggested between SLC24A5 and MC1R. SLC24A5 transcript was reduced in Japanese quail *yellow* compared to wild type (Nadeau et al. 2008). This observation was ascribed to overexpression of ASIP as a result of the *yellow* mutation and implies that SLC24A5 expression is negatively regulated by ASIP. Additional support for a functional link was derived from gene expression microarray analyses of murine melan-a melanocytes, dosed with either αMSH or ASIP (Le Pape et al. 2009). SLC24A5 was increased by αMSH and decreased by ASIP, highlighting its involvement in melanogenic processes.

Because the specific function of SLC24A5/NCKX5 in relation to skin colour variation has still not been determined, we have examined factors controlling its expression and intracellular localization and conducted gene expression and pathway analysis arising from gene knockdown. We find surprisingly that SLC24A5 knockdown markedly perturbs the expression of genes involved in sterol and cholesterol metabolism.

9.2 NCKX5 Is Localized Within the *trans*-Golgi Network

Immunofluorescence confocal microscopy (IFM) studies using polyclonal anti-NCKX5 antibodies have previously suggested a TGN location for NCKX5 (Ginger et al. 2008). These studies did not exclude the possibility that NCKX5 was present within TGN-associated vesicles or that its abundance at another localization (e.g. within melanosomes or at the plasma membrane) was below the detection sensitivity of the antibodies. To further clarify the location of NCKX5, we treated the cells with several agents that disrupt the TGN prior to fixation and IFM analysis.

Nocodazole disrupts the microtubule network, causing the Golgi apparatus to fragment into vesicles that are redistributed to random sites throughout the cytoplasm (Freyberg et al. 2002). Brefeldin A (BFA) disrupts endoplasmic reticulum (ER) to Golgi trafficking. As a result, the TGN fragments into vesicles and tubules which

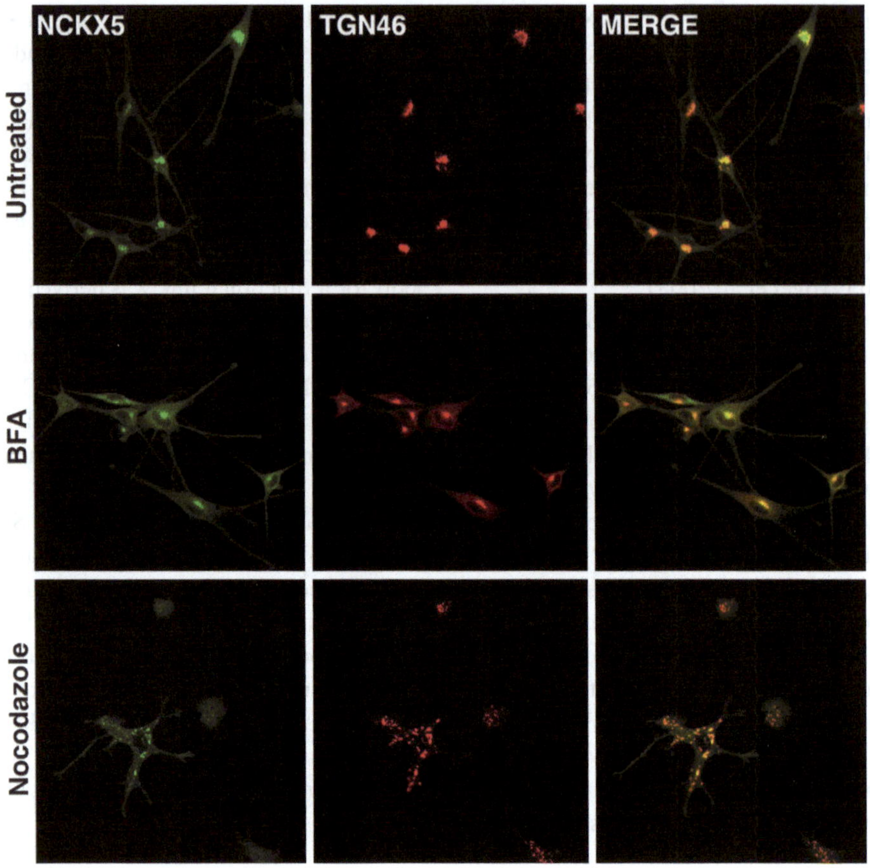

Fig. 9.1 Localization of NCKX5 in NHM. Protein localization of NCKX5 (*green*) and TGN46 (*red*) was determined by IFM, after 2-h incubation with DMSO vehicle (*top panels*), 1μg/mL BFA (*middle panels*) or 5 μg/mL nocodazole (*bottom panels*). Image width = 160μm. Merged images highlighting the closeness of association between the two antibodies are also shown (*right-hand panels*)

localize to a pericentriolar region of the cell (Radulescu et al. 2007). In the absence of agents that interfere with the organization of the TGN, NCKX5-specific antibody was detected coincident with the TGN marker TGN46 (Fig. 9.1, top panels). Some background fluorescence was present throughout the cytosol indicating that some primary antibody binding had occurred outside of the TGN. Non-specific binding has routinely been observed with this anti-NCKX5 polyclonal antibody and persists even after siRNA-mediated SLC24A5 knockdown for 10 days (data not shown). As expected after incubation of NHM with nocodazole, the TGN was fragmented and dispersed randomly throughout the cell (Fig. 9.1, middle panels). A high proportion of NCKX5 remained co-localized with the TGN marker antibody, although some background fluorescence was again observed. BFA pretreatment of NHM also resulted in redistribution of the TGN, with a high proportion localizing to a pericentriolar region of the cell (Fig. 9.1, bottom panels). There was coincident localization of the two antibodies within this region of the cell. These data confirm that the TGN is the major location of NCKX5 in the melanocyte.

9.3 SLC24A5/NCKX5 Expression in NHM Is Regulated by MITF and αMSH

We conducted a 72-h siRNA-mediated knockdown of microphthalmia-associated transcription factor (MITF) in order to determine whether it is a controlling factor in the regulation of SLC24A5 gene expression. Knockdown using multiple MITF-specific siRNA duplexes resulted in a significant depletion of SLC24A5 mRNA (Fig. 9.2). Only one of the duplexes (SI00005369) markedly reduced tyrosinase mRNA abundance, highlighting the limitations of using transient knockdown technologies to investigate gene function and perhaps demonstrating the complexity of MITF transcriptional regulation (Schepsky et al. 2006). Examination of 5-Kb upstream genomic sequence of SLC24A5 for putative MITF promoter sites identified nine E-box motifs (CAYRTG) and two M-box motifs (TCAYRTGA), as has been shown for MITF-regulated genes such as TYR and TYRP1 (Park et al. 2006).

In order to further characterize factors controlling the expression status of SLC24A5/NCKX5, NHM were cultured in the presence of the MC1R agonist αMSH and also in the presence of the adenylate cyclase activator forskolin and the phosphodiesterase inhibitor 3-isobutyl-1-methylxanthine (IBMX), both of which increase intracellular cAMP (D'Orazio et al. 2006; Koo et al. 2002). Culture of NHM for 24 h in the presence of 10–50 nM αMSH and 1–10μM forskolin resulted in a marked increase of SLC24A5 mRNA transcript (Fig. 9.3a). These changes in the transcriptional status of SLC24A5 were replicated at the protein level. 10 nM αMSH increased the apparent abundance of NCKX5 in NHM when evaluated by IFM, without altering its localization within the TGN (Fig. 9.3b). Likewise, Western blot evaluation of NCKX5 abundance showed that 10nM αMSH, 20μM forskolin and also 1μM IBMX increased NCKX5 abundance within 24h of the start of treatment (Fig. 9.3c). The impact of 10nM endothelin-1, 100nM endothelin-3 and 0.53nM keratinocyte growth factor on NCKX5 protein distribution and abundance was examined

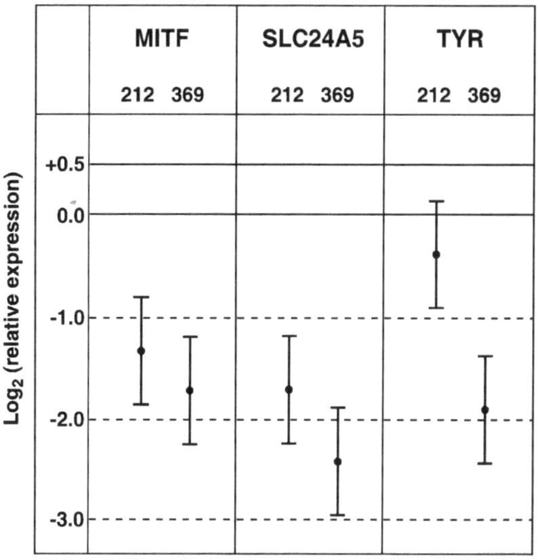

Fig. 9.2 SLC24A5 and TYR gene expression after MITF knockdown for 3d. MITF, SLC24A5 and TYR mRNA transcript abundance (relative to 18S) was determined in NHM using semi-quantitative PCR after 72-h siRNA-mediated knockdown of MITF. 212=MITF-specific duplex SI02781212 and 369=MITF-specific duplex SI00005369. Values presented as $\log_2$ relative expression to the non-targeting siRNA duplex treatment (All*). Mean values and 95% CIs from pooled estimates of model error of three independent experiments (with biological triplicates) are plotted

in similar fashion, but did not have a discernible effect in NHM (data not shown).

9.4 SLC24A5/NCKX5 Knockdown Does Not Alter TYR or TYRP1 Gene Expression

In order to investigate whether SLC24A5/NCKX5 might be involved in the transcriptional regulation of the key melanin biosynthesis enzymes TYR and TYRP1, semi-quantitative PCR was used to examine the impact of SLC24A5 knockdown on their abundance. TYR and TYRP1 mRNA abundance was not reproducibly altered after 5d SLC24A5 knockdown in NHM, although a fairly small decrease was observed in the pooled data of three independent experiments (Fig. 9.4a). We have previously shown that TYR and TYRP1 protein abundance is markedly reduced at this time point (Ginger et al. 2008). TYR and TYRP1 mRNAs were elevated by treating the cells with

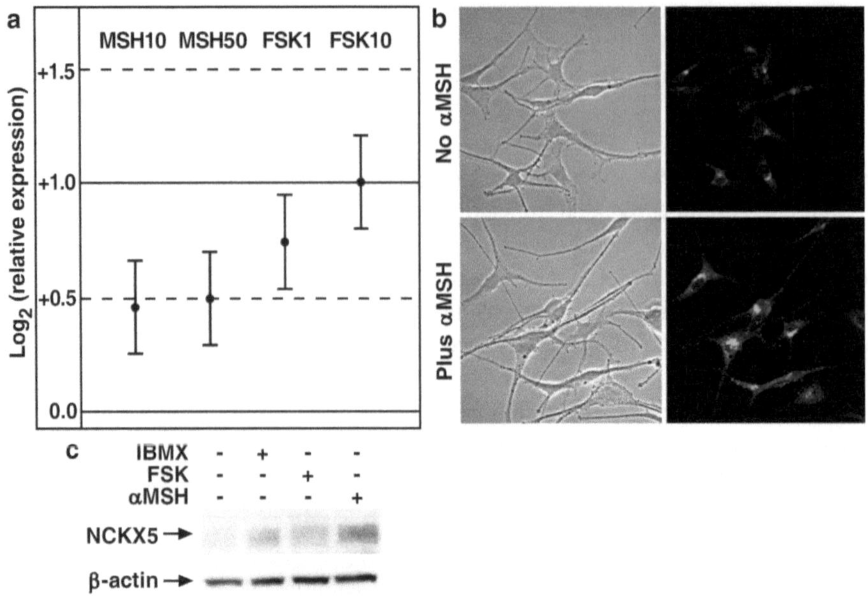

Fig. 9.3 αMSH, forskolin and IBMX increase SLC24A5/NCKX5 abundance. The abundance of SLC24A5/NCKX5 after 24-h incubation with αMSH, forskolin (fsk) or IBMX was determined in NHM. (**a**) SLC24A5 mRNA abundance (relative to 18S) after 24-h incubation with αMSH or forskolin at the concentrations indicated. Values presented as $\log_2$ relative expression to the DMSO-treated vehicle control. Mean values and 95% CIs obtained from pooled estimates of model error from three independent experiments (with biological duplicates) are plotted. (**b**) IFM of NHM labelled using anti-NCKX5 antibody, after 24-h incubation with 10nM αMSH or PBS. Images were captured under identical conditions and are representative of those obtained on three separate occasions. Image width = 160μm. (**c**) Detection of NCKX5 by Western blot (25μg protein/lane) after 24-h incubation with 50nM αMSH, 1μM forskolin, 1μM IBMX or DMSO. Equal loading was verified by reprobing for β-actin

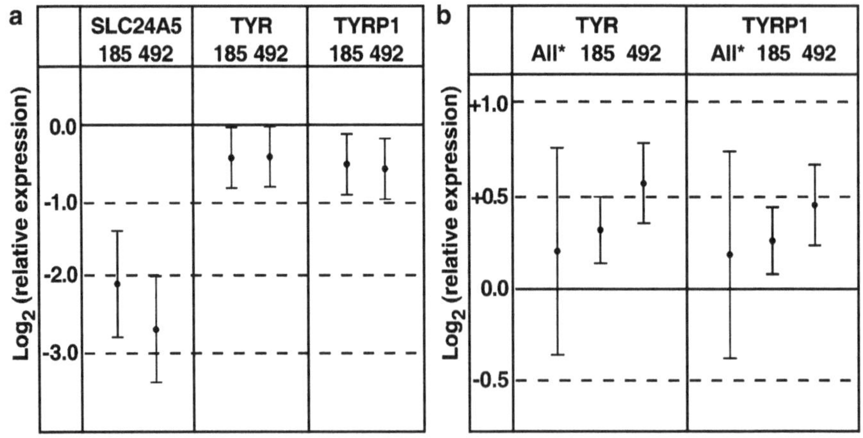

Fig. 9.4 SLC24A5 knockdown does not change the pattern of TYR or TYRP1 mRNA expression with or without αMSH. (**a**) The mRNA abundances of TYR and TYRP1 (relative to 18S) were not affected in NHM after 6d SLC24A5 knockdown. (**b**) TYR and TYRP1 abundances were also determined after addition of 50nM αMSH for the last 24h of the 6d SLC24A5 knockdown period. The elevated gene expression of TYR and TYRP1 in response to αMSH was not markedly altered by the knockdown. All values are presented as $\log_2$ relative expression to the non-stimulated All* control. Mean values and 95% CIs from pooled estimates of model error from three independent experiments (with biological triplicates) are plotted. All* = non-targeting siRNA, 185 = SLC24A5-specific siRNA 1, 492 = SLC24A5-specific siRNA 2

50 nM αMSH for 24 h (<1.5-fold). This elevated gene expression status was not altered by 5d SLC24A5 knockdown (Fig. 9.4b). Our data implies that NCKX5 activity does not influence mRNA expression of melanogenesis-related genes in a coordinated manner and that NCKX5 is not a major contributor to transcriptional regulation involving MC1R and MITF.

9.5 SLC24A5/NCKX5 Knockdown Does Not Alter ER Processing of Melanogenesis Related Enzymes

Although SLC24A5 knockdown did not alter the mRNA abundance of the key melanogenesis-related enzymes TYR and TYRP1, we have previously demonstrated that their protein levels are reduced after SLC24A5 knockdown. In order to determine whether this protein reduction was due to aberrant processing and maturation, EndoH and PNGaseF glycosidase sensitivity was determined for TYR and TYRP1 following 5d SLC24A5 knockdown. Oligosaccharide cleavage by EndoH is indicative of endoplasmic reticulum (ER) limited glycosylation. Glycosylation that has taken place within the Golgi lumen is resistant to EndoH but not to PNGaseF, which is able to cleave oligosaccharides from both ER and TGN processed proteins (Newton et al. 2007). Examination of EndoH- and PNGaseF-digested NHM protein extracts via Western blot demonstrated that although both endoglycosidase enzymes reduced the apparent molecular mass of TYR and TYRP1, no differences between controls and knockdowns were observed. This suggests that the observed post-transcriptional effects of SLC24A5 knockdown on melanogenesis-related proteins occur at a point after maturation within the Golgi apparatus.

9.6 Microarray Analysis of SLC24A5 Knockdown in NHM

To help gain new insights into the regulatory role of SLC24A5 within melanocytes, an mRNA microarray analysis was conducted after SLC24A5 knockdown for five days, in both lightly and darkly pigmented NHM. Global mRNA profiles were compared between treatments comprising two different SLC24A5-specific siRNA duplexes against three separate control treatments.

A total of 604 gene expression changes were identified between the controls and SLC24A5-specific siRNA treatments after selection using ANOVA, plus Bonferroni multiple testing correction ($p<0.05$). Very few of these changes were represented within pigmentation-related GO: processes. For instance, TYR, TYRP1, PMEL17 and MART-1 mRNA levels were unaltered. Of the pigmentation genes that did have altered mRNA expression after the knockdown, MC1R stood out as being reduced approximately two-fold. For the other genes, variations in mRNA abundance were modest, all being under 1.5-fold different from the normalized control samples.

9.7 Validation of Key Gene Expression Changes Following SLC24A5 Knockdown

Based on insights obtained from SLC24A5 siRNA knockdown microarray data, 45 genes were selected for qPCR validation by TaqMan gene expression assays. Samples used for qPCR validation were generated from a SLC24A5 siRNA knockdown time-course experiment independent of the samples used for microarray analysis. Only the 120-h time point post-transfection with siRNA constructs is reported here. Data were normalized using the geometric mean of three reference genes (B2M, PP1A and TBP) to minimize bias.

This confirmed that SLC24A5, which is not annotated on the Agilent microarray, was over fivefold downregulated by 24 h (data not shown) and ninefold downregulated by 120h. Over 90% of genes selected for qPCR validation confirmed previous microarray results from independent SLC24A5 siRNA knockdown experiment, indicating reproducibility and robustness of siRNA experiments. Interestingly, GREM2 and AREG gene expression appeared to oscillate with time, which could explain the difference in magnitude of upregulation of these genes between the two experiments. SLC24A4 expression was downregulated by SLC24A5 siRNA (construct 185) but upregulated by SLC24A5 siRNA (construct 492), indicating that differential expression of SLC24A4 could be a non-specific effect caused by the different siRNA constructs. This qPCR result also validated the microarray results showing downregulation of cholesterol metabolism genes HMGCS1, HMGCR, MVK, MVD, IDI1, SC5DL, SC4MOL and CYP51A1.

9.8 Pathway Analysis of SLC24A5 Knockdown Gene Expression Data

To identify cellular pathways impacted by SLC24A5 knockdown, a GO-based analysis was carried out on the 604 genes impacted by SLC24A5 knockdown (Table 9.1). The classifications most over-represented within the 604 genes were GO:16235 (sterol metabolism), GO:16126 (sterol biosynthesis), GO:6694 (steroid biosynthesis), GO:8202 (steroid metabolism), GO:8610 (lipid biosynthesis) and GO:8203 (cholesterol metabolism). GO:9650 (UV protection) was the most significant pigmentation-relevant process identified by the analysis, being the 21st most impacted pathway, solely represented by MCIR.

In addition, Ingenuity Pathway Analysis was carried out using the same gene list. A number of gene networks were identified as being over-represented by genes identified in the microarray analysis. The results support the GO analysis that lipid and cholesterol-associated cellular pathways and networks are the most impacted by SLC24A5 knockdown.

9.9 SLC24A5 Knockdown Is Associated with Increased Cellular Cholesterol

SLC24A5 knockdown for 5 days resulted in altered expression of a number of genes associated with sterol and cholesterol-related processes. These included the ATP-binding cassette transporter A1 (ABCA1) and the low-density lipoprotein receptor (LDLR), both of which have important roles in cellular cholesterol homeostasis and metabolism (Chen et al. 2007; Zhou et al. 2009). To evaluate whether these gene expression changes were sufficient to alter the cholesterol/cholesterol ester content of NHM, quantitative determination of total cellular cholesterol (and its esters) was conducted following a 5-day siRNA-mediated knockdown of SLC24A5. A 24% increase in cholesterol/cholesterol ester content was observed after knockdown using the siRNA duplex targeting nucleotides 185–210 of SLC24A5 ($p<0.0001$). Compared to control treatments, no reproducible increase in cholesterol abundance was observed after SLC24A5 knockdown using a second duplex (to nucleotides 492–517) at 5 days (Fig. 9.5).

In total, these observations strongly indicate that SLC24A5 knockdown strongly perturbs sterol and cholesterol metabolism in NHM.

Table 9.1 GO: processes impacted by SLC24A5 knockdown in NHM. The 10 most affected biological processes (selected by significance) are shown, using the 604 genes identified in the microarray analysis as input

Category	Genes in list in category	*p* value
GO:16125: sterol metabolism	14	$6.7E^{-09}$
GO:16126: sterol biosynthesis	9	$5.2E^{-08}$
GO:6694: steroid biosynthesis	12	$3.7E^{-07}$
GO:8202: steroid metabolism	17	$5.4E^{-07}$
GO:8610: lipid biosynthesis	21	$7.3E^{-07}$
GO:8203: cholesterol metabolism	10	$9.4E^{-06}$
GO:6066: alcohol metabolism	20	$2.9E^{-05}$
GO:41: transition metal ion transport	8	$1.1E^{-04}$
GO:6825: copper ion transport	4	$1.4E^{-04}$
GO:44255: cellular lipid metabolism	27	$2.3E^{-04}$

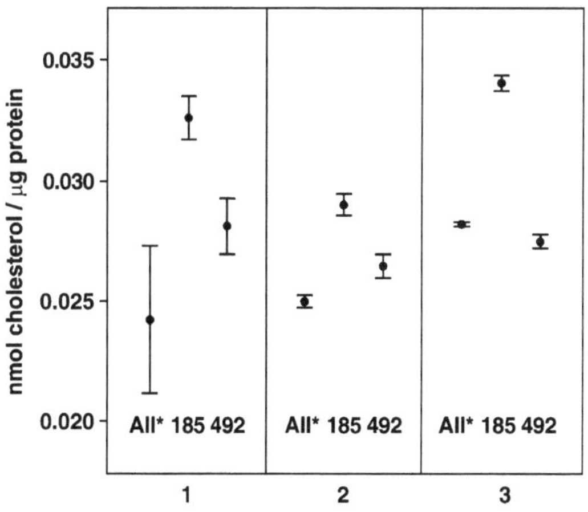

Fig. 9.5 Cellular cholesterol abundance after siRNA-mediated SLC24A5 knockdown. Total cholesterol was measured in NHM cellular extracts 5d after SLC24A5 knockdown. Data presented as nmol cholesterol/µg protein. Mean and SEM of the raw data from three independent experiments (with biological triplicates) are plotted. A two-way ANOVA and Dunnett's adjustment for multiple comparisons to a common control was applied. The p value was <0.0001 for cells challenged with duplex 185. All* = non-targeting siRNA control treatment, 185 = SLC24A5-specific siRNA 1, 492 = SLC24A5-specific siRNA 2

9.10 Conclusions and Discussion

Calcium exchangers of the NCKX family are primarily found in excitable tissues where they are thought to handle large and dynamic Ca^{2+} fluxes, although experimental data as to their physiological roles is still somewhat limited (Altimimi and Schnetkamp 2007b). Unlike the other exchangers of the SLC24 (and the related SLC8) family, SLC24A5 is expressed as an intracellular protein, and functional assays have not yet been developed to directly measure its biological activity. As such, experimental data which clarify its intracellular localization are particularly valuable, providing insights into its possible function within the pigmentary system. It has been proposed that NCKX5 functions to modulate melanosomal ionic equilibrium and pH (Lamason et al. 2005; Cheli et al. 2009), thus altering melanogenic processes within that organelle. An alternative hypothesis is that NCKX5 influences calcium gradients within the TGN and in this way modulates calcium-mediated intracellular signal transduction and/or protein sorting and trafficking to endosomes (Ginger et al. 2008).

In this study, NCKX5 was shown to co-localize with an antibody specific to TGN46, even after significant disruption of the TGN by BFA and nocodazole. If NCKX5 was not part of the resident TGN machinery, then the two marker antibodies would not have remained in such close proximity after the chemical treatments. These data support IFM-based evidence that NCKX5 is primarily localized within the TGN. As such, hypotheses suggesting that NCKX5 might function to influence either protein sorting events in the TGN or modulate Ca^{2+}-mediated signal transduction from its location within the TGN are the most plausible. Several intracellular proteins have been shown to have functions within multiple cellular organelles, including OA1 (Palmisano et al. 2008) and ATP7A (Setty et al. 2008). That NCKX5 is active at additional sites to the TGN, e.g. the plasma membrane or melanosomes, but at abundances below the detection threshold of the antibodies used should not be discounted. Nevertheless, the maturation, sorting and transport of the melanosomal enzymes TYR and TYRP1 in and from the TGN have been investigated over a number of years (e.g. Jimbow et al. 2000), and NCKX5 does not appear to follow a similar secretory pathway.

The Golgi complex has a critical function in the modification, sorting and transport of proteins. After processes such as glycosylation and proteolytic cleavage, "cargo" proteins are delivered to their various destinations by dynamic membrane-bound organelles, which lack resident Golgi enzymes (Emr et al. 2009; Luini et al. 2008). Alteration of the Ca^{2+} microenvironment either within the Golgi lumen or in the adjacent cytosol can regulate its function, and the requirement for Ca^{2+} gradients in the TGN for such processes is well established (Dolman and Tepikin 2006). Because SLC24A5 knockdown has been shown to reduce the protein abundance of TYR and TYRP1, we wanted to determine whether this reduction was due to aberrant post-translational processing within the ER or Golgi apparatus.

The apparent molecular weight of TYR and TYRP1 after endoglycosidase digestion suggested that the overall glycosylation status of each protein was largely unchanged by SLC24A5 knockdown, although their overall abundances were reduced. Increased EndoH sensitivity of TYR and/or TYRP1, especially, would have suggested disruption of ER to Golgi trafficking and retention of immature protein within the ER (Olivares et al. 2003). These data indicate that the observed post-transcriptional effects of SLC24A5 knockdown on melanogenesis-related proteins occur at a point after maturation within the Golgi apparatus.

Here, we have shown that knockdown of MITF results in a significant depletion of SLC24A5 transcript. This is the first experimental evidence to suggest that SLC24A5 is a target of MITF regulation. Knockdown of SLC24A5, however, had little or no effect on transcript abundance of most melanogenesis-related genes, many of which are directly regulated by MITF (Chiaverini et al. 2008), the exception being MC1R. It is therefore unlikely that SLC24A5 is a direct effector of MC1R and/or MITF function. Interestingly, SLC24A5 was not identified in a comprehensive microarray analysis of MITF knockdown, conducted by Hoek et al. (2008). However, another aspect of their study was identification of gene expression increases as a response to heterologous over-expression of MITF in SK-MEL-28 cells. Here, SLC24A5 transcript was elevated approximately 200-fold. Furthermore, the presence of several E-box and M-box motifs (9 and 2, respectively) in the genomic region upstream of SLC24A5 further supports SLC24A5 as being a novel transcriptional target of MITF.

Recent studies have also implicated the MC1R pathway in the control of SLC24A5 gene expression. Cheli et al. (2009) have demonstrated that the mRNA transcripts of both SLC24A5 and its close relative SLC24A4 are increased by the cAMP-elevating agent forskolin. In mouse melan-a cells, SLC24A5 transcript is increased by exogenous αMSH and decreased by ASIP (Le Pape et al. 2009). Furthermore, in Japanese yellow quail, SLC24A5 transcript is significantly decreased compared to that of wild-type birds (Nadeau et al. 2008). Yellow quail carry a genetic mutation in a regulatory region of the ASIP gene, presumed to cause increased ASIP expression. In support of these findings, we have demonstrated that αMSH and chemical agents that elevate cAMP increase SLC24A5 mRNA abundance in NHM. We have additionally shown that these agents increase the expression of the protein product NCKX5, with no apparent impact on its TGN localization. SLC24A5 knockdown did not appear to directly impact basal mRNA levels of TYR and TYRP1 (although gene expression of MC1R was reduced twofold), nor did it impair the increased expression of TYR and TYRP1 mRNA in response to exogenous addition of αMSH. Taken together, these data suggest that SLC24A5/NCKX5 does not function to modulate either constitutive or facultative gene expression of melanogenesis-related enzymes. In summary, transcription of SLC24A5 is controlled by MITF, and, additionally, its expression (and presumably its activity) is influenced by αMSH.

Using microarray analysis, we have demonstrated that the role of SLC24A5 in NHM appears unrelated to the transcriptional regulation of genes involved in melanogenesis, with the exception of MC1R. Instead, the main processes impacted by the knockdown at the transcriptional level were sterol and cholesterol metabolism and biosynthesis. Within these processes, the ABCA1 and LDLR genes stood out being key mediators of intracellular cholesterol homeostasis. ABCA1 is the primary mechanism of cholesterol efflux from cells (Hozoji et al. 2008), and LDLR is the primary means of cellular exogenous cholesterol update. This elevation suggests that NCKX5 might function within the TGN to control intracellular cholesterol levels, perhaps via calcium-mediated regulation such as the liver X receptor, activator protein 2α or protein kinase D (Iwamoto et al. 2008). Import and biosynthesis of cholesterol is tightly controlled by feedback regulation of LDLR in response to elevated sterol levels and cholesterol depletion (Sekar and Veldhuis 2001). The expression of LDLR was twofold lower after SLC24A5 knockdown, which implies that the requirement for exogenous cholesterol in NHM was reduced as a response to the knockdown.

Cholesterol is an essential structural component of mammalian cell membranes, having a major impact on their permeability and fluidity (Chen and Resh 2002). It is also a precursor for steroid hormone synthesis and influences many intracellular processes such as formation of lipid rafts, protein transport, endocytosis and signal transduction (Rodriguez-Agudo et al. 2008; Burgos et al. 2004; Chen et al. 2008). The role of cholesterol in melanocyte function has not been extensively studied. Depletion of cellular cholesterol using methyl-β-cyclodextrin has been shown to inhibit melanogenesis in NHM, via an ERK-dependent mechanism (Jin et al. 2008). Schallreuter et al. (2009) demonstrated that addition of exogenous cholesterol to NHM and melanoma cells stimulated melanogenesis in vitro. In this case, cAMP elevation (with subsequent increases in protein expression of CREB, MITF, tyrosine hydroxylase and TYR) was demonstrated as a response to increased cholesterol availability. Surprisingly, the total cholesterol content of an amelanotic melanoma cell line used in the study was markedly elevated compared to other cell lines examined, as was the melanosomal cholesterol content, suggestive that it is a requirement for early melanosome biogenesis even if pigment is not made. The oxysterol 25-hydroxycholesterol, which enhances the degradation of 3-hydroxy-3-methylglutaryl CoA reductase (HMGCoAR), the rate-limiting enzyme in cholesterol biosynthesis (Liscum and Faust 1989), has been shown to induce the degradation of TYR after its maturation in the TGN (Hall et al. 2004) presumably in a manner similar to that observed after SLC24A5 knockdown. These researchers also identified esterified cholesterol as being required for melanosomal maturation. In our study, knockdown using the siRNA duplex targeting nucleotides 185–210 of SLC24A5 increased the total cholesterol content of NHM by 24% ($p<0.0001$). The other duplex used in this study did not increase the cholesterol/cholesterol ester content in a significant or reproducible manner. This result suggests that the observed effects on ABCA1, LDLR and other genes involved in cholesterol homeostasis might have been an adaptive response to the cholesterol elevation. However, the ratio of cholesterol to its esters, as well as cholesterol abundance within organelles such as melanosomes, is important for normal organelle function (Chen et al. 2008; Hall et al. 2004). A more detailed examination of the effect of SLC24A5 knockdown on these factors is required.

Limitations to these studies include potential off-target effects of siRNA-mediated knockdown and the lack of time-course experiments to pinpoint the epistatic nature of the identified gene regulatory effects. Despite this, we contend and have provided evidence that NCKX5 functions within the TGN to modulate intracellular cholesterol levels (and/or its esters), and we propose that this is a primary role of this Ca^{2+}-exchange protein within melanocytes. It has been predicted that the mechanisms controlling the maintenance of cholesterol homeostasis within melanocytes must diverge from other cell types in order to protect melanogenic processes from fluctuations in dietary cholesterol (Hall et al. 2004). SLC24A5 gene expression appears restricted to pigmented cells such as melanocytes (Ginger et al. 2008), and we suggest that this provides the cell-specific influence over cholesterol homeostasis required for normal melanogenesis.

Acknowledgements We thank Carl Jarman, Wendy Filsell, Amelia Jarman, Gordon James and Dawn Mazzatti for their support and helpful discussions.

References

Z.A. Abdel-Malek, M.C. Scott, M. Furumura, M.L. Lamoreux, M. Ollmann, G.S. Barsh, V.J. Hearing, The melanocortin 1 receptor is the principal mediator of the effects of agouti signaling protein on mammalian melanocytes. J. Cell Sci. **114**(Pt 5), 1019–1024 (2001)

H.F. Altimimi, P.P. Schnetkamp, Na^+-dependent inactivation of the retinal cone/brain Na^+/Ca^{2+}-K^+ exchanger NCKX2. J. Biol. Chem. **282**, 3720–3729 (2007a)

H.F. Altimimi, P.P.M. Schnetkamp, Na^+/Ca^{2+}-K^+ exchangers (NCKX): functional properties and physiological roles. Channels **1**(2), 62–69 (2007b)

K.A. Beaumont, S.N. Shekar, A.L. Cook, D.L. Duffy, R.A. Sturm, Red hair is the null phenotype of MC1R. Hum. Mutat. **29**(8), E88–E94 (2008)

P.V. Burgos, C. Klattenhoff, E. de la Fuente, A. Rigotti, A. González, Cholesterol depletion induces PKA-mediated basolateral-to-apical transcytosis of the scavenger receptor class B type I in MDCK cells. Proc. Natl. Acad. Sci. U. S. A. **101**(11), 3845–3850 (2004)

Y. Cheli, F. Luciani, M. Khaled, L. Beuret, K. Bille, P. Gounon, J.P. Ortonne, C. Bertolotto, R. Ballotti, αMSH and Cyclic AMP elevating agents control melanosome pH through a protein kinase A-independent mechanism. J. Biol. Chem. **284**(28), 18699–18706 (2009)

H. Chen, J. Yang, P.S. Low, J.X. Cheng, Cholesterol level regulates endosome motility via Rab proteins. Biophys. J. **94**(4), 1508–1520 (2008)

X. Chen, M.D. Resh, Cholesterol depletion from the plasma membrane triggers ligand-independent activation of the epidermal growth factor receptor. J. Biol. Chem. **277**(51), 49631–49637 (2002)

Y. Chen, X.Z. Ruan, Q. Li, A. Huang, J.F. Moorhead, S.H. Powis, Z. Varghese, Inflammatory cytokines disrupt LDL-receptor feedback regulation and cause statin resistance: a comparative study in human hepatic cells and mesangial cells. Am. J. Physiol. Renal Physiol. **293**(3), F680–F687 (2007)

A. Chi, J.C. Valencia, Z.Z. Hu, H. Watabe, H. Yamaguchi, N.J. Mangini, H. Huang, V.A. Canfield, K.C. Cheng, F. Yang, R. Abe, S. Yamagishi, J. Shabanowitz, V.J. Hearing, C. Wu, E. Appella, D.F. Hunt, Proteomic and bioinformatic characterization of the biogenesis and function of melanosomes. J. Proteome Res. **5**(11), 3135–3144 (2006)

C. Chiaverini, L. Beuret, E. Flori, R. Busca, P. Abbe, K. Bille, P. Bahadoran, J.P. Ortonne, C. Bertolotto, R. Ballotti, Microphthalmia-associated transcription factor regulates RAB27A gene expression and controls melanosome transport. J. Biol. Chem. **283**(18), 12635–12642 (2008)

J.A. D'Orazio, T. Nobuhisa, R. Cui, M. Arya, M. Spry, K. Wakamatsu, V. Igras, T. Kunisada, S.R. Granter, E.K. Nishimura, S. Ito, D.E. Fisher, Topical drug rescue strategy and skin protection based on the role of Mc1r in UV-induced tanning. Nature **443**(7109), 340–344 (2006)

N.J. Dolman, A.V. Tepikin, Calcium gradients and the Golgi. Cell Calcium **40**(5–6), 505–512 (2006)

S. Emr, B.S. Glick, A.D. Linstedt, J. Lippincott-Schwartz, A. Luini, V. Malhotra, B.J. Marsh, A. Nakano, S.R. Pfeffer, C. Rabouille, J.E. Rothman, G. Warren, F.T. Wieland, Journeys through the Golgi – taking stock in a new era. J. Cell Biol. **187**(4), 449–453 (2009)

Z. Freyberg, S. Bourgoin, D. Shields, Phospholipase D2 is localized to the rims of the Golgi apparatus in mammalian cells. Mol. Biol. Cell **13**(11), 3930–3942 (2002)

R.S. Ginger, S.E. Askew, R.M. Ogborne, S. Wilson, D. Ferdinando, T. Dadd, A.M. Smith, S. Kazi, R.T. Szerencsei, R.J. Winkfein, P.P. Schnetkamp, M.R. Green, SLC24A5 encodes a *trans*-Golgi network protein with potassium-dependent sodium-calcium exchange activity that regulates human epidermal melanogenesis. J. Biol. Chem. **283**(9), 5486–5495 (2008)

A.M. Hall, L. Krishnamoorthy, S.J. Orlow, 25-hydroxycholesterol acts in the Golgi compartment to induce degradation of tyrosinase. Pigment Cell Res. **17**(4), 396–406 (2004)

T. Hida, K. Wakamatsu, E.V. Sviderskaya, A.J. Donkin, L. Montoliu, M. Lynn Lamoreux, B. Yu, G.L. Millhauser, S. Ito, G.S. Barsh, K. Jimbow, D.C. Bennett, Agouti protein, mahogunin, and attractin in pheomelanogenesis and melanoblast-like alteration of melanocytes: a cAMP-independent pathway. Pigment Cell Melanoma Res. **22**(5), 623–634 (2009)

K.S. Hoek, N.C. Schlegel, O.M. Eichhoff, D.S. Widmer, C. Praetorius, S.O. Einarsson, S. Valgeirsdottir, K. Bergsteinsdottir, A. Schepsky, R. Dummer, E. Steingrimsson, Novel MITF targets identified using a two-step DNA microarray strategy. Pigment Cell Melanoma Res. **21**, 665–676 (2008)

M. Hozoji, Y. Munehira, Y. Ikeda, M. Makishima, M. Matsuo, N. Kioka, K. Ueda, Direct interaction of nuclear liver X receptor-beta with ABCA1 modulates cholesterol efflux. J. Biol. Chem. **283**(44), 30057–30063 (2008)

International HapMap Consortium, A haplotype map of the human genome. Nature **437**(7063), 1299–1320 (2005)

N. Iwamoto, S. Abe-Dohmae, R. Lu, S. Yokoyama, Involvement of protein kinase D in phosphorylation and increase of DNA binding of activator protein 2 alpha to downregulate ATP-binding cassette transporter A1. Arterioscler. Thromb. Vasc. Biol. **28**(12), 2282–2287 (2008)

K. Jimbow, J.S. Park, F. Kato, K. Hirosaki, K. Toyofuku, C. Hua, T. Yamashita, Assembly, target-signaling and intracellular transport of tyrosinase gene family proteins in the initial stage of melanosome biogenesis. Pigment Cell Res. **13**(4), 222–229 (2000)

S.H. Jin, Y.Y. Lee, H.Y. Kang, Methyl-beta-cyclodextrin, a specific cholesterol-binding agent, inhibits melanogenesis in human melanocytes through activation of ERK. Arch. Dermatol. Res. **300**(8), 451–454 (2008)

R.W. Klemm, C.S. Ejsing, M.A. Surma, H.J. Kaiser, M.J. Gerl, J.L. Sampaio, Q. de Robillard, C. Ferguson, T.J. Proszynski, A. Shevchenko, K. Simons, Segregation of sphingolipids and sterols during formation of secretory vesicles at the *trans*-Golgi network. J. Cell Biol. **185**(4), 601–612 (2009)

H.M. Koo, M. VanBrocklin, M.J. McWilliams, S.H. Leppla, N.S. Duesbery, G.F. Woude, Apoptosis and melanogenesis in human melanoma cells induced by anthrax lethal factor inactivation of mitogen-activated protein kinase kinase. Proc. Natl. Acad. Sci. U. S. A. **99**(5), 3052–3057 (2002)

R.L. Lamason, M.A. Mohideen, J.R. Mest, A.C. Wong, H.L. Norton, M.C. Aros, M.J. Jurynec, X. Mao, V.R. Humphreville, J.E. Humbert, S. Sinha, J.L. Moore, P. Jagadeeswaran, W. Zhao, G. Ning, I. Makalowska, P.M. McKeigue, D. O'donnell, R. Kittles, E.J. Parra, N.J. Mangini, D.J. Grunwald, M.D. Shriver, V.A. Canfield, K.C. Cheng, SLC24A5, a putative cation

exchanger, affects pigmentation in zebrafish and humans. Science **310**(5755), 1782–17866 (2005)
E. Le Pape, T. Passeron, A. Giubellino, J.C. Valencia, R. Wolber, V.J. Hearing, Microarray analysis sheds light on the dedifferentiating role of agouti signal protein in murine melanocytes via the Mc1r. Proc. Natl. Acad. Sci. U. S. A. **106**(6), 1802–1807 (2009)
L. Liscum, J.R. Faust, The intracellular transport of low density lipoprotein-derived cholesterol is inhibited in Chinese hamster ovary cells cultured with 3-beta-[2-(diethylamino)ethoxy]androst-5-en-17-one. J. Biol. Chem. **264**(20), 11796–11806 (1989)
A. Luini, A.A. Mironov, E.V. Polishchuk, R.S. Polishchuk, Morphogenesis of post-Golgi transport carriers. Histochem. Cell Biol. **129**(2), 153–161 (2008)
N.J. Nadeau, F. Minvielle, S. Ito, M. Inoue-Murayama, D. Gourichon, S.A. Follett, T. Burke, N.I. Mundy, Characterization of Japanese quail yellow as a genomic deletion upstream of the avian homolog of the mammalian ASIP (agouti) gene. Genetics **178**(2), 777–786 (2008)
R.A. Newton, A.L. Cook, D.W. Roberts, J.H. Leonard, R.A. Sturm, Post-transcriptional regulation of melanin biosynthetic enzymes by cAMP and resveratrol in human melanocytes. J. Invest. Dermatol. **127**(9), 2216–2227 (2007)
C. Olivares, F. Solano, J.C. García-Borrón, Conformation-dependent post-translational glycosylation of tyrosinase. Requirement of a specific interaction involving the CuB metal binding site. J. Biol. Chem. **278**(18), 15735–15743 (2003)
I. Palmisano, P. Bagnato, A. Palmigiano, G. Innamorati, G. Rotondo, D. Altimare, C. Venturi, E.V. Sviderskaya, R. Piccirillo, M. Coppola, V. Marigo, B. Incerti, A. Ballabio, E.M. Surace, C. Tacchetti, D.C. Bennett, M.V. Schiaffino, The ocular albinism type 1 protein, an intracellular G protein-coupled receptor, regulates melanosome transport in pigment cells. Hum. Mol. Genet. **17**(22), 3487–3501 (2008)
H.Y. Park, C. Wu, L. Yonemoto, M. Murphy-Smith, H. Wu, C.M. Stachur, B.A. Gilchrest, MITF mediates cAMP-induced protein kinase C-beta expression in human melanocytes. Biochem. J. **395**(3), 571–578 (2006)
A.E. Radulescu, A. Siddhanta, D. Shields, A role for clathrin in reassembly of the Golgi apparatus. Mol. Biol. Cell **18**(1), 94–105 (2007)
D. Rodriguez-Agudo, S. Ren, E. Wong, D. Marques, K. Redford, G. Gil, P. Hylemon, W.M. Pandak, Intracellular cholesterol transporter StarD4 binds free cholesterol and increases cholesteryl ester formation. J. Lipid Res. **49**(7), 1409–1419 (2008)
K.U. Schallreuter, S. Hasse, H. Rokos, B. Chavan, M. Shalbaf, J.D. Spencer, J.M. Wood, Cholesterol regulates melanogenesis in human epidermal melanocytes and melanoma cells. Exp. Dermatol. **18**(8), 680–688 (2009)
A. Schepsky, K. Bruser, G.J. Gunnarsson, J. Goodall, J.H. Hallsson, C.R. Goding, E. Steingrimsson, A. Hecht, The microphthalmia-associated transcription factor Mitf interacts with beta-catenin to determine target gene expression. Mol. Cell. Biol. **26**(23), 8914–8927 (2006)
P.P. Schnetkamp, The SLC24 Na^{+}/Ca^{2+}-K^{+} exchanger family: vision and beyond. Pflugers Arch. **447**(5), 683–688 (2004)
M.C. Scott, I. Suzuki, Z.A. Abdel-Malek, Regulation of the human melanocortin 1 receptor expression in epidermal melanocytes by paracrine and endocrine factors and by ultraviolet radiation. Pigment Cell Res. **15**(6), 433–439 (2002)
N. Sekar, J.D. Veldhuis, Concerted transcriptional activation of the low density lipoprotein receptor gene by insulin and luteinizing hormone in cultured porcine granulosa-luteal cells: possible convergence of protein kinase a, phosphatidylinositol 3-kinase, and mitogen-activated protein kinase signaling pathways. Endocrinology **142**(7), 2921–2928 (2001)
S.R. Setty, D. Tenza, E.V. Sviderskaya, D.C. Bennett, G. Raposo, M.S. Marks, Cell-specific ATP7A transport sustains copper-dependent tyrosinase activity in melanosomes. Nature **454**(7208), 1142–1146 (2008)
R.P. Stokowski, P.V. Pant, T. Dadd, A. Fereday, D.A. Hinds, C. Jarman, W. Filsell, R.S. Ginger, M.R. Green, F.J. van der Ouderaa, D.R. Cox, A genomewide association study of skin pigmentation in a South Asian population. Am. J. Hum. Genet. **81**(6), 1119–1132 (2007)
P. Vogel, R.W. Read, R.B. Vance, K.A. Platt, K. Troughton, D.S. Rice, Ocular albinism and hypopigmentation defects in Slc24a5-/- mice. Vet. Pathol. **45**(2), 264–279 (2008)
Y. Yamaguchi, V.J. Hearing, Physiological factors that regulate skin pigmentation. Biofactors **35**(2), 193–199 (2009)
L. Zhou, H.Y. Choi, W.P. Li, F. Xu, J. Herz, LRP1 controls cPLA2 phosphorylation, ABCA1 expression and cellular cholesterol export. PLoS One **4**(8), e6853 (2009)

Expression and Regulation of Sodium/Calcium Exchangers, NCX and NCKX, in Reproductive Tissues: Do They Play a Critical Role in Calcium Transport for Reproduction and Development?

10

Hyun Yang, Kyung-Chul Choi, Eui-Man Jung, Beum-Soo An, Sang-Hwan Hyun, and Eui-Bae Jeung

Abstract

Plasma membrane sodium/calcium (Na^+/Ca^{2+}) exchangers are an important component of intracellular calcium $[Ca^{2+}]_i$ homeostasis and electrical conduction. Na^+/Ca^{2+} exchangers, NCX and NCKX, play a critical role in the transport of one $[Ca^{2+}]_i$ and potassium ion across the cell membrane in exchange for four extracellular sodium ions $[Na^+]_e$. Mammalian plasma membrane Na^+/Ca^{2+} exchange proteins are divided into two families: one in which Ca^{2+} flux is dependent only on sodium (NCX1–3) and another in which Ca^{2+} flux is also dependent on potassium (NCKX1–4). Both molecules are capable of forward- and reverse-mode exchange. In cells and tissues, Na^+/Ca^{2+} (and K^+) gradients localize to the cell membrane; thus, the exchangers transport ions across a membrane potential. Uterine NCKX3 has been shown to be involved in the regulation of endometrial receptivity by $[Ca^{2+}]_i$. In the uterus and placenta, NCKX3 expression is regulated by the sex steroid hormone estrogen (E2) and hypoxia stress, respectively. In this chapter, we described the expression and regulation of these proteins for reproductive functions in various tissues including uterus, placenta, and kidney of humans and rodents. Evidence to date suggests that NCKX3 and NCX1 may be regulated in a tissue-specific manner. In addition, we focused on the molecular mechanism involved in the regulation of NCKX3 and NCX1 in mammals, based upon our recent results and those of others.

H. Yang • K.-C. Choi • E.-M. Jung
B.-S. An • S.-H. Hyun • E.-B. Jeung (✉)
Laboratory of Veterinary Biochemistry and Molecular Biology, College of Veterinary Medicine, Chungbuk National University, Cheongju, Chungbuk 361-763, Republic of Korea
e-mail: ebjeung@chungbuk.ac.kr

L. Annunziato (ed.), *Sodium Calcium Exchange: A Growing Spectrum of Pathophysiological Implications*, Advances in Experimental Medicine and Biology 961, DOI 10.1007/978-1-4614-4756-6_10,

Keywords
Calcium transport • Uterus • Placenta • NCKX • NCX

10.1 Introduction

The actions of calcium ions (Ca^{2+}) in female reproductive organs have been widely studied for several decades. It has been suggested that Ca^{2+} are involved in uterine smooth muscle contraction, fetal implantation, and fetal bone mineralization. The balance between uterine muscle contraction and relaxation is extremely important throughout pregnancy and during labor. A model of Ca^{2+} transport suggests that calcium flows into the cytoplasm via channel proteins. Two pathways are responsible for Ca^{2+} entry into the body (Peng et al. 2003). In the paracellular pathway, occludin and junction adherence molecular A (JAM-A), which construct tight junctions and regulate paracellular transport, are needed for Ca^{2+} absorption. Paracellular Ca^{2+} absorption is highest in the ileum, while dietary Ca^{2+} absorption via transcellular mechanisms occurs predominantly in the duodenum (Khanal and Nemere 2008). Transepithelial Ca^{2+} active transport occurs according to a well-controlled sequence of events consisting of apical Ca^{2+} entry involving channels such as voltage-dependent Ca^{2+} (VDCCs) and the transient receptor potential (TRP) family, TRPV5 and TRPV6 in particular, followed by the use of calcium-binding proteins (Calbindin-D9k or -28 k) and transient receptor potential vanilloid type 5 and 6 (TRPV5 and TRPV6), which are extruded from the cell membrane by plasma membrane Ca^{2+} ATPase and to a lesser extent by Na^+/Ca^{2+} exchanger (NCX) (Opperman et al. 1992; Bindels 1993; Friedman and Gesek 1995; Hong et al. 2004; Lee and Jeung 2007). Ca^{2+} serves as a universal intracellular messenger to modulate processes such as neurotransmission and hormone secretion as well as many biological processes, e.g., cell cycle regulation and programmed cell death (Berridge 1995; Clapham 1995).

In tissues, the plasma membrane Na^+/Ca^{2+} exchange proteins of mammals have been divided into two families, one in which Ca^{2+} action is dependent only on sodium (NCX family 1–3) and one in which Ca^{2+} action is also dependent on potassium (NCKX family 1–6) (Tsoi et al. 1998; Kraev et al. 2001; Li et al. 2002; Lytton et al. 2002; Dong et al. 2006).

Among the different pathways that mediate Ca^{2+} movement, Na^+/Ca^{2+} exchange has emerged as the predominant mechanism for Ca^{2+} efflux across the plasma membrane, particularly when overall Ca^{2+} levels are high (Lee et al. 2002; Wanaverbecq et al. 2003; Kim et al. 2005). Moreover, several recent studies have highlighted the connection between important physiological events and specific Na^+/Ca^{2+} exchange molecules (Jeon et al. 2003; Li et al. 2006; Pott et al. 2007). Extensive studies have demonstrated that Na^+/Ca^{2+} exchange plays a crucial role in Ca^{2+} extrusion and operates with a stoichiometry of three Na^+ ions to one Ca^{2+} ion, indicating that the transport process is electrogenic (Blaustein and Lederer 1999). The development of partially purified exchange preparations, antibody reagents, and expression cloning techniques led to the molecular cloning of the canine cardiac Na^+/Ca^{2+} exchanger, subsequently denoted NCX1 (Blaustein and Lederer 1999). NCX1 cDNA encodes a protein of 970 amino acids predicted to be approximately 110 kDa in size (Lytton 2007). The mature protein is subject to both signal peptide cleavage and glycosylation and, when analyzed on gels, runs with an apparent size of 120 kDa (Lytton 2007). Additional bands observed at 160 and 70 kDa arise by heat-induced aggregation and proteolytic cleavage, respectively. Further, minor variations in apparent size (5–10 kDa) are generated by alternative splicing (Lytton 2007). Parallel studies in retinal rod photoreceptors have established the presence of a mechanistically similar Na^+/Ca^{2+} exchange process that is critical for visual adaptation and represent the predominant means of Ca^{2+} extrusion from rod outer segments (Schnetkamp 1986). The rod exchanger differs from that described in the heart and axon principally due to its absolute transport requirement for K^+ (Lytton 2007). Studies showed that the rod exchanger catalyzed the transport of four Na^+ in exchange for one Ca^{2+} and one K^+ (Cervetto et al. 1989; Schnetkamp et al. 1989). Purification of the bovine rod $Na^+/Ca^{2+}/K^+$ exchanger subsequently led to its molecular cloning (Cook and Kaupp 1988; Reilander et al. 1992). This cDNA, denoted NCKX1, encodes a protein of 1,216 amino acids predicted to be approximately 132 kDa in size (Lytton 2007). Extensive glycosylation of the protein appears to account for the much larger apparent size of the native protein observed on gels (≤220 kDa) (Lytton 2007). Both forward- and reverse-mode exchange occur via NCX and NCKX exchangers depending on the gradients of the Na^+/Ca^{2+} (and K^+) (Blaustein and Lederer 1999; Lytton et al. 2002). NCX1 is abundantly expressed in the heart, brain, kidney, and smooth muscle (Nicoll et al. 1990, 1996). However, NCX2 and NCX3 expression in the brain and skeletal muscle is limited (Longoni and Carafoli 1987; Nicoll et al. 1996; Kraev et al. 2001; Li et al. 2002; Lytton et al. 2002; Dong et al. 2006). NCKX1 is expressed only in retinal rod photoreceptors. NCKX2 in brain neurons and cone photoreceptors shows restricted expression (Kraev et al. 2001; Li et al. 2002; Lytton et al. 2002; Dong et al. 2006). NCKX3 and NCKX4 are expressed not only in the brain but also in many other tissues, including aorta, uterus, and intestine, which are rich in smooth muscle cells (Kraev et al. 2001; Li et al. 2002; Cai and Lytton 2004; Yang et al. 2009). NCKX5 has recently been demonstrated to be expressed in skin and retinal pigmented epithelium, where it is thought to be present on the melanosome membrane and not the plasma membrane (Lamason et al. 2005). NCKX6 has also been characterized; however, the physiologic function of this protein remains controversial (Cai and Lytton 2004; Palty et al. 2004). Studies examining the physiologic role(s) in vascular contraction via controlling Ca^{2+} homeostasis by NCKX and NCX proteins have been performed in blood vessels, and some recent reports describe NCKX and NCX function in brain, spermatozoa, mast cells, and platelets (Kiedrowski et al.. 2002; Aneiros et al. 2005; Kim et al. 2005; Kip et al. 2006). NCX1 and NCKX3 are expressed in the reproductive organs, including the uterus (Yang et al. 2009, 2011; Quednau et al. 1997; Kraev et al. 2001; Li et al. 2002). As mentioned previously, maintenance of the calcium balance of reproductive organs is of crucial importance for many physiologic functions, including smooth muscle contraction, embryo implantation, and placental transport. The regulation of contraction, muscle excitability, and the maternofetal calcium transport system of the placenta in the reproductive organs is important for lessening maternal and fetal mortality caused by pre- and postmature birth or metroparalysis (Wray et al. 2003). In addition, calcium plays an important role in the endometrium and placenta. A number of calcium-related proteins expressed in the uterus mediate muscle functions (Sanborn 2000) and in the placenta (Lafond et al. 1991; Moreau et al. 2003a; Yang et al. 2011) ensure successful implantation.

In this chapter, we summarize current findings related to the molecular mechanisms involved in NCX1 and NCKX3 regulation in mammals and introduce research data from our recent studies and others.

10.2 Na^+/Ca^{2+} Exchangers in Reproductive Tissues

The kidney plays an important role in the maintenance of calcium balance in both the male and female body by regulating calcium reabsorption and excretion. Ca^{2+} are filtered daily by the glomeruli, and <2% are excreted in the urine (Hoenderop et al. 2002b). In contrast, the transcellular pathway allows the body to regulate Ca^{2+} reabsorption independent of Na^+ balance. This pathway is specifically controlled by the calcitropic hormones, including parathyroid hormone, calcitonin, and 1,25-dihydroxy vitamin D_3 ($1,25[OH]_2D_3$) (Bindels 1993; Friedman and

Gesek 1995; Hoenderop et al. 2000, 2002a). Due to this active process, an organism can respond to fluctuations in dietary Ca^{2+} and adapt to changes in demand during certain processes, i.e., growth, pregnancy, lactation, and aging (Hoenderop et al. 2002a). Disturbances in active Ca^{2+} reabsorption are most likely to be accompanied by significant alterations in overall Ca^{2+} homeostasis (14). At the cellular level, active calcium transport can be divided into three functional steps, Ca^{2+} influx into cells, transferring through the cytosol, and extrusion into the bloodstream. These steps are mediated by three types of proteins: (1) the calcium entry channel proteins of the outer membrane, (2) cytosolic buffering or transfer proteins, and (3) excretory pump proteins (Hoenderop et al. 2002b; Van Cromphaut et al. 2003; Diepens et al. 2004; Choi and Jeung 2008). Two highly selective calcium channels on the apical sides of cells, members of the TRP superfamily of ion channels (TRPV6 and TRPV5), are the main Ca^{2+} entry channels (Hoenderop et al. 2002b; Nijenhuis et al. 2003). Calbindin-D9k (CaBP-9k) and Calbindin-D28k (CaBP-28k) are $[Ca^{2+}]_i$-binding proteins that are thought to participate in shuttling Ca^{2+} from the apical to the basolateral membrane, where the Na^+/Ca^{2+} exchanger and plasma membrane Ca^{2+}-ATPase 1b (PMCA1b) mediate Ca^{2+} extrusion (Christakos et al. 2007; Choi and Jeung 2008). There has been great progress in our understanding of the mechanism of active calcium transport and its role in calcium-related disorders, such as hypocalcemia, rickets, and osteomalacia, using vitamin D receptor-null mice and 1α-hydroxylase-deficient mice (Van Cromphaut et al. 2001; Zheng et al. 2004). The role of several calcium-processing proteins in the active calcium transport system, including TRPV5/6, CaBP-9k/28k, PMCA1b, and NCX1, has been verified recently using gene knockout studies.

Previously, we reported on the phenotype of CaBP-9k-null mice and the effect of compensatory gene induction of calcium-related genes in these mice (Lee et al. 2007). We also examined the differential levels of NCKX3 mRNA and protein in the kidneys of male and female mice (Lee et al. 2002). Renal NCKX3 expression was certainly increased in male mice kidneys compared to those of female mice (Lee et al. 2002). Interestingly, several renal calcium-processing genes were highly expressed in both wild-type and CaBP-9k-null female mice in the absence of any treatment. In addition, several calcium-processing genes were altered in the kidneys and duodenum of both male and female mice by the same hormones; however, we did observe differential expression of the calcium-processing gene between males and females (Lee et al. 2002). We hypothesized that NCKX3 plays a more important role in maintaining the calcium balance in the female reproductive system than in Ca^{2+} metabolism in males. We investigated the importance or involvement of NCKX3 in the tissues of female reproductive organs. In several current studies, we examined whether Na^+/Ca^{2+} exchangers are differentially expressed in the reproductive organs of females. We also examined whether the regulation of Na^+/Ca^{2+} exchangers in female reproductive organs correlated with development and identified a potential mechanism involved in the relationship of female-specific diseases caused by expression of these genes.

10.3 Uterine Expression and Regulation of Na^+/Ca^{2+} Exchangers

The primate endometrium undergoes certain hormone-dependent changes, during a particular time window within the preimplantation phase, which prepare it to receive the growing blastocyst (Schlafke and Enders 1975; Yoshinaga 1988; Carson et al. 2000). A complex interaction between effector molecules, including steroid hormones, growth factors, and cytokines, regulates the development of a "receptive" state in the uterine epithelium (Psychoyos 1976, 1986; Sharkey 1998; Carson et al. 2000). The action of Ca^{2+} in female reproductive organs has been widely studied for several decades. It is suggested that Ca^{2+} is involved in uterine smooth muscle contraction and fetal implantation (Salamonsen et al. 2001; Daston and Naciff 2005; Dong et al. 2006). Also, the balance of Ca^{2+} during uterine contraction/relaxation is extremely important throughout pregnancy and during labor. However, the mechanism underlying the regulation of calcium levels within uterine tissue remains largely unknown. While a physiological role for the NCKX and NCX proteins in the regulation of Ca^{2+} homeostasis during vascular and cardiac myocyte contraction has not yet been definitively established, recent reports describe NCKX and NCX function in brain, spermatozoa, mast cells, and platelets (Kiedrowski 2004; Kiedrowski et al. 2004; Aneiros et al. 2005; Kim et al. 2005; Kip et al. 2006). NCKX3 is expressed in the reproductive organs, including the uterus (Kraev et al. 2001; Yang et al. 2009); however, the specific role of NCKX3 within the uterus has not been fully characterized. Maintenance of calcium balance within the uterus is critically important for many physiological functions, including smooth muscle contraction during embryo implantation (Salamonsen et al. 2001; Luu et al. 2004). Thus, it is likely that the NCKX and NCX proteins may be functionally important in female reproductive organs.

In rodents, the expression of NCKX3 mRNA and protein fluctuates in the uterus during the cycle and is regulated by sex steroids (Yang et al. 2009). In mice, the uterine expression of NCKX3 mRNA and protein was highest at estrus, when the levels of E2 (17β-estradiol) and P4 (progesterone) are relatively low (Yang et al. 2009). However, rat uterine NCKX3 mRNA and protein was highly expressed in proestrus (Yang et al. 2011). Also, the levels of NCKX3 mRNA and protein were regulated by distinct sex steroids between mice and rats. In mice, uterine NCKX3 mRNA and protein expression was downregulated by both E2 and P4 when subcutaneously treated with the sex steroids of immature mice (Yang et al. 2009), and pretreatment with estrogen receptor (ER) or progesterone receptor (PR) antagonists completely recovered E2- or P4-mediated decreases in NCKX3 mRNA expression (Yang et al. 2009). In rats, uterine NCKX3 expression was induced by E2 and reduced by P4, and pretreatment with ER or PR antagonists restored the changed NCKX3 mRNA and protein levels to that of the untreated group (Yang et al. 2010).

In humans, endometrial NCKX3 expression is altered during the menstrual cycle and in an ER-positive endometrial cancer cell line (Ishikawa cell) (Boggess et al. 2006; Yang et al. 2009). Expression levels of NCKX3 mRNA change during the menstrual cycle, but that of NCX1 do not (Yang et al. 2011). The expression levels of NCKX3 during the early- and mid-proliferative phases and the early- secretory phase increase during the menstrual cycle; furthermore, endometrial expression of NCKX3 is lower than that in the early- and mid-proliferative phases and the early-secretory phase (Yang et al. 2011). It is known that the expression patterns of non-calcium-related proteins increase in the human endometrium during the early-secretory phase (Surveyor et al.1998; Marions and Danielsson 1999). In osteoblasts, these genes (COX-1 and COX-2) are regulated by extracellular calcium $[Ca^{2+}]_e$ via the extracellular signal-regulated kinase (ERK) signaling pathway (Choudhary et al. 2004). In pigs, significantly increased concentrations of calcium are observed in the uterine lumen during the implantation period (Geisert et al. 1982). Therefore, we speculated that the level of $[Ca^{2+}]_e$ observed in the human endometrium would be higher (Yang et al. 2011). The transporter function of NCKX has been demonstrated in the outer segments of rod photoreceptors, where NCKX1 is the principal mediator of Ca^{2+} extrusion (Schnetkamp 1995). Although the tissue is different, we hypothesized that NCKX3 induces Ca^{2+} extrusion to extracellular regions, that this high $[Ca^{2+}]_e$ concentration induces the expression of COX-1 and 2, and that these events may be involved in implantation. Na^+/Ca^{2+} exchangers are involved in a multitude of key points for the regulation of uterine function. These points of regulation are of fundamental importance to the function of the uterus, as large changes in contractile behavior are required to satisfy the demands on the tissue at different gestational states. Changes may also be expected in the expression of other proteins associated with the role of the sarcoplasmic reticulum (SR), as well as Ca^{2+} efflux mechanisms. Therefore, Ca^{2+} entry, together with SR Ca^{2+} release and efflux, optimizes the Ca^{2+} transient profile to the function of the uterus. These reports showed that expression levels of NCKX3 in human endometrial tissues vary throughout the menstrual cycle. This fluctuation appears to be regulated by sex steroids, indicating that NCKX3 may be a major regulator of uterine function through its effects on fetal implantation and calcium homeostasis.

10.4 Placental Expression and Regulation of Na^+/Ca^{2+} Exchangers

Placental development depends on the transport of oxygenated maternal blood, nutrients, and mineral ions to the fetus. Fetal bone mineralization requires calcium and phosphorus exchange between maternal and fetal blood, and mammalian fetal nutrition is supplied by placental transfer of maternal nutrients during gestation (Hill and Longo 1980). The syncytiotrophoblast (ST) layer of the human placenta transfers as much as 30 g of Ca^{2+} from the mother to the fetus (Belkacemi et al. 2005). Calcium is actively transported across the placenta at an increased rate in late gestation to meet the needs of the rapidly mineralizing skeleton and to maintain a $[Ca^{2+}]_e$ that is physiologically appropriate for fetal tissues and that is higher than the maternal calcium concentration (Kovacs and Kronenberg 1997).

In the human placenta, NCX may play a significant role in trans-syncytial transfer and in regulating intracellular calcium $[Ca^{2+}]_i$ important for a variety of physiological mechanisms (Kamath and Smith 1994). There has been a controversy concerning the precise placental trophoblast membrane location of NCX. A minimal role of NCX in the trans-placental movement of Ca^{2+} from mother to fetus was observed in perfusion studies of human placental lobules (Williams et al. 1991). Human placental NCX expression was not shown in the basal plasma membrane (BPM) of STs (Lafond et al. 1991). The presence of NCX expression was demonstrated in BPM but not in brush-border membrane (BBM) (Kamath and Smith 1994). Expression of the widely distributed isoform NCX1 at the molecular level in whole human placental tissue has been reported (Kofuji et al. 1992). Recently, the presence of both NCX1 and NCX2 isoforms has been demonstrated in BeWo cells and in trophoblasts from human term placenta (Moreau et al. 2003a, b). The NCX3 gene product distribution pattern is normally restricted to the brain and skeletal muscle (Nicoll et al. 1990); its unexpected presence in trophoblasts may point to a specific role of this isoform in placental physiology (Moreau et al. 2003a, b). Under basal conditions, NCX does not have a major role in Ca^{2+} efflux (Moreau et al. 2003a, b). In fact, a minimal role of NCX in the trans-placental movement of Ca^{2+} from the mother to the fetus has been observed with perfusion of placental lobules (Williams et al. 1991). Therefore, NCXs are more likely to be active in cells where $[Ca^{2+}]_i$ levels are exposed to large variations (Moreau et al. 2003a, b). During human pregnancy, the mRNA expressions of NCKX3 and NCX1 were examined in the placenta (Yang et al. 2011). The patterns of human placental NCX1 mRNA expression were altered during the second and third trimester and rapidly increased at 40 weeks (Yang et al. 2011). However, placental NCKX3 mRNA expression did not evidently fluctuate during the late-second and late-third trimesters (Yang et al. 2011). During preterm labor, placental NCKX3 expression was highly expressed in the maternal section of human placenta compared to the fetal and central section; however, human NCX1 expression was not changed among the three sections of placenta (Yang et al. 2011). During term labor, NCKX3 expression was higher in the maternal section of placenta than the fetal and central sections, and the pattern of NCX1 expression was highest in the fetal section of the placenta compared to the other two sections (Yang et al. 2011). Localization of NCKX3 and NCX1 occurred in the cytoplasm of villous ST and inside of fetal and maternal plate layers, but not in extravillous cytotrophoblast and villous blood vessels (Yang et al. 2011). The calcium homeostasis proteins, NCKX3 and NCX1, were highly expressed in the cytoplasm of both ST and giant cells of the maternal plate in the human placenta (Yang et al. 2011). On the basis of these experiments, altered expression of NCKX3 and NCX1 during preterm or term labor may affect maternofetal calcium absorption via syncytia of floating chorionic villi for $[Ca^{2+}]_i$ homeostasis and successful delivery.

In mice, mRNA and protein expression of NCX1 was significantly increased in two sections (central and fetal) of placenta compared to the maternal section (Koo 2012). In CaBP-9k and CaBP-28k knockout (KO) mice, the mRNA and protein expression of NCX1 was higher in all KO mice than in WT (wild-type) mice in all three sections of placenta (Koo 2012). Also, placental NCX1 expression was elevated in maternal to fetal sections of placental in WT, CaBP-9k, and CaBP-28k KO mice (Koo 2012). Localization of NCX1 protein was detected throughout the placenta along with expression in the STs of the labyrinthine zone and in fetal vascular endothelial cells in mice (Koo 2012). The process involves transepithelial Ca^{2+}/Na^{+} transport, which is mediated by the ST (Stulc et al. 1993). The involvement of NCX in Ca^{2+} extrusion in epithelial cells is less well established. It may be that NCX is more important under conditions where $[Ca^{2+}]_i$ levels show large fluctuations, unlike the basal conditions examined in the current study (Moreau et al. 2003a, b). These results suggest that the compensatory expression of NCX1 is increased by ablation of CaBP-9k or CaBP-28k and that expression of NCX1 influences CaBP-9k and CaBP-28k expression. The enhanced expression of NCX1 in CaBP-9k and 28k KO mice may help compensate for CaBP deficiencies in the placenta.

10.5 The Relationship of Na^{+}/Ca^{2+} Exchangers in Reproductive Tissue Diseases

Diseases during pregnancy such as preeclampsia are characterized by maternal syndromes such as gestational hypertension, proteinuria, and in 30% of cases fetal syndromes such as reduced amniotic fluid and abnormal oxygenation. Preeclamptic placental oxidative stress, resulting from deficient remodeling of maternal spiral arteries, is of importance in preeclampsia. It induces the placenta to release various factors, such as inflammatory cytokines, apoptotic wastes, and anti-angiogenic factors, which change the intracellular environment (Redman and Sargent 2009). These secreted soluble factors are then thought to alter the endothelial metabolic status, mitochondrial integrity, and vascular functions including $[Ca^{2+}]_i$ behavior (Seta et al. 2004; Robinson et al. 2008). Increases in $[Ca^{2+}]_i$ are implicated in cell injury and death induced by hypoxic stress (Seta et al. 2004). Calcium entry blockers have been reported to protect against cellular necrosis caused by experimental ischemia in the liver, kidney, and other tissues (Peck and Lefer 1981; Lee and Lum 1986). In human placenta, oxidative stress exposes the placenta to fluctuating oxygen concentrations during preeclampsia (Kingdom and Kaufmann 1999). Effective calcium homeostasis is certainly required in hypoxic cells to maintain healthy cell environments for calcium homeostatic genes, such as Na^{+}/Ca^{2+} exchangers.

The most effective treatment for preeclampsia is delivery itself; however, several randomized trials report the effective use of various methods to reduce the rate or severity of preeclampsia (Sibai et al. 2005), such as Ca^{2+} supplementation. Several alterations in maternal Ca^{2+} homeostasis were identified in preeclampsia, such as low urinary Ca^{2+} excretion and low circulating levels of 1,25-dihydroxy vitamin D3, parathyroid hormone-related peptide, and calcitonin gene-related peptide (Seely et al. 1992; Halhali et al. 2000). Epidemiologic data suggest an inverse correlation between dietary Ca^{2+} uptake and the incidence of hypertensive disorders during pregnancy in diverse populations (Hofmeyr et al. 2007). Even though the results of two large clinical trials demonstrated disparity in the benefits with respect to Ca^{2+} supplementation in prevention of preeclampsia (Belizan et al. 1991; Levine et al. 1997), Ca^{2+} supplementation did reduce the risk of preeclampsia in high-risk pregnancies as well as in women with low baseline dietary Ca^{2+} uptake (Askie et al. 2007). This is a crucial element for adequate fetal development and prenatal programming of future diseases. Approximately 80% of the total fetal Ca^{2+} is accumulated during the last trimester of pregnancy. This Ca^{2+} transfer allows for adequate fetal skeleton mineralization (Pitkin 1983) and various cellular functions.

There are two pathways for Ca^{2+} entry into the fetus. Paracellular diffusion, active in perfused placental cotyledons, enables Ca^{2+} to cross the placental barrier and represents 66% of the total maternofetal Ca^{2+} transfer (Stulc et al. 1994).

Placental calcium transporting systems involve transepithelial transport, which is mediated by ST which is a polynucleated structure (Malassine and Cronier 2002) formed during implantation and represents the most important maternofetal barrier (Rasmussen 1986). Villous trophoblasts are continuously incorporated by syncytial fusion into the ST. Ca^{2+} absorption occurring in the placenta during pregnancy from the maternal blood pool to chorionic fetal arteries is the result of transport by two mechanisms, paracellular and transcellular (Khanal and Nemere 2008). Occludin and junction adherence molecular A (JAM-A), constructing tight junction and regulating paracellular transport, are needed for Ca^{2+} absorption. Paracellular Ca^{2+} absorption is highest in the ileum, while dietary Ca^{2+} absorption through transcellular mechanisms occurs predominantly in the duodenum (Khanal and Nemere 2008). The Ca^{2+} transepithelial transfer through the ST is passive-active transport and requires various proteins (Belkacemi et al. 2002, 2005). Ca^{2+} signaling pathways in the placenta are still under investigation, and there is very little information concerning the expression of these proteins in preeclamptic placental tissues.

NCKX3 mRNA and protein expression was induced in preeclamptic placenta compared to normal placenta in the fetal and maternal sections during preterm labor (Yang et al. 2011). Placental NCX1 mRNA and protein expression was higher in preeclamptic placenta than in normal placenta (Yang et al. 2011). However, during term labor, placental NCKX3 and NCX1 mRNA and protein expression were downregulated in all sections of preeclamptic placenta compared to normal placenta (Yang et al. 2011). These results suggest that altered expression of NCKX3 and NCX1 during preterm or term labor may affect maternofetal calcium absorption via syncytia of floating chorionic villi for $[Ca^{2+}]_i$ homeostasis and successful delivery.

In preeclamptic STs, unbalanced calcium ion gradients may be protected by the influence of oxidative stress on other factors and calcium transport proteins for safe fetal bone mineralization during pregnancy. In a future study, we plan to investigate the regulators of Na^+/Ca^{2+} exchangers and other calcium transporters in the placenta, which are produced from hypoxia-inducible factor-1α (HIF-1α) and unknown factors in the maternofetal calcium transport pathway. By elucidating factors involved in placenta Ca^{2+} transport during pregnancy, we are able to determine one of the possible mechanisms responsible for fetal predisposition to adult diseases related to preeclampsia. Although this fundamental approach will not have an immediate impact in clinical practice, it will help to characterize the Ca^{2+} transfer process from mother to fetus in pregnancy and evaluate whether or not it will ultimately be possible to improve antenatal placental transfer using nutritional intervention with pharmacological or hormonal agents.

10.6 Regulation of Na^+/Ca^{2+} Exchangers in Hypoxic Placental Cell Lines

Oxidative stress, resulting from deficient remodeling of spiral arteries, is an important aspect of preeclampsia. It induces the placenta to release various factors, such as inflammatory cytokines, apoptotic wastes, and anti-angiogenic factors (Redman and Sargent 2009). These secreted soluble factors are then thought to alter endothelial metabolic status, mitochondrial integrity, and vascular functions (Robinson et al. 2008). An increase in $[Ca^{2+}]_i$ is implicated in cell injury and death induced by hypoxic stress (Seta et al. 2004). Calcium entry blockers have been reported to protect against cellular necrosis caused by experimental ischemia in liver, kidney, and other tissues (Peck and Lefer 1981; Lee and Lum 1986). In human placenta, oxidative stress exposes the placenta to fluctuating oxygen concentrations during preeclampsia (Kingdom and Kaufmann 1999).

In hypoxic placental cells, to assess whether Ca^{2+} transport and homeostasis is affected by hypoxia, we examined Na^+/Ca^{2+} exchangers NCKX3 and NCX1 in placental cell lines, i.e., BeWo, JEG3, and human placental primary cells (hPC). The hPC were obtained from normal placenta (7–12 week of gestation) (Yang et al. 2011). The expression levels of Na^+/Ca^{2+} exchanger mRNA and protein were increased in hPC by oxidative stress, and distinct expression patterns in BeWo and JEG3 cells were found during hypoxia (Yang et al. 2011). TRPV6 mRNA expression was not altered by oxidative stress in BeWo cells, but in JEG3 cells, the level of TRPV6 mRNA was decreased during hypoxia. In addition, PMCA1 mRNA expression was downregulated by oxidative stress in both BeWo and JEG3 cells. The expression of NCKX3 and NCX1 mRNA and protein were increased by oxidative stress in all placental cell lines.

In the ER-positive endometrial carcinoma Ishikawa cell line, NCKX3 mRNA and protein expression was elevated by E2, and induced NCKX3 expression was completely inhibited by the ER-specific antagonist, ICI 182780 (Yang et al. 2011). We also examined the effects of E2 on NCKX3 expression in an ER-negative endometrial carcinoma cell line (RL95). We did not observe a differential expression pattern of NCKX3 in E2-treated RL95 cells (not published data). These results suggest that distinct expression patterns of NCKX3 between ER-positive and ER-negative cell lines may be involved in ER-mediated NCKX3 transcripts in human endometrial ER-positive cells (Ishikawa).

In Chinese hamster ovary cells (CHO), Na^+-dependent inactivation of NCX is manifested in transfected CHO cells as a reduced V_{max} for Ca^{2+} uptake with no change in K_h for allosteric Ca^{2+} activation (Chernysh et al. 2008). The WT canine exchanger used in these studies was quite resistant to Na^+-dependent inactivation, even after extensive phosphatidylinositol-4,5-bisphosphate (PIP_2) depletion, but was strongly inactivated when pH_i was reduced. The resistance of the WT exchanger to Na^+-dependent inactivation suggests that this mode of NCX regulation is of little importance under normal physiologic conditions. Na^+-dependent inactivation could be an important protective response during ischemia, when high $[Na^+]_i$, low ATP/PIP_2, and low pH_i would strongly promote inactivation, thereby reducing NCX-mediated Ca^{2+} influx and toxic Ca^{2+} overload (Chernysh et al. 2008).

On the basis of these experiments, we confirmed that the expression level of Na^+/Ca^{2+} exchangers in human, bovine, and CHO cells varied throughout the reproductive cycle and Na^+-dependent inactivation. This universal expression of Na^+/Ca^{2+} exchangers appears to be regulated by sex steroid (E2) extracellular ion activation, indicating that NCKX3 and/or NCX1 may be a major regulator of uterine and placental functions by extracellular and intracellular Na^+, K^+ activation, and plasma membrane ER via effects on fetal implantation and calcium homeostasis.

10.7 Conclusion

Calcium, among all ions, is one of great importance since it is known to be implicated in many physiological processes. This chapter is an overview of most endometrial and placental proteins involved in calcium homeostatic exchangers during pregnancy. Many hormones and growth factors are involved in placental and fetal development and uterine endometrial recovery. The ST and endometrial epithelial cells, specialized cells in calcium metabolism of reproductive tissues involved in calcium regulation and transport, are also greatly influenced by these factors in the uteroplacental environment. However, the complex calcium movement in the ST necessitates many structures that have specific functional roles in this process. To date, much of our knowledge is fragmentary. In the placenta, the physiologic and clinical relevance of NCKX3 and NCX1 are still unclear. Studies at the whole animal level looking at regulatory proteins that interact with these Na^+/Ca^{2+} exchanger channels and control their functions will certainly contribute to our understanding of the pathophysiologic significance of the expression of these channels in the placenta and uterus. It is hypothesized that active transcellular calcium transport proceeds through a well-controlled sequence of events consisting of apical calcium entry via TRPV5 and TRPV6, present only in BBM, its cytosolic diffusion through its binding to CaBPs, and its basolateral extrusion mainly through PMCA and to a lesser extent by NCX. Moreover, the involvement of VDCCs, store-operated calcium entry (SOC) in Ca^{2+} regulation of hormonal secretions in placental trophoblasts and endometrial glands, and epithelial cells certainly represents interesting possibilities and warrants further investigation. Consequently, it is reasonable to speculate that useful areas of future investigations in thc placenta and uterus are likely to include (1) identification of the regulatory domains in these exchangers with the signaling pathways controlling their activity and (2) the use of inhibitors of some channels for pharmacological manipulation in several disorders related to Ca^{2+} homeostasis.

Acknowledgment This work was supported by the National Research Foundation of Korea (NRF) grant funded by the Korea government (MEST) (No. 2011–0017948).

References

E. Aneiros, S. Philipp, A. Lis, M. Freichel, A. Cavalie, Modulation of Ca^{2+} signaling by Na^+/Ca^{2+} exchangers in mast cells. J. Immunol. **174**, 119–130 (2005)

L.M. Askie, L. Duley, D.J. Henderson-Smart, L.A. Stewart, Antiplatelet agents for prevention of pre-eclampsia: a meta-analysis of individual patient data. Lancet **369**, 1791–1798 (2007)

J.M. Belizan, J. Villar, L. Gonzalez, L. Campodonico, E. Bergel, Calcium supplementation to prevent hypertensive disorders of pregnancy. N. Engl. J. Med. **325**, 1399–1405 (1991)

L. Belkacemi, L. Simoneau, J. Lafond, Calcium-binding proteins: distribution and implication in mammalian placenta. Endocrine **19**, 57–64 (2002)

L. Belkacemi, I. Bedard, L. Simoneau, J. Lafond, Calcium channels, transporters and exchangers in placenta: a review. Cell Calcium **37**, 1–8 (2005)

M.J. Berridge, Capacitative calcium entry. Biochem. J. **312**, 1–11 (1995)

R.J. Bindels, Calcium handling by the mammalian kidney. J. Exp. Biol. **184**, 89–104 (1993)

M.P. Blaustein, W.J. Lederer, Sodium/calcium exchange: its physiological implications. Physiol. Rev. **79**, 763–854 (1999)

J.F. Boggess, C. Zhou, V.L. Bae-Jump, P.A. Gehrig, Y.E. Whang, Estrogen-receptor-dependent regulation of telomerase activity in human endometrial cancer cell lines. Gynecol. Oncol. **103**, 417–424 (2006)

X. Cai, J. Lytton, Molecular cloning of a sixth member of the K^+-dependent Na^+/Ca^{2+} exchanger gene family, NCKX6. J. Biol. Chem. **279**, 5867–5876 (2004)

D.D. Carson, I. Bagchi, S.K. Dey, A.C. Enders, A.T. Fazleabas, B.A. Lessey, K. Yoshinaga, Embryo implantation. Dev. Biol. **223**, 217–237 (2000)

L. Cervetto, L. Lagnado, R.J. Perry, D.W. Robinson, P.A. McNaughton, Extrusion of calcium from rod outer segments is driven by both sodium and potassium gradients. Nature **337**, 740–743 (1989)

O. Chernysh, M. Condrescu, J.P. Reeves, Sodium-dependent inactivation of sodium/calcium exchange in transfected Chinese hamster ovary cells. Am. J. Physiol. Cell Physiol. **295**, C872–C882 (2008)

K.C. Choi, E.B. Jeung, Molecular mechanism of regulation of the calcium-binding protein calbindin-D9k, and its physiological role(s) in mammals: a review of current research. J. Cell. Mol. Med. **12**, 409–420 (2008)

S. Choudhary, A. Kumar, R.K. Kale, L.G. Raisz, C.C. Pilbeam, Extracellular calcium induces COX-2 in osteoblasts via a PKA pathway. Biochem. Biophys. Res. Commun. **322**, 395–402 (2004)

S. Christakos, P. Dhawan, B. Benn, A. Porta, M. Hediger, G.T. Oh, E.B. Jeung, Y. Zhong, D. Ajibade, K. Dhawan, S. Joshi, Vitamin D: molecular mechanism of action. Ann. N. Y. Acad. Sci. **1116**, 340–348 (2007)

D.E. Clapham, Intracellular calcium. Replenishing the stores. Nature **375**, 634–635 (1995)

N.J. Cook, U.B. Kaupp, Solubilization, purification, and reconstitution of the sodium-calcium exchanger from bovine retinal rod outer segments. J. Biol. Chem. **263**, 11382–11388 (1988)

G.P. Daston, J.M. Naciff, Gene expression changes related to growth and differentiation in the fetal and juvenile reproductive system of the female rat: evaluation of microarray results. Reprod. Toxicol. **19**, 381–394 (2005)

R.J. Diepens, E. den Dekker, M. Bens, A.F. Weidema, A. Vandewalle, R.J. Bindels, J.G. Hoenderop, Characterization of a murine renal distal convoluted tubule cell line for the study of transcellular calcium transport. Am. J. Physiol. Renal Physiol. **286**, F483–F489 (2004)

H. Dong, Y. Jiang, C.R. Triggle, X. Li, J. Lytton, Novel role for K^+-dependent Na^+/Ca^{2+} exchangers in regulation of cytoplasmic free Ca^{2+} and contractility in arterial smooth muscle. Am. J. Physiol. Heart Circ. Physiol. **291**, H1226–H1235 (2006)

P.A. Friedman, F.A. Gesek, Stimulation of calcium transport by amiloride in mouse distal convoluted tubule cells. Kidney Int. **48**, 1427–1434 (1995)

R.D. Geisert, R.H. Renegar, W.W. Thatcher, R.M. Roberts, F.W. Bazer, Establishment of pregnancy in the pig: I. Interrelationships between preimplantation development of the pig blastocyst and uterine endometrial secretions. Biol. Reprod. **27**, 925–939 (1982)

A. Halhali, A.R. Tovar, N. Torres, H. Bourges, M. Garabedian, F. Larrea, Preeclampsia is associated with low circulating levels of insulin-like growth factor I and 1,25-dihydroxyvitamin D in maternal and umbilical cord compartments. J. Clin. Endocrinol. Metab. **85**, 1828–1833 (2000)

E.P. Hill, L.D. Longo, Dynamics of maternal-fetal nutrient transfer. Fed. Proc. **39**, 239–244 (1980)

J.G. Hoenderop, P.H. Willems, R.J. Bindels, Toward a comprehensive molecular model of active calcium reabsorption. Am. J. Physiol. Renal Physiol. **278**, F352–F360 (2000)

J.G. Hoenderop, O. Dardenne, M. Van Abel, A.W. Van Der Kemp, C.H. Van Os, R. St Arnaud, R.J. Bindels, Modulation of renal Ca^{2+} transport protein genes by dietary Ca^{2+} and 1,25-dihydroxyvitamin D3 in 25-hydroxyvitamin D3-1alpha-hydroxylase knockout mice. FASEB. J. **16**, 1398–1406 (2002a)

J.G. Hoenderop, B. Nilius, R.J. Bindels, Molecular mechanism of active Ca^{2+} reabsorption in the distal nephron. Annu. Rev. Physiol. **64**, 529–549 (2002b)

G.J. Hofmeyr, L. Duley, A. Atallah, Dietary calcium supplementation for prevention of pre-eclampsia and related problems: a systematic review and commentary. BJOG **114**, 933–943 (2007)

E.J. Hong, K.C. Choi, E.B. Jeung, Induction of calbindin-D9k messenger RNA and protein by maternal exposure to alkylphenols during late pregnancy in maternal and neonatal uteri of rats. Biol. Reprod. **71**, 669–675 (2004)

D. Jeon, Y.M. Yang, M.J. Jeong, K.D. Philipson, H. Rhim, H.S. Shin, Enhanced learning and memory in mice lacking Na^+/Ca^{2+} exchanger 2. Neuron **38**, 965–976 (2003)

S.G. Kamath, C.H. Smith, Na^+/Ca^{2+} exchange, Ca^{2+} binding, and electrogenic Ca^{2+} transport in plasma membranes of human placental syncytiotrophoblast. Pediatr. Res. **36**, 461–467 (1994)

R.C. Khanal, I. Nemere, Endocrine regulation of calcium transport in epithelia. Clin. Exp. Pharmacol. Physiol. **35**, 1277–1287 (2008)

L. Kiedrowski, High activity of K^+-dependent plasmalemmal Na^+/Ca^{2+} exchangers in hippocampal CA1 neurons. Neuroreport **15**, 2113–2116 (2004)

L. Kiedrowski, A. Czyz, X.F. Li, J. Lytton, Preferential expression of plasmalemmal K-dependent Na^+/Ca^{2+} exchangers in neurons versus astrocytes. Neuroreport **13**, 1529–1532 (2002)

L. Kiedrowski, A. Czyz, G. Baranauskas, X.F. Li, J. Lytton, Differential contribution of plasmalemmal Na/Ca exchange isoforms to sodium-dependent calcium influx and NMDA excitotoxicity in depolarized neurons. J. Neurochem. **90**, 117–128 (2004)

M.H. Kim, N. Korogod, R. Schneggenburger, W.K. Ho, S.H. Lee, Interplay between Na^+/Ca^{2+} exchangers and mitochondria in Ca^{2+} clearance at the calyx of Held. J. Neurosci. **25**, 6057–6065 (2005)

J.C. Kingdom, P. Kaufmann, Oxygen and placental vascular development. Adv. Exp. Med. Biol. **474**, 259–275 (1999)

S.N. Kip, N.W. Gray, A. Burette, A. Canbay, R.J. Weinberg, E.E. Strehler, Changes in the expression of plasma membrane calcium extrusion systems during the maturation of hippocampal neurons. Hippocampus **16**, 20–34 (2006)

P. Kofuji, R.W. Hadley, R.S. Kieval, W.J. Lederer, D.H. Schulze, Expression of the Na-Ca exchanger in diverse tissues: a study using the cloned human cardiac Na-Ca exchanger. Am. J. Physiol. **263**, C1241–C1249 (1992)

T.H. Koo, H. Yang, B.S. An, K.C. Choi, S.H. Hyun, E.B. Jeung, Calcium transport genes are differently regulated in maternal and fetal placenta in the knockout mice of calbindin-D(9k) and -D(28k). Mol Reprod Dev. **79**(5), 346–355 (2012)

C.S. Kovacs, H.M. Kronenberg, Maternal-fetal calcium and bone metabolism during pregnancy, puerperium, and lactation. Endocr. Rev. **18**, 832–872 (1997)

A. Kraev, B.D. Quednau, S. Leach, X.F. Li, H. Dong, R. Winkfein, M. Perizzolo, X. Cai, R. Yang, K.D. Philipson, J. Lytton, Molecular cloning of a third member of the potassium-dependent sodium-calcium exchanger gene family, NCKX3. J. Biol. Chem. **276**, 23161–23172 (2001)

J. Lafond, M. Leclerc, M.G. Brunette, Characterization of calcium transport by basal plasma membranes from human placental syncytiotrophoblast. J. Cell. Physiol. **148**, 17–23 (1991)

R.L. Lamason, M.A. Mohideen, J.R. Mest, A.C. Wong, H.L. Norton, M.C. Aros, M.J. Jurynec, X. Mao, V.R. Humphreville, J.E. Humbert, S. Sinha, J.L. Moore, P. Jagadeeswaran, W. Zhao, G. Ning, I. Makalowska, P.M. McKeigue, D. O'Donnell, R. Kittles, E.J. Parra, N.J. Mangini, D.J. Grunwald, M.D. Shriver, V.A. Canfield, K.C. Cheng, SLC24A5, a putative cation exchanger, affects pigmentation in zebrafish and humans. Science **310**, 1782–1786 (2005)

G.S. Lee, E.B. Jeung, Uterine TRPV6 expression during the estrous cycle and pregnancy in a mouse model. Am. J. Physiol. Endocrinol. Metab. **293**, E132–E138 (2007)

H.C. Lee, B.K. Lum, Protective action of calcium entry blockers in endotoxin shock. Circ. Shock **18**, 193–203 (1986)

S.H. Lee, M.H. Kim, K.H. Park, Y.E. Earm, W.K. Ho, K^+-dependent Na^+/Ca^{2+} exchange is a major Ca^{2+} clearance mechanism in axon terminals of rat neurohypophysis. J. Neurosci. **22**, 6891–6899 (2002)

G.S. Lee, K.Y. Lee, K.C. Choi, Y.H. Ryu, S.G. Paik, G.T. Oh, E.B. Jeung, Phenotype of a calbindin-D9k gene knockout is compensated for by the induction of other calcium transporter genes in a mouse model. J. Bone Miner. Res. **22**, 1968–1978 (2007)

R.J. Levine, J.C. Hauth, L.B. Curet, B.M. Sibai, P.M. Catalano, C.D. Morris, R. DerSimonian, J.R. Esterlitz, E.G. Raymond, D.E. Bild, J.D. Clemens, J.A. Cutler, Trial of calcium to prevent preeclampsia. N. Engl. J. Med. **337**, 69–76 (1997)

X.F. Li, A.S. Kraev, J. Lytton, Molecular cloning of a fourth member of the potassium-dependent sodium-calcium exchanger gene family, NCKX4. J. Biol. Chem. **277**, 48410–48417 (2002)

X.F. Li, L. Kiedrowski, F. Tremblay, F.R. Fernandez, M. Perizzolo, R.J. Winkfein, R.W. Turner, J.S. Bains, D.E. Rancourt, J. Lytton, Importance of K^+-dependent Na^+/Ca^{2+}-exchanger 2, NCKX2, in motor learning and memory. J. Biol. Chem. **281**, 6273–6282 (2006)

S. Longoni, E. Carafoli, Identification of the Na^+/Ca^{2+} exchanger of calf heart sarcolemma with the help of specific antibodies. Biochem. Biophys. Res. Commun. **145**, 1059–1063 (1987)

K.C. Luu, G.Y. Nie, A. Hampton, G.Q. Fu, Y.X. Liu, L.A. Salamonsen, Endometrial expression of calbindin (CaBP)-d28k but not CaBP-d9k in primates implies evolutionary changes and functional redundancy of calbindins at implantation. Reproduction **128**, 433–441 (2004)

J. Lytton, Na^+/Ca^{2+} exchangers: three mammalian gene families control Ca^{2+} transport. Biochem. J. **406**, 365–382 (2007)

J. Lytton, X.F. Li, H. Dong, A. Kraev, K^+-dependent Na^+/Ca^{2+} exchangers in the brain. Ann. N. Y. Acad. Sci. **976**, 382–393 (2002)

A. Malassine, L. Cronier, Hormones and human trophoblast differentiation: a review. Endocrine **19**, 3–11 (2002)

L. Marions, K.G. Danielsson, Expression of cyclo-oxygenase in human endometrium during the implantation period. Mol. Hum. Reprod. **5**, 961–965 (1999)

R. Moreau, G. Daoud, A. Masse, L. Simoneau, J. Lafond, Expression and role of calcium-ATPase pump and sodium-calcium exchanger in differentiated trophoblasts from human term placenta. Mol. Reprod. Dev. **65**, 283–288 (2003a)

R. Moreau, L. Simoneau, J. Lafond, Calcium fluxes in human trophoblast (BeWo) cells: calcium channels, calcium-ATPase, and sodium-calcium exchanger expression. Mol. Reprod. Dev. **64**, 189–198 (2003b)

D.A. Nicoll, S. Longoni, K.D. Philipson, Molecular cloning and functional expression of the cardiac sarcolemmal Na^+-Ca^{2+} exchanger. Science **250**, 562–565 (1990)

D.A. Nicoll, B.D. Quednau, Z. Qui, Y.R. Xia, A.J. Lusis, K.D. Philipson, Cloning of a third mammalian Na^+-Ca^{2+} exchanger, NCX3. J. Biol. Chem. **271**, 24914–24921 (1996)

T. Nijenhuis, J.G. Hoenderop, A.W. van der Kemp, R.J. Bindels, Localization and regulation of the epithelial Ca^{2+} channel TRPV6 in the kidney. J. Am. Soc. Nephrol. **14**, 2731–2740 (2003)

L.A. Opperman, T.J. Saunders, D.E. Bruns, J.C. Boyd, S.E. Mills, M.E. Bruns, Estrogen inhibits calbindin-D28k expression in mouse uterus. Endocrinology **130**, 1728–1735 (1992)

R. Palty, E. Ohana, M. Hershfinkel, M. Volokita, V. Elgazar, O. Beharier, W.F. Silverman, M. Argaman, I. Sekler, Lithium-calcium exchange is mediated by a distinct potassium-independent sodium-calcium exchanger. J. Biol. Chem. **279**, 25234–25240 (2004)

R.C. Peck, A.M. Lefer, Protective effect of nifedipine in the hypoxic perfused cat liver. Agents Actions **11**, 421–424 (1981)

J.B. Peng, E.M. Brown, M.A. Hediger, Apical entry channels in calcium-transporting epithelia. News Physiol. Sci. **18**, 158–163 (2003)

R.M. Pitkin, Endocrine regulation of calcium homeostasis during pregnancy. Clin. Perinatol. **10**, 575–592 (1983)

C. Pott, S.A. Henderson, J.I. Goldhaber, K.D. Philipson, Na^+/Ca^{2+} exchanger knockout mice: plasticity of cardiac excitation-contraction coupling. Ann. N. Y. Acad. Sci. **1099**, 270–275 (2007)

A. Psychoyos, Hormonal control of uterine receptivity for nidation. J. Reprod. Fertil. **25**(Suppl), 17–28 (1976)

A. Psychoyos, Uterine receptivity for nidation. Ann. N. Y. Acad. Sci. **476**, 36–42 (1986)

B.D. Quednau, D.A. Nicoll, K.D. Philipson, Tissue specificity and alternative splicing of the Na^+/Ca^{2+} exchanger isoforms NCX1, NCX2, and NCX3 in rat. Am. J. Physiol. **272**, C1250–C1261 (1997)

H. Rasmussen, The calcium messenger system (2). N. Engl. J. Med. **314**, 1164–1170 (1986)

C.W. Redman, I.L. Sargent, Placental stress and pre-eclampsia: a revised view. Placenta **30**(Suppl A), S38–S42 (2009)

H. Reilander, A. Achilles, U. Friedel, G. Maul, F. Lottspeich, N.J. Cook, Primary structure and functional expression of the Na/Ca, K-exchanger from bovine rod photoreceptors. EMBO J. **11**, 1689–1695 (1992)

N.J. Robinson, M. Wareing, N.K. Hudson, R.T. Blankley, P.N. Baker, J.D. Aplin, I.P. Crocker, Oxygen and the liberation of placental factors responsible for vascular compromise. Lab. Invest. **88**, 293–305 (2008)

L.A. Salamonsen, G. Nie, E. Dimitriadis, L. Robb, J.K. Findlay, Genes involved in implantation. Reprod. Fertil. Dev. **13**, 41–49 (2001)

B.M. Sanborn, Relationship of ion channel activity to control of myometrial calcium. J. Soc. Gynecol. Investig. **7**, 4–11 (2000)

S. Schlafke, A.C. Enders, Cellular basis of interaction between trophoblast and uterus at implantation. Biol. Reprod. **12**, 41–65 (1975)

P.P. Schnetkamp, Sodium-calcium exchange in the outer segments of bovine rod photoreceptors. J. Physiol. **373**, 25–45 (1986)

P.P. Schnetkamp, Calcium homeostasis in vertebrate retinal rod outer segments. Cell Calcium **18**, 322–330 (1995)

P.P. Schnetkamp, D.K. Basu, R.T. Szerencsei, Na^+-Ca^{2+} exchange in bovine rod outer segments requires and transports K^+. Am. J. Physiol. **257**, C153–C157 (1989)

E.W. Seely, R.J. Wood, E.M. Brown, S.W. Graves, Lower serum ionized calcium and abnormal calciotropic hormone levels in preeclampsia. J. Clin. Endocrinol. Metab. **74**, 1436–1440 (1992)

K.A. Seta, Y. Yuan, Z. Spicer, G. Lu, J. Bedard, T.K. Ferguson, P. Pathrose, A. Cole-Strauss, A. Kaufhold,

D.E. Millhorn, The role of calcium in hypoxia-induced signal transduction and gene expression. Cell Calcium **36**, 331–340 (2004)

A. Sharkey, Cytokines and implantation. Rev. Reprod. **3**, 52–61 (1998)

B. Sibai, G. Dekker, M. Kupferminc, Pre-eclampsia. Lancet **365**, 785–799 (2005)

J. Stulc, B. Stulcova, C.P. Sibley, Evidence for active maternal-fetal transport of Na^+ across the placenta of the anaesthetized rat. J. Physiol. **470**, 637–649 (1993)

J. Stulc, B. Stulcova, M. Smid, I. Sach, Parallel mechanisms of Ca^{++} transfer across the perfused human placental cotyledon. Am. J. Obstet. Gynecol. **170**, 162–167 (1994)

G.A. Surveyor, A.K. Wilson, D.R. Brigstock, Localization of connective tissue growth factor during the period of embryo implantation in the mouse. Biol. Reprod. **59**, 1207–1213 (1998)

M. Tsoi, K.H. Rhee, D. Bungard, X.F. Li, S.L. Lee, R.N. Auer, J. Lytton, Molecular cloning of a novel potassium-dependent sodium-calcium exchanger from rat brain. J. Biol. Chem. **273**, 4155–4162 (1998)

S.J. Van Cromphaut, M. Dewerchin, J.G. Hoenderop, I. Stockmans, E. Van Herck, S. Kato, R.J. Bindels, D. Collen, P. Carmeliet, R. Bouillon, G. Carmeliet, Duodenal calcium absorption in vitamin D receptor-knockout mice: functional and molecular aspects. Proc. Natl. Acad. Sci. U. S. A. **98**, 13324–13329 (2001)

S.J. Van Cromphaut, K. Rummens, I. Stockmans, E. Van Herck, F.A. Dijcks, A.G. Ederveen, P. Carmeliet, J. Verhaeghe, R. Bouillon, G. Carmeliet, Intestinal calcium transporter genes are upregulated by estrogens and the reproductive cycle through vitamin D receptor-independent mechanisms. J. Bone Miner. Res. **18**, 1725–1736 (2003)

N. Wanaverbecq, S.J. Marsh, M. Al-Qatari, D.A. Brown, The plasma membrane calcium-ATPase as a major mechanism for intracellular calcium regulation in neurones from the rat superior cervical ganglion. J. Physiol. **550**, 83–101 (2003)

J.M. Williams, D.R. Abramovich, C.G. Dacke, T.M. Mayhew, K.R. Page, Inhibitor action on placental calcium transport. Calcif. Tissue Int. **48**, 7–12 (1991)

S. Wray, K. Jones, S. Kupittayanant, Y. Li, A. Matthew, E. Monir-Bishty, K. Noble, S.J. Pierce, S. Quenby, A.V. Shmygol, Calcium signaling and uterine contractility. J. Soc. Gynecol. Investig. **10**, 252–264 (2003)

H. Yang, T.H. Kim, B.S. An, K.C. Choi, H.H. Lee, J.M. Kim, E.B. Jeung, Differential expression of calcium transport channels in placenta primary cells and tissues derived from preeclamptic placenta. Mol Cell Endocrinol (in submission)

H. Yang, G.S. Lee, Y.M. Yoo, K.C. Choi, E.B. Jeung, Sodium/potassium/calcium exchanger 3 is regulated by the steroid hormones estrogen and progesterone in the uterus of mice during the estrous cycle. Biochem. Biophys. Res. Commun. **385**, 279–283 (2009)

H. Yang, Y.M. Yoo, E.M. Jung, K.C. Choi, E.B. Jeung, Uterine expression of sodium/potassium/calcium exchanger 3 and its regulation by sex-steroid hormones during the estrous cycle of rats. Mol. Reprod. Dev. **77**, 971–977 (2010)

H. Yang, T.H. Kim, H.H. Lee, K.C. Choi, E.B. Jeung, Distinct expression of the calcium exchangers, NCKX3 and NCX1, and their regulation by steroid in the human endometrium during the menstrual cycle. Reprod. Sci. **18**, 577–585 (2011)

K. Yoshinaga, Uterine receptivity for blastocyst implantation. Ann. N. Y. Acad. Sci. **541**, 424–431 (1988)

W. Zheng, Y. Xie, G. Li, J. Kong, J.Q. Feng, Y.C. Li, Critical role of calbindin-D28k in calcium homeostasis revealed by mice lacking both vitamin D receptor and calbindin-D28k. J. Biol. Chem. **279**, 52406–52413 (2004)

Part IV

Genetic and Epigenetic Regulation

Transcriptional Pathways and Potential Therapeutic Targets in the Regulation of *Ncx1* Expression in Cardiac Hypertrophy and Failure

11

Donald R. Menick, Mona S. Li, Olga Chernysh, Ludivine Renaud, Denise Kimbrough, Harinath Kasiganesan, and Santhosh K. Mani

Abstract

Changes in cardiac gene expression contribute to the progression of heart failure by affecting cardiomyocyte growth, function, and survival. The Na^{+}-Ca^{2+} exchanger gene (*Ncx1*) is upregulated in hypertrophy and is often found elevated in end-stage heart failure. Studies have shown that the change in its expression contributes to contractile dysfunction. Several transcriptional pathways mediate *Ncx1* expression in pathological cardiac remodeling. Both α-adrenergic receptor (α-AR) and β-adrenergic receptor (β-AR) signaling can play a role in the regulation of calcium homeostasis in the cardiomyocyte, but chronic activation in periods of cardiac stress contributes to heart failure by mechanisms which include *Ncx1* upregulation. Our studies have even demonstrated that NCX1 can directly act as a regulator of "activity-dependent signal transduction" mediating changes in its own expression. Finally, we present evidence that histone deacetylases (HDACs) and histone acetyltransferases (HATs) act as master regulators of *Ncx1* expression. We show that many of the transcription factors regulating *Ncx1* expression are important in cardiac development and also in the regulation of many other genes in the so-called fetal gene program, which are activated by pathological stimuli. Importantly, studies have revealed that the transcriptional network regulating *Ncx1* expression is also mediating many of the other changes in genetic remodeling contributing to the development of cardiac dysfunction and revealed potential therapeutic targets for the treatment of hypertrophy and failure.

D.R. Menick (✉) • M.S. Li • O. Chernysh • L. Renaud
D. Kimbrough • H. Kasiganesan • S.K. Mani
Division of Cardiology, Department of Medicine, Gazes Cardiac Research Institute, Medical University of South Carolina, 114 Doughty Street, Box 250773, Charleston, SC 29425, USA
e-mail: menickd@musc.edu

L. Annunziato (ed.), *Sodium Calcium Exchange: A Growing Spectrum of Pathophysiological Implications*, Advances in Experimental Medicine and Biology 961, DOI 10.1007/978-1-4614-4756-6_11,

Keywords

Na^+-Ca^{2+} exchanger • Transcriptional regulation • α-Adrenergic • β-Adrenergic • HDAC

11.1 Introduction

The Na^+-Ca^{2+} exchanger (NCX1) is one of the essential regulators of Ca^{2+} homeostasis within cardiomyocytes and is an important regulator of contractility. NCX1 plays a critical role in maintaining the balance of Ca^{2+} flux across the sarcolemmal membrane in excitation-contraction coupling (Bers 2002). Cardiac muscle contracts in response to the rise in $[Ca^{2+}]_i$ which is released from the sarcoplasmic reticulum (SR) and from influx across the sarcolemma through voltage-sensitive channels. SR Ca^{2+}-ATPase (SERCA) recycles Ca^{2+} from the cytosol into the lumen of the SR, and NCX1 mediates the movement of $[Ca^{2+}]_i$ across the sarcolemma to the extracellular space. The exchanger catalyzes the electrogenic exchange of Ca^{2+} and Na^+ across the plasma membrane in either the Ca^{2+}-influx or Ca^{2+}-efflux mode. NCX1 transports approximately 28% of the cytosolic Ca^{2+} during a contraction-relaxation cycle in large animals and humans, with 70% being reaccumulated in the SR (via SERCA) (Bassani and Bers 1994; Bers and Bridge 1989; Bers et al. 1990). Alterations in any of the activities associated with this complex process cause a corresponding change in the amount of Ca^{2+} released from the SR and the resulting force of cardiac contraction.

Heart disease can arise from either congenital abnormalities or a combination of acquired long-standing disorders such as hypertension, injury resulting from myocardial infarction, or myocarditis due to an infectious agent. Pathological stimuli such as prolonged mechanical stress or abnormal neurohumoral activation result in an increase in ventricular wall stress, necessitating an increase in contractile Ca^{2+} to maintain cardiac output. These stimuli induce a phase of cardiac hypertrophy in which individual cardiomyocytes increase in size and assembly of sarcomere proteins as a mechanism of augmenting cardiac output. But persistent pathological stress on the heart leads to continued hypertrophic growth and remodeling characterized by interstitial fibrosis, reexpression of cardiac embryonic genes, and transition to heart failure, arrhythmia, and sudden death (Molkentin and Dorn 2001). The exchanger is regulated at the transcriptional level in animal models of cardiac hypertrophy (Kent et al. 1993; Menick et al. 1996), ischemia, and failure (Studer et al. 1997; Hobai and O'Rourke 2000; Pogwizd et al. 2001; Sipido et al. 2002; Ahmmed et al. 2000; Litwin and Bridge 1997). Importantly, both NCX1 mRNA and protein levels are significantly upregulated in human end-stage heart failure (Hasenfuss et al. 1994, 1996, 1997; Studer et al. 1994). The diastolic performance of failing human myocardium correlates inversely with protein levels of NCX1 (Hasenfuss et al. 1999), and upregulation of the NCX1 gene (*Ncx1*) alone contributes directly to impaired SR loading and contractile dysfunction (Schillinger et al. 2000; Ranu et al. 2002). Studer et al. have demonstrated that there is an increase of NCX1 and a decrease in SERCA mRNA and protein in patients with dilated cardiomyopathy and coronary artery disease (Studer et al. 1994).

The cardiomyocytes sense many of the pathological stimuli outlined above either by membrane-bound receptors which are targets of the hormones, cytokines, or by growth factors and initiate intracellular signaling cascade in response to their binding. These signal transduction pathways mediate the cardiac growth/disease response affecting nuclear factors and the regulation of gene expression. Most of the efforts of trying to suppress the pathological outcomes of hypertrophy and heart failure have focused on the signaling pathways that alter gene expression. The paracrine/autocrine actions of growth factors such as transforming growth factor-β (TGFβ) and connective tissue growth factor (CTGF), cytokines such as TNF-α, as well as neurohormonal mediators such as norepinephrine, acetylcholine, phenylephrine, endothelin-1, and angiotensin II can all activate cardiac hypertrophy. Treatment of neonatal or adult cardiomyocytes with TGFβ, TNF-α, endothelin-1, or angiotensin II induces profound changes in cardiac gene expression including the reexpression of fetal isoforms of contractile proteins including β-myosin heavy chain, α-skeletal actin, and increased production of "stress markers" such as atrial natriuretic factor (ANF) and B-type natriuretic peptide (BNP) (Sadoshima and Izumo 1993a, b). Although endothelin-1 enhances *Ncx1* expression in renal epithelial cells, angiotensin II upregulates *Ncx1* in vascular smooth muscle cells (Kita et al. 2004) and TNF-α induces significant increases in *Ncx1* expression in human airway smooth muscle; treatment with TGFβ, TNF-α, endothelin-1, or angiotensin II does not alter the expression of NCX1 in adult cardiocytes (Mani, S. and Menick D. R. unpublished). What controls NCX1 expression in the heart? Unraveling these pathways should give us a better understanding of this complex process at the molecular level and reveal novel therapeutic targets for the prevention of adverse cardiac remodeling, cardiac hypertrophy, ischemia-reperfusion injury, and heart failure. In this chapter, we introduce recent findings on the signal transduction pathways, signaling factors, and transcription factors, which regulate *Ncx1* gene and contribute to the pathological process of heart failure.

11.2 α-Adrenergic Receptor-Stimulated *Ncx1* Upregulation

α- and β-adrenergic agonists play a major role in regulating cardiac metabolism and function. α-adrenergic stimulation stimulates chronotropic and inotropic effects on the heart and can activate many of the hypertrophic growth pathways in isolated adult and neonatal cardiomyocytes (Lee et al. 1988). The *Ncx1* gene contains three promoters (H1, K1, and Br1) and multiple 5′-untranslated exons upstream from the coding region. As a result of alternative promoter usage and the resulting alternative splicing, there are multiple tissue-specific variants of NCX1 (Barnes et al. 1997; Lee et al. 1994; Kofuji et al. 1993; Quednau et al. 1997). In our initial characterization of the NCX1 cardiac promoter (Barnes et al. 1996, 1997), we demonstrated that a construct containing the first 250 bp of the 5′-flanking region, H1 exon, and 67 bp of the first intron is sufficient for cardiac-directed expression and α-adrenergic stimulation of the luciferase reporter gene. We have since shown that a construct containing only 184 bases of the 5′-flanking region, H1 exon, and 67 bp of the first intron is not only sufficient for cardiac-directed expression but also for α-adrenergic stimulation of the luciferase reporter gene (Xu et al. 2006). This is also in agreement with what has been reported for the rat *Ncx1* minimal promoter (Nicholas et al. 1998). There are consensus sequences for a number of potential DNA-binding factors in the *Ncx1* cardiac minimal promoter (Fig. 11.1). There are two potential binding sites for the GATA family of zinc-fingered transcription factors (A/T) GATA(A/G) and two CANNTG motifs (E-boxes) which are potential target sites for the basic helix-loop-helix (bHLH) family of transcription factors. This region also contains a single myocyte enhancer factor 2 (MEF-2) element, a CArG element, and a binding site for the cardiogenic homeodomain factor Nkx-2.5. It is of interest to note that the sequence of both GATA elements, CArG element, MEF element, and −153 E-Box are perfectly conserved in the feline and rat *Ncx1* promoter (Nicholas et al. 1998). The mutational analysis revealed that both the CArG box at −80 and the GATA element at −50 were required for expression in rat neonatal cardiomyocytes but were not required for α-adrenergic induction (Cheng et al. 1999). In contrast to what we found in neonatal cardiomyocytes, the −80 CArG element mediates a part of the α-adrenergic-stimulated upregulation and is required for *Ncx1* upregulation in response to p38 stimulation in isolated adult cardiomyocytes (Xu et al. 2005).

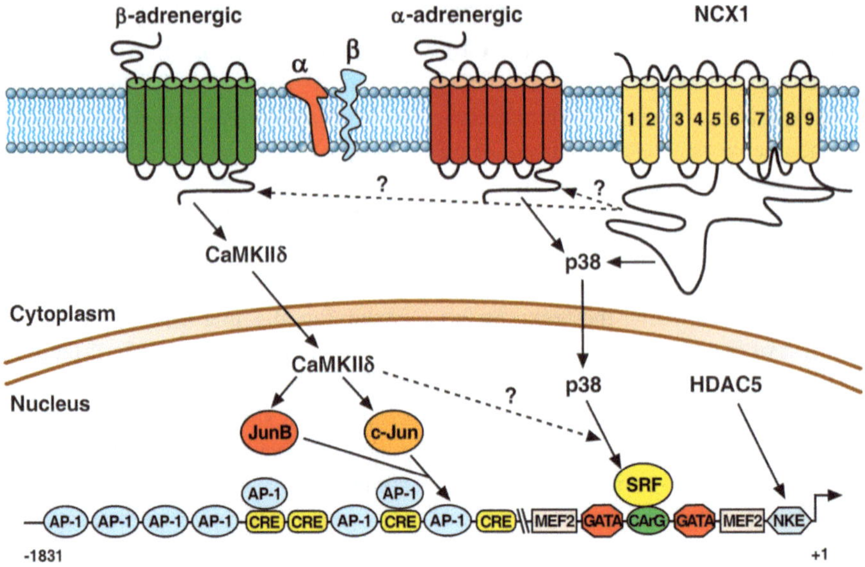

Fig. 11.1 Agonist-specific regulation of *Ncx1*. β-AR agonist isoproterenol (1 μM) or dobutamine (1 μM) activates CaMKinase II which then activates JunB and c-Jun, which sequentially binds to AP-1 elements and activates *Ncx1* expression. α-AR agonist phenylephrine (10 μM) activates SRF via MAP kinase-p38 activation. SRF binds to CArG elements and mediates *Ncx1* upregulation. HDACs regulate *Ncx1* via the NKE element by deacetylating Nkx2.5. Inhibition of NCX1 by its reverse mode inhibitor KB-R7943 activates *Ncx1* upregulation via activating MAP kinase p38 pathway

11.3 β-Adrenergic Receptor-Stimulated *Ncx1* Upregulation

β-adrenergic receptor activation is common during times of cardiac stress. Initially, this leads to increases in heart rate and contractility contributing to increased cardiac output. However, chronic β-AR stimulation leads to changes in cardiac gene expression and eventual heart failure. In congestive heart failure, the heart is under intense sympathetic stimulation with very high levels of circulating norepinephrine (Bristow 2000; Vatner et al. 2000). The changes in gene expression with chronic β-AR stimulation are the same as what is observed in heart failure (Lowes et al. 2002; Rothermel et al. 2001; Sucharov et al. 2006). *Ncx1* is upregulated at both the transcriptional and protein levels with β-AR stimulation in neonatal rat cardiomyocytes and in the adult rat heart (Golden et al. 2000, 2001). β-AR-stimulated upregulation of *Ncx1* is largely dependent on CaMKII activation in the adult heart (Mani et al. 2010). β-AR-stimulated changes in cardiomyocyte gene expression are classically mediated by the cAMP-responsive element-binding protein (CREB) and activating protein-1 (AP-1) transcription factors which bind respectively to CRE and AP-1 promoter elements which are present in the *Ncx1* promoter (Fig. 11.1). Although CREB is phosphorylated by β-AR stimulation in adult cardiomyocytes, it does not mediate *Ncx1* upregulation. Mutation of the *Ncx1* promoter AP-1 elements demonstrates that although no specific AP-1 element is required, retention of a single AP-1 element is sufficient for the majority of the β-AR-stimulated upregulation (Mani et al. 2010). Chromatin immunoprecipitation (ChIP) analysis demonstrates that β-AR stimulation activates an ordered recruitment of JunB, between 30 and 60 min, which then is replaced by c-Jun (between 2 and 6 h) binding to either the −546 AP-1 or −581 AP-1 element of the endogenous *Ncx1* promoter (Mani et al. 2010).

β-AR activation is one of the most important pathways regulating E-C coupling in the heart.

β-AR stimulation results in increased amplitude and rate of cardiomyocyte $[Ca^{2+}]_I$ at each beat resulting in increased contractility. Calcium-/calmodulin-dependent kinase II (CaMKII) is activated by β-AR stimulation in response to the increase in level and frequency of calcium transients (for review (Maier and Bers 2002)). There are four CaMKII isoforms, α, β, γ, and δ. CaMKIIδ is the predominant form in the heart, and the α- and β-isoforms are expressed only in nerve tissue (Tobimatsu and Fujisawa 1989). CaMKII phosphorylates several downstream targets in common with cAMP-activated PKA including the ryanodine receptor, phospholamban, and L-type Ca^{2+} channel complex and thus also plays an important role in regulating E-C coupling in the heart (Lindemann et al. 1983; Takasago, et al. 1989; Karczewski et al. 1997; Maier and Bers 2002). CaMKII has also been shown to phosphorylate Na^+ channels, which may contribute to arrhythmogenesis in heart failure (Wagner et al. 2006). CaMKII is activated in hypertrophy and has been shown to induce dilated cardiomyopathy and heart failure (Zhang et al. 2003; Hoch et al. 1999). In addition to acutely modulating calcium influx, SR calcium release and uptake, chronic activation of CaMKII results in the induction of the fetal gene program, which is expressed in cardiac hypertrophy and failure. Our findings that β-AR-stimulated *Ncx1* upregulation is dependent on CaMKII further illuminate the important role CaMKII has in the chronic dysregulation of cardiac Ca^{2+} homeostasis and E-C coupling. Transgenic overexpression of the cytosolic splice variant, CaMKIIδ_c, induces severe heart failure associated with SR Ca^{2+} leak and reduced SR Ca^{2+} content (Zhang et al. 2003). Correspondingly, the upregulation of *Ncx1* contributes directly to limiting SR loading, contractile dysfunction, and greater potential for delayed afterdepolarizations, which lead to ventricular tachycardia (Pogwizd et al. 1999, 2001; Schillinger et al. 2000; Ranu et al. 2002). Further, numerous studies of human and animal models of heart failure demonstrate that diastolic performance in the failing heart correlates inversely with protein levels of NCX1 (Hasenfuss et al. 1999; Pogwizd et al. 1999; Weisser-Thomas et al. 2005). Importantly, inhibition of CaMKII activity prevents cardiac arrhythmias and suppresses afterdepolarizations (Anderson 2004). Dysregulation of *Ncx1* expression can be added to the list of downstream adverse effects of chronic β-AR stimulation events mediated by the activation of CaMKII.

11.4 Identification of Regulatory Elements Mediating In Vivo Cardiac-Specific Expression and Upregulation

From our in vitro studies, we proposed that *Ncx1 H1* promoter regulates expression in the heart, the *K1* promoter regulates expression in the kidney, and the *Br1* promoter regulates expression in the brain as well as low-level ubiquitous expression. In order to test whether the *H1* promoter directed the correct spatiotemporal pattern of *Ncx1* expression in the developing and adult heart, we engineered a transgenic mouse model with the −1,831 to 67 bp of intron 1 encompassing exon *H1* for the feline *Ncx1* gene. The full-length *1831Ncx1H1* promoter was expressed in a heart-restricted pattern both in early embryos (E7.75–E14) and in late embryos (post-E14) when *Ncx1* is expressed in other tissues (Muller et al. 2002). *Ncx1*-driven reporter gene expression was detected in the cardiogenic plate by E7.75–E8.0, before the first heartbeat. The spatiotemporal expression of the reporter is identical to that previously described for endogenous *Ncx1* (Koushik et al. 1999). High levels of reporter gene expression were restricted to cardiomyocytes in both ventricles and atria in the adult heart. No reporter gene activity was detected in the kidney, liver, spleen, uterus, or skeletal muscle, but trace activity was detected in the brain. Importantly, there was a twofold upregulation of *Ncx1H1* promoter activity in the left ventricle after 7 days of transverse aortic constricted induced pressure overload compared with both sham and control animals (Muller et al. 2002).

The minimal (*184Ncx1*) promoter drives reporter gene expression at levels three- to fourfold greater than the *1831Ncx1* promoter in both neonatal and adult cardiomyocytes (Xu et al. 2006) because of the deletion of putative repressor elements distal to the minimal 184-bp promoter construct (S. Mani, L. Xu, and D. R. Menick, unpublished data). As discussed above, the *1831Ncx1* promoter is upregulated in response to α-adrenergic and β-adrenergic stimulation in both adult and neonatal cardiomyocytes and responsive to pressure-overload hypertrophy in the adult heart. However, the *184Ncx1* promoter is upregulated in response to α-adrenergic stimulation in neonatal cardiomyocytes but is recalcitrant to α-adrenergic and β-adrenergic stimulation in adult cardiomyocytes.

To test whether the *Ncx1* minimal promoter contains sufficient DNA regulatory elements to direct cardiac-specific expression, we established *184Ncx1*-β-galactosidase transgenic mouse lines. The data revealed that the *184Ncx1* minimal promoter retains the necessary enhancer elements to drive the correct spatiotemporal pattern of *Ncx1* expression in development but not for upregulation in response to pressure overload. Our data show that at least a single distal AP-1 element is required for the majority of the β-AR-stimulated upregulation (Mani et al. 2010). This AP-1 element may also be required for upregulation of *Ncx1* expression in response to pressure overload. Mutational analysis revealed that both the −80 CArG and the −50 GATA elements were required for expression in isolated adult cardiomyocytes (Cheng et al. 1999). ChIP assays in adult cardiocytes demonstrate that SRF and GATA4 are associated with the proximal region of the endogenous *Ncx1* promoter. Transgenic lines were established for the *1831Ncx1* promoter-luciferase containing mutations in the −80 CArG or −50 GATA element. No luciferase activity was detected during development, in the adult, or after pressure overload in any of the −80 CArG transgenic lines. Therefore, the −80 CArG element appears to be critical to *Ncx1* cardiac expression and regulation (Xu et al. 2006). The *Ncx1* −50 GATA mutant promoter was sufficient for driving the normal spatiotemporal pattern of *Ncx1* expression in development and for upregulation in response to pressure overload, but importantly, expression was no longer cardiac restricted. Our work demonstrates that the −50 GATA element is critical for cardiac-restricted expression of *Ncx1* (Xu et al. 2006).

11.5 NCX1 Acts as a Regulator of Activity-Dependent Transcription

Many studies have demonstrated that NCX1 inhibitors can act as positive inotropic drugs for the treatment of ischemia-reperfusion injury and congestive heart failure (Hobai et al. 2004; MacDonald and Howlett 2008; Ozdemir et al. 2008; O'Rourke 2008). All three benzyloxyphenyl NCX inhibitors, KB-R7943, SN-6, and SEA-0400, have been reported to confer some cardioprotective effects against ischemia-reperfusion injury and heart failure.

Although KB-R7943, SN-6, and SEA-0400 have been utilized in a variety of animal and cell models, most studies have focused only on the acute effects on I_{NCX1} and Ca^{2+} homeostasis. The potential for modulation of NCX1 activity to correct the impaired contractile properties seen in diseased cardiomyocytes makes it an extremely attractive target for therapeutic intervention. However, these studies primarily focus on acute treatment with NCX1 inhibitors. Interestingly, cardiac NCX1 expression is increased at both the transcriptional and protein levels in response to chronic inhibition of NCX1 activity with KB-R7943. The level of upregulation is similar to what we have observed with pressure-overload hypertrophy (Muller et al. 2002). *Ncx1* upregulation is mediated by p38, which is activated within 5 min of KB-R7943 treatment and persists for more than 72 h (Fig. 11.1). These studies impart compelling insight into the regulation of NCX1 function and expression. Importantly, treatment of adult cardiomyocytes with a second NCX1 inhibitor, SN-6 (Niu et al. 2007), also results in activation of p38.

11.6 Regulation of NCX1 Expression by HDACs and HATs

The reversible acetylation of histones plays a critical role in gene regulation as well as many other nuclear events. Protein acetylation is regulated by histone acetyltransferases (HATs), while protein deacetylation is regulated by histone deacetylases (HDACs). Many transcriptional activators are HATs or recruit HATs, allowing acetylation to be targeted to specific gene promoters. Conversely, HDACs are associated with transcriptional repressor complexes, which are also recruited to specific gene promoters. There is a rapidly growing list of nonhistone nuclear and cytosolic proteins that undergo reversible acetylation (Minucci and Pelicci 2006; Yang and Gregoire 2005). These findings have established that acetylation of nonhistone proteins plays multiple roles in the regulation of many cellular processes.

Interestingly, α-adrenergic, β-adrenergic, and pressure-overload-stimulated *Ncx1* endogenous and reporter gene expression is inhibited in a dose-dependent manner by the class I/IIb HDAC inhibitor trichostatin A (TSA). In addition, only overexpression of the class IIb HDAC, HDAC5, resulted in significant upregulation of both the promoter-luciferase activity as well as the endogenous *Ncx1* gene. Although HDACs are classically regarded as transcriptional repressors, there is mounting evidence that HDACs can serve to activate some genes, often through the direct deacetylation of transcription factors (Zupkovitz et al. 2006; Qiu et al. 2006; Nusinzon and Horvath 2006). Our co-immunoprecipitation data show that HDAC5 is in complex with HDAC1 and HDAC2, and ChIP demonstrates that HDAC1, HDAC2, and HDAC5 are recruited to the *Ncx1* promoter by the Nkx2.5 transcription factor (Fig. 11.1). Overexpression of HDAC5 appears to decrease the level of HDAC1 and HDAC 2, whereas TSA treatment increases the level HDAC1 and HDAC 2 bound to the *Ncx1* promoter. This suggests that association of HDAC5 with HDAC1 facilitates the deacetylation of a transcription factor associated with the *Ncx1* promoter resulting in their dissociation from the promoter, but inhibition of HDAC activity prevents this. Our results demonstrate that Nkx2.5 is acetylated in adult cardiomyocytes and that TSA treatment dramatically increases this acetylation. We demonstrate that when Nkx2.5 is acetylated, it is found associated with HDAC5, whereas deacetylated Nkx2.5 is in complex with the histone acetyltransferase p300. Importantly, TSA treatment prevents p300 from being recruited to the endogenous *Ncx1* promoter resulting in the repression of *Ncx1* expression.

Based on our finding, we speculate that the *Ncx1* promoter cycles through at least four kinetic steps of transcriptional competency, which we are currently experimentally testing (Fig. 11.2). The first step is a low transcriptional activity state (1) where acetylated Nkx2.5 recruits HDAC5 to the promoter. HDAC5 then complexes with HDAC1 and HDAC2, which mediates the deacetylation of Nkx2.5. Nkx2.5 deacetylation triggers the shift to the second kinetic step, the "transition state." Importantly, TSA inhibits the transfer from the low-activity state (1) to the transition state (2). TSA treatment should trap all the cardiomyocyte *Ncx1* promoters in the low transcriptional activity state (1). In the transition state, the HDAC 1/5 complex dissociates from the promoter and p300 in complex with coactivators is recruited to the promoter. The promoter is now in the high transcriptional activity state (3). We speculate that p300 bridges the transcription factors and coactivators with the preinitiation complex, stabilizing it at the initiation site and helps promote Pol II phosphorylation. As we have previously demonstrated, the makeup of the coactivators and transcription factors recruited to the *Ncx1* promoter would depend on which signaling pathways were activated. Therefore, the combinatorial recruitment of coactivators would determine the extent and duration of *Ncx1* transcriptional activation. We predict that the acetylation of Nkx2.5 by p300 results in its dissociation from the promoter and the subsequent recruitment of HDAC5 (transition state 4). This would cycle the promoter back to its low-activity state (1). Our model does not propose that HDAC5 is itself a coactivator but that HDAC5 is required for the recruitment of coactivators to the *Ncx1* promoter in response to hypertrophic stimuli. The HDAC5-dependent recruitment of coactivators to the promoter in state (3) of our model determines the transcriptional productivity and duration of the high transcriptional state. This cycling between low transcriptional activity and high transcriptional activity states allows the cell to continuously regulate transcription in response to physiological and pathophysiological stimuli

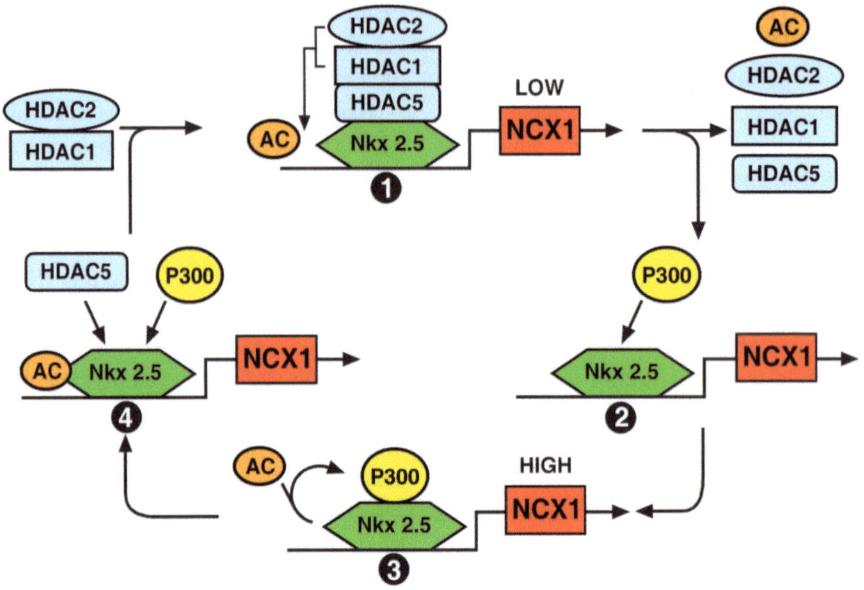

Fig. 11.2 *Regulation of Ncx1 gene by HDAC's.* Model for the role of acetylation in *Ncx1* transcriptional regulation. We speculate that the Ncx1 promoter cycles through at least four kinetic steps. *Step 1:* Acetylated Nkx2.5 recruits HDAC5, HDAC1, and HDAC2 complex to the promoter. The presence of the HDAC complex allows for low levels of *Ncx1* expression. *Step 2:* When the HDAC complex deacetylates Nkx2.5, HDAC5, HDAC1, and HDAC2 dissociate from the promoter and the HAT, p300, is recruited to the promoter. *Step 3:* p300 and its associated coactivators stabilize the Poll II complex to allow a high level of transcription of the *Ncx1* gene. *Step 4:* When p300 acetylates Nkx2.5, it dissociates from the promoter and HDAC5, HDAC1, and HDAC2 are recruited to the promoter. Importantly, TSA treatment traps the *Ncx1* promoters in the low transcriptional state. HDAC5 is not itself a coactivator, but deacetylation of Nkx2.5 is required for the recruitment of activators to the *Ncx1* promoter in response to hypertrophic stimuli

through restricting the duration of the high-activity state (3). Importantly, HDAC and HAT activity appears to act as a dominant regulator over many if not all the pathways that mediate NCX1 expression. Inhibition of HDAC activity trumps both α-adrenergic and β-adrenergic and more importantly prevents *Ncx1* upregulation in the pressure-overloaded ventricle. This is clearly one of the reasons that HDAC inhibitor treatment has been found to be efficacious in several preclinical models of cardiac hypertrophy and failure.

11.7 Conclusions

The process of pathological cardiac hypertrophy involves the change in expression of many genes activated by multiple receptors triggering intracellular signaling cascades that activate transcription factors, enhancers, and repressors. Work over the past 15 years has made major advances in the identification of the molecular regulators involved in *Ncx1* expression in the normal and hypertrophic heart. The MAP kinase, p38, is required for both α-adrenergic-stimulated and NCX1 activity-dependent upregulation. p38 appears to mediate *Ncx1* regulation through SRF activation which binds to the −80 CArG site of the *Ncx1* promoter. Although the −80 CArG element is required for upregulation, the *184Ncx1* minimal promoter is recalcitrant to α-adrenergic-stimulated and NCX1 activity-dependent transcription. Further, the *184Ncx1* minimal promoter does not contain the necessary enhancer elements to drive upregulation in response to pressure overload. Therefore, one or more distal elements, possibly one or more of the AP1 elements, required for β-adrenergic-stimulated upregulation, are requisite with the −80 CArG for hypertrophic upregulation.

HDACs appear to act as master regulators of *Ncx1* expression in the adult heart. Inhibition of class I and class IIb HDACs prevents *Ncx1* upregulation by α-, β-adrenergic stimulation, activity-dependent transcriptional activation and, most importantly, pressure overload. Notably, HDAC inhibition has been shown to attenuate pathological cardiac remodeling in several preclinical models of cardiac hypertrophy and failure. In addition to blocking *ncx1* upregulation, it prevents the dysregulation of many other cardiac genes and highlights the potential for using HDAC inhibitors in the treatment of heart failure (McKinsey 2012).

Acknowledgements This work was supported by NIH R01HL095696 (DRM) and NIH T32HL07260 (MSL, OC, and DK).

References

G.U. Ahmmed, P.H. Dong, G. Song, N.A. Ball, Y. Xu, R.A. Walsh, N. Chiamvimonvat, Changes in Ca^{2+} cycling proteins underlie cardiac action potential prolongation in a pressure-overloaded guinea pig model with cardiac hypertrophy and failure. Circ. Res. **86**, 558–570 (2000)

M.E. Anderson, Calmodulin kinase and L-type calcium channels; a recipe for arrhythmias? Trends Cardiovasc. Med. **14**, 152–161 (2004)

K.V. Barnes, M.M. Dawson, D.R. Menick, Initial characterization of the feline sodium-calcium exchanger gene. Ann. N. Y. Acad. Sci. **779**, 121–125 (1996)

K.V. Barnes, G. Cheng, M.M. Dawson, D.R. Menick, Cloning of cardiac, kidney, and brain promoters of the feline ncx1 gene. J. Biol. Chem. **272**, 11510 11517 (1997)

R.A. Bassani, D.M. Bers, Na-Ca exchange is required for rest-decay but not for rest-potentiation of twitches in rabbit and rat ventricular myocytes. J. Mol. Cell. Cardiol. **26**, 1335–1347 (1994)

D.M. Bers, Cardiac excitation-contraction coupling. Nature **415**, 198–205 (2002)

D.M. Bers, J.H. Bridge, Relaxation of rabbit ventricular muscle by Na-Ca exchange and sarcoplasmic reticulum calcium pump. Ryanodine and voltage sensitivity. Circ. Res. **65**, 334–342 (1989)

D.M. Bers, W.J. Lederer, J.R. Berlin, Intracellular Ca transients in rat cardiac myocytes: role of Na-Ca exchange in excitation-contraction coupling. Am. J. Physiol. **258**, C944–C954 (1990)

M.R. Bristow, Beta-adrenergic receptor blockade in chronic heart failure. Circulation **101**, 558–569 (2000)

G. Cheng, T.P. Hagen, M.L. Dawson, K.V. Barnes, D.R. Menick, The role of GATA, CArG, E-box, and a novel element in the regulation of cardiac expression of the Na^{+}-Ca^{2+} exchanger gene. J. Biol. Chem. **274**, 12819–12826 (1999)

K.L. Golden, Q.I. Fan, B. Chen, J. Ren, J. O'Connor, J.D. Marsh, Adrenergic stimulation regulates Na^{+}/Ca^{2+} exchanger expression in rat cardiac myocytes. J. Mol. Cell. Cardiol. **32**, 611–620 (2000)

K.L. Golden, J. Ren, J. O'Connor, A. Dean, S.E. DiCarlo, J.D. Marsh, In vivo regulation of Na/Ca exchanger expression by adrenergic effectors. Am. J. Physiol. Heart Circ. Physiol. **280**, H1376–H1382 (2001)

G. Hasenfuss, H. Reinecke, R. Studer, M. Meyer, B. Pieske, J. Holtz, C. Holubarsch, H. Posival, H. Just, H. Drexler, Relation between myocardial function and expression of sarcoplasmic reticulum Ca^{2+}-ATPase in failing and nonfailing human myocardium. Circ. Res. **75**, 434–442 (1994)

G. Hasenfuss, M. Preuss, S. Lehnart, J. Prestle, M. Meyer, H. Just, Relationship between diastolic function and protein levels of sodium-calcium-exchanger in end-stage failing human hearts. Circulation **94**, I-443 (1996)

G. Hasenfuss, M. Meyer, W. Schillinger, M. Preuss, B. Pieske, H. Just, Calcium handling proteins in the failing human heart. Basic Res. Cardiol. **92**, 87–93 (1997)

G. Hasenfuss, W. Schillinger, S.E. Lehnart, M. Preuss, B. Pieske, L.S. Maier, J. Prestle, K. Minami, H. Just, Relationship between Na^{+}-Ca^{2+}-exchanger protein levels and diastolic function of failing human myocardium. Circulation **99**, 641–648 (1999)

I.A. Hobai, B. O'Rourke, Enhanced Ca^{2+}-activated Na^{+}-Ca^{2+} exchange activity in canine pacing-induced heart failure. Circ. Res. **87**, 690–698 (2000)

I.A. Hobai, C. Maack, B. O'Rourke, Partial inhibition of sodium/calcium exchange restores cellular calcium handling in canine heart failure. Circ. Res. **95**, 292–299 (2004)

B. Hoch, R. Meyer, R. Hetzer, E.G. Krause, P. Karczewski, Identification and expression of delta-isoforms of the multifunctional Ca^{2+}/calmodulin-dependent protein kinase in failing and nonfailing human myocardium. Circ. Res. **84**, 713–721 (1999)

P. Karczewski, M. Kuschel, L.G. Baltas, S. Bartel, E.G. Krause, Site-specific phosphorylation of a phospholamban peptide by cyclic nucleotide- and Ca^{2+}/calmodulin-dependent protein kinases of cardiac sarcoplasmic reticulum. Basic Res. Cardiol. **92**(Suppl 1), 37–43 (1997)

R.L. Kent, J.D. Rozich, P.L. McCollam, D.E. McDermott, U.F. Thacker, D.R. Menick, P.J. McDermott, G. Cooper IV, Rapid expression of the Na^{+}-Ca^{2+} exchanger in response to cardiac pressure overload. Am. J. Physiol. **265**, H1024–H1029 (1993)

S. Kita, T. Katsuragi, T. Iwamoto, Endothelin-1 enhances the activity of Na^{+}/Ca^{2+} exchanger type 1 in renal epithelial cells. J. Cardiovasc. Pharmacol. **44** (Suppl 1), S239–S243 (2004)

P. Kofuji, W.J. Lederer, D.H. Schulze, Na/Ca exchanger isoforms expressed in kidney. Am. J. Physiol. **265**, F598–F603 (1993)

S.V. Koushik, J. Bundy, S.J. Conway, Sodium-calcium exchanger is initially expressed in a heart-restricted pattern within the early mouse embryo. Mech. Dev. **88**, 119–122 (1999)

H.R. Lee, S.A. Henderson, R. Reynolds, P. Dunnmon, D. Yuan, K.R. Chien, Alpha 1-adrenergic stimulation of cardiac gene transcription in neonatal rat myocardial cells. Effects on myosin light chain-2 gene expression. J. Biol. Chem. **263**, 7352–7358 (1988)

S.L. Lee, A.S. Yu, J. Lytton, Tissue-specific expression of Na^{+}-Ca^{2+} exchanger isoforms. J. Biol. Chem. **269**, 14849–14852 (1994)

J.P. Lindemann, L.R. Jones, D.R. Hathaway, B.G. Henry, A.M. Watanabe, Beta-Adrenergic stimulation of phospholamban phosphorylation and Ca^{2+}-ATPase activity in guinea pig ventricles. J. Biol. Chem. **258**, 464–471 (1983)

S.E. Litwin, J.H. Bridge, Enhanced Na^{+}-Ca^{2+} exchange in the infarcted heart. Implications for excitation-contraction coupling. Circ. Res. **81**, 1083–1093 (1997)

B.D. Lowes, E.M. Gilbert, W.T. Abraham, W.A. Minobe, P. Larrabee, D. Ferguson, E.E. Wolfel, J. Lindenfeld, T. Tsvetkova, A.D. Robertson, R.A. Quaife, M.R. Bristow, Myocardial gene expression in dilated cardiomyopathy treated with beta-blocking agents. N. Engl. J. Med. **346**, 1357–1365 (2002)

A.C. MacDonald, S.E. Howlett, Differential effects of the sodium calcium exchange inhibitor, KB-R7943, on ischemia and reperfusion injury in isolated guinea pig ventricular myocytes. Eur. J. Pharmacol. **580**, 214–223 (2008)

L.S. Maier, D.M. Bers, Calcium, calmodulin, and calcium-calmodulin kinase II: heartbeat to heartbeat and beyond. J. Mol. Cell. Cardiol. **34**, 919–939 (2002)

S.K. Mani, E.A. Egan, B.K. Addy, M. Grimm, H. Kasiganesan, T. Thiyagarajan, L. Renaud, J.H. Brown, C.B. Kern, D.R. Menick, Beta-adrenergic receptor stimulated Ncx1 upregulation is mediated via a CaMKII/AP-1 signaling pathway in adult cardiomyocytes. J. Mol. Cell. Cardiol. **48**, 342–351 (2010)

T.A. McKinsey, Therapeutic potential for HDAC inhibitors in the heart. Annu. Rev. Pharmacol. Toxicol. **10**, 303–319 (2012)

D.R. Menick, K.V. Barnes, U.F. Thacker, M.M. Dawson, D.E. McDermott, J.D. Rozich, R.L. Kent, G. Cooper IV, The exchanger and cardiac hypertrophy. Ann. N. Y. Acad. Sci. **779**, 489–501 (1996)

S. Minucci, P.G. Pelicci, Histone deacetylase inhibitors and the promise of epigenetic (and more) treatments for cancer. Nat. Rev. Cancer **6**, 38–51 (2006)

J.D. Molkentin, G.W. Dorn 2nd, Cytoplasmic signaling pathways that regulate cardiac hypertrophy. Annu. Rev. Physiol. **63**, 391–426 (2001)

J.G. Muller, Y. Isomatsu, S.V. Koushik, M. O'Quinn, L. Xu, C.S. Kappler, E. Hapke, M.R. Zile, S.J. Conway, D.R. Menick, Cardiac-specific expression and hypertrophic upregulation of the feline Na^{+}-Ca^{2+} exchanger gene H1-promoter in a transgenic mouse model. Circ. Res. **90**, 158–164 (2002)

S.B. Nicholas, W. Yang, S.L. Lee, H. Zhu, K.D. Philipson, J. Lytton, Alternative promoters and cardiac muscle cell-specific expression of the Na^{+}/Ca^{2+} exchanger gene. Am. J. Physiol. **274**, H217–H232 (1998)

C.F. Niu, Y. Watanabe, K. Ono, T. Iwamoto, K. Yamashita, H. Satoh, T. Urushida, H. Hayashi, J. Kimura, Characterization of SN-6, a novel Na^{+}/Ca^{2+} exchange inhibitor in guinea pig cardiac ventricular myocytes. Eur. J. Pharmacol. **573**, 161–169 (2007)

I. Nusinzon, C.M. Horvath, Positive and negative regulation of the innate antiviral response and beta interferon gene expression by deacetylation. Mol. Cell. Biol. **26**, 3106–3113 (2006)

B. O'Rourke, The ins and outs of calcium in heart failure. Circ. Res. **102**, 1301–1303 (2008)

S. Ozdemir, V. Bito, P. Holemans, L. Vinet, J.J. Mercadier, A. Varro, K.R. Sipido, Pharmacological inhibition of Na/Ca exchange results in increased cellular Ca^{2+} load attributable to the predominance of forward mode block. Circ. Res. **102**, 1398–1405 (2008)

S.M. Pogwizd, M. Qi, W. Yuan, A.M. Samarel, D.M. Bers, Upregulation of Na^{+}/Ca^{2+} exchanger expression and function in an arrhythmogenic rabbit model of heart failure. Circ. Res. **85**, 1009–1019 (1999)

S.M. Pogwizd, K. Schlotthauer, L. Li, W. Yuan, D.M. Bers, Arrhythmogenesis and contractile dysfunction in heart failure: Roles of sodium-calcium exchange, inward rectifier potassium current, and residual beta-adrenergic responsiveness. Circ. Res. **88**, 1159–1167 (2001)

Y. Qiu, Y. Zhao, M. Becker, S. John, B.S. Parekh, S. Huang, A. Hendarwanto, E.D. Martinez, Y. Chen, H. Lu, N.L. Adkins, D.A. Stavreva, M. Wiench, P.T. Georgel, R.L. Schiltz, G.L. Hager, HDAC1 acetylation is linked to progressive modulation of steroid receptor-induced gene transcription. Mol. Cell **22**, 669–679 (2006)

B.D. Quednau, D.A. Nicoll, K.D. Philipson, Tissue specificity and alternative splicing of the Na^{+}/Ca^{2+} exchanger isoforms NCX1, NCX2, and NCX3 in rat. Am. J. Physiol. **272**, C1250–C1261 (1997)

H.K. Ranu, C.M. Terracciano, K. Davia, E. Bernobich, B. Chaudhri, S.E. Robinson, Z. Bin Kang, R.J. Hajjar, K.T. MacLeod, S.E. Harding, Effects of Na^{+}/Ca^{2+}-exchanger overexpression on excitation-contraction coupling in adult rabbit ventricular myocytes. J. Mol. Cell. Cardiol. **34**, 389–400 (2002)

B.A. Rothermel, T.A. McKinsey, R.B. Vega, R.L. Nicol, P. Mammen, J. Yang, C.L. Antos, J.M. Shelton, R. Bassel-Duby, E.N. Olson, R.S. Williams, Myocyte-enriched calcineurin-interacting protein, MCIP1, inhibits cardiac hypertrophy in vivo. Proc. Natl. Acad. Sci. U. S. A. **98**, 3328–3333 (2001)

J. Sadoshima, S. Izumo, Molecular characterization of angiotensin II–induced hypertrophy of cardiac myocytes and hyperplasia of cardiac fibroblasts. Critical role of the AT1 receptor subtype. Circ. Res. **73**, 413–423 (1993a)

J. Sadoshima, S. Izumo, Signal transduction pathways of angiotensin II–induced c-fos gene expression in cardiac myocytes in vitro. Roles of phospholipid-derived second messengers. Circ. Res. **73**, 424–438 (1993b)

W. Schillinger, P.M. Janssen, S. Emami, S.A. Henderson, R.S. Ross, N. Teucher, O. Zeitz, K.D. Philipson, J. Prestle, G. Hasenfuss, Impaired contractile performance of cultured rabbit ventricular myocytes after adenoviral gene transfer of Na^+-Ca^{2+} exchanger. Circ. Res. **87**, 581–587 (2000)

K.R. Sipido, P.G. Volders, M.A. Vos, F. Verdonck, Altered Na/Ca exchange activity in cardiac hypertrophy and heart failure: a new target for therapy? Cardiovasc. Res. **53**, 782–805 (2002)

R. Studer, H. Reinecke, J. Bilger, T. Eschenhagen, M. Bohm, G. Hasenfuss, H. Just, J. Holtz, H. Drexler, Gene expression of the cardiac Na^+-Ca^{2+} exchanger in end-stage human heart failure. Circ. Res. **75**, 443–453 (1994)

R. Studer, H. Reinecke, R. Vetter, J. Holtz, H. Drexler, Expression and function of the cardiac Na^+/Ca^{2+} exchanger in postnatal development of the rat, in experimental-induced cardiac hypertrophy and in the failing human heart. Basic Res. Cardiol. **92**, 53–58 (1997)

C.C. Sucharov, P.D. Mariner, K.R. Nunley, C. Long, L. Leinwand, M.R. Bristow, A beta1-adrenergic receptor CaM kinase II-dependent pathway mediates cardiac myocyte fetal gene induction. Am. J. Physiol. Heart Circ. Physiol. **291**, H1299–H1308 (2006)

T. Takasago, T. Imagawa, M. Shigekawa, Phosphorylation of the cardiac ryanodine receptor by cAMP-dependent protein kinase. J. Biochem. **106**, 872–877 (1989)

T. Tobimatsu, H. Fujisawa, Tissue-specific expression of four types of rat calmodulin-dependent protein kinase II mRNAs. J. Biol. Chem. **264**, 17907–17912 (1989)

S.F. Vatner, D.E. Vatner, C.J. Homcy, Beta-adrenergic receptor signaling: an acute compensatory adjustment-inappropriate for the chronic stress of heart failure? Insights from Gsalpha overexpression and other genetically engineered animal models. Circ. Res. **86**, 502–506 (2000)

S. Wagner, N. Dybkova, E.C. Rasenack, C. Jacobshagen, L. Fabritz, P. Kirchhof, S.K. Maier, T. Zhang, G. Hasenfuss, J.H. Brown, D.M. Bers, L.S. Maier, Ca^{2+}/calmodulin-dependent protein kinase II regulates cardiac Na^+ channels. J. Clin. Invest. **116**, 3127–3138 (2006)

J. Weisser-Thomas, H. Kubo, C.A. Hefner, J.P. Gaughan, B.S. McGowan, R. Ross, M. Meyer, W. Dillmann, S.R. Houser, The Na^+/Ca^{2+} exchanger/SR Ca^{2+} ATPase transport capacity regulates the contractility of normal and hypertrophied feline ventricular myocytes. J. Card. Fail. **11**, 380–387 (2005)

L. Xu, C.S. Kappler, D.R. Menick, The role of p38 in the regulation of Na^+-Ca^{2+} exchanger expression in adult cardiomyocytes. J. Mol. Cell. Cardiol. **38**, 735–743 (2005)

L. Xu, L. Renaud, J.G. Muller, C.F. Baicu, D.D. Bonnema, H. Zhou, C.S. Kappler, S.W. Kubalak, M.R. Zile, S.J. Conway, D.R. Menick, Regulation of Ncx1 expression. Identification of regulatory elements mediating cardiac-specific expression and up-regulation. J. Biol. Chem. **281**, 34430–34440 (2006)

X.J. Yang, S. Gregoire, Class II histone deacetylases: from sequence to function, regulation, and clinical implication. Mol. Cell. Biol. **25**, 2873–2884 (2005)

T. Zhang, L.S. Maier, N.D. Dalton, S. Miyamoto, J. Ross Jr., D.M. Bers, J.H. Brown, The deltaC isoform of CaMKII is activated in cardiac hypertrophy and induces dilated cardiomyopathy and heart failure. Circ. Res. **92**, 912–919 (2003)

G. Zupkovitz, J. Tischler, M. Posch, I. Sadzak, K. Ramsauer, G. Egger, R. Grausenburger, N. Schweifer, S. Chiocca, T. Decker, C. Seiser, Negative and positive regulation of gene expression by mouse histone deacetylase 1. Mol. Cell. Biol. **26**, 7913–7928 (2006)

Transcriptional Regulation of *ncx1* Gene in the Brain

12

Valeria Valsecchi, Giuseppe Pignataro, Rossana Sirabella, Carmela Matrone, Francesca Boscia, Antonella Scorziello, Maria Josè Sisalli, Elga Esposito, Nicola Zambrano, Mauro Cataldi, Gianfranco Di Renzo, and Lucio Annunziato

Abstract

The ubiquitous sodium–calcium exchanger isoform 1 (NCX1) is a bidirectional transporter that plays a relevant role under physiological and pathophysiological conditions including brain ischemia by regulating intraneuronal Ca^{2+} and Na^{+} homeostasis. Although changes in *ncx1* protein and transcript expression have been detected during stroke, its transcriptional regulation is still largely unexplored. Here, we reviewed our recent findings on several transcription factors including cAMP response element-binding protein (CREB), nuclear factor kappa B (NF-κB), and hypoxia-inducible factor-1 (HIF-1) in the control of the *ncx1* gene expression in neuronal cells.

Keywords

ncx1 • HIF-1 • Ischemic preconditioning • Hypoxia • Transcriptional regulation

V. Valsecchi • G. Pignataro • C. Matrone • F. Boscia
A. Scorziello • M.J. Sisalli • E. Esposito • M. Cataldi
G. Di Renzo
Division of Pharmacology, Department of Neuroscience, School of Medicine, "Federico II" University of Naples, Via S. Pansini 5, Naples, 80131, Italy

R. Sirabella
Fondazione IRCCS SDN, Naples, Italy

N. Zambrano
Department of Biochemistry and Medical Biotechnology, "Federico II" University of Naples, Via S. Pansini 5, Naples 80131, Italy

L. Annunziato (✉)
Division of Pharmacology, Department of Neuroscience, School of Medicine, "Federico II" University of Naples, Via S. Pansini 5, Naples 80131, Italy
e-mail: lannunzi@unina.it

L. Annunziato (ed.), *Sodium Calcium Exchange: A Growing Spectrum of Pathophysiological Implications*, Advances in Experimental Medicine and Biology 961, DOI 10.1007/978-1-4614-4756-6_12,

12.1 Introduction

Three different genes coding for the three different NCX isoforms, NCX1, NCX2, and NCX3, have been identified in mammals (Li et al. 1994; Nicoll et al. 1990, 1996). The *ncx1, ncx2,* and *ncx3* genes are dispersed in the genome, since they map in mouse chromosomes 17, 7, and 12, respectively (Nicoll et al. 1996). The *ncx1* gene is expressed in several tissues, including brain, heart, skeletal muscle, smooth muscle, kidney, eye, secretory, and blood cells, whereas transcripts encoded by the *ncx2* and *ncx3* genes have been found exclusively in neuronal and skeletal muscle tissues (Lee et al. 1994). Both *ncx1* and *ncx3* genes give rise to several splicing variants that appear to be selectively expressed in different regions and cellular populations of the brain (Papa et al. 2003; Quednau et al. 1997; Yu and Colvin 1997).

Although much research has focused on the mechanisms of regulation of ncx1 at the post-translational level, few studies have directly addressed the regulation of *ncx1* expression at the transcriptional level. Specifically, little is known on the mechanisms responsible for the changes in *ncx1* expression caused by a variety of signals such as glucocorticoids, growth factors, depolarization, Ca^{2+} influx, and adrenergic stimulation (Golden et al. 2000; Li et al. 2000; Smith and Smith 1994) or occurring under physiological or pathophysiological conditions such as postnatal development (Gibney et al. 2002; Sakaue et al. 2000) and cerebral ischemia (Boscia et al. 2006; Li et al. 2006; Lu et al. 2002; Pignataro et al. 2004).

ncx1 gene expression appears to be directed by three alternative regulatory regions, named heart (Ht), kidney (Kd), and brain (Br) promoters (Lee et al. 1994). These promoters include different binding sequences for several transcriptional factors that control the tissue-specific expression of ncx1 in the heart, kidney, and brain (Barnes et al. 1997; Lee et al. 1994; Nicholas et al. 1998). Specifically, in cardiomyocytes, the expression of ncx1 is controlled by the Ht promoter that includes serum response factor (SRF), GATA-4, activator protein 1 (AP-1), and Nkx2.5 binding sites (Cheng et al. 1999; Nicholas and Philipson 1999). In particular, the cardiac-specific homeodomain transcription factor Nkx2.5 induces the expression of *ncx1* and enhances SRF-induced *ncx1* gene transcription in neonatal cardiomyocytes (Muller et al. 2002).

In kidney, the ncx1 expression is controlled by the Kd promoter region that includes a typical TATA box preceded by two tandem GATA elements, NF-Y binding site, and two CAAT/enhancer binding protein-like (C/EBP-like) sites. Furthermore, Kd promoter has two putative binding elements for CREB, one for the activating transcription factor (ATF) and one for AP-1. Although each of these transcriptional binding sites is not specific to the kidney, their unique combinations may lead to kidney-specific transcription regulation (Nicholas et al. 1998).

Br *ncx1* promoter is a ubiquitous GC-rich TATA-less promoter, giving rise to the majority of *ncx1* transcripts in the brain (Nicholas et al. 1998). This region contains several putative binding sites for transcriptional factors including specificity protein 1 (Sp-1), AP-1, AP-2, and nuclear factor κB (NF-κB) (Nicholas et al. 1998) (Fig. 12.1). It has been hypothesized that Sp-1 and NF-κB binding sites could be responsible for the ubiquitous transcription of *ncx1* in all tissues, whereas the high level of expression of ncx1 in the brain could be dependent on AP-2 (Nicholas et al. 1998). Consistent with the hypothesis that the constitutive expression depends on Sp-1 and NF-κB, the region spanning from 312 bp upstream the transcription start site (TSS) to 151 bp downstream the TSS of Br promoter, which contains the binding sites for these transcriptional factors, is sufficient to induce the expression outside the brain, in rat vascular smooth muscle cells (VSMC) (Scheller et al. 1998).

To date, the promoter of ncx2 gene has not yet been identified, and moreover, no extensive data are available for its transcriptional regulation. On the other hand, it has been shown that ncx2 expression is rapidly downregulated after membrane depolarization in a calcineurin-dependent manner suggesting the possible involvement of the transcriptional factor CREB (Li et al. 2000).

The promoter of *ncx3* is located immediately upstream of first exon (Gabellini et al. 2002, 2003)

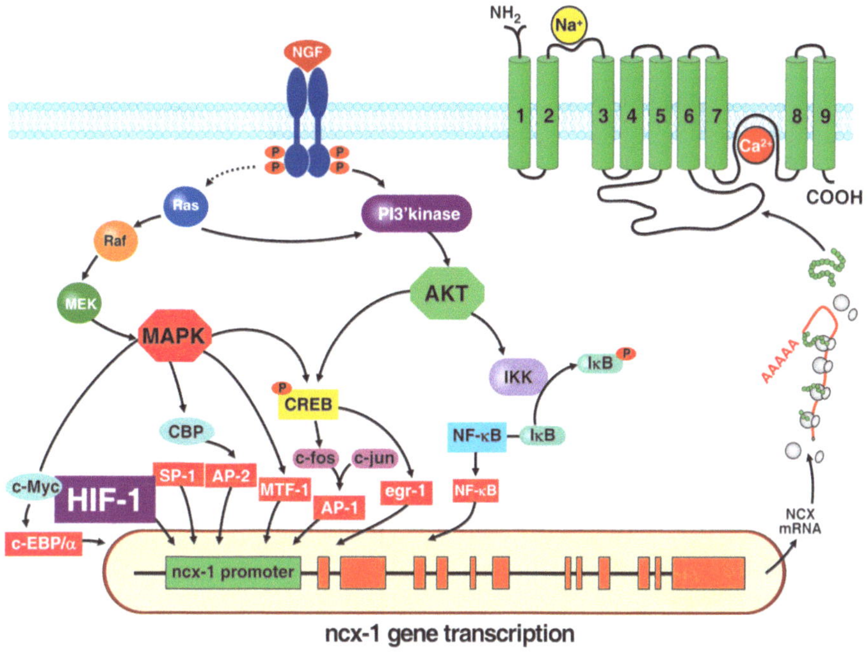

Fig. 12.1 *Cartoon representing the transcription factors regulating ncx1 gene expression in the brain.* The cartoon shows how NGF controls *ncx1* gene expression by promoting CREB phosphorylation by Akt. The regulation of ncx1 by HIF-1 and NF-κB under hypoxic conditions is also reported (Modified from Annunziato et al. 2009)

and includes binding sites for AP-1, AP-2, CREB, downstream regulatory element antagonist modulator (DREAM), Egr-1, KROX24, MyoD, GATA 2/3, and Sp1. Two of these transcription factors, CREB and DREAM, seem to be the most important regulators of *ncx3* gene transcription in response to changes in the intracellular concentrations of Ca^{2+} and cAMP (Gabellini et al. 2003; Gomez Villafuertes et al. 2005).

12.2 *ncx1* Gene Regulation in Response to Neurotrophic Factors and Brain Ischemia

During the last 10 years, our laboratory provided compelling evidence that ncx1 exerts a neuroprotective activity in experimental models of stroke, and we suggested that ischemic neurons could be rescued from cell death by enhancing the expression or the activity of NCX1 isoform with neurotrophic factors or neuroprotective drugs (Annunziato et al. 2004 and 2007; Boscia et al. 2006; Pignataro et al. 2004; Secondo et al. 2007). Such a therapeutic approach appears rational also considering that *ncx1* gene expression was found to be downregulated during stroke by unknown mechanisms (Boscia et al. 2006). In this chapter, we reviewed the work that we performed in the effort to unravel the transcriptional regulation of *ncx1* during brain ischemia and to identify possible tools to increase its transcription.

12.2.1 *ncx1* Is Transcriptionally Upregulated by NF-κB Under Anoxic Conditions

NF-κB is a family of transcription factors that exists as either a hetero- or homodimer of several subunits called RelA (p65), RelB, cRel, p50 and its precursor p105 (NF-κB1), and p52 and its precursor p100 (NF-κB2). Some dimers are more prevalent than others, and they shuttle between the cytoplasm and nucleus but are predominantly sequestered in the inactive state in the cytoplasm

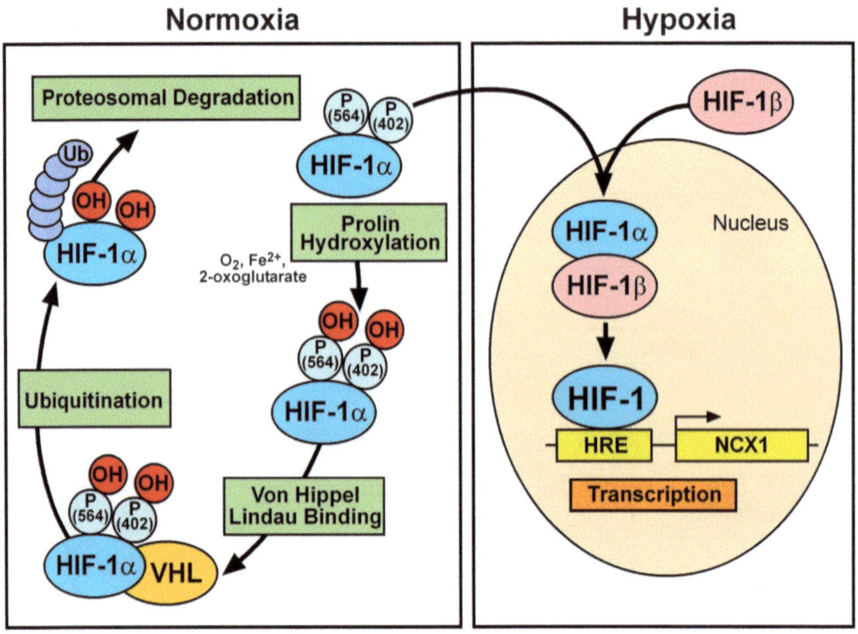

Fig. 12.2 *Regulation and activation pathways of HIF-1α. Left panel* shows how the HIF-1 pathway is regulated under normoxic conditions. The Von Hippel–Lindau (pVHL) tumor suppressor gene product binds the hydroxylated form of HIF-1α and targets it to ubiquitination and proteosomal degradation. *Right panel* shows how the HIF-1 pathway is regulated during hypoxia. In the absence of oxygen, HIF-1α protein escapes proteasomal degradation, accumulates within the nucleus, dimerizes with HIF-1β forming the active complex, translocates into the nucleus, and finally upregulates *ncx1* gene transcription

by members of the inhibitor of NF-κB (IκB) (Hayden and Ghosh 2008). Upon stimulation by tumor necrosis factor-α (TNF-α), oncogenes, or UV light, IκB is firstly phosphorylated by a kinase cascade and then degraded by a ubiquitination-mediated proteasomal pathway. As a result, NF-κB is released and translocates into the nucleus where it binds target DNA sequences and regulates the transcription of several genes (Perkins and Gilmore 2006).

Interestingly, a putative NF-κB binding site was identified on Br ncx1 promoter at -306/-297 bp from the TSS (Nicholas et al. 1998). Recently, we demonstrated that under hypoxic conditions, NF-κB increases the protein expression of NCX1 in cortical neurons (Sirabella et al. 2009). Moreover, the pharmacological blocker SN-50 or the silencing of the NF-κB subunit p65 prevents hypoxia-induced *ncx1* upregulation (Sirabella et al. 2009). This suggests an involvement of the transcriptional factor NF-κB in *ncx1* upregulation in the ischemic brain.

12.2.2 *ncx1* Is a Novel Target Gene for HIF-1

HIF-1 is a nuclear factor required for transcriptional activation in response to hypoxia (Semenza et al. 1991). It consists of two subunits: an oxygen-sensitive HIF-1α and a constitutively expressed HIF-1β subunit (Bruick 2003; Wang and Semenza 1993). Both contain two motifs named Per-ARNT-Sim (PAS) and basic helix-loop-helix (bHLH) (Jiang et al. 1996). Whereas both motifs are required for heterodimer formation between HIF-1α and HIF-1β subunits, only the bHLH region affords specific binding to a region of HIF-1 target genes called hypoxia-responsive element (HRE) (Crews 1998). Under normoxic conditions, HIF-1α protein is rapidly hydroxylated by a proline hydroxylase, and this, in turn, allows the recognition by Von Hippel–Lindau tumor suppressor protein that targets HIF-1α for ubiquitination and proteasomal degradation (Tanimoto et al. 2000) (Fig. 12.2). By contrast,

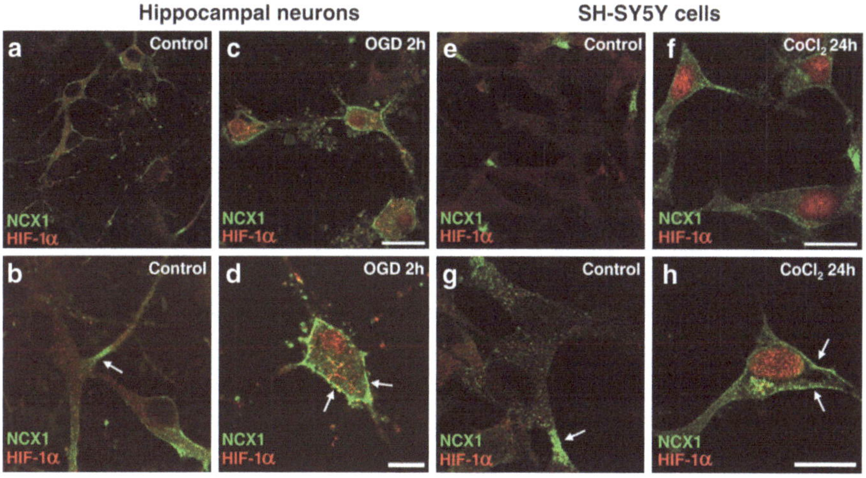

Fig. 12.3 *Confocal images depicting the immunoreactivity of ncx1 and HIF-1α under normoxic and hypoxic conditions.* ncx1 (*green*) and HIF-1α (*red*) immunoreactivity in hippocampal neurons (**a**, **b**) and in SH-SY5Y cells (**e**, **f**) under normoxic conditions. ncx1 and HIF-1α immunoreactivity in hippocampal neurons under hypoxic conditions (**c**, **d**) and in SH-SY5Y cells exposed to $CoCl_2$ (**g**, **h**). Scale bars are 10 μm (**a**–**d**) and 20 μm (**e**–**h**) (Modified from Valsecchi et al. 2011)

during hypoxia or in the presence of iron chelators and divalent cations, such as cobalt, HIF-1α protein escapes proteosomal degradation, dimerizes with HIF-1β, generates HIF-1, and translocates into the nucleus (Kallio et al. 1997) (Fig. 12.2). A large variety of genes involved in erythropoiesis, iron metabolism (Semenza et al. 1991), angiogenesis (Levy et al. 1995), and glucose metabolism (Semenza et al. 1996) are regulated by HIF-1. In addition, HIF-1 is upregulated in rat brain during cerebral ischemia (Bergeron et al. 1999; Chavez and LaManna 2002; Matrone et al. 2004), in repetitive episodes of hypoxia–reoxygenation (Semenza 2009), or in ischemic preconditioning, a condition in which a period of sublethal ischemia is able to protect the brain from a subsequent lethal insult (Pignataro et al. 2009; Taie et al. 2009).

Thus, in an attempt to unravel the mechanisms involved in the regulation of *ncx1* expression at the transcriptional level, we investigated whether *ncx1* gene might be a target of the transcription factor HIF-1 in the brain. We firstly searched for the presence of HRE on the Br *ncx1* promoter. Next, by means of electromobility shift assay (EMSA) and chromatin immunoprecipitation (ChIP), we verified the specific HIF-1 binding to these HREs. Finally, we determined whether ncx1 overexpression induced by HIF-1 in the brain might be partially responsible for neuroprotection in an animal model of cerebral ischemic preconditioning.

Interestingly, primary hippocampal neurons exposed to hypoxia displayed an increase in HIF-1α immunosignal in the nucleus that was accompanied by a remarkable enhancement of *ncx1* expression on the plasma membrane and within the cytoplasm (Fig. 12.3). By contrast, under normoxic conditions, ncx1 immunosignal was mainly confined along the neuropil, whereas HIF-1α immunoreactivity was barely detected in the cytoplasm and in the nucleus (Valsecchi et al. 2011) (Fig. 12.3).

A computational analysis revealed the presence of six putative HRE *consensus* sites for HIF-1 on Br ncx1 promoter. These HREs, including 4 in the forward orientation and 2 in the reverse orientation, are located at -539/-535, -442/-438, -409/-405, -408/-404, -331/-327, and -164/-160 bp referring to the TSS (+1).

By deletion analysis, EMSA and ChIP assays, we showed that the HIF-1-dependent regulation of NCX1 was dependent on the binding of HIF-1

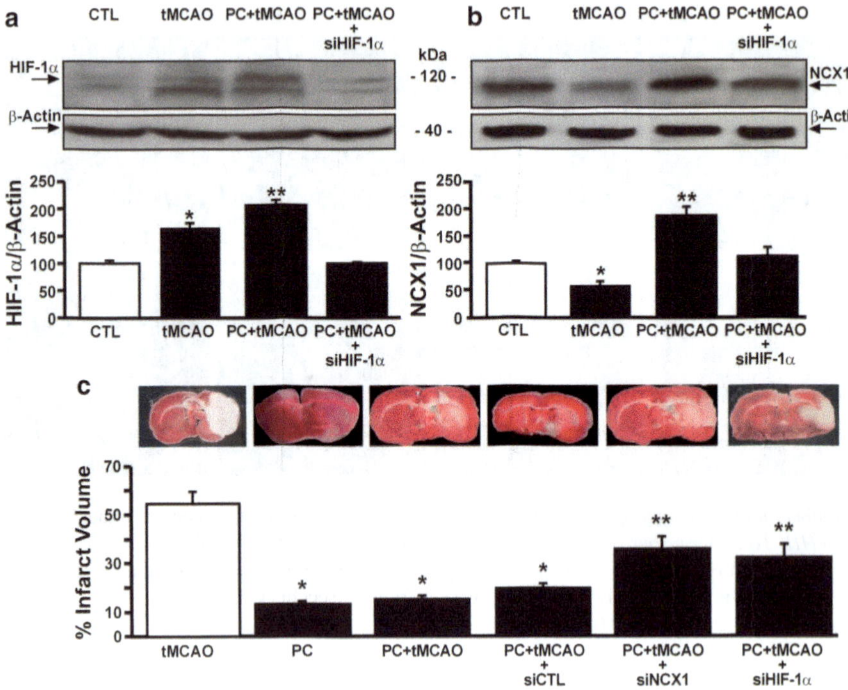

Fig. 12.4 *Silencing of HIF-1α in ischemic preconditioning reverts neuroprotection by preventing ncx1 overexpression.* Panels (**a**) and (**b**) show representative Western blots of HIF-1α and ncx1 protein levels in the cortex of sham-operated animals (CTL), rats subjected to 100 min of tMCAO followed by 24 h of reperfusion, rats subjected to 30 min of tMCAO followed up 72 h later by 100 min of tMCAO and 24 h of reperfusion (PCt+MCAO), and rats intracerebroventricularly treated with siHIF-1α and subjected to 30 min of tMCAO followed up 72 h later by 100 min of tMCAO and 24 h of reperfusion (PC+tMCAO+siHIF-1α). $*p<0.05$ vs CTL; $**p<0.05$ versus all experimental groups. Each column represents the mean±SEM ($n=3$). Panel (**c**) shows the infarct volume in rats subjected to tMCAO, PC, PC+tMCAO, PC+tMCAO+siCTL, PC+tMCAO+sincx1, and PC+tMCAO+siHIF-1α. A representative brain slice from each experimental group is shown on the top of respective column. Ischemic rats were euthanized 24 h after tMCAO. $*p<0.05$ versus tMCAO; $**p<0.05$ versus all experimental groups. Each column represents the mean±SEM of the percentage of the infarct volume compared to the ipsilateral hemisphere (Modified from Valsecchi et al. 2011)

on two HRE sequences: located at -164/-160 bp (HRE1) and -331/-327 bp (HRE2) (Valsecchi et al. 2011).

To evaluate the pathophysiological relevance of the increase in ncx1 transcription and expression induced by HIF-1α, we measured the expression of HIF-1α and NCX1 in the striatum of rats subjected to brain ischemic preconditioning and previously treated with siHIF-1α. In particular, the increase of HIF-1α occurring in the striatum of rats subjected to ischemic preconditioning was higher than that observed in rats subjected to cerebral ischemia alone (Fig. 12.4a). Analogously, NCX1 protein expression was strongly upregulated after ischemic preconditioning and tMCAO compared to cerebral ischemia alone (Fig. 12.4b). Interestingly, NCX1 upregulation induced by ischemic preconditioning and tMCAO was suppressed by silencing HIF-1α. Moreover, the relevant role played by HIF-1α and NCX1 in the neuroprotection induced by ischemic preconditioning and tMCAO was further demonstrated by the evidence that siRNA for NCX1 and for HIF-1α partially reverted the reduction in the infarct volume induced by ischemic preconditioning (Valsecchi et al. 2011) (Fig. 12.4c). Furthermore, immunohistochemical analysis performed in the striatum of preconditioned ischemic rats indicated that either NCX1 or HIF-1α protein expressions were upregulated in this region. In addition, icv infused siRNA for NCX1 and HIF-1α were effective in preventing the upregulation of each protein observed in preconditioned rats (Valsecchi et al. 2011) (Fig. 12.4c).

Collectively, the results that we reviewed demonstrate that *ncx1* gene can be included among the members of the growing family of genes activated by the transcription factor HIF-1.

12.2.3 *ncx1* Is Transcriptionally Activated by Akt1 via CREB

In the attempt to find a neuroprotective strategy to increase *ncx1* gene transcription, we investigated whether the transcriptional factor CREB, which is activated by the neurotrophic factor NGF, could play a role in the regulation of *ncx1* gene. CREB is a member of a large multigene family encompassing at least ten different structurally related cAMP-responsive transcription factors (the ATF/CREB family). All these bZip class transcription factors form homo- or heterodimers that could bind, through their leucine zipper domain, to CRE elements present on the gene promoters. In particular, the transcriptional factor CREB is activated by the phosphorylation of a serine residue, stimulated by a cyclic AMP-dependent kinase, Ca^{2+}, growth factors, and stress signals (De Cesare et al. 1999). More important, CREB has a distinct prosurvival activity and mediates the neuroprotective effect of NGF. Indeed, this neurotrophin, upon binding to its trkA receptors, causes the activation of the Ser/Thr kinase Akt which ultimately phosphorylates CREB and causes its translocation in the nucleus (Autio et al. 2011) (Fig. 12.1). Interestingly, Nicholas et al. in 1998 reported the presence of a putative CRE binding site in Kd *ncx1* promoter, whereas its existence in the Br *ncx1* promoter remained unaddressed. Recently, we found another CREB binding site on the Br *ncx1* promoter that is responsible for the increase in ncx1 expression and activity upon the activation of the neuroprotective kinase Akt1 (Formisano et al. 2008). In particular, the activation of the Akt1 by the nerve growth factor (NGF) leads to the phosphorylation of CREB and causes its translocation into the nucleus (Fukunaga and Kawano 2003), that, in turn, binds the CRE sequence at the level of the Br *ncx1* promoter increasing the transcription of this gene. Furthermore, the silencing of CREB prevents the Akt1-induced increase in ncx1 protein expression (Formisano et al. 2008). Our data demonstrated that CREB controls the gene expression of NCX1 and that NCX1 could participate to the multiple prosurvival actions of this transcriptional factor.

12.3 Conclusions

In this chapter, we reviewed the contribution that our recent research work gave to the understanding of *ncx1* gene regulation during brain ischemia. In particular, we provided evidence that NF-κB and HIF-1 could all drive an increase in *ncx1* gene expression in neurons under anoxic conditions, whereas CREB is responsible for the increase of the expression of this antiporter in response to the neurotrophic factor NGF. Further studies will be necessary to clarify how these different transcriptional factors may interact during stroke and whether their modulation by pharmacological or molecular biology tools could represent a useful strategy to decrease ischemic brain damage.

Acknowledgments This work was supported by COFIN2008, Ricerca-Sanitaria RF-FSL352059 Ricerca finalizzata 2006, Ricerca-Oncologica 2006, Progetto-Strategico 2007, Progetto Ordinario 2007, Ricerca finalizzata 2009, Ricerca-Sanitaria Progetto Ordinario by Ministero della Salute 2008, and MIUR, Progetto Operativo Nazionale (PON) 01602 all to LA.

References

L. Annunziato, G. Pignataro, G.F. Di Renzo, Pharmacology of brain Na^+/Ca^{2+} exchanger: from molecular biology to therapeutic perspectives. Pharmacol. Rev. **56**, 633–654 (2004)

L. Annunziato, G. Pignataro, F. Boscia, R. Sirabella, L. Formisano, M. Saggese, O. Cuomo, R. Gala, A. Secondo, D. Viggiano, P. Molinaro, V. Valsecchi, A. Tortiglione, A. Adornetto, A. Scorziello, M. Cataldi, GF. Di Renzo

L. Annunziato, P. Molinaro, A. Secondo, A. Panaccione, A. Scorziello, G. Pignataro, O. Cuomo, R. Sirabella, F.

Boscia, A. Spinali, G. Di Renzo, The Na^+/Ca^{2+} exchanger: a target for therapeutic intervention in cerebral ischemia, in *New Strategies in Stroke Intervention*, ed. by L. Annunziato (Springer, Totowa, 2009), pp. 65–87
H. Autio, K. Matlik, T. Rantamaki, L. Lindemann, M.C. Hoener, M. Chao, U. Arumae, E. Castren, Acetylcholinesterase inhibitors rapidly activate Trk neurotrophin receptors in the mouse hippocampus. Neuropharmacology **61**, 1291–1296 (2011)
K.V. Barnes, G. Cheng, M.M. Dawson, D.R. Menick, Cloning of cardiac, kidney, and brain promoters of the feline NCX1 gene. J. Biol. Chem. **272**, 11510–11517 (1997)
M. Bergeron, A.Y. Yu, K.E. Solway, G.L. Semenza, F.R. Sharp, Induction of hypoxia-inducible factor-1 (HIF-1) and its target genes following focal ischaemia in rat brain. Eur. J. Neurosci. **11**, 4159–4170 (1999)
F. Boscia, R. Gala, G. Pignataro, A. de Bartolomeis, M. Cicale, A. Ambesi-Impiombato, G. Di Renzo, L. Annunziato, Permanent focal brain ischemia induces isoform-dependent changes in the pattern of Na^+/Ca^{2+} exchanger gene expression in the ischemic core, peri-infarct area, and intact brain regions. J. Cereb. Blood Flow Metab. **26**, 502–517 (2006)
R.K. Bruick, Oxygen sensing in the hypoxic response pathway: regulation of the hypoxia-inducible transcription factor. Genes Dev. **17**, 2614–2623 (2003)
J.C. Chavez, J.C. LaManna, Activation of hypoxia-inducible factor-1 in the rat cerebral cortex after transient global ischemia: potential role of insulin-like growth factor-1. J. Neurosci. **22**, 8922–8931 (2002)
G. Cheng, T.P. Hagen, M.L. Dawson, K.V. Barnes, D.R. Menick, The role of GATA, CArG, E-box, and a novel element in the regulation of cardiac expression of the Na^+-Ca^{2+} exchanger gene. J. Biol. Chem. **274**, 12819–12826 (1999)
S.T. Crews, Control of cell lineage-specific development and transcription by bHLH-PAS proteins. Genes Dev. **12**, 607–620 (1998)
D. De Cesare, G.M. Fimia, P. Sassone-Corsi, Signaling routes to CREM and CREB: plasticity in transcriptional activation. Trends Biochem. Sci. **24**, 281–285 (1999)
L. Formisano, M. Saggese, A. Secondo, R. Sirabella, P. Vito, V. Valsecchi, P. Molinaro, G. Di Renzo, L. Annunziato, The two isoforms of the Na^+/Ca^{2+} exchanger, NCX1 and NCX3, constitute novel additional targets for the prosurvival action of Akt/protein kinase B pathway. Mol. Pharmacol. **73**, 727–737 (2008)
K. Fukunaga, T. Kawano, Akt is a molecular target for signal transduction therapy in brain ischemic insult. J. Pharmacol. Sci. **92**, 317–327 (2003)
N. Gabellini, S. Bortoluzzi, G.A. Danieli, E. Carafoli, The gene promoter of human Na^+/Ca^{2+} exchanger isoform 3 (SLC8A3) is controlled by cAMP and calcium. Ann. N. Y. Acad. Sci. **976**, 282–284 (2002)
N. Gabellini, S. Bortoluzzi, G.A. Danieli, E. Carafoli, Control of the Na^+/Ca^{2+} exchanger 3 promoter by cyclic adenosine monophosphate and Ca^{2+} in differentiating neurons. J. Neurochem. **84**, 282–293 (2003)
G.T. Gibney, J.H. Zhang, R.M. Douglas, G.G. Haddad, Y. Xia, Na^+/Ca^{2+} exchanger expression in the developing rat cortex. Neuroscience **112**, 65–73 (2002)
K.L. Golden, Q.I. Fan, B. Chen, J. Ren, J. O'Connor, J.D. Marsh, Adrenergic stimulation regulates Na^+/Ca^{2+}exchanger expression in rat cardiac myocytes. J. Mol. Cell. Cardiol. **32**, 611–620 (2000)
R. Gomez-Villafuertes, B. Torres, J. Barrio, M. Savignac, N. Gabellini, F. Rizzato, B. Pintado, A. Gutierrez-Adan, B. Mellstrom, E. Carafoli, J.R. Naranjo, Downstream regulatory element antagonist modulator regulates Ca^{2+} homeostasis and viability in cerebellar neurons. J. Neurosci. **25**, 10822–10830 (2005)
M.S. Hayden, S. Ghosh, Shared principles in NF-kappaB signaling. Cell **132**, 344–362 (2008)
B.H. Jiang, E. Rue, G.L. Wang, R. Roe, G.L. Semenza, Dimerization, DNA binding, and transactivation properties of hypoxia-inducible factor 1. J. Biol. Chem. **271**, 17771–17778 (1996)
P.J. Kallio, I. Pongratz, K. Gradin, J. McGuire, L. Poellinger, Activation of hypoxia-inducible factor 1alpha: posttranscriptional regulation and conformational change by recruitment of the ARNT transcription factor. Proc. Natl. Acad. Sci. U. S. A. **94**, 5667–5672 (1997)
S.L. Lee, A.S. Yu, J. Lytton, Tissue-specific expression of Na^+-Ca^{2+} exchanger isoforms. J. Biol. Chem. **269**, 14849–14852 (1994)
A.P. Levy, N.S. Levy, S. Wegner, M.A. Goldberg, Transcriptional regulation of the rat vascular endothelial growth factor gene by hypoxia. J. Biol. Chem. **270**, 13333–13340 (1995)
Z. Li, S. Matsuoka, L.V. Hryshko, D.A. Nicoll, M.M. Bersohn, E.P. Burke, R.P. Lifton, K.D. Philipson, Cloning of the NCX2 isoform of the plasma membrane Na^+-Ca^{2+} exchanger. J. Biol. Chem. **269**, 17434–17439 (1994)
L. Li, D. Guerini, E. Carafoli, Calcineurin controls the transcription of Na^+/Ca^{2+} exchanger isoforms in developing cerebellar neurons. J. Biol. Chem. **275**, 20903–20910 (2000)
L.L. Li, L.N. Sun, H.Y. Zhou, Z.B. Li, X.L. Wang, Selective alteration of expression of Na^+/Ca^{2+} exchanger isoforms after transient focal cerebral ischemia in rats. Neurosci. Lett. **404**, 249–253 (2006)
J. Lu, X.Y. Tong, X.L. Wang, Altered gene expression of Na^+/Ca^{2+} exchanger isoforms NCX1, NCX2 and NCX3 in chronic ischemic rat brain. Neurosci. Lett. **332**, 21–24 (2002)
C. Matrone, G. Pignataro, P. Molinaro, C. Irace, A. Scorziello, G.F. Di Renzo, L. Annunziato, HIF-1alpha reveals a binding activity to the promoter of iNOS gene after permanent middle cerebral artery occlusion. J. Neurochem. **90**, 368–378 (2004)
J.G. Muller, J.T. Thompson, A.M. Edmonson, M.S. Rackley, H. Kasahara, S. Izumo, T.C. McQuinn, D.R. Menick, T.X. O'Brien, Differential regulation of the cardiac sodium calcium exchanger promoter in adult

and neonatal cardiomyocytes by Nkx2.5 and serum response factor. J. Mol. Cell. Cardiol. **34**, 807–821 (2002)

S.B. Nicholas, K.D. Philipson, Cardiac expression of the Na^+/Ca^{2+} exchanger NCX1 is GATA factor dependent. Am. J. Physiol. **277**, H324–H330 (1999)

S.B. Nicholas, W. Yang, S.L. Lee, H. Zhu, K.D. Philipson, J. Lytton, Alternative promoters and cardiac muscle cell-specific expression of the Na^+/Ca^{2+} exchanger gene. Am. J. Physiol. **274**, H217–H232 (1998)

D.A. Nicoll, S. Longoni, K.D. Philipson, Molecular cloning and functional expression of the cardiac sarcolemmal Na^+-Ca^{2+} exchanger. Science **250**, 562–565 (1990)

D.A. Nicoll, B.D. Quednau, Z. Qui, Y.R. Xia, A.J. Lusis, K.D. Philipson, Cloning of a third mammalian Na^+-Ca^{2+} exchanger, NCX3. J. Biol. Chem. **271**, 24914–24921 (1996)

M. Papa, A. Canitano, F. Boscia, P. Castaldo, S. Sellitti, H. Porzig, M. Taglialatela, L. Annunziato, Differential expression of the Na^+-Ca^{2+} exchanger transcripts and proteins in rat brain regions. J. Comp. Neurol. **461**, 31–48 (2003)

N.D. Perkins, T.D. Gilmore, Good cop, bad cop: the different faces of NF-kappaB. Cell Death Differ. **13**, 759–772 (2006)

G. Pignataro, R. Gala, O. Cuomo, A. Tortiglione, L. Giaccio, P. Castaldo, R. Sirabella, C. Matrone, A. Canitano, S. Amoroso, G. Di Renzo, L. Annunziato, Two sodium/calcium exchanger gene products, NCX1 and NCX3, play a major role in the development of permanent focal cerebral ischemia. Stroke **35**, 2566–2570 (2004)

G. Pignataro, A. Scorziello, G. Di Renzo, L. Annunziato, Post-ischemic brain damage: effect of ischemic preconditioning and postconditioning and identification of potential candidates for stroke therapy. FEBS J. **276**, 46–57 (2009)

B.D. Quednau, D.A. Nicoll, K.D. Philipson, Tissue specificity and alternative splicing of the Na^+/Ca^{2+} exchanger isoforms NCX1, NCX2, and NCX3 in rat. Am. J. Physiol. **272**, C1250–C1261 (1997)

M. Sakaue, H. Nakamura, I. Kaneko, Y. Kawasaki, N. Arakawa, H. Hashimoto, Y. Koyama, A. Baba, T. Matsuda, Na^+-Ca^{2+} exchanger isoforms in rat neuronal preparations: different changes in their expression during postnatal development. Brain Res. **881**, 212–216 (2000)

T. Scheller, A. Kraev, S. Skinner, E. Carafoli, Cloning of the multipartite promoter of the sodium-calcium exchanger gene NCX1 and characterization of its activity in vascular smooth muscle cells. J. Biol. Chem. **273**, 7643–7649 (1998)

A. Secondo, R.I. Staiano, A. Scorziello, R. Sirabella, F. Boscia, A. Adornetto, V. Valsecchi, P. Molinaro, L.M. Canzoniero, G. Di Renzo, L. Annunziato, BHK cells transfected with NCX3 are more resistant to hypoxia followed by reoxygenation than those transfected with NCX1 and NCX2: possible relationship with mitochondrial membrane potential. Cell Calcium **42**, 521–535 (2007)

G.L. Semenza, Involvement of oxygen-sensing pathways in physiologic and pathologic erythropoiesis. Blood **114**, 2015–2019 (2009)

G.L. Semenza, M.K. Nejfelt, S.M. Chi, S.E. Antonarakis, Hypoxia-inducible nuclear factors bind to an enhancer element located 3′ to the human erythropoietin gene. Proc. Natl. Acad. Sci. U. S. A. **88**, 5680–5684 (1991)

G.L. Semenza, B.H. Jiang, S.W. Leung, R. Passantino, J.P. Concordet, P. Maire, A. Giallongo, Hypoxia response elements in the aldolase A, enolase 1, and lactate dehydrogenase A gene promoters contain essential binding sites for hypoxia-inducible factor 1. J. Biol. Chem. **271**, 32529–32537 (1996)

R. Sirabella, A. Secondo, A. Pannaccione, A. Scorziello, V. Valsecchi, A. Adornetto, L. Bilo, G. Di Renzo, L. Annunziato, Anoxia-induced NF-kappaB-dependent upregulation of NCX1 contributes to Ca^{2+} refilling into endoplasmic reticulum in cortical neurons. Stroke **40**, 922–929 (2009)

L. Smith, J.B. Smith, Regulation of sodium-calcium exchanger by glucocorticoids and growth factors in vascular smooth muscle. J. Biol. Chem. **269**, 27527–27531 (1994)

S. Taie, J. Ono, Y. Iwanaga, S. Tomita, T. Asaga, K. Chujo, M. Ueki, Hypoxia-inducible factor-1 alpha has a key role in hypoxic preconditioning. J. Clin. Neurosci. **16**, 1056–1060 (2009)

K. Tanimoto, Y. Makino, T. Pereira, L. Poellinger, Mechanism of regulation of the hypoxia-inducible factor-1 alpha by the von Hippel-Lindau tumor suppressor protein. EMBO J. **19**, 4298–4309 (2000)

V. Valsecchi, G. Pignataro, A. Del Prete, R. Sirabella, C. Matrone, F. Boscia, A. Scorziello, M.J. Sisalli, E. Esposito, N. Zambrano, G. Di Renzo, L. Annunziato, NCX1 is a novel target gene for hypoxia-inducible factor-1 in ischemic brain preconditioning. Stroke **42**, 754–763 (2011)

G.L. Wang, G.L. Semenza, Characterization of hypoxia-inducible factor 1 and regulation of DNA binding activity by hypoxia. J. Biol. Chem. **268**, 21513–21518 (1993)

L. Yu, R.A. Colvin, Regional differences in expression of transcripts for Na^+/Ca^{2+} exchanger isoforms in rat brain. Brain Res. Mol. Brain Res. **50**, 285–292 (1997)

Part V

Regulatory Mechanisms of NCX

Metabolic Regulation of the Squid Nerve Na^+/Ca^{2+} Exchanger: Recent Developments

13

Luis Beaugé, Reinaldo DiPolo, Mariana Bollo, Alexandra Cousido, Graciela Berberián, and Alberto Podjarny

Abstract

In squid nerves, MgATP modulation of the Na^+/Ca^{2+} exchanger requires the presence of a cytosolic protein which becomes phosphorylated during the process. This factor has been recently identified. Mass spectroscopy and Western blot analysis established that it is a member of the lipocalin superfamily of lipid-binding proteins (LBP or FABP) of 132 amino acids. We called it regulatory protein of squid nerve sodium/calcium exchanger (ReP1-NCXSQ, access to GenBank EU981897).

ReP1-NCXSQ was cloned, expressed, and purified. Circular dichroism, far-UV, and infrared spectroscopy suggest a secondary structure, predominantly of beta-sheets. The tertiary structure prediction provides ten beta-sheets and two alpha-helices, characteristic of most of LPB. Functional experiments showed that, to be active, ReP1-NCXSQ must be phosphorylated by MgATP, through the action of a kinase present in the plasma membrane. Moreover, PO4-ReP1-NCXSQ can stimulate the exchanger in the absence of ATP. An additional crucial observation was that, in proteoliposomes containing only the purified Na^+/Ca^{2+} exchanger, PO4-ReP1-NCXSQ promotes activation; therefore, this upregulation has no other requirement than a lipid membrane and the incorporated exchanger protein.

L. Beaugé (✉) • M. Bollo • G. Berberián
Laboratorio de Biofísica, Instituto de Investigación Médica "Mercedes y Martín Ferreyra" (INIMEC-CONICET), Córdoba, Argentina
e-mail: lbeauge@immf.uncor.edu

R. DiPolo
Laboratorio de Fisiología Celular, Centro de Biofísica, Instituto Venezolano de Investigaciones Científicas (IVIC), Caracas, Venezuela

A. Cousido • A. Podjarny
Department of Structural Biology and Genomics, IGBMC, CNRS, INSERM, Université de Strasbourg, Illkirch, France

L. Annunziato (ed.), *Sodium Calcium Exchange: A Growing Spectrum of Pathophysiological Implications*, Advances in Experimental Medicine and Biology 961, DOI 10.1007/978-1-4614-4756-6_13,

Recently, we solved the crystal structure of ReP1-NCXSQ which was as predicted: a "barrel" consisting of ten beta-sheets and two alpha-helices. Inside the barrel is the fatty acid coordinated by hydrogen bonds with Arg126 and Tyr128. Point mutations showed that neither Tyr20Ala, Arg58Val, Ser99Ala, nor Arg126Val is necessary for protein phosphorylation or activity. On the other hand, Tyr128 is essential for activity but not for phosphorylation. We can conclude that (1) for the first time, a role of an LBP is demonstrated in the metabolic regulation of an ion exchanger; (2) phosphorylation of this LBP can be separated from the activation capacity; and (3) Tyr128, a candidate to coordinate lipid binding inside the barrel, is essential for activity.

Keywords

Na^+/Ca^{2+} exchanger • Metabolic regulation • Regulatory proteins • Lipid-binding protein • Structure-function • Squid nerve • Membrane transport

13.1 Introduction

Sodium/calcium exchangers comprise a large family of constitutive membrane proteins in charge of extruding Ca^{2+} from the cells in exchange for extracellular Na^+. This system works reversibly and is electrogenic (three Na^+ ions are transported for one Ca^{2+} ion); therefore, the transport rate is affected by the transmembrane potential. They are widely distributed in the organism, appear in practically all animal cells, and constitute a main mechanism for cytosolic Ca^{2+} clearance (see Blaustein and Lederer 1999; DiPolo and Beaugé 2006).

Cation movement has been demonstrated to have four types of modes: the forward Ca_i^{2+}/Na_o^+ and reverse Na_i^+/Ca_o^{2+} and the Na_i^+/Na_o^+ and Ca_i^{2+}/Ca_o^{2+} exchanges. The forward mode is responsible for Ca^{2+} clearance; the reverse seems to work during the depolarizing plateau of cardiac action potential, and it is also important during pathological conditions such as anoxia and ischemia. The Na_i^+/Na_o^+ exchange has been extremely useful in the study of Ca_i^{2+} regulation (see below).

An important property of the exchangers is that they are highly regulated by intracellular ions (ionic regulation) and the metabolic state of the cell (metabolic regulation), including levels of ATP, hormones, and activation-inhibition of the signaling pathway. These regulations imply, in some cases, modifications of the exchange activity and, in others, alterations of the protein expression and docking into the membrane where, associated with other transporters (Na^+/K^+-ATPase and Na^+/H^+ exchanger) and enzymes (kinases and phosphatases), they form functional supramolecular complexes (Bers and Despa 2009; Schulze et al. 2003; Hilgemann 2007; McLaughlin et al. 2002; Berberián et al. 2009b). The personal data presented here deal with the metabolic regulation of the squid Na^+/Ca^{2+} exchanger, a process that is intimately related to ionic modulation. Therefore, before presenting the data, we will review the most important features of these two regulations, emphasizing the comparison between the squid and mammalian exchangers.

13.2 Ionic Regulation

13.2.1 Ca_i^{2+} Activation

In this section, we will review the main, most up-to-date aspects of the interaction of the Na^+/Ca^{2+} exchanger with cytosolic Ca^{2+}, Na^+, and H^+. In Na^+/Ca^{2+} exchangers, the binding of Ca^{2+} to intracellular activating site/s, located in the large intracellular regulatory loop, is essential for function.

This was discovered in dialyzed squid axons (DiPolo 1979) and later found in all species and members of the exchanger family; this Ca^{2+} is not transported across the membrane and is required for all modes of translocation. Interestingly, there is one exchanger, in *Drosophila*, where the effects of cytosolic Ca^{2+} are actually inhibitory (Hryshko et al. 1996). In fact, it is now known that all exchangers have not just one but two Ca^{2+} regulatory binding domains, CBD1 and CBD2, each binding more than one Ca^{2+} ion with different affinities and upregulation abilities, and that includes *Drosophila* (Hilge et al. 2007; Ottolia et al. 2009). In fact, it seems that there is interplay between CBD1 and CBD2, leading to the structural changes influencing transport activity; a different type of conformational change could explain the opposite effects observed in *Drosophila* compared to the other exchangers (Wu et al. 2009). In other words, in general, without cytosolic Ca^{2+}, the loop would be in an "inhibitory conformation," whereas in *Drosophila*, it would be in a "stimulatory conformation." This hypothesis agrees with the observation that removal of the loop in NCX1 results in marked activation of the exchanger. Further support may be found in experiments removing the loop in *Drosophila*, which has not yet been done.

The overall apparent affinities of Ca^{2+} regulatory sites for Ca^{2+} go from below 100 nM in mammals to about 400 nM in squid (DiPolo and Beaugé 2006). It must be stressed that estimation of those affinities is not easy, particularly when exchange currents are measured, because in some instances the intracellular transport and regulatory Ca^{2+} sites coexist (forward mode) and, in others, $[Ca^{2+}]$ building up near the intracellular sites of the exchanger cannot be ruled out (reverse mode). One way around this problem is to look at an exchange mode involving only transport of Na^+, that is, the electro-silent Na^+/Na^+ exchange estimated from isotopic fluxes, as is done in dialyzed squid axons (DiPolo and Beaugé 2006). In this case, Ca_i^{2+} activation would represent only Ca_i^{2+} binding to the regulatory sites. It must be pointed out here that the Ca^{2+}-binding domains of the squid nerve exchanger have not yet been identified.

13.2.2 Na_i^+ Inhibition

The countertransport system is based on competition of the transported species for the same binding site or for the same conformation of the carrier. As expected, in the Na^+/Ca^{2+} exchangers, intracellular Na^+ inhibits transport by displacing Ca_i^{2+} from its transport sites but, unexpectedly, also by acting on other site/s. The latter effect, named Na_i^+ inactivation (Hilgemann et al. 1992a), is observed both in pre-steady- and steady-state conditions (Blaustein and Lederer 1999; DiPolo and Beaugé 2006). Na_i^+ inactivation disappears when the cytoplasmic side of the membrane is treated with chymotrypsin, indicating that it takes place at the large intracellular loop. The N-terminus of the loop contains a region called "XIP" (exchanger inhibitory peptide) that is related to Na_i^+ inactivation. XIP is a synthetic peptide with an amino acid sequence similar to the XIP region, which completely inactivates the exchanger when applied intracellularly (Li et al. 1991). Mutations of NCX1 that eliminate Na_i^+ inactivation produce stronger interactions between PtdIns-4,5-P_2 (an up-regulator of NCX1) and XIP. Conversely, those interactions are weakened by mutations that promote Na_i^+ inactivation (Matsuoka et al. 1997; He et al. 2000); that is, this region functions as an auto-inhibitory domain that shifts the active-inactive states of the Na^+/Ca^{2+} exchanger. Calcium ions antagonize Na_i^+ inactivation, and all the experimental data available point to the idea that inhibition by Na_i^+ responds to a reduction of the affinity for Ca^{2+} at its regulatory sites (DiPolo and Beaugé 2006).

13.2.3 Intracellular Proton Inhibition

Inhibition of the Na^+/Ca^{2+} exchanger by intracellular protons in the presence of internal Na^+ was first described in injected axons (Baker and McNaughton 1977), later confirmed in axons under internal dialysis by DiPolo and Beaugé (1982). This inhibition was further characterized in mouse heart (Wakabayashi and Goshima 1981), dog cardiac sarcolemmal vesicles (Philipson et al. 1982), reconstituted Na^+/Ca^{2+}

exchanger in liposomes (Khananshvili et al. 1996), and myocyte excised patches (Doering and Lederer 1993, 1994). There is a remarkable similarity in all these preparations regarding the profound dependency of exchange activity on intracellular protons near physiological pH_i. Also, chymotrypsin digestion of the regulatory loop dramatically reduces the effects of H_i^+, indicating that these interactions take place on, or are related to, that loop. The exquisite sensitivity of the exchanger to pH_i's around 7.2–7.4 suggests that H_i^+ could play a significant role in Ca^{2+} transport modulation under physiopathological conditions. On the other hand, intracellular protons also inhibit the Na^+/Ca^{2+} exchanger in the complete absence of Na_i^+ (proton inhibition per se); this was first shown in giant excised membrane patches from cardiac guinea pig myocytes (Doering and 1993) and confirmed later in dialyzed squid axons (DiPolo and Beaugé 2002). In Na_i^+-free conditions, intracellular Ca^{2+} competitively antagonizes H_i^+ inhibition, and the experiments strongly indicate that the Ca_i^{2+} regulatory sites are involved (Hilgemann et al. 1992b; DiPolo and Beaugé 2002).

13.2.4 ($H_i^+ + Na_i^+$) Synergic Inhibition

The inhibition of Na^+/Ca^{2+} exchangers by H_i^+ is enhanced by Na_i^+, while Na_i^+ inhibition, both at pre-steady (Na_i^+ inactivation) and steady state, is increased by H_i^+. In other words, there is a synergic ($H_i^+ + Na_i^+$) impairment of the exchanger function. Also, there are similarities between Na_i^+-dependent inactivation and H_i^+ inhibition: both develop with a slow time course (seconds) and both are prevented by chymotrypsin digestion. A Na_i^+-H_i^+ synergism predicts that internal alkalinization should reduce Na_i^+ inhibition. This was confirmed in cardiac excised giant patches (Hilgemann et al. 1992b), where the fast Na_i^+ inactivation and the reduction in the steady-state exchange currents seen at pH_i 6.8 are completely overcome by raising internal pH to 7.8. Intracellular ($H_i^+ + Na_i^+$) synergism in the inhibition of the Na^+/Ca^{2+} exchanger in steady state has also been observed in dialyzed squid axons, where the $K_{0.5}$ for Na_i^+ inhibition goes from 10 mM at pH_i 6.9 to 90 mM at pH_i 8.8 (DiPolo and Beaugé 2002). There is no question that all these effects are produced via a reduction in the affinity of the Ca_i^{2+} regulatory sites for Ca^{2+} and, as expected, are antagonized by Ca_i^{2+}. In summary, intracellular ionic regulation of the Na^+/Ca^{2+} exchangers can be reduced to a Ca^{2+} stimulation at the Ca_i^{2+} regulatory sites and a H^+ plus synergic ($H^+ + Na^+$) inhibition. Regardless of where H^+ and Na^+ ions bind, the final target is the Ca_i^{2+} regulatory sites (DiPolo and Beaugé 2006).

13.3 Metabolic Regulation

An energy-dependent upregulation of the Na^+/Ca^{2+} exchanger has been known since the early work of Baker and Glitsch (1973) in injected squid axons, where reduction in the levels of ATP inhibited a large fraction of the Na^+-dependent Ca^{2+} efflux. This observation was extended to axons subjected to internal dialysis, showing that inhibition of the exchanger following ATP removal was reversed by adding the nucleotide to the dialysis solution (DiPolo 1973, 1974; Blaustein 1977). At pH_i 7.3, 40 mM Na_i^+ and 0.8 μM Ca_i^{2+}, the ATP-dependent forward Na^+/Ca^{2+} exchange is about 80% of the total flux, and the $K_{1/2}$ for ATP about 250 μM (DiPolo 1974; DiPolo and Beaugé 1979). In addition, the simultaneous presence of Mg^{2+} was required (DiPolo and Beaugé 1984) while non-phosphorylating ATP analogues were ineffective (DiPolo 1976; DiPolo and Beaugé 1999). On the other hand, the phosphorylating analogue ATP-γ-S, substrate for kinases but not for ATPases, is an even better up-regulator than ATP. This may be explained because a protein subjected to plain phosphorylation is more sensitive to phosphatases than when thio-phosphorylated (DiPolo and Beaugé 1999, 2006).

MgATP stimulation was also seen in the reverse Na_i^+/Ca_o^{2+} exchanger (DiPolo 1979; DiPolo and Beaugé 1984) and Ca^{2+}/Ca^{2+} (Blaustein and Hodgkin 1969; Baker et al. 1969), and Na^+/Na^+ exchange modes (DiPolo and Beaugé 1984, 1987). An important aspect of MgATP modulation of the squid nerve exchanger is the absolute requirement of a cytosolic soluble protein (CSRP from cytosolic soluble regulatory protein). For modulation to be effective, the SCRP must become phosphorylated; actually, the phosphorylated CSRP can upregulate NCXSQ1 even in the absence of ATP (DiPolo and Beaugé 2006). Phosphorylation was obtained by incubating SCRP with squid nerve vesicles and MgATP. Also important when comparing metabolic modulation in the squid with other isoforms, like NCX1, is that, in the squid, it is not associated to phosphoinositides (DiPolo et al. 2000).

The other preparation where the metabolic regulation of the Na^+/Ca^{2+} exchanger was widely explored is the mammalian heart (NCX1 isoform). This was done by using heart cells or NCX1 overexpressed in alien cells. The approaches consisted in measuring isotopic fluxes in inside-out cardiac microsomal vesicles, liposomes with purified NCX1 incorporated, or whole alien cells expressing NCX1 and also with the powerful giant patch technique, both in heart cells, and in alien cells overexpressing NCX1 (Philipson et al. 1982; Hilgemann et al. 1992b; Luciani et al. 1991, 1995; Condrescu et al. 1995; Hilgemann 1996; Hilgemann and Ball 1996; Berberián and Beaugé 1996; Hilgemann 1997; Berberián et al. 1998; He et al. 2000; Asteggiano et al. 2001; Posada et al. 2007; Hilgemann 2007; Yaradanakul et al. 2007).

MgATP modulation of NCX1 has two important properties not found in squid nerve: (1) it does not require a cytosolic regulatory protein, and (2) it is exerted through the production of PtdInl,4,5-P2. Actually, PtdInl,4,5-P2, in the absence of the nucleotide, mimics the effects of the nucleotide in all details (Hilgemann and Ball 1996; Asteggiano et al. 2001; Posada et al. 2007). Furthermore, the inositide responsible for upregulation is not the total present in the membrane, but just that bound to NCX1 (Asteggiano et al. 2001; He et al. 2000; Ottolia et al. 2007). It is likely that PtdInl,4,5-P2 binds into the XIP region of the "regulatory loop" (He et al. 2000; Ottolia et al. 2007), although there are regions in that loop that also qualify to bind the phosphoinositide (Berberián et al. 2009b). In addition, and concurring with these observations, a supramolecular complex is formed between at least PtdInl,4,5-P2, a PtdInl,4,5-P2-kinase, and NCX1. This complex is so important that when the PtdInl,4,5-P2-kinase is detached from NCX1, as what happens with dithiothreitol treatment, optimal stimulation of the exchanger is impaired (Forcato et al. 2010). A supramolecular complex between NCX1 and other enzymes and proteins related to regulation and signal transduction, including PKC, PKA, a protein kinase A-anchoring protein (mAKAP), and two serine/threonine protein phosphatases (PP1 and PP2A), has already been described (Schulze et al. 2003). The latter observation may be related to a new role of the exchanger acting on the signaling cascade; however, the evidence for that role is still controversial (Zhang and Hancox 2009).

As can be seen from the above, the metabolic pathways for MgATP upregulation of the squid (NCXSQ1) and mammalian heart (NCX1) exchangers go through different metabolic pathways. However, the remarkable observation is that in both cases the final effects concur: both protect the Ca_i^{2+} regulatory sites against H_i^+ and ($H_i^+ + Na_i^+$). This takes place in steady-state and pre-steady-state conditions, for example, the transient Na_i^+-dependent inactivation and the H_i^+ and Na_i^+ inhibition of steady fluxes. This is supported by the fact that, at alkaline pH_i, Na_i^+ inhibition does not arise, MgATP (NCXSQ1 and NCX1) and PtdIns-4,5-P_2 (NCX1) do not modify the exchange fluxes, and the overall affinity of the Ca_i^{2+} regulatory sites remains high regardless of the $[Na^+]_i$ (Blaustein and Lederer 1999; DiPolo and Beaugé 2006).

13.4 Identification and Properties of the Soluble Cytosolic Regulatory Protein (SCRP) in Squid Nerve

13.4.1 Identification

The fact that, during upregulation of NCXSQ1, the SCRP must become phosphorylated, enabled it to be identified. Peptide sequences, obtained by mass spectroscopy of a ~13-kDa phosphoprotein

protein band isolated from the 30 to 10 kDa cytosolic fraction of squid optic ganglia, were matched against a squid expressed sequence tag (EST) database containing approximately 23,000 nucleotide sequences (see Berberián et al. 2009a). Four peptides matched a predicted amino acid sequence of 396 cases, encoding 132 amino acids. A polyclonal antibody, based on the amino acid sequence obtained from the squid gene database, recognized both ReP1-NCXSQ and phosphorylated SCRP, while it failed to detect any protein in nerve membrane vesicles. This is further evidence that the gene and the protein are one and the same. Initial alignment by BLAST search indicated that this amino acid sequence matched members of the lipocalin superfamily of lipid-binding proteins; the theoretical molecular weight and pI were 14.8 kDa and 5.85, respectively (Berberián et al. 2009a); both predictions agree with those of SCRP.

The new protein was named regulatory protein of the squid Na^+/Ca^{2+} exchanger (ReP1-NCXSQ; GenBank accession number EU981897); the number 1 implies the possibility that other proteins may have a similar function. When it was expressed in BL21 *E. coli,* purified and checked for function, it was observed that ReP1-NCXSQ behaved exactly as SCRP (Berberián et al. 2009a; Raimunda et al. 2009). Further sequence alignments showed a strong similarity, both in sequence and structure, to "myelin P2 proteins" from vertebrates (58% identity and 25% similarity); on the other hand, both parameters were lower with invertebrate FABPs: 23% and 7.5%, respectively. Alignment of ReP1-NCXSQ to CRABPs (cellular retinoic acid-binding proteins) had a 70% similarity and 37% identity with invertebrates and 66% and 30% with vertebrates. The closest sequence found corresponded to a CRABP from shrimp.

13.4.2 Structure of ReP1-NCXSQ

The data obtained from far- and near-UV CD and infrared spectra analysis, the latter both original and Fourier's self-deconvoluted, indicated a highly structured protein with a well-defined tertiary structure in the native state, containing a sizable number of β-strands in the secondary structure. On the whole, the data agree with those obtained for the intestinal fatty acid-binding protein (I-FABP) (Clérico and Ermácora 2001), human liver and heart (Tanfani et al. 1995), mammalian adipocytes (Gericke et al. 1997), and avian liver (Nolan et al. 2003). Thus, the infrared data suggested 63% of beta-sheets and 18% of alpha-helices. Taking together, the spectral data coincide with that expected from the structure predicted on the basis of the primary amino acid sequence, using as template the myelin P2 LBP from vertebrates (Protein Data Bank entry 1YIV). The new cartoon representation is illustrated in Fig. 13.1a.

The crystallization of ReP1-NCXSQ was performed, screening initially with the robot at a protein concentration of 10 mg/ml in a solution containing 10 mM Tris (pH 7.5 at RT) and 150 mM NaCl. A total of seven screens yielded crystals in six conditions.

The final condition, after a manual screen of the best condition (INDEX n°96) obtained with the robot, had 200 mM KCl and 35% PEG 2000 MME in the well. The structure was solved by molecular replacement using the Protein Data Bank entry 1YIV. A first data set, collected at 1.8-Å resolution, was fitted with the ARP/wARP program and refined with the REFMAC5 program; a second set of data was collected and refined at 1.28-Å resolution (REFMAC5 and PHENIX programs) to an R-factor of 15% and an R-free of 19%. A new cartoon representation of the crystal structure is shown in Fig. 13.1b, which looks very much like that predicted (Fig. 13.1a), folding in a β-sandwich, with ten β-sheets and two alpha-coils, typical of lipid-binding proteins.

Mass spectrometry identified the 16 carbons fatty acid in the barrel of the *E. coli*-expressed ReP1-NCXSQ; tentative either palmitic or palmitoleic acid. This coincides with previous observations in a human muscle FABP expressed in similar conditions (Young et al. 1994). In the crystal, this lipid is bent in a U-shape conformation, like that found in palmitic acid found in FABPs from human heart, adipocytes, epidermis, ileum, and brain (Furuhashi and Hotamisligil

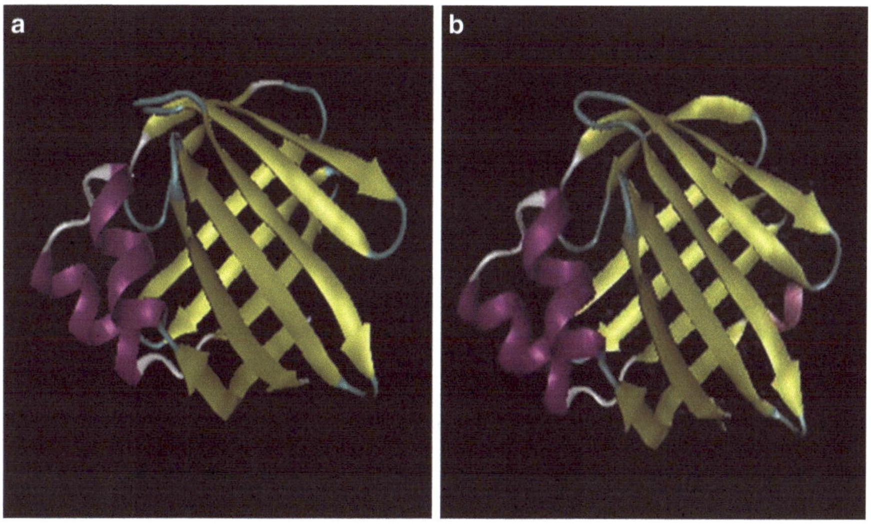

Fig. 13.1 *Predicted and solved structure of ReP1-NCXSQ*. (**a**) Predicted structure of ReP1-NCXSQ obtained with the HERMES Swiss Project (hermes@isb.admin.ch) on the basis of the amino acid sequence. (**b**) Solved crystal structure. Note that both are similar and correspond to a fatty acid-binding protein (see text for details)

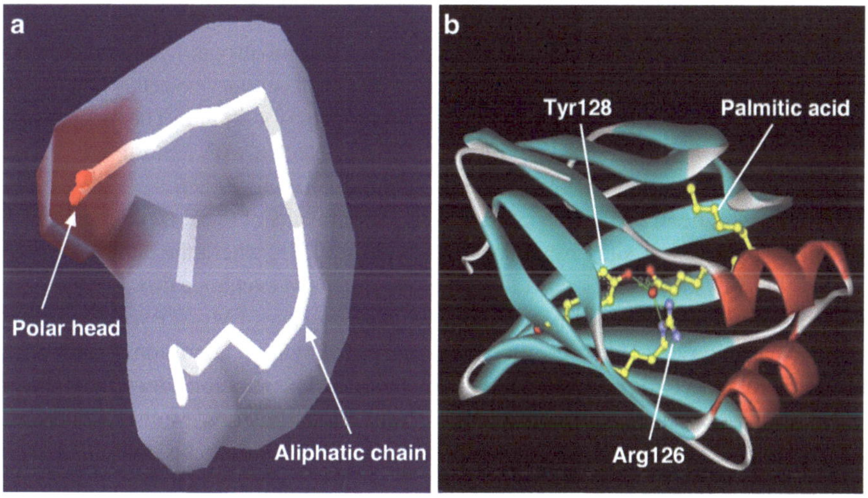

Fig. 13.2 (**a**) *Crystal structure of palmitic acid inserted in ReP1-NCXSQ*. The structure was solved with the HERMES Swiss Project (hermes@isb.admin.ch) by isolating the amino acids corresponding to palmitic acid from the overall structure of ReP1-NCXSQ, with the ligand inserted in the barrel. The picture shows the backbone structure covered by the molecular surface. (**b**) *Anchorage of palmitic acid inside the barrel of ReP-NCXSQ*. The figure shows the polar head of a palmitic acid making hydrogen bond contact with Arg126 (2.67) and Tyr128 (2.62) of ReP1-NCXSQ. Note that no covalent bonds are present

2008). The isolated crystal structure of palmitic acid inside ReP1-NCSQ is depicted in Fig. 13.2a, showing the backbone and the energy surfaces. This lipid fits tightly in the barrel where the CO_2 head of the lipid makes contact, through short hydrogen bonds, with the amino acids Arg126 (2.67) and Tyr128 (2.62) (Fig. 13.2b), forming an ordered structure up to C12 (not shown). However, no covalent bonds were observed.

Based on the stabilization of the fatty acid by proton bonds connecting with Tyr128 and Arg126, it was decided to investigate the relevance of these

Table 13.1 Effect of WT and mutated ReP1-NCXSQ on the MgATP stimulation of a Na^+ gradient-dependent Ca^{2+} uptake and MgATP-dependent ReP1-NCXSQ phosphorylation

Addition ReP1-NCXSQ	MgATP stimulation of a Na^+ gradient-dependent Ca^{2+} uptake (nmol/mg.10s)	Phosphorylation from [^{32}P]-γ-ATP (arbitrary units)
WT (no vesicles)[a]	0.51 ± 0.01	–
WT	1.25 ± 0.03	1
Arg126Ala	1.34 ± 0.06	1.08 ± 0.16
Tyr128Phe	0.53 ± 0.025	1.05 ± 0.15
Arg126Ala-Tyr128Phe	0.61 ± 0.05	0.98 ± 0.10
Tyr20Phe	1.16 ± 0.07	0.94 ± 0.12
Ser99Ala	1.20 ± 0.10	0.99 ± 0.13
Phe58Val	1.32 ± 0.045	0.97[b]

Stimulation of the squid nerve Na^+/Ca^{2+} exchanger and ReP1-NCXSQ phosphorylation by MgATP with the wild-type regulatory protein (WT, row 2) and with the following mutations: Arg26Ala (row 3), Tyr128Phe (row 4), Arg26Ala-Tyr128Phe (row 5), Tyr20Phe (row 6), Ser99Ala (row 7), and Phe58Glu (row 8). Row 1 refers to an experiment in the absence of the cytosolic factor (see text for technical details)
[a]ReP1-NCXSQ was incubated in the absence of squid nerve vesicles
[b]Duplicate determinations

bonds for the activity of ReP1-NCSQ. To that aim, two single mutations, Arg126Ala and Tyr128Phe, and the double mutation Arg126Ala-Tyr128Phe, were performed. On the other hand, looking for similarities of amino acid residues between ReP1-NCXSQ and other FABPs, we found that Tyr20, part of the potential phosphorylation sites, is conserved among many of these. NetPhos analysis of the potential phosphorylation sites by using the primary ReP1-NCXSQ structure shows a total of eight: five for serine, two for threonine, and one for tyrosine (actually Tyr20). However, by using the crystal structure, only two potential phosphorylating sites, located on the surface of the barrel, are obtained: Tyr20 and Ser99. We therefore added two other mutations: Tyr20Phe and Arg99Ala. The activity of the exchanger was estimated by measuring the Na^+ gradient-dependent $45Ca^{2+}$ uptake in squid membrane vesicles. These vesicles, prepared from squid optical nerve by differential centrifugation, are about 40% inside out. The uptake was measured at RT by incubating the vesicles (25–30 μg protein) for 10s in media with high (300 mM) or low (30 mM) Na^+ in a total volume of 100 μl (Berberián et al. 2009a). Phosphorylation of the WT and different mutants was carried out with 0.5 mM (500 cpm/pmol) [^{32}P]-γ-ATP (PerkinElmer, USA), in the same solution used for Na^+/Ca^{2+} transport experiments. The reaction was terminated by adding 5× Laemmli sample buffer (Laemmli 1970), run in SDS-PAGE (4–20% gradient gels, Invitrogen), and then transferred to PVDF by semidry electrophoretic transfer at 2.5 mA/cm^2 for 40 min. The incorporation of [^{32}P] Pi was detected with a Storm 840 image analyzer (Berberián et al. 2009a).

The results of these experiments are illustrated in Table 13.1, where the central column refers to the effects of exchange fluxes. Row 1 corresponds to the presence of MgATP without ReP1-NCXSQ. The following conditions indicate that the Arg126Ala mutation does not interfere with MgATP stimulation of the exchanger (row 3), while that stimulation is completely absent in the Tyr128Phe mutant (row 4). The double mutant (row 5) just expresses the expected results due to the mutation of Tyr128. These results are extremely interesting since, in the human muscle fatty acid-binding protein, the mutation Arg126Phe induced a large decrease or complete loss of oleic acid binding, whereas the Tyr128Phe mutation had no great effect. Similarly, in adipocyte fatty acid- (A-FABP) and intestinal fatty acid-binding proteins (I-FABP), which interact with the fatty acid carboxylate group, thermodynamic analysis of Arg126Ala mutations predicted a large reduction of the affinities for lipid binding (Richieri et al. 1998). This is intriguing but at the

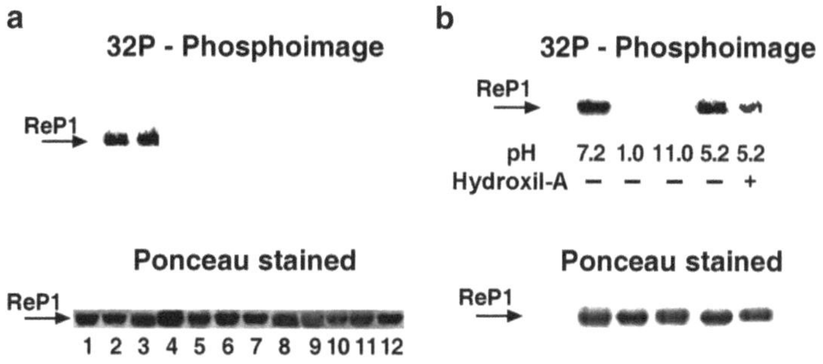

Fig. 13.3 (**a**) *Controls of ReP1-NCXSQ phosphorylation.* The possibility of artifacts was explored by incubating ReP1-NCSQ under the following conditions: *lanes 1–3*: 0.5 mM [^{32}P]-γ-ATP (note that no membrane vesicles were included in 1); *lane 4*: MW markers without radioactive ATP; *lanes 5 and 6*: 05 mM [^{32}P]-α-ATP; *lanes 7 and 8*: 0.5 mM [^{32}P]Pi; *lanes 9 and 10*: 0.5 mM cold ATP plus 0.5 mM [^{32}P]Pi; and *lanes 11 and 12*: 0.5 mM cold ATP plus carrier-free [^{32}P]Pi. Note: (1) a radioactive band was only detected with [^{32}P]-γ-ATP; (2) the protein content was similar under all experimental conditions. (**b**) *Stability of phosphorylated ReP1-NCXSQ.* Note that phosphorylation completely disappears at extreme acid and alkaline pHs and at the same time that it is sensitive to hydroxylamine at pH 5.2 (see text for details). Note that the protein content was similar under all experimental conditions

same time shows an important difference between squid and human FABPs that may even be a fingerprint of marine invertebrates.

Finally, neither Tyr20Phe nor Ser99Ala mutations affect stimulation. On the other hand, the results on ReP1-NCXSQ phosphorylation (right column) show no difference between the diverse mutants. This could be interpreted as an indication that, although phosphorylation of the protein is essential for activity, it is not the only requirement. An interesting result was that mass spectrometry analysis of phosphorylated ReP1-NCXSQ did not show phosphate incorporation in serines, threonines, or tyrosines.

This observation led to two additional experiments. In the first group, we explored the possibility of an artifact due to [32P]-γ-ATP merely trapped in the precipitates without actual protein phosphorylation.

To that aim, ReP1-NCXSQ was incubated in otherwise the same conditions but in the presence of the following ligands: 0.5 mM [^{32}P]-γ-ATP, 0.5 mM [^{32}P]-α-ATP, 0.5 mM [^{32}P]Pi, 0.5 mM cold ATP plus 0.5 mM [^{32}P]Pi, and 0.5 mM cold ATP plus carrier-free [^{32}P]Pi. The results, seen in Fig. 13.3a, rule out any artifact, clearly indicating the existence of protein phosphorylation. In a second group, the experiments explored the stability of phosphorylated ReP1-NCXSQ at extremes and its sensitivity to hydroxylamine. Figure 13.3b shows that phosphorylation disappears at both extreme alkaline and acid pHs; this concurs with what is seen in the phosphorylation of the small protein in the two-component signaling systems (Janiak-Spens et al. 1999), where the residue incorporating Pi is an aspartate. Actually, the fact that at pH 5.2 phosphorylated ReP1-NCXSQ is sensitive to hydroxylamine points to that residue, and this coincides with the fact that phosphorylated aspartic acid is not detected by mass spectrometry. These results do not necessarily mean that we are in the presence of a two-component system with a lipid-binding protein as the small partner (Beaugé et al. 1996; Shinpei et al. 2007), although this mechanism cannot be completely excluded, but, at the same time, it strongly suggests that the responsible kinase is of histidine type.

A study was made of the crystal structures of the WT and ReP1-NCXSQ mutated at the lipid-anchoring residues. Unfortunately, although crystallization of the Tyr128Phe mutant was straightforward, that of the Arg126Ala was elusive. Therefore, at this point the only data correspond to the former. Two results seem relevant: (1) The overall structure of the WT and the Tyr128Phe mutant were quite similar, except for a displacement in Phe58 near the barrel mouth;

the displacement is large enough to modify the mouth, opening a channel leading into the cavity. This channel is closed in the WT, where Phe58 is stabilized by a hydrophobic contact with palmitic acid (not shown here). In other words, in the WT, the Phe58 is stabilized by hydrophobic contact with palmitic acid (3.84 Å), while in the Tyr128Phe mutant, the disordered palmitic acid tail leads to a double conformation of the Phe58 loop. (2) Regarding the lipid structure, the difference maps in Tyr128Phe show a disordered signal for palmitic acid, both at the carboxylate head and of carbons C7–C16, where atoms C7–C12, clearly seen in the WT, are not seen here. Also, the carboxylate head is significantly more disordered than in the WT. Regarding occupancy, mass spectrometry shows that it is lower in the Tyr128Phe mutant, with only 13% of the detected species corresponding to a ReP1-NCXSQ-palmitic acid complex (not shown).

There was a prospect that the displacement of Phe58 observed in Try128Phe, opening the mouth, is responsible for the loss of activity. This was checked by following MgATP stimulation with a new point mutation, Phe58Glu. However, the results seen in the last row of Table 13.1 rule out this possibility.

13.5 Conclusions

This chapter presents results of crystallographic and mass spectrometry analysis of the ReP1-NCXSQ protein, essential for MgATP upregulation of the squid nerve Na^+/Ca^{2+} exchanger, and of a ligand present in the protein when expressed in *E. coli.* It also shows the changes induced both in ReP1-NCXSQ and in the inserted ligand, following the Tyr128Phe mutation that abolishes activity. It may be that palmitic acid is not the actual ligand related to activation of this exchanger, although it must not be a priori disregarded since it belongs to the three most abundant fatty acids, present in Mediterranean cuttlefish, octopus, and squid (Zlatanos et al. 2006). At any rate, the drastic changes observed in the bound palmitic acid structure when comparing WT with the mutated Tyr128Phe are likely to occur also with the fatty acids related to these effects.

Despite the advances presented here, the actual mechanism by which ReP1-NCXSQ participates in the MgATP regulation of the squid Na^+/Ca^{2+} exchanger remains unknown. Several possibilities are illustrated in Fig. 13.4.

At the top of the figure, the phosphorylation and overall final effects on the exchanger have five

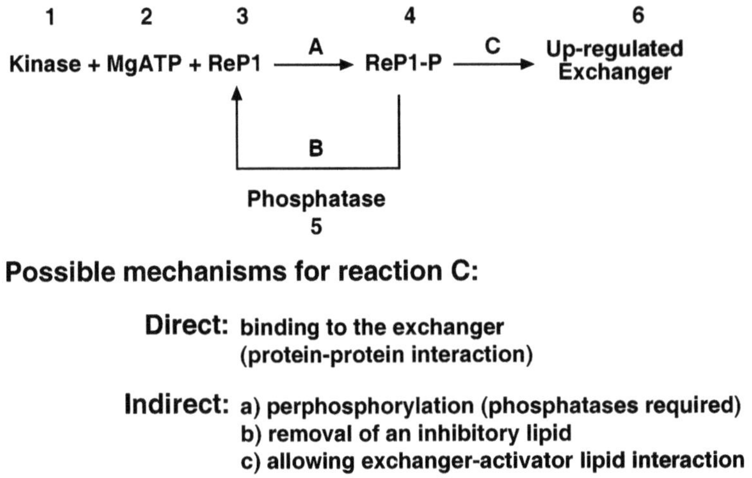

Fig. 13.4 *Possible biochemical reactions and mechanisms associated to ReP1-NCXSQ required MgATP stimulation of the squid nerve Na^+/Ca^{2+} exchanger.* The top part displays possible reaction mechanisms including structures and partial reactions involved. The bottom part summarizes possible mechanisms for the final steps of exchanger activation (see text for details)

structures/molecules and three reactions involved. These are a kinase (1) which by using MgATP (2) acts upon the substrate ReP1-NCXSQ (3) which becomes phosphorylated (reaction A) producing P-ReP1-NCXSQ (4). Also, as much experimental evidence indicates (see DiPolo and Beaugé 2006), a phosphatase (5) must come into play (reaction B). Finally, reaction C covers the mechanism/s leading to upregulation of NCXSQ (6).

The bottom of Fig. 13.4 suggests possible intimate mechanisms for reaction C. The actual modulation of the exchanger (protection of the Ca_i^{2+} regulatory sites) may be direct or indirect through other structure/s. A direct ReP1-NCXSQ-exchanger (protein-protein) interaction has several possibilities, particularly considering that the C-terminus part of the large intracellular loop is markedly hydrophobic; it can also be of the type seen in the attachment of human lipocalins to macromolecules (Skerra 2008). The requirement for phosphorylation of ReP1-NCXSQ could be accounted for, if incorporation of inorganic phosphate is essential to produce a configuration able to bind to the target. Also, a kind of two-component system modulation (see Shinpei et al. 2007) cannot be completely ruled out; if this is the case, one would be in the presence of a novel system where the regulator is a lipid-binding protein. In the indirect mechanism/s, the exchanger protein is not the target of phosphorylated ReP1-NCXSQ. These would offer several possibilities, including the following: (1) Perphosphorylation, where the phosphate group is transferred from the phosphorylated ReP1-NCXSQ to another structure/s that eventually interacts with the exchanger. This would justify the need for Mg^{2+} and may resemble that seen in the immunoglobulin E receptor signaling system (Holowka et al. 2007). (2) A P-Rep1-NCXSQ removal of inhibitory fatty acid/s from the membrane, as is done by a FABP with the Na^+-dependent amino acid uptake in rat brain synaptosomes (Bass et al. 1984). (3) Opposite to the latter, by allowing an interaction between the exchanger and an activator lipid. The last two possibilities are supported by the fact that, in plastic strips, ReP1-NCXSQ can bind polyphosphoinositides and phosphatidic acid, whereas it does not bind cholesterol, sphingomyelin, phosphoserine, phosphoethanolamine, or ptdIns (Berberián et al. 2009a).

Acknowledgments The work was supported by grants from the US National Science Foundation (MCB 0444598); Fondo Nacional para Investigaciones Científicas y Tecnológicas (PICT-05-38073) and Consejo Nacional de Investigaciones Científicas y Técnicas (PIP 2010–2012: GI11220090100063), Argentina; Fondo Nacional para Ciencia y Técnica (S1-9900009046 and G-2001000637 FONACIT) and Fundación Polar, Venezuela; and CNRS and INSERM, Strasbourg, France.

References

C. Asteggiano, G. Berberián, L. Beaugé, Phosphatidyl inositol-4,5-bisphosphate bound to bovine cardiac Na^+/Ca^{2+} exchanger displays a MgATP regulation similar to that of the exchange fluxes. Eur. J. Biochem. **268**, 437–442 (2001)

P.F. Baker, H.G. Glitsch, Does metabolic energy participate directly in the Na^+-dependent extrusion of Ca^{2+} ions from squid giant axons? J. Physiol. **233**, 44–46 (1973)

P.F. Baker, P.A. McNaughton, Selective inhibition of the Ca-dependent Na efflux from intact squid axons by a fall in intracellular pH. J. Physiol. **269**, 78–79 (1977)

P.F. Baker, M.P. Blaustein, A.L. Hodgkin, R.A. Steinhardt, The influence of calcium ions on sodium efflux in squid axons. J. Physiol. **200**, 431–458 (1969)

R. Bass, D.E. Raghupathy, J.A. Rhoads, R. Manning, R. Ockner, Partial purification of molecular weight 12.000 fatty acid binding proteins from rat brain and their effect on synaptosomal Na^+-dependent amino acid uptake. Biochemistry **23**, 6539–6544 (1984)

L. Beaugé, D. Girardi, H. Rojas, G. Berberian, R. Di Polo, A nerve cytosolic factor is required for MgATP stimulation of Na^+ gradient- dependent Ca^+ uptake in plasma membrane vesicles from squid optic nerve. Ann. N.Y. Acad. Sci. **779**, 208–216 (1996)

G. Berberián, C. Hidalgo, R. DiPolo, L. Beaugé, ATP stimulation of Na^+/Ca^{2+} exchange in cardiac sarcolemmal vesicles. Am. J. Physiol. **274**, C724–C733 (1998)

G. Berberián, M. Bollo, G. Montich, G. Roberts, J. De Giorgis, R. DiPolo, L. Beaugé, A novel lipid binding protein is a factor required for MgATP stimulation of the squid Na/Ca exchanger. Biochim. Biophys. Acta Biomembr. **1788**, 1255–1262 (2009a)

G. Berberián, D. Forcato, L. Beaugé, Key role of PTDIns-4,5P2 micro domain in ionic regulation of the mammalian heart Na/Ca exchanger. Cell Calcium **45**, 546–553 (2009b)

D. Bers, S. Despa, Na^+/K^+-ATPase, an integral player in the adrenergic fight-or-flight response. Trends Cardiovasc. Med. **19**, 111–118 (2009)

M.P. Blaustein, Effects of internal and external cations and ATP on sodium-calcium exchange and calcium-calcium exchange in squid axons. Biophys. J. **20**, 79–111 (1977)

M.P. Blaustein, A.L. Hodgkin, The effect of cyanide on the efflux of calcium from squid axons. J. Physiol. **200**, 497–527 (1969)

M. Blaustein, W.J. Lederer, Sodium/calcium exchange: its physiological implication. Physiol. Rev. **79**, 763–854 (1999)

E.M. Clérico, M.R. Ermácora, Tryptophan mutants of intestinal fatty acid-binding protein: ultraviolet absorption and circular dichroism studies. Arch. Biochem. Biophys. **395**, 215–224 (2001)

M. Condrescu, J.P. Gardner, G. Chernaya, J.F. Aceto, C. Kroupis, J.P. Reeves, ATP-dependent regulation of sodium-calcium exchange in Chinese hamster ovary cells transfected with the bovine cardiac sodium-calcium exchanger. J. Biol. Chem. **270**, 9137–9146 (1995)

R. DiPolo, Calcium efflux from internally dialyzed squid giant axons. J. Gen. Physiol. **62**, 575–589 (1973)

R. DiPolo, The effect of ATP on the calcium efflux in dialyzed squid axons. J. Gen. Physiol. **64**, 503–517 (1974)

R. DiPolo, The influence of nucleotides on calcium fluxes. Fed. Proc. **35**, 2579–2582 (1976)

R. DiPolo, Calcium influx in internally dialyzed squid giant axons. J. Gen. Physiol. **73**, 91–113 (1979)

R. DiPolo, L. Beaugé, Physiological role of ATP-driven calcium pump in squid axons. Nature **278**, 271–273 (1979)

R. DiPolo, L. Beaugé, The effect of pH on Ca extrusion mechanisms in dialyzed squid axons. Biochim. Biophys. Acta **688**, 237–245 (1982)

R. DiPolo, L. Beaugé, Interactions of ligands with the Ca pump and Na/Ca exchange. J. Gen. Phys. **84**, 895–914 (1984)

R. DiPolo, L. Beaugé, In squid axons, ATP modulates Na-Ca exchange by Ca(i)-dependent phosphorylation. Biochim. Biophys. Acta **897**, 347–354 (1987)

R. DiPolo, L. Beaugé, Metabolic pathways in the regulation of invertebrate and vertebrate Na^+/Ca^{2+} exchange. Biochim. Biophys. Acta **1422**, 57–71 (1999)

R. DiPolo, L. Beaugé, MgATP counteracts intracellular proton inhibition of the sodium-calcium exchanger in dialyzed squid axons. J. Physiol. **539**, 791–803 (2002)

R. DiPolo, L. Beaugé, Sodium/calcium exchanger: influence of metabolic regulation on ion carrier interactions. Physiol. Rev. **86**, 155–203 (2006)

R. DiPolo, G. Berberián, L. Beaugé, In squid nerves intracellular Mg promotes deactivation of the ATP-upregulated Na/Ca exchanger. Am. J. Physiol. **279**, C1631–C1639 (2000)

A.E. Doering, W.J. Lederer, The mechanism by which cytoplasmic protons inhibit the sodium-calcium exchanger in guinea pig heart cells. J. Physiol. **466**, 481–499 (1993)

A.E. Doering, W.J. Lederer, The action of Na^+ as a cofactor in the inhibition by cytoplasmic protons of the cardiac Na^+-Ca^{2+} exchanger in the guinea pig. J. Physiol. **480**, 9–20 (1994)

D. Forcato, V. Posada, L. Beaugé, G. Berberián, Optimal metabolic regulation of the mammalian heart Na/Ca exchanger requires a special arrangement with a PtdIns(4)-5 kinase. Biochem. Biophys. Res. Comm. **402**, 147–152 (2010)

A. Furuhashi, G.S. Hotamisligil, Fatty acid-binding proteins: role in metabolic diseases and potential as drug targets. Nat. Rev. Drug Discov. **7**, 489–503 (2008)

E.R. Gericke, D.J. Smith, R. Moore, J. Mendelsohn, J. Storch, Adipocyte fatty acid-binding protein: interaction with phospholipid membranes and thermal stability studied by FT-IR spectroscopy. Biochemistry **36**, 8311–8317 (1997)

Z. He, S. Feng, Q. Tong, D.W. Hilgemann, K.D. Phipipson, Interaction of PIP2 with the XIP region of the cardiac Na/Ca exchanger. Am. J. Physiol. **278**, C661–C666 (2000)

M. Hilge, J. Aelen, A. Perrakis, G.W. Vuister, Structural basis for Ca^{2+} regulation in the Na^+/Ca^{2+} exchanger. Ann. N. Y. Acad. Sci. **1099**, 7–15 (2007)

D. Hilgeman, The cardiac Na^+/Ca^{2+} exchanger in giant membrane patches. Ann. N. Y. Acad. Sci. **779**, 136–158 (1996)

D.W. Hilgemann, Cytoplasmic ATP-dependent regulation of ion transporters and channels: mechanisms and messengers. Annu. Rev. Physiol. **59**, 193–220 (1997)

D.W. Hilgemann, On the physiological roles of PIP_2 at cardiac Na^+/Ca^{2+} exchangers and K(ATP) channels: a long journey from membrane biophysics into cell biology. J. Physiol. **582**, 903–909 (2007)

D.W. Hilgemann, R. Ball, Regulation of cardiac Na^+, Ca^{2+} exchange and KATP potassium channels by PIP2. Science **273**, 956–959 (1996)

D.W. Hilgemann, A. Collins, S. Matsuoka, Steady-state and dynamic properties of cardiac sodium-calcium exchange. Secondary modulation by cytoplasmic calcium and ATP. J. Gen. Physiol. **100**, 933–961 (1992a)

D.W. Hilgemann, S. Matsuoka, G.A. Nagel, A. Collins, Steady-state and dynamic properties of cardiac sodium-calcium exchanger. Sodium-dependent inactivation. J. Gen. Physiol. **100**, 905–932 (1992b)

D. Holowka, D. Sil, C. Torigoe, B. Baird, Insights into immunoglobulin E receptor signaling from structurally defined ligands. Immunol. Rev. **217**, 269–279 (2007)

L.V. Hryshko, S. Matsuoka, D.A. Nicoll, J.N. Weiss, E.M. Schwarz, S. Benzer, K.D. Philipson, Anomalous regulation of the *Drosophila* Na^+/Ca^{2+} exchanger by Ca^{2+}. J. Gen. Physiol. **108**, 67–74 (1996)

F. Janiak-Spens, J.M. Sparling, M. Gurfinkel, A.H. West, Differential stabilities of phosphorylated response regulator domains reflect functional roles of the yeast osmoregulatory SLN1 and SSK1 proteins. J. Bacteriol. **181**, 411–417 (1999)

D. Khananshvili, E. Weil-Maslansky, D. Baazou, Kinetics and mechanism; Modulation of ion transport in the cardiac sarcolemma sodium-calcium exchanger by protons, monovalent ions and temperature. Ann. N. Y. Acad. Sci. **779**, 217–235 (1996)

U.K. Laemmli, Cleavage of structural proteins during the assembly of the head of bacteriophage T4. Nature **227**, 680–685 (1970)

Z. Li, D.A. Nicoll, A. Collins, D.W. Hilgemann, A.G. Filoteo, J.T. Penniston, Identification of a peptide inhibitor of the cardiac sarcolemmal Na^+-Ca^{2+} exchanger. J. Biol. Chem. **266**, 1014–1020 (1991)

S. Luciani, S. Bova, G. Cargnelli, F. Cusinato, P. Debetto, Modulation of sodium-calcium exchange by lipids. Ann. N. Y. Acad. Sci. **639**, 156–165 (1991)

S. Luciani, M. Antonini, S. Bova, G. Cargnelli, F. Cusinato, P. Debetto, L. Trevisi, R. Varotto, Inhibition of cardiac sarcolemmal sodium-calcium exchanger by glycerophosphoinositol 4-phosphate and glycerophosphoinositol 4-5-bisphosphate. Biochem. Biophys. Res. Commun. **206**, 674–680 (1995)

S. Matsuoka, D.A. Nicoll, Z. He, K.D. Philipson, Regulation of cardiac Na^+-Ca^{2+} exchanger by the endogenous XIP region. J. Gen. Physiol. **109**, 273–286 (1997)

S . McLaughlin, J. Wang, A. Gambhir, D. Murray, PIP(2) and proteins: interactions, organization, and information flow. Annu. Rev. Biophys. Biomol. Struct. **31**, 151–175 (2002)

V. Nolan, M. Parduca, H.L. Monaco, B. Maggio, G.G. Montich, Interactions of chicken liver basic fatty acid-binding protein with lipid membranes. Biochim. Biophys. Acta. **1611**, 98–106 (2003)

M. Ottolia, S. John, X. Ren, K.D. Philipson, Fluorescent Na^+/Ca^{2+} exchangers: electrophysiological and optical characterization. J. Biol. Chem. **282**, 3695–3701 (2007)

M. Ottolia, D.A. Nicoll, K.D. Philipson, Roles of two Ca^{2+}-binding domains in regulation of the cardiac Na^+-Ca^{2+} exchanger. J. Biol. Chem. **284**, 32735–32741 (2009)

K.D. Philipson, M.M. Bersohn, Y. Nishimoto, Effects of pH on Na^+-Ca^{2+} exchange in canine cardiac sarcolemmal vesicles. Circ. Res. **50**, 287–293 (1982)

V. Posada, L. Beaugé, G. Berberián, Maximal Ca^{2+}i stimulation of cardiac Na^+/Ca^{2+} exchange requires simultaneous alkalinization and binding of the PtdIns-4,5-P2to the exchanger. Biol. Chem. **388**, 281–288 (2007)

D. Raimunda, M. Bollo, L. Beaugé, G. Berberián, Squid nerve Na/Ca exchanger expressed in *Saccharomyces cerevisiae*: Up-regulation by a phosphorylated cytosolic protein (ReP1-NCXSQ) is identical to that of native exchanger in situ. Cell Calcium **45**, 499–508 (2009)

D.H. Schulze, M. Muqhal, W.J. Lederer, A.M. Ruknudin, Sodium/calcium exchanger (NCX1) macromolecular complex. J. Biol. Chem. **278**, 28849–28855 (2003)

B. Shinpei, R. Noguchi, K. Yamashita, F. Fukumori, M. Kimura, I. Yamaguchi, M. Fujimura, Roles of putative His-to-Asp signaling modules HPT-1 and RRG-2, on viability and sensitivity to osmotic and oxidative stresses in *Neurospora crassa*. Curr. Genet. **51**, 197–208 (2007)

A. Skerra, Alternative binding proteins: anticalins-harnessing the structural plasticity of the lipocalin ligand pocked of engineer novel binding activities. FEBS J. **275**, 2677–2683 (2008)

F. Tanfani, Z. Kochan, J. Swierczynski, M.M. Zydowo, E. Bertoli, Structural properties and thermal stability of human liver and heart fatty acid binding proteins: a Fourier transform IR spectroscopy study. Biopolymers **36**, 569–577 (1995)

S. Wakabayashi, K. Goshima, Comparison of kinetic characteristics of Na-Ca exchange in sarcolemma vesicles and cultured cells from chick heart. Biochim. Biophys. Acta **645**, 311–317 (1981)

M. Wu, M. Wang, J. Nix, L.V. Hryshko, L.J. Zheng, Crystal structure of CBD2 from the *Drosophila* Na^+/Ca^{2+} exchanger: diversity of Ca^{2+} regulation and its alternative splicing modification. Mol. Biol. **387**, 104–112 (2009)

A. Yaradanakul, S. Feng, C. Shen, V. Lariccia, M.J. Lin, J. Yang, T.M. Kang, P. Dong, H.L. Yin, J.P. Albanesi, D.W. Hilgemann, Dual control of cardiac Na^+ Ca^{2+} exchange by PIP(2): electrophysiological analysis of direct and indirect mechanism. J. Physiol. **582**, 991–1010 (2007)

A.C. Young, G. Scapin, A. Kromminga, S.B. Patel, J.H. Veerkamp, J.C. Sacchettini, Structural studies on human muscle fatty acid binding protein at 1.4 Å resolution: binding interactions with three C18 fatty acids. Structure **15**, 523–534 (1994)

S. Zlatanos, K. Laskaridis, C. Feist, A. Sagredos, Proximate composition, fatty acid analysis and protein digestibility-corrected amino acids core of three Mediterranean cephalopods. Mol. Nutr. Food Res. **50**, 967–970 (2006)

Y.H. Zhang, J.C. Hancox, Regulation of cardiac Na^+-Ca^{2+} exchanger activity by protein kinase phosphorylation still a paradox? Cell Calcium **45**, 1–10 (2009)

Regulation of Sodium-Calcium Exchanger Activity by Creatine Kinase

14

Ya-Chi Yang and Lung-Sen Kao

Abstract

It has been shown that in rat heart NCX1 exists in a macromolecular complex including PKA, PKA-anchoring protein, PKC, and phosphatases PP1 and PP2A. In addition, several lines of evidence suggest that the interactions of the exchanger with other molecules are closely associated with its function in regulation of $[Ca^{2+}]_i$. NCX contains a large intracellular loop (NCXIL) that is responsible for regulating NCX activity. We used the yeast two-hybrid method to screen a human heart cDNA library and found that the C-terminal region of sarcomeric mitochondrial creatine kinase (sMiCK) interacted with NCX1IL. Among the four creatine kinase (CK) isozymes, both sMiCK and the muscle-type cytosolic creatine kinase (CKM) co-immunoprecipitated with NCX1. Both sMiCK and CKM were able to produce a recovery in the decreased NCX1 activity that was lost under energy-compromised conditions. This regulation is mediated through a putative PKC phosphorylation site of sMiCK and CKM. The catalytic activity of sMiCK and CKM is not required for their regulation of NCX1 activity. Our results suggest a novel mechanism for the regulation of NCX1 activity and a novel role for CK.

Keywords

Sodium-calcium exchanger1 (NCX1) • Creatine kinase • Energy-compromised conditions • NCX1 macromolecular complex

Y.-C. Yang
Institute of Biochemistry and Molecular Biology, National Yang-Ming University, Taipei, Taiwan, 112, Republic of China

L.-S. Kao, Ph.D. (✉)
Department of Life Sciences and Institute of Genome Sciences, National Yang-Ming University, Taipei, Taiwan, 112, Republic of China
e-mail: lskao@ym.edu.tw

L. Annunziato (ed.), *Sodium Calcium Exchange: A Growing Spectrum of Pathophysiological Implications*, Advances in Experimental Medicine and Biology 961, DOI 10.1007/978-1-4614-4756-6_14,

14.1 Introduction

Na^+/Ca^{2+} exchanger (NCX) plays an important role in maintaining intracellular Ca^{2+} homeostasis in cells which handle a large flux of Ca^{2+} across their plasma membrane, such as secretory, cardiac, and neuronal cells (Blaustein et al. 1991). The NCX has attracted many attentions in recent years because it is present in the plasma membrane of virtually every cell type examined and involved in many physiological and pathological processes including, in particular, (1) embryonic heart development (Cho et al. 2000); (2) normal physiological, such as action potential and E-C coupling, and pathological function, such as cardiac arrhythmias and reperfusion injury (Eigel et al. 2004) of cardiac muscle; (3) the development of hypertension in smooth muscle cells; and (4) neurotransmitter or hormone secretion in central and peripheral nervous system and secretory cells (Blaustein and Lederer 1999).

A large intracellular loop of NCX (NCXIL), which is located between the fifth and sixth transmembrane segments, is critical for the regulation of NCX activity (DiPolo and Beauge 2006). It has been shown that NCX1 exists in a large macromolecular complex including several kinases and phosphatases. This chapter describes the search for molecules that interact with NCXIL and provide evidence to show that creatine kinase interacts with NCX1 and regulates its activity under energy-compromised conditions.

14.2 Molecular Characterization of NCX

Three mammalian NCX genes, *ncx1*, *ncx2*, and *ncx3*, have been identified (Komuro et al. 1992; Li et al. 1994; Nicoll et al. 1996). There is about 70 % of identity in the amino acid sequences of the three NCX isoforms (Quednau et al. 2004). Expression of the NCX1 gene is controlled by alternative promoters in a tissue-specific and transcription factor-specific manner. NCX1 is expressed at high levels in the heart but is observed in most other tissues in varying amounts (Quednau et al. 1997). NCX2 and NCX3 are expressed primarily in brain and skeletal muscle (Li et al. 1994).

We have studied the role of NCX in excitable cells using bovine chromaffin cells as a model system and found that NCX plays a preeminent role in maintaining the Ca^{2+} homeostasis both at resting and stimulated states (Chern et al. 1992; Kao 1988; Kao and Cheung 1990; Liu and Kao 1990). We have isolated three alternative spliced variants of *ncx* clones from a bovine chromaffin cell cDNA library (Pan et al. 1998). In addition, another type of the exchanger that depends on K^+ for its activity, NCKX, was also present in the chromaffin cells, and the two types of exchanger were copresent in the same cell (Pan et al. 2008). Similar phenomenon was also observed in primary cultured rat cortical neurons (Wu et al. 2008). The physiological significance of the presence of the various types of the exchanger in the same cell requires further study.

NCX1 has 938 amino acids and has been modeled to contain nine transmembrane segments (Ren et al. 2006). Each NCX contains two pairs of internal repeats, designated α- and β-repeats (Quednau et al. 2004). The α-repeats consist of two groups of highly conserved residues separated by a short unconserved linker and are located in the groups of transmembrane segments. The α-repeat regions have been implicated in ion binding and transport and may form membrane reentrant segments. The α-repeat regions interact with one another in the tertiary structure of the protein. The α-2 domain is involved in determining the sensitivity of the exchanger to the inhibitor KBR7943 which represses the reverse-mode NCX1 activity (Iwamoto et al. 2001). The β-repeats are in the intracellular loop of NCX and share sequence similarity with a motif found in β4 integrin (Schwarz and Benzer 1997). The function of this motif is unknown.

A large intracellular loop of NCX (NCXIL), which is located between the fifth and sixth transmembrane segments, is critical for the regulation of NCX activity (DiPolo and Beauge 2006). The activity of NCX is regulated by Ca^{2+} through a large intracellular loop that contains two Ca^{2+}-binding domains, calcium-binding domain 1 (CBD1) and calcium-binding domain 2 (CBD2). CBD1 binds Ca^{2+} with much higher affinity than CBD2 and is considered to be the primary Ca^{2+} sensor (Besserer et al. 2007; Ottolia et al. 2009). Near the N terminus of NCX1 large intracellular loop is a 20-amino acid region-designated exchanger inhibitory peptide (XIP). NCX is inhibited by the synthetic peptides XIP (Li et al. 1991). Mutations in the XIP region (Matsuoka et al. 1997) or in the intracellular loop between transmembrane segments 1 and 2 (Doering et al. 1998) alter the properties of Na^+-dependent inactivation.

14.3 Regulation of NCX Activity by MgATP

The cardiac NCX1 is positively regulated by intracellular Ca^{2+} and MgATP and inactivated by cytoplasmic Na^+. Several possible regulatory mechanisms on NCX by MgATP have been proposed (DiPolo and Beauge 1999). First, MgATP binds to the protein kinase near the exchanger. The phosphate group transfers from MgATP by the kinase to the exchanger. The activity of NCX exchanges because of the phosphorylation. Second, ATP involves in the synthesis of PIP_2 from the phosphorylation cascade of phosphatidylinositol by endogenous lipids kinases, phosphoinositide 4-kinase and phosphoinositide 4-phosphate 5 kinase. The PIP_2 directly activates the exchanger by binding to a positively charged cytoplasmic domain on the exchanger (Hilgemann 1997). Third, protein kinase transfers the phosphate group of the MgATP to a regulatory protein, and phosphorylated regulatory protein interacts with NCX to alter the exchanger activity. This model is developed based on a finding that a novel 13-kDa cytoplasmic soluble protein is required for the MgATP-dependent modulation of NCX in squid axons (DiPolo et al. 1997).

Reversible phosphorylation of proteins is widely considered as an important mechanism for the control of many cellular processes. PKC has been shown to be involved in the phosphorylation-dephosphorylation of NCX. Different isoforms of the exchanger are regulated differently by protein kinase A and protein kinase C (Pan et al. 1998). However, regulation of NCX activity by PKC remains controversial: in most cases, PKC upregulates the NCX. For instance, PKC upregulates NCX activity in rat aortic smooth muscle cells (Iwamoto et al. 1995), rat neonatal cardiomyocytes (Iwamoto et al. 1996), and rat hepatocytes (Ikari et al. 1998). However, PKC downregulates the NCX activity in other cell types, such as bovine chromaffin cells (Tokumura et al. 1998). Macromolecular complexes have been found to regulate specific K^+ channels (Marx et al. 2002) and cardiac ryanodine receptors (RyR2). In general, these complexes are composed of kinases, phosphatases, and kinase-anchoring proteins. It is possible that some unfound factors participate in the regulation of NCX activity in different tissues. Several proteins, including protein kinase A (PKA), protein kinase C (PKC), muscle protein kinase A-anchoring protein, and the phosphatases PP1/PP2A, have been shown to exist within the NCX1 macromolecular complex (Schulze et al. 2003).

Several lines of evidence suggest that the interaction of NCX with other molecules is closely associated with its physiological function. It has been shown that 14-3-3 protein interacts with the three NCX isoforms and inhibits NCX activity (Pulina et al. 2006). In smooth muscle cells, NCX co-localizes with Na^+/K^+-ATPase in regions closed to the sarcoplasmic reticulum that store Ca^{2+}, which suggests a connection between NCX with its associated Na^+/K^+-ATPase and Ca^{2+} release from the sarcoplasmic reticulum (Moore et al. 1993). Our previous results from a functional study of NCX1 in relation to the regulation of $[Ca^{2+}]_i$ and catecholamine secretion in bovine chromaffin cells suggested that there is a close association between the exchanger, the Ca^{2+} channels, the Ca^{2+} pump in the endoplasmic reticulum, the mitochondria, and the exocytotic sites (Pan and Kao 1997).

14.4 Interaction of NCX with Creatine Kinase

It appears that regulation of NCX activity and the physiological function of NCX require the interaction of NCX with other molecules. Therefore, we used the intracellular loop of NCX1 as a bait to screen a human heart cDNA library by yeast two-hybrid system to search for the molecules that interact with NCX. Among the 52 clones with sequences of known proteins obtained, 32 clones were identified as the sarcomeric mitochondrial creatine kinase (sMiCK). Alignment of the DNA sequences of these sMiCK clones shows that the shortest region of sMiCK cDNA that contributed to the interaction between sMiCK and NCX1IL was the C terminus of sMiCK corresponding to amino acids 226–380. The interaction between full-length sMiCK and NCX1IL was confirmed by GST pull-down assay. Our results obtained by yeast two-hybrid assay show that the entire NCX1IL including the exchanger inhibitory peptide (XIP), Ca^{2+}-binding domain, and alternative splicing regions (Kofuji et al. 1994; Levitsky et al. 1994; Xu et al. 1997) was required for the interaction.

The possibility that NCX interacts with sMiCK is very intriguing. Creatine kinase (CK) catalyzes the reversible transfer of a phosphate group from phosphocreatine to ADP to yield ATP and creatine. CK plays important roles in the cells where rapid regeneration of ATP is required such as muscle cells and neurons. There are four creatine kinase (CK) isozymes in mammals, ubiquitous mitochondrial CK (uMiCK), sarcomeric mitochondrial CK (sMiCK), cytoplasmic brain type CK (CKB), and cytoplasmic muscle-type CK (CKM); these isozymes are found in different tissues and have different subcellular localizations. Both mitochondrial forms of CK, sMiCK and uMiCK, exist in dimeric and octameric forms at the contact sites between inner and outer membranes of the mitochondria (Wallimann et al. 1992). Octameric CKMT, together with porin and adenine nucleotide translocase, forms a microcompartment at contact sites between inner and outer mitochondrial membranes and facilitates the production and export into the cytosol of phosphocreatine. Octamer-dimer transitions of CKMT as well as different creatine kinase substrates have a profound influence on controlling mitochondrial permeability transition pore (PTP).

Despite the high sequence similarity among the various CK isoforms (Fig. 14.1a), the interaction between NCX and CK is highly specific for sMiCK and CKM. The C-terminal regions of sMiCK and CKM were found to be responsible for their interaction with NCX1 and the effect on the NCX1 activity. The results are in accord with the three-dimensional structures of CKs, which show that the C-terminal region is located on the outside of the octameric mitochondrial (Fig. 14.1b, c) and dimeric cytosolic CK isozymes and that the N-terminal regions of CK isozymes are known to be critical for CK dimerization.

14.5 CK Recovers NCX Activity Under Energy-Compromised Conditions

We have carried out a series of experiments to study the physiological significance of the interactions between NCX1 and CK. CK is known to be important for the maintenance of cellular energy homeostasis. It has been shown that NCX activity is stimulated by ATP and is inhibited under an ATP-depleted condition in rat cardiomyocytes (Haworth and Biggs 1997) and in bovine cardiac NCX-expressing Chinese hamster ovary cells (Condrescu et al. 1995). In NCX1 and CK-co-expressing HEK293T cells, the NCX1 activity was decreased under energy-compromised conditions. The decreased NCX1 activity can be reversed by sMiCK and CKM, but not

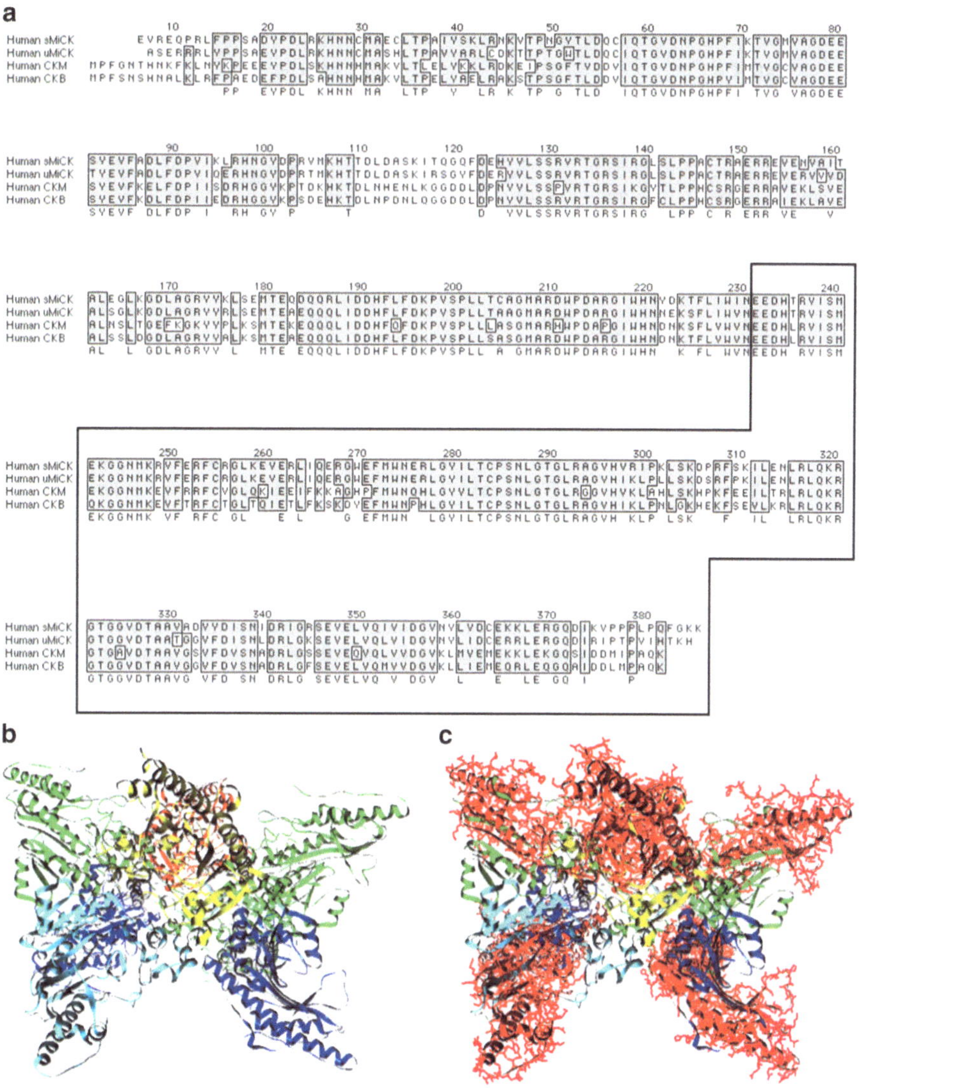

Fig. 14.1 *Alignment of the amino acid sequence of human uMiCK, sMiCK, CKB, and CKM.* (**a**) The amino acid residues in the frame are the shortest region of the human sMiCK required to interact with NCX1IL based on the results obtained from yeast two-hybrid assays (From Yang et al. 2010. Reprinted by permission from the *Journal of Biological Chemistry*). (**b**) The three-dimensional structure of octameric uMiCK (human uMiCK: PDB code 1qk1) (Eder et al. 2000a). (**c**) The C-terminal region (*red*) is located on the outside of the octameric mitochondrial uMiCK

uMiCK and CKB (Fig. 14.2). The results are compatible with that only sMiCK and CKM are co-immunoprecipitated with NCX1. The C termini of sMiCK and CKM are critical for the recovery of decreased NCX1 activity under the energy-compromised conditions because the chimeric CK, sMiCK-CKB and CKM-CKB, in which the C termini of sMiCK and CKM were replaced by the corresponding region of CKB, a.a. 231–381, showed no effects on NCX1 activity.

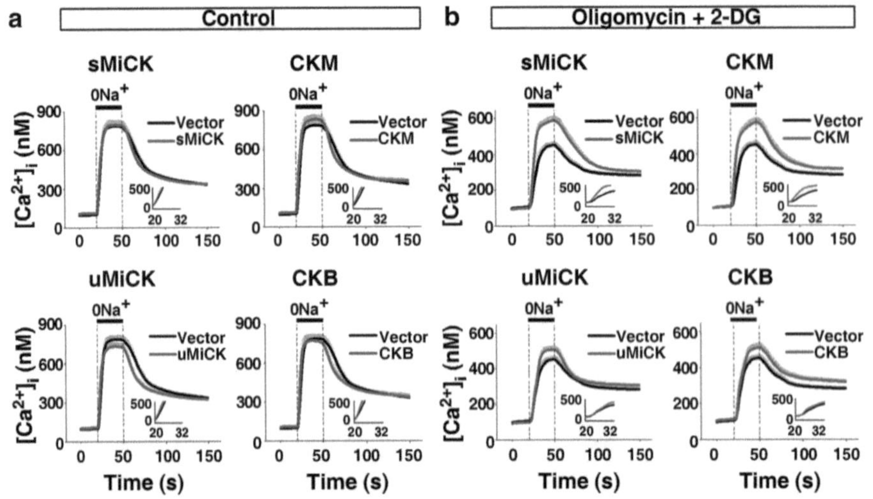

Fig. 14.2 *sMiCK and CKM maintain NCX1 activity under energy-compromised conditions.* HEK293T cells were co-transfected with pGFP, pNCX1, and plasmids encoding various CK isozymes as indicated. Cells with GFP fluorescence were selected to measure the reverse-mode NCX activity by puffing with a Na^+-free buffer for 30s as indicated. (**a**) Reverse-mode NCX activity under control conditions. (**b**) Reverse-mode NCX activity under energy-compromised conditions induced by treatment of oligomycin and 2-DG (From Yang et al. 2010. Reprinted by permission from the *Journal of Biological Chemistry*)

14.6 Phosphorylation of CK Is Critical but Catalytic Activity of CK Is Not Required

It has been shown that CKM is phosphorylated by PKC, possibly at Ser128 (Lin et al. 2009). This putative PKC phosphorylation site of sMiCK and CKM is critical to their regulation of NCX1 activity because the Ser128-equivalent mutants of sMiCK and CKM, sMiCK-S123A and CKM-S128A, failed to produce a recovery in the decreased NCX1 activity under the energy-compromised conditions. In contrast, the autophosphorylation of sMiCK and CKM is not required for the regulation of NCX1 activity. Interestingly, the putative PKC phosphorylation serine residue, which is required for the effects of CK on NCX activity, is located adjacent to the C terminus based on the three-dimensional structure. It is possible that phosphorylation of the serine residue is critical for maintaining the C terminus in a proper position for maintaining NCX activity.

It has been shown in squid nerve that NCX is upregulated by phosphoarginine in addition to ATP (DiPolo and Beauge 1995). Phosphoarginine in invertebrates, such as mollusks, is functionally equivalent to phosphocreatine in vertebrates; both compounds are "high-energy" compound, and together with the enzyme responsible for their formation, they facilitate the communication between ATP generation and ATP utilization. In analogous to squid nerve, it is possible that CK provides NCX with phosphocreatine or ATP for regulation of NCX activity. However, it is disappointing to find that the enzyme activity of creatine kinase is not required for the regulation of NCX1 activity. It has been shown that in the active site of CK isozymes, a conserved negatively charged amino acid cluster, EED, is essential for CK enzyme activity (Eder et al. 2000b). Activity-devoid mutants at the catalytic site of sMiCK and CKM, sMiCK-E226Q, sMiCK-E227L, CKM-E231Q, or CKM-E232L, have the same ability as the wild-type sMiCK and CKM for the recovery of decreased NCX1 activity under energy-compromised conditions.

14.7 Subcellular Localization of NCX1 and CK Under Energy-Compromised Conditions

It is puzzling, however, how NCX1 and sMiCK physically interact with each other in vivo when the two molecules are thought to locate in different subcellular compartments; NCX1 is known to locate in the plasma membrane, and sMiCK is in the intermembrane space of mitochondria. Under energy-compromised conditions, the subcellular localizations of NCX1, uMiCK, sMiCK, and CKB show a similar pattern to that found under the control conditions; NCX1 was localized to the plasma membrane, uMiCK and sMiCK were localized to the mitochondria (Fig. 14.3a, b), and CKM and CKB were distributed in the cytosol (Fig. 14.3c, d). The subcellular localization of CKM is interesting; CKM, localized in the cytosol under control condition, was recruited to the plasma membrane and co-localized with NCX1 under energy-compromised conditions (Fig. 14.3d). Most cells showed preferential localization of CKM to the plasma membrane, but a few cells (~10 %) showed predominant localization of CKM to the plasma membrane.

We could not observe the co-localization of sMiCK and NCX1. One possibility is that the amount of sMiCK that interacts with NCX1 is too small to be detected and that sMiCK is highly concentrated in the mitochondria, which renders the detection of cytosolic sMiCK more difficult. It has previously been shown that sMiCK is present in a large proteolipid complex composed of adenine translocator, voltage-dependent anion channels, and cytochrome C in the intermembrane space of mitochondria (Schlattner et al. 2006). It is also possible that sMiCK is only released into the cytosol under pathological conditions and that is when it interacts with NCX1.

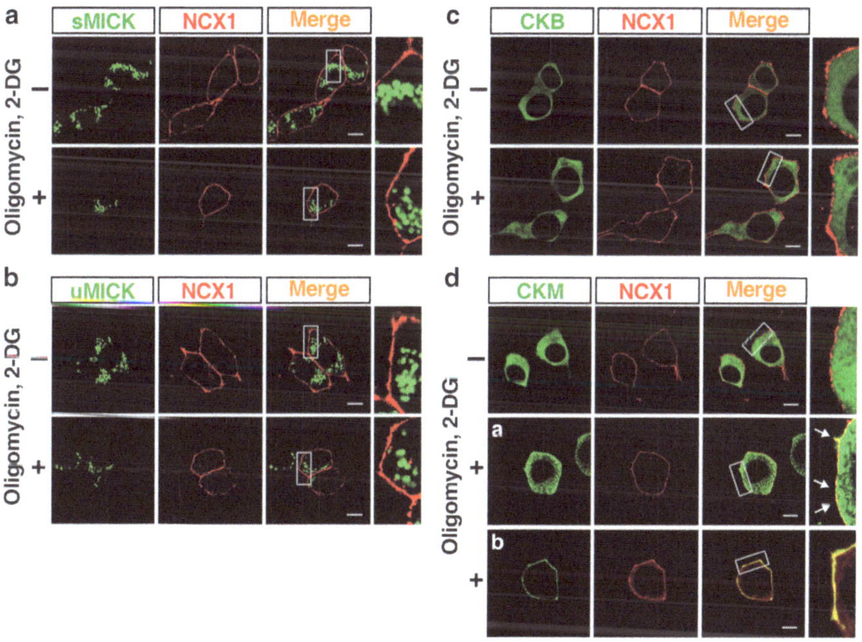

Fig. 14.3 *Subcellular localization of NCX1 and the CK isozymes in HEK293T cells under control and energy-compromised conditions.* NCX1 and the four CK isozymes were transiently expressed in HEK293T cells, which were treated with (+) or without (–) oligomycin and 2-DG. Immunocytochemistry was performed with rabbit anti-NCX1IL antibodies and mouse anti-myc antibody. The subcellular localization of NCX1 (*red*) and the four CK isozymes (*green*), (**a**) sMiCK, (**b**) uMiCK, (**c**) CKB, and (**d**) CKM, were examined by confocal microscopy. (**d**, *a* and *b*) Two different patterns of CKM distribution under the energy-compromised conditions. Scale bars are 5 μm in all images (From Yang et al., 2010. Reprinted by permission from the *Journal of Biological Chemistry*)

The subcellular localizations of NCX1, sMiCK, and CKM in the membrane and cytosolic fractions isolated from mouse cardiac tissues support the possible interaction between the two CKs and NCX1. NCX1 was detected only in the membrane fraction, while CKM was present in both membrane and cytosolic fractions. Surprisingly, although not detected in the immunostaining images, sMiCK was detected not only in mitochondrial fraction but also present in the cytosolic fraction. The presence of sMiCK in the cytosol suggests that some sMiCK may be directly interacting with NCX in the plasma membrane.

The fact that NCX1 is predominantly expressed in cardiac myocytes, where sMiCK and CKM are also present, supports the physiological implications of the interaction between CK and NCX1. NCX is known to play a critical role in the maintenance of Ca^{2+} homeostasis in cardiomyocytes (Reppel et al. 2007; Wakimoto et al. 2000) and thus in the modulation of cardiac contractility. It has been shown that NCX activity is altered during cardiac remodeling in the hypertrophic heart and during heart failure (Mattiello et al. 1998; Weber et al. 2003). The hearts from CKM and sMiCK double-knockout mice have been demonstrated to show increased sensitivity to ischemia-reperfusion injury and show a greater increase in diastolic Ca^{2+} concentration during ischemia (Spindler et al. 2004); therefore, the interaction between NCX1 and CKs may be important to prolonging normal heart functioning when a pathological condition such as ischemia starts to develop. A significant change in muscle force generation and cell morphology is found in CKM and sMiCK double-knockout mice (Steeghs et al. 1997). This supports the hypothesis that there is a specific and direct coupling between the cytosolic and mitochondrial CKs that helps to maintain the normal physiology and functioning of the heart muscles.

For CKM, the immunoprecipitation results showed that CKM interacted with NCX1 in HEK293T cells and cardiac myocytes. Moreover, under energy-compromised conditions, CKM translocated from the cytosol to the plasma membrane, where it co-localized with NCX1. Moreover, CKM mutants without the catalytic enzyme activity were also able, after energy depletion, to recover the reduced reverse-mode NCX1 activity to the same degree as wild-type CKM. Thus, a direct physical interaction between CKM and NCX1 is able to account for the recovery of the decreased reverse-mode NCX1 activity under the energy-compromised conditions.

Our results clearly show that the activity of CK is not required for the effect of CK on NCX activity, and the evidence supports the idea that phosphorylation of sMiCK at Serine 123 and CKM at Serine 128, probably by PKC, is critical for the effect of these CKs on NCX. It has been shown that NCX1IL is phosphorylated by PKC; however, the regulation of NCX activity by PKC remains controversial. PKC upregulates NCX1 activity in rat aortic smooth muscle cells (Iwamoto et al. 1995), neonatal cardiomyocytes (Iwamoto et al. 1996), and hepatocytes (Ikari et al. 1998). In contrast, PKC downregulates NCX activity in other cell types such as bovine chromaffin cells (DiPolo and Beauge 1995; Tokumura et al. 1998). It is possible that some as yet unknown factors participate in the regulation of the NCX activity in these different tissues and that this contributes to these contradictory results. In this context, the interaction of NCX1 with CK represents an indirect regulation of NCX activity by PKC. It is possible that CK interacts with NCX via its C terminus under control conditions, and under energy-compromised conditions, CK is phosphorylated by PKC or other kinases, which further induces the C terminus of CK to take up a position for maintaining NCX activity.

14.8 Conclusions

We discovered a novel way by which NCX1 activity is regulated that involves two CK isozymes, sMiCK and CKM, which are able to recover the decreased NCX1 activity under energy-compromised conditions. The C-terminal region and the putative PKC phosphorylation Ser residue of sMiCK and CKM are critical for the effect of these CKs on NCX1 activity, but catalytic activity and autophosphorylation of sMiCK and CKM are not required. Our results show not only a novel mechanism for regulation of NCX1 activity by CK but also a novel function for CK. Our results provide new direction for the understanding of the role of NCX in the regulation of cardiac physiology.

Acknowledgments This research was originally published in the Journal of Biological Chemistry; Yang, Y.-C., Fann, M.-J., Chang, W.-H., Tai, L.-H., Jiang, J.-H., and Kao, L.-S. (2010) Regulation of sodium-calcium exchanger activity by creatine kinase under energy-compromised conditions. J. Biol. Chem. Vol. 285, 28275–28285 © the American Society for Biochemistry and Molecular Biology.

References

G.M. Besserer, M. Ottolia, D.A. Nicoll, V. Chaptal, D. Cascio, K.D. Philipson, J. Abramson, The second Ca^{2+}-binding domain of the Na^+ Ca^{2+} exchanger is essential for regulation: crystal structures and mutational analysis. Proc. Natl. Acad. Sci. U. S. A. **104**, 18467–18472 (2007)

M.P. Blaustein, W.J. Lederer, Sodium/calcium exchange: its physiological implications. Physiol. Rev. **79**, 763–854 (1999)

M.P. Blaustein, W.F. Goldman, G. Fontana, B.K. Krueger, E.M. Santiago, T.D. Steele, D.N. Weiss, P.J. Yarowsky, Physiological roles of the sodium-calcium exchanger in nerve and muscle. Ann. N. Y. Acad. Sci. **639**, 254–274 (1991)

Y.J. Chern, S.H. Chueh, Y.J. Lin, C.M. Ho, L.S. Kao, Presence of Na^+/Ca^{2+} exchange activity and its role in regulation of intracellular calcium concentration in bovine adrenal chromaffin cells. Cell Calcium **13**, 99–106 (1992)

C.H. Cho, S.S. Kim, M.J. Jeong, C.O. Lee, H.S. Shin, The Na^+-Ca^{2+} exchanger is essential for embryonic heart development in mice. Mol. Cells **10**, 712–722 (2000)

M. Condrescu, J.P. Gardner, G. Chernaya, J.F. Aceto, C. Kroupis, J.P. Reeves, ATP-dependent regulation of sodium-calcium exchange in Chinese hamster ovary cells transfected with the bovine cardiac sodium-calcium exchanger. J. Biol. Chem. **270**, 9137–9146 (1995)

R. DiPolo, L. Beauge, Phosphoarginine stimulation of Na+-Ca2+ exchange in squid axons–a new pathway for metabolic regulation? J. Physiol. **487**(Pt 1), 57–66 (1995)

R. DiPolo, L. Beauge, Metabolic pathways in the regulation of invertebrate and vertebrate Na^+/Ca^{2+} exchange. Biochim. Biophys. Acta **1422**, 57–71 (1999)

R. DiPolo, L. Beauge, Sodium/calcium exchanger: influence of metabolic regulation on ion carrier interactions. Physiol. Rev. **86**, 155–203 (2006)

R. DiPolo, G. Berberian, D. Delgado, H. Rojas, L. Beauge, A novel 13 kDa cytoplasmic soluble protein is required for the nucleotide (MgATP) modulation of the Na^+/Ca^{2+} exchange in squid nerve fibers. FEBS Lett. **401**, 6–10 (1997)

A.E. Doering, D.A. Nicoll, Y. Lu, L. Lu, J.N. Weiss, K.D. Philipson, Topology of a functionally important region of the cardiac Na^+/Ca^{2+} exchanger. J. Biol. Chem. **273**, 778–783 (1998)

M. Eder, K. Fritz-Wolf, W. Kabsch, T. Wallimann, U. Schlattner, Crystal structure of human ubiquitous mitochondrial creatine kinase. Proteins **39**, 216–225 (2000a)

M. Eder, M. Stolz, T. Wallimann, U. Schlattner, A conserved negatively charged cluster in the active site of creatine kinase is critical for enzymatic activity. J. Biol. Chem. **275**, 27094–27099 (2000b)

B.N. Eigel, H. Gursahani, R.W. Hadley, Na^+/Ca^{2+} exchanger plays a key role in inducing apoptosis after hypoxia in cultured guinea pig ventricular myocytes. Am. J. Physiol. Heart Circ. Physiol. **287**, H1466–H1475 (2004)

R.A. Haworth, A.V. Biggs, Effect of ATP depletion on kinetics of Na^+/Ca^{2+} exchange-mediated Ca^{2+} influx in Na^+-loaded heart cells. J. Mol. Cell. Cardiol. **29**, 503–514 (1997)

D.W. Hilgemann, Cytoplasmic ATP-dependent regulation of ion transporters and channels: mechanisms and messengers. Annu. Rev. Physiol. **59**, 193–220 (1997)

A. Ikari, H. Sakai, N. Takeguchi, Protein kinase C-mediated up-regulation of Na^+/Ca^{2+}-exchanger in rat hepatocytes determined by a new Na^+/Ca^{2+}-exchanger inhibitor, KB-R7943. Eur. J. Pharmacol. **360**, 91–98 (1998)

T. Iwamoto, S. Wakabayashi, M. Shigekawa, Growth factor-induced phosphorylation and activation of aortic smooth muscle Na^+/Ca^{2+} exchanger. J. Biol. Chem. **270**, 8996–9001 (1995)

T. Iwamoto, Y. Pan, S. Wakabayashi, T. Imagawa, H.I. Yamanaka, M. Shigekawa, Phosphorylation-dependent regulation of cardiac Na^+/Ca^{2+} exchanger via protein kinase C. J. Biol. Chem. **271**, 13609–13615 (1996)

T. Iwamoto, S. Kita, A. Uehara, Y. Inoue, Y. Taniguchi, I. Imanaga, M. Shigekawa, Structural domains influencing sensitivity to isothiourea derivative inhibitor KB-R7943 in cardiac Na^+/Ca^{2+} exchanger. Mol. Pharmacol. **59**, 524–531 (2001)

L.S. Kao, Calcium homeostasis in digitonin-permeabilized bovine chromaffin cells. J. Neurochem. **51**, 221–227 (1988)

L.S. Kao, N.S. Cheung, Mechanism of calcium transport across the plasma membrane of bovine chromaffin cells. J. Neurochem. **54**, 1972–1979 (1990)

P. Kofuji, W.J. Lederer, D.H. Schulze, Mutually exclusive and cassette exons underlie alternatively spliced

isoforms of the Na/Ca exchanger. J. Biol. Chem. **269**, 5145–5149 (1994)
I. Komuro, K.E. Wenninger, K.D. Philipson, S. Izumo, Molecular cloning and characterization of the human cardiac Na^{+}/Ca^{2+} exchanger cDNA. Proc. Natl. Acad. Sci. U. S. A. **89**, 4769–4773 (1992)
D.O. Levitsky, D.A. Nicoll, K.D. Philipson, Identification of the high affinity Ca^{2+}-binding domain of the cardiac Na^{+}-Ca^{2+} exchanger. J. Biol. Chem. **269**, 22847–22852 (1994)
Z. Li, D.A. Nicoll, A. Collins, D.W. Hilgemann, A.G. Filoteo, J.T. Penniston, J.N. Weiss, J.M. Tomich, K.D. Philipson, Identification of a peptide inhibitor of the cardiac sarcolemmal Na^{+}-Ca^{2+} exchanger. J. Biol. Chem. **266**, 1014–1020 (1991)
Z. Li, S. Matsuoka, L.V. Hryshko, D.A. Nicoll, M.M. Bersohn, E.P. Burke, R.P. Lifton, K.D. Philipson, Cloning of the NCX2 isoform of the plasma membrane Na^{+}-Ca^{2+} exchanger. J. Biol. Chem. **269**, 17434–17439 (1994)
G. Lin, Y. Liu, K.M. MacLeod, Regulation of muscle creatine kinase by phosphorylation in normal and diabetic hearts. Cell. Mol. Life Sci. **66**, 135–144 (2009)
P.S. Liu, L.S. Kao, Na^{+}-dependent Ca^{2+} influx in bovine adrenal chromaffin cells. Cell Calcium **11**, 573–579 (1990)
S.O. Marx, J. Kurokawa, S. Reiken, H. Motoike, J. D'Armiento, A.R. Marks, R.S. Kass, Requirement of a macromolecular signaling complex for beta adrenergic receptor modulation of the KCNQ1-KCNE1 potassium channel. Science **295**, 496–499 (2002)
S. Matsuoka, D.A. Nicoll, Z. He, K.D. Philipson, Regulation of cardiac Na^{+}-Ca^{2+} exchanger by the endogenous XIP region. J. Gen. Physiol. **109**, 273–286 (1997)
J.A. Mattiello, K.B. Margulies, V. Jeevanandam, S.R. Houser, Contribution of reverse-mode sodium-calcium exchange to contractions in failing human left ventricular myocytes. Cardiovasc. Res. **37**, 424–431 (1998)
E.D. Moore, E.F. Etter, K.D. Philipson, W.A. Carrington, K.E. Fogarty, L.M. Lifshitz, F.S. Fay, Coupling of the Na^{+}/Ca^{2+} exchanger, Na^{+}/K^{+} pump and sarcoplasmic reticulum in smooth muscle. Nature **365**, 657–660 (1993)
D.A. Nicoll, B.D. Quednau, Z. Qui, Y.R. Xia, A.J. Lusis, K.D. Philipson, Cloning of a third mammalian Na^{+}-Ca^{2+} exchanger, NCX3. J. Biol. Chem. **271**, 24914–24921 (1996)
M. Ottolia, D.A. Nicoll, K.D. Philipson, Roles of two Ca^{2+}-binding domains in regulation of the cardiac Na^{+}-Ca^{2+} exchanger. J. Biol. Chem. **284**, 32735–32741 (2009)
C.Y. Pan, L.S. Kao, Catecholamine secretion from bovine adrenal chromaffin cells: the role of the Na^{+}/Ca^{2+} exchanger and the intracellular Ca^{2+} pool. J. Neurochem. **69**, 1085–1092 (1997)
C.Y. Pan, Y.S. Chu, L.S. Kao, Molecular study of the Na^{+}/Ca^{2+} exchanger in bovine adrenal chromaffin cells. Biochem. J. **336**(Pt 2), 305–310 (1998)
C.Y. Pan, L.L. Tsai, J.H. Jiang, L.W. Chen, L.S. Kao, The co-presence of Na^{+}/Ca^{2+}-K^{+} exchanger and Na^{+}/Ca^{2+} exchanger in bovine adrenal chromaffin cells. J. Neurochem. **107**, 658–667 (2008)
M.V. Pulina, R. Rizzuto, M. Brini, E. Carafoli, Inhibitory interaction of the plasma membrane Na^{+}/Ca^{2+} exchangers with the 14-3-3 proteins. J. Biol. Chem. **281**, 19645–19654 (2006)
B.D. Quednau, D.A. Nicoll, K.D. Philipson, Tissue specificity and alternative splicing of the Na^{+}/Ca^{2+} exchanger isoforms NCX1, NCX2, and NCX3 in rat. Am. J. Physiol. **272**, C1250–C1261 (1997)
B.D. Quednau, D.A. Nicoll, K.D. Philipson, The sodium/calcium exchanger family-SLC8. Pflugers Arch. **447**, 543–548 (2004)
X. Ren, D.A. Nicoll, K.D. Philipson, Helix packing of the cardiac Na^{+}-Ca^{2+} exchanger: proximity of transmembrane segments 1, 2, and 6. J. Biol. Chem. **281**, 22808–22814 (2006)
M. Reppel, B.K. Fleischmann, H. Reuter, P. Sasse, H. Schunkert, J. Hescheler, Regulation of the Na^{+}/Ca^{2+} exchanger (NCX) in the murine embryonic heart. Cardiovasc. Res. **75**, 99–108 (2007)
U. Schlattner, M. Tokarska-Schlattner, T. Wallimann, Mitochondrial creatine kinase in human health and disease. Biochim. Biophys. Acta **1762**, 164–180 (2006)
D.H. Schulze, M. Muqhal, W.J. Lederer, A.M. Ruknudin, Sodium/calcium exchanger (NCX1) macromolecular complex. J. Biol. Chem. **278**, 28849–28855 (2003)
E.M. Schwarz, S. Benzer, Calx, a Na-Ca exchanger gene of *Drosophila* melanogaster. Proc. Natl. Acad. Sci. U. S. A. **94**, 10249–10254 (1997)
M. Spindler, K. Meyer, H. Stromer, A. Leupold, E. Boehm, H. Wagner, S. Neubauer, Creatine kinase-deficient hearts exhibit increased susceptibility to ischemia-reperfusion injury and impaired calcium homeostasis. Am. J. Physiol. Heart Circ. Physiol. **287**, H1039–H1045 (2004)
K. Steeghs, A. Benders, F. Oerlemans, A. de Haan, A. Heerschap, W. Ruitenbeek, C. Jost, J. van Deursen, B. Perryman, D. Pette, M. Bruckwilder, J. Koudijs, P. Jap, J. Veerkamp, B. Wieringa, Altered Ca^{2+} responses in muscles with combined mitochondrial and cytosolic creatine kinase deficiencies. Cell **89**, 93–103 (1997)
A. Tokumura, M. Okuno, K. Fukuzawa, H. Houchi, K. Tsuchiya, M. Oka, Positive and negative controls by protein kinases of sodium-dependent Ca^{2+} efflux from cultured bovine adrenal chromaffin cells stimulated by lysophosphatidic acid. Biochim. Biophys. Acta **1389**, 67–75 (1998)
K. Wakimoto, K. Kobayashi, O.M. Kuro, A. Yao, T. Iwamoto, N. Yanaka, S. Kita, A. Nishida, S. Azuma,

Y. Toyoda, K. Omori, H. Imahie, T. Oka, S. Kudoh, O. Kohmoto, Y. Yazaki, M. Shigekawa, Y. Imai, Y. Nabeshima, I. Komuro, Targeted disruption of Na^+/Ca^{2+} exchanger gene leads to cardiomyocyte apoptosis and defects in heartbeat. J. Biol. Chem. **275**, 36991–36998 (2000)

T. Wallimann, M. Wyss, D. Brdiczka, K. Nicolay, H.M. Eppenberger, Intracellular compartmentation, structure and function of creatine kinase isoenzymes in tissues with high and fluctuating energy demands: the 'phosphocreatine circuit' for cellular energy homeostasis. Biochem. J. **281**(Pt 1), 21–40 (1992)

C.R. Weber, V. Piacentino 3rd, S.R. Houser, D.M. Bers, Dynamic regulation of sodium/calcium exchange function in human heart failure. Circulation **108**, 2224–2229 (2003)

M.P. Wu, L.S. Kao, H.T. Liao, C.Y. Pan, Reverse mode Na^+/Ca^{2+} exchangers trigger the release of Ca^{2+} from intracellular Ca^{2+} stores in cultured rat embryonic cortical neurons. Brain Res. **1201**, 41–51 (2008)

W. Xu, H. Denison, C.C. Hale, C. Gatto, M.A. Milanick, Identification of critical positive charges in XIP, the Na/Ca exchange inhibitory peptide. Arch. Biochem. Biophys. **341**, 273–279 (1997)

Coordinated Regulation of Cardiac Na^+/Ca^{2+} Exchanger and Na^+-K^+-ATPase by Phospholemman (FXYD1)

15

Joseph Y. Cheung, Xue-Qian Zhang, Jianliang Song, Erhe Gao, Tung O. Chan, Joseph E. Rabinowitz, Walter J. Koch, Arthur M. Feldman, and JuFang Wang

Abstract

Phospholemman (PLM) is the founding member of the FXYD family of regulators of ion transport. PLM is a 72-amino acid protein consisting of the signature PFXYD motif in the extracellular N terminus, a single transmembrane (TM) domain, and a C-terminal cytoplasmic tail containing three phosphorylation sites. In the heart, PLM co-localizes and co-immunoprecipitates with Na^+-K^+-ATPase, Na^+/Ca^{2+} exchanger, and L-type Ca^{2+} channel. The TM domain of PLM interacts with TM9 of the α-subunit of Na^+-K^+-ATPase, while its cytoplasmic tail interacts with two small regions (spanning residues 248–252 and 300–304) of the proximal intracellular loop of Na^+/Ca^{2+} exchanger. Under stress, catecholamine stimulation phosphorylates PLM at serine68, resulting in relief of inhibition of Na^+-K^+-ATPase by decreasing Km for Na^+ and increasing Vmax, and simultaneous inhibition of Na^+/Ca^{2+} exchanger. Enhanced Na^+-K^+-ATPase activity lowers intracellular Na^+, thereby minimizing Ca^{2+} overload and risks of arrhythmias. Inhibition of Na^+/Ca^{2+} exchanger reduces Ca^{2+} efflux, thereby preserving contractility. Thus, the coordinated actions of PLM during stress serve to minimize arrhythmogenesis and maintain inotropy. In acute cardiac ischemia and chronic heart failure, either expression or phosphorylation of PLM or both are altered. PLM regulates important ion transporters in the heart and offers a tempting target for development of drugs to treat heart failure.

J.Y. Cheung, M.D., Ph.D. (✉) • X.-Q. Zhang • J. Song
E. Gao • T.O. Chan • J.E. Rabinowitz • W.J. Koch
A.M. Feldman • J. Wang
Center of Translational Medicine, Temple University School of Medicine, 3500 N. Broad Street, MERB-960, Philadelphia, PA 19140, USA
e-mail: joseph.cheung@tuhs.temple.edu

L. Annunziato (ed.), *Sodium Calcium Exchange: A Growing Spectrum of Pathophysiological Implications*, Advances in Experimental Medicine and Biology 961, DOI 10.1007/978-1-4614-4756-6_15,

Keywords
FXYD proteins • Ion transport • Cardiac excitation-contraction coupling • Inotropy • Arrhythmias • Catecholamines • Stress protein • Intracellular Na^+ and Ca^{2+}

15.1 Introduction

The cardiac Na^+/Ca^{2+} exchanger (NCX1) is an important sarcolemmal ion transporter in the heart (Bers 2002). During most of the cardiac excitation-contraction (EC) cycle, NCX1 operates in the forward or Ca^{2+} efflux (3 Na^+ in: 1 Ca^{2+} out) mode. During systole, when the membrane potential (E_m) exceeds the equilibrium potential of NCX1 (E_{NaCa}), Ca^{2+} influx (3 Na^+ out: 1 Ca^{2+} in) is favored. This unique capability of NCX1 to modulate both Ca^{2+} influx and efflux during EC confers upon it the ability to regulate both contraction and relaxation. Altered expression and activity of NCX1 has been postulated to account for contractile dysfunction in many cardiac disease models, including human cardiomyopathy (Sipido et al. 2002). In addition, elevated NCX1 activity has been proposed as a mechanism for arrhythmogenesis in heart failure (Pogwizd et al. 2001). NCX1 therefore is a tempting target for drug development in therapy of heart failure, arrhythmias, and cardiomyopathy (Sipido et al. 2002; Hasenfuss and Schillinger 2004).

NCX1 is a 938-amino acid (939 amino acid in the rat) protein consisting of an extracellular N-terminal domain comprising the first five transmembrane (TM) segments, a large intracellular loop (residues 218–764), and an intracellular C-terminal domain comprising the last four TM segments (Fig. 15.1) (Nicoll et al. 1999; Philipson and Nicoll 2000). The α-repeats in TM segments 2, 3, and 7 of NCX1 are important in ion transport activity (Nicoll et al. 1996; Iwamoto et al. 2000), while the large intracellular loop contains the regulatory domains of the exchanger (Li et al. 1991; Levitsky et al. 1994; Maack et al. 2005). Specifically, the exchange inhibitory peptide (XIP) region (residues 219–238) (Li et al. 1991), the proximal linker domain (residues 259–370), the Ca^{2+}-binding domains (CBD) 1 (residues 371–508) (Levitsky et al. 1994) and 2 (residues 501–689) (Hilge et al. 2006), and the interaction site for endogenous XIP (residues 562–679) (Maack et al. 2005) all reside in the intracellular loop. Although the structures of CBD1 and CBD2 have been elucidated (Hilge et al. 2006, 2009; Nicoll et al. 2006; Ottolia et al. 2009), the three-dimensional structures of proximal linker domain and the distal intracellular loop remain unsolved.

Although NCX1 is a key Ca^{2+} transporter in the heart, remarkably little is known about its functional regulation (Blaustein and Lederer 1999; Philipson and Nicoll 2000; DiPolo and Beauge 2006). In 2002, we first speculated that phospholemman (PLM), the founding member of the FXYD family of small regulators of ion transport (Sweadner and Rael 2000), modulates NCX1 activity in the heart (Song et al. 2002). In 2003, we demonstrated that overexpression of PLM in adult rat cardiac myocytes inhibits Na^+/Ca^{2+} exchange current (I_{NaCa}) (Zhang et al. 2003). In 2005, we confirmed in transfected HEK293 cells that PLM is indeed the first reported endogenous regulator of NCX1 (Ahlers et al. 2005). In 2006, we showed that despite similar NCX1 protein levels (Tucker et al. 2006), I_{NaCa} is higher in PLM-knockout (KO) myocytes (Zhang et al. 2006a). Our results provide unambiguous proof that PLM is the first endogenous regulator of NCX1. This chapter reviews the discovery, structure, and known functions of PLM; the molecular sites of interaction between PLM and NCX1 and between PLM and Na^+-K^+-ATPase; and finally the role of PLM in maintenance of cardiac contractility and decrease in risks of arrhythmogenesis under stressful conditions.

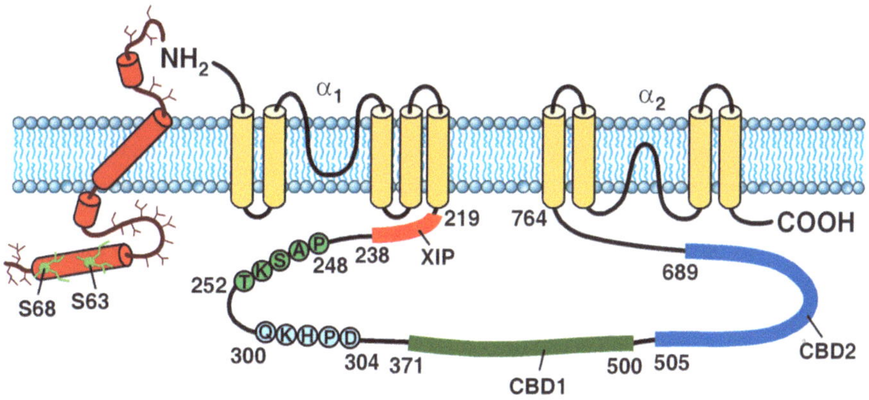

Fig. 15.1 *Molecular models of phospholemman and Na^{+}/Ca^{2+} exchanger. Left*: nuclear magnetic resonance studies of highly purified phospholemman in micelles reveal four helices of the protein with a single transmembrane (TM) domain. The FXYD motif is in the extracellular domain, and the physiologically important phosphorylation sites serine[63] and serine[68] are in the cytoplasm. *Right*: the mature Na^{+}/Ca^{2+} exchanger is modeled to consist of nine TM segments with two re-entrant loops (between TM2 and TM3 and between TM7 and TM8) as part of the conserved α-repeat motifs that are important in ion transport activity. The N terminus is extracellular and the C terminus is intracellular. Between TM5 and TM6 is a large intracellular loop (residues 218–764) which contains the regulatory domains of the exchanger. Specifically, the proximal linker domain (residues 218–358) which interacts with phospholemman, the exchange inhibitory peptide (XIP) region (residues 219–238), the two calcium-binding domains 1 (residues 371–500) and 2 (residues 505–689) connected in tandem by a short linker (residues 501–504), and the interaction site for endogenous XIP (residues 562–679) all reside within the intracellular loop. The two specific segments [residues 248–252 (PASKT) and residues 300–304 (QKHPD)] in the proximal linker domain responsible for inhibition of Na^{+}/Ca^{2+} exchanger by phospholemman are shown

15.2 Phospholemman: History

PLM was identified in 1985 as a 15-kDa sarcolemmal (SL) protein that is phosphorylated in response to isoproterenol and is distinct from the sarcoplasmic reticulum (SR) protein phospholamban (PLB) (Presti et al. 1985a). PLM is also phosphorylated by α-adrenergic agonists (Lindemann 1986) and protein kinase (PK) C (Presti et al. 1985b). The complete protein sequence of PLM has been determined and the cDNA cloned (Palmer et al. 1991). In humans, PLM gene is localized to chromosome 19q13.1 (Chen et al. 1997).

Early studies based on PLM overexpression in Xenopus oocytes suggest that PLM is a hyperpolarization-activated anion-selective channel (Moorman et al. 1995). In lipid bilayers, PLM forms a channel that is highly selective for taurine (Chen et al. 1998). In noncardiac tissues, PLM may function as a regulator of cell volume (Morales-Mulia et al. 2000; Davis et al. 2004). The function of PLM in the heart remained unknown until this century.

15.3 Phospholemman: Molecular Structure and Phosphorylation

PLM is synthesized as a 92-amino acid peptide with a 20-amino acid signal peptide at the N terminus. The mature PLM consists of 72 amino acids with a calculated molecular weight of 8,409 although the protein migrates at 15 kDa on sodium dodecyl sulfate polyacrylamide gel electrophoresis (SDS-PAGE) gels. The extracellular N terminus contains 17 amino acids which includes the PFXYD motif; a single transmembrane (TM) domain comprises of 20 amino acids, while the remaining 35 amino acids form the cytoplasmic tail. The C-terminal cytoplasmic tail of dog, human, rabbit, and rat PLM contains three serines (at residues 62, 63, and 68) and one threonine (at residue 69), but threonine[69] is replaced by serine in mouse PLM (Sweadner and Rael 2000). When reconstituted in liposomes, the TM domain of PLM is an α-helix with a maximum tilt of 15–17° (Beevers and Kukol 2006; Franzin et al. 2007). Highly purified PLM in model micelles consists of four α-helices: H1 (residues 12–17) is in the extracellular domain, H2 (residues 22–38) is the main TM helix followed by a short H3 (residues 39–45), and H4 (residues 60–68) in the C terminus is connected to H3 by a flexible linker (Fig. 15.1) (Teriete et al. 2007). Synthetic peptide encompassing the TM domain of PLM forms tetramers when reconstituted in liposomes (Beevers and Kukol 2006). In transfected human embryonic kidney (HEK) 293 cells overexpressing PLM, fluorescence resonance energy transfer (FRET) studies suggest PLM forms oligomers consisting of 3–4 PLM molecules (Bossuyt et al. 2006, 2009; Song et al. 2011).

In isolated rat diaphragm, PLM is phosphorylated at serine^{68} by PKA and at both serine^{63} and serine^{68} by PKC (Waalas et al. 1994). In vitro studies using synthetic PLM fragments suggest that serine^{63} and threonine^{69} may be additional phosphorylation targets for PKA and PKC, respectively (Fuller et al. 2009). By assessing the effects of serine mutants of PLM on I_{NaCa}, ~46% of serine^{68} and ~16% of serine^{63} in adult rat cardiac myocytes are estimated to be phosphorylated in the resting state (Song et al. 2005). When evaluated with phospho-specific anti-PLM antibodies (Rembold et al. 2005; Fuller et al. 2009), ~30–40% of PLM in adult rat cardiac myocytes (Zhang et al. 2006b; Fuller et al. 2009) and ~25% of PLM in guinea pig myocytes (Silverman et al. 2005) are phosphorylated under resting conditions. In transfected HEK293 cells overexpressing PLM, ~30–45% of PLM is phosphorylated in the basal state (Zhang et al. 2009).

15.4 Phospholemman and the FXYD Family of Regulators of Ion Transport

Members in the FXYD family are small, single membrane-spanning proteins involved in regulation of ion transport (Sweadner and Rael 2000). PLM (FXYD1) is the first cloned member. All FXYD family members contain the signature PFXYD motif in the N terminus. With the exception of γ-subunit of Na^+-K^+-ATPase (FXYD2), all other FXYD proteins contain potential phosphorylation sites in the C terminus. As a family, FXYD proteins are expressed in tissues involved in fluid and solute transport (kidney, colon, pancreas, mammary gland, liver, lung, prostate, and placenta) or are electrically excitable (cardiac and skeletal muscle, neural tissues). To date, there are at least 12 known FXYD proteins including mammary-associated tumor 8 kDa (MAT-8 or FXYD3), channel-inducing factor (CHIF or FXYD4), dysadherin (FXYD5; also known as RIC [related to ion channel]), phosphohippolin (FXYD6), FXYD7, and phospholemman shark (PLM-S or FXYD10). Notably, FXYD3 and FXYD5 are overexpressed in a variety of tumors and may be associated with tumor progression (Nam et al. 2007; Yamamoto et al. 2009).

15.5 Phospholemman: Regulator of Na^+-K^+-ATPase

PLM co-immunoprecipitates (Crambert et al. 2002; Fuller et al. 2004; Bossuyt et al. 2005; Wang et al. 2010a) and co-localizes (Silverman et al. 2005; Cheung et al. 2010) with α-subunits of Na^+-K^+-ATPase in mammalian hearts. When co-expressed with α- and β-subunits of Na^+-K^+-ATPase in Xenopus oocytes, PLM increases K_m for Na^+ and K^+ without affecting V_{max} of the enzyme (Crambert et al. 2002). In cardiac myocytes and homogenates, PLM reduces its apparent affinities for intracellular Na^+ (Despa et al. 2005; Bossuyt et al. 2009) and extracellular K^+ (Han et al. 2009) as well as decreases V_{max} (Fuller et al. 2004; Zhang et al. 2006a; Bell et al. 2008; Wang et al. 2010a). Phosphorylation of PLM at serine^{68} results in increasing V_{max} (Fuller et al. 2004; Silverman et al. 2005; Wang et al. 2010a) while reducing the apparent K_m for Na^+ (Despa et al. 2005) but not for K^+ (Han et al. 2009) of cardiac Na^+-K^+-ATPase.

There are four isoforms of the catalytic α-subunits of Na^+-K^+-ATPase, and expression of each isoform is tissue and species dependent (Blanco and Mercer 1998). Whereas human (Zahler et al. 1993; McDonough et al. 1996) and rabbit (Bossuyt et al. 2005) hearts express α1- (ouabain-resistant), α2-, and α3-isoforms, rodent hearts express only α1- and α2-isoforms of Na^+-K^+-ATPase. PLM co-immunoprecipitates all three α-subunits in human and rabbit (Bossuyt et al. 2005) and α1- and α2-subunits in mouse (Wang et al. 2010a) and bovine (Crambert et al. 2002), but only α1-subunit of Na^+-K^+-ATPase in rat (Fuller et al. 2004) and guinea pig (Silverman et al. 2005) hearts. It is important to note that co-immunoprecipitation of PLM with α-subunits of Na^+-K^+-ATPase does not require the presence of NCX1 in cardiac membranes (Wang et al. 2010a).

Exploiting the differential ouabain sensitivity between α1- and α2-subunits of Na^+-K^+-ATPase, currents due to Na^+-K^+-ATPase (I_{pump}) can be separated into that due to α1- ($I_{\alpha 1}$) and α2-subunit ($I_{\alpha 2}$) activities. $I_{\alpha 1}$ represents ~82–88% and ~73% of I_{pump} in mouse (Berry et al. 2007; Wang et al. 2010a) and guinea pig cardiac myocytes (Silverman et al. 2005), respectively. By measuring I_{pump} in adult mouse and guinea pig cardiac myocytes at baseline and after isoproterenol stimulation, it appears that PLM regulates the activity of α1- but not α2-subunit of Na^+-K^+-ATPase. This conclusion is based on derivation of $I_{\alpha 2}$ from the difference of two large numbers ($I_{pump} - I_{\alpha 1}$), and the method may not have the requisite sensitivity to detect small changes in $I_{\alpha 2}$ in response to isoproterenol stimulation. Using "SWAP" mouse (Dostanic et al. 2004) in which the ouabain affinities of the α-subunits are reversed, PLM regulates the apparent K_m for Na^+ of both α1- and α2-subunits of Na^+-K^+-ATPase (Bossuyt et al. 2009). It is therefore plausible but not definitively proven that in wild-type (WT) hearts, PLM regulates the activities of both α1- and α2-subunits of Na^+-K^+-ATPase.

15.6 Molecular Interactions Between Phospholemman and Na^+-K^+-ATPase

Early studies using mutational analysis suggest that FXYD proteins (FXYD2, 4 and 7) interact with TM9 segment of Na^+-K^+-ATPase (Li et al. 2004). Covalent cross-linking and co-immunoprecipitation studies demonstrate that the TM segment of PLM is close to TM2 of Na^+-K^+-ATPase (Lindzen et al. 2006). Based on crystal structure of Ca^{2+}-ATPase in the E_1-ATP bound conformation, the TM segment of PLM is modeled to dock in the groove between TM segments 2, 6, and 9 of the α-subunit of Na^+-K^+-ATPase. High-resolution (2.4 Å) crystal structure of shark rectal gland Na^+-K^+-ATPase in the E2.2 K^+. P_i state indicates that FXYD proteins interact almost exclusively with the outside of TM9 of the α-subunit (Shinoda et al. 2009). The FXYD motif stabilizes interactions between α- and β-subunits of the Na^+ pump.

Phosphorylation of PLM-S causes it to dissociate from shark Na^+-K^+-ATPase (Mahmmoud et al. 2000). By contrast, phosphorylated PLM remains associated with the α-subunit of Na^+-K^+-ATPase (Silverman et al. 2005; Bossuyt et al. 2009). Nuclear magnetic resonance studies of highly purified PLM reconstituted in micelles show no major conformational changes on phosphorylation of serine^{68} (Teriete et al. 2009). By contrast, in transfected HEK293 cells co-expressing PLM and Na^+-K^+-ATPase, FRET demonstrates decrease in interaction between phosphorylated PLM and Na^+-K^+-ATPase (Bossuyt et al. 2006; 2009). The conclusion that PLM phosphorylation alters its conformation is given additional support by FRET which shows phosphomimetic mutants of PLM reduce the apparent affinity of interaction between Na^+-K^+-ATPase and PLM (Song et al. 2011). It is likely that phosphorylation of PLM results in subtle changes in its conformation and alters its physical interaction with Na^+-K^+-ATPase.

15.7 Functional Significance of Na^+-K^+-ATPase Regulation by PLM in Resting Heart

A major contribution to the elucidation of the physiological function of PLM in the heart is the generation of PLM-KO mice (Jia et al. 2005). Compared to WT hearts, congenic PLM-KO hearts express similar levels of α1-, α2-, β1-, and β2-subunits of Na^+-K^+-ATPase, NCX1, SR Ca^{2+}-ATPase (SERCA2), PLB, and calsequestrin

(Tucker et al. 2006; Bell et al. 2008). However, Na^+-K^+-ATPase enzymatic activity (Bell et al. 2008) and I_{pump} (Song et al. 2008; Wang et al. 2010a) are higher in PLM-KO hearts, as expected from relief of tonic inhibition by PLM. Higher Na^+-K^+-ATPase activity in PLM-KO hearts would be expected to lower intracellular Na^+ concentration ($[Na^+]_i$), thereby thermodynamically favoring Ca^{2+} efflux via NCX1 and resulting in lower cardiac contractility (Bell et al. 2008). However, basal $[Na^+]_i$ was similar between WT and PLM-KO cardiac myocytes (Despa et al. 2005; Wang et al. 2010a). In isolated cardiac myocytes paced to contract at 1 Hz and extracellular Ca^{2+} concentration ($[Ca^{2+}]_o$) of 1.8 mM, intracellular Ca^{2+} concentration ($[Ca^{2+}]_i$) transient and myocyte contraction amplitudes are similar between WT and PLM-KO myocytes (Tucker et al. 2006). Finally, in vivo myocardial contractility assessed by echocardiography and cardiac catheterization shows no differences in ejection fraction, $^+dP/dt$ and $-dP/dt$ between WT and PLM-KO hearts (Bell et al. 2008; Wang et al. 2010a) in the resting state. These observations indicate that under resting conditions, the regulatory effects of PLM on Na^+-K^+-ATPase in the heart are not discernible and PLM is functionally quiescent.

The paradox that Na^+-K^+-ATPase activity is higher but cardiac contractility is similar between WT and PLM-KO hearts may be reconciled by the following considerations: First, there may be a distinct pool of Na^+-K^+-ATPase not regulated by PLM but intimately involved in regulation of cardiac contractility. For example, in adult mouse and rat cardiac myocytes, the α2-subunit of Na^+-K^+-ATPase is preferentially distributed in the t-tubules (Berry et al. 2007; Swift et al. 2007) and is involved with regulation of Ca^{2+} (James et al. 1999) and contractility (Swift et al. 2007). PLM appears not to regulate α2-subunit of Na^+-K^+-ATPase (Silverman et al. 2005; Wang et al. 2010a) although this point is controversial (Bossuyt et al. 2009). Second, maintenance of resting $[Na^+]_i$ under basal conditions may only require a fraction of Na^+-K^+-ATPase activity. Assuming the 1:1 stoichiometry of PLM:Na^+-K^+-ATPase observed in transfected HEK293 cells (Bossuyt et al. 2009), the 25–40% of PLM phosphorylated in resting cardiac myocytes (Silverman et al. 2005; Zhang et al. 2006b; Fuller et al. 2009) may unencumber enough Na^+-K^+-ATPase for basal $[Na^+]_i$ to be maintained at similar levels between WT and PLM-KO myocytes. Third, resting $[Na^+]_i$ in mouse myocytes of 6–12 mM (Despa et al. 2005; Wang et al. 2010a, 2011) is close to the K_m for Na^+ (~10 mM) of Na^+-K^+-ATPase (Despa et al. 2002; Zhang et al. 2006b). Since PLM regulates Na^+-K^+-ATPase by both K_m for Na^+ (Despa et al. 2005) and V_{max} (Zhang et al. 2006b; Bell et al. 2008) mechanisms, the effects of PLM on Na^+-K^+-ATPase activity at resting $[Na^+]_i$ may be rather small. This hypothesis is supported by the observation that I_{pump} is higher in PLM-KO myocytes under high but not low pipette $[Na^+]$ conditions (Wang et al. 2010a).

15.8 Phospholemman: Endogenous Regulator of Na^+/Ca^{2+} Exchanger

When dog PLM is overexpressed (1.4- to 3.5-fold) in adult rat cardiac myocytes by adenovirus-mediated gene transfer, expression of α1- and α2-subunits of Na^+-K^+-ATPase, NCX1, SERCA2, and calsequestrin is not affected (Song et al. 2002; Zhang et al. 2003, 2006a). As expected, I_{pump} is decreased in rat myocytes overexpressing PLM, primarily due to reduction in V_{max} rather than changes in apparent K_m for Na^+ or K^+ (Zhang et al. 2006b). Inhibition of Na^+-K^+-ATPase activity would be expected to enhance myocyte contractility, similar to the mechanism of action of digitalis glycosides (Grupp et al. 1985). A totally unexpected but critical observation is that both contraction and $[Ca^{2+}]_i$ transient amplitudes (1 Hz, 5 mM $[Ca^{2+}]_o$) are lower, rather than higher, in myocytes overexpressing PLM (Song et al. 2002). By manipulating $[Ca^{2+}]_o$ and thus the thermodynamic driving force favoring Ca^{2+} influx or efflux via NCX1, the contraction phenotype of myocytes overexpressing PLM is reminiscent of that of myocytes in which NCX1 is downregulated (Tadros et al. 2002), leading us to speculate that PLM may directly inhibit NCX1 (Song et al. 2002). Follow-up studies demonstrate that PLM co-localizes with NCX1 in the sarcolemma and t-tubules (Zhang et al. 2003);

that PLM co-immunoprecipitates with NCX1 in rat, pig, and guinea pig cardiac membranes (Mirza et al. 2004; Ahlers et al. 2005; Wang et al. 2010b); that PLM overexpression inhibits I_{NaCa} (Zhang et al. 2003; Song et al. 2005) and PLM downregulation increases I_{NaCa} in rat cardiac myocytes (Mirza et al. 2004); and that I_{NaCa} is higher in PLM-KO compared to WT mouse myocytes despite no differences in NCX1 protein levels (Zhang et al. 2006a). The conclusion that PLM directly inhibits NCX1 is further supported by observations in transfected HEK293 cells co-expressing PLM and NCX1, in which both I_{NaCa} and Na^+-dependent Ca^{2+} uptake are depressed by the presence of PLM (Ahlers et al. 2005).

15.9 PLM Regulation of NCX1 Is not Mediated by Ca^{2+}-Dependent Activation

Regulation of NCX1 by PLM does not depend on synergistic interactions between CBD1 and CBD2 (Hilge et al. 2009; Ottolia et al. 2009; Giladi et al. 2010; John et al. 2011), thereby modulating Ca^{2+}-dependent activation of NCX1 (Matsuoka et al. 1995). First, I_{NaCa} measured in HEK293 cells expressing split NCX1 exchangers (comprising of N- or C-terminal domains with varying lengths of the intracellular loop) (Ottolia et al. 2001) in which both CBD1 and CBD2 are absent is still inhibited by PLM (Wang et al. 2006). Second, CBD1 (spanning residues 371–508) has no physical association with PLM (Wang et al. 2006). Finally, I_{NaCa} measured in HEK293 cells expressing NCX1-G503P mutant which lacks Ca^{2+}-dependent activation of NCX1 (Matsuoka et al. 1995) is still inhibited by PLM (Zhang et al. 2011).

15.10 Molecular Interactions Between PLM and Na^+/Ca^{2+} Exchanger

Glutathione S-transferase (GST) pull-down assay demonstrates that the intracellular loop but not the N or C terminus of NCX1 associates with PLM (Wang et al. 2006). Specifically, PLM binds to GST constructs fused to NCX1 loop segments encompassing residues 218–371 and 508–764 but not 371–508 (CBD1). Co-expressing split NCX1 exchangers (Ottolia et al. 2001) with PLM in HEK293 cells, PLM inhibits I_{NaCa} only when split exchangers contain residues 218–358 of the intracellular loop (Wang et al. 2006). PLM co-immunoprecipitates with N-terminal split exchanger only when it contains the intracellular loop segment spanning residues 218–358. These observations strongly indicate that PLM physically associates with the proximal linker domain (encompassing residues 218–358) of NCX1 and regulates I_{NaCa} via interacting with this region.

To further refine the region of NCX1 required for regulation by PLM, a family of overlapping NCX1 deletion mutants spanning the proximal linker domain is constructed and expressed in HEK293 cells. By mapping whether I_{NaCa} inhibition by PLM is preserved in overlapping NCX1 deletion mutants, it was deduced that two regions encompassing residues 238–270 and 300–328 must be present in order for PLM to regulate NCX1 (Zhang et al. 2009). Indeed, NCX1 mutants in which either residues 238–270 or residues 300–328 are deleted lose their ability to be regulated by PLM. Finally, GST constructs fused to residues 218–270 and 300–373 but not residues 250–300 physically associate with PLM. These observations indicate that two discrete regions (spanning residues 238–270 and 300–328) in the proximal linker domain of NCX1 interact with and are responsible for its regulation by PLM.

Alanine linker scanning was next used to pinpoint the residues in the proximal linker domain responsible for regulation of NCX1 by PLM. Mutating residues 248–252 (PASKT) or 300–304 (QKHPD) to alanine results in loss of I_{NaCa} inhibition by PLM in transfected HEK293 cells (Zhang et al. 2011). While mutating residue 301 alone completely abolishes PLM inhibition, single alanine mutation of residues 250–252, 300, or 302–304 results in partial reduction in inhibition. In addition, mutating residues 248–252 to alanine weakens the association with PLM, as demonstrated by GST pull-down assays. Taken together, the results indicate that the cytoplasmic tail of PLM physically and functionally interacts with residues 248–252 (PASKT) and 300–304 (QKHPD) of NCX1 (Fig. 15.1).

Although the structure of proximal linker domain is unknown, one hypothetical model of the region encompassing residues 259–370 suggests that Val^{261} is in close proximity with Ala^{314} (Hilge et al. 2006). Consistent with this model and assuming one PLM molecule interacts with one NCX1 molecule, the two small regions spanning residues 248–252 and 300–324 required for I_{NaCa} inhibition by PLM are close to each other and suggest some "structure" of the proximal linker domain. Alternatively, if, as observed in model lipid membranes and heterologous expression systems, both NCX1 (Ren et al. 2008; John et al. 2011) and PLM (Beevers and Kukol 2007; Song et al. 2011) exist as dimers and oligomers, then one PLM molecule interacts with the region encompassing residues 248–252 of one NCX1 molecule while a second PLM molecule interacts with the region spanning residues 300–304 of the other NCX1 molecule. Whether NCX1 exists as dimers and oligomers in native cardiac membranes remains to be established.

15.11 Differences Between PLM Regulation of Na^+-K^+-ATPase and Na^+/Ca^{2+} Exchanger

There are major mechanistic differences between regulation of Na^+-K^+-ATPase and NCX1 by PLM. First, inhibition of Na^+-K^+-ATPase is relieved by PLM phosphorylation. By contrast, PLM phosphorylated at $serine^{68}$ inhibits NCX1 in transfected HEK293 cells (Zhang et al. 2006a). When expressed in WT rat (Song et al. 2005) or PLM-KO mouse myocytes (Song et al. 2008), the phosphomimetic PLM S68E mutant inhibits I_{NaCa} with no effect on I_{pump}. By contrast, the non-phosphorylable PLM S68A mutant inhibits I_{pump} but not I_{NaCa} in PLM-KO mouse myocytes. Second, the single TM segment of FXYD proteins interacts with TM segments of Na^+-K^+-ATPase. By contrast, association between PLM and NCX1 does not appear to involve the TM segments (Wang et al. 2006). Rather, regulation of NCX1 is mediated through the interaction between the cytoplasmic tail of PLM and intracellular loop of NCX1 (Fig. 15.1) (Song et al. 2005; Wang et al. 2006).

15.12 Regulation of Myocyte Contraction by PLM: Na^+-K^+-ATPase vs. NCX1

In PLM-KO myocytes expressing the phosphomimetic PLM S68E mutant, contraction and $[Ca^{2+}]_i$ transient amplitudes are decreased at 5.0 but not 1.8 mM $[Ca^{2+}]_o$ (Song et al. 2008). When the non-phosphorylable PLM S68A mutant is expressed, no changes in $[Ca^{2+}]_i$ transient and contraction amplitudes are observed at either $[Ca^{2+}]_o$. Therefore, under conditions in which $[Ca^{2+}]_o$ is varied to manipulate the thermodynamic driving force for NCX1, regulation of single cardiac myocyte contractility by PLM is mediated by its inhibitory effects of NCX1 rather than Na^+-K^+-ATPase.

On the other hand, when WT myocytes are Na^+ loaded by rapid pacing and isoproterenol exposure, relief of inhibition of Na^+-K^+-ATPase associated with PLM phosphorylation results in lowering of $[Na^+]_i$, $[Ca^{2+}]_i$ transient and contraction amplitudes (Fig. 15.2) (Despa et al. 2008; Wang et al. 2010a). In PLM-KO myocytes subjected to similar treatment, no enhancement of

before and after isoproterenol (1 μM) addition. There are eight WT (o) and six PLM-KO (•) myocytes. Note $\Delta[Na^+]_i$ (increase of $[Na^+]_i$ above baseline) reaches a peak after isoproterenol followed by a decline in WT myocytes. The time-dependent decline in $\Delta[Na^+]_i$ is due to relief of inhibition of Na^+-K^+-ATPase upon phosphorylation of phospholemman. The time-dependent decline in $[Na^+]_i$ in WT myocytes promotes increased Ca^{2+} efflux via Na^+/Ca^{2+} exchanger, leading to decreased $[Ca^{2+}]_i$ transient (**e**) and contraction (**d**) amplitudes. In PLM-KO myocytes, no increase Na^+-K^+-ATPase activity occurs following isoproterenol stimulation and $\Delta[Na^+]_i$ continues its monotonous rise, without any decrease in $[Ca^{2+}]_i$ transient and contraction amplitudes

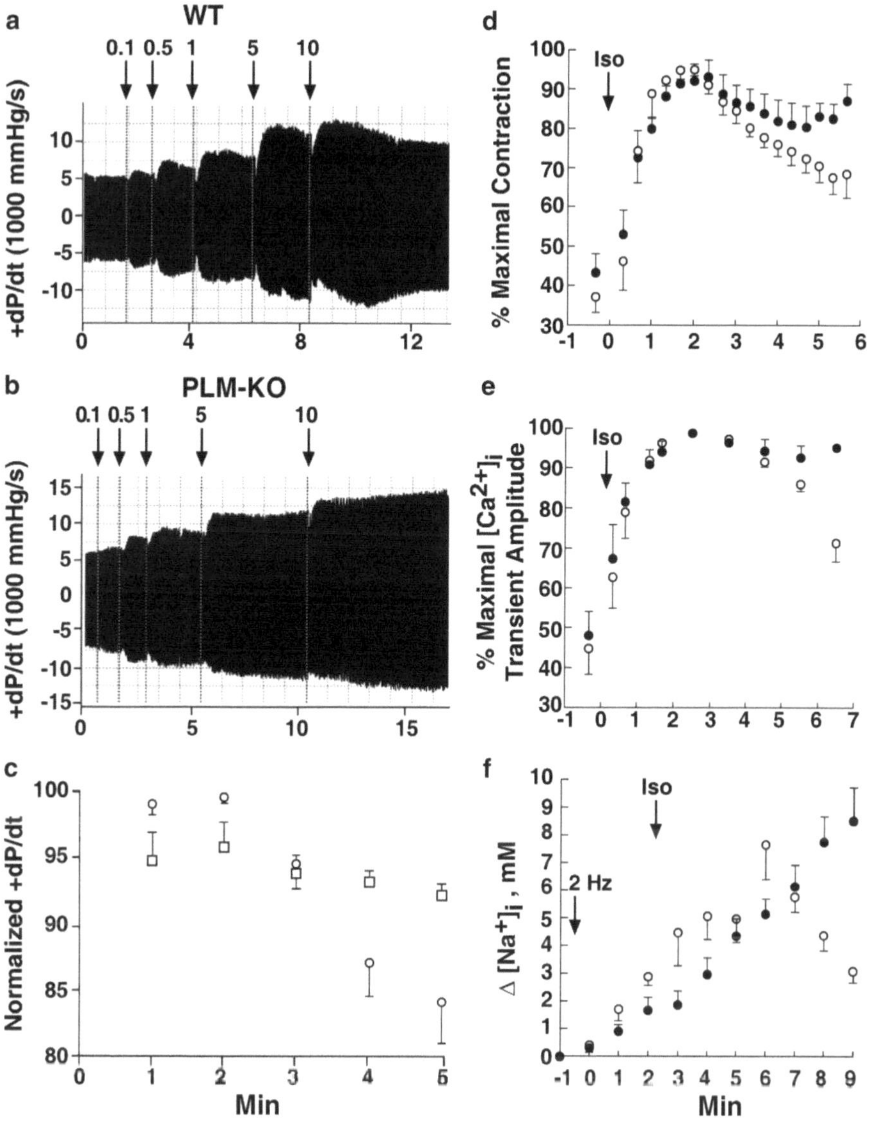

Fig. 15.2 *Disinhibiting Na^+-K^+-ATPase by phosphorylated phospholemman minimizes $[Ca^{2+}]_i$ and $[Na^+]_i$ overload and reduces risks of arrhythmogenesis in hearts under stress but at the apparent expense of reduced inotropy.* (**a**) First-time derivatives of left ventricular pressure rise ($^+dP/dt$) measured in an anesthetized, closed-chest wild-type (WT) mice are continuously measured, both at baseline and with increasing doses of isoproterenol (arrows; ng). Note $^+dP/dt$ increases to a maximum followed by time-dependent decline. (**b**) Absence of time-dependent decline in $^+dP/dt$ in phospholemman knockout (PLM-KO) after isoproterenol addition. (**c**) Time course of normalized $+dP/dt$ (means ± SE) following addition of maximal doses (10 or 25 ng) of isoproterenol to five WT (o) and eight PLM-KO (□) mice. (**d**) To elucidate the cellular mechanisms responsible for the time-dependent decline in $+dP/dt$ in WT hearts following isoproterenol stimulation, isolated myocytes are paced at 2 Hz, 37°C, and 1.8 mM $[Ca^{2+}]_o$. After reaching steady-state contraction amplitudes, isoproterenol (Iso; 1 μM) is added. There are six WT (o) and six PLM-KO (•) myocytes. Similar to intact hearts, in isolated myocytes stimulated with isoproterenol, contraction amplitude reaches a peak followed by rapid decline in WT but not PLM-KO myocytes. (**e**) $[Ca^{2+}]_i$ transient amplitudes are measured in paced, fura-2 loaded myocytes, before and after isoproterenol (1 μM) addition. There are ten WT (o) and six PLM-KO (•) myocytes. Note $[Ca^{2+}]_i$ transient amplitude reaches a peak after isoproterenol stimulation, followed by decline in WT but not PLM-KO myocytes. (**f**) $[Na^+]_i$ is measured in paced, sodium-binding benzofuran isophthalate (SBFI)-loaded myocytes,

Na^+-K^+-ATPase activity is observed, $[Na^+]_i$ continues to increase, and $[Ca^{2+}]_i$ transient and contraction amplitudes remain stable (Fig. 15.2). Thus, at the level of a single myocyte, PLM can be manipulated to regulate Na^+ and Ca^{2+} fluxes (and therefore $[Ca^{2+}]_i$ transients and contractility) by either Na^+/Ca^{2+} exchanger or Na^+-K^+-ATPase, depending on experimental conditions.

15.13 Functional Significance of Na^+/Ca^{2+} Exchanger Regulation by PLM at Rest

Inhibition of NCX1 activity by the ~30–40% phosphorylated PLM in resting hearts would be predicted to reduce Ca^{2+} efflux and secondarily increase SR Ca^{2+} load, resulting in enhanced contractility. However, in vivo cardiac function (Wang et al. 2010a) and in vitro myocyte contractility (Tucker et al. 2006) are similar between WT and PLM-KO mice, indicating that the regulatory effects of PLM on NCX1 are not detectable in hearts under basal conditions and PLM is functionally quiescent.

15.14 Phospholemman: A Novel Cardiac Stress Protein

Under stressful conditions when catecholamine levels are high, increased heart rate results in increased Na^+ entry. In addition, β-adrenergic agonists also enhance Ca^{2+} channel and SERCA2 activity, resulting in increased Ca^{2+} entry and SR Ca^{2+} loading. Increased Ca^{2+} entry must be balanced by greater Ca^{2+} efflux via NCX1, thereby bringing more Na^+ into the myocyte. Therefore, stress results in elevations in $[Na^+]_i$ and $[Ca^{2+}]_i$, with consequent increased risks of arrhythmogenesis. Phosphorylation of PLM by PKA relieves tonic inhibition of Na^+-K^+-ATPase, thereby lowering $[Na^+]_i$. The lower $[Na^+]_i$ promotes Ca^{2+} efflux via NCX1, thereby minimizing $[Ca^{2+}]_i$ overload. However, increased Ca^{2+} efflux results in decreased SR Ca^{2+} load, leading to reduced $[Ca^{2+}]_i$ transient and contraction amplitudes. Indeed, when stimulated with isoproterenol, $[Na^+]_i$, $[Ca^{2+}]_i$ transient and contraction amplitudes initially rise followed by rapid decline in WT but not PLM-KO myocytes (Fig. 15.2) (Despa et al. 2008; Wang et al. 2010a). Similarly, when hearts in vivo are stressed with maximal doses of isoproterenol, $^+dP/dt$ rises to a peak within 2 min followed by decline in WT but not PLM-KO hearts (Fig. 15.2). Thus, one of the major physiological functions of PLM (mediated by relief of inhibition of Na^+-K^+-ATPase) is to limit Na^+ and Ca^{2+} overload in hearts under duress, thereby minimizing risks of arrhythmogenesis but at the apparent expense of reduced contractility.

Decreased inotropy in fight or flight circumstances is clearly not in the best interests of the animal attempting to survive. What, if any, are the effects of PLM regulation of NCX1 in hearts under stress?

To eliminate the effects of Na^+-K^+-ATPase regulation by PLM, a model of "pure" NCX1 inhibition is engineered to evaluate the functional significance of NCX1 regulation by PLM in the heart. Recombinant adeno-associated virus, serotype 9 (rAAV9) expressing either the phosphomimetic PLM S68E mutant (rAAV9-S68E) or control green fluorescent protein (rAAV9-GFP) is injected into left ventricles (LV) of PLM-KO hearts (Fig. 15.3). After 5–6 weeks, ~40% of LV myocytes express the exogenous gene (Wang et al. 2011). Since S68E mutant inhibits I_{NaCa} but not I_{pump} (Song et al. 2008), and since PLM-KO myocytes do not exhibit regulation of Na^+-K^+-ATPase (Despa et al. 2005; Wang et al. 2010a), the effects of PLM regulation of NCX1 can be evaluated without the confounding influence of changes in Na^+-K^+-ATPase activity. Baseline cardiac output as evaluated by echocardiography is similar between PLM-KO hearts expressing S68E mutant or GFP (Wang et al. 2011), in agreement with previous conclusion that under resting conditions, PLM is functionally quiescent. When isoproterenol is added to simulate stressful conditions, PLM-KO hearts expressing S68E mutant have significantly higher contractility than those expressing GFP, despite <40% of LV expresses

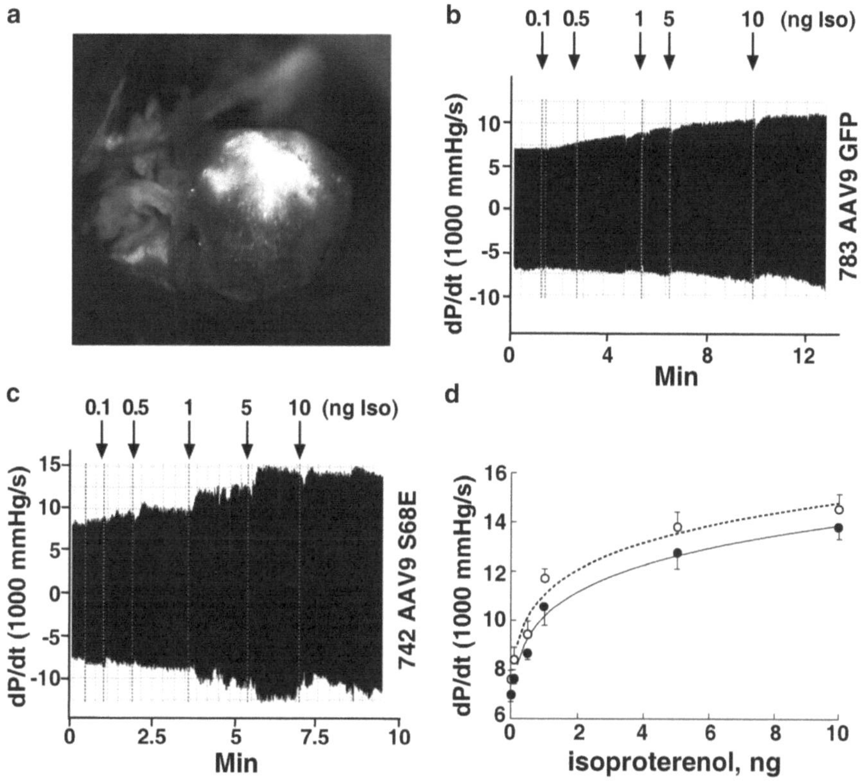

Fig. 15.3 *Inhibiting Na^+/Ca^{2+} exchanger by phosphorylated phospholemman preserves inotropy in hearts under stress.* (**a**) The apex, anterior, and posterior left ventricular (LV) wall of phospholemman knockout (PLM-KO) mice are injected with recombinant adeno-associated virus, serotype 9 (rAAV9), expressing either control green fluorescent protein (GFP) or GFP+phosphomimetic phospholemman S68E mutant (inhibits Na^+/Ca^{2+} exchanger but not Na^+-K^+-ATPase). The exogenous genes are expressed 5–6 weeks after injection as indicated by GFP fluorescence. About 40% of LV myocytes are successfully infected by rAAV9 injection. (**b**) First-time derivatives of LV pressure rise (+d*P*/d*t*) measured in an anesthetized, closed-chest PLM-KO mice previously injected with rAAV9-GFP, both at baseline and at increasing doses of isoproterenol (*arrows*, 0.1–10 ng). (**c**) +d*P*/d*t* in PLM-KO mice previously injected with rAAV9-S68E, both at baseline and at increasing doses of isoproterenol (arrows). Note there is no time-dependent decline in +d*P*/d*t* following isoproterenol addition, in contrast to the progressive decline in +d*P*/d*t* in WT hearts (Fig. 15.2a). This is consistent with lack of effect on Na^+-K^+-ATPase by S68E mutant. Note also +d*P*/d*t* is higher in PLM-KO heart expressing S68E mutant compared to that expressing control GFP (b). (**d**) Averaged maximal +d*P*/d*t* (means ± SE) achieved with each dose of isoproterenol in five PLM-KO hearts expressing S68E (o) and six PLM-KO hearts expressing GFP (•). Two-way ANOVA indicates significant ($P<0.047$) differences in +d*P*/d*t* between the two groups

the mutant (Fig. 15.3). Consistent with NCX1 inhibition in PLM-KO myocytes expressing S68E, increase in $[Na^+]_i$ after isoproterenol stimulation is smaller (when compared to PLM-KO myocytes expressing GFP), due to decreased Na^+ entry via forward NCX1. Therefore, a second major function of PLM (mediated by inhibition of NCX1) is to maintain inotropy under stress.

15.15 Phospholemman in Cardiac Diseases

In the rat myocardial infarction (MI) model, PLM is one of 19 genes (in a cDNA microarray containing 86 known genes and 989 unknown cDNAs) to increase after MI (Sehl et al. 2000).

PLM protein levels increase 2.4- and 4-fold at 3 and 7 days post-MI, respectively, in the rat (Zhang et al. 2006b). PLM overexpression may explain decreased NCX1 (Dixon et al. 1992a; Zhang et al. 1996) and Na^+-K^+-ATPase (Dixon et al. 1992b) activities in the post-MI rat model. In rat hearts subjected to acute ischemia, PLM is phosphorylated with resultant increase in Na^+-K^+-ATPase activity (Fuller et al. 2004). In mouse hearts subjected to ischemia/reperfusion, protection against infarction by sildenafil is associated with increased PLM phosphorylation at serine^{69} that enhances Na^+-K^+-ATPase activity during reperfusion (Madhani et al. 2010). Maintaining $[Na^+]_i$ homeostasis during ischemia and reperfusion by enhanced Na^+-K^+-ATPase activity minimizes the deleterious effects of elevated $[Na^+]_i$ on contractility and arrhythmogenesis. In a rabbit model of volume overload heart failure that is prone to arrhythmias, PLM expression is reduced by 42–48%, but serine^{68} phosphorylation is dramatically increased (Bossuyt et al. 2005). In human heart failure, PLM in left ventricles is reduced by 24% (Bossuyt et al. 2005). Thus, both expression and phosphorylation state of PLM are altered in various cardiac disease models, making PLM a rational therapeutic target. In this context, it is very relevant to note that the two classes of drugs that are clinically efficacious in the treatment of human heart failure, β-adrenergic blockers (targeting PKA) and angiotensin-converting-enzyme inhibitors (targeting PKC), both have PLM as a common target.

15.16 Future Directions

Regulation of L-type Ca^{2+} channel by PLM has been demonstrated in transfected HEK293 cells (Wang et al. 2010b). The possibility that PLM regulates L-type Ca^{2+} channel in the heart needs to be unequivocally proved since this adds tremendous complexity to the design and interpretation of experiments. The stoichiometry of interaction between PLM and Na^+-K^+-ATPase, PLM and NCX1, and PLM and L-type Ca^{2+} channel in cardiac tissues needs to be determined. The structure of the proximal linker domain of NCX1 crucial for its regulation by PLM needs to be elucidated. Whether PLM exists as monomers or oligomers in cardiac membranes requires clarification. The effects of oxidative stress (Figtree et al. 2009), NO (William et al. 2005), and glutathionylation (Bibert et al. 2011) on Na^+-K^+-ATPase and FXYD proteins are just beginning to emerge. Methods needed to be developed to measure $[Na^+]_i$ in vivo to prove that the time-dependent decrease in contractility in WT hearts stimulated with isoproterenol is associated with decreased $[Na^+]_i$ due to enhanced Na^+-K^+-ATPase activity. Finally, the role of PLM in regulating in vivo contractility, both in health and disease, needs to be further delineated and will likely require novel genetic models.

15.17 Conclusion

Phospholemman, the inaugural member of FXYD family, regulates Na^+-K^+-ATPase, Na^+/Ca^{2+} exchanger, and possibly L-type Ca^{2+} channel in the heart. In the resting state, phospholemman is functionally quiescent in that its regulatory effects on Na^+-K^+-ATPase and Na^+/Ca^{2+} exchanger on cardiac contractility are not manifested. Under stress when catecholamine levels are high, the coordinated actions of phospholemman on Na^+-K^+-ATPase and Na^+/Ca^{2+} exchanger minimize the risks of arrhythmogenesis and preserve inotropy, respectively. Phospholemman expression and phosphorylation are altered in ischemic heart disease and heart failure. Phospholemman is likely a useful target for drug therapy.

15.18 Disclosures

No conflicts of interest, financial or otherwise, are declared by the authors.

Acknowledgments This work was supported in part by National Institutes of Health grants RO1-HL58672 and RO1-HL74854 (JYC); RO1-HL91096 (JER); RO1-HL56205, RO1-HL-61690, RO1-HL85503, PO1-HL-75443, and PO1-HL-91799 (WJK); and PO1-HL-91799 (AMF) and by American Heart Association Scientist Development grant F64702 (TOC).

References

B.A. Ahlers, X.Q. Zhang, J.R. Moorman, L.I. Rothblum, L.L. Carl, J. Song, J. Wang, L.M. Geddis, A.L. Tucker, J.P. Mounsey, J.Y. Cheung, Identification of an endogenous inhibitor of the cardiac Na^+/Ca^{2+} exchanger, phospholemman. J. Biol. Chem. **280**, 19875–19882 (2005)

A.J. Beevers, A. Kukol, Secondary structure, orientation, and oligomerization of phospholemman, a cardiac transmembrane protein. Protein Sci. **15**, 1127–1132 (2006)

A.J. Beevers, A. Kukol, Phospholemman transmembrane structure reveals potential interactions with Na^+/K^+-ATPase. J. Biol. Chem. **282**, 32742–32748 (2007)

J.R. Bell, E. Kennington, W. Fuller, K. Dighe, P. Donoghue, J.E. Clark, L.G. Jia, A.L. Tucker, J.R. Moorman, M.S. Marber, P. Eaton, M.J. Dunn, M.J. Shattock, Characterisation of the phospholemman knockout mouse heart: depressed left ventricular function with increased Na/K ATPase activity. Am. J. Physiol. Heart Circ. Physiol. **294**, H613–H621 (2008)

R.G. Berry, S. Despa, W. Fuller, D.M. Bers, M.J. Shattock, Differential distribution and regulation of mouse cardiac Na^+/K^+-ATPase alpha1 and alpha2 subunits in T-tubule and surface sarcolemmal membranes. Cardiovasc. Res. **73**, 92–100 (2007)

D.M. Bers, Cardiac excitation-contraction coupling. Nature **415**, 198–205 (2002)

S. Bibert, C.C. Liu, G.A. Figtree, A. Garcia, E.J. Hamilton, F.M. Marassi, K.J. Sweadner, F. Cornelius, K. Geering, H.H. Rasmussen, FXYD proteins reverse inhibition of the Na^+-K^+ pump mediated by glutathionylation of its beta1 subunit. J. Biol. Chem. **286**, 18562–18572 (2011)

G. Blanco, R.W. Mercer, Isozymes of the Na-K-ATPase: heterogeneity in structure, diversity in function. Am. J. Physiol. **275**, F633–F650 (1998)

M. Blaustein, W. Lederer, Sodium/calcium exchange: its physiological implications. Physiol. Rev. **79**, 763–854 (1999)

J. Bossuyt, X. Ai, J.R. Moorman, S.M. Pogwizd, D.M. Bers, Expression and phosphorylation of the Na-pump regulatory subunit phospholemman in heart failure. Circ. Res. **97**, 558–565 (2005)

J. Bossuyt, S. Despa, J.L. Martin, D.M. Bers, Phospholemman phosphorylation alters its fluorescence resonance energy transfer with the Na/K-ATPase pump. J. Biol. Chem. **281**, 32765–32773 (2006)

J. Bossuyt, S. Despa, F. Han, Z. Hou, S.L. Robia, J.B. Lingrel, D.M. Bers, Isoform-specificity of the Na/K-ATPase association and regulation by phospholemman. J. Biol. Chem. **284**, 26749–26757 (2009)

L.S. Chen, C.F. Lo, R. Numann, M. Cuddy, Characterization of the human and rat phospholemman (PLM) cDNAs and localization of the human PLM gene to chromosome 19q13.1. Genomics **41**, 435–443 (1997)

Z. Chen, L.R. Jones, J.J. O'Brian, J.R. Moorman, S.E. Cala, Structural domains in phospholemman: a possible role for the carboxyl terminus in channel inactivation. Circ. Res. **82**, 367–374 (1998)

J. Cheung, X. Zhang, J. Song, E. Gao, J. Rabinowitz, T. Chan, J. Wang, Phospholemman: a novel cardiac stress protein. Clin. Transl. Sci. **3**, 189–196 (2010)

G. Crambert, M. Fuzesi, H. Garty, S. Karlish, K. Geering, Phospholemman (FXYD1) associates with Na,K-ATPase and regulates its transport properties. Proc. Natl. Acad. Sci. U. S. A. **99**, 11476–11481 (2002)

C.E. Davis, M.K. Patel, J.R. Miller, J.E. John 3rd, L.R. Jones, A.L. Tucker, J.P. Mounsey, J.R. Moorman, Effects of phospholemman expression on swelling-activated ion currents and volume regulation in embryonic kidney cells. Neurochem. Res. **29**, 177–187 (2004)

S. Despa, M.A. Islam, S.M. Pogwizd, D.M. Bers, Intracellular $[Na^+]$ and Na^+ pump rate in rat and rabbit ventricular myocytes. J. Physiol. **539**, 133–143 (2002)

S. Despa, J. Bossuyt, F. Han, K.S. Ginsburg, L.G. Jia, H. Kutchai, A.L. Tucker, D.M. Bers, Phospholemman-phosphorylation mediates the beta-adrenergic effects on Na/K pump function in cardiac myocytes. Circ. Res. **97**, 252–259 (2005)

S. Despa, A. Tucker, D. Bers, PLM-mediated activation of Na/K-ATPase limits [Na]i and inotropic state during Β-adrenergic stimulation in mouse ventricular myocytes. Circulation **117**, 1849–1855 (2008)

R. DiPolo, L. Beauge, Sodium/calcium exchanger: influence of metabolic regulation on ion carrier interactions. Physiol. Rev. **86**, 155–203 (2006)

I.M.C. Dixon, T. Hata, N.S. Dhalla, Sarcolemmal calcium transport in congestive heart failure due to myocardial infarction in rats. Am. J. Physiol. Heart Circ. Physiol. **262**, H1387–H1394 (1992a)

I.M.C. Dixon, T. Hata, N.S. Dhalla, Sarcolemmal Na^+-K^+-ATPase activity in congestive heart failure due to myocardial infarction. Am. J. Physiol. Cell Physiol. **262**, C664–C671 (1992b)

I. Dostanic, J. Schultz Jel, J.N. Lorenz, J.B. Lingrel, The alpha 1 isoform of Na, K-ATPase regulates cardiac contractility and functionally interacts and co-localizes with the Na/Ca exchanger in heart. J. Biol. Chem. **279**, 54053–54061 (2004)

G.A. Figtree, C.C. Liu, S. Bibert, E.J. Hamilton, A. Garcia, C.N. White, K.K. Chia, F. Cornelius, K. Geering, H.H. Rasmussen, Reversible oxidative modification: a key mechanism of Na^+-K^+ pump regulation. Circ. Res. **105**, 185–193 (2009)

C.M. Franzin, X.M. Gong, K. Thai, J. Yu, F.M. Marassi, NMR of membrane proteins in micelles and bilayers: the FXYD family proteins. Methods (San Diego, Calif.) **41**, 398–408 (2007)

W. Fuller, P. Eaton, J.R. Bell, M.J. Shattock, Ischemia-induced phosphorylation of phospholemman directly activates rat cardiac Na/K-ATPase. FASEB J. **18**, 197–199 (2004)

W. Fuller, J. Howie, L. McLatchie, R. Weber, C.J. Hastie, K. Burness, D. Pavlovic, M.J. Shattock, FXYD1 phosphorylation in vitro and in adult rat cardiac myocytes:

threonine 69 is a novel substrate for protein kinase C. Am. J. Physiol. Cell Physiol. **296**, C1346–C1355 (2009)
M. Giladi, L. Boyman, H. Mikhasenko, R. Hiller, D. Khananshvili, Essential role of the CBD1-CBD2 linker in slow dissociation of Ca^{2+} from the regulatory two-domain tandem of NCX1. J. Biol. Chem. **285**, 28117–28125 (2010)
I. Grupp, W.-B. Im, C.O. Lee, S.-W. Lee, M.S. Pecker, A. Schwartz, Regulation of sodium pump inhibition to positive inotrophy at low concentrations of ouabain in rat heart muscle. J. Physiol. **360**, 149–160 (1985)
F. Han, A.L. Tucker, J.B. Lingrel, S. Despa, D.M. Bers, Extracellular potassium dependence of the Na^+-K^+-ATPase in cardiac myocytes: isoform specificity and effect of phospholemman. Am. J. Physiol. Cell Physiol. **297**, C699–C705 (2009)
G. Hasenfuss, W. Schillinger, Is modulation of sodium-calcium exchange a therapeutic option in heart failure? Circ. Res. **95**, 225–227 (2004)
M. Hilge, J. Aelen, G.W. Vuister, Ca^{2+} regulation in the Na^+/Ca^{2+} exchanger involves two markedly different Ca^{2+} sensors. Mol. Cell **22**, 15–25 (2006)
M. Hilge, J. Aelen, A. Foarce, A. Perrakis, G.W. Vuister, Ca^{2+} regulation in the Na^+/Ca^{2+} exchanger features a dual electrostatic switch mechanism. Proc. Natl. Acad. Sci. U. S. A. **106**, 14333–14338 (2009)
T. Iwamoto, A. Uehara, I. Imanaga, M. Shigekawa, The Na^+/Ca^{2+} exchanger NCX1 has oppositely oriented reentrant loop domains that contain conserved aspartic acids whose mutation alters its apparent Ca^{2+} affinity. J. Biol. Chem. **275**, 38571–38580 (2000)
P.F. James, I.L. Grupp, G. Grupp, A.L. Woo, G.R. Askew, M.L. Croyle, R.A. Walsh, J.B. Lingrel, Identification of a specific role for the Na, K-ATPase alpha 2 isoform as a regulator of calcium in the heart. Mol. Cell **3**, 555–563 (1999)
L.G. Jia, C. Donnet, R.C. Bogaev, R.J. Blatt, C.E. McKinney, K.H. Day, S.S. Berr, L.R. Jones, J.R. Moorman, K.J. Sweadner, A.L. Tucker, Hypertrophy, increased ejection fraction, and reduced Na-K-ATPase activity in phospholemman-deficient mice. Am. J. Physiol. Heart Circ. Physiol. **288**, H1982–H1988 (2005)
S.A. John, B. Ribalet, J.N. Weiss, K.D. Philipson, M. Ottolia, Ca^{2+}-dependent structural rearrangements within Na^+-Ca^{2+} exchanger dimers. Proc. Natl. Acad. Sci. U. S. A. **108**, 1699–1704 (2011)
D.O. Levitsky, D.A. Nicoll, K.D. Philipson, Identification of the high affinity Ca^{2+}-binding domain of the cardiac Na^+-Ca^{2+} exchanger. J. Biol. Chem. **269**, 22847–22852 (1994)
Z.P. Li, D.A. Nicoll, A. Collins, D.W. Hilgemann, A.G. Filoteo, J.T. Penniston, J.N. Weiss, J.M. Tomich, K.D. Philipson, Identification of a peptide inhibitor of the cardiac sarcolemmal Na^+-Ca^{2+} exchanger. J. Biol. Chem. **266**, 1014–1020 (1991)
C. Li, A. Grosdidier, G. Crambert, J.D. Horisberger, O. Michielin, K. Geering, Structural and functional interaction sites between Na, K-ATPase and FXYD proteins. J. Biol. Chem. **279**, 38895–38902 (2004)
J.P. Lindemann, Alpha-adrenergic stimulation of sarcolemmal protein phosphorylation and slow responses in intact myocardium. J. Biol. Chem. **261**, 4860–4867 (1986)
M. Lindzen, K.E. Gottschalk, M. Fuzesi, H. Garty, S.J. Karlish, Structural interactions between FXYD proteins and Na^+, K^+-ATPase: alpha/beta/FXYD subunit stoichiometry and cross-linking. J. Biol. Chem. **281**, 5947–5955 (2006)
C. Maack, A. Ganesan, A. Sidor, B. O'Rourke, Cardiac sodium-calcium exchanger is regulated by allosteric calcium and exchanger inhibitory peptide at distinct sites. Circ. Res. **96**, 91–99 (2005)
M. Madhani, A.R. Hall, F. Cuello, R.L. Charles, J.R. Burgoyne, W. Fuller, A.J. Hobbs, M.J. Shattock, P. Eaton, Phospholemman Ser-69 phosphorylation contributes to sildenafil-induced cardioprotection against reperfusion injury. Am. J. Physiol. Heart Circ. Physiol. **299**, H827–H836 (2010)
Y.A. Mahmmoud, H. Vorum, F. Cornelius, Purification of a phospholemman-like protein from shark rectal glands. J. Biol. Chem. **275**, 35969–35977 (2000)
S. Matsuoka, D.A. Nicoll, L.V. Hryshko, D.O. Levitsky, J.N. Weiss, K.D. Philipson, Regulation of the cardiac Na^+-Ca^{2+} exchanger by Ca^{2+}. Mutational analysis of the Ca^{2+}-binding domain. J. Gen. Physiol. **105**, 403–420 (1995)
A.A. McDonough, Y. Zhang, V. Shin, J.S. Frank, Subcellular distribution of sodium pump isoform subunits in mammalian cardiac myocytes. Am. J. Physiol. Cell Physiol. **270**, C1221–C1227 (1996)
M.A. Mirza, X.Q. Zhang, B.A. Ahlers, A. Qureshi, L.L. Carl, J. Song, A.L. Tucker, J.P. Mounsey, J.R. Moorman, L.I. Rothblum, T.S. Zhang, J.Y. Cheung, Effects of phospholemman downregulation on contractility and $[Ca^{2+}]i$ transients in adult rat cardiac myocytes. Am. J. Physiol. Heart Circ. Physiol. **286**, H1322–H1330 (2004)
J.R. Moorman, S.J. Ackerman, G.C. Kowdley, M. Griffin, J.P. Mounsey, Z. Chen, S.E. Cala, J.J. O'Brian, G. Szabo, L.R. Jones, Unitary onion currents through phospholemman channel molecules. Nature **377**, 737–740 (1995)
M. Morales-Mulia, H. Pasantes-Morales, J. Moran, Volume sensitive efflux of taurine in HEK 293 cells overexpressing phospholemman. Biochim. Biophys. Acta **1496**, 252–260 (2000)
J.S. Nam, S. Hirohashi, L.M. Wakefield, Dysadherin: a new player in cancer progression. Cancer lett. **255**, 161–169 (2007)
D.A. Nicoll, L.V. Hryshko, S. Matsuoka, J.S. Frank, K.D. Philipson, Mutation of amino acid residues in the putative transmembrane segments of the cardiac sarcolemmal Na^+-Ca^{2+} exchanger. J. Biol. Chem. **271**, 13385–13391 (1996)
D.A. Nicoll, M. Ottolia, L. Lu, Y. Lu, K.D. Philipson, A new topological model of the cardiac sarcolemmal

Na^+-Ca^{2+} exchanger. J. Biol. Chem. **274**, 910–917 (1999)

D.A. Nicoll, M.R. Sawaya, S. Kwon, D. Cascio, K.D. Philipson, J. Abramson, The crystal structure of the primary Ca^{2+} sensor of the Na^+/Ca^{2+} exchanger reveals a novel Ca^{2+} binding motif. J. Biol. Chem. **281**, 21577–21581 (2006)

M. Ottolia, S. John, Z. Qiu, K.D. Philipson, Split Na^+-Ca^{2+} exchangers. Implications for function and expression. J. Biol. Chem. **276**, 19603–19609 (2001)

M. Ottolia, D.A. Nicoll, K.D. Philipson, Roles of two Ca^{2+}-binding domains in regulation of the cardiac Na^+-Ca^{2+} exchanger. J. Biol. Chem. **284**, 32735–32741 (2009)

C.J. Palmer, B.T. Scott, L.R. Jones, Purification and complete sequence determination of the major plasma membrane substrate for cAMP-dependent protein kinase and protein kinase C in myocardium. J. Biol. Chem. **266**, 11126–11130 (1991)

K.D. Philipson, D.A. Nicoll, Sodium-calcium exchange: a molecular perspective. Annu. Rev. Physiol. **62**, 111–133 (2000)

S.M. Pogwizd, K. Schlotthauer, L. Li, W. Yuan, D.M. Bers, Arrhythmogenesis and contractile dysfunction in heart failure: roles of sodium-calcium exchange, inward rectifier potassium current, and residual beta-adrenergic responsiveness. Circ. Res. **88**, 1159–1167 (2001)

C.F. Presti, L.R. Jones, J.P. Lindemann, Isoproterenol-induced phosphorylation of a 15-kilodalton sarcolemmal protein in intact myocardium. J. Biol. Chem. **260**, 3860–3867 (1985a)

C.F. Presti, B.T. Scott, L.R. Jones, Identification of an endogenous protein kinase C activity and its intrinsic 15-kilodalton substrate in purified canine cardiac sarcolemmal vesicles. J. Biol. Chem. **260**, 13879–13889 (1985b)

C.M. Rembold, M.L. Ripley, M.K. Meeks, L.M. Geddis, H.C. Kutchai, F.M. Marassi, J.Y. Cheung, J.R. Moorman, Serine 68 phospholemman phosphorylation during forskolin-induced swine carotid artery relaxation. J. Vasc. Res. **42**, 483–491 (2005)

X. Ren, D.A. Nicoll, G. Galang, K.D. Philipson, Intermolecular cross-linking of Na^+-Ca^{2+} exchanger proteins: evidence for dimer formation. Biochemistry **47**, 6081–6087 (2008)

P.D. Sehl, J.T. Tai, K.J. Hillan, L.A. Brown, A. Goddard, R. Yang, H. Jin, D.G. Lowe, Application of cDNA microarrays in determining molecular phenotype in cardiac growth, development, and response to injury. Circulation **101**, 1990–1999 (2000)

T. Shinoda, H. Ogawa, F. Cornelius, C. Toyoshima, Crystal structure of the sodium-potassium pump at 2.4 A resolution. Nature **459**, 446–450 (2009)

B.D. Silverman, W. Fuller, P. Eaton, J. Deng, J.R. Moorman, J.Y. Cheung, A.F. James, M.J. Shattock, Serine 68 phosphorylation of phospholemman: acute isoform-specific activation of cardiac Na/K ATPase. Cardiovasc. Res. **65**, 93–103 (2005)

K.R. Sipido, P.G.A. Volders, M.A. Vos, F. Verdonck, Altered Na/Ca exchange activity in cardiac hypertrophy and heart failure: a new target for therapy? Cardiovasc. Res. **53**, 782–805 (2002)

J. Song, X.Q. Zhang, L.L. Carl, A. Qureshi, L.I. Rothblum, J.Y. Cheung, Overexpression of phospholemman alter contractility and $[Ca^{2+}]$i transients in adult rat myocytes. Am. J. Physiol. Heart Circ. Physiol. **283**, H576–H583 (2002)

J. Song, X.Q. Zhang, B.A. Ahlers, L.L. Carl, J. Wang, L.I. Rothblum, R.C. Stahl, J.P. Mounsey, A.L. Tucker, J.R. Moorman, J.Y. Cheung, Serine 68 of phospholemman is critical in modulation of contractility, $[Ca^{2+}]$i transients, and Na^+/Ca^{2+} exchange in adult rat cardiac myocytes. Am. J. Physiol. Heart Circ. Physiol. **288**, H2342–H2354 (2005)

J. Song, X.Q. Zhang, J. Wang, E. Cheskis, T.O. Chan, A.M. Feldman, A.L. Tucker, J.Y. Cheung, Regulation of cardiac myocyte contractility by phospholemman: Na^+/Ca^{2+} exchange vs. Na^+-K^+-ATPase. Am. J. Physiol. Heart Circ. Physiol. **295**, H1615–H1625 (2008)

Q. Song, S. Pallikkuth, J. Bossuyt, D.M. Bers, S.L. Robia, Phosphomimetic mutations enhance oligomerization of phospholemman and modulate its interaction with the Na/K-ATPase. J. Biol. Chem. **286**, 9120–9126 (2011)

K.J. Sweadner, E. Rael, The FXYD gene family of small ion transport regulators or channels: cDNA sequence, protein signature sequence, and expression. Genomics **68**, 41–56 (2000)

F. Swift, N. Tovsrud, U.H. Enger, I. Sjaastad, O.M. Sejersted, The Na^+/K^+-ATPase alpha2-isoform regulates cardiac contractility in rat cardiomyocytes. Cardiovasc. Res. **75**, 109–117 (2007)

G.M. Tadros, X.Q. Zhang, J. Song, L.L. Carl, L.I. Rothblum, Q. Tian, J. Dunn, J. Lytton, J.Y. Cheung, Effects of Na^+/Ca^{2+} exchanger downregulation on contractility and $[Ca^{2+}](i)$ transients in adult rat myocytes. Am. J. Physiol. Heart Circ. Physiol. **283**, H1616–H1626 (2002)

P. Teriete, C.M. Franzin, J. Choi, F.M. Marassi, Structure of the Na, K-ATPase regulatory protein FXYD1 in micelles. Biochemistry **46**, 6774–6783 (2007)

P. Teriete, K. Thai, J. Choi, F.M. Marassi, Effects of PKA phosphorylation on the conformation of the Na, K-ATPase regulatory protein FXYD1. Biochim. Biophys. Acta **1788**, 2462–2470 (2009)

A.L. Tucker, J. Song, X.Q. Zhang, J. Wang, B.A. Ahlers, L.L. Carl, J.P. Mounsey, J.R. Moorman, L.I. Rothblum, J.Y. Cheung, Altered contractility and $[Ca^{2+}]$i homeostasis in phospholemman-deficient murine myocytes: Role of Na^+/Ca^{2+} exchange. Am. J. Physiol. Heart Circ. Physiol. **291**, H2199–H2209 (2006)

S.I. Waalas, A.J. Czernik, O.K. Olstad, K. Sletten, O. Walaas, Protein kinase C and cyclic AMP-dependent protein kinase phosphorylate phospholemman, an insulin and adrenaline-regulated membrane phosphoprotein, at specific sites in the carboxy terminal domain. Biochem. J. **304**(Pt 2), 635–640 (1994)

J. Wang, X.Q. Zhang, B.A. Ahlers, L.L. Carl, J. Song, L.I. Rothblum, R.C. Stahl, D.J. Carey, J.Y. Cheung, Cytoplasmic tail of phospholemman interacts with the intracellular loop of the cardiac Na^+/Ca^{2+} exchanger. J. Biol. Chem. **281**, 32004–32014 (2006)

J. Wang, E. Gao, J. Song, X.Q. Zhang, J. Li, W.J. Koch, A.L. Tucker, K.D. Philipson, T.O. Chan, A.M. Feldman, J.Y. Cheung, Phospholemman and {beta}-adrenergic stimulation in the heart. Am. J. Physiol. Heart Circ. Physiol. **298**, H807–H815 (2010a)

X. Wang, G. Gao, K. Guo, V. Yarotskyy, C. Huang, K.S. Elmslie, B.Z. Peterson, Phospholemman modulates the gating of cardiac L-type calcium channels. Biophys. J. **98**, 1149–1159 (2010b)

J. Wang, E. Gao, J. Rabinowitz, J. Song, X.Q. Zhang, W.J. Koch, A.L. Tucker, T.O. Chan, A.M. Feldman, J.Y. Cheung, Regulation of in vivo cardiac contractility by phospholemman: role of Na^+/Ca^{2+} exchange. Am. J. Physiol. Heart Circ. Physiol. **300**, H859–H868 (2011)

M. William, J. Vien, E. Hamilton, A. Garcia, H. Bundgaard, R.J. Clarke, H.H. Rasmussen, The nitric oxide donor sodium nitroprusside stimulates the Na^+-K^+ pump in isolated rabbit cardiac myocytes. J. Physiol. **565**, 815–825 (2005)

H. Yamamoto, K. Okumura, S. Toshima, K. Mukaisho, H. Sugihara, T. Hattori, M. Kato, S. Asano, FXYD3 protein involved in tumor cell proliferation is overproduced in human breast cancer tissues. Biol. Pharm. Bull. **32**, 1148–1154 (2009)

R. Zahler, M. Gilmore-Hebert, J.C. Baldwin, K. Franco, E.J. Benz Jr., Expression of alpha isoforms of the Na, K-ATPase in human heart. Biochim. Biophys. Acta **1149**, 189–194 (1993)

X. Zhang, D. Tillotson, R. Moore, R. Zelis, J. Cheung, Na^+/Ca^{2+} exchange currents and SR Ca^{2+} contents in postinfarction myocytes. Am. J. Physiol. **271**, C1800–C1807 (1996)

X.Q. Zhang, A. Qureshi, J. Song, L.L. Carl, Q. Tian, R.C. Stahl, D.J. Carey, L.I. Rothblum, J.Y. Cheung, Phospholemman modulates Na^+/Ca^{2+} exchange in adult rat cardiac myocytes. Am. J. Physiol. Heart Circ. Physiol. **284**, H225–H233 (2003)

X.Q. Zhang, B.A. Ahlers, A.L. Tucker, J. Song, J. Wang, J.R. Moorman, J.P. Mounsey, L.L. Carl, L.I. Rothblum, J.Y. Cheung, Phospholemman inhibition of the cardiac Na^+/Ca^{2+} exchanger. Role of phosphorylation. J. Biol. Chem. **281**, 7784–7792 (2006a)

X.Q. Zhang, J.R. Moorman, B.A. Ahlers, L.L. Carl, D.E. Lake, J. Song, J.P. Mounsey, A.L. Tucker, Y.M. Chan, L.I. Rothblum, R.C. Stahl, D.J. Carey, J.Y. Cheung, Phospholemman overexpression inhibits Na^+-K^+-ATPase in adult rat cardiac myocytes: relevance to decreased Na^+ pump activity in post-infarction myocytes. J. Appl. Physiol. **100**, 212–220 (2006b)

X.Q. Zhang, J. Wang, L.L. Carl, J. Song, B.A. Ahlers, J.Y. Cheung, Phospholemman regulates cardiac Na^+/Ca^{2+} exchanger by interacting with the exchanger's proximal linker domain. Am. J. Physiol. Cell Physiol. **296**, C911–C921 (2009)

X.Q. Zhang, J. Wang, J. Song, A.M. Ji, T.O. Chan, J.Y. Cheung, Residues 248–252 and 300–304 of the cardiac Na^+/Ca^{2+} exchanger are involved in its regulation by phospholemman. Am. J. Physiol. Cell Physiol. **301**, C833–C840 (2011)

Part VI

Subcellular Localization and Function of NCX in Ca^{2+}-Storing Organelles and Mitochondria

Mitochondria Na^+-Ca^{2+} Exchange in Cardiomyocytes and Lymphocytes

16

Bongju Kim, Ayako Takeuchi, Orie Koga, Masaki Hikida, and Satoshi Matsuoka

Abstract

Mitochondria Na^+-Ca^{2+} exchange (NCX_{mit}) was first discovered by Carafoli et al. in 1974. Thereafter, the mechanisms and roles of NCX_{mit} have been extensively studied. We review NCX_{mit} in cardiomyocytes and lymphocytes by presenting our recent studies on it. Studies of NCX_{mit} in rat ventricular cells demonstrated that NCX_{mit} is voltage dependent and electrogenic. A targeted knockdown and knockout of NCLX in HL-1 cardiomyocytes and B lymphocytes, respectively, significantly reduced the NCX_{mit} activity, indicating that NCLX is a major component of NCX_{mit} in these cells. The store-operated Ca^{2+} entry was greatly attenuated in NCLX knockout lymphocytes, suggesting that substantial amount of Ca^{2+} enters into mitochondria and is released to cytosol via NCX_{mit}. NCX_{mit} or NCLX has pivotal roles in Ca^{2+} handling in mitochondria and cytoplasm.

Keywords

Mitochondrial NCX • Calcium • Voltage dependence • Electrogenicity • NCLX • Lymphocytes • Cardiomyocytes • Store-operated Ca^{2+} entry

16.1 Introduction

Mitochondria are pivotal organelle as master switches of cell fate upon exposure of various stimuli, modulators of cell excitability, as well as ATP producing factories (Bernardi 1999; Murgia et al. 2009; Celsi et al. 2009). Mitochondrial Ca^{2+} dynamically changes upon stimulation by various ligands, and several enzymes in mitochondria are regulated by Ca^{2+}. Therefore, Ca^{2+} has been considered as one of the important regulators of mitochondria function.

B. Kim • O. Koga • M. Hikida • S. Matsuoka (✉)
Center for Innovation in Immunoregulative Technology and Therapeutics, Graduate School of Medicine, Kyoto University, Yoshida-konoe, Sakyo-ku, Kyoto 606-8501, Japan
e-mail: smatsuoka@ak.med.kyoto-u.ac.jp

A. Takeuchi
Department of Physiology and Biophysics, Graduate School of Medicine, Kyoto University, Yoshida-konoe, Sakyo-ku, Kyoto 606-8501, Japan

L. Annunziato (ed.), *Sodium Calcium Exchange: A Growing Spectrum of Pathophysiological Implications*, Advances in Experimental Medicine and Biology 961, DOI 10.1007/978-1-4614-4756-6_16,

Carafoli et al. (1974) first discovered Na^+-dependent Ca^{2+} release from mitochondria. Thereafter, the mechanisms and roles of mitochondrial Ca^{2+} handling have been extensively studied (Carafoli 2010). In many cells, Ca^{2+} mainly enters into mitochondria through Ca^{2+} uniporter (CaUni) and is released via Na^+-Ca^{2+} exchange (NCX_{mit}) and H^+-Ca^{2+} exchange (HCX_{mit}) (Bernardi 1999). In the last few years, genes responsible for these mitochondrial Ca^{2+} carriers were discovered, and our understanding of mitochondrial Ca^{2+} dynamics is being updated drastically. Here, we review roles and biophysical properties of NCX_{mit}, especially focusing on NCX_{mit} in cardiomyocytes and lymphocytes.

16.2 Ca^{2+} and Mitochondria Function

Ca^{2+} uptake by mitochondria was first discovered in the 1960s (DeLuca and Engstrom 1961). Although the Ca^{2+} accumulation by mitochondria is probably general phenomenon in all types of cell, a volume occupied by mitochondria differs markedly from cell to cell. Figure 16.1a presents images of mitochondria stained with a mitochondria-specific dye in a rat ventricular cell and a chicken DT40 B lymphocyte, representative example of excitable and non-excitable cells, respectively. Cardiomyocytes are one of the mitochondria-rich cells, and the mitochondria volume ranges between 22.0 % and 37.0 % of total cell volume (Barth et al. 1992). Smaller animals tend to have larger mitochondria fraction in cardiomyocytes. On the other hand, the number of mitochondria in lymphocytes is small and the occupied volume is 4~7 % (Petrzilka and Schroeder 1979). In excitable cells such as cardiomyocytes, cytoplasmic Ca^{2+} continuously changes during excitation-contraction coupling. Mitochondrial Ca^{2+} is influenced by the amplitude and the cycle of cytoplasmic Ca^{2+} transient. In non-excitable cells such as lymphocytes, cytoplasmic Ca^{2+} level is kept low and various external stimuli provoke a rise of cytoplasmic Ca^{2+}. In lymphocytes, antigen binding to the surface receptor induces an oscillatory Ca^{2+} rise which lasts for hours. Therefore, mitochondrial Ca^{2+} dynamics and its role may be different between excitable and non-excitable cells.

16.2.1 Mitochondrial Ca^{2+} Pathways

Figure 16.1b summarizes mitochondrial Ca^{2+} carriers. Ca^{2+} enters mitochondria through a Ca^{2+}-selective channel, CaUni, according to a large negative membrane potential of −150~−180 mV (Kirichok et al. 2004). The molecule was identified recently by Baughman et al. (2011) as MCU or CCDC109A. A regulator of CaUni was also discovered (MICU1 or CBARA1; Perocchi et al. 2010). A mitochondria ryanodine receptor (RyR) and a rapid mode of Ca^{2+} uptake (RaM) have been suggested as additional Ca^{2+} influx pathways (Ryu et al. 2010). Mitochondrial Ca^{2+} is extruded by HCX_{mit} and/or NCX_{mit}. NCX_{mit} exists in many tissues such as brain, adrenal cortex, parotid gland, skeletal muscle, and heart (Crompton et al. 1978). However, its activity is absent or weak in liver, kidney, uterus muscle, and ileum muscle (Crompton et al. 1978), where HCX_{mit} has a dominant role in extruding mitochondrial Ca^{2+} (Bernardi 1999). Letm1 and NCLX were recently identified as a responsible gene for HCX_{mit} (Jiang et al. 2009) and NCX_{mit} (Palty et al. 2010), respectively.

The change in cytoplasmic Ca^{2+} influences mitochondria Ca^{2+}. It is well known in heart that the activation of sympathetic nervous system increases heart rate and strengthens heart contraction while it increases mitochondrial Ca^{2+} besides cytosolic Ca^{2+} (Bell et al. 2006). Also in lymphocytes, the activation of antigen receptor increases cytoplasmic Ca^{2+} followed by mitochondrial Ca^{2+} increase (Kim et al. 2012). Ca^{2+} entered in mitochondria exerts a great impact on mitochondria metabolisms, such as NADH and ATP synthesis.

16.2.2 Mitochondria Enzyme Activation by Ca^{2+}

It has been demonstrated that three dehydrogenases in mitochondria are activated by Ca^{2+}; pyruvate dehydrogenase (PDHC), isocitrate

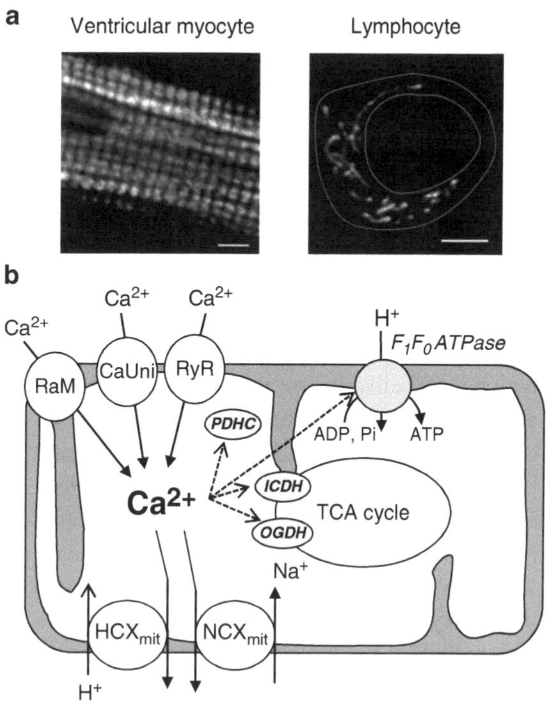

Fig. 16.1 *Mitochondria and Ca^{2+} pathways in mitochondria.* (**a**) Images of mitochondria in a rat ventricular cell (*left*) and a chicken DT40 B lymphocyte (*right*). Mitochondria were stained with a mitochondria-specific dye, MitoTracker Green. *Bars* indicate 5 μm. (**b**) Ca^{2+} pathways in mitochondria. *Dotted lines* indicate Ca^{2+} activation processes

dehydrogenase (ICDH), and 2-oxoglutarate dehydrogenase (OGDH) (McCormack et al. 1990). The increase in mitochondrial Ca^{2+} following the rise of cytoplasmic Ca^{2+} augments NADH content in isolated mitochondria (Territo et al. 2001). The rise of NADH amount is probably caused by the Ca^{2+} activation of the dehydrogenases. The Ca^{2+} activation of the dehydrogenases is suggested to be also related to the transient change of mitochondria NADH upon the workload transition (Cortassa et al. 2006; Jo et al. 2006). Several studies demonstrated that Ca^{2+} activates ATP synthase (F_1F_0-ATPase) (Territo et al. 2001), although the mechanism has not been clarified. Therefore, mitochondrial Ca^{2+} and Ca^{2+} transporters play pivotal roles in regulating cellular energy metabolisms.

16.2.3 Ca^{2+} Communication Between SR/ER and Mitochondria

Sarcoplasmic reticulum (SR) and endoplasmic reticulum (ER) are major Ca^{2+} stores within the cells. Mitochondria locate closely to SR/ER, and narrow interorganellar spaces exist between the organelles. Interestingly, a voltage-dependent anion channel of outer mitochondrial membrane and an inositol1,4,5-trisphosphate ($InsP_3$) receptor (IP3R) of ER are suggested to locate in the interorganellar regions and create a Ca^{2+} pathway from ER to mitochondria (Mendes et al. 2005; Szabadkai et al. 2006; Csordás and Hajnóczky 2009). Ca^{2+} released from RyR on SR also accumulates in mitochondria (Sharma et al. 2000). Thus, SR/ER and mitochondria are functionally related with each other via Ca^{2+}.

Several studies suggested that reverse Ca^{2+} movement, from mitochondria to ER/SR, is also important. It was reported that an inhibition of NCX_{mit} by CGP-37157 attenuates ER Ca^{2+} uptake in HeLa cells, endothelial cells, and vascular smooth muscle cells (Arnaudeau et al. 2001; Malli et al. 2005; Poburko et al. 2009), though the specificity of the drug is uncertain. In our studies on DT40 and A20 B lymphocytes (Kim et al. 2012), the targeted knockout or knockdown of NCLX caused a drastic decrease of ER Ca^{2+} content, demonstrating the essential role of NCX_{mit} (NCLX) in mitochondria-ER communication.

Therefore, considerable amount of Ca^{2+} probably recycles between mitochondria and ER, and the inter-organelle Ca^{2+} recycling may be pivotal for maintaining ER Ca^{2+} content.

16.2.4 Store-Operated Ca^{2+} Entry (SOCE) and Mitochondria in Lymphocytes

Ca^{2+} is an important second messenger in the lymphocyte activation by antigen. Upon the antigen binding to the surface receptor of lymphocytes, $InsP_3$ increases and facilitates Ca^{2+} release from the IP3R on the ER membrane. The subsequent Ca^{2+} depletion of ER causes translocation of the stromal interaction molecule 1 to the vicinity of plasmalemma, inducing sustained and oscillatory cytoplasmic Ca^{2+} increase by the activation of SOCE through Ca^{2+} release-activated Ca^{2+} channel encoded by ORAI1 (Feske 2007; Vig and Kinet 2009). The mitochondrial Ca^{2+} sequestration and/or the mitochondria metabolites have been reported to fine-tune the amplitude of SOCE (Hoth et al. 1997; Parekh 2008). Interestingly, mitochondria accumulate in the vicinity of immunological synapse in Jurkat T cells upon the T cell receptor activation (Quintana et al. 2007), which may support the sustainment of cytoplasmic Ca^{2+} elevation by facilitating SOCE.

16.3 Mitochondria Na^+-Ca^{2+} Exchange

16.3.1 Biophysical Property of NCX_{mit}

Crompton et al. (1977) demonstrated that Li^+ can substitute for Na^+ in promoting mitochondrial Ca^{2+} efflux. This property is distinctly different from plasma membrane Na^+-Ca^{2+} exchange (NCX) which cannot mediate Li^+-Ca^{2+} exchange, but its reverse exchange is stimulated by external Li^+ (Matsuoka and Hilgemann 1994; Blaustein and Lederer 1999; DiPolo and Beaugé 2006). Interestingly, a point mutation of threonine 103 to valine or cysteine in canine NCX1 increases an apparent affinity of the exchanger for cytoplasmic Na^+ and produces a significant Li^+-Ca^{2+} exchange (Doering et al. 1998). The threonine residue is conserved among all members of NCX, NCKX, and NCLX, whereas only in NCLX the amino acid preceding threonine is valine (Palty et al. 2004). Mechanisms for transporting Li^+ may be similar to those for Na^+.

Figure 16.2a demonstrates NCX_{mit} activity in permeabilized rat ventricular myocytes (Kim and Matsuoka 2008). After loading the mitochondria with Ca^{2+}, Ca^{2+} efflux from mitochondria via NCX_{mit} was activated by the Ca^{2+} removal from and the Na^+ addition to the bath solution. The decay of mitochondrial Ca^{2+} monitored by Rhod-2 was clearly dependent on cytoplasmic Na^+, having a half maximum concentration ($K_{1/2}$) of cytoplasmic Na^+ of ~1 mM and Hill coefficient of 3.8 (Kim and Matsuoka 2008). It has been reported that the $K_{1/2}$ of Na^+ for the mitochondrial Ca^{2+} efflux was 2–12 mM in the isolated mitochondria (Crompton et al. 1976, 1978; Hayat and Crompton 1982; Wingrove and Gunter 1986; Cox and Matlib 1993) and reconstituted mitochondrial Na^+-Ca^{2+} exchange (Paucek and Jab rek 2004). Our data show higher Na^+ affinity of NCX_{mit}. Despa et al. (2002) demonstrated that resting cytoplasmic Na^+ in the rat ventricular myocyte is ~10 mM and that Na^+ increased linearly to 15 mM when beating frequency was increased to 2.5 Hz. Therefore, under the physiological concentration, Na^+ binding to NCX_{mit} is probably saturated, and further increase in cytoplasmic Na^+ will not activate turnover of NCX_{mit}.

The dependence of NCX_{mit} on mitochondrial membrane potential ($\Delta\Psi$) has been controversial. In early studies by Crompton et al. (1976, 1977), electrogenic or voltage-dependent Na^+-Ca^{2+} exchange was suggested, considering the Hill coefficient for cytoplasmic Na^+ of ~3 and the attenuation of Na^+-dependent Ca^{2+} efflux by membrane depolarization induced by an uncoupler. A later study by Jung et al. (1995) further supported their conclusion by measuring matrix pH and Ca^{2+} with fluorescence probes. To the contrary, Affolter and Carafoli (1980) demonstrated that $\Delta\Psi$ did not change when the Ca^{2+} efflux via NCX_{mit} was induced and suggested NCX_{mit} is electroneutral. Brand (1985) and

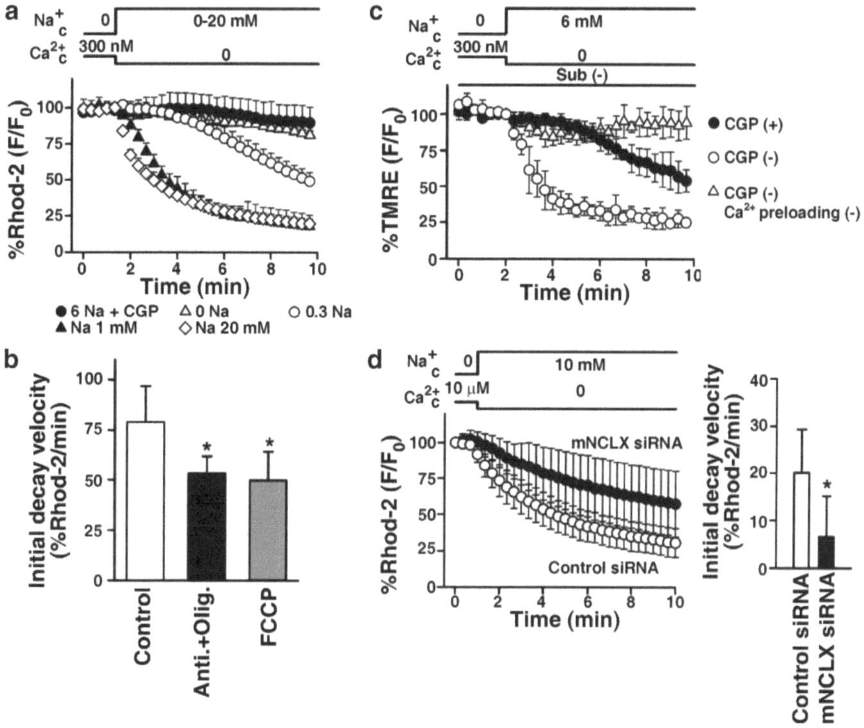

Fig. 16.2 *NCX_{mit} in cardiomyocytes.* (**a**) Cytoplasmic Na^+-dependent Ca^{2+} efflux from mitochondria (Modified from Kim and Matsuoka (2008)). (**b**) Attenuation of forward mode of NCX_{mit} by depolarization of mitochondria membrane (Modified from Kim and Matsuoka (2008)). (**c**) $\Delta\Psi$ changes induced by activating forward mode of NCX_{mit} (Modified from Kim and Matsuoka (2008)). (**d**) Attenuation of mitochondrial Ca^{2+} efflux via NCX_{mit} by knocking down mNCLX with siRNA in HL-1 atrial myocytes

Wingrove and Gunter (1986) also suggested the voltage-independent exchange.

Our study on NCX_{mit} using the permeabilized rat ventricular cells demonstrated that NCX_{mit} is voltage dependent and electrogenic (Kim and Matsuoka 2008). First, when $\Delta\Psi$ was intact, an addition of cytoplasmic Ca^{2+} promoted Ca^{2+} influx through CaUni and increased mitochondrial Ca^{2+}. The increase of mitochondrial Ca^{2+} was augmented by a blocker of NCX_{mit}, CGP-37157, suggesting the operation of forward mode of NCX_{mit}. Contrarily, when $\Delta\Psi$ was decreased (depolarization) by a mitochondrial uncoupler FCCP, the addition of cytoplasmic Ca^{2+} increased mitochondrial Ca^{2+} and the Ca^{2+} increase was suppressed by CGP-37157, suggestive of reverse mode of NCX_{mit}. Ruthenium red (a blocker of CaUni) or cyclosporin A (a blocker of permeability transition pore) did not affect the Ca^{2+} increase. Namely, mitochondrial depolarization changes the exchange mode from forward to reverse. Secondly, the Ca^{2+} efflux via the forward mode of NCX_{mit} was significantly reduced by depolarization of mitochondria membrane as shown in Fig. 16.2b. Thus, NCX_{mit} is voltage dependent. Lastly, when the respiratory chain was compromised, the induction of the reverse mode of NCX_{mit} hyperpolarized mitochondria membrane, while mitochondria membrane depolarized upon inducing the forward mode of NCX_{mit}. Both changes in $\Delta\Psi$ were remarkably inhibited by CGP-37157. The experimental data of forward mode of NCX_{mit} were presented in Fig. 16.2c. These findings indicate that NCX_{mit} is voltage dependent and electrogenic.

16.3.2 NCLX in Cardiomyocytes and Lymphocytes

The gene-encoding NCX_{mit} has not been found for a long time. Gobbi et al. (2007) indicated by immunocytochemistry that NCX1 ~ 3 localize in mitochondria of neurons and astrocytes of adult rat brain. However, functional evidence for the localization of NCX1–3 in mitochondria has not been demonstrated yet. In 2010, Palty et al. elegantly demonstrated that NCLX or NCKX6, which were previously cloned as an isoform of NCX or NCKX (Cai and Lytton 2004; Palty et al. 2004), is a gene-encoding NCX_{mit}.

16.3.2.1 NCLX in Cardiomyocytes

We investigated whether NCLX is functional in the cardiomyocytes using the HL-1 cells, a spontaneously beating cardiomyocyte cell line originated from mouse atrial myocytes (Claycomb et al. 1998). Although they can be serially passaged, it retains phenotypes of differentiated adult atrial cardiomyocytes (White et al. 2004). First, we confirmed the expression of mouse NCLX (mNCLX) mRNA in HL-1 cells (data not shown). Then we examined the contribution of NCLX on the NCX_{mit} activity by knocking down the mNCLX. The activity of NCX_{mit}, which was measured in a similar manner to Fig. 16.2a, was significantly slower in the mNCLX siRNA-transfected cells compared with that observed in the control siRNA-transfected cells (Fig. 16.2d). These results suggest that NCX_{mit} activity, which had been demonstrated in ventricular myocytes, is preserved also in atrial myocytes and NCLX has a major role in mediating it.

16.3.2.2 NCLX in Lymphocytes

NCLX also functions in non-excitable cell, lymphocytes. We studied NCX_{mit} function by creating heterozygous NCLX knockout of chicken DT40 B lymphocytes ($NCLX^{+/-}$) (Kim et al. 2012). As shown in Fig. 16.3a, cytoplasmic Na^+-dependent Ca^{2+} efflux from mitochondria was greatly suppressed in $NCLX^{+/-}$ cells, indicating that NCLX is a major component of NCX_{mit}, and also in lymphocytes. Mitochondrial Ca^{2+}

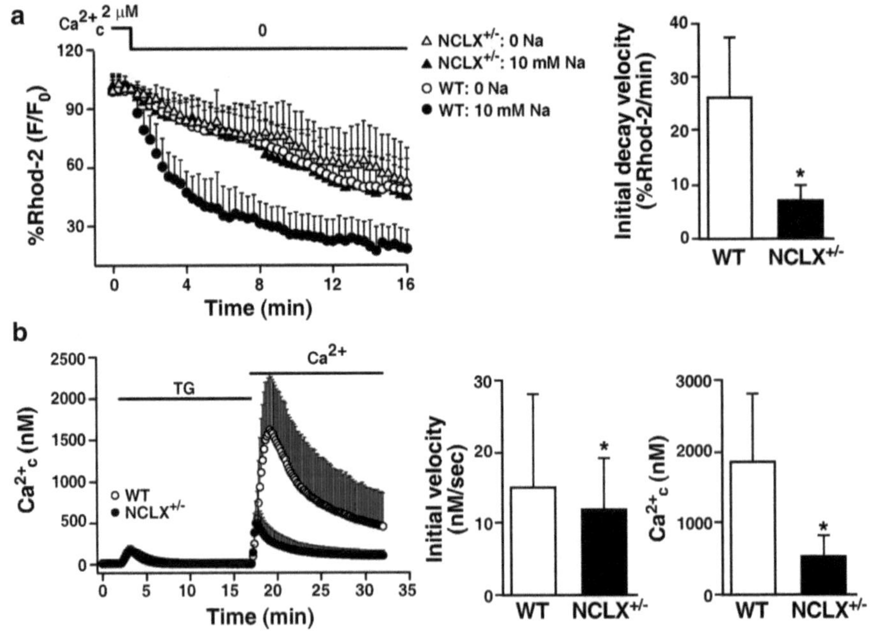

Fig. 16.3 *NCX_{mit} in lymphocytes.* (**a**) Cytoplasmic Na^+-dependent Ca^{2+} efflux from mitochondria in control and $NCLX^{+/-}$ DT40 cells. Similar protocol to Fig. 16.2a was used. A bar graph shows summary of the initial velocity of WT and $NCLX^{+/-}$ cells. (**b**) SOCE in control and $NCLX^{+/-}$ DT 40 cells. Fura-2-loaded DT40 cells were superfused with a nominally Ca^{2+}-free solution for 10 min, followed by that containing 0.1 μM thapsigargin (TG) to induce store depletion. Then 2 mM Ca^{2+} was added to activate SOCE. Bar graphs show summary of the initial velocity (*left*) and cytoplasmic Ca^{2+} peak (*right*). The initial velocity of cytoplasmic Ca^{2+} rise was obtained by fitting a linear function to initial three data (10 s)

spontaneously and slowly decayed even without cytoplasmic Na^{+}. The slow spontaneous decay is probably mediated by HCX_{mit}, whose activity is about one-third of NCX_{mit}.

The Ca^{2+} accumulation in and the NCLX-mediated Ca^{2+} release from mitochondria are involved in SOCE as suggested by previous studies (Hoth et al. 1997; Parekh 2008; Quintana et al. 2007). As shown in Fig. 16.3b, the initial velocity of cytoplasmic Ca^{2+} rise induced by SOCE was not much affected in the $NCLX^{+/-}$ cells, suggesting Ca^{2+} influx pathways were not compromised. To the contrary, the peak amplitude of cytoplasmic Ca^{2+} was greatly reduced. The reduction is probably caused by Ca^{2+} trap in mitochondria of $NCLX^{+/-}$ cells, and this indicates that the majority of Ca^{2+} that entered via SOCE subsequently goes into mitochondria and is released to cytosol via NCX_{mit}. In $NCLX^{+/-}$ cells, cytoplasmic Ca^{2+} rise induced by stimulation of B cell antigen receptor was also suppressed, indicating that NCX_{mit} or NCLX plays pivotal role in ER-mitochondrial Ca^{2+} communication (Kim et al. 2012).

16.4 Conclusions

NCX_{mit} is voltage dependent and electrogenic. NCLX is a major component of NCX_{mit} in cardiomyocytes and B lymphocytes and has a pivotal role in Ca^{2+} handling in mitochondria and cytoplasm in these cells.

References

H. Affolter, E. Carafoli, The Ca^{2+}-Na^{+} antiporter of heart mitochondria operates electroneutrally. Biochem. Biophys. Res. Commun. **95**, 193–196 (1980)

S. Arnaudeau, W.L. Kelley, J.V. Walsh Jr., N. Demaurex, Mitochondria recycle Ca^{2+} to the endoplasmic reticulum and prevent the depletion of neighboring endoplasmic reticulum regions. J. Biol. Chem. **276**, 29430–29439 (2001)

E. Barth, G. Stämmler, B. Speiser, J. Schaper, Ultrastructural quantitation of mitochondria and myofilaments in cardiac muscle from 10 different animal species including man. J. Mol. Cell. Cardiol. **24**, 669–681 (1992)

J.M. Baughman, F. Perocchi, H.S. Girgis, M. Plovanich, C.A. Belcher-Timme, Y. Sancak, X.R. Bao, L. Strittmatter, O. Goldberger, R.L. Bogorad, V. Koteliansky, V.K. Mootha, Integrative genomics identifies MCU as an essential component of the mitochondrial calcium uniporter. Nature **476**, 341–345 (2011)

C.J. Bell, N.A. Bright, G.A. Rutter, E.J. Griffiths, ATP regulation in adult rat cardiomyocytes: time-resolved decoding of rapid mitochondrial calcium spiking imaged with targeted photoproteins. J. Biol. Chem. **281**, 28058–28067 (2006)

P. Bernardi, Mitochondrial transport of cations: channels, exchangers, and permeability transition. Physiol. Rev. **79**, 1127–1155 (1999)

M.P. Blaustein, W.J. Lederer, Sodium/calcium exchange: its physiological implications. Physiol. Rev. **79**, 763–854 (1999)

M.D. Brand, The stoichiometry of the exchange catalysed by the mitochondrial calcium/sodium antiporter. Biochem. J. **229**, 161–166 (1985)

X. Cai, J. Lytton, Molecular cloning of a sixth member of the K^{+}-dependent Na^{+}/Ca^{2+} exchanger gene family, NCKX6. J. Biol. Chem. **279**, 5867–5876 (2004)

E. Carafoli, The fateful encounter of mitochondria with calcium: how did it happen? Biochim. Biophys. Acta **1797**, 595–606 (2010)

E. Carafoli, R. Tiozzo, G. Lugli, F. Crovetti, C. Kratzing, The release of calcium from heart mitochondria by sodium. J. Mol. Cell. Cardiol. **6**, 361–371 (1974)

F. Celsi, P. Pizzo, M. Brini, S. Leo, C. Fotino, P. Pinton, R. Rizzuto, Mitochondria, calcium and cell death: a deadly triad in neurodegeneration. Biochim. Biophys. Acta **1787**, 335–344 (2009)

W.C. Claycomb, N.A. Lanson Jr., B.S. Stallworth, D.B. Egeland, J.B. Delcarpio, A. Bahinski, N.J. Izzo Jr., HL-1 cells: a cardiac muscle cell line that contracts and retains phenotypic characteristics of the adult cardiomyocyte. Proc. Natl. Acad. Sci. U. S. A. **95**, 2979–2984 (1998)

S. Cortassa, M.A. Aon, B. O'Rourke, R. Jacques, H.J. Tseng, E. Marbán, R.L. Winslow, A computational model integrating electrophysiology, contraction, and mitochondrial bioenergetics in the ventricular myocyte. Biophys. J. **91**, 1564–1589 (2006)

D.A. Cox, M.A. Matlib, A role for the mitochondrial Na^{+}-Ca^{2+} exchanger in the regulation of oxidative phosphorylation in isolated heart mitochondria. J. Biol. Chem. **268**, 938–947 (1993)

M. Crompton, M. Capano, E. Carafoli, The sodium induced efflux of calcium from heart mitochondria. A possible mechanism for the regulation of mitochondrial calcium. Eur. J. Biochem. **69**, 453–462 (1976)

M. Crompton, M. Künzi, E. Carafoli, The calcium-induced and sodium-induced effluxes of calcium from heart mitochondria. Evidence for a sodium-calcium carrier. Eur. J. Biochem. **79**, 549–558 (1977)

M. Crompton, R. Moser, H. Lüdi, E. Carafoli, The interrelations between the transport of sodium and calcium in mitochondria of various mammalian tissues. Eur. J. Biochem. **82**, 25–31 (1978)

G. Csordás, G. Hajnóczky, SR/ER-mitochondrial local communication: calcium and ROS. Biochim. Biophys. Acta **1787**, 1352–1362 (2009)

H.F. DeLuca, G.W. Engstrom, Calcium uptake by rat kidney mitochondria. Proc. Natl. Acad. Sci. U. S. A. **47**, 1744–1750 (1961)

S. Despa, M.A. Islam, S.M. Pogwizd, D.M. Bers, Intracellular [Na^+] and Na^+ pump rate in rat and rabbit ventricular myocytes. J. Physiol. **539**, 133–143 (2002)

R. DiPolo, L. Beaugé, Sodium/calcium exchanger: influence of metabolic regulation on ion carrier interactions. Physiol. Rev. **86**, 155–203 (2006)

A.E. Doering, D.A. Nicoll, Y. Lu, L. Lu, J.N. Weiss, K.D. Philipson, Topology of a functionally important region of the cardiac Na^+/Ca^{2+} exchanger. J. Biol. Chem. **273**, 778–783 (1998)

S. Feske, Calcium signalling in lymphocyte activation and disease. Nat. Rev. Immunol. **7**, 690–702 (2007)

P. Gobbi, P. Castaldo, A. Minelli, S. Salucci, S. Magi, E. Corcione, S. Amoroso, Mitochondrial localization of Na^+/Ca^{2+} exchangers NCX1-3 in neurons and astrocytes of adult rat brain in situ. Pharmacol. Res. **56**, 556–565 (2007)

L.H. Hayat, M. Crompton, Evidence for the existence of regulatory sites for Ca^{2+} on the Na^+/Ca^{2+} carrier of cardiac mitochondria. Biochem. J. **202**, 509–518 (1982)

M. Hoth, C.M. Fanger, R.S. Lewis, Mitochondrial regulation of store-operated calcium signaling in T lymphocytes. J. Cell Biol. **137**, 633–648 (1997)

D. Jiang, L. Zhao, D.E. Clapham, Genome-wide RNAi screen identifies Letm1 as a mitochondrial Ca^{2+}/H^+ antiporter. Science **326**, 144–147 (2009)

H. Jo, A. Noma, S. Matsuoka, Calcium-mediated coupling between mitochondrial substrate dehydrogenation and cardiac workload in single guinea-pig ventricular myocytes. J. Mol. Cell. Cardiol. **40**, 394–404 (2006)

D.W. Jung, K. Baysal, G.P. Brierley, The sodium-calcium antiport of heart mitochondria is not electroneutral. J. Biol. Chem. **270**, 672–678 (1995)

B. Kim, S. Matsuoka, Cytoplasmic Na^+-dependent modulation of mitochondrial Ca^{2+} via electrogenic mitochondrial Na^+-Ca^{2+} exchange. J. Physiol. **586**, 1683–1697 (2008)

B. Kim, A. Takeuchi, O. Koga, M. Hikida, S. Matsuoka, Pivotal role of mitochondrial Na^+-Ca^{2+} exchange in antigen receptor mediated Ca^{2+} signalling in DT40 and A20 B lymphocytes. J. Physiol. (2012 in press)

Y. Kirichok, G. Krapivinsky, D.E. Clapham, The mitochondrial calcium uniporter is a highly selective ion channel. Nature **427**, 360–364 (2004)

R. Malli, M. Frieden, M. Trenker, W.F. Graier, The role of mitochondria for Ca^{2+} refilling of the endoplasmic reticulum. J. Biol. Chem. **280**, 12114–12122 (2005)

S. Matsuoka, D.W. Hilgemann, Inactivation of outward Na^+-Ca^{2+} exchange current in guinea-pig ventricular myocytes. J. Physiol. **476**, 443–458 (1994)

J.G. McCormack, A.P. Halestrap, R.M. Denton, Role of calcium ions in regulation of mammalian intramitochondrial metabolism. Physiol. Rev. **70**, 391–425 (1990)

C.C. Mendes, D.A. Gomes, M. Thompson, N.C. Souto, T.S. Goes, A.M. Goes, M.A. Rodrigues, M.V. Gomez, M.H. Nathanson, M.F. Leite, The type III inositol 1,4,5-trisphosphate receptor preferentially transmits apoptotic Ca^{2+} signals into mitochondria. J. Biol. Chem. **280**, 40892–40900 (2005)

M. Murgia, C. Giorgi, P. Pinton, R. Rizzuto, Controlling metabolism and cell death: at the heart of mitochondrial calcium signalling. J. Mol. Cell. Cardiol. **46**, 781–788 (2009)

R. Palty, E. Ohana, M. Hershfinkel, M. Volokita, V. Elgazar, O. Beharier, W.F. Silverman, M. Argaman, I. Sekler, Lithium-calcium exchange is mediated by a distinct potassium-independent sodium-calcium exchanger. J. Biol. Chem. **279**, 25234–25240 (2004)

R. Palty, W.F. Silverman, M. Hershfinkel, T. Caporale, S.L. Sensi, J. Parnis, C. Nolte, D. Fishman, V. Shoshan-Barmatz, S. Herrmann, D. Khananshvili, I. Sekler, NCLX is an essential component of mitochondrial Na^+/Ca^{2+} exchange. Proc. Natl. Acad. Sci. U. S. A. **107**, 436–441 (2010)

A.B. Parekh, Mitochondrial regulation of store-operated CRAC channels. Cell Calcium **44**, 6–13 (2008)

P. Paucek, M. Jab rek, Kinetics and ion specificity of Na^+/Ca^{2+} exchange mediated by the reconstituted beef heart mitochondrial Na^+/Ca^{2+} antiporter. Biochim. Biophys. Acta **1659**, 83–91 (2004)

F. Perocchi, V.M. Gohil, H.S. Girgis, X.R. Bao, J.E. McCombs, A.E. Palmer, V.K. Mootha, MICU1 encodes a mitochondrial EF hand protein required for Ca^{2+} uptake. Nature **467**, 291–296 (2010)

G.E. Petrzilka, H.E. Schroeder, Activation of human T-lymphocytes. A kinetic and stereological study. Cell Tissue Res. **201**, 101–127 (1979)

D. Poburko, C.H. Liao, C. van Breemen, N. Demaurex, Mitochondrial regulation of sarcoplasmic reticulum Ca^{2+} content in vascular smooth muscle cells. Circ. Res. **104**, 104–112 (2009)

A. Quintana, C. Schwindling, A.S. Wenning, U. Becherer, J. Rettig, E.C. Schwarz, M. Hoth, T cell activation requires mitochondrial translocation to the immunological synapse. Proc. Natl. Acad. Sci. U. S. A. **104**, 14418–14423 (2007)

S.Y. Ryu, G. Beutner, R.T. Dirksen, K.W. Kinnally, S.S. Sheu, Mitochondrial ryanodine receptors and other mitochondrial Ca^{2+} permeable channels. FEBS Lett. **584**, 1948–1955 (2010)

V.K. Sharma, V. Ramesh, C. Franzini-Armstrong, S.S. Sheu, Transport of Ca^{2+} from sarcoplasmic reticulum to mitochondria in rat ventricular myocytes. J. Bioenerg. Biomembr. **32**, 97–104 (2000)

G. Szabadkai, K. Bianchi, P. Várnai, D. De Stefani, M.R. Wieckowski, D. Cavagna, A.I. Nagy, T. Balla, R. Rizzuto, Chaperone-mediated coupling of endoplasmic reticulum and mitochondrial Ca^{2+} channels. J. Cell Biol. **175**, 901–911 (2006)

P.R. Territo, S.A. French, M.C. Dunleavy, F.J. Evans, R.S. Balaban, Calcium activation of heart mitochondrial oxidative phosphorylation: rapid kinetics of mVO2, NADH, AND light scattering. J. Biol. Chem. **276**, 2586–2599 (2001)

M. Vig, J.P. Kinet, Calcium signaling in immune cells. Nat. Immunol. **10**, 21–27 (2009)

S.M. White, P.E. Constantin, W.C. Claycomb, Cardiac physiology at the cellular level: use of cultured HL-1 cardiomyocytes for studies of cardiac muscle cell structure and function. Am. J. Physiol. Heart Circ. Physiol. **286**, H823–H829 (2004)

D.E. Wingrove, T.E. Gunter, Kinetics of mitochondrial calcium transport. II. A kinetic description of the sodium-dependent calcium efflux mechanism of liver mitochondria and inhibition by ruthenium red and by tetraphenylphosphonium. J. Biol. Chem. **261**, 15166–15171 (1986)

New Insights in Mitochondrial Calcium Handling by Sodium/ Calcium Exchanger

17

Antonella Scorziello, Claudia Savoia, Agnese Secondo, Francesca Boscia, Maria Josè Sisalli, Alba Esposito, Annalisa Carlucci, Pasquale Molinaro, Luca Lignitto, Gianfranco Di Renzo, Antonio Feliciello, and Lucio Annunziato

Abstract

Mitochondria are now recognized as one of the main intracellular calcium-storing organelles which play a key role in the intracellular calcium signalling. Indeed, besides performing oxidative phosphorylation, mitochondria are able to sense and shape calcium (Ca^{2+}) transients, thus controlling cytosolic Ca^{2+} signals and Ca^{2+}-dependent protein activity. It has been well established for many years that mitochondria have a huge capacity to accumulate calcium. While the physiological significance of this pathway was hotly debated until relatively recently, it is now clear that the ability of mitochondria in calcium handling is a ubiquitous phenomenon described in every cell system in which the issue has been addressed.

A. Scorziello • C. Savoia • A. Secondo • F. Boscia
M.J. Sisalli • A. Esposito • P. Molinaro • G. Di Renzo
Division of Pharmacology, Department of Neuroscience, School of Medicine, Federico II University of Naples-National Institute of Neuroscience, Via S. Pansini, 5, Naples, 80131, Italy

A. Carlucci • L. Lignitto • A. Feliciello
Department of Molecular and Cellular Biology and Pathology "L. Califano", Federico II University of Naples, Via S. Pansini, 5, Naples 80131, Italy

L. Annunziato, M.D. (✉)
Division of Pharmacology, Department of Neuroscience, School of Medicine, Federico II University of Naples-National Institute of Neuroscience, Via S. Pansini, 5, Naples 80131, Italy
e-mail: lannunzi@unina.it

L. Annunziato (ed.), *Sodium Calcium Exchange: A Growing Spectrum of Pathophysiological Implications*, Advances in Experimental Medicine and Biology 961, DOI 10.1007/978-1-4614-4756-6_17,

In this chapter, we will review the molecular mechanisms involved in the regulation of mitochondrial calcium cycling in physiological conditions with particular regard to the role played by the mitochondrial Na+/Ca^{2+} exchanger.

Keywords
Mitochondria • Calcium • NCX3 • AKAP • MCU

17.1 Introduction

Mitochondria are highly specialized organelles and major players in fundamental aspects of cell physiology. In eukaryotic cells, energy production is functionally coupled to metabolic demands, and the cells efficiently adapt oxidative respiration in response to changes in extracellular microenvironment and metabolic nutrient availability.

These organelles produce approximately 15 times more ATP from glucose than the glycolytic pathway in eukaryotic cells by coupling electron transport to the generation of proton gradients for oxidative phosphorylation. Cells of highly metabolic tissues such as muscle, liver and brain are therefore particularly dependent on mitochondria. In particular, in the central nervous system, mitochondria produce over 95% of ATP utilized by the brain (Erecinska and Silver 1994), and within the brain, they are distributed to regions of high metabolic demand, including synapses, nodes of Ranvier and myelination/demyelination interfaces (Berthold et al. 1993; Bristow et al. 2002; Kageyama and Wong-Riley 1982; Rowland et al. 2000). Generation, processing and transmission of neural impulses rely heavily on sodium (Na^+), potassium (K^+) and calcium (Ca^{2+}) ion gradients across the plasma membrane. In fact, 50–60% of total brain ATP is used to maintain these gradients, especially through Na^+/K^+ pumps (Erecinska and Silver 1994). Mitochondria also play an important role in regulating intracellular calcium homeostasis (Babcock et al. 1997; Nicholls 1978; Jouaville et al. 1995; Werth and Thayer 1994). Mitochondrial calcium overload and subsequent dysfunction are thought to be critically important for triggering the cell death that follows ischemic and traumatic brain injury as well as in several neurodegenerative disorders including Alzheimer's, Parkinson's, Huntington's diseases and amyotrophic lateral sclerosis (ALS).

17.2 Relevance of Mitochondrial Calcium Handling Under Physiological and Pathophysiological Conditions

Besides performing oxidative phosphorylation, mitochondria are able to sense and shape Ca^{2+} transients, thus controlling cytosolic Ca^{2+} signals and Ca^{2+}-dependent proteins. Indeed, it has been well established for many years that mitochondria have a huge capacity to accumulate calcium. In particular, the contribution of mitochondrial Ca^{2+} uptake becomes crucial for Ca_i^{2+} homeostasis when $[Ca^{2+}]_i$ exceeds levels of 400–500 nM (Nicholls and Cromton 1980). While the physiological significance of this pathway was debated until relatively recently, it is now clear that the ability of mitochondria in calcium handling is a ubiquitous phenomenon described in every cell system in which the issue has been addressed. Therefore, mitochondria are now recognized as one of the main intracellular calcium storing organelles which play a key role in the intracellular calcium signalling (Rizzuto et al. 2000). The maintenance of membrane potential ($\Delta\Psi_m$), strictly dependent by the holding of mitochondrial Ca^{2+} and Na^+ within a narrow range of concentrations, is an essential requirement for calcium accumulation into mitochondria (Murgia et al. 2009). This process has enormous functional consequences both for cell physiology and for pathophysiology.

Indeed, the presence of high levels of Ca^{2+} inside the mitochondrial matrix is necessary for the right functioning of mitochondrial enzymes (Nicholls 2004). Nevertheless, when mitochondrial calcium concentration overcomes the storage capability of these organelles, as it happens in neuronal anoxia, a subsequent increase in free radical production (ROS) occurs. This phenomenon results in a damage of the inner mitochondrial membrane and in the oxidation of the proteins involved in the electron transport, in proton pumping and in ATP production (Dugan and Choi 1994). A further consequence of $[Ca^{2+}]_m$ increase is the alteration of inner mitochondrial membrane permeability that leads to the opening of mitochondrial permeability transition pore (mPTP). This causes the release from mitochondria to cytosol of molecules potentially harmful for the cell such as cytochrome *c* (Atlante et al. 2003; Petrosillo et al. 2004; Scorziello et al. 2007), apoptosis inducing factor (AIF), and Smac/Diablo (Du et al. 2000; Verhagen et al. 2000; Susin et al. 1999; Li et al. 1997).

17.3 Ion Channels and Transporters Controlling the Mitochondrial Calcium Handling

The ionic homeostasis of mitochondria is largely maintained by mechanisms regulating the efflux or the influx of Ca^{2+} (Fig. 17.1). Mitochondrial calcium uptake is primarily driven by the electrochemical potential gradient and by a relatively low $[Ca^{2+}]_m$. Ca^{2+} is taken up through the inner mitochondrial membrane (IMM) by a uniporter (MCU) whose molecular identity and functional activity have been recently demonstrated (De Stefani et al. 2011; Baughman et al. 2011). An uptake pathway with properties distinct from those of the uniporter has also been described (Sparagna et al. 1995; Buntinas et al. 2001). This has been referred as the rapid uptake mode (RaM). This pathway has the capacity to transfer Ca^{2+} very rapidly into the mitochondria during the rising phase of a Ca^{2+} pulse. On the other

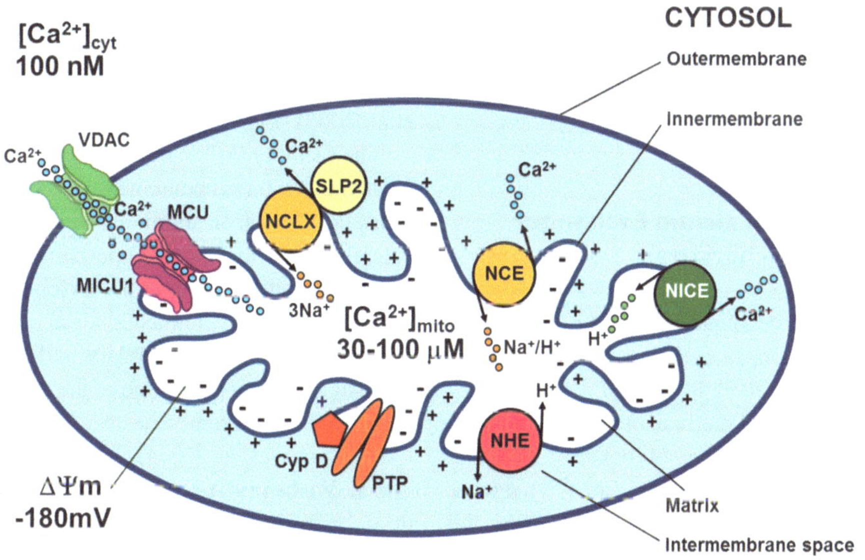

Fig. 17.1 *Schematic model of mitochondrial influx and efflux pathways.* In mammalian mitochondria, the uptake of Ca^{2+} into the matrix is mediated by the mitochondrial Ca^{2+} uniporter (MCU), a calcium-sensitive channel, regulated by a calcium-sensing accessory subunit (MICU1). The efflux is mediated by the Na^+-Li^+/Ca^{2+} exchanger NCLX whose activity is downregulated by the protein SLP-2. High levels of Ca^{2+} into the matrix trigger the opening of the permeability transition pore, PTP, responsible for mitochondrial membrane permeabilization and neuronal cell death. *VDAC*, voltage-dependent anionic channel; *NICE*, sodium-independent calcium exchanger; *NCE*, sodium-dependent calcium exchange

hand, compared to the MCU and the others Ca^{2+} influx mechanism, the proteins that catalyze the efflux of this ion from mitochondria have received much less attention. Ca^{2+} efflux is catalyzed by antiporters that drive Ca^{2+} out of the mitochondrial matrix in exchange with either Na^+ or H^+ (Nicholls and Crompton 1980). Two types of exchangers have been functionally characterized in the 1970s: the Na^+/Ca^{2+} and the H^+/Ca^{2+} exchangers (Carafoli 2003). These two pathways have been defined Na^+-independent pathway for Ca^{2+} efflux ("NICE") and Na^+-dependent pathway for Ca^{2+} efflux ("NCE"), respectively. They have different kinetics of activation and calcium affinity (Harris 1979; Lehninger et al. 1978; Ramachandran and Bygrave 1978). The Na^+-independent Ca^{2+} efflux is the main mitochondrial Ca^{2+} efflux system in non-excitable cells, and since no specific cations have been found to be exchanged with Ca^{2+}, it is believed to be a Ca^{2+}-H^+ exchanger (Saris and Carafoli 2005). A characteristic of this transporter is that it saturates at low calcium loads and its kinetics is extremely slow (Bernardi 1999). This emphasizes a feature of the mitochondrial Ca^{2+} machinery: it is equipped with high V_{max} uptake transport systems coupled to slow and easily saturable efflux system, increasing the risk of Ca^{2+} overload (Murgia et al. 2009).

17.3.1 Sodium Calcium Exchanger in Mitochondria

The first evidence suggesting that mitochondria can efflux Ca^{2+} ions in exchange for Na^+ ions was reported by Carafoli et al. in 1974. Later, in 1988 the identification of the benzothiazepine derivative CGP37157 (CGP) as a blocker of the mitochondrial Na^+/Ca^{2+} exchanger (mNCX) opened the way to pharmacologically characterize this exchanger (Chiesi et al. 1988). However, its identity and localization have been a matter of debate in the last three decades. Recently, among the calcium efflux mechanisms, NCLX, a member of the CCX superfamily, has been found on the cristae of the inner mitochondrial membrane (IMM), where it mediates sodium- and lithium-dependent calcium efflux from the mitochondrial matrix (Palty et al. 2010). This antiporter protein is both phylogenetically and functionally distinct from NCX and NCKX family members (Palty et al. 2010). However, the molecular mechanisms involved in NCLX expression, transport, tissue distribution, as well as its role in mitochondrial activity regulation still remain unidentified. On the other hand, mounting evidence has suggested that the outer mitochondrial membrane (OMM), where NCLX has never been found, is not passively permeable to calcium fluxes into the cytoplasm, but rather plays a key role in controlling mitochondrial function and Ca^{2+} cycling (Szabadkai and Duchen 2008). More specifically, the OMM serves as a significant permeability barrier not only to Ca^{2+} influx but also to Ca^{2+} efflux (Crompton et al. 2002). It has been demonstrated that the voltage-dependent anion channel (VDAC), located on the OMM, plays a relevant role in the regulation of Ca^{2+}permeability across this external membrane, thus favouring the activity of the specific transport systems of the inner membrane (Crompton et al. 2002). Moreover, VDAC, together with the adenine nucleotide translocase (ANT) and cyclophilin-D, might also elicit mitochondrial Ca^{2+}efflux since it is a part of the mitochondrial PTP (Bernardi 1999; Crompton et al. 2002) (Fig. 17.1).

The fact that Ca^{2+} is transported from the cytoplasm to the intermembrane space of mitochondria suggests that an additional sodium-dependent calcium influx mechanism operating between the intermembrane space and the cytoplasmic compartment could affect mitochondrial calcium buffering capacity. Our preliminary and unpublished data suggest that NCX3, one of the three isoforms of the plasma membrane Na^+/Ca^{2+} exchanger (Annunziato et al. 2004), is localized on the OMM. We have some evidence that the mitochondrial NCX3 works by promoting mitochondrial Ca^{2+} efflux from the intermembrane space to the cytosol. Interestingly, this Ca^{2+}-buffering activity seems to be modulated by AKAP121 (A-kinase-anchoring protein) that we previously showed to play a critical role in controlling mitochondrial membrane potential and oxidative phosphorylation (Livigni et al. 2006). Another evidence that

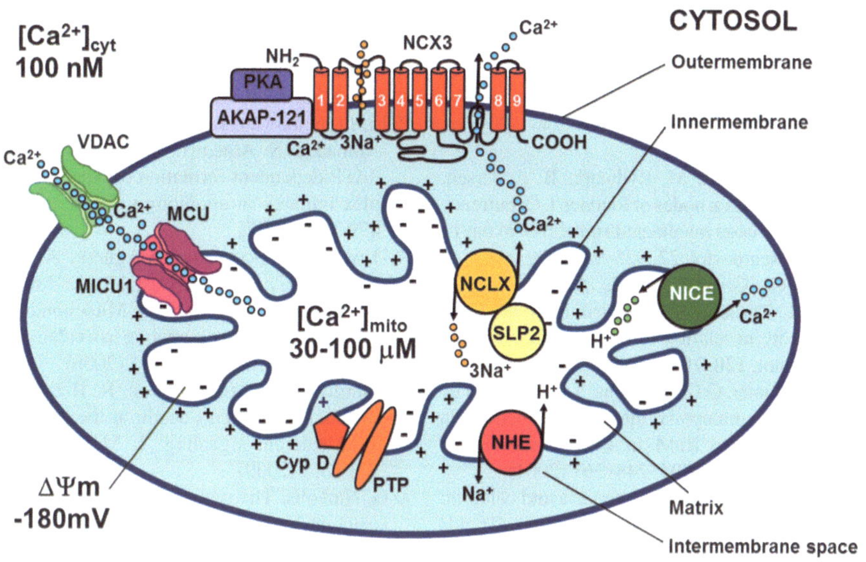

Fig. 17.2 *Schematic model of NCX3 and AKAP121 interaction on the outer mitochondrial membrane.* NCX3 on the outer mitochondrial membrane works in concert with NCLX to allow mitochondrial calcium efflux. This effect is promoted by the interaction with AKAP121 on the outer mitochondrial membrane and is dependent on PKA

mNCX3 could be involved in Ca^{2+} extrusion is that the Ca^{2+} lowering effect found in cells transfected with NCX3 was completely prevented by the selective mitochondrial Na^+/Ca^{2+}exchanger inhibitor benzothiazepine compound CGP-37157 (Cox et al. 1993; Nicolau 2009).

Therefore, we propose a model in which NCX3 interacting with AKAP121 on the OMM could promote mitochondrial calcium efflux working in concert with NCLX that, on the IMM, promotes calcium efflux from the mitochondrial matrix to the intermembrane space (Fig. 17.2).

17.4 Conclusions

Here we reviewed the recent work that is starting to shed light on the molecular identity and functional properties of the mitochondrial Na^+/Ca^{2+} exchanger. The picture that is emerging is more complex that initially believed since different forms of the exchanger seems to exist in different compartments of the mitochondria where they subserve different physiological and possibly pathophysiological roles. The new data that we reviewed will presumably open the way to new intense investigation aimed to further clarify how the different mitochondrial NCX are regulated and how they interact each other.

Acknowledgments This work was supported by COFIN2008; Ricerca-Sanitaria RF-FSL352059 Ricerca finalizzata 2006; Ricerca-Oncologica 2006; Progetto-Strategico 2007; Progetto Ordinario 2007; Ricerca finalizzata 2009; Ricerca-Sanitaria progetto Ordinario by Ministero della Salute 2008 all to LA

References

L. Annunziato, G. Pignataro, G.F. Di Renzo, Pharmacology of brain Na^+/Ca^{2+} exchanger: from molecular biology to therapeutic perspectives. Pharmacol. Rev. **56**, 633–654 (2004)

A. Atlante, A. Bobba, P. Calissano, S. Passarella, E. Marra, The apoptosis/necrosis transition in cerebellar granule cells depends on the mutual relationship of the antioxidant and the proteolytic systems which regulate ROS production and cytochrome c release en route to death. J. Neurochem. **84**, 960–971 (2003)

D.F. Babcock, J. Herrington, P.C. Goodwin, Y.B. Park, B. Hille, Mitochondrial participation in the intracellular Ca^{2+} network. J. Cell Biol. **136**, 833–844 (1997)

J.M. Baughman, F. Perocchi, H.S. Girgis, M. Plovanich, C.A. Belcher-Timme, Y. Sancak, X.R. Bao, L. Strittmatter, O. Goldberger, R.L. Bogorad, V. Koteliansky,

V.K. Mootha, Integrative genomics identifies MCU as an essential component of the mitochondrial calcium uniporter. Nature **476**, 341–345 (2011)

P. Bernardi, Mitochondrial transport of cations: channels, exchangers, and permeability transition. Physiol. Rev. **79**, 1127–1155 (1999)

C.H. Berthold, C. Fabricius, M. Rydmark, B. Andersen, Axoplasmic organelles at nodes of Ranvier. I. Occurrence and distribution in large myelinated spinal root axons of the adult cat. J. Neurocytol. **22**, 925–940 (1993)

E.A. Bristow, P.G. Griffiths, R.M. Andrews, M.A. Johnson, D.M. Turnbull, The distribution of mitochondrial activity in relation to optic nerve structure. Arch. Ophthalmol. **120**, 791–796 (2002)

L. Buntinas, K.K. Gunter, G.C. Sparagna, T.E. Gunter, The rapid mode of calcium uptake into heart mitochondria (RaM): comparison to RaM in liver mitochondria. Biochim. Biophys. Acta **1504**, 248–261 (2001)

E. Carafoli, Historical review: mitochondria and calcium: ups and downs of an unusual relationship. Trends Biochem. Sci. **28**, 175–181 (2003)

E. Carafoli, R. Tiozzo, G. Lugli, F. Crovetti, C. Kratzing, The release of calcium from heart mitochondria by sodium. J. Mol. Cell. Cardiol. **6**, 361–371 (1974)

M. Chiesi, R. Schwaller, K. Eichenberger, Structural dependency of the inhibitory action of benzodiazepines and related compounds on the mitochondrial Na^+-Ca^{2+} exchanger. Biochem. Pharmacol. **22**, 4399–4403 (1988)

M. Crompton, E. Barksby, N. Johnson, M. Capano, Mitochondrial intermembrane junctional complexes and their involvement in cell death. Biochimie **84**, 143–152 (2002)

D.A. Cox, L. Conforti, N. Sperelakis, M.A. Matlib. Selectivity of inhibition of $Na^{(+)}$-Ca^{2+} exchange of heart mitochondria by benzothiazepine CGP-37157. J. Cardiovasc. Pharmacol. **21**, 595–599 (1993)

D. De Stefani, A. Raffaello, E. Teardo, I. Szabo, R. Rizzuto, A forty-kilodalton protein of the inner membrane is the mitochondrial calcium uniporter. Nature **476**, 336–340 (2011)

C. Du, M. Fang, Y. Li, L. Li, X. Wang, Smac, a mitochondrial protein that promotes cytochrome c-dependent caspase activation by eliminating IAP inhibition. Cell **102**, 33–42 (2000)

L.L. Dugan, D.W. Choi, Excitotoxicity, free radicals, and cell membrane changes. Ann. Neurol. **35**, S17–S21 (1994)

M. Erecinska, I.A. Silver, Ions and energy in mammalian brain. Prog. Neurobiol. **43**, 37–71 (1994)

E.J. Harris, Modulation of Ca^{2+} efflux from heart mitochondria. Biochem. J. **178**, 673–680 (1979)

L.S. Jouaville, F. Ichas, E.L. Holmuhamedov, P. Camacho, J.D. Lechleiter, Synchronization of calcium waves by mitochondrial substrates in *Xenopus laevis* oocytes. Nature **377**, 438–441 (1995)

G.H. Kageyama, M.T. Wong-Riley, Histochemical localization of cytochrome oxidase in the hippocampus: correlation with specific neuronal types and afferent pathways. Neuroscience **7**, 2337–2361 (1982)

A.L. Lehninger, A. Vercesi, E.A. Bababunmi, Regulation of Ca^{2+} release from mitochondria by the oxidation-reduction state of pyridine nucleotides. Proc. Natl. Acad. Sci. U. S. A. **75**, 1690–1694 (1978)

P. Li, D. Nijhawan, I. Budihardjo, S.M. Srinivasula, M. Ahmad, E.S. Alnemri, X. Wang, Cytochrome c and dATP-dependent formation of Apaf-1/caspase-9 complex initiates an apoptotic protease cascade. Cell **91**, 479–489 (1997)

A. Livigni, A. Scorziello, S.Agnese, A. Adornetto, A. Carlucci, C. Garbi, I. Castaldo, L. Annunziato, E.V. Avvedimento, A. Feliciello, Mitochondrial AKAP121 links cAMP and src signaling to oxidative metabolism. Mol. Biol. Cell **17**, 263–271 (2006)

C. Murgia, P. Giorgi, P. Pinton, R. Rizzuto, Controlling metabolism and cell death: at the heart of mitochondrial calcium signaling. J. Mol. Cell. Cardiol. **46**, 781–788 (2009)

D.G. Nicholls, The regulation of extramitochondrial free calcium ion concentration by rat liver mitochondria. Biochem. J. **176**, 463–474 (1978)

D.G. Nicholls, Mitochondrial membrane potential and aging. Aging Cell **3**, 35–40 (2004)

D.G. Nicholls, M. Crompton, Mitochondrial calcium transport. FEBS Lett. **111**, 261–268 (1980)

S.M. Nicolau, Mitochondrial Na^+/Ca^{2+}-exchanger blocker CGP37157 protects against chromaffin cell death elicited by veratridine. J. Pharmacol. Exp. Ther. **330**(3), 844–854 (2009)

R. Palty, W.F. Silverman, M. Hershfinkel, T. Caporale, S.L. Sensi, J. Parnis, C. Nolte, D. Fishman, V. Shoshan-Barmatz, S. Herrmann, D. Khananshvili, I. Sekler, NCLX is an essential component of mitochondrial Na^+/Ca^{2+} exchange. Proc. Natl. Acad. Sci. U. S. A. **107**, 436–441 (2010)

G. Petrosillo, F.M. Ruggiero, M. Pistolese, G. Paradies, Ca^{2+}-induced reactive oxygen species production promotes cytochrome c release from rat liver mitochondria via mitochondrial permeability transition (MPT)-dependent and MPT-independent mechanisms: role of cardiolipin. J. Biol. Chem. **279**, 53103–53108 (2004)

C. Ramachandran, F.L. Bygrave, Calcium ion cycling in rat liver mitochondria. Biochem. J. **174**, 613–620 (1978)

R. Rizzuto, P. Bernardi, T. Pozzan, Mitochondria as all-round players of the calcium game. J. Physiol. **1**, 37–47 (2000)

K.C. Rowland, N.K. Irby, G.A. Spirou, Specialized synapse-associated structures within the calyx of Held. J. Neurosci. **20**, 9135–9144 (2000)

N.E. Saris, E. Carafoli, A historical review of cellular calcium handling, with emphasis on mitochondria. Biochemistry (Mosc.) **70**, 187–194 (2005)

A. Scorziello, M. Santillo, A. Adornetto, C. Dell'Aversano, R. Sirabella, S. Damiano, L.M. Canzoniero, G.F. Renzo, L. Annunziato, NO-induced neuroprotection in ischemic preconditioning stimulates mitochondrial Mn-SOD activity and expression via Ras/ERK1/2 pathway. J. Neurochem. **103**, 1472–80 (2007)

G.C. Sparagna, K.K. Gunter, S.S. Sheu, T.E. Gunter, Mitochondrial calcium uptake from physiological-type pulses of calcium. A description of the rapid uptake mode. J. Biol. Chem. **270**, 27510–27515 (1995)

S.A. Susin, H.K. Lorenzo, N. Zamzami, I. Marzo, B.E. Snow, G.M. Brothers, J. Mangion, E. Jacotot, P. Costantini, M. Loeffler, N. Larochette, D.R. Goodlett, R. Aebersold, D.P. Siderovski, J.M. Penninger, G. Kroemer, Molecular characterization of mitochondrial apoptosis-inducing factor. Nature **397**, 441–446 (1999)

G. Szabadkai, M.R. Duchen, Mitochondria: the hub of cellular Ca^{2+} signaling. Physiology (Bethesda) **23**, 84–94 (2008)

A.M. Verhagen, P.G. Ekert, M. Pakusch, J. Silke, L.M. Connolly, G.E. Reid, R.L. Moritz, R.J. Simpson, D.L. Vaux, Identification of DIABLO, a mammalian protein that promotes apoptosis by binding to and antagonizing IAP proteins. Cell **102**, 43–53 (2000)

J.L. Werth, S.A. Thayer, Mitochondria buffer physiological calcium loads in cultured rat dorsal root ganglion neurons. J. Neurosci. **14**, 348–356 (1994)

Part VII

NCX in Neurodegenerative Diseases

Genetically Modified Mice as a Strategy to Unravel the Role Played by the Na^+/Ca^{2+} Exchanger in Brain Ischemia and in Spatial Learning and Memory Deficits

18

Pasquale Molinaro, Mauro Cataldi, Ornella Cuomo, Davide Viggiano, Giuseppe Pignataro, Rossana Sirabella, Agnese Secondo, Francesca Boscia, Anna Pannaccione, Antonella Scorziello, Sophie Sokolow, André Herchuelz, Gianfranco Di Renzo, and Lucio Annunziato

Abstract

Because no isoform-specific blocker of NCX has ever been synthesized, a more selective strategy to identify the role of each antiporter isoform in the brain was represented by the generation of knockout and knockin mice for the different isoforms of the antiporter.

Experiments performed in NCX2 and NCX3 knockout mice provided evidence that these two isoforms participate in spatial learning and memory consolidation, although in an opposite manner. These new data from *ncx2−/−* and *ncx3−/−* mice may open new experimental avenues for the development of effective therapeutic compounds that, by selectively inhibiting or activating these molecular targets, could treat patients affected by cognitive impairment including Alzheimer's, Parkinson's, Huntington's diseases, and infarct dementia.

P. Molinaro • M. Cataldi • O. Cuomo • D. Viggiano
G. Pignataro • A. Secondo • F. Boscia • A. Pannaccione
A. Scorziello • G. Di Renzo
Division of Pharmacology, Department of Neuroscience, School of Medicine, "Federico II" University of Naples, Via S. Pansini, 5, Naples 80131, Italy

R. Sirabella
Fondazione IRCCS SDN, Naples, Italy

S. Sokolow • A. Herchuelz
Laboratory of Pharmacology and Therapeutics, Universitè Libre de Bruxelles, Brussels, Gosselies 6041, Belgium

L. Annunziato, M.D. (✉)
Division of Pharmacology, Department of Neuroscience, School of Medicine, "Federico II" University of Naples
Via S. Pansini, 5, Naples 80131, Italy
e-mail: lannunzi@unina.it

L. Annunziato (ed.), *Sodium Calcium Exchange: A Growing Spectrum of Pathophysiological Implications*, Advances in Experimental Medicine and Biology 961, DOI 10.1007/978-1-4614-4756-6_18,

More importantly, knockout and knockin mice also provided new relevant information on the role played by NCX in maintaining the intracellular Na^+ and Ca^{2+} homeostasis and in protecting neurons during brain ischemia. In particular, both *ncx2−/−* and *ncx3−/−* mice showed an increased neuronal vulnerability after the ischemic insult induced by transient middle cerebral artery occlusion.

As the ubiquitous deletion of NCX1 brings about to an early death of embryos because of a lack of heartbeat, this strategy could not be successfully pursued. However, information on the role of NCX1 in normal and ischemic brain could be obtained by developing conditional knockout mice lacking NCX1 in the brain. Preliminarily results obtained in these conditional mice suggest that also NCX1 protects neurons from ischemic cell death.

Overall, the use of genetic-modified mice for NCX1, NCX2, and NCX3 represents a fruitful strategy to characterize the physiological role exerted by NCX in CNS and to identify the isoforms of the antiporter as potential molecular targets for therapeutic intervention in cerebral ischemia.

Keywords
NCX1, NCX2, NCX3 • Brain ischemia • Learning and memory • Knockout mice

18.1 Introduction

Three genes coding for the three different Na^+/Ca^{2+} exchanger (NCX), NCX1 (Nicoll et al. 1990), NCX2 (Li et al. 1994), and NCX3 (Nicoll et al. 1996), have been identified in neurons, astrocytes, microglia, and oligodendrocytes, where they play a relevant role in maintaining Na^+ and Ca^{2+} homeostasis under different neurophysiological and neuropathological conditions. In neurons, the level of expression of NCX is particularly high in those sites where a large movement of Ca^{2+} ions occurs across the plasma membrane, as it happens at the level of synapses (Canitano et al. 2002; Juhaszova et al. 1996). In fact, during an action potential or after glutamate-activated channel activity, Ca^{2+} massively enters the plasma membrane and is rapidly extruded by the plasma membrane Ca^{2+} ATPase (PMCA) and by NCX. However, when $[Ca^{2+}]_i$ raises to a value higher than 500 nM, as it happens when a train of action potentials reaches the nerve terminals, NCX becomes the dominant Ca^{2+} extrusion mechanism in neurons.

While the role of NCX in glial cells remained obscure for years (Finkbeiner 1993; Goldman et al. 1994; Holgado and Beauge 1996; Takuma et al. 1994), recent studies suggest that this antiporter could be involved in the activation of microglia (Boscia et al. 2009) and in the differentiation of oligodendrocytes (Boscia et al. 2011).

Although the three NCX isoforms showed remarkably similar biophysical properties in expression systems, studies performed in genetic-modified mice provided evidence that the three NCX isoforms actually subserve different functional roles as we will review in the following sections.

18.2 Genetic-Modified Mice as a Model to Dissect the Role Played by NCX Isoforms in Brain Functions

In the last three decades, an increasing number of compounds inhibiting NCX have been synthesized in order to identify the physiological role played by NCX and/or its isoforms, NCX1,

NCX2, and NCX3; however, at present, no isoform-specific compounds have been yet identified (Annunziato et al. 2004). Moreover, these drugs, despite their potency, possess some nonspecific actions against several ion channels and receptors (Pintado et al. 2000; Reuter et al. 2002a). As a result, it is difficult to discriminate the direct effects of NCX inhibition from the inhibition/activation of channels, receptors, or enzymes. Thus, the use of these nonspecific inhibitors to characterize the role of each NCX isoform should be interpreted with caution (Reuter et al. 2002a). Furthermore, although more than 16 chemical classes of NCX inhibitors have been synthesized (Annunziato et al. 2004; Hasegawa et al. 2003; Iwamoto et al. 2004; Iwamoto and Kita 2006; Matsuda et al. 2001; Secondo et al. 2009; Watano et al. 1999), no compounds provided with stimulatory effects on NCX isoform activity have been generated. Because of the limitations of the currently available pharmacological tools acting on NCX, knockout/knockin and transgenic mice for NCX1, NCX2, and NCX3 were generated in order to identify the role of each antiporter isoform under physiological and pathophysiological conditions.

In 2000, four independent laboratories generated knockout mice for NCX1. Unfortunately, these mice die in utero at 9.5 days postcoitum because the lack of cardiac NCX1 in the homozygote state causes the absence of a spontaneous heartbeat (Cho et al. 2000; Koushik et al. 2001; Reuter et al. 2002b; Wakimoto et al. 2000). Initially, these results slowed down the study of NCX1 by a classic knockout strategy; however, later, Henderson et al. (2004) showed how to overcome this limitation by using the Cre/loxP technique in the heart, demonstrating that NCX1 gene ablation in adult mice is compatible with life.

As far as the other isoform NCX2 is concerned, its knocking-out is compatible with life and survival until adulthood. Studying these mice, the role of NCX2 in neuronal Ca^{2+} homeostasis and synaptic plasticity has been determined. Specifically, it was shown that hippocampal neurons from *ncx2−/−* mice display a significant delay in Ca_i^{2+} clearance following neuronal depolarization, thus enhancing both paired-pulse facilitation (PPF) and post-tetanic potentiation. Furthermore, *ncx2−/−* hippocampal slices also show a decreased frequency threshold for long-term potentiation (LTP) and long-term depression (LTD) (Jeon et al. 2003) that, in turn, enhances synaptic plasticity at the presynaptic level. As a result, *ncx2−/−* mice show an improved performance in several hippocampus-dependent learning and memory tasks, including water-maze and fear-conditioning tests (Jeon et al. 2003). This evidence seems to suggest that the specific inhibition of NCX2 could improve learning and memory consolidation.

Also NCX3 knockout mice survive until the adulthood and display relevant alterations in Ca_i^{2+} homeostasis and synaptic plasticity. Studies performed in *ncx3−/−* mice show that, besides contributing to control Ca_i^{2+} clearance in hippocampal neurons like NCX2, NCX3 also plays a role in controlling basal $[Ca^{2+}]_i$ (Molinaro et al. 2011). In particular, *ncx3−/−* hippocampal neurons show an increased basal level of $[Ca^{2+}]_i$ and a slow decline in $[Ca^{2+}]_i$ after depolarization. However, the lack of NCX3 does not affect neither peak values of $[Ca^{2+}]_i$ after neuronal depolarization nor Ca^{2+} storage in the ER. More importantly, in *ncx3−/−* hippocampal neurons, the moderate elevation of Ca_i^{2+} basal levels activates a phosphatase cascade, rather than a kinase cascade, with a consequent decrease in the phosphorylated form of Ca^{2+}-calmodulin Kinase IIα (CaMKIIα) (Molinaro et al. 2011) that has a crucial role in controlling LTP at postsynaptic level (Silva et al. 1992a, b). Moreover, in the CA1 area of hippocampus, *ncx3−/−* mice also show a reduction in the expression of CaMKIIα. Accordingly, *ncx3−/−* hippocampal slices show a marked impairment of LTP that occurs predominantly at postsynaptic level, as suggested by paired-pulse facilitation experiments (Molinaro et al. 2011) (Fig. 18.1). LTP is conventionally considered as an in vitro equivalent of learning and memory, and alterations in this neuronal plasticity usually mirror alterations in the whole animal. This also happened in our *ncx3−/−* mice where we found a significant reduction in hippocampus-dependent spatial learning and memory performance,

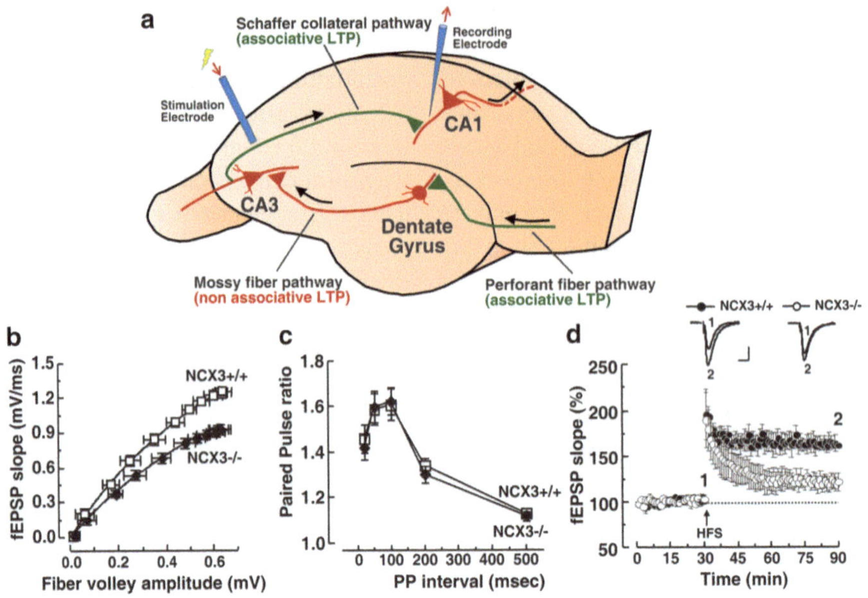

Fig. 18.1 *Synaptic function in ncx3−/− hippocampal slices.* (**a**) Cartoon showing the stimulation of hippocampal LTP in the Schaffer collateral axons given off by CA3 pyramidal cells. (**b**) Input/output curves for *ncx3−/−* and *ncx3+/+* mice measured by plotting the fEPSP slopes and their corresponding presynaptic fiber volley amplitudes at increasing stimulus strengths. Basal synaptic transmission was impaired in *ncx3* knockout mice compared with wild-type littermates. (**c**) Paired-pulse (PP) facilitation between the two experimental groups. The facilitation ratio (slope of second EPSP/slope of first EPSP) was plotted as a function of interpulse intervals of 20, 50, 100, 200, and 500 ms. For each group, the mean ± SEM is indicated. (**d**) Superimposed pooled data showing the normalized changes in field potential slope (±SEM) induced by HFS in *ncx3−/−* and *ncx3+/+* mice. fEPSP slopes were recorded and were expressed as the percentage of the pretetanus baseline. Representative fEPSP traces, before and 60 min after the induction of LTP, are shown. Calibration bars, 0.5 mV, 10 ms (Modified from Molinaro et al. (2011))

measured by Barnes maze, fear conditioning, and novel object recognition tests (Molinaro et al. 2011) (Fig. 18.2). Consistent with our results, an increased level of $[Ca^{2+}]_i$ (Gomez-Villafuertes et al. 2005), a reduction in basal synaptic transmission, and a reduction in long-term depression (LTD) (Wu et al. 2010) were reported in transgenic mice overexpressing a constitutive active mutant of the NCX3 repressor, named DREAM (downstream regulatory element antagonist modulator). Conversely, knockout mice of the NCX3 transcriptional repressor DREAM display enhanced learning and memory performance (Fontan-Lozano et al. 2009).

Another role emerging from *ncx3−/−* mice is that this isoform also plays a role in motor activity. In fact, 6-month-old mice lacking the NCX3 isoform exhibit reduced motor activity, weakness of forelimb muscles, and fatigability (Sokolow et al. 2004). However, since NCX3 is expressed in the CNS, in peripheral nervous system, and in skeletal muscles, it cannot be established whether these symptoms can be attributed to CNS defects or to alterations at the neuronal muscular junctions and skeletal fibers levels (Sokolow et al. 2004).

Interestingly, NCKX2, another antiporter belonging to the Ca^{2+} exchanger superfamily, that transports and requires extracellular potassium for its exchange activity, is involved in LTP and LTD. Indeed, the analysis of knockout mouse for *nckx2−/−* revealed a profound loss of LTP similar to *ncx3−/−* mice, and an increase in LTD, similar to *ncx2−/−* mice, at the level of hippocampal Schaffer/CA1 synapses (Li et al. 2006). Moreover, *nckx2−/−* mice show deficits

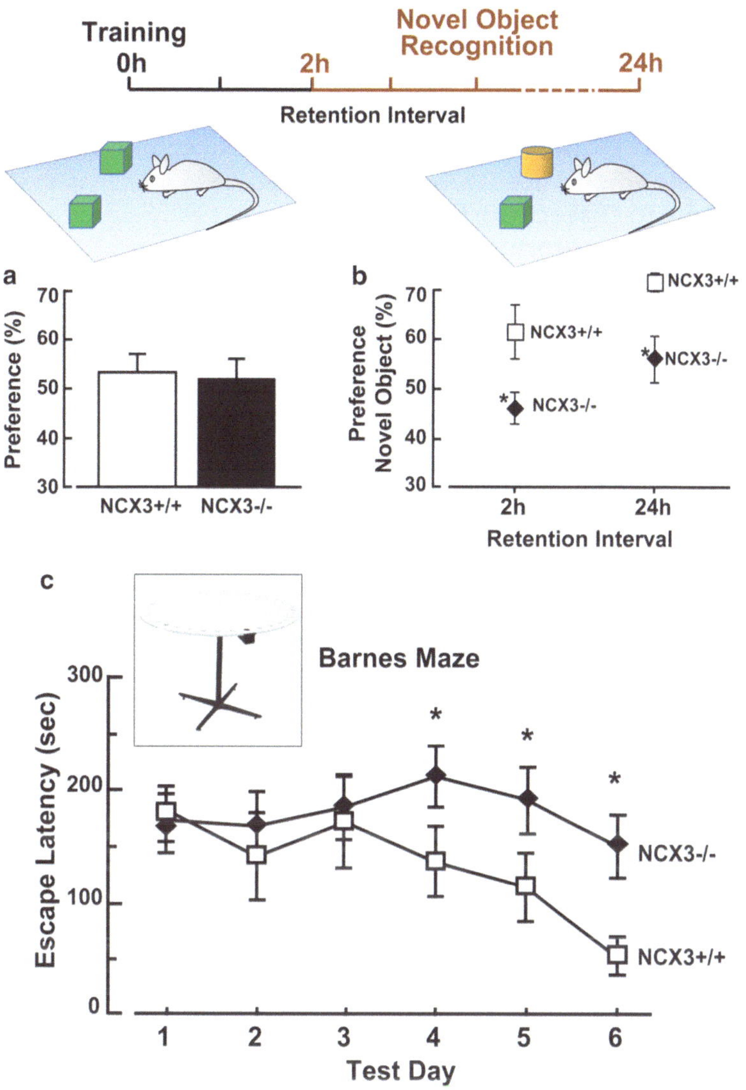

Fig. 18.2 *Hippocampal-dependent long-term memory tests in ncx3+/+ and ncx3−/− mice.* (**a**) Percentage of preference between two objects in the training phase of the novel object test performed in *ncx3−/−* and *ncx3+/+* mice. (**b**) Percentage of preference in the trial phase of the novel object recognition test 2 and 24 h after the training phase. $^*p < 0.05$ versus *ncx3+/+* mice. (**c**) Results of *ncx3−/−* and *ncx3+/+* mice subjected to Barnes maze test during 6 days of training as revealed by two-way ANOVA with repeated measures across training days and as assessed by one-way Dunnett's test for repeated comparisons. $^*p < 0.05$ (Modified from Molinaro et al. (2011))

in specific tests of motor learning and spatial working memory.

Collectively, the evidence emerging from the study of knockout mice for NCX2 and NCX3 showing that NCX2 and NCX3 have a critical role in controlling learning and memory suggests that these two exchangers could be a potential target for diseases causing a significant cognitive impairment such as Alzheimer's, Parkinson's, Huntington's diseases, and infarct dementia. Until now, neither activators of NCX3 nor specific inhibitor for NCX2 are available. However, since several inhibitors have been synthesized in the last few years, these compounds could be used as scaffold molecules for the development of selective NCX2 inhibitors or NCX3 activators able to improve learning and memory deficits.

18.3 Genetic-Modified Mice as a Model to Dissect the Role Played by NCX Isoforms in Brain Ischemia

NCX plays a pivotal role in maintaining intracellular Na^+ and Ca^{2+} homeostasis during pathophysiological conditions in the brain. However, it still remains to be fully clarified whether it is the suppression or the activation of the exchanger to yield potentially beneficial effects on a number of neurodegenerative diseases, such as Alzheimer's disease, aging, and white matter trauma. In particular, in stroke models, a number of conflicting results have been obtained using drugs that inhibit NCX, probably because they lack selectivity and some of them possess a remarkable and long-lasting hypothermic effect. By using the more selective approach of antisense oligodeoxynucleotides to selectively inhibit the translation of specific NCX isoforms, we demonstrated that NCX1 and NCX3 limit brain damage in experimental model of permanent occlusion of middle cerebral artery occlusion (Annunziato et al. 2007; Pignataro et al. 2004).

The potential role of NCX3 in brain ischemia is of particular interest since this isoform, unlike NCX1 and NCX2, displays the peculiar property of being able to operate in the absence of ATP (Linck et al. 1998; Secondo et al. 2007), whereas NCX1 and NCX2 require elevated ATP levels for their activity. This ATP independence of NCX3 is suggestive of its role under conditions in which there is an energy deprivation such as during brain ischemia. In fact, primary cortical neurons and organotypic hippocampal cultures from *ncx3−/−* mice display an enhanced vulnerability to oxygen and glucose deprivation (OGD) followed by reoxygenation (Molinaro et al. 2008). Intriguingly, a more severe cell death occurs in the CA3 and DG hippocampal subregions as compared to CA1 upon *ncx3* gene deletion; this effect is probably caused by the removal of this isoform in those hippocampal subregions where it is more intensely expressed (Papa et al. 2003). Consistent with studies in vitro, the transient middle cerebral artery occlusion (tMCAO) in *ncx3−/−* mice causes a remarkable increase in the ischemic volume (Molinaro et al. 2008) (Fig. 18.3). Another interesting aspect deriving from genetic-modified mice is that *ncx3* gene

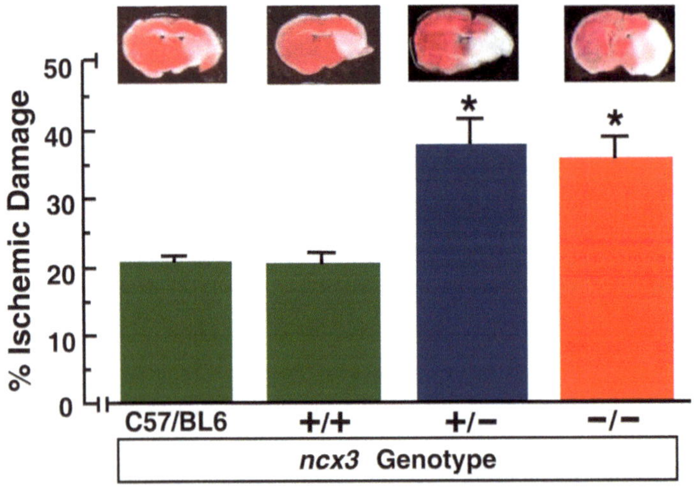

Fig. 18.3 *Effect of NCX3 ablation on brain ischemia.* Effect of *ncx3* knocking-out on infarct volume in C57BL/6 wild-type, *ncx3+/+*, *ncx3+/−*, and *ncx3−/−* mice subjected to tMCAO. Each column represents the mean ± SEM of the percentage of the infarct volume compared with the ipsilateral hemisphere. Ischemic mice were killed 24 h after tMCAO. *p <0.05 versus C57BL/6 wild-type and *ncx3+/+* ischemic mice. A representative brain slice from each ischemic experimental group is shown on the top of each column (Modified from Molinaro et al. (2008))

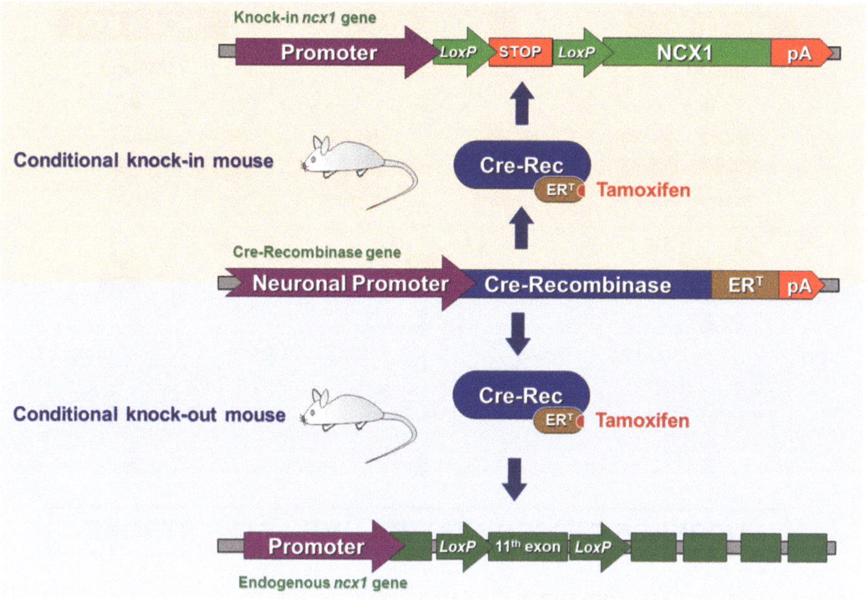

Fig. 18.4 *Cartoon of conditional knock-in and knockout strategy for NCX1*. The *upper panel* shows a schematic representation of a conditional knock-in mouse for NCX1 in the brain. A Cre-recombinase gene is expressed in neurons, and upon tamoxifen administration, it excises a STOP cassette signal in a new NCX1 gene. The *lower panel* shows a schematic representation of a conditional knockout mouse for NCX1 in the brain. Also in this case a Cre-recombinase gene is expressed in neurons, and upon tamoxifen administration, it excises a crucial exon of the endogenous *ncx1* gene

displays haploinsufficiency during the ischemic brain damage; in fact, no significant differences are detected between *ncx3+/−* and *ncx3−/−* ischemic groups (Molinaro et al. 2008). A possible explanation for the higher susceptibility to ischemic damage of neurons from *ncx3−/−* mice could rely in disturbance in intracellular Ca^{2+} homeostasis observed upon the deletion of NCX3.

Interestingly, similar in vitro and in vivo results were obtained by Jeon et al. (2008) and Cuomo et al. (2008) with *ncx2−/−* and *nckx2−/−* mice, respectively. In particular, NCX2 gene deletion results in a sustained elevation of $[Ca^{2+}]_i$, an increased membrane depolarization in hippocampus, and a worsening of the ischemic brain damage (Jeon et al. 2008). Similarly, the genetic ablation of NCKX2, the most expressed isoform in neurons among NCKX members (Tsoi et al. 1998), dramatically increases the infarct volume following tMCAO and increases the neuronal vulnerability of primary cortical neurons exposed to OGD (Cuomo et al. 2008).

As the ubiquitous deletion of NCX1 brings about to NCX1-null embryos that die in utero, this strategy could not be successfully pursued to characterize NCX1 role during brain ischemia. On the other hand, studies conducted in *ncx1+/−* mice, which survive until adulthood, showed that this condition does not cause an increased susceptibility to ischemic brain damage despite the reduction of NCX1 protein levels by almost 50 % and of NCX activity by 60 % (Luo et al. 2007).

A more rational and realizable strategy to obtain adult mice lacking neuronal NCX1 could be the generation of a conditional NCX1 knockout mouse by crossbreeding two genetic-modified mice: the first one carrying a floxed NCX1 gene and the other one expressing a conditional Cre recombinase in a specific subpopulation of neurons. This technique allows to activate upon tamoxifen administration the Cre recombinase specifically in those neurons that express this enzyme and to remove the 11th exon that codifies for the α2 repeat of NCX1 gene (Fig. 18.4). Our preliminary and unpublished results suggest that the conditional

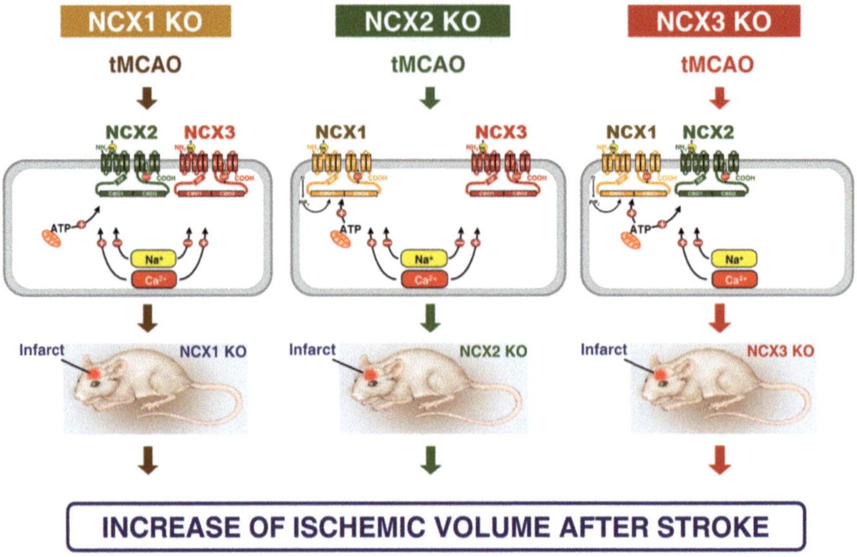

Fig. 18.5 *Summary of the effect of NCX1, NCX2, or NCX3 knocking-out in mice subjected to brain ischemia.* Disruption of one of the three NCX genes aggravates brain damage induced by the transient middle cerebral artery occlusion

removal of NCX1 in neurons causes an increase in the infarct volume upon tMCAO.

As NCX1 knocking-down leads to a worsening of ischemic damage, a strategy aimed to increase the expression and the activity of this isoform should bring about to an amelioration of the ischemic brain damage (Molinaro et al. 2012). The development of a conditional knockin mouse that, upon tamoxifen administration, overexpresses the most abundant splicing form of NCX1 in selective neurons should cause a reduction of ischemic volume following tMCAO (Fig. 18.4). Preliminary and unpublished results seem to be in accordance with this hypothesis, since the overexpression of NCX1 in neurons renders mice more resistant to the ischemic damage induced by tMCAO.

18.4 Conclusions

Overall, the use of knockout and knock-in mice for NCX1, NCX2, and NCX3 proved to be a fruitful strategy to characterize the physiological role exerted by NCX in CNS. Thanks to experimental work performed on these animals, the role of NCX2 and NCX3 in learning and memory has been discovered, and the involvement of all the NCX isoforms in brain ischemia has been better characterized (Fig. 18.5). These studies could pave the way to the development of selective drugs acting on specific NCX isoforms for the treatment of cognitive disorders or brain ischemia.

Acknowledgments This work was supported by COFIN2008; Ricerca-Sanitaria RF-FSL352059 Ricerca finalizzata 2006; Ricerca-Oncologica 2006; Progetto-Strategico2007; Progetto Ordinario 2007; Ricerca finalizzata 2009; Ricerca-Sanitaria progetto Ordinario by Ministero della Salute 2008 all to LA

References

L. Annunziato, G. Pignataro, G.F. Di Renzo, Pharmacology of brain Na^+/Ca^{2+} exchanger: from molecular biology to therapeutic perspectives. Pharmacol. Rev. **56**, 633–654 (2004)

L. Annunziato, G. Pignataro, F. Boscia, R. Sirabella, L. Formisano, M. Saggese, O. Cuomo, R. Gala, A. Secondo, D. Viggiano, P. Molinaro, V. Valsecchi, A. Tortiglione, A. Adornetto, A. Scorziello, M. Cataldi, GF. Di Renzo ncx1, ncx2, and ncx3 gene product expression and function in neuronal anoxia and brain ischemia. Ann. N. Y. Acad. Sci. 1099, 413-426 (2007)

F. Boscia, R. Gala, A. Pannaccione, A. Secondo, A. Scorziello, G. Di Renzo, L. Annunziato, NCX1 expression and functional activity increase in microglia invading the infarct core. Stroke **40**, 3608–3617 (2009)

F. Boscia, C. D'Avanzo, A. Pannaccione, A. Secondo, A. Casamassa, L. Formisano, N. Guida, L. Annunziato, Silencing or knocking out the Na^+/Ca^{2+} exchanger-3 (NCX3) impairs oligodendrocyte differentiation. Cell. Death. Differ. **19**, 562–572 (2011)

A. Canitano, M. Papa, F. Boscia, P. Castaldo, S. Sellitti, M. Taglialatela, L. Annunziato, Brain distribution of the Na^+/Ca^{2+} exchanger-encoding genes NCX1, NCX2, and NCX3 and their related proteins in the central nervous system. Ann. N. Y. Acad. Sci. **976**, 394–404 (2002)

C.H. Cho, S.S. Kim, M.J. Jeong, C.O. Lee, H.S. Shin, The Na^+ -Ca^{2+} exchanger is essential for embryonic heart development in mice. Mol. Cells **10**, 712–722 (2000)

O. Cuomo, R. Gala, G. Pignataro, F. Boscia, A. Secondo, A. Scorziello, A. Pannaccione, D. Viggiano, A. Adornetto, P. Molinaro, X.F. Li, J. Lytton, G. Di Renzo, L. Annunziato, A critical role for the potassium-dependent sodium-calcium exchanger NCKX2 in protection against focal ischemic brain damage. J. Neurosci. **28**, 2053–2063 (2008)

S.M. Finkbeiner, Glial calcium. Glia **9**, 83–104 (1993)

A. Fontan-Lozano, R. Romero-Granados, Y. del-Pozo-Martin, I. Suarez-Pereira, J.M. Delgado-Garcia, J.M. Penninger, A.M. Carrion, Lack of DREAM protein enhances learning and memory and slows brain aging. Curr. Biol. **19**, 54–60 (2009)

W.F. Goldman, P.J. Yarowsky, M. Juhaszova, B.K. Krueger, M.P. Blaustein, Sodium/calcium exchange in rat cortical astrocytes. J. Neurosci. **14**, 5834–5843 (1994)

R. Gomez-Villafuertes, B. Torres, J. Barrio, M. Savignac, N. Gabellini, F. Rizzato, B. Pintado, A. Gutierrez-Adan, B. Mellstrom, E. Carafoli, J.R. Naranjo, Downstream regulatory element antagonist modulator regulates Ca^{2+} homeostasis and viability in cerebellar neurons. J. Neurosci. **25**, 10822–10830 (2005)

H. Hasegawa, M. Muraoka, K. Matsui, A. Kojima, Discovery of a novel potent Na^+/Ca^{2+} exchanger inhibitor: design, synthesis and structure-activity relationships of 3,4-dihydro-2(1H)-quinazolinone derivatives. Bioorg. Med. Chem. Lett. **13**, 3471–3475 (2003)

S.A. Henderson, J.I. Goldhaber, J.M. So, T. Han, C. Motter, A. Ngo, C. Chantawansri, M.R. Ritter, M. Friedlander, D.A. Nicoll, J.S. Frank, M.C. Jordan, K.P. Roos, R.S. Ross, K.D. Philipson, Functional adult myocardium in the absence of Na^+-Ca^{2+} exchange: cardiac-specific knockout of NCX1. Circ. Res. **95**, 604–611 (2004)

A. Holgado, L. Beauge, Effects of external monovalent cations on Na^+-Ca^{2+} exchange in cultured rat glial cells. Ann. N. Y. Acad. Sci. **779**, 279–281 (1996)

T. Iwamoto, S. Kita, YM-244769, a novel Na^+/Ca^{2+} exchange inhibitor that preferentially inhibits NCX3, efficiently protects against hypoxia/reoxygenation-induced SH-SY5Y neuronal cell damage. Mol. Pharmacol. **70**, 2075–2083 (2006)

T. Iwamoto, Y. Inoue, K. Ito, T. Sakaue, S. Kita, T. Katsuragi, The exchanger inhibitory peptide region-dependent inhibition of Na^+/Ca^{2+} exchange by SN-6 [2-[4-(4-nitrobenzyloxy)benzyl]thiazolidine-4-carboxylic acid ethyl ester], a novel benzyloxyphenyl derivative. Mol. Pharmacol. **66**, 45–55 (2004)

D. Jeon, Y.M. Yang, M.J. Jeong, K.D. Philipson, H. Rhim, H.S. Shin, Enhanced learning and memory in mice lacking Na^+/Ca^{2+} exchanger 2. Neuron **38**, 965–976 (2003)

D. Jeon, K. Chu, K.H. Jung, M. Kim, B.W. Yoon, C.J. Lee, U. Oh, H.S. Shin, Na^+/Ca^{2+} exchanger 2 is neuroprotective by exporting Ca^{2+} during a transient focal cerebral ischemia in the mouse. Cell Calcium **43**, 482–491 (2008)

M. Juhaszova, H. Shimizu, M.L. Borin, R.K. Yip, E.M. Santiago, G.E. Lindenmayer, M.P. Blaustein, Localization of the Na^+-Ca^{2+} exchanger in vascular smooth muscle, and in neurons and astrocytes. Ann. N. Y. Acad. Sci. **779**, 318–335 (1996)

S.V. Koushik, J. Wang, R. Rogers, D. Moskophidis, N.A. Lambert, T.L. Creazzo, S.J. Conway, Targeted inactivation of the sodium-calcium exchanger (Ncx1) results in the lack of a heartbeat and abnormal myofibrillar organization. FASEB J. **15**, 1209–1211 (2001)

Z. Li, S. Matsuoka, L.V. Hryshko, D.A. Nicoll, M.M. Bersohn, E.P. Burke, R.P. Lifton, K.D. Philipson, Cloning of the NCX2 isoform of the plasma membrane Na^+-Ca^{2+} exchanger. J. Biol. Chem. **269**, 17434–17439 (1994)

X.F. Li, L. Kiedrowski, F. Tremblay, F.R. Fernandez, M. Perizzolo, R.J. Winkfein, R.W. Turner, J.S. Bains, D.E. Rancourt, J. Lytton, Importance of K^+-dependent Na^+/Ca^{2+}-exchanger 2, NCKX2, in motor learning and memory. J. Biol. Chem. **281**, 6273–6282 (2006)

B. Linck, Z. Qiu, Z. He, Q. Tong, D.W. Hilgemann, K.D. Philipson, Functional comparison of the three isoforms of the Na^+/Ca^{2+} exchanger (NCX1, NCX2, NCX3). Am. J. Physiol. **274**, C415–C423 (1998)

J. Luo, Y. Wang, X. Chen, H. Chen, D.B. Kintner, G.E. Shull, K.D. Philipson, D. Sun, Increased tolerance to ischemic neuronal damage by knockdown of Na^+-Ca^{2+} exchanger isoform 1. Ann. N. Y. Acad. Sci. **1099**, 292–305 (2007)

T. Matsuda, N. Arakawa, K. Takuma, Y. Kishida, Y. Kawasaki, M. Sakaue, K. Takahashi, T. Takahashi, T. Suzuki, T. Ota, A. Hamano-Takahashi, M. Onishi, Y. Tanaka, K. Kameo, A. Baba, SEA0400, a novel and selective inhibitor of the Na^+-Ca^{2+} exchanger, attenuates reperfusion injury in the in vitro and in vivo cerebral ischemic models. J. Pharmacol. Exp. Ther. **298**, 249–256 (2001)

P. Molinaro, O. Cuomo, G. Pignataro, F. Boscia, R. Sirabella, A. Pannaccione, A. Secondo, A. Scorziello, A. Adornetto, R. Gala, D. Viggiano, S. Sokolow, A.

Herchuelz, S. Schurmans, G. Di Renzo, L. Annunziato, Targeted disruption of Na^{+}/Ca^{2+} exchanger 3 (NCX3) gene leads to a worsening of ischemic brain damage. J. Neurosci. **28**, 1179–1184 (2008)

P. Molinaro, D. Viggiano, R. Nistico, R. Sirabella, A. Secondo, F. Boscia, A. Pannaccione, A. Scorziello, B. Mehdawy, S. Sokolow, A. Herchuelz, G.F. Di Renzo, L. Annunziato, Na^{+}-Ca^{2+} exchanger (NCX3) knock-out mice display an impairment in hippocampal long-term potentiation and spatial learning and memory. J. Neurosci. **31**, 7312–7321 (2011)

P. Molinaro, M. Cantile, O. Cuomo, A. Secondo, A. Pannaccione, P. Ambrosino, G. Pignataro, F. Fiorino, B. Severino, E. Gatta, M.J. Sisalli, M. Milanese, A. Scorziello, G. Bonanno, M. Robello, V. Santagada, G. Caliendo, G. Di Renzo, L. Annunziato, Neurounina-1, a Novel Compound that Increases Na+/Ca2+ Exchanger Activity, Effectively Protects Against Stroke Damage. Mol Pharmacol. doi:10.1124/mol.112.080986 (2012)

D.A. Nicoll, S. Longoni, K.D. Philipson, Molecular cloning and functional expression of the cardiac sarcolemmal Na^{+}-Ca^{2+} exchanger. Science **250**, 562–565 (1990)

D.A. Nicoll, B.D. Quednau, Z. Qui, Y.R. Xia, A.J. Lusis, K.D. Philipson, Cloning of a third mammalian Na^{+}-Ca^{2+} exchanger, NCX3. J. Biol. Chem. **271**, 24914–24921 (1996)

M. Papa, A. Canitano, F. Boscia, P. Castaldo, S. Sellitti, H. Porzig, M. Taglialatela, L. Annunziato, Differential expression of the Na^{+}-Ca^{2+} exchanger transcripts and proteins in rat brain regions. J. Comp. Neurol. **461**, 31–48 (2003)

G. Pignataro, R. Gala, O. Cuomo, A. Tortiglione, L. Giaccio, P. Castaldo, R. Sirabella, C. Matrone, A. Canitano, S. Amoroso, G. Di Renzo, L. Annunziato, Two sodium/calcium exchanger gene products, NCX1 and NCX3, play a major role in the development of permanent focal cerebral ischemia. Stroke **35**, 2566–2570 (2004)

A.J. Pintado, C.J. Herrero, A.G. Garcia, C. Montiel, The novel Na^{+}/Ca^{2+} exchange inhibitor KB-R7943 also blocks native and expressed neuronal nicotinic receptors. Br. J. Pharmacol. **130**, 1893–1902 (2000)

H. Reuter, S.A. Henderson, T. Han, T. Matsuda, A. Baba, R.S. Ross, J.I. Goldhaber, K.D. Philipson, Knockout mice for pharmacological screening: testing the specificity of Na^{+}-Ca^{2+} exchange inhibitors. Circ. Res. **91**, 90–92 (2002a)

H. Reuter, S.A. Henderson, T. Han, R.S. Ross, J.I. Goldhaber, K.D. Philipson, The Na^{+}-Ca^{2+} exchanger is essential for the action of cardiac glycosides. Circ. Res. **90**, 305–308 (2002b)

A. Secondo, R.I. Staiano, A. Scorziello, R. Sirabella, F. Boscia, A. Adornetto, V. Valsecchi, P. Molinaro, L.M. Canzoniero, G. Di Renzo, L. Annunziato, BHK cells transfected with NCX3 are more resistant to hypoxia followed by reoxygenation than those transfected with NCX1 and NCX2: Possible relationship with mitochondrial membrane potential. Cell Calcium **42**, 521–535 (2007)

A. Secondo, A. Pannaccione, P. Molinaro, P. Ambrosino, P. Lippiello, A. Esposito, M. Cantile, P.R. Khatri, D. Melisi, G. Di Renzo, L. Annunziato, Molecular pharmacology of the amiloride analog 3-amino-6-chloro-5-[(4-chloro-benzyl)amino]-n-[[(2,4-dimethy lbenzyl)-amino] iminomethyl]-pyrazinecarboxamide (CB-DMB) as a pan inhibitor of the Na^{+}-Ca^{2+} exchanger isoforms NCX1, NCX2, and NCX3 in stably transfected cells. J. Pharmacol. Exp. Ther. **331**, 212–221 (2009)

A.J. Silva, R. Paylor, J.M. Wehner, S. Tonegawa, Impaired spatial learning in alpha-calcium-calmodulin kinase II mutant mice. Science **257**, 206–211 (1992a)

A.J. Silva, C.F. Stevens, S. Tonegawa, Y. Wang, Deficient hippocampal long-term potentiation in alpha-calcium-calmodulin kinase II mutant mice. Science **257**, 201–206 (1992b)

S. Sokolow, M. Manto, P. Gailly, J. Molgo, C. Vandebrouck, J.M. Vanderwinden, A. Herchuelz, S. Schurmans, Impaired neuromuscular transmission and skeletal muscle fiber necrosis in mice lacking Na/Ca exchanger 3. J. Clin. Invest. **113**, 265–273 (2004)

K. Takuma, T. Matsuda, H. Hashimoto, S. Asano, A. Baba, Cultured rat astrocytes possess Na^{+}-Ca^{2+} exchanger. Glia **12**, 336–342 (1994)

M. Tsoi, K.H. Rhee, D. Bungard, X.F. Li, S.L. Lee, R.N. Auer, J. Lytton, Molecular cloning of a novel potassium-dependent sodium-calcium exchanger from rat brain. J. Biol. Chem. **273**, 4155–4162 (1998)

K. Wakimoto, K. Kobayashi, O.M. Kuro, A. Yao, T. Iwamoto, N. Yanaka, S. Kita, A. Nishida, S. Azuma, Y. Toyoda, K. Omori, H. Imahie, T. Oka, S. Kudoh, O. Kohmoto, Y. Yazaki, M. Shigekawa, Y. Imai, Y. Nabeshima, I. Komuro, Targeted disruption of Na^{+}/Ca^{2+} exchanger gene leads to cardiomyocyte apoptosis and defects in heartbeat. J. Biol. Chem. **275**, 36991–36998 (2000)

T. Watano, Y. Harada, K. Harada, N. Nishimura, Effect of Na^{+}/Ca^{2+} exchange inhibitor, KB-R7943 on ouabain-induced arrhythmias in guinea-pigs. Br. J. Pharmacol. **127**, 1846–1850 (1999)

L.J. Wu, B. Mellstrom, H. Wang, M. Ren, S. Domingo, S.S. Kim, X.Y. Li, T. Chen, J.R. Naranjo, M. Zhuo, DREAM (downstream regulatory element antagonist modulator) contributes to synaptic depression and contextual fear memory. Mol. Brain **3**, 3 (2010)

NCX as a Key Player in the Neuroprotection Exerted by Ischemic Preconditioning and Postconditioning

19

Giuseppe Pignataro, Ornella Cuomo, Antonio Vinciguerra, Rossana Sirabella, Elga Esposito, Francesca Boscia, Gianfranco Di Renzo, and Lucio Annunziato

Abstract

Ischemic preconditioning is a neuroprotective mechanism in which a brief non-injurious episode of ischemia protects the brain from a subsequent lethal insult. Recently, it has been reported that modified reperfusion subsequent to a prolonged ischemic episode may also confer neuroprotection, a phenomenon termed postconditioning. Mitogen-activated protein kinases (MAPK) play a key role in these two neuroprotective mechanisms. The aim of this study was to evaluate whether Na^+/Ca^{2+} exchangers (NCXs), a family of ionic transporters that contribute to the maintenance of intracellular ionic homeostasis, contribute to the neuroprotection elicited by ischemic preconditioning and postconditioning.

Results of this study indicated that (1) NCX1 and NCX3 are upregulated in those brain regions protected by preconditioning, while (2) postconditioning treatment induces an upregulation only in NCX3 expression. (3) NCX1 upregulation and NCX3 upregulation are mediated by p-AKT since its inhibition reverted the neuroprotective effect of preconditioning and postconditioning and prevented NCXs overexpression. (4) The involvement of NCX in preconditioning and postconditioning neuroprotection is further supported by the results of experiments showing that a partial reversion of the protective effect induced by preconditioning was obtained by silencing NCX1 or NCX3, while the silencing of NCX3 was able to mitigate the protection induced by ischemic postconditioning.

G. Pignataro • O. Cuomo • A. Vinciguerra • E. Esposito • F. Boscia • G. Di Renzo
Division Pharmacology, Department Neuroscience, School of Medicine, Federico II University of Naples, Via S. Pansini 5, Naples, 80131, Italy
e-mail: lannunzi@unina.it

R. Sirabella
Fondazione IRCCS SDN, Naples, Italy

L. Annunziato (✉)
Division Pharmacology, Department Neuroscience, School of Medicine, Federico II University of Naples, Via S. Pansini 5, Naples 80131, Italy
e-mail: lannunzi@unina.it

L. Annunziato (ed.), *Sodium Calcium Exchange: A Growing Spectrum of Pathophysiological Implications*, Advances in Experimental Medicine and Biology 961, DOI 10.1007/978-1-4614-4756-6_19,

Altogether, the data presented here suggest that NCX1 and NCX3 represent two promising druggable targets for setting on new strategies in stroke therapy.

Keywords
Preconditioning • Postconditioning • NCX1 • NCX3 • p-AKT

19.1 Introduction

Due to the failure of clinical trials for pharmacological neuroprotective strategies in stroke (Gladstone et al. 2002), attention of researchers has been recently focused to the identification of additional transductional and transcriptional pathways eventually activated by endogenous neuroprotective mechanisms. Interestingly, it has been recently reported that ischemic preconditioning, a sublethal ischemic episode applied before a longer harmful ischemia (Kirino 2002; Dirnagl et al. 2003; Gidday 2006), and ischemic postconditioning, a sublethal ischemia subsequent to a prolonged harmful ischemic episode (Burda et al. 1991, 1995, 2006; Hausenloy et al. 2005a, b; Yellon and Hausenloy 2005; Danielisova et al. 2006; Pignataro et al. 2006, 2008; Zhao et al. 2006a, b; Zhao 2007; Scartabelli et al. 2008), are both able to exert a remarkable neuroprotection. Therefore, ischemic preconditioning and postconditioning represent useful tools to identify the transductional and transcriptional factors activated during these two experimental conditions that can be targeted in the attempt to identify new neuroprotective molecular targets. The molecular mechanisms contributing to preconditioning- and postconditioning-mediated tissue protection have been classified as (1) triggers, such as adenosine (ADO), opioids, erythropoietin (EPO), nitric oxide, reactive oxygen species, cytokines, and bradykinin; (2) transductors, such as reperfusion injury salvage kinase (RISK) pathways and other protein kinases; and finally (3) effectors, such as mitochondrial permeability transition pore and mitochondrial potassium ATP channels (Zhao 2007, 2009; Pignataro et al. 2009).

Among the several transductors, the well-known families of mitogen-activated protein kinases (MAPK) and phosphatidylinositol 3-kinase (PI3-K) have been proposed as important factors in mediating ischemic postconditioning neuroprotection (Zhao et al. 2006a, b; Pignataro et al. 2008). Interestingly, we have previously demonstrated that NCX1 and NCX3, two of the three brain isoforms of the plasma membrane Na^+/Ca^{2+} exchanger, are novel additional targets for the survival action of the (PI3-K)/Akt pathway (Formisano et al. 2008). In fact, AKT functions as a major downstream target of phosphatidylinositol 3-kinase (PI3-K), and after phosphorylation, it phosphorylates some substrates on the serine or threonine residues, including glycogen synthase kinase-3, *Caenorhabditis elegans* DAF-16 transcription factor, BAD, phosphodiesterase 3B, and the tuberous sclerosis complex-2 tumor suppressor gene product tuberin (Chan 2004a, b). In particular, it has been proposed that PI3-K/Akt signaling pathway by phosphorylating specific substrates is determinant for the control of cell death in ischemic neurons during stroke (Chan 2004a, b). In addition, NCX isoforms play a fundamental role in regulating and maintaining cellular calcium and sodium homeostasis (Annunziato et al. 2004) and are involved in the pathophysiology of stroke damage. In particular, it has been shown that NCX gene expression after permanent middle cerebral artery occlusion (MCAO) in rats is regulated in a differential manner, depending on the exchanger isoform (NCX1, NCX2, or NCX3) and on the region involved in the insult (Pignataro et al. 2004; Boscia et al. 2006). Furthermore, NCX1 and NCX3 downregulation or genetic ablation worsens the experimentally induced ischemic damage in mice and rats (Pignataro et al. 2004; Molinaro et al. 2008).

In this chapter, we will discuss evidence showing that the two NCX isoforms, NCX1 and NCX3, might take part as effectors in the neuroprotection evoked by preconditioning and postconditioning.

In the next sections we will analyze:

(a) The main features of ischemic preconditioning and postconditioning as neuroprotective strategies
(b) The pattern of expression of the three NCX isoforms in preconditioned and postconditioned brains
(c) The effect of NCX silencing induced by siRNA on neuroprotection induced by ischemic preconditioning and postconditioning
(d) The effect of p-AKT inhibition on NCX expression during ischemic preconditioning and postconditioning

19.2 NCX in Ischemic Preconditioning and Postconditioning

Stroke is the third most common cause of death after heart attack and cancer and has profound negative social and economic effects. The failure over the past decades of multiple clinical trials of exogenously administered drugs as potential stroke neuroprotectants has enhanced the ongoing effort to identify endogenously modulated mechanisms activated after cerebral ischemia, which might be harnessed as neuroprotectants in stroke (Kirino 2002; Dirnagl et al. 2003; Gidday 2006). Such molecular mechanisms might be those evolutionarily conserved to counteract the damage induced by a disruption to the cerebral blood supply (Gladstone et al. 2002).

In this attempt, ischemic preconditioning and ischemic postconditioning represent two promising strategies in modulating ischemic damage. The neuroprotective concept of ***preconditioning*** is based on the observation that a brief non-injurious episode of ischemia is able to protect the brain from a subsequent longer ischemic insult (Dirnagl et al. 2003). A major goal of this research is to understand the mechanisms involved in the preconditioning-induced neuroprotection in the attempt to identify genes and proteins increased in abundance during preconditioning as potential drug targets for stroke therapy. In fact, such preconditioning as a strategy to attenuate the pathophysiological consequences of ischemia–reperfusion injury would be focused on pretreatment situations, such as protection before cardiac bypass surgery. A non-pharmacological neuroprotective strategy for administration after ischemia onset, however, remains elusive.

Based on recent studies on the heart (Hausenloy et al. 2005a, b; Yellon and Hausenloy 2005) and proof of principle experiments in the brain (Burda et al. 2006; Danielisova et al. 2006; Zhao et al. 2006a, b), a hypothesis has been offered that a modified reperfusion subsequent to a prolonged ischemic episode may also confer ischemic neuroprotection, a phenomenon termed ***postconditioning*** (Zhao et al. 2006a, b; Pignataro et al. 2008). In the published studies in myocardium describing postconditioning, a number of issues are presented. In 2005, Yellon and coworkers have suggested that pre- and postconditioning are similar phenomena with the effectors being a similar group of downstream signaling cascades (Yellon and Hausenloy 2005). Alternately, the mechanisms regulating postconditioning might be entirely different than preconditioning, as the rapidity of onset of postconditioning-induced neuroprotection contrasts with a significant temporal delay for (protein synthesis dependent) preconditioning-induced neuroprotection. Furthermore, postconditioning may not involve the activation of endogenous neuroprotection, but rather merely attenuate the burst of free radicals occurring with reperfusion. Indeed, it has been suggested that the protection of postconditioning could be accomplished by a gradual increase in the reperfusion rate (Burda et al. 1991, 1995). Other hypotheses have been offered, such as that postconditioning results from intravascular adenosine washout (Kin et al. 2005) or is affected by intravascular pressures associated with reperfusion (Allen et al. 2000). However, the major focus of postconditioning mechanisms is the role of the effector protein kinases and the question of the similarity or difference of these effectors versus those thought to be involved in preconditioning (Hausenloy et al. 2005a, b). Although the time window of effectiveness of postconditioning-induced neuroprotection is narrow, postconditioning may have translational relevance to reperfusion and thrombolytic treatments in acute brain ischemia.

19.2.1 Ischemic Preconditioning

Many studies in vivo and in vitro have demonstrated that neurons exposed to brief periods of sublethal anoxia develop resistance to subsequent, more prolonged, and lethal anoxic insults (Kitagawa et al. 1990, 1991; Simon et al. 1993; Gidday et al. 1994; Glazier et al. 1994; Miyashita et al. 1994). This phenomenon known as anoxic preconditioning (APC) was firstly described in the myocardium (Murry et al. 1986; Meldrum et al. 1997) and only recently in the brain (Kitagawa et al. 1991; Dawson and Dawson 2000; Kirino 2002; Schaller and Graf 2002; Meller et al. 2005). Consequently, over the last decades, many efforts have been addressed to identify the molecular mechanisms involved in this phenomenon in order to open up therapeutic avenues for the treatment of cerebral ischemia.

19.2.1.1 Ischemic Preconditioning Induction

It is generally accepted that preconditioning requires small doses of an otherwise harmful stimulus to induce protection against subsequent injurious challenge (Dirnagl et al. 2003). Several distinct preconditioning stimuli can induce tolerance to ischemic brain injury; among them are non-injurious ischemia, cortical spreading depression, brief episode of seizure, exposure to anesthetics inhalants, and low doses of endotoxin, hyperthermia, or heat shock (Kobayashi et al. 1995; Plamondon et al. 1999).

Differences in the intensity, duration, and/or frequency of a particular stimulus potentially able to induce protection determine whether that stimulus is too weak to elicit a response, of sufficient intensity to serve as a preconditioning trigger, or too robust to be harmful (Dirnagl et al. 2003).

In general, it is widely accepted that immediate acquisition of protein-synthesis-independent tolerance is mediated by posttranslational modification, and its effective duration is brief. Conversely, there is general agreement that delayed induction of ischemic tolerance requires new protein synthesis and is sustained for a time interval comprised between few days and weeks (Dirnagl et al. 2003). In the brain, the time course of ischemic tolerance apparently follows the delayed pattern, suggesting that synthesis of active proteins may be necessary for full development of ischemic tolerance (Barone et al. 1998). Once induced, the ischemic tolerance is believed to last for a few days and to diminish gradually few weeks after acquisition (Kirino et al. 1991).

19.2.1.2 Mechanisms Involved in Ischemic Preconditioning Neuroprotection

The molecular mechanisms responsible for the induction and maintenance of ischemic tolerance in the brain are complex and remain largely undefined. In this context, some studies have demonstrated that several events such as the activation of protein kinases (Lange-Asschenfeldt et al. 2004; Perez-Pinzon et al. 2005), the induction of transcription factors (Digicaylioglu and Lipton 2001; Jones and Bergeron 2001; Ginis et al. 2002), and of immediate early genes (Patel et al. 2004; Rybnikova et al. 2005) play pivotal roles in the development of ischemic tolerance. Moreover, oxygen free radicals, generated during the preconditioning stimuli, may also elicit ischemic tolerance (Ravati et al. 2000) through the neo-synthesis of neuroprotective proteins such as the heat shock protein 72 (HSP72) (Massa et al. 1996; Rejdak et al. 2001), through anti-apoptotic proteins such as Bcl2 (Brambrink et al. 2000; Meller et al. 2005), and through antioxidant enzymes (Toyoda et al. 1997; Arthur and Wilkins 2004). Finally, it has also been hypothesized that nitric oxide (NO) could bolster intracellular signaling, which is activated in neurons during APC (Gonzalez-Zulueta et al. 2000; Huang 2004). In this contest, NO acting as an important transducer would activate Ras/extracellular-regulated kinase (ERK) ERK1/2 pathway thus eliciting beneficial effects (Dawson and Dawson 2000; Gonzalez-Zulueta et al. 2000; Scorziello et al. 2007).

The early phase of ischemic preconditioning is characterized by rapid posttranslational modification of pre-existing proteins through signaling pathways which involve protein kinase C (PKC) (Speechly-Dick et al. 1994) and mitogen-activated protein (MAP) kinases. The late preconditioning, PC, is mediated by protective gene expression and by the synthesis of new protective proteins. This mechanism involves redox-sensitive activation of transcriptional factors through PKC and tyrosine kinase signaling pathways that are in common with the early phase of preconditioning.

Important steps in the signaling pathway of ischemic preconditioning are represented by the activation of phospholipase D, tyrosine kinase, and MAP kinase with interactions among these pathways not yet entirely elucidated (Brooks and Hearse 1996). In particular, MAP kinases, a serine/threonine protein kinase family, play a crucial role in triggering of the intracellular events leading to the activation of adaptive response observed in ischemic preconditioning both in the heart (Maulik et al. 1996) and in the brain (Shamloo and Wieloch 1999; Pignataro et al. 2008). Interestingly, postischemic activation of Akt/protein kinase B may contribute to the induction of ischemic tolerance (Yano et al. 2001). In fact, Akt is activated after sublethal ischemia, and its inhibition resulted in attenuation of ischemic tolerance (Yano et al. 2001).

19.2.1.3 Cellular Ionic Homeostasis and Ischemic Preconditioning: Role of Na^+/Ca^{2+} Exchanger

Since ischemic preconditioning activates intracellular biological responses prior to a potential lethal insult, it is expected that an improvement of energy metabolism or a latency in anoxic depolarization after the onset of ischemic insult might represent the mechanisms by which organs strengthen their tolerance when exposed to a sublethal insult. In this regard, several experiments have been performed both in vivo and in vitro in order to demonstrate that a reduction in energy demand and in the activity of ion channels represents determinant factors for ischemic tolerance (Stenzel-Poore et al. 2003). In fact, an impairment in voltage-gated potassium channels has been observed in cortical neurons exposed to brief non-injurious oxygen and glucose deprivation. Similarly, in vivo experiments demonstrated that ischemic preconditioning prevented the inhibition of Na^+-K^+-ATPase activity after brain ischemia in hippocampal and cortical neurons of rats exposed to global forebrain ischemia (de Souza Wyse et al. 2000). As far as calcium homeostasis is concerned, during ischemic preconditioning, the results of in vivo experiments in gerbils showed an increase in Ca^{2+}-ATPase activity and an enhancement in mitochondrial calcium sequestration in CA1 hippocampal neurons after preconditioning (Ohta et al. 1996). In line with this result, intracellular calcium imaging performed in hippocampal neurons of preconditioned gerbils showed that the increase in $[Ca^{2+}]i$ occurring after anoxic and aglycemic episode was markedly inhibited in the ischemic-tolerant animals (Shimazaki et al. 1998). The molecular mechanisms underlying this effect are still under investigation. A possible explanation could be the increased expression of Ca^{2+}-ATPases isoform 1 (PMCA-1) as recently demonstrated by Kato and coworkers (Kato et al. 2005). Furthermore, the hypothesis that a modulation of expression and activity of sodium calcium exchanger (NCX) might play role in the regulation of calcium and sodium homeostasis during ischemic tolerance is likely. In this respect, it is relevant to mention that NCX gene expression was reduced during cerebral ischemia in rats in a different manner depending on the exchanger isoforms and on region involved in the insult (Pignataro et al. 2004; Boscia et al. 2006).

In our recent study, we demonstrated that among the three NCX brain isoforms, NCX1 and NCX3 represent two additional new molecular effectors involved in the neuroprotective mechanisms elicited by ischemic preconditioning (Pignataro et al. 2012).

In fact, whereas NCX1 and NCX3 silencing partially prevented ischemic preconditioning neuroprotection, the prosurvival factor p-AKT mediated NCX1 and NCX3 upregulation

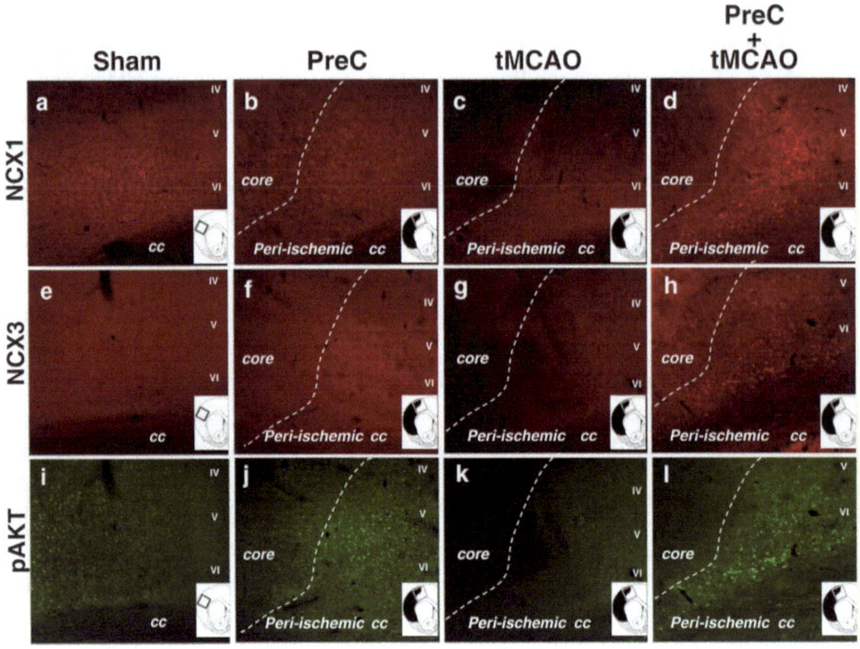

Fig. 19.1 *Distribution of NCX1, NCX3, and p-AKT immunoreactivity in the rat cerebral cortex.* Single staining of NCX1 (**a–d**), NCX3 (**e–h**), and p-Akt (**i–l**) in brain sections from sham-operated (**a**, **e**, **i**, respectively), preconditioned (**b**, **f**, **j**, respectively), ischemic (**c**, **g**, **k**, respectively), and ischemic preconditioned (**d**, **h**, **l**, respectively) animals (Reprinted with permission from Pignataro et al. (2012))

during ischemic preconditioning (Figs. 19.1, 19.2, and 19.3).

Our results were obtained in a rat model of ischemic preconditioning in which the preconditioning stimulus is represented by a transient occlusion of the middle cerebral artery (MCA) for 30 min, a time interval that is not able to induce a brain damage but is capable of protecting the brain from a subsequent harmful stimulus obtained through a longer occlusion of the MCA, 100 min, applied 72 h after the preconditioning stimulus. Data obtained in these experimental conditions do indeed support the importance of NCX1 and NCX3 in the pathogenesis of ischemic lesion and, most important, offer a new possible interpretation of the neuroprotective mechanism elicited by ischemic preconditioning. In addition, it has been shown that p-AKT, by acting on NCX1 and NCX3, represents a fundamental transducer of the neuroprotection exerted by preconditioning. That a tight relationship between NCX and p-AKT exists was already demonstrated in our previous studies in which NCX1 and NCX3 emerged as novel additional targets for the survival action of the phosphatidylinositol 3-kinase (PI3-K)/Akt pathway (Formisano et al. 2008). In addition, it is known that p-AKT takes part in the preconditioning-induced neuroprotection since a persistent overexpression of p-AKT occurs after ischemic preconditioning, and the use of the PI3K inhibitor LY294002 is able to revert the ameliorative effect induced by ischemic preconditioning (Endo et al. 2006; Pignataro et al. 2008).

The overexpression of NCX1 and NCX3 during preconditioning may be related to their ability to counteract the dysregulation of intracellular Na^+, ($[Na^+]_i$) and Ca^{2+}, ($[Ca^{2+}]_i$) homeostasis occurring in the brain under anoxic conditions. More important, the increased expression of NCX1 and NCX3 observed at early time points does not necessarily implicate that the neuroprotection occurs at the same time points. Rather, it is possible to hypothesize that the increased

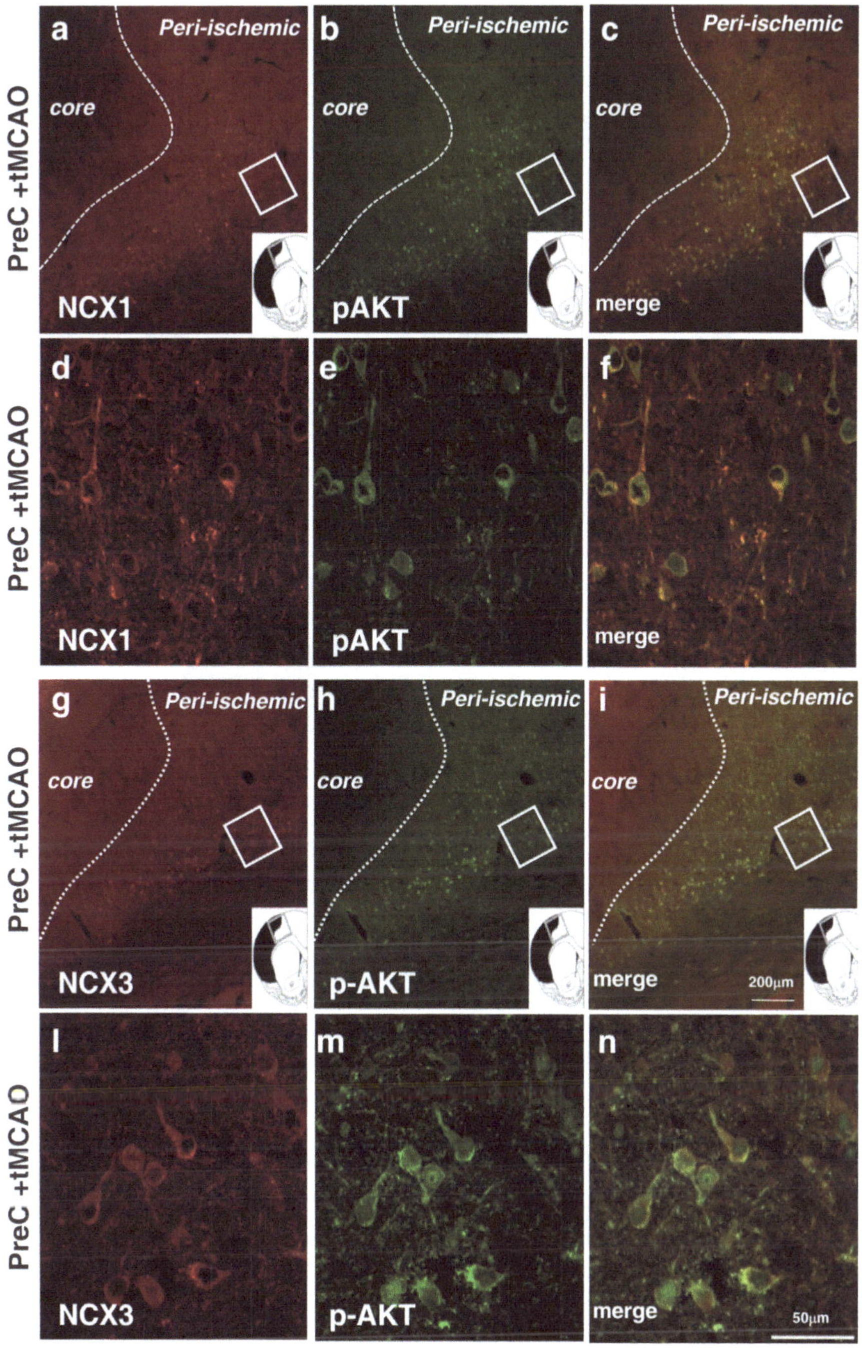

Fig. 19.2 *Co-localization of p-AKT with NCX1 or NCX3 in the peri-ischemic region of ischemic preconditioned rats*. Double-staining of NCX1 (**a–c**) and NCX3 (**g–i**) with p-AKT in the peri-ischemic region of ischemic preconditioned rats. Overview pictures were taken in the deep cortical layers on the border between the ischemic core and the peri-ischemic region (*dashed lines*). High-magnification photomicrographs depicting NCX1/p-AKT or NCX3/p-AKT double-labeled cells are shown in panels **d–f** and **j–l**, respectively. Overview pictures were taken in the deep cortical layers on the border between the ischemic core and the peri-ischemic region (*dashed lines*). Corpus callosum (cc); scale bars 200 μm, except for *Panel B*, **d–f,** and **j–l**, 50 μm (Reprinted with permission from Pignataro et al. (2012))

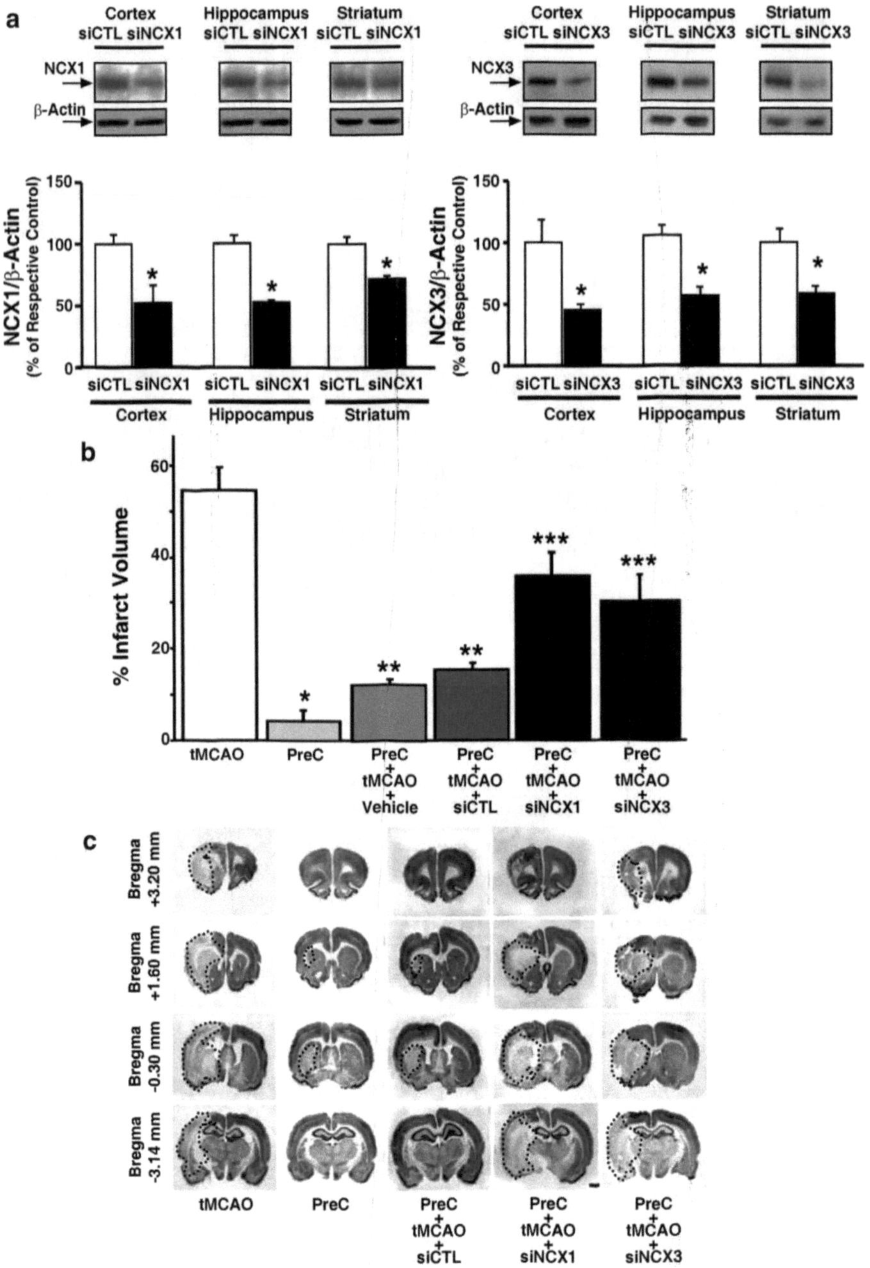

Fig. 19.3 (**a**) *Effect of siNCX1 and siNCX3 on NCX1 and NCX3 expression evaluated in cortex, hippocampus, and striatum of rats icv treated with silencing.* (**b**) *Effect of siNCX1 and siNCX3 on neuroprotection mediated by ischemic preconditioning*; (**c**) *distribution of NeuN immunoreactivity in adult rat brain sections of ischemic, preconditioned, and ischemic preconditioned animals. Panel A left side*, representative Western blots of NCX1 protein levels in the temporoparietal cortex, hippocampus, and striatum of rats treated icv with siNCX1. *Right side*, representative Western blots of NCX3 protein levels in the temporoparietal cortex, hippocampus, and striatum of rats treated icv with siNCX1. All animals were euthanized 24 h after siRNA injection. Values are mean ± SEM. *p <0.05, compared with control group (siCTL). n = 6–8 animals for each column. *Panel B* infarct volume in rats subjected to tMCAO, PreC, PreC + tMCAO + vehicle, PreC + tMCAO + siCTL, PreC + tMCAO + siNCX1, and PreC + tMCAO + siNCX3. Rats were euthanized 24 h after tMCAO. *p <0.05 versus control group; **p <0.05 versus control and preconditioned groups; ***p <0.05 versus all experimental groups, p <0.05. Each column represents the

expression of certain proteins induced by a neuroprotective strategy like preconditioning could render the brain tissue ready to withstand subsequent, more severe brain conditions. On the other hand, NCX1 and NCX3 neuroprotective role in ischemic preconditioning is in accordance with previous research. For instance, in homozygous *ncx3–/–* mice subjected to MCAO, an increased brain damage occurs (Molinaro et al. 2008). In addition, NCX1 and NCX3 silencing by RNA interference increases cerebellar granule neurons vulnerability to Ca^{2+} overload and excitotoxicity (Bano et al. 2005; Secondo et al. 2007). Moreover, the vulnerability to chemical hypoxia of BHK cells overexpressing NCX1 or NCX3 considerably increases when either NCX1 or NCX3 is silenced (Bano et al. 2005; Secondo et al. 2007). Finally, ischemic rats treated with NCX1 or NCX3 antisense display a remarkable enlargement of the infarct volume (Pignataro et al. 2004).

Consistently, by means of confocal fluorescence experiments, we observed that the temporoparietal cortex of preconditioned rats displayed a greater increase in p-AKT, NCX1, and NCX3 immunofluorescent signal if compared to the same brain region of rats subjected only to 100 min of tMCAO. More interestingly, the tight relationship existing between NCX1, NCX3, and p-AKT was further demonstrated by co-localization experiments, in which the increased expression of these three proteins occurred in the same brain cells. Furthermore, since the downregulation of NCX1 and NCX3 expression induced by siRNA did not modify p-AKT expression, the effect of p-AKT on NCX can be considered unidirectional.

Altogether, these data support the importance of p-AKT in mediating preconditioning neuroprotection and suggest that NCX1 and NCX3 are indeed two additional signals downstream of p-AKT that are involved in the neuroprotective process of ischemic preconditioning. On the other hand, p-AKT should not be considered the only transducer able to activate NCX1 and NCX3. In fact, numerous other cellular factors are most likely released even in earlier stages, as for instance right after preconditioning induction, and can therefore control the levels of NCX expression. In this respect, we have recently demonstrated that after ischemic preconditioning, HIF-1α is strongly augmented. This increase, in turn, is accompanied by an increase in NCX1 expression, which contributes to brain preconditioning neuroprotection (Valsecchi et al. 2011). These results demonstrate that NCX1 gene is a novel HIF-1 target and that HIF-1 exerts its pro-survival role also through NCX1 upregulation during brain preconditioning. Therefore, HIF-1, at least in part, exerts its neuroprotective effect by inducing an overexpression of NCX1 that is mediated by the interaction between HIF-1 and NCX1 promoter (Valsecchi et al. 2011).

The overexpression of the two isoforms NCX1 and NCX3 can be interpreted as a compensatory mechanism activated by preconditioning in neurons and glial cells to counteract the dysregulation of intracellular Na^{+} and Ca^{2+} homeostasis occurring after harmful ischemia. Interestingly, the activation of these mechanisms appears to be long-lasting, as the upregulation of NCX1 and NCX3 was still present even 72 h after preconditioning induction, thus suggesting that both NCX1 and NCX3 might be considered as two possible effectors of delayed preconditioning.

In conclusion, the results above discussed suggest that the enhancement of NCX1 and NCX3 expression and/or activity might be a reasonable pharmacological strategy to reduce the extension of the infarct volume after a harmful ischemic insult.

Fig. 19.3 (continued) mean ± S.E.M. (n = 5–7) of the percentage of the infarct volume compared to the ipsilateral hemisphere. *Panel C* representative brain NeuN-immuno histochemistry-processed coronal sections deriving from tMCAO, preconditioned, and ischemic preconditioned-bearing rats treated with siCTL, siNCX1, or siNCX3. All sections corresponded approximately to the bregma levels of +3.20, +1.60, −0.30, and -3.14 mm, respectively. Scale bar = 2 mm. The infarct area is circled in *dotted lines* (Reprinted from Pignataro et al. (2011))

19.2.2 Ischemic Postconditioning

Unlike ischemic preconditioning, the neuroprotective strategy named ischemic postconditioning is a relatively novel concept (Pignataro et al. 2006; Zhao et al. 2006a, b).

Rapid revascularization of the occluded vessels and timely reperfusion are among the most effective approaches currently used for acute ischemic stroke. However, it has been repeatedly demonstrated that during the early reperfusion phase, reactive oxygen species are generated, and intracellular free Ca^{2+} overload may occur, potentially leading to additional injury (Kuroda and Siesjo 1997). In an attempt to attenuate the injurious early hyperemic response after reperfusion, a novel neuroprotective procedure termed ischemic "postconditioning" has been reported. This neuroprotective strategy is defined as a repetitive series of brief interruptions of reperfusion applied immediately after ischemia. Repeated cycles of brief reperfusion and re-occlusion were initially demonstrated to reduce the infarct size after cardiac ischemia both in the experimental (Zhao et al. 2003) and in the clinical (Staat et al. 2005) setting. More recently, ischemic postconditioning has also been shown to attenuate neuronal damage in rodent models of spinal cord (Jiang et al. 2006), focal (Pignataro et al. 2006, 2008; Zhao et al. 2006a, b) and global (Burda et al. 2006; Wang et al. 2008) ischemic injury.

19.2.2.1 Ischemic Postconditioning Induction

Brain neuroprotection induced by postconditioning has been achieved by subjecting the brain to different cycles of short non-dangerous ischemia applied after harmful ischemia. In focal ischemia, two models have been used so far to induce postconditioning: in the first model, permanent distal occlusion of MCA was followed by a series of occlusion of both common carotid arteries (CCA) (Zhao et al. 2006a, b); in the other method, harmful transient MCAO was followed by a series of brief non-injurious MCA occlusions and reperfusions (Pignataro et al. 2006, 2008). Ischemic postconditioning was achieved also in animals subjected to global ischemia induced by occlusion of CCAs and of the two vertebral arteries, 4-vessel occlusion, 4-VO, by subjecting the animals to different cycles of non-injurious CCAs occlusion.

Recently, this neuroprotective strategy has been reproduced also in in vitro preparations, i.e., hippocampal organotypic slice cultures and primary neurons subjected to harmful oxygen glucose deprivation, OGD, followed by non-injurious cycles of oxygen glucose deprivation and reoxygenation (Pignataro et al. 2006, 2008; Scartabelli et al. 2008).

Interestingly, the cross-tolerance phenomenon, in which one stressor induces protection against a different stressor, occurs even in postconditioning. In fact, it has been shown that not only brief periods of ischemia produced by repeated interruptions of reperfusion but also other pharmacological strategies that have been previously used as preconditioning stimuli can be used for postconditioning (Burda et al. 2006). Collectively, it is possible to state that several treatments can be used as postconditioning stimuli and are effective for ischemic tolerance induction, i.e., short ischemia, hypoxia, isoflurane, norepinephrine, and 3-nitropropionic acid.

19.2.2.2 Mechanisms Involved in Ischemic Postconditioning Neuroprotection

Little is known about the postconditioning protective mechanisms against cerebral ischemia. However, because postconditioning, by definition, is performed after the insult, it is essential to better characterize this phenomenon in hopes to develop an effective clinical approach to treat stroke. It has been hypothesized that ischemic postconditioning reduces infarct size after focal stroke as a function of stroke severity, probably by reducing apoptosis and free radical products (Zhao 2007). In fact, the expression and the activity of endogenous antioxidant enzymes such as manganese superoxide dismutase are increased in the brain of animal subjected to postconditioning treatment, thus playing an important role in this neuroprotective strategy (Nemethova et al. 2008). Additionally, it has been shown that postconditioning enhances ERK1/2 expression and promotes increase in phosphorylated Akt expression and activity (Gao et al. 2008; Pignataro et al. 2008). In particular, the role of Akt pathway in postconditioning has probably received the major attention in different animal models of brain ischemic postconditioning (Rehni and Singh 2007; Gao et al. 2008; Pignataro et al. 2008). In summary, although several groups have reported that postconditioning reduces ischemic damage in the brain, many outstanding issues remain elusive, such as number of ischemia/reperfusion cycles as well as durations of reperfusion and occlusion for postconditioning, therapeutic time windows, and, more importantly, the underlying protective mechanisms of postconditioning. Future studies should address these and other missing points in the attempt to better clarify this important neuroprotective strategy.

In the cardiac model of postconditioning, the signaling pathways involved in protection include the activation of prosurvival protein kinases, and these same kinases may be effectors of cardiac preconditioning as well (Hausenloy et al. 2005a, b). Recent studies suggest that postconditioning in the brain may also involve the activation of the family of protein kinases called reperfusion injury salvage kinases, RISK (Pignataro et al. 2006, 2008; Gao et al. 2008). Among them, Akt and ERK have been shown to regulate ischemic preconditioning-induced neuroprotection both in the heart and in the brain (Shamloo and Wieloch 1999; Meller et al. 2005). Interestingly, during harmful ischemia, Akt is transiently phosphorylated, and consequently activated, only for a short interval of time after reperfusion, whereas after postconditioning, the phosphorylation of Akt persists longer, being still present in the phosphorylated form even 24 h later. Erk and P38 MAPK also show similar activation profiles after postconditioning.

The pharmacological postconditioning effect of the Akt activation after harmful ischemia may account for the delayed activation of cell death pathways, as inhibition of AKT has been shown to speed cell death after ischemia (Noshita et al. 2001). Specific targets of AKT have not been shown in postconditioning, and potential effectors include (1) cAMP-responsive element binding protein (CRE) and (2) bcl-2 upregulation (Willaime-Morawek et al. 2005), both involved in preconditioning neuroprotection (Chen and Xia 2000; Meller et al. 2005); (3) activation of the Akt/GSK3 (glycogen synthase kinase 3 beta) signaling pathway (Endo et al. 2006); or (4) prosurvival effect on BAD and caspase 9 (Chan 2004a, b).

From these results, postconditioning stimulus results in the prolonged activation of AKT, which has also been shown to mediate intrinsic protection to ischemia in models of ischemic preconditioning. In addition, recent studies showed that postconditioning-mediated protection is correlated with inhibition of ERK 1/2 and JNK activities, promotion of εPKC phosphorylation, and reduction of δPKC cleavage.

19.2.2.3 Cellular Ionic Homeostasis and Ischemic Postconditioning: Role of Na^+/Ca^{2+} Exchanger

Given the potentially lethal consequences of intracellular Na^+ and Ca^{2+} overload, it is relevant to examine whether Na^+ and Ca^{2+} homeostasis is altered during postconditioning.

As previously discussed, the role of proteins involved in Ca^{2+} homeostasis during cerebral ischemia has been recently highlighted in terms of expression, activity, and pharmacological relevance. Na^+/Ca^{2+} exchangers and plasma membrane Ca^{2+} pumps are crucial for intracellular Ca^{2+} homeostasis and Ca^{2+} signaling. Different neurotoxic stimuli are able to modify expression and the function of these two families of transporters. In fact, 2–3-h exposure to 300 microM H_2O_2 induces a significantly downregulation of all NCXs and PMCAs at the RNA and protein level (Kip and Strehler 2007). In addition, in previous works, it has been shown that NCX gene expression after permanent MCAO in rats is regulated in a differential manner, depending on the exchanger isoform (NCX1, NCX2, or NCX3) and the region involved in the insult, i.e., ischemic core, peri-infarct areas, or spared regions (Pignataro et al. 2004; Boscia et al. 2006).

As far as the role played by NCX during brain postconditioning is concerned, we demonstrated that among the three NCX isoforms expressed in

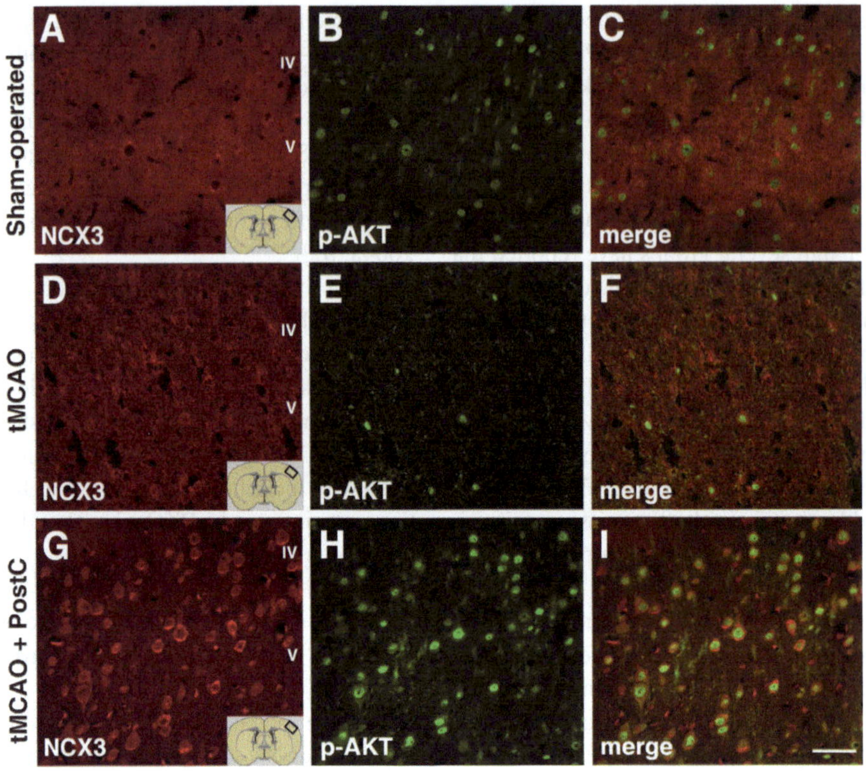

Fig. 19.4 *Co-localization of NCX3 and p-Akt in the temporoparietal cortex of sham-operated, ischemic, and ischemic postconditioned rats 5 h after sham surgery, ischemia, or postconditioning. Panels A–C* confocal microscopic images displaying both NCX3 (*red*) and p-AKT (*green*) immunoreactivity in IV and V cortical layers of sham animals. *Panels D–E* confocal microscopic images displaying both NCX3 (*red*) and p-AKT (*green*) immunoreactivity in IV and V cortical layers of ischemic animals. *Panels G–I* confocal microscopic images displaying both NCX3 (*red*) and p-AKT (*green*) immunoreactivity in IV and V cortical layers of ischemic postconditioned rats. A representative brain slice cartoon indicating the area of interest is on the left top of the figure. Scale bars in *A–I*, 50 μm (Reprinted from Pignataro et al. (2011))

the CNS, NCX3 represents an additional new molecular effector involved in the neuroprotection exerted by ischemic postconditioning. In particular, in our experimental model of ischemic postconditioning, obtained by subjecting adult male rats to 10 min of subliminal tMCAO applied 10 min after 100 min of tMCAO, we provided solid evidence showing that p-AKT is the mediator of this action since (1) p-AKT expression after postconditioning increased, and its increase timely mirrors that of NCX3 (Fig. 19.4); (2) NCX3 downregulation, induced by siRNA, reverts the neuroprotection induced by ischemic postconditioning (Fig. 19.5); and (3) the selective p-AKT inhibition prevents NCX3 upregulation thus reverting the postconditioning-induced neuroprotection (Pignataro et al. 2011).

That NCX3 is overexpressed during postconditioning may be related to its ability to counteract the dysregulation of intracellular Na^+, ($[Na^+]_i$) and Ca^{2+}, ($[Ca^{2+}]_i$) homeostasis occurring in the brain under anoxic conditions. This peculiar capability of NCX3 isoform to maintain $[Ca^{2+}]_i$ and $[Na^+]_i$ homeostasis in anoxic conditions might be correlated to its ability to operate, unlike the other two NCX isoforms, NCX1 and NCX2, even when ATP levels are reduced (Secondo et al. 2007). As a matter of fact, the three NCX isoforms display a different sensitivity to ATP levels (Linck et al. 1998; Secondo et al. 2007). In particular, during ATP depletion, NCX1 and NCX2 isoform activity is reduced, whereas NCX3 is still operative (Linck et al. 1998; Secondo et al. 2007). On the other hand, the

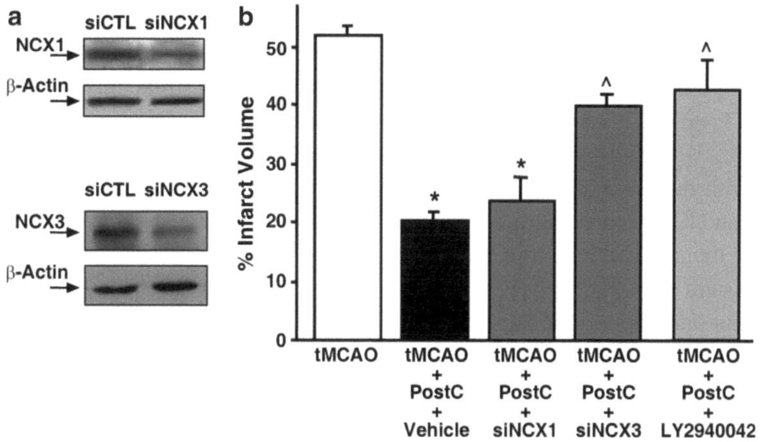

Fig. 19.5 (**a**) *NCX3 protein expression in rat brain after NCX3 siRNA treatment.* (**b**) *Effect of the p-AKT inhibitor LY2940002, siNCX1, and siNCX3 on the neuroprotective effect mediated by ischemic postconditioning. Panel A*, representative Western blots of NCX1 and NCX3 protein levels in the temporoparietal cortex of rats icv treated with siNCX1 or siNCX3, respectively. *Panel B*, infarct volume in rats subjected to tMCAO, tMCAO+PostC+siCTL, tMCAO+PostC+siNCX1, tMCAO+PostC+siNCX3, and tMCAO+PostC+LY2940042. Rats were euthanized 24 h after tMCAO. *p <0.05 versus tMCAO; ^p <0.05 versus all experimental groups. Each column represents the mean±S.E.M. (n=5–7) of the percentage of the infarct volume compared to the ipsilateral hemisphere

NCX3 neuroprotective role in ischemic postconditioning is in accordance with the results demonstrating that in homozygous *ncx3–/–* mice subjected to MCAO, an increased brain damage occurs (Molinaro et al. 2008). In addition, the silencing of NCX3 expression by RNA interference increases cerebellar granule neurons' vulnerability to Ca^{2+} overload and excitotoxicity and renders BHK cells transfected with NCX3 extremely vulnerable to chemical hypoxia (Bano et al. 2005; Secondo et al. 2007).

It is important to underline that NCX1 downregulation induced by siRNA is not able to revert the postconditioning-induced neuroprotection, thus showing that, differently from what occurs in ischemic preconditioning, NCX1 does not play a relevant role in this phenomenon. This result can be explained by taking into account the above-mentioned different sensitivity of NCX1 and NCX3 to ATP levels (Secondo et al. 2007). In addition, NCX1 and NCX3 promoters show structural differences that render NCX3 a better target for the prosurvival kinase CREB, an Akt downstream player (Gabellini et al. 2003). In fact, previous results showed that the stimulation of NCX3 promoter is mediated by the CRE sequence, which binds transcription factors of the ATF/CREB family, an AKT downstream pathway (Gabellini et al. 2003).

We previously demonstrated that during ischemia, Akt is transiently phosphorylated, and consequently activated, only for a short interval of time after reperfusion (Pignataro et al. 2008), whereas after postconditioning, the phosphorylation of Akt persists longer, being still present in the phosphorylated form even 24 h later (Pignataro et al. 2008). In addition, we showed that after ischemic postconditioning Akt phosphorylation is greater than that observed after ischemia alone. The time course of the increase of AKT phosphorylation after postconditioning parallels the same time interval at which an NCX3 upregulation occurs. Furthermore, double-staining experiments in temporoparietal cortex of postconditioned rats further confirm the greater increase in p-AKT and NCX3 expression if compared to ischemic rats. The tight relationship existing between NCX3 and p-AKT was further demonstrated by confocal microscopy results showing that the increased expression of these two proteins occurs in the same cells. Since the downregulation of NCX3 expression induced by siRNA did not modify p-AKT expression, the effect of p-AKT on NCX3 can be considered unidirectional.

Consistent with the present results, we previously showed that NCX3 represents a novel additional target for the survival action of the p-Akt pathway (Formisano et al. 2008).

Altogether, our data support the importance of p-AKT in mediating postconditioning neuroprotection and suggest NCX3 as one of the additional signals downstream to p-AKT strongly involved in the neuroprotective effect of ischemic postconditioning. The results of the present study support the idea that the enhancement of NCX3 expression and activity might be a reasonable strategy to reduce the infarct extension after a harmful ischemic insult. However, at present time, compounds able to selectively enhance NCX3 expression or activity are not available.

Although the results obtained in models of cerebral ischemia are encouraging, several other issues need to be addressed in order to better clarify the role of ionic homeostasis in the course of ischemic postconditioning.

19.3 Conclusions

Studies on postconditioning and preconditioning as neuroprotective strategy are currently being published at what appears to be a near exponential rate. The bursting of studies on this issue should help to improve our understanding of their mechanistic basis and, in turn, their clinical relevance. An important point to be underlined is that the endogenous survival mechanisms activated in response to postconditioning and preconditioning do not depend on differences in drug pharmacokinetics or administration protocols that can confound the translation of neuroprotective strategies from rodents to humans. Therefore, the identification of intrinsic cell survival pathways should provide more direct opportunities for translational neuroprotection trials. Data presented at the 6th International Conference on NCX highlighted the role of NCX1 and NCX3 in mediating neuroprotection elicited by preconditioning and postconditioning, thus suggesting that an effective stroke therapy could be designed by inducing an overexpression of these two NCX isoforms or by increasing their activity.

Acknowledgments This work was supported by COFIN 2008; Ricerca-Sanitaria RF-FSL352059 Ricerca finalizzata 2006; Ricerca-Oncologica 2006; Progetto-Strategico 2007; Progetto Ordinario 2007; Ricerca finalizzata 2009; Ricerca-Sanitaria Progetto Ordinario by Ministero della Salute 2008 all to LA.

The authors thank:

1. Elsevier for permission of using Figs. 19.1 and 19.2, license number 2821360173915, and three excerpts, license number 2821360363363, from the article that appeared in Neurobiology of Disease, 2012, by Pignataro G. et al.
2. John Wiley and Sons for permission of using text extracts from the article that appeared in FEBS Journal, 2009, by Pignataro G. et al.
3. Nature Publishing Group for permission of using text extracts and Figs. 19.3 and 19.4 from the article that appeared in Journal of Cerebral Blood Flow and Metabolism, 2011, by Pignataro G. et al.

References

B.S. Allen, A.O. Halldorsson, M.J. Barth, M.N. Ilbawi, Modification of the subclavian patch aortoplasty for repair of aortic coarctation in neonates and infants. Ann. Thorac. Surg. **69**, 877–880 (2000). discussion 881

L. Annunziato, G. Pignataro, G.F. Di Renzo, Pharmacology of brain Na^+/Ca^{2+} exchanger: from molecular biology to therapeutic perspectives. Pharmacol. Rev. **56**, 633–654 (2004)

J.W. Arthur, M.R. Wilkins, Using proteomics to mine genome sequences. J. Proteome Res. **3**, 393–402 (2004)

D. Bano, K.W. Young, C.J. Guerin, R. Lefeuvre, N.J. Rothwell, L. Naldini, R. Rizzuto, E. Carafoli, P. Nicotera, Cleavage of the plasma membrane Na^+/Ca^{2+} exchanger in excitotoxicity. Cell **120**, 275–285 (2005)

F.C. Barone, R.F. White, P.A. Spera, J. Ellison, R.W. Currie, X. Wang, G.Z. Feuerstein, Ischemic preconditioning and brain tolerance: temporal histological and functional outcomes, protein synthesis requirement, and interleukin-1 receptor antagonist and early gene expression. Stroke **29**, 1937–1950 (1998). discussion 1950–1931

F. Boscia, R. Gala, G. Pignataro, A. de Bartolomeis, M. Cicale, A. Ambesi-Impiombato, G. Di Renzo, L. Annunziato, Permanent focal brain ischemia induces isoform-dependent changes in the pattern of Na^+/Ca^{2+} exchanger gene expression in the ischemic core, peri-infarct area, and intact brain regions. J. Cereb. Blood Flow Metab. **26**, 502–517 (2006)

A.M. Brambrink, A. Schneider, H. Noga, A. Astheimer, B. Gotz, I. Korner, A. Heimann, M. Welschof, O. Kempski, Tolerance-Inducing dose of 3-nitropropionic acid modulates bcl-2 and bax balance in the rat brain: a potential mechanism of chemical preconditioning. J. Cereb. Blood Flow Metab. **20**, 1425–1436 (2000)

G. Brooks, D.J. Hearse, Role of protein kinase C in ischemic preconditioning: player or spectator? Circ. Res. **79**, 627–630 (1996)

J. Burda, M. Marsala, J. Radonak, J. Marsala, Graded postischemic reoxygenation ameliorates inhibition of cerebral cortical protein synthesis in dogs. J. Cereb. Blood Flow Metab. **11**, 1001–1005 (1991)

J. Burda, M. Gottlieb, I. Vanicky, M. Chavko, J. Marsala, Short-term postischemic hypoperfusion improves recovery of protein synthesis in the rat brain cortex. Mol. Chem. Neuropathol. **25**, 189–198 (1995)

J. Burda, V. Danielisova, M. Nemethova, M. Gottlieb, M. Matiasova, I. Domorakova, E. Mechirova, M. Ferikova, M. Salinas, R. Burda, Delayed postconditionig initiates additive mechanism necessary for survival of selectively vulnerable neurons after transient ischemia in rat brain. Cell. Mol. Neurobiol. **26**, 1141–1151 (2006)

P.H. Chan, Future targets and cascades for neuroprotective strategies. Stroke **35**, 2748–2750 (2004a)

P.H. Chan, Mitochondria and neuronal death/survival signaling pathways in cerebral ischemia. Neurochem. Res. **29**, 1943–1949 (2004b)

Y.Y. Chen, Q. Xia, Evaluation of G(i/o) protein signal transduction pathway in cardioprotection of hypoxic preconditioning. Sheng Li Xue Bao **52**, 93–97 (2000)

V. Danielisova, M. Nemethova, M. Gottlieb, J. Burda, The changes in endogenous antioxidant enzyme activity after postconditioning. Cell. Mol. Neurobiol. **26**, 1181 1191 (2006)

V.L. Dawson, T.M. Dawson, Neuronal ischaemic preconditioning. Trends Pharmacol. Sci. **21**, 423–424 (2000)

A.T. de Souza Wyse, E.L. Streck, P. Worm, A. Wajner, F. Ritter, C.A. Netto, Preconditioning prevents the inhibition of Na^+, K^+-ATPase activity after brain ischemia. Neurochem. Res. **25**, 971–975 (2000)

M. Digicaylioglu, S.A. Lipton, Erythropoietin-mediated neuroprotection involves cross-talk between Jak2 and NF-kappaB signalling cascades. Nature **412**, 641–647 (2001)

U. Dirnagl, R.P. Simon, J.M. Hallenbeck, Ischemic tolerance and endogenous neuroprotection. Trends Neurosci. **26**, 248–254 (2003)

H. Endo, C. Nito, H. Kamada, T. Nishi, P.H. Chan, Activation of the Akt/GSK3beta signaling pathway mediates survival of vulnerable hippocampal neurons after transient global cerebral ischemia in rats. J. Cereb. Blood Flow Metab. **26**, 1479–1489 (2006)

L. Formisano, M. Saggese, A. Secondo, R. Sirabella, P. Vito, V. Valsecchi, P. Molinaro, G. Di Renzo, L. Annunziato, The two isoforms of the Na^+/Ca^{2+} exchanger, NCX1 and NCX3, constitute novel additional targets for the prosurvival action of Akt/protein kinase B pathway. Mol. Pharmacol. **73**, 727–737 (2008)

N. Gabellini, S. Bortoluzzi, G.A. Danieli, E. Carafoli, Control of the Na^+/Ca^{2+} exchanger 3 promoter by cyclic adenosine monophosphate and Ca^{2+} in differentiating neurons. J. Neurochem. **84**, 282–293 (2003)

X. Gao, H. Zhang, T. Takahashi, J. Hsieh, J. Liao, G.K. Steinberg, H. Zhao, The Akt signaling pathway contributes to postconditioning's protection against stroke; the protection is associated with the MAPK and PKC pathways. J. Neurochem. **105**, 943–955 (2008)

J.M. Gidday, Cerebral preconditioning and ischaemic tolerance. Nat. Rev. Neurosci. **7**, 437–448 (2006)

J.M. Gidday, J.C. Fitzgibbons, A.R. Shah, T.S. Park, Neuroprotection from ischemic brain injury by hypoxic preconditioning in the neonatal rat. Neurosci. Lett. **168**, 221–224 (1994)

I. Ginis, R. Jaiswal, D. Klimanis, J. Liu, J. Greenspon, J.M. Hallenbeck, TNF-alpha-induced tolerance to ischemic injury involves differential control of NF-kappaB transactivation: the role of NF-kappaB association with p300 adaptor. J. Cereb. Blood Flow Metab. **22**, 142–152 (2002)

D.J. Gladstone, S.E. Black, A.M. Hakim, Toward wisdom from failure: lessons from neuroprotective stroke trials and new therapeutic directions. Stroke **33**, 2123–2136 (2002)

S.S. Glazier, D.M. O'Rourke, D.I. Graham, F.A. Welsh, Induction of ischemic tolerance following brief focal ischemia in rat brain. J. Cereb. Blood Flow Metab. **14**, 545–553 (1994)

M. Gonzalez-Zulueta, A.B. Feldman, L.J. Klesse, R.G. Kalb, J.F. Dillman, L.F. Parada, T.M. Dawson, V.L. Dawson, Requirement for nitric oxide activation of p21(ras)/extracellular regulated kinase in neuronal ischemic preconditioning. Proc. Natl. Acad. Sci. U. S. A. **97**, 436–441 (2000)

D.J. Hausenloy, A. Tsang, M.M. Mocanu, D.M. Yellon, Ischemic preconditioning protects by activating prosurvival kinases at reperfusion. Am. J. Physiol. Heart Circ. Physiol. **288**, H971–H976 (2005a)

D.J. Hausenloy, A. Tsang, D.M. Yellon, The reperfusion injury salvage kinase pathway: a common target for both ischemic preconditioning and postconditioning. Trends Cardiovasc. Med. **15**, 69–75 (2005b)

P.L. Huang, Nitric oxide and cerebral ischemic preconditioning. Cell Calcium **36**, 323–329 (2004)

X. Jiang, E. Shi, Y. Nakajima, S. Sato, Postconditioning, a series of brief interruptions of early reperfusion, prevents neurologic injury after spinal cord ischemia. Ann. Surg. **244**, 148–153 (2006)

N.M. Jones, M. Bergeron, Hypoxic preconditioning induces changes in HIF-1 target genes in neonatal rat brain. J. Cereb. Blood Flow Metab. **21**, 1105–1114 (2001)

K. Kato, K. Shimazaki, T. Kamiya, S. Amemiya, T. Inaba, K. Oguro, Y. Katayama, Differential effects of sublethal ischemia and chemical preconditioning with 3-nitropropionic acid on protein expression in gerbil hippocampus. Life Sci. **77**, 2867–2878 (2005)

H. Kin, A.J. Zatta, M.T. Lofye, B.S. Amerson, M.E. Halkos, F. Kerendi, Z.Q. Zhao, R.A. Guyton, J.P. Headrick, J. Vinten-Johansen, Postconditioning reduces infarct size via adenosine receptor activation by endogenous adenosine. Cardiovasc. Res. **67**, 124–133 (2005)

S.N. Kip, E.E. Strehler, Rapid downregulation of NCX and PMCA in hippocampal neurons following H2O2 oxidative stress. Ann. N. Y. Acad. Sci. **1099**, 436–439 (2007)

T. Kirino, Ischemic tolerance. J. Cereb. Blood Flow Metab. **22**, 1283–1296 (2002)

T. Kirino, Y. Tsujita, A. Tamura, Induced tolerance to ischemia in gerbil hippocampal neurons. J. Cereb. Blood Flow Metab. **11**, 299–307 (1991)

K. Kitagawa, M. Matsumoto, M. Tagaya, R. Hata, H. Ueda, M. Niinobe, N. Handa, R. Fukunaga, K. Kimura, K. Mikoshiba et al., 'Ischemic tolerance' phenomenon found in the brain. Brain Res. **528**, 21–24 (1990)

K. Kitagawa, M. Matsumoto, K. Kuwabara, M. Tagaya, T. Ohtsuki, R. Hata, H. Ueda, N. Handa, K. Kimura, T. Kamada, 'Ischemic tolerance' phenomenon detected in various brain regions. Brain Res. **561**, 203–211 (1991)

S. Kobayashi, V.A. Harris, F.A. Welsh, Spreading depression induces tolerance of cortical neurons to ischemia in rat brain. J. Cereb. Blood Flow Metab. **15**, 721–727 (1995)

S. Kuroda, B.K. Siesjo, Reperfusion damage following focal ischemia: pathophysiology and therapeutic windows. Clin. Neurosci. **4**, 199–212 (1997)

C. Lange-Asschenfeldt, A.P. Raval, K.R. Dave, D. Mochly-Rosen, T.J. Sick, M.A. Perez-Pinzon, Epsilon protein kinase C mediated ischemic tolerance requires activation of the extracellular regulated kinase pathway in the organotypic hippocampal slice. J. Cereb. Blood Flow Metab. **24**, 636–645 (2004)

B. Linck, Z. Qiu, Z. He, Q. Tong, D.W. Hilgemann, K.D. Philipson, Functional comparison of the three isoforms of the Na^+/Ca^{2+} exchanger (NCX1, NCX2, NCX3). Am. J. Physiol. **274**, C415–C423 (1998)

S.M. Massa, R.A. Swanson, F.R. Sharp, The stress gene response in brain. Cerebrovasc. Brain Metab. Rev. **8**, 95–158 (1996)

N. Maulik, M. Watanabe, Y.L. Zu, C.K. Huang, G.A. Cordis, J.A. Schley, D.K. Das, Ischemic preconditioning triggers the activation of MAP kinases and MAPKAP kinase 2 in rat hearts. FEBS Lett. **396**, 233–237 (1996)

D.R. Meldrum, J.C. Cleveland Jr., R.T. Rowland, A. Banerjee, A.H. Harken, X. Meng, Early and delayed preconditioning: differential mechanisms and additive protection. Am. J. Physiol. **273**, H725–H733 (1997)

R. Meller, M. Minami, J.A. Cameron, S. Impey, D. Chen, J.Q. Lan, D.C. Henshall, R.P. Simon, CREB-mediated Bcl-2 protein expression after ischemic preconditioning. J. Cereb. Blood Flow Metab. **25**, 234–246 (2005)

K. Miyashita, H. Abe, T. Nakajima, A. Ishikawa, M. Nishiura, T. Sawada, H. Naritomi, Induction of ischaemic tolerance in gerbil hippocampus by pretreatment with focal ischaemia. Neuroreport **6**, 46–48 (1994)

P. Molinaro, O. Cuomo, G. Pignataro, F. Boscia, R. Sirabella, A. Pannaccione, A. Secondo, A. Scorziello, A. Adornetto, R. Gala, D. Viggiano, S. Sokolow, A. Herchuelz, S. Schurmans, G. Di Renzo, L. Annunziato, Targeted disruption of Na^+/Ca^{2+} exchanger 3 (NCX3) gene leads to a worsening of ischemic brain damage. J. Neurosci. **28**, 1179–1184 (2008)

C.E. Murry, R.B. Jennings, K.A. Reimer, Preconditioning with ischemia: a delay of lethal cell injury in ischemic myocardium. Circulation **74**, 1124–1136 (1986)

M. Nemethova, V. Danielisova, M. Gottlieb, J. Burda, Post-conditioning exacerbates the MnSOD immunereactivity after experimental cerebral global ischemia and reperfusion in the rat brain hippocampus. Cell Biol. Int. **32**, 128–135 (2008)

N. Noshita, T. Sugawara, M. Fujimura, Y. Morita-Fujimura, P.H. Chan, Manganese superoxide dismutase affects cytochrome c release and caspase-9 activation after transient focal cerebral ischemia in mice. J. Cereb. Blood Flow Metab. **21**, 557–567 (2001)

S. Ohta, S. Furuta, I. Matsubara, K. Kohno, Y. Kumon, S. Sakaki, Calcium movement in ischemia-tolerant hippocampal CA1 neurons after transient forebrain ischemia in gerbils. J. Cereb. Blood Flow Metab. **16**, 915–922 (1996)

A. Patel, M.C. van de Poll, J.W. Greve, W.A. Buurman, K.C. Fearon, S.J. McNally, E.M. Harrison, J.A. Ross, O.J. Garden, C.H. Dejong, S.J. Wigmore, Early stress protein gene expression in a human model of ischemic preconditioning. Transplantation **78**, 1479–1487 (2004)

M.A. Perez-Pinzon, K.R. Dave, A.P. Raval, Role of reactive oxygen species and protein kinase C in ischemic tolerance in the brain. Antioxid. Redox Signal. **7**, 1150–1157 (2005)

G. Pignataro, R. Gala, O. Cuomo, A. Tortiglione, L. Giaccio, P. Castaldo, R. Sirabella, C. Matrone, A. Canitano, S. Amoroso, G. Di Renzo, L. Annunziato, Two sodium/calcium exchanger gene products, NCX1 and NCX3, play a major role in the development of permanent focal cerebral ischemia. Stroke **35**, 2566–2570 (2004)

G. Pignataro, Z. Xiong, R.P. Simon, *Ischemic post-conditioning: a new neuroprotective strategy* (Society for Neuroscience, Atlanta, 2006)

G. Pignataro, R. Meller, K. Inoue, A.N. Ordonez, M.D. Ashley, Z. Xiong, R. Gala, R.P. Simon, In vivo and in vitro characterization of a novel neuroprotective strategy for stroke: ischemic postconditioning. J. Cereb. Blood Flow Metab. **28**, 232–241 (2008)

G. Pignataro, A. Scorziello, G. Di Renzo, L. Annunziato, Post-ischemic brain damage: effect of ischemic preconditioning and postconditioning and identification of potential candidates for stroke therapy. FEBS J. **276**, 46–57 (2009)

G. Pignataro, E. Esposito, O. Cuomo, R. Sirabella, F. Boscia, N. Guida, G. Di Renzo, L. Annunziato, The NCX3 isoform of the Na^+/Ca^{2+} exchanger contributes to neuroprotection elicited by ischemic postconditioning. J. Cereb. Blood Flow Metab. **31**, 362–370 (2011)

G. Pignataro, F. Boscia, E. Esposito, R. Sirabella, O. Cuomo, A. Vinciguerra, G. Di Renzo, L. Annunziato, NCX1 and NCX3: Two new effectors of delayed preconditioning in brain ischemia. Neurobiol. Dis. **45**, 616–623 (2012)

H. Plamondon, N. Blondeau, C. Heurteaux, M. Lazdunski, Mutually protective actions of kainic acid epileptic preconditioning and sublethal global ischemia on hippocampal neuronal death: involvement of adenosine A1 receptors and K(ATP) channels. J. Cereb. Blood Flow Metab. **19**, 1296–1308 (1999)

A. Ravati, B. Ahlemeyer, A. Becker, J. Krieglstein, Preconditioning-induced neuroprotection is mediated by reactive oxygen species. Brain Res. **866**, 23–32 (2000)

A.K. Rehni, N. Singh, Role of phosphoinositide 3-kinase in ischemic postconditioning-induced attenuation of cerebral ischemia-evoked behavioral deficits in mice. Pharmacol. Rep. **59**, 192–198 (2007)

R. Rejdak, K. Rejdak, M. Sieklucka-Dziuba, Z. Stelmasiak, P. Grieb, Brain tolerance and preconditioning. Pol. J. Pharmacol. **53**, 73–79 (2001)

E. Rybnikova, L. Vataeva, E. Tyulkova, T. Gluschenko, V. Otellin, M. Pelto-Huikko, M.O. Samoilov, Mild hypoxia preconditioning prevents impairment of passive avoidance learning and suppression of brain NGFI-A expression induced by severe hypoxia. Behav. Brain Res. **160**, 107–114 (2005)

T. Scartabelli, E. Gerace, E. Landucci, F. Moroni, D.E. Pellegrini-Giampietro, Neuroprotection by group I mGlu receptors in a rat hippocampal slice model of cerebral ischemia is associated with the PI3K-Akt signaling pathway: a novel postconditioning strategy? Neuropharmacology **55**, 509–516 (2008)

B. Schaller, R. Graf, Cerebral ischemic preconditioning. An experimental phenomenon or a clinical important entity of stroke prevention? J. Neurol. **249**, 1503–1511 (2002)

A. Scorziello, M. Santillo, A. Adornetto, C. Dell'Aversano, R. Sirabella, S. Damiano, L.M. Canzoniero, G.F. Renzo, L. Annunziato, NO-induced neuroprotection in ischemic preconditioning stimulates mitochondrial Mn-SOD activity and expression via Ras/ERK1/2 pathway. J. Neurochem. 103, 1472–80 (2007)

A. Secondo, R.I. Staiano, A. Scorziello, R. Sirabella, F. Boscia, A. Adornetto, V. Valsecchi, P. Molinaro, L.M. Canzoniero, G. Di Renzo, L. Annunziato, BHK cells transfected with NCX3 are more resistant to hypoxia followed by reoxygenation than those transfected with NCX1 and NCX2: Possible relationship with mitochondrial membrane potential. Cell Calcium **42**, 521–535 (2007)

M. Shamloo, T. Wieloch, Changes in protein tyrosine phosphorylation in the rat brain after cerebral ischemia in a model of ischemic tolerance. J. Cereb. Blood Flow Metab. **19**, 173–183 (1999)

K. Shimazaki, T. Nakamura, K. Nakamura, K. Oguro, T. Masuzawa, Y. Kudo, N. Kawai, Reduced calcium elevation in hippocampal CA1 neurons of ischemia-tolerant gerbils. Neuroreport **9**, 1875–1878 (1998)

R.P. Simon, M. Niiro, R. Gwinn, Prior ischemic stress protects against experimental stroke. Neurosci. Lett. **163**, 135–137 (1993)

M.E. Speechly-Dick, M.M. Mocanu, D.M. Yellon, Protein kinase C. Its role in ischemic preconditioning in the rat. Circ. Res. **75**, 586–590 (1994)

P. Staat, G. Rioufol, C. Piot, Y. Cottin, T.T. Cung, I. L'Huillier, J.F. Aupetit, E. Bonnefoy, G. Finet, X. Andre-Fouet, M. Ovize, Postconditioning the human heart. Circulation **112**, 2143–2148 (2005)

M.P. Stenzel-Poore, S.L. Stevens, Z. Xiong, N.S. Lessov, C.A. Harrington, M. Mori, R. Meller, H.L. Rosenzweig, E. Tobar, T.E. Shaw, X. Chu, R.P. Simon, Effect of ischaemic preconditioning on genomic response to cerebral ischaemia: similarity to neuroprotective strategies in hibernation and hypoxia-tolerant states. Lancet **362**, 1028–1037 (2003)

T. Toyoda, N.F. Kassell, K.S. Lee, Induction of ischemic tolerance and antioxidant activity by brief focal ischemia. Neuroreport **8**, 847–851 (1997)

V. Valsecchi, G. Pignataro, A. Del Prete, R. Sirabella, C. Matrone, F. Boscia, A. Scorziello, M.J. Sisalli, E. Esposito, N. Zambrano, G. Di Renzo, L. Annunziato, NCX1 is a novel target gene for hypoxia-inducible factor-1 in ischemic brain preconditioning. Stroke **42**, 754–763 (2011)

J.Y. Wang, J. Shen, Q. Gao, Z.G. Ye, S.Y. Yang, H.W. Liang, I.C. Bruce, B.Y. Luo, Q. Xia, Ischemic postconditioning protects against global cerebral ischemia/reperfusion-induced injury in rats. Stroke **39**, 983–990 (2008)

S. Willaime-Morawek, N. Arbez, J. Mariani, B. Brugg, IGF-I protects cortical neurons against ceramide-induced apoptosis via activation of the PI-3K/Akt and ERK pathways; is this protection independent of CREB and Bcl-2? Brain Res. Mol. Brain Res. **142**, 97–106 (2005)

S. Yano, M. Morioka, K. Fukunaga, T. Kawano, T. Hara, Y. Kai, J. Hamada, E. Miyamoto, Y. Ushio, Activation of Akt/protein kinase B contributes to induction of ischemic tolerance in the CA1 subfield of gerbil hippocampus. J. Cereb. Blood Flow Metab. **21**, 351–360 (2001)

D.M. Yellon, D.J. Hausenloy, Realizing the clinical potential of ischemic preconditioning and postconditioning. Nat. Clin. Pract. Cardiovasc. Med. **2**, 568–575 (2005)

H. Zhao, The protective effect of ischemic postconditioning against ischemic injury: from the heart to the brain. J. Neuroimmune Pharmacol. **2**, 313–318 (2007)

H. Zhao, Ischemic postconditioning as a novel avenue to protect against brain injury after stroke. J. Cereb. Blood Flow Metab. **29**, 873–885 (2009)

Z.Q. Zhao, J.S. Corvera, M.E. Halkos, F. Kerendi, N.P. Wang, R.A. Guyton, J. Vinten-Johansen, Inhibition of myocardial injury by ischemic postconditioning during reperfusion: comparison with ischemic preconditioning. Am. J. Physiol. Heart Circ. Physiol. **285**, H579–H588 (2003)

H. Zhao, R.M. Sapolsky, G.K. Steinberg, Interrupting reperfusion as a stroke therapy: ischemic postconditioning reduces infarct size after focal ischemia in rats. J. Cereb. Blood Flow Metab. **26**, 1114–1121 (2006a)

H. Zhao, R.M. Sapolsky, G.K. Steinberg, Phosphoinositide-3-kinase/akt survival signal pathways are implicated in neuronal survival after stroke. Mol. Neurobiol. **34**, 249–270 (2006b)

The Role of the Mitochondrial NCX in the Mechanism of Neurodegeneration in Parkinson's Disease

20

Alison Wood-Kaczmar, Emma Deas,
Nicholas W. Wood, and Andrey Y. Abramov

Abstract

Mitochondrial Na^+/Ca^{2+} exchange (NCXmito) is critical for neuronal Ca^{2+} homeostasis and prevention of cell death from excessive mitochondrial Ca^{2+} ($m[Ca^{2+}]$) accumulation. The mitochondrial kinase PINK1 appears to regulate the mCa^{2+} efflux from dopaminergic (DAergic) neurons, possibly by directly regulating NCXmito. However, the precise molecular identity of NCXmito is unknown and has been the subject of great controversy. Here we propose that the previously characterised plasmalemmal NCX isoforms (NCX2, NCX3) contribute to mitochondrial Na^+/Ca^{2+} exchange in human DAergic neurons and may act downstream of PINK1 in the prevention of neurodegeneration by $m[Ca^{2+}]$ accumulation. Firstly, we definitively show the existence of a mitochondrial pool of endogenous plasmalemmal NCX isoforms in human DAergic neurons and cell lines using immunolocalisation and fluorescence-assisted organelle sorting (FAOS). Secondly, we demonstrate reduced mitochondrial Ca^{2+} efflux occurs following inhibition of NCX2 or NCX3 (but not NCX1) using siRNA or antibody blocking. This study has potentially revealed a new molecular target in Parkinson's disease pathology which ultimately may open up new avenues for future therapeutic intervention.

Keywords

PINK1 • Mitochondria • NCX • Calcium • Parkinson's disease

A. Wood-Kaczmar • E. Deas • N.W. Wood
A.Y. Abramov (✉)
Department of Molecular Neuroscience,
UCL Institute of Neurology, Queen Square, London,
WC1N 3BG, UK
e-mail: a.abramov@ucl.ac.uk

L. Annunziato (ed.), *Sodium Calcium Exchange: A Growing Spectrum of Pathophysiological Implications*, Advances in Experimental Medicine and Biology 961, DOI 10.1007/978-1-4614-4756-6_20,

20.1 Introduction

Parkinson's disease (PD) is a debilitating and incurable movement disorder and is the second most common age-dependent neurodegenerative disease in the Western world. It affects ~1% of the population over 65 years of age, rising to 5 % over 80 years of age. PD is characterised clinically by motor symptoms including bradykinesia (slow movement), rigidity, resting tremor and postural instability. The observed motor symptoms are caused by the progressive loss of DAergic (DA) neurons located within a pigmented region of the brain called the substantia nigra (SN). The pathological hallmarks of PD in post-mortem tissue are pallor of the SN and the presence of proteinaceous inclusions (Lewy bodies and Lewy neurites) in surviving neurons. In the majority of cases, the underlying cause of PD is unknown. However, research in the last few decades has highlighted the interconnected roles of oxidative stress, protein mishandling and mitochondrial dysfunction in the aetiology of PD (Cali et al. 2011). Furthermore, the quest for a deeper understanding of the molecular causes of disease has been greatly aided by the identification of 'PD genes' which give rise to heritable forms of parkinsonism. These genes include α-synuclein, leucine-rich repeat kinase 2 (LRRK2), DJ-1, ATP13A2, parkin and PTEN-induced kinase 1 (PINK1). Autosomal recessive mutations in the PINK1 gene cause early onset parkinsonism and the PINK1 protein was characterised as a serine-threonine kinase which localises to mitochondria (Valente et al. 2004). Recent work in our laboratory has focused on understanding the normal function of PINK1 by creating loss-of-function cellular models *in vitro*, including in human primary DAergic neurons (Wood-Kaczmar et al. 2008). DAergic neurons deficient in PINK1 expression have significantly lower long-term survival rates, with surviving neurons exhibiting signs of acute oxidative stress and degeneration, including impaired respiratory chain activity, decreased glutathione levels and abnormal mitochondrial morphology. Live cell imaging studies show that in the absence of PINK1, the basal mitochondrial membrane potential ($\Delta\Psi m$) is significantly lowered, and there is enhanced mitochondrial and cytoplasmic production of reactive oxygen species. Further investigation into the causes of these physiological abnormalities revealed an underlying impairment in neuronal calcium homeostasis (Gandhi et al. 2009). Transient elevations of intracellular calcium occur following by influx of extracellular calcium through ion channels or transporters or a release of calcium from intracellular stores such as the endoplasmic reticulum. In neurons, calcium transients play a critical role in the generation of an action potential, neurotransmitter release and intracellular signalling. However, to prevent calcium overload and excitotoxicity, calcium levels must be restored to basal levels through extrusion and buffering (reviewed in Szydlowska and Tymianski (2010)). Mitochondria are important players in this process. Calcium enters mitochondria passively via the calcium uniporter (MCU), its accumulation into the matrix driven by the $\Delta\Psi m$ across the inner membrane (Baughman et al. 2011). Accumulation of calcium has two effects on mitochondria – the activation of certain oxidative enzymes and the lowering of $\Delta\Psi m$, which in turn reduces ATP production driven by ATP synthetase (complex V). Normally, calcium is extruded back into the cytoplasm down its concentration gradient via ionic antiporters. In excitable cells, this is predominantly via the Na^+/Ca^{2+} exchanger, NCX_{mito}.

20.2 The Identity of Mitochondrial NCX in Dopaminergic Neurons

In neurons, NCX_{mito} is the primary mechanism by which mCa^{2+} is returned to the cytoplasm and therefore is critical to a multitude of Ca^{2+}-dependent processes including neurotransmitter release, synaptic plasticity, bioenergetics and mitochondrial NO and free radical generation (reviewed comprehensively in Castaldo et al. (2009)). It was found that in the absence of PINK1, NCX_{mito} activity was severely impaired, leading to mitochondrial calcium overload, permeability transition pore opening and cell death (see Fig. 4, Gandhi et al. (2009)). Despite being extensively characterised in terms of ionic selectivity, stoichiometry and pharmacological sensitivity, the precise molecular identity of the NCX_{mito} in human neurons remains unclear. In contrast, genes for NCX at the plasma membrane are well characterised (Lytton 2007). In mammals, three NCX genes have been cloned, NCX1, NCX2 and NCX3, which show striking sequence homology and have similar ionic and pharmacological properties (Linck et al. 1998). It was proposed that NCX_{mito} is entirely distinct from the characterised plasmalemmal NCX isoforms, due to the specific sensitivity of NCX_{mito} to the inhibitor CGP-37157 (Czyz and Kiedrowski 2003). In purification assays using reconstituted bovine cardiac mitochondrial microliposomes, a 110-kDa protein with demonstrable Na^+/Ca^{2+} exchange activity was isolated. However, this candidate exchanger was not identified by peptide sequencing (Li et al. 1992). Moreover, an antibody raised to the purified exchanger did not cross-react with brain mitochondria. This raises the possibility of molecular heterogeneity between tissue types, and characterisation of NCX_{mito} in a particular tissue type may not necessarily inform on the identity of the exchanger elsewhere. Furthermore, it has been demonstrated that in neurons, plasmalemmal NCX *is* inhibited by CGP-37157 within the micromolecular range typically used to define NCX_{mito} in cardiac sarcolemma (2). In a study aimed at identifying NCX_{mito} in rodent neurons, Gobbi et al. used antibodies raised against NCX1, NCX2 and NCX3 in immunogold labelling and subcellular fractionation experiments (Gobbi et al. 2007). All three isoforms were found expressed in neuronal mitochondria. As we had previously identified a striking NCX_{mito} impairment in PINK1-deficient neurons, we sought the molecular target of PINK1 in the regulation of mitochondrial Ca^{2+} efflux. With an absence of any cloned human NCX_{mito}, we decided to test the hypothesis that NCX1–NCX3 were also expressed in the mitochondria of human midbrain neurons and were interacting with PINK1 to regulate Ca^{2+} efflux.

20.2.1 NCX Expression in Human DAergic Neurons

In order to study NCX_{mito} in the most appropriate cell models for Parkinson's disease, we used primary human DAergic neurons derived from a novel embryonic stem cell line, ReNCell VM (Donato et al. 2007; Wood-Kaczmar et al. 2008). We also used SH-SY5Y cells engineered to constitutively express a fluorescent mitochondrial marker (DsRed2-Mito (Deas et al. 2011)). First, we determined which NCX isoforms were expressed in human DAergic neurons. The relative expression levels of all three genes NCX1–NCX3 were investigated using Western blotting of human neuronal cell extracts (see Fig. 20.1a). Three developmental time points were assessed: neural stem cells (NSCs), young differentiated DAergic neurons (day of differentiation 5 (dd5)) and aged DAergic neurons (dd30). We also compared NCX expression in cultures enriched in DAergic neurons (PreD) with those containing predominantly other subtypes (StdD). For detailed culturing procedures, see Wood-Kaczmar et al. (2008). NCX1 is predicted to be expressed as a 120-kDa protein, with cleaved forms appearing around 60 kDa. Using a monoclonal anti-NCX1 antibody (Swant), we found that NCX1 expression was barely detectable in NSCs and newly differentiated neurons (data not shown). NCX2 is detected as a 60-kDa form using the monoclonal antibody W1C3 (Thurneysen et al. 2002). NCX2 is expressed in NSCs and in young (dd5) and aged (dd30) neurons. Using the polyclonal antibody anti-NCX3 (Swant), we were able to detect the predicted 120 kDa species of NCX3, as well as additional bands at slightly lighter (~80 and ~110 kDa) and heavier (140, 180 and >260 kDa) molecular weights using longer exposures.

In whole cell lysates from SH-SY5Y cells, we occasionally detected the cleaved form of NCX1 only (~60 kDa). Using the mAb W1C3, we detected the 54-kDa NCX2 form and a doublet at high molecular weight between the 148-kDa and 250-kDa markers. This may represent heat-induced aggregated forms of NCX2 or non-denatured exchanger complexes. We detected the NCX3 as a predominant band at the

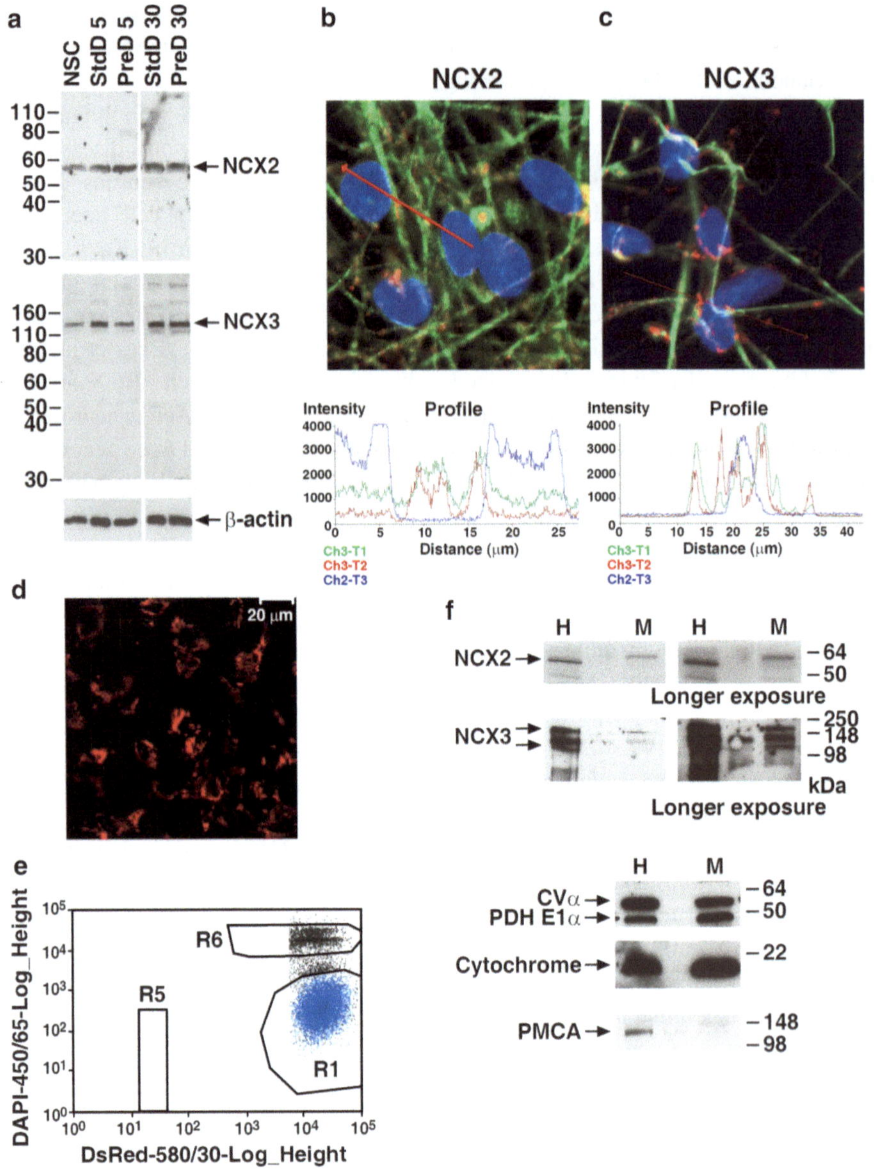

Fig. 20.1 *Expression of NCX2 and NCX3 in the mitochondria of human DAergic neurons.* **(a)** Western blots showing expression of NCX2 and NCX3 in neural stem cells (NSCS), newly differentiated neurons (PreD5 and StdD5) and aged neurons (PreD30 and StdD30). Neuronal cultures were either enriched in DAergic neurons using a pre-aggregation differentiation procedure (PreD) or were composed of other non-DAergic neuronal subtypes from midbrain (StdD). (b) Top: Micrograph showing co-localisation of NCX2 expression (*green*) with COXIV (mitochondrial marker, *red*) in human DAergic neurons (PreD5). Nuclear staining with Hoeschst is shown in *blue*. Bottom: Transectional fluorescence intensity plots showing regions of co-occurrence of NCX2 signal with COXIV. Y-axis = Fluorescence intensity (Arbitrary Units). X-axis = Distance (μM). (c) Top: Micrograph showing co-localisation of NCX3 expression with mitochondrial marker Complex V in human DAergic neurons. Nuclear staining with Hoeschst is shown in *blue*. Bottom: Transectional fluorescence intensity plots showing regions of co-occurrence of NCX3 signal with Complex V. Y-axis = Fluorescence intensity (Arbitrary Units). X-axis = Distance (μM). (d) Micrograph showing SH-SY5Y cells expressing DsRed2 Mito. (e) Scatter plots showing FS and SSC profiles of DsRed2-Mito-positive, DAPI-positive mitochondrial populations (R1) collected during FAOS from stably transfected DsRed2-Mito SH-SY5Y cells. (f) Expression of NCX2 and NCX3 in mitochondrial fractions purified by FACS from DsRed2-Mito SH-SY5Y. The purity of mitochondrial fractions was assessed by Western blot against compartmental markers for mitochondria and plasma membrane (PMCA)

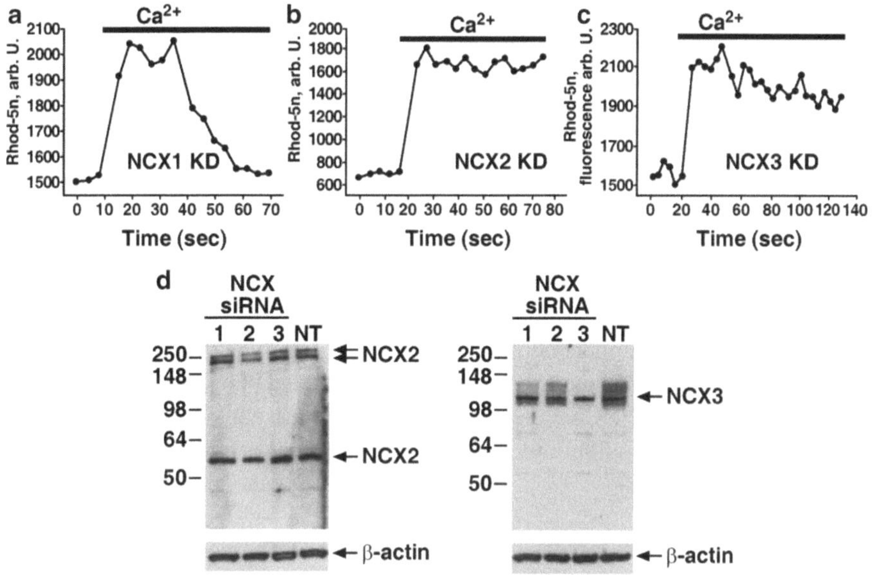

Fig. 20.2 *Acute siRNA-mediated knockdown of NCX2 and NCX3 impairs Ca^{2+} efflux from isolated mitochondria.* (a–c) Ca^{2+} efflux from mitochondria within intact SH-SY5Y cells was measured in response to 5-mM Ca^{2+} following siRNA-mediated knockdown of NCX1–NCX3. Mitochondrial Ca^{2+} efflux was reduced in SH-SY5Y cells with NCX2 and NCX3 siRNA (b and c) but not in controls or NCX1 siRNA (**a**). (d) Western blots showing levels of NCX gene expression in SH-SY5Y cells following transient transfection with siRNA pools targeting each NCX gene individually (NCX1–3). Control cultures were transfected with a pool of non-targeting siRNA (NT)

Table 20.1 Primary antibodies

Antibody	Host	Subtype	Supplier	[Western blot]	[IF]
Anti-β-actin	m	IgG	Sigma	1:5,000	–
Anti-PINK1 505A	rb	IgG	Novus	1:500	1:50
Anti-NCX1	m	IgG	Swant	1:500	1:50
W1C3	m	IgM	Hybridoma cells (H. Porzig)	1:50–1:100	1:5–1:10
Anti-NCX3	rb	IgG	Swant	1:2,000	1:200
Total OXPHOS mAb cocktail	m	–	Mitosciences	1:1,000	–
ApoTrack	m	–	Mitosciences	1:1,000	–

predicted 120 kDa, within a distinctive smear, extending from ~140 kDa to 100 kDa. This pattern is presumably due to the glycosylated forms of NCX3. For SH-SY5Y band patterns, see Fig. 20.2d.

20.2.2 NCX Genes Are Expressed in Neuronal Mitochondria

20.2.2.1 Immunofluorescence

To assess mitochondrial localisation of the three NCX genes, ReNCell VM NSCs differentiated to DAergic neurons (dd5) or SH-SY5Y plated the day previously were fixed in warm 4 % paraformaldehyde in PBS. Immunostaining of coverslips was carried out as described previously (Wood-Kaczmar et al. 2008). Details of all antibodies are available in Table 20.1. Confocal images were obtained using a Zeiss 510 uv-vis CLSM equipped with a META detection system and a 40× oil immersion objective. The 488-nm Argon laser line was used to excite fluo-4 fluorescence which was measured using a bandpass filter from 505 to 550 nm. Illumination intensity was kept to a minimum (at 0.1–0.2 % of laser output) to avoid phototoxicity and the pinhole set to give an optical slice of ~2 μm.

We saw no signal with anti-NCX1 on either cell line (data not shown). This is likely due to the low abundance of NCX1 protein expressed in these cells. However, NCX2 and NCX3 were highly expressed on the plasma membrane and in mitochondria (Fig. 20.1b, c) and DsRed2-Mito SH-SY5Y. Using cross-sectional fluorescence profile plots, the co-occurrence of an NCX signal with mitochondrial markers could clearly be seen. Interestingly, NCX2 and NCX3 immunostaining was not uniform along mitochondrial networks, but rather appeared brighter in certain regions, suggesting NCX2 and NCX3 are localised to specific microdomains within mitochondrial membranes.

20.2.2.2 Fluorescence-Activated Organelle Sorting

Subcellular fractionation by differential centrifugation was initially used to isolate mitochondria from differentiated ReNCell VM humans and wt SH-SY5Y. We found NCX1, NCX2 and NCX3 within the mitochondrial fraction, but these fractions also contained contamination from plasma membrane, assessed by Western blotting for plasma membrane calcium ATPase (PMCA) and potassium channel Kv1.1.

To obtain mitochondrial fractions free of plasma membrane contamination, we utilised flow cytometry to detect fluorescently labelled mitochondria from a crude cell extract of DsRed2-Mito SH-SY5Y followed by fluorescence-assisted organelle sorting (FAOS). DsRed2-Mito SH-SY5Y cells (Fig. 20.1d) were grown in T175 flasks to 60%–80% confluency. On the day of cell sorting, cells were trypsinised and pelleted at 1,200 rpm for 4 min in 15-mL Falcon tubes. Cells were washed in PBS and repelleted before being resuspended in chilled 500-mL mitochondrial isolation buffer (Gandhi et al. 2009) containing 1× protease inhibitors (Roche). Cells were transferred to a 1.5-mL microcentrifuge tube and lysed on ice using a plastic hand-held homogeniser (30 passes). Crude homogenates were kept on ice prior to flow cytometry. Immediately prior to sorting, homogenates (0.5 mL) were spiked with 10-μL DAPI [200 μg/mL] or 7.5-μL Hoechst 33342 [1 mg/mL]. Organelles were sorted on a MoFlo XDP cell sorter (Beckman Coulter, USA). DAPI or Hoechst (UV 350 nm excitation) was detected in FL-6 (450/65) channel and DsRed-Mito (Argon 488-nm excitation) was detected in FL-2 (580/30) channel. We sorted DAPI/Hoechst bright – DsRed-Mito-positive nuclei and intact cells and DAPI/Hoechst dim – DsRed-Mito-positive mitochondria. Both gates were set on DsRed vs DAPI/Hoechst dot plot (Fig. 20.1e). The MoFlo was triggered on FL-2 fluorescence signal to exclude small debris. Mitochondria were collected into 1.5-mL centrifuge tubes containing protease inhibitors. A small volume (50–100 μL) crude homogenate was retained for analysis. Mitochondrial fractions were sonicated on ice using 3×10-s pulses at frequency 14–18 with 10 s between pulses. Proteins were concentrated using 10-k-MW spin columns (VivaSpin). Protein concentrations of non-sorted and concentrated mitochondrial fractions were assessed using BCA colorimetric assay (Pierce) and gel samples made by boiling fractions in 2× sample buffer for 3 min. Equal amounts of protein were loaded onto gels for SDS-PAGE.

Using this novel method, we confirmed expression of all three NCX isoforms in isolated mitochondria (Fig. 20.1f). Sorted mitochondrial fractions were assessed for purity against non-sorted homogenate, and we confirmed that whilst containing multiple mitochondrial markers, they did not contain plasma membrane markers including PMCA.

20.2.3 NCX2 and NCX3 Contribute to Mitochondrial Na^+/Ca^{2+} Exchanger Activity

20.2.3.1 siRNA Knockdown of NCX2 and NCX3 Causes NCX_{mito} Impairment

To assess the functional role of NCX in mitochondria, we used siRNA to silence expression of each NCX gene individually, followed by live imaging to measure mitochondrial calcium dynamics (see Fig. 20.2). The following Dharmacon siGENOME SMARTpool siRNAs were used to silence NCX1–NCX3 in SH-SY5Y cells: Human SLC8A1 (NM_001112801),

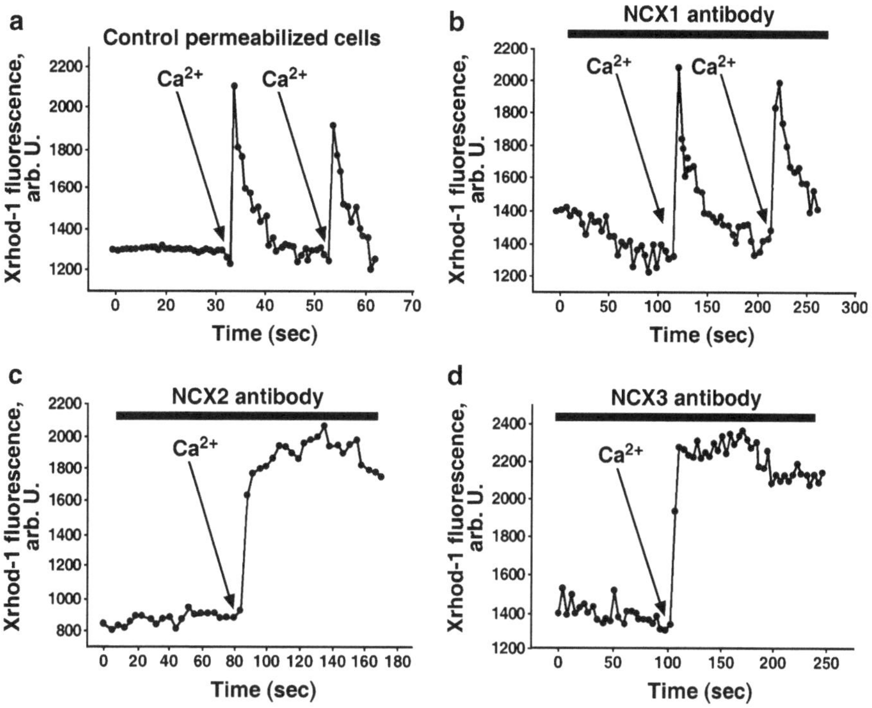

Fig. 20.3 *Ca^{2+} efflux from mitochondria can be blocked by anti-NCX2 and anti-NCX3 antibodies*. (a–**d**) Flash photolysis of permeabilised ReNCell VM neurons loaded with Rhod-5N demonstrated a flash-induced increase in $[Ca^{2+}]_m$ followed by Ca^{2+} efflux from mitochondria pre-incubated with control antibodies (**a**) or anti-NCX1 (**b**). In permeabilised neurons incubated with anti-NCX2 (**c**) or anti-NCX3 (**d**), there was no recovery of the $[Ca^{2+}]_m$ signal

Human SLC8A2 (NM_015063), Human SLC8A3 (NM_182936) and Human siGENOME non-targeting siRNA pool #1. Mitochondrial calcium uptake in SH-SY5Y cells was stimulated by the addition of 5 μm calcium to the medium. In control cells transfected with the non-targeting siRNA pool or in NCX-1 siRNA cultures, cytosolic transience caused a rapid accumulation of calcium in the mitochondria. This was observed as a sharp increase in Rhod-5N fluorescence, followed by a steady decrease back to basal levels. After approximately one minute, mitochondrial calcium had returned to normal levels (see Fig. 20.2a). Following transient NCX2 or NCX3 knockdown, no decline in Rhod-5N signal occurred following a calcium spike, indicating a severe impairment of NCX_{mito} activity (see Fig. 20.2b). Silencing NCX-3 expression impaired NCX_{mito} in a similar manner. To investigate the specificity of NCX siRNA pools, Western blot analysis was performed on transfected cultures. NCX2 siRNA reduced NCX2 expression and had a minor silencing effect on NCX3. However, NCX3 siRNA did not affect NCX2 expression. It is therefore likely that both exchangers contribute to NCX_{mito} activity in this cell model.

20.2.3.2 Antibody-Mediated Blocking of NCX_{mito} Activity

We also blocked NCX function in mitochondria using specific anti-NCX antibodies (Fig. 20.3). Calcium transients in digitonin-permeabilised DAergic neurons were elicited using flash photolysis of caged calcium (described in Gandhi et al. (2009)). Mitochondrial calcium levels were quantified over time using Xrhod-1 fluorescence. Pre-incubation with either control antibodies (anti-COXIV, antiNDUFS-4) or anti-NCX1 at 1 μg/mL did not perturb normal calcium uptake and extrusion into mitochondria; see Fig. 20.3. However, pre-incubation with anti-NCX2 or anti-NCX3 antibodies severely impaired NCX_{mito} activity. We found that NCX2 and NCX3, but not NCX1, significantly contribute to mitochondrial calcium efflux following stimulation and therefore play a crucial role in calcium homeostasis in neurons.

20.3 Conclusions

The role of calcium in the aetiology of Parkinson's disease has recently come to light following the discovery that PINK1 loss-of-function causes dysregulated calcium homeostasis in DAergic neurons (Abramov et al. 2011; Gandhi et al. 2009). Specifically, we have found that PINK1 is crucial for the correct functioning of the mitochondrial Na^+/Ca^{2+} exchanger (NCX_{mito}) which extrudes calcium back to the cytoplasm following intracellular transients. However, in order to determine the exact mode of regulation by PINK1, it is necessary to know the precise molecular identity of NCX_{mito}. Up until recently, this was unknown and a source of controversy in the field (Drago et al. 2011). We and others postulated that in the absence of any known mitochondrial isoforms of NCX, perhaps the plasmalemmal forms of the exchanger were performing a dual role (Gobbi et al. 2007). Our study is the first to report expression of NCX2 and NCX3 in human neuronal mitochondria, and the first to demonstrate that NCX2 and 3 contribute to NCX_{mito} activity in DAergic neurons. Other groups have reported mitochondrial localisation of an additional calcium antiporter – NCLX – previously thought to localise only to the plasma membrane (Palty et al. 2010). We did not investigate expression of this isoform in our models, so we cannot rule out the possibility of a contribution of NCLX to NCX_{mito} activity. Indeed, it is likely that NCX_{mito} activity is not dependent solely on one gene product. We found that blocking activity of either NCX2 or NCX3 virtually abolished NCX_{mito} activity. These non-additive effects are puzzling. One explanation may be that all NCX isoforms expressed in mitochondria somehow act co-operatively, such that blocking activity or expression of one gene product interferes with the activity of the others. This could be due to their functioning as a heteromeric complex. NCLX is detected as a 50–70 kDa species and an SDS-resistant 100 kDa species (Palty et al. 2010), and it is well documented that NCX proteins form dimers. It would be intriguing to perform co-immunoprecipitation experiments with NCX subunits using mitochondrial lysates to test for heteromeric interactions.

Further questions remain unanswered: How do NCX proteins become targeted to mitochondria as none contain a recognisable mitochondrial localisation sequence? And how does PINK1 regulate NCX_{mito} activity? PINK1 is a serine-threonine kinase that localises to mitochondrial inner membranes (Gandhi et al. 2006) and several putative substrates have been proposed which also localise to mitochondria, including parkin and TRAP1 (Pilsl and Winklhofer 2012). It would be interesting to discover how PINK1 interacts with NCX_{mito} to protect DAergic neurons from degeneration in order for us to better understand the pathobiology of Parkinson's disease and bring us one step closer to real therapeutic intervention for patients.

References

A.Y. Abramov, M. Gegg, A. Grunewald, N.W. Wood, C. Klein, A.H. Schapira, Bioenergetic consequences of PINK1 mutations in Parkinson disease. PLoS One **6**, e25622 (2011)

J.M. Baughman, F. Perocchi, H.S. Girgis, M. Plovanich, C.A. Belcher-Timme, Y. Sancak, X.R. Bao, L. Strittmatter, O. Goldberger, R.L. Bogorad, V. Koteliansky, V.K. Mootha, Integrative genomics identifies MCU as an essential component of the mitochondrial calcium uniporter. Nature **476**, 341–345 (2011)

T. Cali, D. Ottolini, M. Brini, Mitochondria, calcium, and endoplasmic reticulum stress in Parkinson's disease. Biofactors **37**, 228–240 (2011)

P. Castaldo, M. Cataldi, S. Magi, V. Lariccia, S. Arcangeli, S. Amoroso, Role of the mitochondrial sodium/calcium exchanger in neuronal physiology and in the pathogenesis of neurological diseases. Prog. Neurobiol. **87**, 58–79 (2009)

A. Czyz, L. Kiedrowski, Inhibition of plasmalemmal Na^+/Ca^{2+} exchange by mitochondrial Na^+/Ca^{2+} exchange inhibitor 7-chloro-5-(2-chlorophenyl)-1,5-dihydro-4,1-benzothiazepin-2(3 H)-one (CGP-37157) in cerebellar granule cells. Biochem. Pharmacol. **66**, 2409–2411 (2003)

E. Deas, H. Plun-Favreau, S. Gandhi, H. Desmond, S. Kjaer, S.H. Loh, A.E. Renton, R.J. Harvey, A.J. Whitworth, L.M. Martins, A.Y. Abramov, N.W. Wood, PINK1 cleavage at position A103 by the mitochondrial protease PARL. Hum. Mol. Genet. **20**, 867–879 (2011)

R. Donato, E.A. Miljan, S.J. Hines, S. Aouabdi, K. Pollock, S. Patel, F.A. Edwards, J.D. Sinden, Differential development of neuronal physiological responsiveness in two human neural stem cell lines. BMC Neurosci. **8**, 36 (2007)

I. Drago, P. Pizzo, T. Pozzan, After half a century mitochondrial calcium in- and efflux machineries reveal themselves. EMBO J. **30**, 4119–4125 (2011)

S. Gandhi, M.M. Muqit, L. Stanyer, D.G. Healy, P.M. Abou-Sleiman, I. Hargreaves, S. Heales, M. Ganguly, L. Parsons, A.J. Lees, D.S. Latchman, J.L. Holton, N.W. Wood, T. Revesz, PINK1 protein in normal human brain and Parkinson's disease. Brain **129**, 1720–1731 (2006)

S. Gandhi, A. Wood-Kaczmar, Z. Yao, H. Plun-Favreau, E. Deas, K. Klupsch, J. Downward, D.S. Latchman, S.J. Tabrizi, N.W. Wood, M.R. Duchen, A.Y. Abramov, PINK1-associated Parkinson's disease is caused by neuronal vulnerability to calcium-induced cell death. Mol. Cell **33**, 627–638 (2009)

P. Gobbi, P. Castaldo, A. Minelli, S. Salucci, S. Magi, E. Corcione, S. Amoroso, Mitochondrial localization of Na^+/Ca^{2+} exchangers NCX1-3 in neurons and astrocytes of adult rat brain in situ. Pharmacol. Res. **56**, 556–565 (2007)

W. Li, Z. Shariat-Madar, M. Powers, X. Sun, R.D. Lane, K.D. Garlid, Reconstitution, identification, purification, and immunological characterization of the 110-kDa Na^+/Ca^{2+} antiporter from beef heart mitochondria. J. Biol. Chem. **267**, 17983–17989 (1992)

B. Linck, Z. Qiu, Z. He, Q. Tong, D.W. Hilgemann, K.D. Philipson, Functional comparison of the three isoforms of the Na^+/Ca^{2+} exchanger (NCX1, NCX2, NCX3). Am. J. Physiol. **274**, C415–C423 (1998)

J. Lytton, Na^+/Ca^{2+} exchangers: three mammalian gene families control Ca^{2+} transport. Biochem. J. **406**, 365–382 (2007)

R. Palty, W.F. Silverman, M. Hershfinkel, T. Caporale, S.L. Sensi, J. Parnis, C. Nolte, D. Fishman, V. Shoshan-Barmatz, S. Herrmann, D. Khananshvili, I. Sekler, NCLX is an essential component of mitochondrial Na^+/Ca^{2+} exchange. Proc. Natl. Acad. Sci. U. S. A. **107**, 436–441 (2010)

A. Pilsl, K.F. Winklhofer, Parkin, PINK1 and mitochondrial integrity: emerging concepts of mitochondrial dysfunction in Parkinson's disease. Acta Neuropathol. **123**, 173–188 (2012)

K. Szydlowska, M. Tymianski, Calcium, ischemia and excitotoxicity. Cell Calcium **47**, 122–129 (2010)

T. Thurneysen, D.A. Nicoll, K.D. Philipson, H. Porzig, Immunohistochemical detection of the sodium-calcium exchanger in rat hippocampus cultures using subtype-specific antibodies. Ann. N. Y. Acad. Sci. **976**, 367–375 (2002)

E.M. Valente, P.M. Abou-Sleiman, V. Caputo, M.M. Muqit, K. Harvey, S. Gispert, Z. Ali, D. Del Turco, A.R. Bentivoglio, D.G. Healy, A. Albanese, R. Nussbaum, R. Gonzalez-Maldonado, T. Deller, S. Salvi, P. Cortelli, W.P. Gilks, D.S. Latchman, R.J. Harvey, B. Dallapiccola, G. Auburger, N.W. Wood, Hereditary early-onset Parkinson's disease caused by mutations in PINK1. Science **304**, 1158–1160 (2004)

A. Wood-Kaczmar, S. Gandhi, Z. Yao, A.Y. Abramov, E.A. Miljan, G. Keen, L. Stanyer, I. Hargreaves, K. Klupsch, E. Deas, J. Downward, L. Mansfield, P. Jat, J. Taylor, S. Heales, M.R. Duchen, D. Latchman, S.J. Tabrizi, N.W. Wood, PINK1 is necessary for long term survival and mitochondrial function in human dopaminergic neurons. PLoS One **3**, e2455 (2008)

The Contribution of the Sodium-Calcium Exchanger (NCX) and Plasma Membrane Ca^{2+} ATPase (PMCA) to Cerebellar Synapse Function

21

Chris J. Roome and Ruth M. Empson

Abstract

The cerebellum, a part of the brain critically involved in motor learning and sensory adaptation, expresses high levels of the sodium-calcium exchanger (NCX) and the plasma membrane calcium ATPase (PMCA). Both these transporters control calcium dynamics at a variety of synapses, and here, we draw upon the available literature to discuss how NCX and PMCA work together to shape pre-synaptic calcium dynamics at cerebellar synapses.

Keywords

Cerebellum • Parallel fibre • PMCA • NCX • Synapse

21.1 Introduction

In this chapter, we look at how the sodium-calcium exchanger (NCX) acts to control synaptic cerebellar calcium (Ca^{2+}) dynamics alongside the plasma membrane calcium ATPase (PMCA), another important calcium transporter.

We will first explain the different cerebellar synapse connections and then explain why Ca^{2+} is critical at the pre-synaptic side of the functionally important parallel fibre to Purkinje neuron synapse. We will then go on to compare the different contributions of PMCA and NCX to the function of this synapse, drawing upon the available literature and the known similarities and differences between the expression levels and functional dynamics of PMCA and NCX. This will lead to some testable hypotheses as to how these transporters work together to control Ca^{2+} in small compartments such as pre-synaptic terminals and a discussion of how new emerging tools can advance the field.

21.2 What Is the Cerebellum?

21.2.1 The Role of the Cerebellum

The cerebellum is a fascinating region of the brain located in the inferior posterior portion of the head, whose primary function is the integration of motor learning and sensory perception.

C.J. Roome • R.M. Empson (✉)
Department of Physiology, Brain Health Research Centre, University of Otago, PO Box 913 Dunedin, 9054, New Zealand
e-mail: ruth.empson@otago.ac.nz

L. Annunziato (ed.), *Sodium Calcium Exchange: A Growing Spectrum of Pathophysiological Implications*, Advances in Experimental Medicine and Biology 961, DOI 10.1007/978-1-4614-4756-6_21,

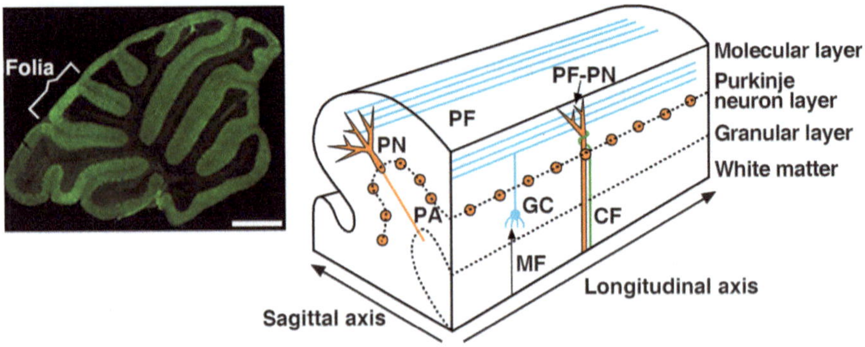

Fig. 21.1 *Left* is a sagittal slice of mouse cerebellum labelled with fluorescent parvalbumin antibodies, to illustrate repeated 'folia' structures (scale bar is ~1 mm). *Right* shows a representation of a cerebellar folia section showing major cellular architecture in sagittal and longitudinal planes. *PN* Purkinje neurons and *PA* Purkinje neuron axons are each labelled in the sagittal plane. *GC* granule cell, *MF* mossy fibres, *CF* climbing fibre, *PF* parallel fibres, and *PF-PN* parallel fibre-Purkinje neuron synapses are each labelled in the longitudinal plane

Well conserved throughout all vertebrate species and often referred to as the 'neuronal machine', the cerebellum is notorious for its beautifully crystalline architecture and machine-like neuronal circuitry (Eccles 1967; Ito 2006).

The cerebellum is built upon repeated modules or 'folia' containing identical cell types, each connected in a similar manner and organised in a distinctly planar fashion (Fig. 21.1). Folia are primarily composed of three cortical layers: the outermost molecular layer, the Purkinje neuron layer and the innermost granular layer (Andersen et al. 1964). The sensorimotor information that bypasses the cerebellum to the deep cerebellar nuclei is carried via mossy fibres and climbing fibres and relayed to various pre-motor areas. These fibres link a feedback-type side loop via the cerebellar Purkinje neurons. Purkinje neurons are then connected through inhibitory synapses back to the deep cerebellar nuclei and vestibular nuclei in the brainstem, thus completing a cerebellar inhibitory feedback loop (Jorntell and Hansel 2006). The synaptic strength of almost every synapse in the cerebellar circuitry has been shown to undergo synaptic plasticity. This plasticity is thought to regulate the extent to which the mossy and climbing fibres activate the deep cerebellar nuclei and in doing so contributes to the continuous adjustment and fine-tuning that is critical for controlling motor learning in the brain (Matsukawa et al. 2003; Boyden et al. 2004; Ito 2006; Qiu and Knopfel 2009; Bao et al. 2010).

21.2.2 Components of the Cerebellar Circuitry

The function of the cerebellum as a 'neuronal machine' relies upon clockwork precision in timing and activity between individual neuronal components within the cerebellar microcircuitry (Eccles 1967; Ito 2006). Most of the synapses within this microcircuitry are highly plastic and believed to be instrumental in cerebellar function as a whole (Boyden et al. 2004). Constituting the cerebellar microcircuitry, Purkinje neurons receive two primary excitatory inputs and two primary inhibitory inputs. Purkinje neuron excitation originates from granule cell parallel fibre (PF) inputs, and the inferior olive climbing fibre input, each performing very different roles in Purkinje neuron excitation (Eccles et al. 1966a, b; Llinas and Sugimori 1980). The inhibitory inputs are received from basket and stellate neurons located in the molecular layer that also receive input from parallel fibres and so create the important feedforward inhibition onto Purkinje neurons as a vital counterbalance to their two excitatory pathways (Bao et al. 2010).

Mossy fibres extend from pre-cerebellar nuclei into the granular layer that is composed of numerous tiny granule cells (Andersen et al. 1964) and also the larger Golgi cells (Eccles et al. 1964; Ito 2006). Here, excitatory synapses are formed with granule cells and deep cerebellar nuclei. Granule cells send characteristic T-shaped parallel fibres, each 0.2–0.3 μm in diameter, that run for ~2–4 mm in both directions along the longitudinal axis of the folia (Andersen et al. 1964) up into the superficial molecular layer forming hundreds of thousands of glutamatergic synapses on Purkinje neuron dendrites (PF-PN synapses). The Purkinje neuron dendritic arbours are planar and lie orientated with each neighbouring Purkinje neuron arbour at a right angle to the longitudinal axis (and PFs) within the folium.

Stellate and basket inhibitory interneurons are located in the molecular layer and form strong GABAergic synapses onto the Purkinje neuron (Andersen et al. 1964; Ito 2006; Bao et al. 2010). Basket cells are located closer to the Purkinje neuron soma and form a large complex synaptic structure called a ‘pinceau’. The basket cell to Purkinje neuron synapse exerts a sevenfold stronger inhibitory effect than stellate cell to Purkinje neuron synapse which is located on the Purkinje neuron distal dendrites (Sakaba 2008; Bao et al. 2010).

21.2.3 A Key Synapse: The Parallel Fibre to Purkinje Neuron Synapse

21.2.3.1 Properties

Approximately 100,000–200,000 parallel fibres run orthogonally through the Purkinje neuron (PN) dendritic plane (in the rat cerebellum), and each PF creates one or two synapses as they transverse the PN dendrites (Palay and Chan-Palay 1974; Barbour 1993). It is estimated that only fifty PF-PN synapses need to be simultaneously activated to maintain the simple spike activity of the Purkinje neuron (which fire at an intrinsic background rate of ~50 Hz). Given that each Purkinje neuron receives so many PF-PN inputs, individual granule cells are expected to fire at a normally very low rate (<0.5 Hz) in between high-frequency (250 Hz) ‘bursts’ of activity (Eccles 1967; Barbour 1993; Chadderton et al. 2004; Ito 2006; De Schutter and Steuber 2009).

The PF-PN synapse is considered a low probability synapse whereby only one or two quanta of glutamate neurotransmitter is released upon stimulation (Barbour 1993). Each granule cell firing event creates an excitatory post-synaptic current (EPSC) of ~2–60 pA at the Purkinje neuron (Barbour 1993) by activating glutamatergic 2-amino-3-(5-methyl-3-oxo-1,2-oxazol-4-yl) propanoic acid (AMPA)-type post-synaptic receptors (Konnerth et al. 1990; Barbour 1993). In stark contrast, each Purkinje neuron receives only one input via a climbing fibre extending from the inferior olive nuclei. The climbing fibre winds around the PN dendrite creating a large number (several hundred) of synaptic contacts, which collectively are capable of producing a massive and powerful input of complex Ca^{2+} spike activity within the PN soma and dendrites (by a single incoming action potential (AP)), that temporarily suppresses simple spike activity induced by the PF inputs (Eccles et al. 1966a; Palay and Chan-Palay 1974; Miyakawa et al. 1992; De Schutter and Bower 1994; Schmidt et al. 2003; Ito 2006).

This weakening of PF-PN synaptic strength by climbing fibre activity is thought to play a key role during cerebellar motor learning (Boyden et al. 2004; Jorntell and Hansel 2006). Cerebellar motor learning is a process of improving the smoothness and accuracy of movements, and the Marr-Albus model proposes that modulation of PF-PN synaptic strength via concomitant climbing fibre input is pivotal for this process (Marr 1969; Ito 2006). Accordingly, it is not surprising that mouse models in which PF-PN synaptic plasticity is dysfunctional suffer from severe forms of cerebral ataxia and impairment of motor coordination (Matsukawa et al. 2003; Empson et al. 2007).

21.2.3.2 Importance of Pre-synaptic Calcium for Pre-synaptic Function

Early studies that sought to characterise the mechanism underlying synaptic transmitter release used the frog neuromuscular junction and squid giant synapse axon terminal to identify Ca^{2+} as essential for initiating exocytosis (Katz and Miledi 1965). These studies inspired a hypothesis that proposed the amount of transmitter release (and strength of the post-synaptic response) should be related to Ca^{2+} concentration upon influx, which was later confirmed and described by a non-linear relationship between Ca^{2+} concentration and the synaptic end plate potential (Dodge and Rahamimoff 1967a, b). These studies also indicated that multiple 'Ca^{2+} sensor' sites need to be accommodated as part of the mechanism to trigger neurotransmitter release. It is now known that during a precisely tuned rise and fall of pre-synaptic Ca^{2+}, called a 'Ca^{2+} transient', intracellular Ca^{2+} ions bind to multiple pre-synaptic 'Ca^{2+} sensors' within the pre-synaptic terminal which consequently trigger synaptic vesicle fusion and initiate neurotransmitter release (Burgoyne and Weiss 2001; Atwood and Karunanithi 2002; Stevens and Sullivan 2003; Burgoyne 2007; Neher and Sakaba 2008).

More recent studies have utilised the large calyx of Held synapse by combining fluorimetric Ca^{2+} measurement and caged-Ca^{2+} stimulation with pre- and post-synaptic voltage clamp electrophysiology to further probe the critical relationship between the pre-synaptic Ca^{2+} transient and neurotransmitter release (Schneggenburger and Neher 2000). It was found that a large elevation of pre-synaptic Ca^{2+} concentration of ~10 μM was required to evoke transmitter release, equivalent to that evoked by a single AP. It is now well established that a low-affinity transmitter release 'Ca^{2+} sensor' is common across many central nervous system (CNS) synapse types (Augustine 2001; Neher and Sakaba 2008).

Further characterisation of the pre-synaptic Ca^{2+} transient found that a single AP evoked Ca^{2+} transient often exceeds concentrations of ~50 μM or more within microseconds and then recovers back to resting levels (~50 nM) within a few hundred milliseconds (Helmchen et al. 1997; Koester and Sakmann 2000; Augustine 2001; Brenowitz and Regehr 2007). A recent study has also shown that to produce a typical post-synaptic current waveform, the pre-synaptic Ca^{2+} transient must be very rapid indeed (with a half-width <0.5 ms) (Bollmann and Sakmann 2005), and it was well known that the delay between the Ca^{2+} transient and post-synaptic current is also very rapid (<100 μs) (Llinas et al. 1981; Sabatini and Regehr 1996). Thus, all lines of evidence indicate that the processes governing transmitter release function over very short (microsecond) timescales, during which time diffusion of Ca^{2+} ions within the local environment of the pre-synaptic membrane and subsequent interaction with numerous synaptic vesicle associated proteins in close vicinity are expected to be the determining factors of transmitter release (Sudhof 2006; Neher and Sakaba 2008).

Accordingly, the predominant mechanisms that control initiation of synchronous transmitter release are believed to reside in highly localised sub-domains in very close vicinity to Ca^{2+} channels and synaptic vesicle release machinery, in so-called micro- or nanodomains of the pre-synaptic terminal (Llinas et al. 1992; Stanley 1997; Augustine 2001). Since these domains lie well below the diffraction limit of light microscopy (micrometre–nanometre), fluorimetric Ca^{2+} imaging techniques are limited to recording only bulk Ca^{2+} levels within the pre-synaptic terminal and are unable to probe beyond the dynamics of these global Ca^{2+} signals (Neher and Sakaba 2008). This is an important consideration since localised pre-synaptic Ca^{2+} signals within micro- or nanodomains of the release machinery are likely to have critical impact on synaptic transmitter release that differs from that of the global pre-synaptic Ca^{2+} signal (Myoga and Regehr 2011).

Global pre-synaptic Ca^{2+} signals measured experimentally are limited by diffusion of Ca^{2+} within the terminal and by the Ca^{2+} indicator binding kinetics. Consequently, fluorimetric Ca^{2+} imaging techniques currently being used provide only a rough indicator of Ca^{2+} signals at the pre-synaptic release site, which are at very least expected to have faster dynamics and attain higher peak Ca^{2+} concentrations than those measured experimentally. As yet only biophysical modelling studies have been able to explore how local Ca^{2+} diffusion and the properties of pre-synaptic nanodomains might uniquely influence transmitter release (Neher 1998).

21.2.3.3 Importance of Pre-Synaptic Calcium at the PF-PN Synapse

At synapses with low initial release probability of neurotransmitter, such as the parallel fibre to Purkinje neuron synapses in the cerebellum (Dittman et al. 2000), a form of short-term synaptic enhancement called 'facilitation' occurs over timescales of milliseconds, which, by acting as a high-pass filter, allows these synapses to maintain reliable transmission during high-frequency input activity (Eccles et al. 1966b; Dittman et al. 2000).

At PF-PN synapses, facilitation can be measured experimentally using paired pulse stimulation, called 'paired pulse facilitation' (PPF) (Atluri and Regehr 1996). During PPF, the ratio of paired excitatory post-synaptic potentials (EPSC) amplitudes measured from a voltage-clamped Purkinje neuron during paired stimulation of PFs, called the paired pulse ratio (PPR), is calculated to measure the extent of facilitation. Recovery of facilitation (i.e. the time taken for the synapse to return to a non-facilitated state) can be measured by varying the interstimulus interval between the stimulating pulses (ISI). Recovery of facilitation is intimately related to the recovery of pre-synaptic residual Ca^{2+} (i.e. the time taken for intracellular Ca^{2+} to return to basal levels) which follows a similar time course (Atluri and Regehr 1996). Accordingly, this technique also provides an indirect measure for the recovery dynamics of pre-synaptic residual Ca^{2+} (Parnas and Segel 1980, 1989; Empson et al. 2007).

Recovery of pre-synaptic residual Ca^{2+} is dependent upon the cellular mechanisms that control Ca^{2+} within the pre-synaptic terminal (or interact with free Ca^{2+} within the terminal) following synaptic transmission. It is expected that recovery of facilitation is indirectly influenced by such 'Ca^{2+} signalling' cellular mechanisms through their control over residual Ca^{2+} (Parnas and Segel 1989). This supposition has been confirmed by studies that identify modification in the recovery of facilitation following an alteration of the recovery of residual Ca^{2+} as determined by Ca^{2+} signalling mechanisms (Atluri and Regehr 1996; Matsukawa et al. 2003; Empson et al. 2007). Thus, this PPF experimental approach has successfully been used to provide a meaningful functional readout of how the properties and behaviour of the PF-PN synapse are influenced when the dynamics of pre-synaptic (residual) Ca^{2+} are altered.

21.3 The Sodium-Calcium Exchanger (NCX)

21.3.1 Structure and Function

Sodium/calcium exchangers (NCXs) are expressed in the plasma membrane of virtually all mammalian cells and function by utilising sodium ion (Na^+) concentration gradient to transport Ca^{2+} out of the cell. In most cells, three Na^+ ions are exchanged for one Ca^{2+} ion by what is referred to as 'forward mode' NCX activity. Under certain physiological conditions, it is possible for NCX to transport Ca^{2+} into cells (and Na^+ out), via so-called reverse mode NCX activity that is largely dependent on the Na^+ and Ca^{2+} electrochemical gradients and the membrane potential (DiPolo and Beauge 1983; Blaustein et al. 2002; Doi et al. 2002). In select neurons and certain other cells, potassium ions (K^+) are also transported in the same direction as calcium, with a coupling ratio of four Na^+ to one Ca^{2+} plus one K^+, by the potassium-dependent NCX (NCKX) (Blaustein and Lederer 1999; Sheng et al. 2000; Lee et al. 2002).

A schematic of the NCX protein is shown in Fig. 21.2. An amino terminus is located in extracellular space, following an initial leader peptide (0). Five transmembrane segments (1–5) precede a large intracellular loop region, followed by an additional four transmembrane segments (6–9) at the COOH-terminal. Although the intracellular loop is thought not to play a direct role in Ca^{2+} transportation, it is involved in the modulation of NCX function predominantly via Ca^{2+} interaction with two Ca^{2+}-binding domains located in the middle of the intracellular loop (Hilge et al. 2009; Ottolia et al. 2009; Wu et al. 2009). The NCX activity is also modulated by intracellular kinases,

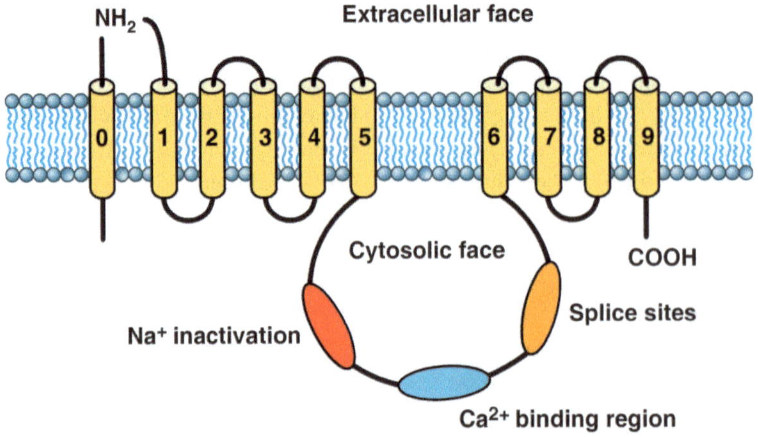

Fig. 21.2 A schematic of the sodium/calcium exchanger protein (NCX). A NH_2 terminus is located in extracellular space, following an initial leader peptide (0). Five transmembrane segments (1–5) prelude a large intracellular loop region containing modulating interaction sites, including calcium binding, sodium inactivation and alternative splice site domains, followed by four additional transmembrane segments (6–9) and a cytosolic COOH-terminal (Figure modified from Blaustein and Lederer (1999))

Na^+ and H^+ intracellular ions (Blaustein and Lederer 1999) that interact with regions of the intracellular loop and alternatively spliced NCX1 variants (of which at least 32 distinctive isoforms have been identified) arise from alterations in a limited region of the intracellular loop (Blaustein and Lederer 1999; Wu et al. 2009).

Early studies proposing the existence of a plasma membrane sodium-calcium exchanger identified a coupling of Na^+ and Ca^{2+} currents across a squid axon membrane by altering Na^+ and Ca^{2+} concentration in the intracellular and extracellular medium (Baker et al. 1969; Blaustein and Hodgkin 1969) whereby replacement of external Na^+ with lithium, choline or dextrose reduced Ca^{2+} efflux from the axon. It was later confirmed that the NCX utilises the Na^+ gradient by allowing Na^+ to be transported into the cell in exchange for Ca^{2+} transport out of the cell (Blaustein 1974; DiPolo and Beauge 1983). As such, exchanger activity is predominantly dependent on both Ca^{2+} and Na^+ concentration gradients; furthermore, it is electrogenic (one positive ion is transported into the cell per cycle) and is therefore dependent on the membrane potential (Blaustein and Lederer 1999). This dependence on the electrochemical gradients of both Ca^{2+} and Na^+ has allowed for an effective method of experimentally manipulating the activity of the NCX, by changing extracellular Na^+ concentrations. This method has been used extensively to study exchanger kinetics and determine its importance for controlling Ca^{2+} dynamics within axons and synaptic terminals (DiPolo and Beauge 1983; Fontana et al. 1995; Reuter and Porzig 1995; Fierro et al. 1998).

21.3.2 Isoforms

Nine members of the sodium-calcium exchanger family have been identified in mammals: three isoforms in the NCX family (NCX1, NCX2 and NCX3) and six in the potassium-dependent NCKX family (NCKX1–NCKX6) (Linck et al. 1998; Blaustein and Lederer 1999; Canitano et al. 2002; Thurneysen et al. 2002; Altimimi and Schnetkamp 2007). Although the primary NCX protein appears ubiquitously expressed across many tissue types including heart, nerve and kidney, nonradioactive in situ hybridisation and immunohistochemical analysis of NCX/NCKX isoforms and alternatively spliced variants of NCX1 and NCX3 (but not NCX2) have revealed highly specific expression patterns for each variant (Porzig et al. 1993; Thurneysen et al. 2002; Papa et al. 2003).

These expression patterns indicate that variation in structural and functional properties of each NCX member is important for cell-specific Ca^{2+} signalling. Expression levels of NCX in neurons are very high, especially at synapses where transmembrane Ca^{2+} flux is high, suggesting an important role of the exchanger at these junctions (Baker et al. 1969; Juhaszova et al. 2000; Blaustein et al. 2002). NCX1 is the most dominant exchanger transcript in the brain and is strongly expressed in the cerebellum (Papa et al. 2003) and other brain regions. NCX2 transcripts are not so easily recognised in cerebellum, but NCX3 mRNA and protein have been detected in the cerebellar molecular layer and cerebellar granule cells (Papa et al. 2003; Sokolow et al. 2004).

21.3.3 Functional Expression at Synapses

Studies at the model calyx of Held synapse revealed a clear role for NCX during Ca^{2+} extrusion from the pre-synaptic terminal (Kim et al. 2005) such that inhibition of NCX in its forward mode led to a prolonged pre-synaptic Ca^{2+} transient. At the post-synaptic site in the hippocampus, NCX controls the temporal and spatial spread of Ca^{2+} from the restricted volume of the dendritic spines to the dendrite (Lörincz et al. 2007).

NCX2 and 3 and NCKX2 'knockout' mice models have also been used to examine NCX/NCKX function, specifically for synaptic Ca^{2+} control (Jeon et al. 2003; Sokolow et al. 2004; Li et al. 2006). Electrophysiological studies on NCX2 'knockout' mice revealed that NCX2 is an important regulator of pre-synaptic Ca^{2+} control and is essential for regulating synaptic plasticity at hippocampal pyramidal neurons in vitro (Jeon et al. 2003). In vivo studies on the NCX3 'knockout' mouse indicate a similarly important role for NCX3 at the neuromuscular junction, which was demonstrated through a delay in pre-synaptic Ca^{2+} recovery and subsequent physiological dysfunction of the NCX2(–/–) mouse neuromuscular junction (Sokolow et al. 2004). The NCKX2 knockout mouse also exhibits altered synaptic plasticity at hippocampal Schaffer/CA1 synapses and concurrent deficits in motor learning and spatial working memory (Li et al. 2006).

21.3.4 Functional Expression of NCX in the Cerebellum

Despite the clear evidence for expression of NCX in the cerebellum, there is as yet little available functional evidence for a contribution by NCX to cerebellar synapse Ca^{2+} dynamics, although post-synaptic metabotropic glutamate receptor-mediated Ca^{2+} transients on Purkinje neurons are reduced by pharmacological NCX inhibitors bepridil and KB-R7943 and also by siRNA-mediated knockdown of NCX (Kim et al. 2007). At the Purkinje neuron soma, NCX removes Ca^{2+} in response to a depolarisation and contributes approximately 40 % of the Ca^{2+} recovery (Fierro et al. 1998). Furthermore, it is tempting to speculate that the non-inactivating Na^+ conductance in Purkinje neuron dendrites (Llinas and Sugimori 1980) could be driven by forward mode NCX.

21.4 The Plasma Membrane Calcium ATPase PMCA

21.4.1 PCMA Structure and Function

The PMCAs belong to a family of P-type primary ion transport ATPases characterised by the formation of an aspartyl phosphate intermediate during an ATP hydrolysis reaction cycle. PMCAs consist of 10 transmembrane (TM) spanning segments with amino (NH_2) and carboxyl (COOH) cytosolic termini. There are three major parts to the protein, all of which are contained on the cytosolic side of the membrane: an intracellular loop between TM segments 2 and 3, a large unit between membrane-spanning domains 4 and 5 and the extended C-terminal containing "tail" following the last transmembrane domain (Strehler and Zacharias 2001).

An extended carboxyl (COOH)-terminal tail creates a major regulatory domain of the PMCA; Ca^{2+}-bound calmodulin proteins (Ca^{2+}-CaM) bind to a region in the COOH-terminal portion located

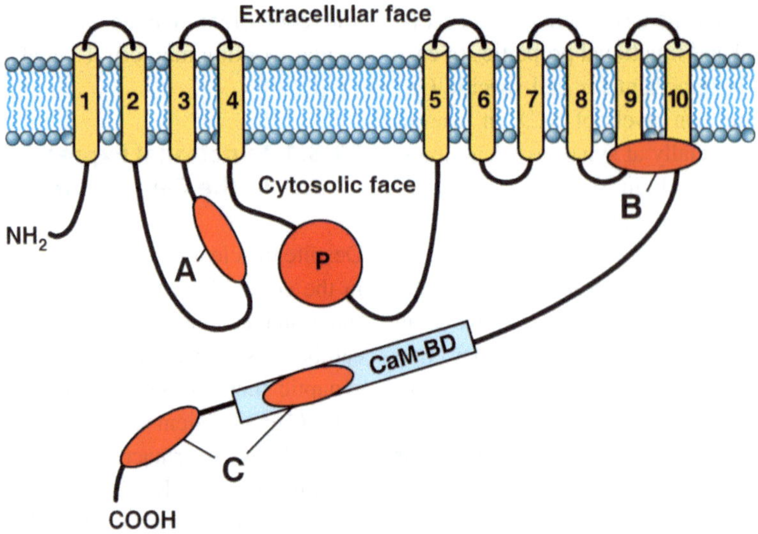

Fig. 21.3 A schematic of the plasma membrane calcium ATPase protein (PMCA). The 10 transmembrane segments have been numbered beginning at the amino (NH_2) terminus (1) and with the extended carboxyl (COOH)-terminal 'tail' (10). Here, the PMCA is in an auto-inhibited state, with Ca^{2+}-CaM-binding domain (CaM-BD) interacting with the intracellular loops. P is the obligatory aspartyl phosphate formed during pump function. Sites where alternative RNA splicing may occur are labelled *A*, *B* and *C* (Figure modified from Strehler and Zacharias (2001))

~40 residues downstream of the last transmembrane domain to activate catalytic turnover (Caride et al. 2001; Strehler and Zacharias 2001). In the absence of Ca^{2+}-CaM, catalytic turnover is prevented, while the pump is kept in an 'autoinhibitory' inactive state. Elevations in cytosolic Ca^{2+} result in an increase in Ca^{2+}-CaM that binds with high affinity to PMCA autoinhibitory domains, thereby releasing inhibition and stimulating pump activity (Strehler et al. 2007).

21.4.2 PMCA Isoforms

Mammalian PMCAs are encoded by a gene family composed of at least four members (PMCA1, 2, 3 and 4). Further PMCA isoform variants are generated via alternative RNA splicing of the primary gene transcripts. Sites where alternative RNA splicing may occur in the protein-coding sequence of the PMCA transcripts are shown in Fig. 21.3 and are labelled A, B and C.

Studies that revealed a tissue- and developmental-specific expression pattern of several PMCA isoforms contradicted the initial assumption that PMCAs function primarily as ubiquitous Ca^{2+} homeostasis regulators and suggested that individual isoforms may play unique Ca^{2+} signalling roles (Burette et al. 2003; Strehler et al. 2007). While the brain expresses a great abundance and diversity of PMCA isoforms and splice variants, the PMCA2 and 3 isoforms in particular are expressed in a very restricted manner (Zacharias and Kappen 1999; Burette et al. 2003). Furthermore, studies of the kinetics of activation of PMCA2 and 3 isoforms reveal significant differences in their Ca^{2+} extrusion characteristics (Caride et al. 2001). For example, the isoform variants, PMCA3f and PMCA2a, respond more rapidly to Ca^{2+} elevations and are referred to as the 'fast' PMCA isoforms (Caride et al. 2001; Jensen et al. 2007). The different time course of inactivation also results in striking differences in the 'memory' of certain PMCA isoforms following recent activation. Thus, in frequently excited cells, PMCA2b will remain 'pre-activated' and will respond almost immediately to successive Ca^{2+} signals (Caride et al. 2001; Brini et al. 2003).

21.4.3 Functional Expression at Synapses

There are several pieces of strong evidence to support a contribution from PMCAs to Ca^{2+} extrusion during synaptic activity. Some of this evidence comes from cerebellar synapses (see below Sect. 21.4.4), but other evidence comes from hippocampal and calyx of Held synapses (Kim et al. 2005). The fast isoform of PMCA2, PMCA2a, is critical for the control of glutamate release from hippocampal CA3 terminals (Jensen et al. 2007), and the same isoform uniquely changes its expression levels in a model of homeostatic plasticity where glutamate release is also altered (Jensen et al. 2009). Other studies in the hippocampus have identified how PMCA controls Ca^{2+} in dendritic spines of hippocampal CA2 pyramidal neurons to regulate long-term synaptic plasticity (Simons et al. 2009).

21.4.4 Functional Expression in the Cerebellum

The PMCA2 isoform is highly expressed within the cerebellum (Filoteo et al. 1997), particularly in cerebellar Purkinje neurons (Hillman et al. 1996) and its synaptic connections where it is critical for pre-synaptic Ca^{2+} control and short-term synaptic plasticity at PF-PN synapses (Empson et al. 2007). This finding is further supported by behavioural studies using genetically engineered PMCA2 'knockout' mice that exhibit severe ataxia and motor coordination deficits. Such phenotypes are normally associated with a dysfunction of cerebellar circuitry (Kozel et al. 1998; Empson et al. 2007; Empson et al. 2010). Like NCX, PMCA makes a rather modest contribution to the clearance of calcium from Purkinje neuron soma (Fierro et al. 1998), but in the more restricted space of the Purkinje neuron dendrite, removal of only 50 % of the PMCA2 protein in PMCA2 heterozygous mice leads to a significant slowing of the recovery of dendritic calcium.

21.5 Dynamic Cooperation of NCX and PMCA at Synapses

Given that both PMCA and NCX clear Ca^{2+} from synaptic compartments, we would predict that when expressed in the same compartment they should work together. As described above, the evidence suggests that both mechanisms are present at the parallel fibre to Purkinje neuron terminals but their relative functional contributions are not understood.

21.5.1 Complimentary Calcium Affinity and Extrusion Rate

Under normal physiological conditions, NCX has approximately 10-fold lower affinity for Ca^{2+} than the PMCA ($k_{0.5} > 1$ μM) but has a 10- to 50-fold higher turnover rate than the PMCA (which is typically ~100 s^{-1} for PMCA) (Fontana et al. 1995; Blaustein et al. 2002). Consequently, only a small fraction of the exchangers are expected to be active at normal resting Ca^{2+} concentrations, yet when Ca^{2+} is elevated, such as within pre-synaptic terminals during synaptic transmission, it is expected that the NCX-mediated Ca^{2+} efflux would become increasingly important (DiPolo and Beauge 1983; Blaustein et al. 2002). In contrast, the PMCAs have a high affinity for Ca^{2+} by virtue of their need for activation (or release of autoinhibition; see above Sect. 21.3.1) as a consequence of Ca^{2+}-binding calmodulin. Furthermore, the PMCAs have a far lower capacity for Ca^{2+} transport when compared with the exchanger. This means that PMCAs are responsive to small changes in intracellular Ca^{2+} concentration (half-maximal activation ($k_{0.5}$) ~100 nM) and are quick to bind excess Ca^{2+}, while net transportation rates can be relatively slow compared to that of NCX (Blaustein and Lederer 1999; Strehler et al. 2007).

Based upon this information, we can hypothesise that NCX and PMCA should work together to control the recovery of pre-synaptic Ca^{2+}. In the context of a large rise in pre-synaptic Ca^{2+} (several μM; see Sect. 12.2.3.2 above), NCX will be most suited to contribute very soon after the peak of the Ca^{2+} transient (when the high-affinity PMCA will be saturated), but that as Ca^{2+} levels recover, PMCA, with its higher affinity for Ca^{2+}, takes over to extrude the remaining Ca^{2+}.

21.5.2 Distinct Sub-compartment Locations

However, the above hypothesis does not take into account the many other mechanisms that operate in Ca^{2+} signalling compartments. These include Ca^{2+} buffer proteins and uptake of Ca^{2+} into other compartments such as the endoplasmic reticulum and mitochondria. Furthermore, the relative contributions of NCX and PMCA will be influenced by their location with respect to the point of entry of Ca^{2+} and the speed of diffusion of Ca^{2+}, as well as the shape and size of the compartment. Interestingly, it is compartments that have a restricted cytosolic space and large membrane surface area, such as long, thin pre-synaptic varicosities (such as the parallel fibre terminal Palay and Chan-Palay 1974) and small post-synaptic spine heads that have restricted access to the larger volume of the dendrite, where membrane transport proteins will have a major impact.

It is interesting to speculate that PMCA2 and NCX are uniquely segregated within the pre-synaptic parallel fibre to Purkinje neuron synapse compartment. Recent findings suggest that PMCA2 makes molecular interactions with the pre-synaptic protein syntaxin in the cerebellum as well as with the post-synaptic protein PSD95 (Garside et al. 2009). Both these molecular interactions would place PMCA very close to the site of calcium entry, and indeed, recent electron microscopy (EM) evidence places PMCAs close to the pre-synaptic active zone and post-synaptic density (Burette and Weinberg 2007). Within the literature, there is little evidence of close molecular interactions between NCX and synaptic proteins. Furthermore, in the ciliary ganglion of the chick, immunocytochemistry and light microscopy revealed this segregation. PMCAs were located close to the release sites, while NCX was located further away at non-synaptic sites (Juhaszova et al. 2000). Under these conditions, the diffusional distance will prevent NCX from 'seeing' high peak Ca^{2+} during synaptic transmission; instead, PMCA ensures efficient re-priming of the Ca^{2+} sensor by keeping local Ca^{2+} low. Similar complexities may operate at the parallel fibre to Purkinje neuron synapse.

21.5.3 Unique Properties

Despite their shared responsibility to act to clear Ca^{2+} from synaptic compartments, the PMCA and NCX are very different proteins.

Whilst the PMCA is heavily and directly dependent upon ATP and glycolysis (Ivannikov et al. 2010), the NCX is less so and, instead, indirectly uses the energy within the Na^+ gradient. This means that the two transporters could behave differently depending upon the metabolic conditions.

Perhaps the most important functional distinction between the two proteins is the fact that NCX, unlike PMCA, can switch its behaviour from a Ca^{2+} extrusion protein to a Ca^{2+} entry protein when it switches into reverse mode. This means NCX has a greater scope to influence synapse function, where the timing of Ca^{2+} entry is so critical. Although reverse mode NCX has been associated with pathological arrhythmia (Woodcock et al. 2001) when the cytosolic compartment becomes overloaded with Na^+, it is possible that even brief rises in intracellular Na^+ especially in diffusion-restricted spaces, such as the heart dyadic cleft (Pott et al. 2005; Larbig et al. 2010), or even a synaptic terminal could switch the mode of action of NCX.

21.6 Future Directions

The fascinating world of NCX in the nervous system began with its original discovery in the squid giant axon. Today, there are tools available that mean we can begin to uncover the functional role of this important Ca^{2+} handling molecule and particularly its significance at synapses; for synaptic facilitation, plasticity and in the cerebellum for motor control. To help this, the development of better pharmacological tools must continue and the application of cell-type and even synapse-specific tools to knockout, or overexpress NCX must also progress.

References

H.F. Altimimi, P.P. Schnetkamp, Na^+/Ca^{2+}-K^+ exchangers (NCKX): functional properties and physiological roles. Channels (Austin) **1**, 62–69 (2007)

P. Andersen, J.C. Eccles, P.E. Voorhoeve, Postsynaptic inhibition of cerebellar Purkinje cells. J. Neurophysiol. **27**, 1138–1153 (1964)

P.P. Atluri, W.G. Regehr, Determinants of the time course of facilitation at the granule cell to Purkinje cell synapse. J. Neurosci. **16**, 5661–5671 (1996)

H.L. Atwood, S. Karunanithi, Diversification of synaptic strength: presynaptic elements. Nat. Rev. Neurosci. **3**, 497–516 (2002)

G.J. Augustine, How does calcium trigger neurotransmitter release? Curr. Opin. Neurobiol. **11**, 320–326 (2001)

P.F. Baker, M.P. Blaustein, A.L. Hodgkin, R.A. Steinhardt, The influence of calcium on sodium efflux in squid axons. J. Physiol. **200**, 431–458 (1969)

J. Bao, K. Reim, T. Sakaba, Target-dependent feed forward inhibition mediated by short-term synaptic plasticity in the cerebellum. J. Neurosci. **30**, 8171–8179 (2010)

B. Barbour, Synaptic currents evoked in Purkinje cells by stimulating individual granule cells. Neuron **11**, 759–769 (1993)

M.P. Blaustein, The interrelationship between sodium and calcium fluxes across cell membranes. Rev. Physiol. Biochem. Pharmacol. **70**, 33–82 (1974)

M.P. Blaustein, A.L. Hodgkin, The effect of cyanide on the efflux of calcium from squid axons. J. Physiol. **200**, 497–527 (1969)

M.P. Blaustein, W.J. Lederer, Sodium/calcium exchange: its physiological implications. Physiol. Rev. **79**, 763–854 (1999)

M.P. Blaustein, M. Juhaszova, V.A. Golovina, P.J. Church, E.F. Stanley, Na^+/Ca^{2+} exchanger and PMCA localization in neurons and astrocytes: functional implications. Ann. N. Y. Acad. Sci. **976**, 356–366 (2002)

J.H. Bollmann, B. Sakmann, Control of synaptic strength and timing by the release-site Ca^{2+} signal. Nat. Neurosci. **8**, 426–434 (2005)

E.S. Boyden, A. Katoh, J.L. Raymond, Cerebellum-dependent learning: the role of multiple plasticity mechanisms. Annu. Rev. Neurosci. **27**, 581–609 (2004)

S.D. Brenowitz, W.G. Regehr, Reliability and heterogeneity of calcium signaling at single presynaptic boutons of cerebellar granule cells. J. Neurosci. **27**, 7888–7898 (2007)

M. Brini, L. Coletto, N. Pierobon, N. Kraev, D. Guerini, E. Carafoli, A comparative functional analysis of plasma membrane Ca^{2+} pump isoforms in intact cells. J. Biol. Chem. **278**, 24500–24508 (2003)

A. Burette, R.J. Weinberg, Perisynaptic organization of plasma membrane calcium pumps in cerebellar cortex. J. Comp. Neurol. **500**, 1127–1135 (2007)

A. Burette, J.M. Rockwood, E.E. Strehler, R.J. Weinberg, Isoform-specific distribution of the plasma membrane Ca^{2+} ATPase in the rat brain. J. Comp. Neurol. **467**, 464–476 (2003)

R.D. Burgoyne, Neuronal calcium sensor proteins: generating diversity in neuronal Ca^{2+} signalling. Nat. Rev. Neurosci. **8**, 182–193 (2007)

R.D. Burgoyne, J.L. Weiss, The neuronal calcium sensor family of Ca^{2+}-binding proteins. Biochem. J. **353**, 1–12 (2001)

A. Canitano, M. Papa, F. Boscia, P. Castaldo, S. Sellitti, M. Taglialatela, L. Annunziato, Brain distribution of the Na^+/Ca^{2+} exchanger-encoding genes NCX1, NCX2, and NCX3 and their related proteins in the central nervous system. Ann. N. Y. Acad. Sci. **976**, 394–404 (2002)

A.J. Caride, A.G. Filoteo, A.R. Penheiter, K. Pászty, A. Enyedi, J.T. Penniston, Delayed activation of the plasma membrane calcium pump by a sudden increase in Ca^{2+}: fast pumps reside in fast cells. Cell Calcium **30**, 49–57 (2001)

P. Chadderton, T.W. Margrie, M. Häusser, Integration of quanta in cerebellar granule cells during sensory processing. Nature **428**, 856–860 (2004)

E. De Schutter, J.M. Bower, An active membrane model of the cerebellar Purkinje cell II. Simulation of synaptic responses. J. Neurophysiol. **71**, 401–419 (1994)

E. De Schutter, V. Steuber, Patterns and pauses in Purkinje cell simple spike trains: experiments, modeling and theory. Neuroscience **162**, 816–826 (2009)

R. DiPolo, L. Beauge, The calcium pump and sodium-calcium exchange in squid axons. Annu. Rev. Physiol. **45**, 313–324 (1983)

J.S. Dittman, A.C. Kreitzer, W.G. Regehr, Interplay between facilitation, depression, and residual calcium at three presynaptic terminals. J. Neurosci. **20**, 1374–1385 (2000)

F.A. Dodge Jr., R. Rahamimoff, Co-operative action a calcium ions in transmitter release at the neuromuscular junction. J. Physiol. **193**, 419–432 (1967a)

F.A. Dodge Jr., R. Rahamimoff, On the relationship between calcium concentration and the amplitude of the end-plate potential. J. Physiol. **189**, 90P–92P (1967b)

A. Doi, Y. Kakazu, N. Akaike, Na^+/Ca^{2+} exchanger in GABAergic presynaptic boutons of rat central neurons. J. Neurophysiol. **87**, 1694–1702 (2002)

J.C. Eccles, Circuits in the cerebellar control of movement. Proc. Natl. Acad. Sci. U. S. A. **58**, 336–343 (1967)

J. Eccles, R. Llinas, K. Sasaki, Golgi cell inhibition in the cerebellar cortex. Nature **204**, 1265–1266 (1964)

J.C. Eccles, R. Llinas, K. Sasaki, The excitatory synaptic action of climbing fibres on the Purkinje cells of the cerebellum. J. Physiol. **182**, 268–296 (1966a)

J.C. Eccles, R. Llinas, K. Sasaki, Parallel fibre stimulation and the responses induced thereby in the Purkinje cells of the cerebellum. Exp. Brain Res. **1**, 17–39 (1966b)

R.M. Empson, M.L. Garside, T. Knöpfel, Plasma membrane Ca^{2+} ATPase 2 contributes to short-term synapse plasticity at the parallel fiber to Purkinje neuron synapse. J. Neurosci. **27**, 3753–3758 (2007)

R.M. Empson, P.R. Turner, R.Y. Nagaraja, P.W. Beesley, T. Knöpfel, Reduced expression of the Ca^{2+} transporter protein PMCA2 slows Ca^{2+} dynamics in mouse cerebellar Purkinje neurones and alters the precision of motor coordination. J. Physiol. **588**(Pt 6), 907–922 (2010)

L. Fierro, R. DiPolo, I. Llano, Intracellular calcium clearance in Purkinje cell somata from rat cerebellar slices. J. Physiol. **510**(Pt 2), 499–512 (1998)

A.G. Filoteo, N.L. Elwess, A. Enyedi, A. Caride, H.H. Aung, J.T. Penniston, Plasma membrane Ca^{2+} pump in rat brain. Patterns of alternative splices seen by isoform-specific antibodies. J. Biol. Chem. **272**, 23741–23747 (1997)

G. Fontana, R.S. Rogowski, M.P. Blaustein, Kinetic properties of the sodium-calcium exchanger in rat brain synaptosomes. J. Physiol. **485**(Pt 2), 349–364 (1995)

M.L. Garside, P.R. Turner, B. Austen, E.E. Strehler, P.W. Beesley, R.M. Empson, Molecular interactions of the plasma membrane calcium ATPase 2 at pre- and postsynaptic sites in rat cerebellum. Neuroscience **162**, 383–395 (2009)

F. Helmchen, J.G. Borst, B. Sakmann, Calcium dynamics associated with a single action potential in a CNS presynaptic terminal. Biophys. J. **72**, 1458–1471 (1997)

M. Hilge, J. Aelen, A. Foarce, A. Perrakis, G.W. Vuister, Ca^{2+} regulation in the Na^{+}/Ca^{2+} exchanger features a dual electrostatic switch mechanism. Proc. Natl. Acad. Sci. U. S. A. **106**, 14333–14338 (2009)

D.E. Hillman, S. Chen, R. Bing, J.T. Penniston, R. Llinas, Ultrastructural localization of the plasmalemmal calcium pump in cerebellar neurons. Neuroscience **72**, 315–324 (1996)

M. Ito, Cerebellar circuitry as a neuronal machine. Prog. Neurobiol. **78**, 272–303 (2006)

M.V. Ivannikov, M. Sugimori, R.R. Llinás, Calcium clearance and its energy requirements in cerebellar neurons. Cell Calcium **47**, 507–513 (2010)

T.P. Jensen, A.G. Filoteo, T. Knopfel, R.M. Empson, Presynaptic plasma membrane Ca^{2+} ATPase isoform 2a regulates excitatory synaptic transmission in rat hippocampal CA3. J. Physiol. **579**(Pt 1), 85–99 (2007)

T.P. Jensen, L.E. Buckby, R.M. Empson, Reduced expression of the "fast" calcium transporter PMCA2a during homeostatic plasticity. Mol. Cell. Neurosci. **41**, 364–372 (2009)

D. Jeon, Y.M. Yang, M.J. Jeong, K.D. Philipson, H. Rhim, H.S. Shin, Enhanced learning and memory in mice lacking Na^{+}/Ca^{2+} exchanger 2. Neuron **38**, 965–976 (2003)

H. Jorntell, C. Hansel, Synaptic memories upside down: bidirectional plasticity at cerebellar parallel fiber-Purkinje cell synapses. Neuron **52**, 227–238 (2006)

M. Juhaszova, P. Church, M.P. Blaustein, E.F. Stanley, Location of calcium transporters at presynaptic terminals. Eur. J. Neurosci. **12**, 839–846 (2000)

B. Katz, R. Miledi, The effect of calcium on acetylcholine release from motor nerve terminals. Proc. R. Soc. Lond. B Biol. Sci. **161**, 496–503 (1965)

M.H. Kim, N. Korogod, R. Schneggenburger, W.K. Ho, S.H. Lee, Interplay between Na^{+}/Ca^{2+} exchangers and mitochondria in Ca^{2+} clearance at the calyx of Held. J. Neurosci. **25**, 6057–6065 (2005)

Y.T. Kim, Y.L. Namkung, J. Kwak, C.K. Suh, Involvement of Na^{+}-Ca^{2+} exchanger on metabotropic glutamate receptor 1-mediated $[Ca^{2+}]$i transients in rat cerebellar Purkinje neurons. Neuroscience **146**, 170–177 (2007)

H.J. Koester, B. Sakmann, Calcium dynamics associated with action potentials in single nerve terminals of pyramidal cells in layer 2/3 of the young rat neocortex. J. Physiol. **529**(Pt 3), 625–646 (2000)

A. Konnerth, I. Llano, C.M. Armstrong, Synaptic currents in cerebellar Purkinje cells. Proc. Natl. Acad. Sci. U. S. A **87**, 2662–2665 (1990)

P.J. Kozel, R.A. Friedman, L.C. Erway, E.N. Yamoah, L.H. Liu, T. Riddle, J.J. Duffy, T. Doetschman, M.L. Miller, E.L. Cardell, G.E. Shull, Balance and hearing deficits in mice with a null mutation in the gene encoding plasma membrane Ca^{2+}-ATPase isoform 2. J. Biol. Chem. **273**, 18693–18696 (1998)

R. Larbig, N. Torres, J.H. Bridge, J.I. Goldhaber, K.D. Philipson, Activation of reverse Na^{+}-Ca^{2+} exchange by the Na^{+} current augments the cardiac Ca^{2+} transient: evidence from NCX knockout mice. J. Physiol. **588**(Pt 17), 3267–3276 (2010)

S.H. Lee, M.H. Kim, K.H. Park, Y.E. Earm, W.K. Ho, K^{+}-dependent Na^{+}/Ca^{2+} exchange is a major Ca^{2+} clearance mechanism in axon terminals of rat neurohypophysis. J. Neurosci. **22**, 6891–6899 (2002)

X.F. Li, L. Kiedrowski, F. Tremblay, F.R. Fernandez, M. Perizzolo, R.J. Winkfein, R.W. Turner, J.S. Bains, D.E. Rancourt, J. Lytton, Importance of K^{+}-dependent Na^{+}/Ca^{2+}-exchanger 2, NCKX2, in motor learning and memory. J. Biol. Chem. **281**, 6273–6282 (2006)

B. Linck, Z. Qiu, Z. He, Q. Tong, D.W. Hilgemann, K.D. Philipson, Functional comparison of the three isoforms of the Na^{+}/Ca^{2+} exchanger (NCX1, NCX2, NCX3). Am. J. Physiol. **274**(2 Pt 1), C415–C423 (1998)

R. Llinas, M. Sugimori, Electrophysiological properties of in vitro Purkinje cell somata in mammalian cerebellar slices. J. Physiol. **305**, 171–195 (1980)

R. Llinas, I.Z. Steinberg, K. Walton, Relationship between presynaptic calcium current and postsynaptic potential in squid giant synapse. Biophys. J. **33**, 323–351 (1981)

R. Llinas, M. Sugimori, R.B. Silver, Microdomains of high calcium concentration in a presynaptic terminal. Science **256**, 677–679 (1992)

A. Lörincz, B. Rózsa, G. Katona, E.S. Vizi, G. Tamás, Differential distribution of NCX1 contributes to spine-dendrite compartmentalization in CA1 pyramidal cells. Proc. Natl. Acad. Sci. U. S. A. **104**, 1033–1038 (2007)

D. Marr, A theory of cerebellar cortex. J. Physiol. **202**, 437–470 (1969)

H. Matsukawa, A.M. Wolf, S. Matsushita, R.H. Joho, T. Knöpfel, Motor dysfunction and altered synaptic transmission at the parallel fiber-Purkinje cell synapse in mice lacking potassium channels Kv3.1 and Kv3.3. J. Neurosci. **23**, 7677–7684 (2003)

H. Miyakawa, V. Lev-Ram, N. Lasser-Ross, W.N. Ross, Calcium transients evoked by climbing fiber and parallel fiber synaptic inputs in guinea pig cerebellar Purkinje neurons. J. Neurophysiol. **68**, 1178–1189 (1992)

M.H. Myoga, W.G. Regehr, Calcium microdomains near R-type calcium channels control the induction of presynaptic long-term potentiation at parallel fiber to purkinje cell synapses. J. Neurosci. **31**, 5235–5243 (2011)

E. Neher, Usefulness and limitations of linear approximations to the understanding of Ca^{2+} signals. Cell Calcium **24**, 345–357 (1998)

E. Neher, T. Sakaba, Multiple roles of calcium ions in the regulation of neurotransmitter release. Neuron **59**, 861–872 (2008)

M. Ottolia, D.A. Nicoll, K.D. Philipson, Roles of two Ca^{2+}-binding domains in regulation of the cardiac Na^+-Ca^{2+} exchanger. J. Biol. Chem. **284**, 32735–32741 (2009)

S.L. Palay, V. Chan-Palay, *Cerebellar cortex: cytology and organization* (Springer, New York/Heidelberg, 1974)

M. Papa, A. Canitano, F. Boscia, P. Castaldo, S. Sellitti, H. Porzig, M. Taglialatela, L. Annunziato, Differential expression of the Na^+-Ca^{2+} exchanger transcripts and proteins in rat brain regions. J. Comp. Neurol. **461**, 31–48 (2003)

H. Parnas, L.A. Segel, A theoretical explanation for some effects of calcium on the facilitation of neurotransmitter release. J. Theor. Biol. **84**, 3–29 (1980)

H. Parnas, L.A. Segel, Facilitation as a tool to study the entry of calcium and the mechanism of neurotransmitter release. Prog. Neurobiol. **32**, 1–9 (1989)

H. Porzig, Z. Li, D.A. Nicoll, K.D. Philipson, Mapping of the cardiac sodium-calcium exchanger with monoclonal antibodies. Am. J. Physiol. **265**, C748–C756 (1993)

C. Pott, K.D. Philipson, J.I. Goldhaber, Excitation-contraction coupling in Na^+-Ca^{2+} exchanger knockout mice: reduced transsarcolemmal Ca^{2+} flux. Circ. Res. **97**, 1288–1295 (2005)

D.L. Qiu, T. Knopfel, Presynaptically expressed long-term depression at cerebellar parallel fiber synapses. Pflugers Arch. **457**, 865–875 (2009)

H. Reuter, H. Porzig, Localization and functional significance of the Na^+/Ca^{2+} exchanger in presynaptic boutons of hippocampal cells in culture. Neuron **15**, 1077–1084 (1995)

B.L. Sabatini, W.G. Regehr, Timing of neurotransmission at fast synapses in the mammalian brain. Nature **384**, 170–172 (1996)

T. Sakaba, Two Ca^{2+}-dependent steps controlling synaptic vesicle fusion and replenishment at the cerebellar basket cell terminal. Neuron **57**, 406–419 (2008)

H. Schmidt, K.M. Stiefel, P. Racay, B. Schwaller, J. Eilers, Mutational analysis of dendritic Ca^{2+} kinetics in rodent Purkinje cells: role of parvalbumin and calbindin D28k. J. Physiol. **551**, 13–32 (2003)

R. Schneggenburger, E. Neher, Intracellular calcium dependence of transmitter release rates at a fast central synapse. Nature **406**, 889–893 (2000)

J.Z. Sheng, C.F. Prinsen, R.B. Clark, W.R. Giles, P.P. Schnetkamp, Na^+-Ca^{2+}-K^+ currents measured in insect cells transfected with the retinal cone or rod Na^+-Ca^{2+}-K^+ exchanger cDNA. Biophys. J. **79**, 1945–1953 (2000)

S.B. Simons, Y. Escobedo, R. Yasuda, S.M. Dudek, Regional differences in hippocampal calcium handling provide a cellular mechanism for limiting plasticity. Proc. Natl. Acad. Sci. U. S. A. **106**, 14080–14084 (2009)

S. Sokolow, M. Manto, P. Gailly, J. Molgó, C. Vandebrouck, J.M. Vanderwinden, A. Herchuelz, S. Schurmans, Impaired neuromuscular transmission and skeletal muscle fiber necrosis in mice lacking Na/Ca exchanger 3. J. Clin. Invest. **113**, 265–273 (2004)

E.F. Stanley, The calcium channel and the organization of the presynaptic transmitter release face. Trends Neurosci. **20**, 404–409 (1997)

C.F. Stevens, J.M. Sullivan, The synaptotagmin C2A domain is part of the calcium sensor controlling fast synaptic transmission. Neuron **39**, 299–308 (2003)

E.E. Strehler, D.A. Zacharias, Role of alternative splicing in generating isoform diversity among plasma membrane calcium pumps. Physiol. Rev. **81**, 21–50 (2001)

E.E. Strehler, A.G. Filoteo, J.T. Penniston, A.J. Caride, Plasma-membrane Ca^{2+} pumps: structural diversity as the basis for functional versatility. Biochem. Soc. Trans. **35**, 919–922 (2007)

T.C. Sudhof, Synaptic vesicles: an organelle comes of age. Cell **127**, 671–673 (2006)

T. Thurneysen, D.A. Nicoll, K.D. Philipson, H. Porzig, Sodium/calcium exchanger subtypes NCX1, NCX2 and NCX3 show cell-specific expression in rat hippocampus cultures. Brain Res. Mol. Brain Res. **107**, 145–156 (2002)

E.A. Woodcock, J.F. Arthur, S.N. Harrison, X.M. Gao, X.J. Du, Reperfusion-induced Ins(1,4,5)P(3) generation and arrhythmogenesis require activation of the Na^+/Ca^{2+} exchanger. J. Mol. Cell. Cardiol. **33**, 1861–1869 (2001)

M. Wu, M. Wang, J. Nix, L.V. Hryshko, L. Zheng, Crystal structure of CBD2 from the Drosophila Na^+/Ca^{2+} exchanger: diversity of Ca^{2+} regulation and its alternative splicing modification. J. Mol. Biol. **387**, 104–112 (2009)

D.A. Zacharias, C. Kappen, Developmental expression of the four plasma membrane calcium ATPase (PMCA) genes in the mouse. Biochim. Biophys. Acta **1428**, 397–405 (1999)

Part VIII

Emerging Role of NCX Activity in Immune and Glial Cells

Sodium-Calcium Exchanger Modulates the L-Glutamate Ca_i^{2+} Signalling in Type-1 Cerebellar Astrocytes

22

Héctor Rojas, Claudia Colina, Magaly Ramos, Gustavo Benaim, Erica Jaffe, Carlo Caputo, and Reinaldo DiPolo

Abstract

We have previously demonstrated that rat type-1 cerebellar astrocytes express a very active Na^+/Ca^{2+} exchanger which accounts for most of the total plasma membrane Ca^{2+} fluxes and for the clearance of Ca_i^{2+} induced by physiological agonist. In this chapter, we have explored the mechanism by which the reverse Na^+/Ca^{2+} exchange is involved in agonist-induced Ca^{2+} signalling in rat cerebellar astrocytes. Laser-scanning confocal microscopy experiments using immunofluorescence labelling of Na^+/Ca^{2+} exchanger and RyRs demonstrated that they are highly co-localized. The most important finding presented in this chapter is that L-glutamate activates the reverse mode of the Na^+/Ca^{2+} exchange by inducing a Na^+ entry through the electrogenic Na^+-glutamate co-transporter and not through the ionophoric L-glutamate receptors as confirmed by pharmacological experiments with specific blockers of ionophoric L-glutamate receptors, electrogenic glutamate transporters and the Na/Ca exchange.

Keywords

Na^+/Ca^{2+} exchange • CICR • Glutamate • Glutamate transporter

22.1 Introduction

The Na^+/Ca^{2+} exchanger a plasma membrane counter-transport system plays a critical role in the control of intracellular calcium. In its forward mode (Ca^{2+} efflux), the exchanger has an important physiological role for the rapid extrusion of large amounts of Ca^{2+} from the cell. However, the physiological role of the exchanger, working in its reverse mode (Ca^{2+} entry), is still

H. Rojas • C. Colina • M. Ramos • E. Jaffe • C. Caputo
R. DiPolo (✉)
Laboratorio de Fisiología Celular, Centro de Biofísica, Instituto Venezolano de Investigaciones Científicas (IVIC), Caracas, Venezuela
e-mail: rdipolo@gmail.com

G. Benaim
Laboratorio de Señalización Celular, Instituto de Estudios Avanzados (IDEA), Caracas, Venezuela

L. Annunziato (ed.), *Sodium Calcium Exchange: A Growing Spectrum of Pathophysiological Implications*, Advances in Experimental Medicine and Biology 961, DOI 10.1007/978-1-4614-4756-6_22,

controversial (Blaustein and Lederer 1999). In principle, there are two non-exclusive possibilities for the involvement of the reverse exchange in Ca_i^{2+} signalling in type-1 cerebellar astrocytes: (1) Ca_i^{2+} entering by the reverse exchanger directly triggers calcium-dependent processes and/or (2) Ca_i^{2+} entering through the exchanger serves as messenger for a Ca_i^{2+} signal amplification through a Ca_i^{2+}-induced-Ca_i^{2+}-release (CICR) mechanism. In favour of the first possibility are the reports that L-glutamate (L-Glu) through activation of kainate receptor channels leads to the influx of Na^+ ions which activates the reverse Na^+/Ca^{2+} exchange, thus leading to $[Ca^{2+}]_i$ increase (Goldman et al. 1994; Takuma et al. 1996). A similar mechanism has been proposed to explain the glutamate-induced homocysteic acid release from cortical astrocytes (Benz et al. 2004). On the other hand, the existence and functional significance of CICR coupled to ryanodine receptors (RyRs) is well documented in astrocytes (Verkhratsky and Kettenmann 1994). Nevertheless, the existence and functional relevance of RyRs in type-1 cerebellar astrocytes, if any, have not been demonstrated.

The experiments reported here examine the role of Ca^{2+} entry through reverse Na^+/Ca^{2+} exchange as a mechanism for inducing amplification of Ca_i^{2+} signals that occur during conditions of agonist activation. Using microspectrofluorometric measurements, pharmacological tools, immunofluorescence labelling and laser-scanning confocal microscopy (LSCM) analyses, we present for the first time evidences that in rat type-1 cerebellar astrocytes, (1) Ca^{2+} entry during operation of reverse Na^+/Ca^{2+} markedly increases $[Ca^{2+}]_i$ by a CICR mechanism, followed by the opening of store-operated Ca^{2+} channels (SOCC); (2) immunofluorescence labelling of both Na^+/Ca^{2+} exchanger and RyRs using confocal microscopy demonstrates that they are highly co-localized; and (3) unexpectedly, physiological agonist concentrations of L-Glu increase $[Ca^{2+}]_i$ through activation of the reverse exchange as a result of Na^+ entry through the electrogenic glutamate transporters.

22.2 Role of the Sodium-Calcium Exchanger in the Control of L-Glutamate Ca_i^{2+} Signalling in Type-1 Cerebellar Astrocytes

Figure 22.1a shows a run in which a cell was exposed to a long 70-s pulse to a 0NaCa solution, producing a larger increase in intracellular Ca^{2+}. Readmission of external Na causes the Ca_i^{2+} to drop to a sloping plateau which was cut short by superfusing the cell with a Na^+-containing Ca^{2+}-free medium (Na_0Ca) lowering the Ca_i^{2+} to nearly resting values. This experiment suggests that in this preparation, substantial Ca^{2+} entry through the reverse exchange may activate the release of Ca^{2+} from intracellular Ca^{2+} stores, and the opening of the SOCC (store-operated calcium channels). The protocol of Fig. 22.1b was designed to disable the forward Na^+/Ca^{2+} exchange with a 50-s pulse of 0NaCa (Ca^{2+} entry mode, horizontal slash) and then rapidly enable the forward exchange (Ca^{2+} extrusion mode) for about 40 s by rapidly superfusing with the test Na_0Ca medium (vertical arrows). This protocol was repeated during nine consecutive pulses. The results of Fig. 22.1b indicate that in the absence of external Ca^{2+}, the forward mode of the exchange lowers the Ca_i^{2+} faster and to a greater extent than in its presence. More importantly, they also show that the peak of the Ca_i^{2+}-dependent fluo-3 signal induced by the reverse exchange decreases progressively after each period of activation of the forward exchange mode. The fact that this decrease is due to depletion of ryanodine-sensitive Ca_i^{2+} stores is confirmed by the absence of Ca_i^{2+} release by the ryanodine receptor agonist 4-C*m*C (end of experiment).

The pharmacological experiments of Fig. 22.1 indicate that ryanodine receptors are somehow involved in the amplification of the Ca^{2+} signal during reverse operation of exchanger in this glial cell type. Therefore, the next step was focused on the immunolocalization of this transporter in the plasma membrane of these cells as well as on the spatial relationship between the plasmalemmal Na^+/Ca^{2+} exchanger and the

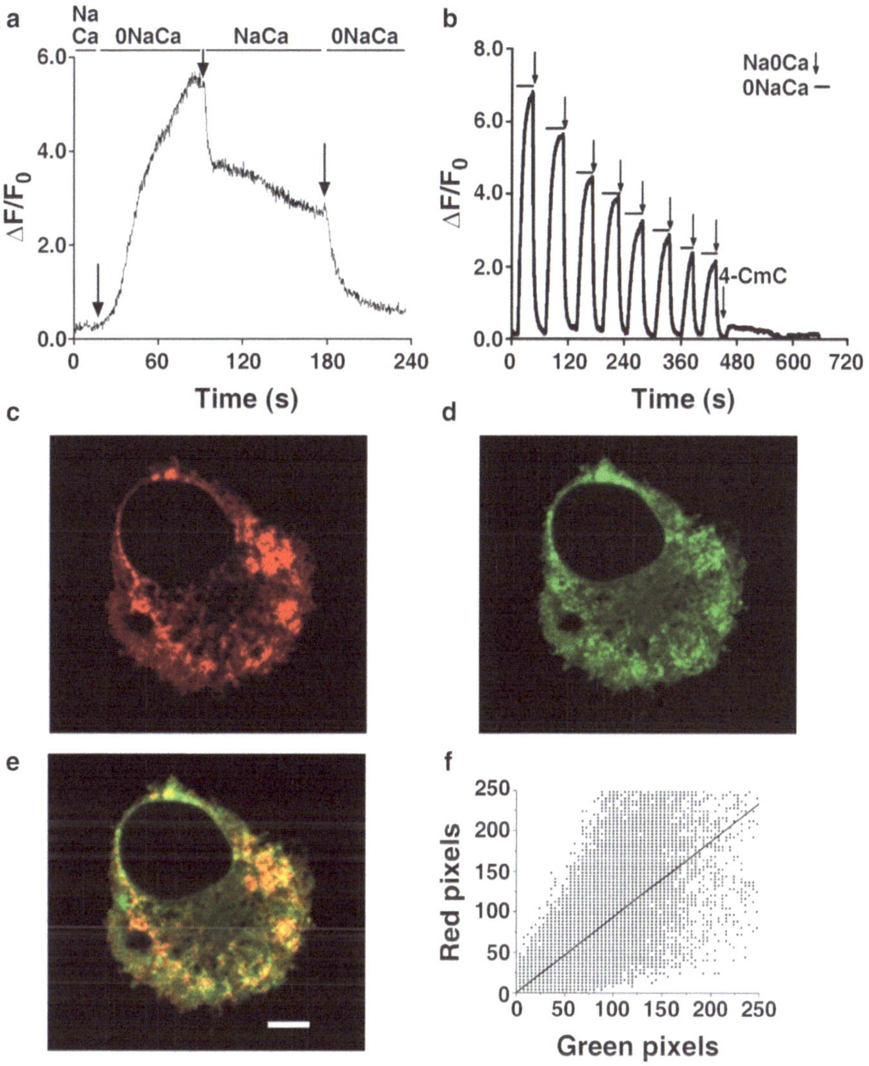

Fig. 22.1 (**a**) Effect of a 0NaCa pulse induces a large Ca_i^{2+} signal, which, upon re-exposure to the normal medium, (NaCa) is partially reversed reaching a sloping plateau value. Additions of a solution containing Na^+ but no Ca^{2+} bring the signal to background levels. (**b**) The effect of nine consecutive reverse (0NaCa) and forward (Na_0Ca) short pulses causes a progressive decrease in the Ca_i^{2+} signal until it reaches a constant small Ca^{2+} value. At the end of experiment, the ryanodine agonist 4-C*m*C fails to release Ca^{2+} from ryanodine-sensitive Ca^{2+} stores, thus indicating that activation of the reverse exchange empty the calcium accumulated in the endoplasmic reticulum. (**C**) and (**D**) show the immunofluorescent labelling of Na^+/Ca^{2+} exchanger (in *red*) with affinity-purified antibodies raised against the cardiac sarcolemmal Na^+/Ca^{2+} exchanger Alexa Fluor 546-Na/Ca and ryanodine receptors (in *green*) with Bodipy-FL-ryanodine, respectively. (**e**) and (**f**) show the merge of the images obtained with the two different labels (*orange* colour corresponding to regions of overlap) and the mathematical analysis of co-localization (Pearson's correlation of 0.89) (*white bar* indicate 10 μm)

underlying endoplasmic reticulum (ER), in particular the ryanodine receptors. For this, cells were incubated first with a purified canine cardiac Na^+/Ca^{2+} exchanger mouse monoclonal antibody and second with a secondary labelled goat anti-mouse Igm antibody Na^+/Ca^{2+} exchange Alexa Fluor 546. The immunofluorescence of the Na^+/Ca^{2+} exchanger (red colour) in a representative cell is presented in Fig. 22.1c. In all cells studied ($n=6$), the labelling was punctual suggesting a

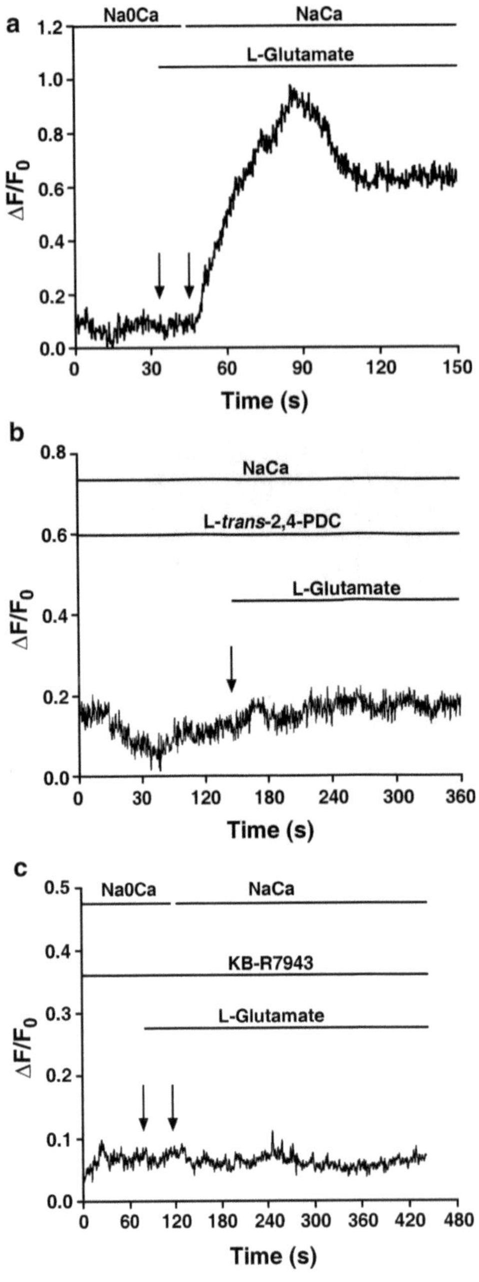

Fig. 22.2 *The effect of low (<30 μM)* L-*Glu on the* $Ca^{2+}{}_i$*-dependent fluo-3 signal in the presence and absence of extracellular* Ca^{2+}. (**a**) This astrocyte was perfused from the beginning with a medium containing no external Ca^{2+} (Na_0Ca). Notice that no effect of L-Glu is observed under this condition. Addition of 2 mM external Ca^{2+} rapidly induces a biphasic response. (**b**) The effect of L-*trans*-2,4-PDC (100 μM), a blocker of the electrogenic glutamate transporter, completely eliminates the L-Glu-induced increase in Ca_i^{2+}.(**c**) The effect of the Na^+/Ca^{2+} exchange inhibitor the KB-R7943 (10 μM) over the L-Glu-induced Ca_i^{2+} rise even at 1 mM [L-Glu]

cluster of Na^+/Ca^{2+} exchange molecules. Labelling of the exchanger was more intense at cell edges suggesting that the exchanger is distributing in an organized manner in the astrocyte plasmalemma. Figure 22.1d shows the localization of ryanodine receptors (green colour) in the same cell using Bodipy-FL-ryanodine, a specific ryanodine receptor marker (Hua et al. 2004). Figure 22.1e shows the co-localization (orange colour) of the Na^+/Ca^{2+} exchanger and the ryanodine receptors. Figure 22.1 shows that from co-localization analysis, the observed overlap was found to be highly significant with a Pearson`s correlation of about 0.89. This indicates that the Na^+/Ca^{2+} exchanger is indeed co-localized with some of the ER, in particular with the ryanodine receptors.

Figure 22.2a shows that in the absence of external Ca, 30-μM L-Glu does not modify the calcium fluorescence signal. Following addition of external Ca^{2+} in the continuous presence of L-Glu, a substantial increase in the $[Ca^{2+}]_i$ was observed, thus indicating that the fluo-3 signal induced by L-Glu is mediated by Ca^{2+} entering from the extracellular medium. On the other hand, Fig. 22.2b shows that the specific blocker of the electrogenic Na^+-glutamate co-transporter, L-*trans*-2,4,-PDC, completely eliminates the L-Glu induced increase in Ca_i^{2+}. Figure 22.2c shows that in presence of a potent inhibitor of the Na^+/Ca^{2+} exchanger, KB-R7943 (Iwamoto et al. 1998) and a preincubation (60 s) with 10 μM of the inhibitor completely block the L-Glu effect.

22.3 Conclusion

The present work demonstrates that in type-1 cerebellar astrocytes in culture, the Ca^{2+} signal generated by Ca^{2+} entry through the reverse Na^+/Ca^{2+} exchange is greatly amplified by a Ca_i^{2+}-induced Ca^{2+} release mechanism which involves ryanodine receptors and ryanodine-sensitive Ca^{2+} stores. While the presence of RyRs has been demonstrated in this preparation, their physiological significance was not clear (Langley and Pearce 1994; Simpson et al. 1998; Matyash et al. 2002; Golovina and Blaustein 2000; Beck et al.

2004; Aley et al. 2006). Caffeine may induce Ca^{2+} release from RyRs-operated Ca^{2+} stores in different neurons (Uneyama et al. 1993; Usachev et al. 1993; Kano et al. 1995; Llano et al. 2000) and glia preparations, (Verkhratsky and Shmilgol 1996; Beck et al. 2004).

In addition, and most importantly, for the first time, we provide evidence that the intracellular Ca^{2+} signal induced by physiological concentrations of the excitatory amino acid L-glutamate is the result of Na^+ entry through the electrogenic glutamate transporter that activates the reverse Na^+/Ca^{2+} exchange and leads to Ca^{2+} entry, with a concomitant increase in $[Ca^{2+}]_i$. The finding of a functional co-expression of Na^+/Ca^{2+} exchangers with ryanodine receptors strongly supports the idea that the original Ca^{2+} signal due to Ca_i^{2+} entry through the exchanger is largely amplified by a CICR process.

Previous studies have shown that the Na^+/Ca^{2+} exchanger working in its reverse mode can induce Ca^{2+} entry in cultured astrocytes (Goldman et al. 1994; Takuma et al. 1994; Blaustein and Lederer 1999). Moreover, Ca^{2+} influx via the exchanger may be responsible for $[Ca^{2+}]_i$ increases under certain pathological conditions (Kin-Lee et al. 1992; Matsuda et al. 1996). Type-1 cerebellar astrocytes express a highly active Na^+/Ca^{2+}exchanger responsible for the balance of the plasma membrane Ca^{2+} fluxes under resting physiological conditions (Rojas et al. 2004). In different preparations, there is evidence of an intimate association between the Na^+/Ca^{2+} exchanger and internal Ca^{2+} stores (Juhaszova et al. 1996). Such association is well established in smooth muscle cells where the exchanger is in close proximity to the sarcoplasmic reticulum (SR) so that Ca^{2+} release from the SR through RyRs is closely coupled to its extrusion by the exchanger (Nazer and van Breemen 1998). Furthermore, in neurons, there is evidence for a functional (Hurtado et al. 2002) and spatial association of the exchanger with the intracellular Ca^{2+} stores (Juhaszova et al. 1996). Micci and Cristensen (1998) working in catfish retinal neurons have studied the interaction between the exchanger and caffeine-sensitive Ca^{2+} stores showing that reverse operation of the Na^+/Ca^{2+} exchanger refills Ca^{2+}-depleted ER. For the case of astrocytes, however, the relationship between the Na^+/Ca^{2+} exchanger and the RyRs is unknown.

One of the aims of the present work was to investigate whether the magnitude of the increase in $[Ca^{2+}]_i$ observed when the operation of the Na^+/Ca^{2+} exchanger was reversed was due solely to Ca^{2+} entry or whether this entry could trigger further Ca^{2+} release from RyRs-operated intracellular Ca^{2+} stores. During long (>60 s) Na^+ gradient reversal pulses, the increase in $[Ca^{2+}]_i$ is much larger and leads to depletion of RyRs-operated intracellular Ca^{2+} store, indicating the presence of a CICR mechanism. Furthermore, depletion of intracellular Ca^{2+} stores causes the activation of SOCC, as confirmed by the extracellular Ca^{2+} dependency (Fig. 22.1a) and sensitivity to 2-APB (Lo et al. 2002; Rojas et al. 2007) of a late, residual component of the Ca^{2+} signal. The presence of ryanodine receptors in type-1 cerebellar astrocytes has been confirmed using conventional Ca^{2+} imaging confocal microscopy and immunocytochemistry techniques. The close proximity of the Na^+/Ca^{2+} exchanger to the ER membranes, where the RyRs are localized, allows the former to rapidly extrude Ca^{2+} ions released from the ER before their recapture by the ER Ca^{2+}-ATPase. This leads to depletion of the ER Ca^{2+} stores as demonstrated by the consecutive reverse-forward pulse experiments. The fact that no release of Ca^{2+} is observed at the end of the run in the presence of the ryanodine agonist 4-C*m*C or a combination (Fig. 22.1b) demonstrates that the exchanger is capable of depleting the ER.

An important discovery in glial cell research is that $[Ca^{2+}]_i$ increase may trigger glutamate release from astrocytes which then mediates Ca_i^{2+} increases in nearby neurons, thus indicating a crosstalk between neurons and astrocytes (Parpura et al. 1994; Jeftinija et al. 1997; Pasti et al. 1995; Calegari et al. 1999; Araque et al. 2000; Fellin and Carmignoto 2004). Benz et al. (2004) have demonstrated the importance of the Na^+/Ca^{2+} exchanger in the glutamate response in cortical astrocytes from mice. Their experiments show that 500-μM L-Glu induces a Ca_i^{2+}-dependent release of homocysteic acid from astrocytes through activation of glutamate receptors, leading to an

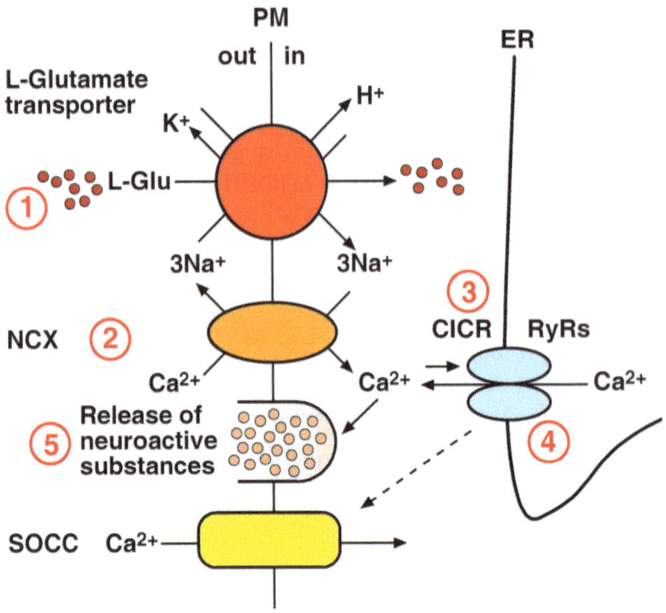

Fig. 22.3 *Role of Na^+/Ca^{2+} exchanger in glutamate-induced rise of intracellular Ca^{2+} in rat cerebellar type 1 astrocytes.* The events that lead to glutamate-induced rise in intracellular Ca^{2+} involve (*1*) Na^+ entry through the electrogenic glutamate co-transporter, (*2*) activation of the reverse Na^+/Ca^{2+} exchange (NCX) by the rise in intracellular Na^+ (Na^+ inward current through the glutamate transporter), (*3*) rise in the $[Ca^{2+}]_i$ near the RyRs to trigger a CICR from the ER, (*4*) activation by Ca^{2+} of RyRs followed by Ca^{2+} release from ryanodine channels leading to an amplification of the original Ca^{2+} entry through the exchanger and (*5*) opening of the store-operated calcium channels and release of neuroactive substrates

influx of Na^+ and to an increase in Ca^{2+} entry through the reverse Na^+/Ca^{2+} exchange (Benz et al. 2004). Previous electrophysiological studies in rat type-1 cerebellar astrocytes show that application of as low as 30-μM L-Glu produced large inward currents which remains inward going at potentials up to $^+80$ mV being the result of the presence of an electrogenic glutamate uptake carrier (Wyllie et al. 1991). In cells kept up to 4 days in culture, quisqualate, kainate and NMDA failed to produce any current indicating the absence at this early stage of glutamate ionotropic receptors in rat type-1 cerebellar astrocytes (Wyllie et al. 1991). These authors showed that even in older cultures, in which ionotropic glutamate receptors are well expressed, most of the L-Glu-induced inward current can be ascribed to the Na^+-glutamate co-transporter (Wyllie et al. 1991). Based on these findings, we considered the possibility that the electrogenic Na^+-glutamate transporter might be involved in the L-Glu-dependent $[Ca^{2+}]_i$ increase in type-1 cerebellar astrocytes through an increase in $[Na^+]_i$.

The major finding in the present work is that activation of the reverse Na^+/Ca^{2+} exchange by physiological [L-Glu] is not the consequence of Na^+ entry through ionotropic receptors as occurs in other astrocyte preparations (Benz et al. 2004) but the result of Na^+ entry through the electrogenic glutamate transporter (see the scheme of Fig. 22.3). An important role of the electrogenic glutamate transporter in the L-Glu-induced Ca_i^{2+} increase and its relationship with the reverse Na^+/Ca^{2+} exchange are supported by the demonstration that (1) no effect of L-Glu is observed in the absence of external Ca^{2+}, (2) inhibition of the ionotropic glutamate receptors does not impair the Ca_i^{2+} rise induced by L-Glu, (3) inhibition of the Na^+/Ca^{2+} exchanger completely blocks the L-Glu effect, (4) L-Glu effect is abolished by depletion of the ryanodine-sensitive intracellular stores (by 4-C*m*C) and (5) specific inhibition of the electrogenic Na^+-glutamate co-transporter completely eliminates the L-Glu effect.

Considering that the transport current generated by the glutamate transporter is evoked by the inward movement of two positive charges per transported glutamate (1Glu:1H^+:3Na^+ entering vs. 1K^+ moving outward; Greever and Rauen 2005) and an average inward current of 800 pA/cm^2 for a 30-μM L-Glu (Wyllie et al. 1991), then for a hypothetical type-1 astrocyte resembling a rectangular triangle of 25 μm in the base and an approximate astrocyte volume of 1.2×10^{-6} μl, enough Na^+ will enter the astrocyte during L-Glu activation as to induce increases of the intracellular [Na^+] in tens of millimolar in less than 10 s, sufficient to greatly activate the reverse mode of the Na^+/Ca^{2+}.

Finally, an interesting recent finding is an acute up-regulation of the Na^+-glutamate transporter mediated by metabotropic glutamate receptors in rat cortical astrocytes, in which activation of mGluR5a induces a PKC-dependent up-regulation of GLT-1 activity (Vermeiren et al. 2005). Further experiments are necessary to link this cross regulation with our proposed model.

Acknowledgements This work was supported by FONACIT-Venezuela (G-0010000637).

References

P.K. Aley, H.J. Murray, J.P. Boyle, H.A. Pearson, C. Peers, Hypoxia stimulates Ca^{2+} release from intracellular stores in astrocytes via cyclic ADP ribose-mediated activation of ryanodine receptors. Cell Calcium **39**, 95–100 (2006)

A. Araque, N. Li, R.T. Doyle, P.G. Haydon, SNARE protein-dependent glutamate release from astrocytes. J. Neurosci. **20**, 666–673 (2000)

A. Beck, R.Z. Nieden, H.P. Scheideer, J.W. Ditmer, Calcium release from intracellular stores in rodent astrocytes and neurons in situ. Cell Calcium **35**, 47–58 (2004)

B. Benz, G. Grima, K.Q. Do, Glutamate-induced homocysteic acid release from astrocytes: possible implication in glia-neuron signalling. Neuroscience **124**, 377–386 (2004)

M.P. Blaustein, J. Lederer, Sodium/calcium exchange: its physiological implication. Physiol. Rev. **79**, 763–854 (1999)

F. Calegari, S. Coco, E. Taverna, M. Bassetti, C. Verderio, N. Corradi, M. Matteoti, P. Rosa, A regulated secretory pathway in cultured hippocampal astrocytes. J. Biol. Chem. **274**, 22539–22547 (1999)

T. Fellin, G. Carmignoto, Neurone-to-astrocyte signalling in the brain represents a distinct multifunctional unit. J. Physiol. **559**, 3–15 (2004)

W.F. Goldman, P.J. Yarowsky, M. Juhaszova, B.K. Krueger, M.P. Blaustein, Sodium/calcium exchange in rat cortical astrocytes. J. Neurosci. **14**, 5834–5843 (1994)

V.A. Golovina, M.P. Blaustein, Unloading and refilling of two classes of spatially resolved endoplasmic reticulum Ca^{2+} stores in astrocytes. Glia **31**, 15–28 (2000)

C. Greever, T. Rauen, Electrogenic glutamate transporters in the CNS: molecular mechanism, pre-steady state kinetics, and their impact on synaptic signaling. J. Memb. Biol. **203**, 1–20 (2005)

X. Hua, E.B. Melarkey, V. Sunjara, S.E. Rosenwald, W.H. Li, V. Parpura, Ca^{2+}-dependent glutamate release involves two classes of endoplasmic reticulum Ca^{2+} stores in astrocytes. J. Neurosci. Res. **76**, 86–97 (2004)

J. Hurtado, S. Borges, M. Wilson, Na^+-Ca^{2+} exchanger controls the gain of the Ca^{2+} amplifier in the dendrites of amacrine cells. J. Neurophysiol. **88**, 2765–2777 (2002)

S.D. Jeftinija, K.V. Jeftinija, G. Stefanovic, Cultured astrocytes express proteins involved in vesicular glutamate release. Brain Res. **750**, 41–47 (1997)

M. Juhaszova, H. Shimizu, M.L. Borin, R.K. Yip, E.M. Santiago, G.E. Lindenmayer, M.P. Blaustein, Localization of the Na^+-Ca^{2+} exchanger in vascular smooth muscle, and in neurons and astrocytes. Ann. N. Y. Acad. Sci. **779**, 318–335 (1996)

M. Kano, O. Garaschuk, A. Verkhratsky, A. Konnerth, Ryanodine receptor-mediated intracellular calcium release in rat cerebellar Purkinje neurones. J. Physiol. **487**, 1–16 (1995)

M.H. Kin-Lee, B.T. Stoke, A.J. Yates, Reperfusion paradox: a novel mode of glial cell injury. Glia **5**, 56–64 (1992)

D. Langley, B. Pearce, Ryanodine-induce intracellular calcium mobilization in culture astrocytes. Glia **12**, 128–134 (1994)

I. Llano, J. Gonzalez, C. Caputo, F.A. Lai, L.M. Blaynet, Y.P. Tan, A. Marty, Presynaptic calcium stores underlie large-amplitude miniature IPSCs and spontaneous calcium transients. Nat. Neurosci. **3**, 1256–1265 (2000)

K. Lo, H.N. Luk, T.Y. Chin, S.H. Chueh, Store depletion-induced calcium influx in rat cerebellar astrocytes. Br. J. Pharmacol. **135**, 1383–1392 (2002)

T. Iwamoto, T. Watano, M. shigegawa, A novel isothiourea frrivative selectivele inhibits de reverse mode of Na/Ca exchange in cells expresing NCX1. J Biol Biochem. **271**, 22391–23397 (1998)

T. Matsuda, K. Takuma, E. Nishiguchi, H. Hashimoto, J. Azuma, A. Baba, Involvement of Na^+-Ca^{2+} exchanger in reperfusion-induced delayed cell death of cultured rat astrocytes. Eur. J. Neurosci. **8**, 951–958 (1996)

M. Matyash, V. Matyash, C. Nolte, V. Sorrentino, H. Kettenmann, Requirement of functional ryanodine receptor type 3 for astrocyte migration. FASEB. J. **16**, 84–86 (2002)

M.A. Micci, B.N. Cristensen, Na^+/Ca^{2+} exchange in catfish retina horizontal cells: regulation of intracellular Ca^{2+} store function. Am. J. Physiol. **274**, C1625–C1633 (1998)

M.A. Nazer, C. van Breemen, Functional linkage of Na^+-Ca^{2+} exchange and sarcoplasmic reticulum Ca^{2+} release mediates Ca^{2+} cycling in vascular smooth muscle. Cell Calcium **24**, 275–283 (1998)

V. Parpura, T.A. Basarsky, F. Liu, S. Jeftinija, P.G. Haydon, Glutamate-mediated astrocyte-neuron signalling. Nature **369**, 744–747 (1994)

L. Pasti, T. Pozzan, G. Carmignoto, Long-lasting changes of calcium oscillations in astrocytes. A new form of glutamate-mediated plasticity. J. Biol. Chem. **270**, 15203–15210 (1995)

H. Rojas, M. Ramos, R. DiPolo, A genistein-sensitive Na^+/Ca^{2+} exchange is responsible for the resting $[Ca^{2+}]i$ and most of the Ca^{2+} plasma membrane fluxes in stimulated rat cerebellar type 1 astrocytes. Jpn. J. Physiol. **54**, 249–262 (2004)

H. Rojas, C. Colina, M. Ramos, G. Benaim, H. Jaffe, C. Caputo, R. Dipolo, Na^+ entry via glutamate transporter activates the reverse Na^+/Ca^{2+} exchange and triggers Cai^{2+}-induced Ca^{2+} release in rat cerebellar Type-1 astrocytes. J. Neurochem. **100**, 1188–1202 (2007)

P.B. Simpson, S. Mehotra, D. Langley, C.A. Sheppard, J.T. Russell, Specialized distribution of mitochondria and endoplasmic reticulum proteins define Ca^{2+} wave amplification sites in cultured astrocytes. J. Neurosci. Res. **52**, 672–683 (1998a)

P.B. Simpson, L.A. Holtzclaw, D.B. Langley, J.T. Russel, Characterization of ryanodine receptors in oligodendrocytes, type 2 astrocytes, and O-2A progenitors. J. Neurosci. Res. **52**, 468–482 (1998b)

K. Takuma, T. Matsuda, H. Hashimoto, S. Asano, A. BaBa, Culture rat astrocytes posses Na^+-Ca^{2+} exchanger. Glia **12**, 336–342 (1994)

K. Takuma, T. Matsuda, U. Hashimoto, J. Kitanaka, S. Asano, A. Baba, Role of Na^+-Ca^{2+} exchanger in agonist-induce Ca^{2+} signalling in cultured rat astrocytes. J. Neurochem. **67**, 1840–1845 (1996)

H. Uneyama, M. Munakata, N. Akaike, Caffeine response in pyramidal neurones freshly dissociated from rat hippocampus. Brain Res. **604**, 24–31 (1993)

Y. Usachev, A. Shmigol, P. Pronchuk, A. Kostyuk, A. Verkhratsky, Caffeine-induce calcium release from internal stores in cultured rat sensory neurones. Neuroscience **57**, 845–859 (1993)

A. Verkhratsky, H. Kettenmann, Calcium signalling in glial cells. TINS **19**, 346–352 (1994)

A. Verkhratsky, A. Shmilgol, Calcium-induce calcium release in neurons. Cell Calcium **19**, 1–14 (1996)

C. Vermeiren, M. Najimi, M. Vanhoutte, S. Tilleux, L. de Hemptinne, J.M. Maloteaux, E. hermans, Acute up-regulation of glutamate uptake mediated by mGluR5a in reactive astrocytes. J. Neurochem. **94**, 405–416 (2005)

D.J.A. Wyllie, A. Mathie, C.D. Symonds, S.G. Cull-Candy, Activation of glutamate receptors and mGluR5a Glutamate uptake in identified macroglial cells in rat cerebellar culture. J. Physiol. **432**, 235–258 (1991)

Immunosuppressive Drugs, Immunophilins, and Functional Expression of NCX Isoforms

23

Hannah Rahamimoff, Benayahu Elbaz, Michael Valitsky, Mahdi Khatib, Marina Eskin-Schwartz, and Daniela Elmaz

Abstract

Although the three mammalian Na^+–Ca^{2+} exchangers share considerable amino acid sequence homology, they exhibit substantial immunosuppressive drug specificity. We have shown that cyclosporin A (CsA) treatment of *NCX*1-, *NCX2*-, or *NCX3*-transfected HEK 293 cells and non-transfected H9c2, L6, and aortic smooth muscle cells, which express NCX1 protein naturally, reduces NCX surface expression and transport activity but has no impact on total cell NCX protein. Similar effect on functional expression of NCX1 protein can be obtained also without CsA treatment by knockdown of cell cyclophilin A (CypA), one of the cellular receptor of CsA. This suggests that CypA has a role in acquisition of function competence of NCX1 protein.

Unlike CsA treatment, which affects the functional expression of all three mammalian NCX proteins similarly, FK506 and rapamycin treatment modulates only the functional expression of NCX2 and NCX3 proteins. FK506 reduces NCX2 and NCX3 surface expression and transport activity without affecting cell NCX protein. Rapamycin reduces NCX2 and NCX3 transport activity but has no effect on their surface expression or total cell NCX protein expression suggesting that, although it shares a common receptor FKBP with FK506, its mode of action follows a different pathway.

We are showing now that the large cytosolic loop of NCX1, NCX2, and NCX3 is involved in acquisition of immunosuppressive drug specificity: truncation of the large cytosolic loop of NCX1 renders the protein sensitive to FK506. Exchange of the large cytosolic loop of NCX3 with that of NCX1 renders the mutant protein insensitive to FK506.

H. Rahamimoff(✉) • B. Elbaz • M. Valitsky • M. Khatib
M. Eskin-Schwartz • D. Elmaz
Department of Biochemistry and Molecular Biology,
Hebrew University-Hadassah Medical School,
Jerusalem, Israel
e-mail: Hannah.Rahamimoff@Huji.ac.il

L. Annunziato (ed.), *Sodium Calcium Exchange: A Growing Spectrum of Pathophysiological Implications*, Advances in Experimental Medicine and Biology 961, DOI 10.1007/978-1-4614-4756-6_23,

Keywords

Na^+–Ca^{2+} exchanger • Immunosuppressive drugs • Cyclosporin A • FK506 (Tacrolimus) • Rapamycin (Sirolimus) • PSC833 (Valspodar) • Cyclophilin A • Cyclophilin B

23.1 Introduction

In the late 1970s, two parallel developments revolutionized medicine: the discovery of immunosuppressive drugs and improved surgical procedures of organ transplantation (Hariharan et al. 2000). These two developments increased considerably human life expectancy. Yet the prolonged treatment with immunosuppressive drugs led also to clinical complications. Among these were hypertension, nephrotoxicity, neurotoxicity, bone loss, and others (Bechstein 2000). Taken together, many of these complications could potentially be linked to impaired Ca^{2+} homeostasis.

The Na^+–Ca^{2+} exchanger plays a major role in regulation of cell Ca^{2+} (Carafoli 1987). Therefore, we started to study the connection between the Na^+–Ca^{2+} exchanger, immunosuppression, and impaired cell Ca^{2+}.

Our studies focused on three major aspects: first, we established that there is a link between the functional expression of NCX1 protein and immunosuppressive drugs (Kimchi-Sarfaty et al. 2002). Next, we studied the specificity and selectivity of individual NCX isoforms and different immunosuppressive drugs (Elbaz et al. 2008). Recently, we established that cyclophilin A was linked to the functional expression of NCX1 protein (Elbaz et al. 2010), and by manipulating cell cyclophilin A levels using targeting siRNA, we could demonstrate changes in the functional expression of the Na^+–Ca^{2+} exchanger.

Two experimental systems were used in our studies: heterologous expression system in which HEK 293 cells were transfected with FN-(amino terminal Flag epitope tagged) *NCX1.5, FN-NCX2*, or *FN-NCX3* – Elbaz et al. 2008, 2010; Kimchi-Sarfaty et al. 2002) and H9c2 cells that express NCX1 naturally. Detailed experimental protocols are published (Elbaz et al. 2008, 2010; Kimchi-Sarfaty et al. 2002). Introduction of the Flag epitope into the extracellular amino terminal of NCX protein did not change its functional expression (Cook et al. 1998; Kasir et al. 1999).

Our current research focuses on identification of the molecular entities that determine the immunosuppressive drug specificity of Na^+–Ca^{2+} exchanger isoforms. Some of these results that are described in this chapter form part of an M. Sc. thesis (Elmaz 2011) that was submitted and approved by the committee of advanced studies, The Hebrew University Medical School. In this chapter, we are summarizing the main aspect of our research.

23.2 Background

Three major drugs are clinically used to suppress the immune system and are routinely employed after organ transplantation: cyclosporin A (Sandimmune), FK506 (Tacrolimus), and rapamycin (Sirolimus) (First 2004; Hariharan et al. 2000; Levy 2000; Meier-Kriesche et al. 2004, 2006; Tsang et al. 2007). The drugs repress the immune system by binding to their respective immunophilin receptors, the cyclophilins or FKBPs, that are highly conserved families of proteins present in all cells and compartments (Barik 2006; Ivery 2000). The immunosuppressive action of CsA and FK506 is based on the interaction of the complex immunophilin-CsA/FK506 with the Ca^{2+} and calmodulin-dependent phosphatase calcineurin (Liu et al. 1992), followed by inhibition of the dephosphorylation of NFAT (nuclear factor of activated T cells) and its translocation to the nucleus, which leads to subsequent suppression of the immune reaction. Rapamycin is also a potent inhibitor of the immune system (Schreiber 1991). It interacts with FKBPs, but the complex FKBP-rapamycin does not inhibit calcineurin but the mTOR (target of rapamycin) protein which is a cell-cycle-specific serine/threonine kinase involved in cell growth, proliferation, protein transcription, initiation, and translation (Proud 2007).

In addition to their involvement in suppression of the immune reaction, the cellular role of immunophilins is their involvement in protein folding. This is mediated by two different activities localized within separate immunophilin domains: PPIase (peptidyl-prolyl cis-trans isomerase) domain, which is a rate-limiting enzyme activity related to acquisition of functional conformation of proteins (Lang et al. 1987) and their chaperone activity (Barik 2006; Galat 2003). The PPIase domain of both immunophilin receptors and its role in acquisition of functional conformation of proteins was established (Barik 2006; Fischer et al. 1989; Galat 2003; Pirkl et al. 2001; Zydowsky et al. 1992). Likewise, the domains responsible for the chaperone activity (Mok et al. 2006; Pirkl et al. 2001) were mapped in some immunophilins. Binding of immunosuppressive drugs to their immunophilin receptors inhibits both PPIase and chaperone activity (Barik 2006; Galat 2003).

CsA-dependent reduction in surface expression has been demonstrated for several membrane proteins: the homo-oligomeric acetylcholine receptor, the homo-oligomeric type 3 hydroxytryptamine receptor, for Kir2.1 potassium channel, and the creatinine transporter (Chen et al. 1998; Helekar et al. 1994; Helekar and Patrick 1997; Tran et al. 2000).

Four different genes code for the Na^+–Ca^{2+} exchangers (Cai and Lytton 2004; On et al. 2009). Three of the gene products NCX1, NCX2, and NCX3 were detected in mammalian genomes, whereas NCX4 was detected in teleost genomes. They all share high amino acid sequence and structural homology. The presence of intramolecular homology and relative conservation of transmembrane segments suggests that the various members arose by gene duplication of an ancient gene (Schwartz and Benzer 1997).

Many studies were carried out to detect functional differences among the mammalian NCX proteins in order to understand the importance of different gene products, multiple splice isoforms, and tissue-specific expression. Among these were differences in ionic sensitivity (Iwamoto et al. 1999b), isoform-specific pharmacological differences (Iwamoto and Kita 2006), isoform-specific neuroprotection (Pignataro et al. 2011), functional differences (Omelchenko et al. 1998), changes in expression during oligodendrocyte differentiation (Boscia et al. 2009, 2011), cell death, and many more.

The immunosuppressive drugs – CsA, FK506, and rapamycin – were shown to cross the blood–brain barrier (Hsiao et al. 2006; Kochi et al. 1999; Lemaire et al. 1996; Pong and Zaleska 2003; Shirai et al. 1994; Tai 2000). This can potentially expose one, two or all NCX protein isoforms in the brain to modulation by a particular immunosuppressive drug- depending on the drug specificity and the appropriate isoform. Therefore and due to the tissue-specific expression of the *NCX* genes (Annunziato et al. 2004; Quednau et al. 2004), it is of specific interest to explore the specificity of immunosuppressive drugs and their interaction with NCX1, NCX2, and NCX3 proteins.

23.3 Immunosuppressive Drugs and Na^+–Ca^{2+} Exchanger Isoforms

The three mammalian Na^+–Ca^{2+} exchangers NCX1, NCX2, and NCX3 share about 75 % overall amino acid sequence identity (Nicoll et al. 1996). Substantial amount of research was done to elucidate the spatial organization of NCX1 protein. This was done by hydropathy analysis, mutagenesis, epitope tagging, NMR, and X-ray crystallography (Cook et al. 1998; Hilge et al. 2006; Iwamoto et al. 1999a; Nicoll et al. 1990, 1999, 2006). It is generally assumed that NCX2 and NCX3 proteins are organized similarly.

Treatment of HEK 293 cells expressing each one of the mammalian NCX isoforms with CsA results in dose-dependent reduction of NCX surface expression, parallel reduction in Na^+-dependent Ca^{2+} transport activity, and no change in total cell NCX protein expression (Kimchi-Sarfaty

Table 23.1 Effect of CsA, FK506, rapamycin, and PSC833 on functional expression of NCX1, NCX2, and NCX3 proteins

	Surface expression	Transport activity	Total cell NCX protein	Drug
NCX1	↓	↓	–	CsA
NCX2	↓	↓	–	
NCX3	↓	↓	–	
NCX1	–	–	–	FK506
NCX2	↓	↓	–	
NCX3	↓	↓	–	
NCX1	–	–	–	Rapamycin
NCX2	–	↓	–	
NCX3	–	↓	–	
NCX1	↓	↓	–	PSC833
NCX2	↓	↓	↓	
NCX3	↓	↓	↓	

This table compiles schematically the results of experiments in which the effect of immunosuppressive and non-immunosuppressive drugs on functional expression of NCX isoforms was examined. HEK 293 cells were transfected with cloned (in pcDNA3.1) N-terminal Flag tagged (FN) *NCX1.5*, *NCX2*, or *NCX3*. Four hours posttransfection, different amounts of each drug as specified were added to the transfected cells. The following concentrations of each drug were used: CsA to transfected cells 0–50 μM; naturally expressing cells, H9c2, L6, or SMC, 0–10 μM; FK506 0–70 μM; rapamycin 0–30 μM; and PSC833 0-30 μM. Na^+-dependent Ca^{2+} uptake, surface-expressed NCX protein, and total cell NCX protein were determined as described in detail (Elbaz et al. 2008; Kimchi-Sarfaty et al. 2002). Downregulation is marked by vertical down pointing arrow; no change is shown by horizontal line

et al. 2002). Similar results are obtained also when H9c2, L6, or smooth muscle aortic cells, all of which express NCX1 protein naturally, are treated with CsA (Rahamimoff et al. 2007).

Unlike CsA, which modulates the expression of all three mammalian NCX proteins similarly, FK506 and rapamycin modulate the functional expression of NCX2 and NCX3 proteins only (Elbaz et al. 2008). Neither FK506 nor rapamycin modulates the functional expression of NCX1 protein (see Table 23.1).

Although FK506 and rapamycin bind to the same cellular receptors from the FKBP family, they modulate the expression of NCX2 and NCX3 proteins differently. Treatment of HEK 293 cells transfected with plasmids encoding *FN-NCX2* or *FN-NCX3* with FK506 reduces the surface expression and transport activity of both proteins, with no change in total cell NCX protein. In this respect, both FK506 and CsA modulate the functional expression of NCX proteins similarly in a posttranslational manner.

Rapamycin, treatment of transfected HEK 293 cells, however, reduces NCX2 and NCX3 transport activity but has no effect on their surface expression or total cell NCX protein expression. This suggests that impaired FN-NCX2 and FN-NCX3 proteins that are formed by a yet unknown pathway bypass the ER-related quality control mechanism and reach the surface membrane (Elbaz et al. 2008). It is possible that the differences in modulation of NCX functional expression between CsA, FK506, and rapamycin are related to the difference between the calcineurin pathway (Klee et al. 1998), which both complexes, CsA/cyclophilins and FK506/FKBPs, follow, and the rapamycin/FKBP–mTOR pathway (Wang and Proud 2006).

We have also examined the effect of PSC833, a non-immunosuppressive chemical analogue of CsA (Boesch et al. 1991) on functional expression of NCX1, NCX2, and NCX2 proteins (Elbaz et al. 2008; Kimchi-Sarfaty et al. 2002). PSC833 treatment of *NCX1*-transfected HEK293 cells, similarly to CsA, reduced surface expression and Na^{+}–Ca^{2+} exchange activity of NCX1 protein and had no impact on total cell NCX1 protein. Treatment of HEK293 cells transfected with *NCX2* or *NCX3* with PSC833 reduced surface expression and Na^{+}–Ca^{2+} exchange activity, but unlike its effect on the expression of NCX1, the drug reduced also total cell NCX2 and NCX3 protein expression (Elbaz et al. 2008). To elucidate the mode of action of PSC833 on NCX2 and NCX3 expression, we have measured the corresponding mRNA levels by quantitative PCR. The PSC833-dependent reduction in mRNA levels suggests that the drug acted at the transcriptional level.

The concentrations of the drugs that we have used in our research (Elbaz et al. 2008; Kimchi-Sarfaty et al. 2002; Rahamimoff et al. 2007), CsA, FK506, and rapamycin, are similar to those that are used clinically in transplant patients to protect against rejection (Crespo et al. 2009; Holt et al. 2000; Kramer et al. 2010; Oellerich et al. 1995; Trunecka et al. 2010). Clinical trials and different surveys compare the advantages and adverse effects including clinical complications that the different immunosuppressive drugs cause (First 2004; Halloran 2004; Kramer et al. 2005). Several protocols employ combinations of two different drugs – reporting lower rejection and lesser complications. Based on the diversity of the interaction between different immunosuppressive drugs and different NCX isoform and the possibility that additional players take part in the various clinical complications, this approach seems to offer advantages.

23.4 Cyclophilins and NCX1 Expression

Downregulation of NCX surface expression and transport activity by CsA treatment of cells expressing the protein could have potentially resulted via two alternative pathways: (1) by binding of the drug to its cyclophilin receptor and resulting in inhibition of the enzyme activity of PPIase and/or the chaperone activity (Barik 2006; Galat 2003; Mok et al. 2006). Inhibition of these activities could potentially impair NCX protein folding and reduce its trafficking to its cellular destination (Kopito 1997), the surface membrane. (2) Alternatively, CsA was shown to generate reactive oxygen species that could in turn damage the surface-expressed NCX protein (Ahmed et al. 1993, 1996; Huschenbett et al. 1998; Iwamoto et al. 1999a) and result in downregulation of Na^{+}–Ca^{2+} exchange activity. Experiments that we did, however, to test this hypothesis suggested (Elbaz et al. 2008; Kimchi-Sarfaty et al. 2002) that this possibility was unlikely. Generation of CsA-derived reactive oxygen species was shown to occur rapidly, and detection of its effects was noted within 1–15 min. When we added CsA 24-h posttransfection directly to the buffered Na^{+}–Ca^{2+} exchange reaction mixture for 20 min, no effect on reduced NCX surface expression and transport activity could be detected.

We examined, therefore, the possible involvement of cyclophilins in functional expression of NCX protein directly, without the involvement of CsA (Elbaz et al. 2010). To do so, we have knocked down cell cyclophilin A or cell cyclophilin B using targeting siRNA and examined the results of the knockdown by two experimental approaches: (1) biochemical methodology (Elbaz et al. 2010) to measure surface-expressed NCX protein, Na^{+}-dependent Ca^{2+} transport activity, and total cell NCX1 protein expression in comparison with cells that were transfected with nontargeting siRNA. (2) We measured Na^{+}-dependent Ca^{2+} fluxes in Fluo-4 AM-loaded single cells. Two expression systems were used, siRNA *and FN NCX1.5* transfected HEK 293 cells and siRNA transfected H9c2 cells, that express NCX1 naturally. The repertoire of siRNAs that were transfected into the cells included cyclophilin A or B targeting and nontargeting siRNAs. The amounts to be used for transfection were precalibrated in preliminary experiments (Elbaz et al. 2010). The results of these experiments are summarized in Table 23.2 schematically.

The experimental protocols that were used in these experiments demanded two separate transfections (siRNA and *FN-NCX1.5*) 24 h apart when

Table 23.2 The effect of cyclophilin A or cyclophilin B knockdown on NCX1 surface expression, Na^+-dependent Ca^{2+} uptake, and total cell NCX1 protein

	Nontargeting siRNA	Cyclophilin A siRNA	Cyclophilin B siRNA
NCX1 surface expression	–	↓	–
Na^+-dependent Ca^{2+} uptake	–	↓	–
Total cell NCX1 protein	–	–	–

This table summarizes experiments that were done in HEK 293 cells or in H9c2 cells. HEK 293 cells were transfected with nontargeting siRNA (20 nM) or siRNA-targeting CypA (10 nM) or CypB (20 nM). Twenty-four hours later, the cells were transfected with a plasmid encoding the Na^+–Ca^{2+} exchanger *FN-NCX1.5* (1 μg per well of 1/12 wells). Twenty-four hours posttransfection with *FN-NCX1.5,* Na^+-dependent Ca^{2+} uptake, surface expression, and total cell NCX protein were determined. H9c2 cells were transfected with nontargeting siRNA (20 nM) or siRNA-targeting CypA (10 nM) or CypB (20 nM). Eight days later, Na^+-dependent Ca^{2+} uptake, surface expression, and total cell NCX1 protein were determined as described in detail in Elbaz et al. (2010). Downregulation is marked by vertical down pointing arrow and no change by a horizontal line

HEK 293 cells were used and a single transfection step (siRNA) for H9c2 cells (Elbaz et al. 2010). Therefore, although our transfection procedures of the Na^+–Ca^{2+} exchanger using DharmaFECT reagent for siRNA and PEI (polyethyleneimine) reagent (von Harpe et al. 2000) for *FN-NCX1.5* were quite efficient (transfection rates were 71.6 %; S.D. of 8.7), not all cells expressed both siRNA and NCX protein. By using optical methods, we could select cells that expressed FN-NCX1.5 (immunostained with M2 anti-Flag antibody), stained for siRNA with siGLO red transfection indicator, and were Fluo-4 AM loaded.

Figure 23.1 summarizes the properties of the Na^+–Ca^{2+} exchange reaction in transfected HEK 293 cells. Panel A left shows mock-transfected (with *pcDNA3.1*) HEK 293 cells that were loaded with 5-μM Fluo-4 AM in a buffered Na^+-loading solution.

Exchange of the bathing solution of the cells to a buffered K^+ solution has no impact on the Fluo-4 AM fluorescence (panel A – right). Following transfection of the cells with *FN-NCX1.5* (panel B), the cells that express the protein can be stained with M2, the anti-Flag antibody (marked by arrows). Exchange of the bathing solution to a buffered K^+ solution results in an increase in Fluo-4 AM fluorescence (panel B – right) of the FN-NCX1.5 expressing cells. Panel C shows that *FN-NCX1.5*-transfected cells carry out bidirectional Na^+–Ca^{2+} exchange, depending on the direction of the Na^+ gradient. Figure 23.1d shows cells that are Fluo-4 AM loaded (green), express FN-NCX1 (blue), and contain siRNA (red).

The combination of biochemical and optical measurements showed that cyclophilin A but not B was involved in functional expression of NCX1 protein (Elbaz et al. 2010).

Fig. 23.1 (continued) FN-NCX1.5 expressing cells. *Panel C* shows that the direction of the calcium flux depends on the sodium gradient. FN-NCX1.5-transfected HEK 293 cells were loaded with 5-μM Fluo-4 AM (*green*) in a Na^+-loading solution. Exchange from Na^+- to K^+-loading solution resulted in a rapid elevation of Fluo-4 AM fluorescence indicating Ca^{2+} influx. Exchange of the bathing solution from K^+ to Na^+ resulted in a rapid reduction of Fluo-4 AM fluorescence, indicating Ca^{2+} efflux. $n=x$ shows the number of cells from which the data were compiled. (**d**) HEK 293 cells were transfected with siRNAs (nontargeting, targeting cyclophilin A or B) and 24 h later with FN-NCX1.5. They were loaded with 5-μM Fluo-4 AM in a Na^+-loading solution. To detect FN-NCX1.5 expression, the Fluo-4 AM-loaded cells were labeled with M2 anti-Flag monoclonal antibodies and Cy5 conjugated anti-mouse secondary Ab. siRNA transfection reagent contained siGLO red transfection indicator. *Panel D* shows an overlay of Fluo-4 AM-loaded cells expressing FN-NCX1.5 (*blue*) that contain siRNA (*red*) (methods are described in Elbaz et al. (2010)

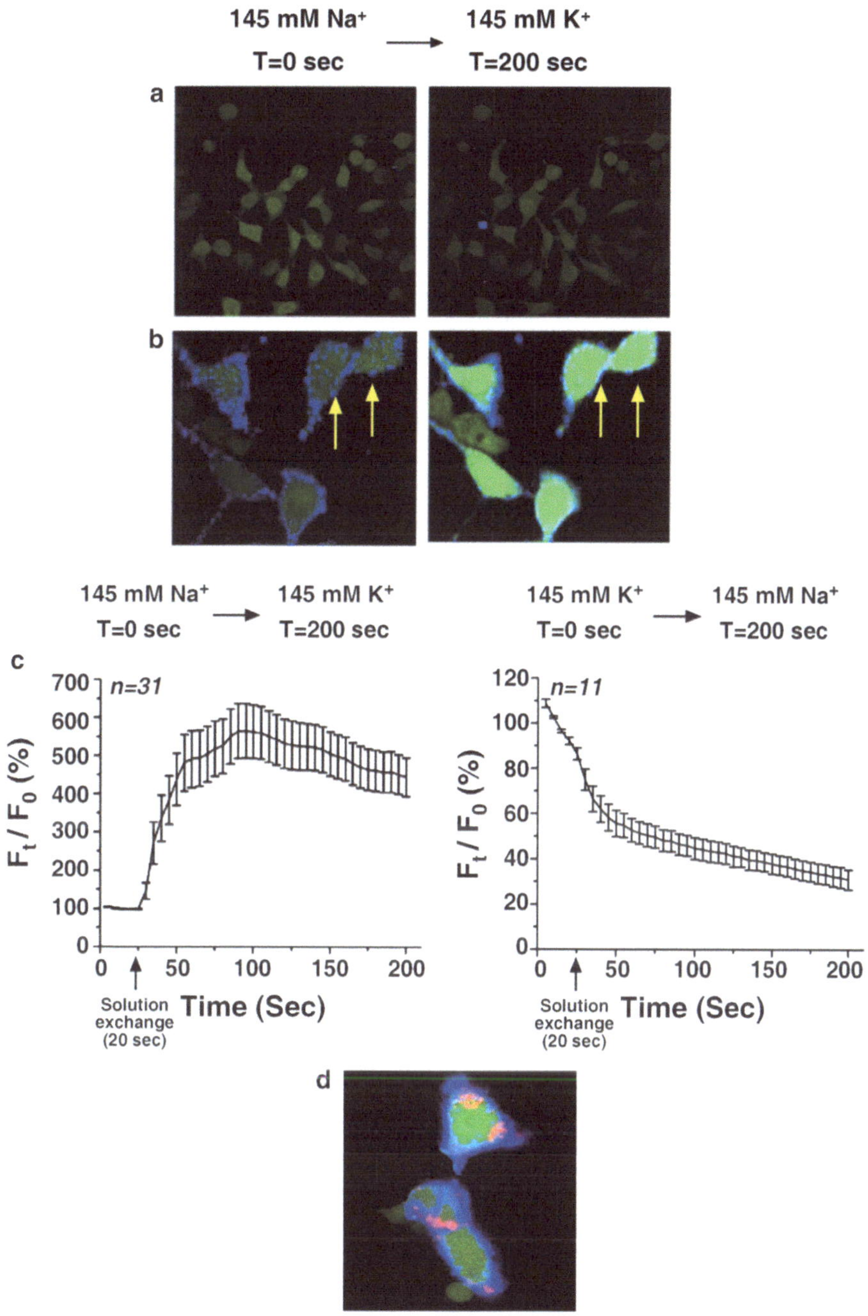

Fig. 23.1 *Na^+-dependent Ca^{2+} fluxes in transfected HEK 293 cells.* (**a**) HEK 293 cells were transfected with *pcDNA3.1* (mock-transfected) *panel A* or with *FN-NCX1.5* (in *pcDNA3.1*) *panel B*. The cells were loaded with 5-μM Fluo-4 AM (*green*) in a buffered Na^+-loading solution. FN-NCX1.5 expression was determined by staining the cells with the M2 primary anti-Flag monoclonal antibodies and Cy5 conjugated anti-mouse secondary antibodies (*blue*). Exchange of cell bathing solution from Na^+-(*left panel*) to K^+- (*right panel*) loading solution was used to initiate Na^+-dependent Ca^{2+} fluxes. *Panel B* is a 2× fold magnification compared to *A*. *Arrows* indicate two of the

23.5 Overexpression of Cyclophilins and NCX Expression

Transfected cells produce large amounts of the expressed protein, most of which is cell retained and only small amounts reach the surface membrane. To study the involvement of cyclophilins in functional expression of NCX1 protein, we have carried out experiments to examine whether overexpression of cyclophilin A or of cyclophilin B can mobilize some of the cell-retained NCX1 and increase its expression in the cell surface. Preliminary experiments were done to ensure that both naturally expressed and transfected Cyps (6xHis-tagged) were expressed. Our results, however, suggested that overexpression of cyclophilin A or of cyclophilin B did not increase the cell surface-expressed NCX protein nor its transport activity (see Table 23.3).

Surprisingly, both cyclophilins A and B alleviated the CsA-related downregulation of surface expression and reduction of transport activity of NCX1 protein (Elbaz et al. 2010). Similar results were also obtained when instead of cyclophilin A, co-transfection was carried out with R55A mutant of cyclophilin A (Helekar and Patrick 1997) in which the PPIase activity was substantially reduced (not shown). This suggests that the PPIase domain of cyclophilin A was probably not mandatory for NCX1 functional expression.

Price et al. (1994) have shown that CsA mobilized cyclophilin B from the endoplasmic reticulum and promoted its secretion from HeLa and BHK cells. If this was the case also in transfected HEK 293 cells, this finding could explain why overexpression of cyclophilin B, which had no impact on NCX1 surface expression or transport activity, did relieve substantially the effect of CsA.

Figure 23.2 shows the cell content and the extracellular medium content of cyclophilin A and cyclophilin B determined by Western analysis in HEK 293 cells that were co-transfected with *FN-NCX1.5* and cloned *cyclophilin A or FN-NCX1* and cloned *cyclophilin B* in the absence (control cells) or in the presence of CsA. It can be seen that although cyclophilin B is released from HEK 293 cells also in the absence of CsA (left panel vs. right panel), addition of the drug to the transfected cells reduces substantially the cell content of cyclophilin B and increases dramatically its content in the extracellular medium. The cell content of cyclophilin A is almost not altered by treatment of the transfected HEK 293 cells with CsA, and only trace amounts are released to the external medium. This suggests that cyclophilin A and cyclophilin B relieve the CsA-dependent downregulation of NCX1 surface expression by different modes of action. Overexpression of cyclophilin A probably reduces the effective cell concentration of CsA whereas in cyclophilin B overexpressing cells, the drug promotes the release of the complex.

We have also examined whether overexpression of cyclophilin 60 or cyclophilin 40 together

Table 23.3 The effect of overexpression of cyclophilin A and cyclophilin B on Na^+-dependent Ca^{2+} transport activity and surface expression of NCX1 without and with CsA treatment

	Relative Na^+-dependent Ca^{2+} uptake (%)	Surface expression (%) (MFI)
NCX1	100	100
NCX1 + cyclophilin A	128 (S.D. = 26.5)	89 (S.D. = 13.4)
NCX1 + CsA (10 μM)	57 (S.D. = 22)	57.4 (S.D. = 11.8)
NCX1 + CsA + cyclophilin A	92 (S.D. = 31)	84.6 (S.D. = 15.6)
NCX1 + cyclophilin B	120 (S.D. = 23)	101.2 (S.D. = 1.06)
NCX1 + CsA + cyclophilin B	96 (S.D. = 23)	86 (S.D. = 21.53)

HEK 293 cells were transfected with a plasmid encoding the Na^+–Ca^{2+} exchanger *FN-NCX1.5* or co-transfected also with 6xHis-tagged cyclophilin A or 6×His-tagged cyclophilin B. Na^+-dependent Ca^{2+} transport activities were measured without and with 10-μM CsA treatment of the transfected cells. The drug was dissolved in DMSO, and an equal amount of DMSO was added to non-treated cells. The volume of DMSO was kept to 0.1 % of the medium. The drug was added to the cell medium 3 h after transfection and kept till 24 h. The transport activities of *FN-NCX1.5*-transfected cells were taken as 100 %, and all other transport activities were calculated in relative values

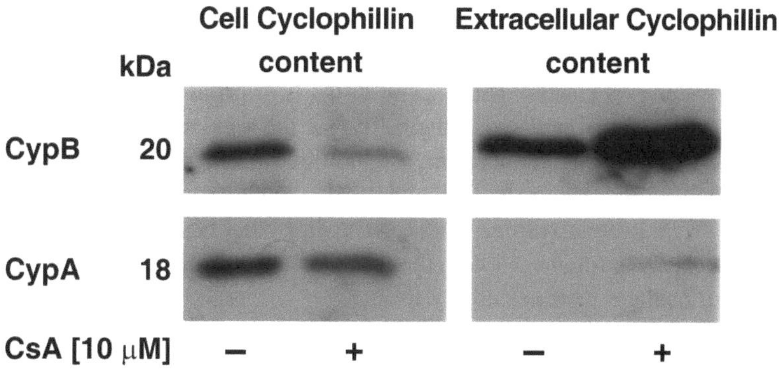

Fig. 23.2 *Distribution of cyclophilins in FN-NCX1.5 transfected cell interior and extracellular medium without and with CsA treatment.* HEK 293 cells were co-transfected with plasmids encoding *FN-NCX1* and 6xHis-tagged cyclophilin A or B without and with CsA treatment. The transfected cells were separated from the extracellular growth medium and lysed, and the membrane-free fraction was analyzed by SDS PAGE. The cell-free extracellular medium was concentrated and analyzed by SDS PAGE

with FN-NCX1.5 in HEK 293 cells without and with CsA had any effect on surface expression, transport activity, or total FN-NCX1.5 protein. No change in any of these parameters was detected when compared to cells transfected with *FN-NCX1.5* alone (not shown).

23.6 The Role of the Large Cytosolic Loop of the Na^+–Ca^{2+} Exchangers NCX1, NCX2, and NCX3 in Acquisition of Immunosuppressive Drug Specificity

Cloning of the first Na^+–Ca^{2+} exchanger – NCX1 (Nicoll et al. 1990) – and elucidation of its primary amino acid sequence suggested that the protein has a cleavable signal peptide and the mature protein starts with an extracellular amino terminal tail followed by a cluster of five transmembrane segments. These are separated from the four C-terminal hydrophobic segments (Nicoll et al. 1999) by a large cytosolic loop of about 500 amino acids. The hydrophobic transmembrane segments are involved in Na^+–Ca^{2+} exchange whereas the cytosolic loop regulates the transport process (Levitsky et al. 1994; Matsuoka et al. 1993). Truncation of most of the large cytosolic loop of NCX1 protein leaves a functional backbone that carries out Na^+–Ca^{2+} exchange. The relationship, however, between the exchange activity, remaining after truncation of the loop, to that of the WT parent exchanger varies and differs in the different expression system used (Eskin-Schwartz and Rahamimoff 1999; Matsuoka et al. 1993; Pan et al. 2000).

Our studies show that the three mammalian NCX isoforms respond in a different way to treatment with immunosuppressive and non-immunosuppressive drugs (Elbaz et al. 2008; Kimchi-Sarfaty et al. 2002). Since in every other respect, except their individual structural differences, all other experimental conditions were identical, we have started a study designed to identify the protein domains that could be involved in acquisition of immunosuppressive drug specificity. First, we have truncated parallel segments from within the large cytosolic loop of NCX2 and NCX3 to those that were previously published for NCX1 protein (Eskin-Schwartz and Rahamimoff 1999). The loop-truncated mutants were expressed in HEK 293 cells in parallel with their WT parent exchangers. Next, the constructs were exposed to immunosuppressive drugs as described previously (Elbaz et al. 2008; Kimchi-Sarfaty et al. 2002), and their Na^+–Ca^{2+} exchange activity, surface expression, and total cell NCX protein expression were determined. These studies showed that the large cytosolic loop-truncated mutants of FN-NCX1, FN-NCX2, and FN-NCX3 responded to CsA treatment similarly to their WT parent exchangers: Their surface expression was downregulated, their Na^+–Ca^{2+} exchange activity was reduced in a dose-dependent manner, and the drug had no effect on total cell NCX protein (Elmaz 2011). Similarly, NCX2 and NCX3 proteins, without the large cytosolic loop, retained their sensitivity to FK506 as their parent WT exchangers.

Interestingly, however, the WT NCX1 protein, which was not sensitive to FK506 or to rapamycin (Kimchi-Sarfaty et al. 2002), became FK506 and rapamycin sensitive when its large cytosolic loop was truncated (Elmaz 2011). The importance of the large cytosolic loop in acquisition of immunosuppressive drug sensitivity is further demonstrated when some NCX chimeras are constructed: for example, WT NCX3 protein, which is FK506 sensitive (Elbaz et al. 2008), loses its FK506 sensitivity when its large cytosolic loop is replaced by the large cytosolic loop of NCX1, which is FK506 insensitive. These results suggest that the large cytosolic loop (and its absence) plays a role in acquisition of immunosuppressive drug specificity.

23.7 Conclusions

While studying the involvement of cyclophilin A in functional expression of NCX1 protein and alleviation of the inhibitory effect of CsA by both cyclophilin A and B overexpression (Elbaz et al. 2010), we have attempted to co-immunoprecipitate the NCX1/cyclophilin complex. Yet attempts to do so were unsuccessful. Likewise, attempts to pull down the homo-oligomeric α7 nicotinic acetylcholine receptor with cyclophilin A (that was shown to participate in functional expression of the receptor) (Helekar and Patrick 1997) were unsuccessful. It is quite possible that an additional protein/s are participating in formation and/or stabilizing the complex. Several cytoskeletal proteins were shown to co-localize with NCX protein (Condrescu and Reeves 2006) or cyclophilin A (Galigniana et al. 2004a, b). Elucidation of the mode of interaction between NCX proteins, their immunophilin receptors, and the domains involved in interaction with immunosuppressive drugs is of prime importance in our understanding of the regulation of functional expression of the three mammalian NCX isoforms and hence also provides some clues toward understanding the molecular basis of the clinical complications related to immunosuppression.

Acknowledgments This research was supported in part by the Israel Science Foundation and the Israel Ministry of Health.

References

S.S. Ahmed, H.W. Strobel, K.L. Napoli, J. Grevel, Adrenochrome reaction implicates oxygen radicals in metabolism of cyclosporine A and FK-506 in rat and human liver microsomes. J. Pharmacol. Exp. Ther. **265**, 1047–1054 (1993)

S.S. Ahmed, K.L. Napoli, H.W. Strobel, Oxygen radical formation due to the effect of varying hydrogen ion concentrations on cytochrome P450-catalyzed cyclosporine metabolism in rat and human liver microsomes. Adv. Exp. Med. Biol. **387**, 135–139 (1996)

L. Annunziato, G. Pignataro, G.F. Di Renzo, Pharmacology of brain Na^+/Ca^{2+} exchanger: from molecular biology to therapeutic perspectives. Pharmacol. Rev. **56**, 633–654 (2004)

S. Barik, Immunophilins: for the love of proteins. Cell. Mol. Life Sci. **31**, 1–12 (2006)

W.O. Bechstein, Neurotoxicity of calcineurin inhibitors: impact and clinical management. Transpl. Int. **13**, 313–326 (2000)

D. Boesch, C. Gaveriaux, B. Jachez, A. Pourtier-Manzanedo, P. Bollinger, F. Loor, In vivo circumvention of P-glycoprotein-mediated multidrug resistance of tumor cells with SDZ PSC 833. Cancer Res. **51**, 4226–4233 (1991)

F. Boscia, R. Gala, A. Pannaccione, A. Secondo, A. Scorziello, G. Di Renzo, L. Annunziato, NCX1 expression and functional activity increase in microglia invading the infarct core. Stroke **40**, 3608–3617 (2009)

F. Boscia, C. D'Avanzo, A. Pannaccione, A. Second, A. Casamassa, L. Formisano, N. Guida, L. Annunziato, Silencing or knocking out the Na^+/Ca^{2+} exchanger-3 (NCX3) impairs oligodendrocyte differentiation. Cell Death Differ. (2011). doi:10.1038/cdd.2011.125. Epub ahead of print

X. Cai, J. Lytton, The cation/Ca^{2+} exchanger superfamily: phylogenetic analysis and structural implications. Mol. Biol. Evol. **21**, 1692–1703 (2004)

E. Carafoli, Intracellular calcium homeostasis. Ann. Rev. Biochem. **56**, 395–433 (1987)

H. Chen, Y. Kubo, T. Hoshi, S.H. Heinemann, Cyclosporin A selectively reduces the functional expression of Kir2.1 potassium channels in *Xenopus* oocytes. FEBS Lett. **422**, 307–310 (1998)

M. Condrescu, J.P. Reeves, Actin-dependent regulation of the cardiac Na^+/Ca^{2+} exchanger. Am. J. Physiol. **290**, C691–C701 (2006)

O. Cook, W. Low, H. Rahamimoff, Membrane topology of the rat brain sodium-calcium exchanger. Biochim. Biophys. Acta **1371**, 40–52 (1998)

M. Crespo, M. Mir, M. Marin, S. Hurtado, C. Estadella, X. Guri, O. Rap, R. Moral, J.M. Puig, J. Lloveras, De novo kidney transplant recipients need higher doses of Advagraf compared with Prograf to get therapeutic levels. Transplant. Proc. **41**, 2115–2117 (2009)

B. Elbaz, A. Alperovitch, M.M. Gottesman, C. Kimchi-Sarfaty, H. Rahamimoff, Modulation of Na^+-Ca^{2+} exchanger expression by immunosuppressive drugs is isoform-specific. Mol. Pharmacol. **73**, 1254–1263 (2008)

B. Elbaz, M. Valitsky, G. Davidov, H. Rahamimoff, Cyclophilin A is involved in functional expression of the Na^+-Ca^{2+} exchanger NCX1. Biochemistry **49**, 7634–7642 (2010)

D. Elmaz, The involvement of the large cytoplasmic loop of the Na^+-Ca^{2+} exchanger in functional expression of the protein. M.sc. thesis, Submitted and approved by the Hebrew University -Hadassah Medical School, Jerusalem, 2011

M. Eskin-Schwartz, H. Rahamimoff, Expression of a mutant of the Na/Ca exchanger RBE-1 lacking 442 amino acids in the large cytoplasmic loop of the protein in HEK 293 cells. Neurosci. Lett. **54**, S36 (1999)

M.R. First, Tacrolimus based immunosuppression. J. Nephrol. **17**(Suppl 8), S25–S31 (2004)

G. Fischer, B. Wittmann-Liebold, K. Lang, T. Kiefhaber, F.X. Schmid, Cyclophilin and peptidyl-prolyl cis-trans isomerase are probably identical proteins. Nature **337**, 476–478 (1989)

A. Galat, Peptidylprolyl cis/trans isomerases (immunophilins): biological diversity–targets–functions. Curr. Top. Med. Chem. **3**, 1315–1347 (2003)

M.D. Galigniana, J.M. Harrell, H.M. O'Hagen, M. Ljungman, W.B. Pratt, Hsp90-binding immunophilins link p53 to dynein during p53 transport to the nucleus. J. Biol. Chem. **279**, 22483–22489 (2004a)

M.D. Galigniana, Y. Morishima, P.A. Gallay, W.B. Pratt, Cyclophilin-A is bound through its peptidylprolyl isomerase domain to the cytoplasmic dynein motor protein complex. J. Biol. Chem. **279**, 55754–55759 (2004b)

P.F. Halloran, Immunosuppressive drugs for kidney transplantation. N. Engl. J. Med. **351**, 2715–2729 (2004)

S. Hariharan, C.P. Johnson, B.A. Bresnahan, S.E. Taranto, M.J. McIntosh, D. Stablein, Improved graft survival after renal transplantation in the United States, 1988 to 1996. N. Engl. J. Med. **342**, 605–612 (2000)

S.A. Helekar, J. Patrick, Peptidyl prolyl cis-trans isomerase activity of cyclophilin A in functional homo-oligomeric receptor expression. Proc. Natl. Acad. Sci. U. S. A. **94**, 5432–5437 (1997)

S.A. Helekar, D. Char, S. Neff, J. Patrick, Prolyl isomerase requirement for the expression of functional homo-oligomeric ligand-gated ion channels. Neuron **12**, 179–189 (1994)

M. Hilge, J. Aelen, G.W. Vuister, Ca^{2+} regulation in the Na^+/Ca^{2+} exchanger involves two markedly different Ca^{2+} sensors. Mol. Cell **22**, 15–25 (2006)

D.W. Holt, A. Johnston, B.D. Kahan, R.G. Morris, M. Oellerich, L.M. Shaw, New approaches to cyclosporine monitoring raise further concerns about analytical techniques. Clin. Chem. **46**, 872–874 (2000)

P. Hsiao, L. Sasongko, J.M. Link, D.A. Mankoff, M. Muzi, A.C. Collier, J.D. Unadkat, Verapamil P-glycoprotein transport across the rat blood–brain barrier: cyclosporine, a concentration inhibition analysis, and comparison with human data. J. Pharmacol. Exp. Ther. **317**, 704–710 (2006)

J. Huschenbett, A. Zaidi, M.L. Michaelis, Sensitivity of the synaptic membrane Na^+/Ca^{2+} exchanger and the expressed NCX1 isoform to reactive oxygen species. Biochim. Biophys. Acta **1374**, 34–46 (1998)

M.T. Ivery, Immunophilins: switched on protein binding domains? Med. Res. Rev. **20**, 452–484 (2000)

T. Iwamoto, S. Kita, YM-244769, a novel Na^+/Ca^{2+} exchange inhibitor that preferentially inhibits NCX3, efficiently protects against hypoxia/reoxygenation-induced SH-SY5Y neuronal cell damage. Mol. Pharmacol. **70**, 2075–2083 (2006)

T. Iwamoto, T.Y. Nakamura, Y. Pan, A. Uehara, I. Imanaga, M. Shigekawa, Unique topology of the internal repeats in the cardiac Na^+/Ca^{2+} exchanger. FEBS Lett. **446**, 264–268 (1999a)

T. Iwamoto, A. Uehara, T.Y. Nakamura, I. Imanaga, M. Shigekawa, Chimeric analysis of Na^+/Ca^{2+} exchangers NCX1 and NCX3 reveals structural domains important for differential sensitivity to external Ni^{2+} or Li^+. J. Biol. Chem. **274**, 23094–23102 (1999b)

J. Kasir, X. Ren, I. Furman, H. Rahamimoff, Truncation of the C-terminal of the rat brain Na^+-Ca^{2+} exchanger RBE-1 (NCX1.4) impairs surface expression of the protein. J. Biol. Chem. **274**, 24873–24880 (1999)

C. Kimchi-Sarfaty, J. Kasir, S. Ambudkar, H. Rahamimoff, Transport activity and surface expression of the Na^+-Ca^{2+} exchanger NCX1 is inhibited by the immunosuppressive agent cyclosporin A and the non-immunosuppressive agent PSC833. J. Biol. Chem. **277**, 2505–2510 (2002)

C.B. Klee, H. Ren, X. Wang, Regulation of the calmodulin-stimulated protein phosphatase, calcineurin. J. Biol. Chem. **273**, 13367–13370 (1998)

S. Kochi, H. Takanaga, H. Matsuo, M. Naito, T. Tsuruo, Y. Sawada, Effect of cyclosporin A or tacrolimus on the function of blood–brain barrier cells. Eur. J. Pharmacol. **372**, 287–295 (1999)

R.R. Kopito, ER quality control: the cytoplasmic connection. Cell **88**, 427–430 (1997)

B.K. Kramer, C. Boger, B. Kruger, J. Marienhagen, M. Pietrzyk, A. Obed, L. Paczek, M. Mack, B. Banas, Cardiovascular risk estimates and risk factors in renal transplant recipients. Transplant. Proc. **37**, 1868–1870 (2005)

B.K. Kramer, B. Charpentier, L. Backman, H.T. Silva Jr., G. Mondragon-Ramirez, E. Cassuto-Viguier, G. Mourad, R. Sola, P. Rigotti, J.O. Mirete, Tacrolimus once daily (ADVAGRAF) versus twice daily (PROGRAF) in de novo renal transplantation: a randomized phase III study. Am. J. Transplant. **10**, 2632–2643 (2010)

K. Lang, F.X. Schmid, G. Fischer, Catalysis of protein folding by prolyl isomerase. Nature **329**, 268–270 (1987)

M. Lemaire, A. Bruelisauer, P. Guntz, H. Sato, Dose-dependent brain penetration of SDZ PSC 833, a novel multidrug resistance-reversing cyclosporin, in rats. Cancer Chemother. Pharmacol. **38**, 481–486 (1996)

D.O. Levitsky, D.A. Nicoll, K.D. Philipson, Identification of the high affinity Ca^{2+}–binding domain of the cardiac Na^{+}–Ca^{2+} exchanger. J. Biol. Chem. **269**, 22847–22852 (1994)

G.A. Levy, Neoral use in the liver transplant recipient. Transplant. Proc. **32**(3A Suppl), 2S–9S (2000)

J. Liu, M.W. Albers, T.J. Wandless, S. Luan, D.G. Alberg, P.J. Belshaw, P. Cohen, C. MacKintosh, C.B. Klee, S.L. Schreiber, Inhibition of T cell signaling by immunophilin-ligand complexes correlates with loss of calcineurin phosphatase activity. Biochemistry **31**, 3896–3901 (1992)

S. Matsuoka, D.A. Nicoll, R.F. Reilly, D.W. Hilgemann, K.D. Philipson, Initial localization of regulatory regions of the cardiac sarcolemmal Na^{+}-Ca^{2+} exchanger. Proc. Natl. Acad. Sci. U. S. A. **90**, 3870–3874 (1993)

H.U. Meier-Kriesche, B.J. Steffen, A.H. Chu, J.J. Loveland, R.D. Gordon, J.A. Morris, B. Kaplan, Sirolimus with neoral versus mycophenolate mofetil with neoral is associated with decreased renal allograft survival. Am. J. Transplant. **4**, 2058–2066 (2004)

H.U. Meier-Kriesche, A.H. Chu, K.M. David, K. Chi-Burris, B.J. Steffen, Switching immunosuppression medications after renal transplantation–a common practice. Nephrol. Dial. Transplant. **21**, 2256–2262 (2006)

D. Mok, R.K. Allan, A. Carrello, K. Wangoo, M.D. Walkinshaw, T. Ratajczak, The chaperone function of cyclophilin 40 maps to a cleft between the prolyl isomerase and tetratricopeptide repeat domains. FEBS Lett. **580**, 2761–2768 (2006)

D.A. Nicoll, S. Longoni, K.D. Philipson, Molecular cloning and functional expression of the cardiac sarcolemmal Na^{+}-Ca^{2+} exchanger. Science **250**, 562–565 (1990)

D.A. Nicoll, B.D. Quednau, Z. Qui, Y.-R. Xia, A.J. Lusis, K.D. Philipson, Cloning of a third mammalian Na^{+} -Ca^{2+} exchanger, NCX3. J. Biol. Chem. **271**, 24914–24921 (1996)

D.A. Nicoll, M. Ottolia, L. Lu, Y. Lu, K.D. Philipson, A new topological model of the cardiac sarcolemmal Na^{+}-Ca^{2+} exchanger. J. Biol. Chem. **274**, 910–917 (1999)

D.A. Nicoll, M. Sawaya, S. Kwon, D. Cascio, K.D. Philipson, J. Abramson, The crystal structure of the primary Ca^{2+} sensor of the Na^{+}/Ca^{2+} exchanger reveals a novel Ca^{2+} binding motif. J. Biol. Chem. **281**, 21577–25181 (2006)

M. Oellerich, V.W. Armstrong, B. Kahan, L. Shaw, D.W. Holt, R. Yatscoff, A. Lindholm, P. Halloran, K. Gallicano, K. Wonigeit et al., Lake louise consensus conference on cyclosporin monitoring in organ transplantation: report of the consensus panel. Ther. Drug Monit. **17**, 642–654 (1995)

A. Omelchenko, C. Dyck, M. Hnatowich, J. Buchko, D. Nicol, K. Philipson, L. Hryshko, Functional differences in ionic regulation between alternatively spliced isoforms of the Na^{+}-Ca^{2+} exchanger from *Drosophila* Melanogaster. J. Gen. Physiol. **111**, 691–702 (1998)

C. On, C.R. Marshall, S.F. Perry, H.D. Le, V. Yurkov, A. Omelchenko, M. Hnatowich, L.V. Hryshko, G.F. Tibbits, Characterization of zebrafish (Danio rerio) NCX4: a novel NCX with distinct electrophysiological properties. Am. J. Physiol. **296**, C173–C181 (2009)

Y. Pan, T. Iwamoto, A. Uehara, T.Y. Nakamura, I. Imanaga, M. Shigekawa, Physiological functions of the regulatory domains of the cardiac Na^{+}/Ca^{2+} exchanger NCX1. Am. J. Physiol. **279**, C393–C402 (2000)

G. Pignataro, E. Esposito, O. Cuomo, R. Sirabella, F. Boscia, N. Guida, G. Di Renzo, L. Annunziato, The NCX3 isoform of the Na^{+}/Ca^{2+} exchanger contributes to neuroprotection elicited by ischemic postconditioning. J. Cereb. Blood Flow Metab. **31**, 362–370 (2011)

F. Pirkl, E. Fischer, S. Modrow, J. Buchner, Localization of the chaperone domain of FKBP52. J. Biol. Chem. **276**, 37034–37041 (2001)

K. Pong, M.M. Zaleska, Therapeutic implications for immunophilin ligands in the treatment of neurodegenerative diseases. Curr. Drug Targets CNS Neurol. Disord. **2**, 349–356 (2003)

E.R. Price, M. Jin, D. Lim, S. Pati, C.T. Walsh, F.D. McKeon, Cyclophilin B trafficking through the secretory pathway is altered by binding of cyclosporin A. Proc. Natl. Acad. Sci. U. S. A. **91**, 3931–3935 (1994)

C.G. Proud, Signalling to translation: how signal transduction pathways control the protein synthetic machinery. Biochem. J. **403**, 217–234 (2007)

B.D. Quednau, D.A. Nicoll, K.D. Philipson, The sodium/calcium exchanger family-SLC8. Pflugers Arch. **447**, 543–548 (2004) Epub 2003 May 7

H. Rahamimoff, B. Elbaz, A. Alperovich, C. Kimchi-Sarfaty, M.M. Gottesman, Y. Lichtenstein, M. Eskin-Schwartz, J. Kasir, Cyclosporin A-dependent downregulation of the Na^{+}/Ca^{2+} exchanger expression. Ann. N. Y. Acad. Sci. **1099**, 204–214 (2007)

S.L. Schreiber, Chemistry and biology of the immunophilins and their immunosuppressive ligands. Science **251**, 283–287 (1991)

E.M. Schwarz, S. Benzer, Calx, a Na-Ca exchanger gene of *Drosophila* Melanogaster. Proc. Natl. Acad. Sci. U. S. A. **94**, 10249–10254 (1997)

A. Shirai, M. Naito, T. Tatsuta, J. Dong, K. Hanaoka, K. Mikami, T. Oh-hara, T. Tsuruo, Transport of cyclosporin A across the brain capillary endothelial cell monolayer by P-glycoprotein. Biochim. Biophys. Acta **1222**, 400–404 (1994)

H.L. Tai, Technology evaluation: Valspodar, Novartis AG. Curr. Opin. Mol. Ther. **2**, 459–467 (2000)

T.T. Tran, W. Dai, H.K. Sarkar, Cyclosporin A inhibits creatine uptake by altering surface expression of the creatine transporter. J. Biol. Chem. **275**, 35708–35714 (2000)

P. Trunecka, O. Boillot, D. Seehofer, A.D. Pinna, L. Fischer, B.G. Ericzon, R.I. Troisi, U. Baccarani, J. Ortizde Urbina, W. Wall, Once-daily prolonged-release tacrolimus (ADVAGRAF) versus twice-daily tacrolimus (PROGRAF) in liver transplantation. Am. J. Transplant. **10**, 2313–2323 (2010)

C.K. Tsang, H. Qi, L.F. Liu, X.F. Zheng, Targeting mammalian target of rapamycin (mTOR) for health and diseases. Drug Discov. Today **12**, 112–124 (2007)

A. von Harpe, H. Petersen, Y. Li, T. Kissel, Characterization of commercially available and synthesized polyethylenimines for gene delivery. J. Control. Release **69**, 309–322 (2000)

X. Wang, C.G. Proud, The mTOR pathway in the control of protein synthesis. Physiology (Bethesda, Md.) **21**, 362–369 (2006)

L.D. Zydowsky, F.A. Etzkorn, H.Y. Chang, S.B. Ferguson, L.A. Stolz, S.I. Ho, C.T. Walsh, Active site mutants of human cyclophilin A separate peptidyl-prolyl isomerase activity from cyclosporin A binding and calcineurin inhibition. Protein Sci. **1**, 1092–1099 (1992)

Calcium Influx Through Reversed NCX Controls Migration of Microglia

24

Mami Noda, Masataka Ifuku, Yuki Mori, and Alexei Verkhratsky

Abstract

Microglia, the immune cells of the central nervous system (CNS), are busy and vigilant guards of the adult brain, which scan brain parenchyma for damage and activate in response to lesions. Release of danger signals/ chemoattractants at the site of damage initiates microglial activation and stimulates migration. The main candidate for a chemoattractant sensed by microglia is adenosine triphosphate (ATP); however, many other substances can have similar effects. Some neuropeptides such as angiotensin II, bradykinin, endothelin, galanin and neurotensin are also chemoattractants for microglia. Among them, bradykinin increases microglial migration using mechanism distinct from that of ATP. Bradykinin-induced migration is controlled by a Gi/o-protein-independent pathway, while ATP-induced migration involves Gi/o proteins as well as mitogen-activated protein kinase (MAPK)/extracellular signal-regulated kinase (ERK)-dependent pathway. Galanin was reported to share certain signalling cascades with bradykinin; however, this overlap is only partial. Bradykinin,

M. Noda (✉) • Y. Mori
Laboratory of Pathophysiology, Graduate School of Pharmaceutical Sciences, Kyushu University, Fukuoka, Japan
e-mail: noda@phar.kyushu-u.ac.jp

M. Ifuku
Department of Integrative Physiology, Graduate School of Medical Sciences, Kyushu University, Fukuoka 812-8582, Japan

A. Verkhratsky
Faculty of Life Sciences, The University of Manchester, Manchester, UK

IKERBASQUE, Basque Foundation for Science, Bilbao, Spain

Department of Neurosciences, University of the Basque Country UPV/EHU, Leioa, Spain

L. Annunziato (ed.), *Sodium Calcium Exchange: A Growing Spectrum of Pathophysiological Implications*, Advances in Experimental Medicine and Biology 961, DOI 10.1007/978-1-4614-4756-6_24,

for example, stimulates Ca^{2+} influx through the reversed Na^+/Ca^{2+} exchange (NCX), whereas galanin induces intracellular Ca^{2+} mobilization by inositol-3,4,5-trisphosphate (InsP3)-dependent Ca^{2+} release from the intracellular store. These differences in signal cascades indicate that different chemoattractants such as ATP, bradykinin and galanin control distinct microglial functions in pathological conditions such as lesion and inflammation and NCX contributes to a special case of microglial migration.

Keywords
ATP • Bradykinin • Calcium influx • Galanin • Neuropeptides • Reverse mode • Sodium-calcium exchanger

24.1 Introduction

Sodium-calcium exchanger (NCX; the protein family is represented by three members NCX1, NCX2 and NCX3) is ubiquitously expressed in peripheral tissues and the nervous system where it is expressed in both neurones and neuroglia. In the CNS, neuroglial cells are fundamental for mounting defence in response to pathological insults. Microglial cells represent innate immune system of the mammalian brain and therefore are critically important for progress (Kim and de Vellis 2005; Kreutzberg 1996; Perry et al. 1993; Kettenmann et al., 2011) and outcome of various injuries and diseases. Microglial cells express NCX (Boscia et al. 2009; Matsuda et al. 2006; Nagano et al. 2004, 2005), although its functional role remains hitherto mostly unknown. Evidence begins to accumulate, however, indicating the contribution of Ca^{2+} influx due to reverse mode of NCX to microglial migration induced by certain endogenous peptide, such as bradykinin (Ifuku et al. 2007). So far, none of other peptides were found to activate reverse mode of NCX with subsequent Ca^{2+} influx in microglia, although further studies may reveal additional endogenous stimulators of NCX as well as functional importance of forward versus reverse mode of NCX in microglia. Incidentally, microglial expression of the NCX was reported to be up-regulated by exposure to interferon-gamma in culture (Nagano et al. 2004).

Microglial migration is central for immune defence and wound healing. It is not yet fully explored whether motility, migration and process movement are distinctly regulated (Kettenmann et al., 2011). Nevertheless, there are many candidate molecules, which serve as 'pathological' signals to activate microglia; the molecules include adenosine triphosphate (ATP) (Davalos et al., 2005; Honda et al. 2001), cannabinoids (Walter et al. 2003), morphine (Takayama and Ueda 2005), chemokine CCL21 (Rappert et al. 2002), neurotensin (Martin et al. 2005), lysophosphatidic acid (Schilling et al. 2004), neurotransmitters such as glutamate (Liu et al. 2009) dopamine and adrenaline (Farber et al. 2005), bradykinin (Ifuku et al. 2007), and galanin (Ifuku et al. 2011). Interleukin 1β (IL-1β) was also reported to act as a chemotactic signal guiding microglial cells to the site of injury, simultaneously promoting cells recruitment (Cartier et al. 2005), which is inhibited by neuropeptide Y (Ferreira et al. 2011). Downstream to various receptors, ion channels and transporters control cell migration. These include K^+ channels, Cl^- channels, Na^+/H^+ exchanger, Cl^-/HCO_3^- exchanger and Na^+/HCO_3^- cotransporter, all of them being linked to the actin cytoskeleton (Schwab 2001a, b). The involvement of NCX has been reported specifically for bradykinin-induced microglial migration (Ifuku et al. 2007). The intracellular signalling cascades underlying ATP-induced microglial migration are completely different (Noda et al. 2011) and do not require the involvement of NCX. Another neuropeptide, galanin, having quite similar intracellular signalling in many cell types, has also partially different signalling from that of bradykinin and does not need NCX for microglial migration (Ifuku et al. 2011).

Here we shall overview signalling pathways regulating microglial migration induced by some chemoattractants and critically address the question of the importance of reverse mode of NCX. We shall also focus on Ca^{2+} signalling systems including plasmalemmal ion channels, intracellular Ca^{2+} mobilization and reverse mode of NCX.

24.2 Mechanisms of ATP-, Bradykinin- and Galanin-induced Microglial Migration Are Agonist-Specific

ATP is the most potent microglial chemoattractant (Davalos et al. 2005; Honda et al. 2001). ATP/ADP promotes microglial chemotaxis via the $G_{i/o}$-coupled $P2Y_{12}$ receptor (Honda et al. 2001; Ohsawa and Kohsaka 2011) which is sensitive to pertussis toxin. Phospholipase C (PLC)-mediated increase in intracellular Ca^{2+} and Akt activation are also involved in ATP-induced microglial migration (Irino et al. 2008). It has been additionally reported that integrin-β1 activation regulates extension of microglial process following ATP stimulation (Ohsawa et al. 2010). The ATP action on microglia is also mediated through ionotropic $P2X_4$ purinoceptor and, by the virtue of relatively high Ca^{2+} permeability, contributes to intracellular Ca^{2+} signal (Ohsawa et al. 2007) (Fig. 24.1).

Microglial migration induced by bradykinin is controlled by different mechanisms when compared to that of ATP. Bradykinin-induced migration is resistant to pertussis toxin and is therefore executed through $G_{i/o}$-protein-independent pathway, whereas ATP-induced migration is $G_{i/o}$-protein-dependent (Honda et al. 2001). In the signal cascade followed by activation of G proteins, cell migration needs intracellular Ca^{2+} rise. Application of ATP (100 μM) induces Ca^{2+} signals, mediated by both P2Y and P2X receptors (Light et al. 2006; Moller et al. 2000; Walz et al. 1993), with a larger amplitude than bradykinin (Noda et al. 2003, 2007). ATP evoked an increase in $[Ca^{2+}]_i$ even in the cells which did not respond to bradykinin, suggesting that majority of cultured microglial cells possess ATP receptors whereas functional bradykinin receptors are restricted to a cell subpopulation.

As for the mechanism of intracellular Ca^{2+} rise, bradykinin induces Ca^{2+} influx due to the activation of reverse mode of NCX1. Bradykinin- but not ATP-induced microglial migration is attenuated in microglia from transgenic mice expressing lower levels of NCX1 ($NCX1^{+/-}$ mice) (Ifuku et al. 2007). Operational Na^+-dependent Ca^{2+} uptake was already detected in cultured microglial cells (Matsuda et al. 2001). Subsequent experiments revealed an expression of all three NCX isoforms, NCX1, NCX2 and NCX3, at both translational and protein levels in cultured rat microglia. At the mRNA level, however, expression of NCX1 was predominant and was substantially larger than in neurones from the same animals; expression of NCX2 and NCX3 was weak, being much less than in nerve cells (Nagano et al. 2004). As for the mechanism how ATP induces intracellular Ca^{2+} rise, ionotropic ATP receptors such as $P2X_4$ and $P2X_7$ would presumably contribute to the influx of Ca^{2+}. Particular importance of $P2X_4$ receptors in ATP-induced migration has been reported (Ohsawa et al. 2007).

As a consequence of intracellular Ca^{2+} rise, Ca^{2+}-dependent K^+ ($K_{(Ca)}$) channels are activated. It is reported that charybdotoxin-sensitive $K_{(Ca)}$ channel is important for regulating migratory behaviour of various cell types such as leukocytes and fibroblasts (Schwab 2001a, b), although there is little information on which type of $K_{(Ca)}$ channel is relevant in each cell type. Bradykinin-induced migration of microglia depends on activation of charybdotoxin (ChTX)-sensitive $K_{(Ca)}$ channels, suggesting LK-type (large conductance) and/or IK (intermediate conductance) $K_{(Ca)}$ channels. Since iberiotoxin (IBX) and apamin (APA), a blocker for LK-type and SK-type (small conductance) $K_{(Ca)}$ channels, respectively, have no effect on bradykinin-induced microglial motility (Ifuku et al. 2007), it is likely that IK-type (intermediate

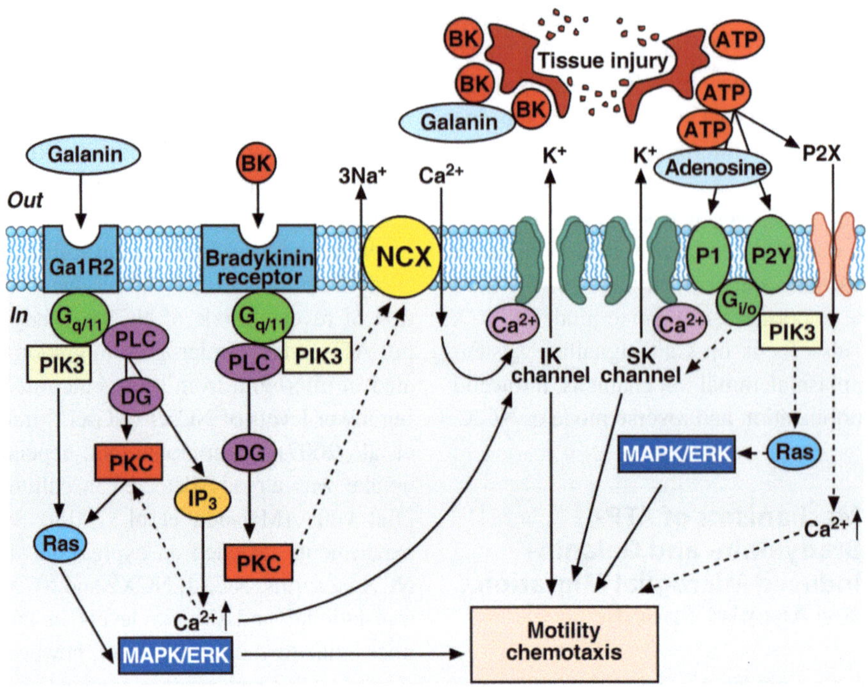

Fig. 24.1 ATP, bradykinin and galanin activate different signalling cascades, leading microglial migration. ATP binds to $P2Y_{12}$ receptor which couples to $G_{i/o}$-protein-dependent signalling, leading activation of PI3K and MAPK/ERK. Activation of $P2X_4$ also contributes to ATP-induced microglial migration, allowing influx of Ca^{2+}, and activates SK-type $K_{(Ca)}$ channel. Bradykinin receptor couples to $G_{q/11}$ protein, activates PLC and subsequent PKC and activates reverse mode of NCX with undetermined mechanism. Influx of extracellular Ca^{2+}, activating IK-type of $K_{(Ca)}$ channels. Another neuropeptide produced at lesion sites, GAL, binds to GalR2 and also activates $G_{q/11}$-protein-dependent pathway. However, GAL does not activate NCX but activates PKC- and $InsP_3$-dependent pathway, leading to activation of $K_{(Ca)}$ channels, presumably IK-type. GAL-induced microglial migration is also dependent on PI3K and MAPK/ERK (Modified from Noda et al. (2011))

conductance) $K_{(Ca)}$ channels are critical for bradykinin-induced migration. Incidentally, ATP-induced microglial migration is not inhibited by ChTX but is markedly inhibited by APA, indicating that ATP-induced microglial migration depends on the activation of SK-type $K_{(Ca)}$ channels. Activation of $I_{K(Ca)}$ causes hyperpolarization, which increases the electro-driving force (the electrochemical gradient) for Ca^{2+} influx and thus enhances the intracellular Ca^{2+} signal to stimulate cell migration.

Galanin is a neuropeptide which is up-regulated following neuronal axotomy or inflammation. One subtype of GAL receptor (specifically GalR2 subtype) is expressed in microglia. Galanin significantly increases the migration of rat-cultured microglia at an extremely low concentration, which can be as low as 0.1 pM. The galanin-induced signal cascade is partially similar to that induced by bradykinin. It does not involve $G_{i/o}$ protein but includes activation of protein kinase C, phosphoinositide 3-kinase and $K_{(Ca)}$ channels. However, Ca^{2+} influx through the reverse mode of NCX1 is not involved in galanin-induced microglial migration. Likewise, nominally free extracellular Ca^{2+} inhibited bradykinin-induced migration but not galanin-induced migration. Rather than extracellular Ca^{2+} influx, inositol-1,4,5-triphosphate ($InsP_3$)-induced intracellular Ca^{2+} mobilization is operationally important for galanin-induced microglial migration (Ifuku et al. 2011).

24.3 Possibility of Reverse NCX in Microglia

NCX is electrogenic, and the NCX currents were initially recorded from cardiac cells using the patch clamp technique (Kimura et al. 1986; Mechmann and Pott 1986). The NCX current is driven by the concentration gradients of both Na^+ and Ca^{2+} across the membrane. The stoichiometry of the exchanger varies between $3Na^+$:$1Ca^{2+}$ in cardiomyocytes and ($3Na^+$ + $1H^+$):($1Ca^{2+}$ + $1K^+$) in retina, which stimulates NCX electrogeneity. Under physiological conditions, when the exchanger operates in the forward mode, the NCX currents are inward, reflecting extrusion of intracellular Ca^{2+} in exchange for Na^+ influx. Depolarization of the membrane or increase in intracellular Na^+ concentration can reverse the NCX, causing Ca^{2+} influx in exchange to Na^+ efflux. When patch pipette is applied to a cardiac cell, loading internal Ca^{2+} of ~500 nM range and application of extracellular Na^+ (Li^+ as a control) induce a large inward shift of the holding current which decayed after a peak. The *I*-*V* relationship shows outward currents at depolarized membrane potential, suggesting the reverse-mode NCX and extracellular Ca^{2+} influx (in this case, with 1 mM-external Ca^{2+}) (Watanabe et al. 2006). Since microglial cells are characterized by high input resistance and little voltage-gated membrane currents and a very low membrane potential in situ, in brain slices (−20 mV, range −2 mV to −40 mV) (Boucsein et al. 2000), it is possible that even small changes in transmembrane ion gradients trigger or increase reverse mode of NCX operation.

24.4 Localization of Ion Channels and Transporters in Microglia

Ion channels and transporters such as K^+ channels, Cl^- channels, Na^+/H^+ exchanger, Cl^-/HCO_3^- exchanger and Na^+/HCO_3^- cotransporter linked to the actin cytoskeleton (Schwab 2001a, b) are critical for regulation of migration. Though transporters are reported to be expressed at protrusion site of the cell, expression pattern of NCX has not been investigated yet. Dynamic change of NCX expression in migrating cells would be extremely interesting to analyze and constitute the essence of our future studies.

24.5 Conclusion

Signalling cascades activated in microglia following arrival of different chemoattractants are agonist-specific. ATP, bradykinin and galanin control distinct microglial functions in pathological conditions such as lesion and inflammation, and contribution of NCX is strictly restricted to bradykinin-induced microglial migration. Functional importance of NCX and its connection with other membrane proteins in migrating cells will help us to understand one of the fundamental mechanisms of cell migration.

References

F. Boscia, R. Gala, A. Pannaccione, A. Secondo, A. Scorziello, G. Di Renzo, L. Annunziato, NCX1 expression and functional activity increase in microglia invading the infarct core. Stroke **40**, 3608–3617 (2009)

C. Boucsein, H. Kettenmann, C. Nolte, Electrophysiological properties of microglial cells in normal and pathologic rat brain slices. Eur. J. Neurosci. **12**, 2049–2058 (2000)

L. Cartier, O. Hartley, M. Dubois-Dauphin, K.H. Krause, Chemokine receptors in the central nervous system: role in brain inflammation and neurodegenerative diseases. Brain Res. Rev. **48**, 16–42 (2005)

D. Davalos, J. Grutzendler, G. Yang, J.V. Kim, Y. Zuo, S. Jung, D.R. Littman, M.L. Dustin, W.B. Gan, ATP mediates rapid microglial response to local brain injury in vivo. Nat. Neurosci. **8**, 752–758 (2005)

K. Farber, U. Pannasch, H. Kettenmann, Dopamine and noradrenaline control distinct functions in rodent microglial cells. Mol. Cell. Neurosci. **29**, 128–138 (2005)

R. Ferreira, T. Santos, L. Cortes, S. Cochaud, F. Agasse, A.P. Silva, S. Xapelli, J.O. Malva, Neuropeptide Y inhibits interleukin-1 beta (IL-1beta)-induced microglia motility. J. Neurochem. **120**, 93–105 (2011)

S. Honda, Y. Sasaki, K. Ohsawa, Imai, Y. Nakamura, K. Inoue, S. Kohsaka, Extracellular ATP or ADP induce chemotaxis of cultured microglia through Gi/o-coupled P2Y receptors. J. Neurosci. **21**, 1975–1982 (2001)

M. Ifuku, K. Farber, Y. Okuno, Y. Yamakawa, T. Miyamoto, C. Nolte, V.F. Merrino, S. Kita, T. Iwamoto, I. Komuro,

B. Wang, G. Cheung, E. Ishikawa, H. Ooboshi, M. Bader, K. Wada, H. Kettenmann, M. Noda, Bradykinin-induced microglial migration mediated by B1-bradykinin receptors depends on Ca^{2+} influx via reverse-mode activity of the Na^+/Ca^{2+} exchanger. J. Neurosci. **27**, 13065–13073 (2007)

M. Ifuku, Y. Okuno, Y. Yamakawa, K. Izumi, H.S. Seifert, Kettenmann, M. Noda, Functional importance of inositol-1,4,5-triphosphate-induced intracellular Ca^{2+} mobilization in galanin-induced microglial migration. J. Neurochem. **117**, 61–70 (2011)

Y. Irino, Y. Nakamura, K. Inoue, S. Kohsaka, K. Ohsawa, Akt activation is involved in P2Y12 receptor-mediated chemotaxis of microglia. J. Neurosci. Res. **86**, 1511–1519 (2008)

H. Kettenmann, U.K. Hanisch, M. Noda, A. Verkhratsky, Physiology of microglia. Physiol. Rev. **91**, 461–553 (2011)

S.U. Kim, J. de Vellis, Microglia in health and disease. J. Neurosci. Res. **81**, 302–313 (2005)

J. Kimura, A. Noma, H. Irisawa, Na-Ca exchange current in mammalian heart cells. Nature **319**, 596–597 (1986)

G.W. Kreutzberg, Microglia: a sensor for pathological events in the CNS. Trends Neurosci. **19**, 312–318 (1996)

A.R. Light, Y. Wu, R.W. Hughen, P.B. Guthrie, Purinergic receptors activating rapid intracellular Ca increases in microglia. Neuron Glia Biol. **2**, 125–138 (2006)

G.J. Liu, R. Nagarajah, R.B. Banati, M.R. Bennett, Glutamate induces directed chemotaxis of microglia. Eur. J. Neurosci. **29**, 1108–1118 (2009)

S. Martin, E. Dicou, J.P. Vincent, J. Mazella, Neurotensin and the neurotensin receptor-3 in microglial cells. J. Neurosci. Res. **81**, 322–326 (2005)

T. Matsuda, N. Arakawa, K. Takuma, Y. Kishida, Y. Kawasaki, M. Sakaue, K. Takahashi, T. Takahashi, T. Suzuki, T. Ota, A. Hamano-Takahashi, M. Onishi, Y. Tanaka, K. Kameo, A. Baba, SEA0400, a novel and selective inhibitor of the Na^+-Ca^{2+} exchanger, attenuates reperfusion injury in the in vitro and in vivo cerebral ischemic models. J. Pharmacol. Exp. Ther. **298**, 249–256 (2001)

T. Matsuda, T. Nagano, M. Takemura, A. Baba, Topics on the Na^+/Ca^{2+} exchanger: responses of Na^+/Ca^{2+} exchanger to interferon-gamma and nitric oxide in cultured microglia. J. Pharm. Sci. **102**, 22–26 (2006)

S. Mechmann, L. Pott, Identification of Na-Ca exchange current in single cardiac myocytes. Nature **319**, 597–599 (1986)

T. Moller, O. Kann, A. Verkhratsky, H. Kettenmann, Activation of mouse microglial cells affects P2 receptor signaling. Brain Res. **853**, 49–59 (2000)

T. Nagano, Y. Kawasaki, A. Baba, M. Takemura, T. Matsuda, Up-regulation of Na^+-Ca^{2+} exchange activity by interferon-gamma in cultured rat microglia. J. Neurochem. **90**, 784–791 (2004)

T. Nagano, M. Osakada, Y. Ago, Y. Koyama, A. Baba, S. Maeda, M. Takemura, T. Matsuda, SEA0400, a specific inhibitor of the Na^+-Ca^{2+} exchanger, attenuates sodium nitroprusside-induced apoptosis in cultured rat microglia. Br. J. Pharmacol. **144**, 669–679 (2005)

M. Noda, Y. Kariura, T. Amano, Y. Manago, K. Nishikawa, S. Aoki, K. Wada, Expression and function of bradykinin receptors in microglia. Life Sci. **72**, 1573–1581 (2003)

M. Noda, K. Sasaki, M. Ifuku, K. Wada, Multifunctional effects of bradykinin on glial cells in relation to potential anti-inflammatory effects. Neurochem. Int. **51**, 185–191 (2007)

M. Noda, M. Ifuku, Y. Okuno, K. Beppu, Y. Mori, S. Naoe. Neuropeptides as attractants of immune cells in the brain and their distinct signaling. Adv. Neuron. Biol. **1**, 53–62 (2011)

K. Ohsawa, S. Kohsaka, Dynamic motility of microglia: Purinergic modulation of microglial movement in the normal and pathological brain. Glia **59**, 1793–1799 (2011)

K. Ohsawa, Y. Irino, Y. Nakamura, C. Akazawa, K. Inoue, S. Kohsaka, Involvement of P2X4 and P2Y12 receptors in ATP-induced microglial chemotaxis. Glia **55**, 604–616 (2007)

K. Ohsawa, Y. Irino, T. Sanagi, Y. Nakamura, E. Suzuki, K. Inoue, S. Kohsaka, P2Y12 receptor-mediated integrin-beta1 activation regulates microglial process extension induced by ATP. Glia **58**, 790–801 (2010)

V.H. Perry, P.B. Andersson, S. Gordon, Macrophages and inflammation in the central nervous system. Trends Neurosci. **16**, 268–273 (1993)

A. Rappert, K. Biber, C. Nolte, M. Lipp, A. Schubel, B. Lu, N.P. Gerard, C. Gerard, H.W. Boddeke, H. Kettenmann, Secondary lymphoid tissue chemokine (CCL21) activates CXCR3 to trigger a Cl- current and chemotaxis in murine microglia. J. Immunol. **168**, 3221–3226 (2002)

T. Schilling, C. Stock, A. Schwab, C. Eder, Functional importance of Ca^{2+}-activated K^+ channels for lysophosphatidic acid-induced microglial migration. Eur. J. Neurosci. **19**, 1469–1474 (2004)

A. Schwab, Function and spatial distribution of ion channels and transporters in cell migration. Am. J. Physiol. Renal Physiol. **280**, F739–F747 (2001a)

A. Schwab, Ion channels and transporters on the move. News Physiol. Sci. **16**, 29–33 (2001b)

N. Takayama, H. Ueda, Morphine-induced chemotaxis and brain-derived neurotrophic factor expression in microglia. J. Neurosci. **25**, 430–435 (2005)

L. Walter, A. Franklin, A. Witting, C. Wade, Y. Xie, G. Kunos, K. Mackie, N. Stella, Nonpsychotropic cannabinoid receptors regulate microglial cell migration. J. Neurosci. **23**, 1398–1405 (2003)

W. Walz, S. Ilschner, C. Ohlemeyer, R. Banati, H. Kettenmann, Extracellular ATP activates a cation conductance and a K^+ conductance in cultured microglial cells from mouse brain. J. Neurosci. **13**, 4403–4411 (1993)

Y. Watanabe, Y. Koide, J. Kimura, Topics on the Na^+/Ca^{2+} exchanger: pharmacological characterization of Na^+/Ca^{2+} exchanger inhibitors. J. Pharm. Sci. **102**, 7–16 (2006)

Sodium Fluxes and Astroglial Function

25

Alexei Verkhratsky, Mami Noda, Vladimir Parpura, and Sergei Kirischuk

Abstract

Astrocytes exhibit their excitability based on variations in cytosolic Ca^{2+} levels, which leads to variety of signalling events. Only recently, however, intracellular fluctuations of more abundant cation Na^{+} are brought in the limelight of glial signalling. Indeed, astrocytes possess several plasmalemmal molecular entities that allow rapid transport of Na^{+} across the plasma membrane: (1) ionotropic receptors, (2) canonical transient receptor potential cation channels, (3) neurotransmitter transporters and (4) sodium-calcium exchanger. Concerted action of these molecules in controlling cytosolic Na^{+} may complement Ca^{2+} signalling to provide basis for complex bidirectional astrocyte-neurone communication at the tripartite synapse.

Keywords

Ionotropic receptors • Sodium-calcium exchanger • Sodium potassium pump • Glutamate transporter • Sodium signalling

A. Verkhratsky (✉)
Faculty of Life Sciences, The University of Manchester, Oxford Road, Manchester, M13 9PT, UK

IKERBASQUE, Basque Foundation for Science, Bilbao, Spain

Department of Neurosciences, University of the Basque Country UPV/EHU, Leioa, Spain
e-mail: alex.verkhratsky@manchester.ac.uk

M. Noda
Laboratory of Pathophysiology, Graduate School of Pharmaceutical Sciences, Kyushu University, Fukuoka, Japan

V. Parpura
IKERBASQUE, Basque Foundation for Science, Bilbao, Spain

Department of Neurosciences, University of the Basque Country UPV/EHU, Leioa, Spain

Department of Neurobiology, Center for Glial Biology in Medicine, Civitan International Research Center, Atomic Force Microscopy & Nanotechnology Laboratories, and Evelyn F. McKnight Brain Institute, University of Alabama, Birmingham, USA

Department of Biotechnology, University of Rijeka, Rijeka, Croatia

S. Kirischuk
Institute of Physiology and Pathophysiology, Universal Medical Center of the Johannes Gutenberg, University Mainz, Mainz, Germany

L. Annunziato (ed.), *Sodium Calcium Exchange: A Growing Spectrum of Pathophysiological Implications*, Advances in Experimental Medicine and Biology 961, DOI 10.1007/978-1-4614-4756-6_25,

25.1 Introduction

Neuroglia represent the main cellular homeostatic system of the brain. Evolution of the central nervous system (CNS) resulted in a high specialisation of elements of brain circuitry: neurones perfected rapidly propagating action potentials and synaptic transmission, whereas glial cells assumed full responsibility for brain homeostasis and defence. Astrocytes, which are the main type of glial cells in the brain and in the spinal cord, have an astonishingly wide array of functions that include regulation of neurogenesis and CNS development, shaping the brain micro-architecture, controlling ion and neurotransmitter homeostasis in the extracellular space, providing activity-dependent metabolic support to neurones and mounting the evolutionary conserved astrogliotic response to CNS lesions (for general review of astroglia in physiology and pathophysiology, see (Heneka et al. 2010; Kettenmann and Ransom 2005; Kimelberg and Nedergaard 2010; Nedergaard et al. 2010; Oberheim et al. 2006; Rodriguez et al. 2009; Verkhratsky 2009, 2011; Verkhratsky and Butt 2007; Verkhratsky et al. 2011)).

Astrocytes are in a possession of several signalling cascades that are activated in response to various physiological and pathological stimuli. These signalling cascades are triggered by activation of numerous plasmalemmal metabotropic and ionotropic receptors (Lalo et al. 2011b; Verkhratsky et al. 2009; Verkhratsky and Steinhauser 2000). The calcium signalling system is of a particular importance for electrically non-excitable astroglial cells, and propagating waves of inositol 1,4,5 trisphosphate ($InsP_3$)-mediated Ca^{2+} release from the endoplasmic reticulum store are considered to be a specific form of astroglial excitability (reviewed in (Agulhon et al. 2008; Parpura et al. 2011; Verkhratsky 2006; Verkhratsky et al. 1998)). Glial calcium signals, however, are rather slow when compared to the speed of synaptic transmission and may not necessarily participate in local neuronal-glial signalling at the level of individual synapses. In the CNS, a substantial proportion of synaptic contacts is closely enwrapped by astroglial membranes (Peters et al. 1991) which effectively shield the synapses preventing neurotransmitter spillover, which ascertains spatial precision of synaptic signalling. In addition to forming physical barrier, perisynaptic processes of astrocytes are endowed with neurotransmitter transporters that provide for neurotransmitter removal and neurotransmitter homeostasis thus contributing to functional isolation of individual synapses. The perisynaptic processes are also rich in ionotropic receptors, cationic channels and sodium-dependent pumps (Na^+/K^+ pump, Na^+/HCO_3^- co-transporters, etc.), the latter being critical for maintaining ion homeostasis in the synaptic cleft.

In this chapter, we shall overview pathways governing sodium fluxes through astroglial plasma membrane and critically address the question of the importance of local sodium fluctuations in the function of astroglial cell. We shall focus on several plasmalemmal Na^+-transporting systems that include ionotropic receptors, canonical transient receptor potential (TRPC) cation channels, Na^+/Ca^{2+} exchanger and Na^+-dependent glutamate transporters.

25.2 Ionotropic Receptors

Astroglial cells are highly heterogeneous in their morphology and function; astrocytes from different brain regions also differ in their assortment of neurotransmitter receptors (Matyash and Kettenmann 2010; Verkhratsky 2011; Verkhratsky and Steinhauser 2000). Astroglial ionotropic receptors are generally represented by receptors for glutamate and adenosine 5′-triphosphate (ATP). Ionotropic glutamate receptors of α-amino-3-hydroxy-5-methyl-isoxazole propionate (AMPA) type are present in astrocytes throughout the CNS, including hippocampus, cerebellum and cortex (Condorelli et al. 1999; Gallo and Ghiani 2000; Seifert and Steinhauser 2001; Steinhäuser and Gallo 1996; Verkhratsky and Kirchhoff 2007a). All four subunits of AMPA receptors have been detected in astroglia, although the actual assembly varies between brain regions. In hippocampal astrocytes, the glutamate receptor (GluR)2 and GluR4 subunits are predominantly expressed, which stipulates specific electrophysiology (linear *I-V* relation and low Ca^{2+} permeability)(Gallo and Ghiani 2000; Seifert and Steinhauser 1995; Steinhäuser and Gallo 1996); in cortical astroglial cells, the GluR1 and GluR4 subunits are the most abundant (Conti et al. 1994). In Bergmann glial cells in situ and in several in vitro astroglial preparations (Geiger et al. 1995; Muller et al. 1992), the AMPA receptors are devoid of GluR2 subunit that makes the receptors moderately permeable to Ca^{2+} ($P_{Ca}/P_{monovalent}$ ~ 1–1.5 (Burnashev et al. 1996; Isa et al. 1996; Itazawa et al. 1997; Pankratov et al. 2009)). Astroglial N-methyl D-aspartate (NMDA) receptors were characterised both in vitro and in situ (Kondoh et al. 2001; Lopez et al. 1997; Nishizaki et al. 1999; Puro et al. 1996), in particular, in astroglial cells from cortex and spinal cord (Lalo et al. 2006; Schipke et al. 2001; Verkhratsky and Kirchhoff 2007b; Ziak et al. 1998). Both NMDA receptor-specific mRNAs and receptor protein were found in cortical astrocytes (Conti et al. 1996; Schipke et al. 2001). In the cortex, the NMDA-mediated astroglial currents are positively potentiated by glycine and are blocked by NMDA antagonists D-2-amino-phosphonopentanoic acid and MK-801 (Lalo et al. 2006; Palygin et al. 2011). The astroglial NMDA receptors have several peculiar features (Lalo et al. 2006; Palygin et al. 2010) which include weak Mg^{2+} block at characteristic levels of astroglial membrane potential of −80 mV (the block develops at V_m values ~ −100 to −120 mV) and moderate Ca^{2+} permeability ($P_{Ca}/P_{monovalent}$ ~3). Incidentally, similar Mg^{2+} sensitivity was determined in oligodendroglial NMDA receptors (Karadottir et al. 2005; Micu et al. 2006; Salter and Fern 2005), which possibly allows classifying a special class of glial NMDA receptors. Based on electrophysiology, Ca^{2+} permeability and sensitivity to NR2C/D subunit-selective antagonist UBP141 the most probable assembly of glial NMDA receptors include two NR1, one NR2C/D and one NR3 subunit (Palygin et al. 2011).

Astroglial functional expression of ionotropic ATP (P2X) receptors remains poorly characterised. The mRNAs specific for various P2X receptors subunits were identified in cultured astrocytes, in freshly isolated retinal Müller cells and in astrocytes in situ (Franke et al. 2001, 2004; Fumagalli et al. 2003; Jabs et al. 2000; Lalo et al. 2008). At the protein level (as determined by immunoreactivity), $P2X_2$, $P2X_3$ and $P2X_4$ receptors were identified in astrocytes from the nucleus accumbens (Franke et al. 2001); the $P2X_1$ and $P2X_2$ receptors were found in astroglial cells in the cerebellum and in the spinal cord (Kanjhan et al. 1996; Loesch and Burnstock 1998). Immunoreactivity for $P2X_4$ receptors was detected in astrocytes from the brainstem (Ashour and Deuchars 2004). In the hippocampus, immunostaining revealed astroglial expression of $P2X_{1-4}$, $P2X_6$ and $P2X_7$ subunits (Kukley et al. 2001).

Functionally $P2X_{1/5}$ heteromeric receptor-mediated currents were identified in cortical astrocytes (Lalo et al. 2008, 2011c). These $P2X_{1/5}$ heteromeric receptors are characterised by special features which include (1) a very high sensitivity to ATP (EC_{50} for current activation of ~40 nM), (2) biphasic kinetics with distinct peak and steady-state components and (3) very little desensitisation in response to the repetitive agonist applications. As a result, the $P2X_{1/5}$ receptors allow cortical astrocytes to detect extremely low levels of extracellular ATP. Astroglial $P2X_{1/5}$ receptors have a moderate Ca^{2+} permeability ($P_{Ca}/P_{monovalent}$ ~ 2 (Palygin et al. 2010)). The $P2X_7$ receptor-mediated currents were also detected in cortical astrocytes in situ (Oliveira et al. 2011), although their low sensitivity to the ATP possibly indicates their pathophysiological importance (Illes et al. 2011). P2X receptor(s)-mediated Ca^{2+} signalling was also described in astroglial cells from acutely isolated optic nerves. These Ca^{2+} signals were inhibited by P2X receptor antagonist NF023 (James and Butt 2001); in addition, astrocytes from the optic nerve seem to express functional $P2X_7$ receptors (Hamilton et al. 2008).

Astroglial ionotropic receptors are activated by endogenous neurotransmitters released in the course of synaptic transmission. In the cortical astrocytes voltage-clamped in the brain slice, both NMDA and $P2X_{1/5}$ receptors mediated the major part of currents triggered by electrical stimulation of neuronal afferents (Lalo et al. 2006, 2011a). The spontaneous ('miniature') currents mediated by AMPA/NMDA glutamate receptors and $P2X_{1/5}$ receptors were also detected in cortical astrocytes indicating close apposition of astroglial membranes bearing these receptors to the presynaptic sites of neurotransmitter release (Lalo et al. 2006, 2011a, b).

Taken together, astrocytes have several types of fast ionotropic receptors, activated by neurotransmitters released to the synaptic cleft. All these receptors, however, have relatively low Ca^{2+} permeability with predicted fractional Ca^{2+} currents in the range of 1–5 %. At the same time, activation of these receptors at resting membrane potential triggers currents mainly carried by Na^+ ions.

25.3 TRP Cationic Channels

The detailed analysis of various types of cationic channels expressed in astroglia is still needed. Nonetheless, these channels are potentially important for controlling cytosolic sodium concentration because negative resting potential of astrocytes makes Na^+ virtually the sole permeating cation. Among many cationic channels, the products of TRP genes have been identified in astrocytes (Golovina 2005; Grimaldi et al. 2003; Malarkey et al. 2008; Pizzo et al. 2001). These TRP channels are reported to be activated following intracellular Ca^{2+} release acting as store-operated channels (Parpura et al. 2011). It was shown that antisense-based inhibition of expression of the TRPC1 gene (Golovina 2005) or occlusion of the same channel by blocking antibodies raised against the TRPC1 protein channel pore (Malarkey et al. 2008) markedly inhibited store-operated Ca^{2+} entry in cultured astrocytes. In addition to TRPC1, acutely isolated astrocytes as well as astrocytes in vitro express TRPC4 and TRPC5 subunits which are needed to form functional TRPC channel (Strubing et al. 2001, 2003). The Na^+ fluxes generated by activation of TRPC channels have not yet been characterised; nonetheless, it is conceivable to speculate that metabotropically induced depletion of the ER Ca^{2+} stores results not only in $[Ca^{2+}]_i$ signalling but also in elevation of $[Na^+]_i$ through the opening of store-operated TRPC channels.

25.4 Neurotransmitter Transporters

Astroglia is central for neurotransmitter homeostasis, turnover and metabolism in the CNS (Danbolt 2001; Verkhratsky and Butt 2007). The action of two most important transmitters in the brain, glutamate and γ–aminobutyric acid (GABA), critically depends on astroglial transporters that remove these transmitters from the cleft thus terminating their action. The subsequent astroglial processing of glutamate through glutamine-glutamate shuttle is fundamental for replenishing glutamatergic terminals, which are incapable of producing glutamate from their own resources (Hertz and Zielke 2004). Glutamate and GABA transport into astrocytes is achieved through Na^+-dependent transporters that utilise energy of transmembrane Na^+ gradient. Astroglial glutamate transporters are represented by excitatory amino acid transporter type 1 and 2 (EAAT1 and EAAT2; analogues of these transporters in rodents are known as glutamate/aspartate transporter, GLAST and glutamate transporter-1, GLT-1 (Danbolt 2001; Gadea and Lopez-Colome 2001)). The stoichiometry of transporting one molecule of glutamate through both transporters involves influx of three Na^+ ions and one H^+ ion and efflux of one K^+ ion (Owe et al. 2006; Zerangue and Kavanaugh 1996). As a result, the transporter generates inward cationic current and produces substantial elevation of cytosolic Na^+ concentration (Kirischuk et al. 2007). The GABA transporters expressed in astrocytes (GAT1-3, (Heja et al. 2009)) are similarly Na^+ dependent with a stoichiometry of $2Na^+/1GABA$.

25.5 Sodium-Calcium Exchanger

Astrocytes express all three types of mammalian Na^+/Ca^{2+} exchangers, namely, NCX1, NCX2 and NCX3, which are primarily localised in perisynaptic processes, in particular those associated with excitatory synapses (Minelli et al. 2007). According to their thermodynamics (NCX stoichiometry is $3Na^+/1Ca^{2+}$), the NCX may operate in both forward (Ca^{2+} extrusion associated with Na^+ influx) and reverse (Ca^{2+} entry associated with Na^+ extrusion) modes. The transition between forward/reverse operations is controlled by transmembrane ion gradients and the level of membrane potential (DiPolo and Beauge 1983). Both modes of NCX activity are present in astroglial cells in vitro and in situ through analysing respective $[Ca^{2+}]_i/[Na^+]_i$ concentrations (Goldman et al. 1994; Kirischuk et al. 1997; Matsuda et al. 1996; Takuma et al. 1994). The NCX dynamically fluctuates between forward/reverse modes; in Bergmann glial cells, the NCX working in reverse mode significantly contributes to the peak $[Ca^{2+}]_i$ elevation following activation of kainate receptors; at the same time, NCX participates in relaxation of kainate-mediated $[Ca^{2+}]_i$ transients by extruding Ca^{2+} in the forward mode (Kirischuk et al. 1997). The reverse mode of NCX is activated following Na^+ entry via glutamate transporter in cultured cerebellar astrocytes (Rojas et al. 2007). Similarly, mild depolarization induced by high extracellular K^+ stimulation of adult rat astrocytes in culture promoted reverse mode of NCX that generated $[Ca^{2+}]_i$ transients (Paluzzi et al. 2007). These NCX-associated transients were specifically blocked by 2-[2-[4-(4-nitrobenzyloxy)phenyl]ethyl]isothiourea (KB-R7943), a drug selectively inhibiting reverse mode of NCX operation (Paluzzi et al. 2007). Treatment of unstimulated cultured astrocytes with KB-R7943 also cased moderate decreases in the resting $[Ca^{2+}]_i$ suggesting that NCX may operate in reverse mode at rest (Reyes et al. 2011). This seems to be a plausible suggestion because reversal potential for NCX calculated from the $[Ca^{2+}]_i$ and $[Na^+]_i$ levels measured from these cells was −98 mV. The resting potential of these cultured astrocytes is ~−70 mV which should set the resting operation mode of NCX as the reversed one.

25.6 Sodium Dynamics in Astrocytes

The resting intracellular Na^+ concentration in astrocytes is generally somewhat higher that in neurones, being ~10 mM in cultured cortical astrocytes (Chatton et al. 2003; Floyd et al. 2005), 15–16 mM in cultured hippocampal astrocytes (Rose and Ransom 1996a), 17 mM in cultured astrocytes from visual cortex (Reyes et al. 2011) and ~20 mM in astrocytes in situ in cortical slices (Kirischuk, unpublished observations). In neurones in contrast, average $[Na^+]_i$ is substantially lower being determined at 4 mM in cultured cerebellar granular cells (Kiedrowski et al. 1994), 9 mM in cultured hippocampal neurones (Rose and Ransom 1996a), ~11 mM in dopaminergic cells in substantia nigra pars compacta (Knopfel et al. 1998) and 10 mM in pyramidal neurones from cortical slices (Pisani et al. 1998). Chemical stimulation of astrocytes triggers spatio-temporally organised $[Na^+]_i$ fluctuations. Exposure of cultured astrocytes to glutamate triggered both $[Na^+]_i$ transients and propagating $[Na^+]_i$ waves (Bernardinelli et al. 2004; Kimelberg et al. 1989; Rose and Ransom 1996b, 1997). Similarly, $[Na^+]_i$ transients occur in situ in Bergmann glial cells and hippocampal astrocytes exposed to exogenous ionotropic glutamate receptor agonists or to electrical stimulation of neuronal afferents (Bennay et al. 2008; Kirischuk et al. 1997, 2007). Stimulation of ionotropic glutamate receptors can elevate $[Na^+]_i$ by 10–25 mM, (Fig. 25.1a) (Deitmer and Rose 2010; Kirischuk et al. 2007). In addition, extracellular glutamate activates glutamate transporters, which also produce substantial Na^+ fluxes elevating $[Na^+]_i$ by 10–20 mM (Fig. 25.1b (Kirischuk et al. 2007)). There are indications that Na^+ can travel between astrocytes via gap junctions, and inhibition of the latter desynchronises $[Na^+]_i$ dynamics in astroglia (Bernardinelli et al. 2004; Rose and Ransom 1996a). Astroglial $[Na^+]_i$ signals are also triggered by stimulation of neuronal afferents; these $[Na^+]_i$ responses develop in parallel with glial synaptic currents mediated by both ionotropic receptors and glutamate transporter (Bennay et al. 2008; Clark and Barbour 1997; Kirischuk et al. 2007). Short bursts of stimuli (5–10 pulses) elevated $[Na^+]_i$ by 5–10 mM. These $[Na^+]_i$ transients last much longer than glutamate-induced $[Ca^{2+}]_i$ responses; the decay time constant of $[Na^+]_i$ transients is about 100 s (Kirischuk et al. 2007). In the cerebellum, electrical stimulation of parallel fibres induces local $[Na^+]_i$ responses in Bergmann glia, whereas activation of climbing

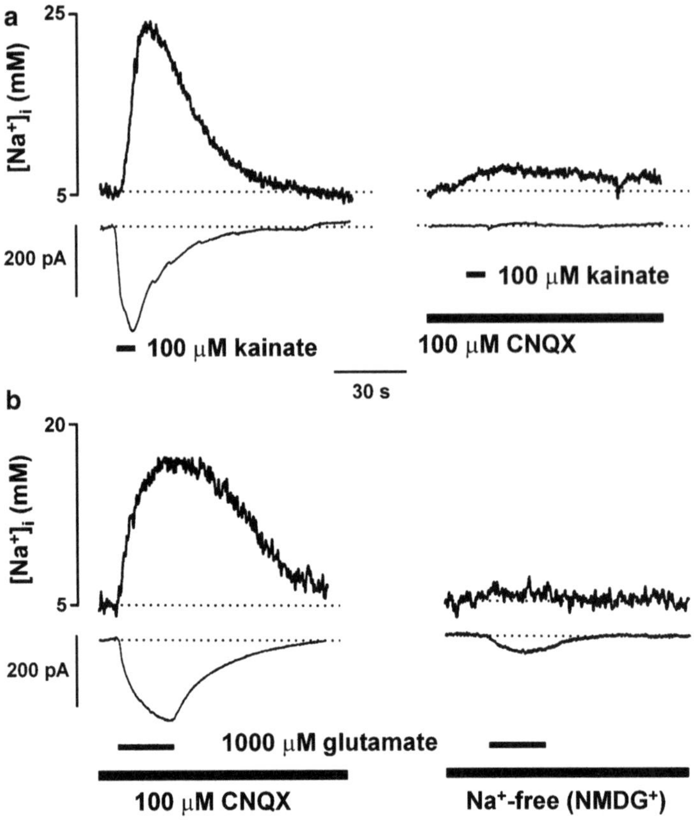

Fig. 25.1 *Glutamatergic stimulation triggers* $[Na^+]_i$ *elevation in Bergmann glial cells in situ in cerebellar slice.* (**a**) Simultaneous recordings of glutamate-induced inward current and $[Na^+]_i$ in response to cell stimulation with 100 µM of kainate, which opens AMPA receptors without triggering their desensitisation and is inactive against glutamate transporters (*left*). Both kainate-induced current and $[Na^+]_i$ transient are blocked by specific antagonist cyano-7-nitroquinoxaline-2,3-dione (CNQX, 100 µM) (*right*). (**b**) Similar to (a) inward currents and $[Na^+]_i$ were measured in Bergmann glial cells stimulated with 1 mM glutamate in the presence of 100 µM CNQX (the latter was added to exclude activation of AMPA ionotropic receptors) (*left*). Replacement of extracellular Na^+ by the organic cation N-methyl-D-glucamine ($NMDG^+$) eliminates both membrane current and $[Na^+]_i$ transient (*right*) (Modified from Kirischuk et al. (2007))

fibres activation triggers global $[Na^+]_i$ rise (Bennay et al. 2008). Thus, synaptic activity-induced intra-glial Na^+ responses show dependency on the synaptic input and significantly outlast the duration of synaptic activity.

25.7 Functional Significance of $[Na^+]_i$ Signalling

Rapid fluctuations of cytosolic Na^+ concentration can regulate numerous astroglial processes, which in turn can provide for local neuronal-glial communication (Fig. 25.2). In particular, elevation of $[Na^+]_i$ is directly coupled with generation of local $[Ca^{2+}]_i$ signals through favouring the reverse mode of NCX; indeed, $[Na^+]_i$ rises were directly demonstrated to induce additional Ca^{2+} influx that contributed to neurotransmitter-evoked $[Ca^{2+}]_i$ transients (Kirischuk et al. 1997). Our own data (Reyes et al. 2011) indicate that in cultured astrocytes, the reversal potential of NCX lies very close to the levels of resting membrane potential, and therefore, even moderate increases in $[Na^+]_i$ may rapidly lead to the NCX reverse operation. The NCX-mediated Ca^{2+} entry can in

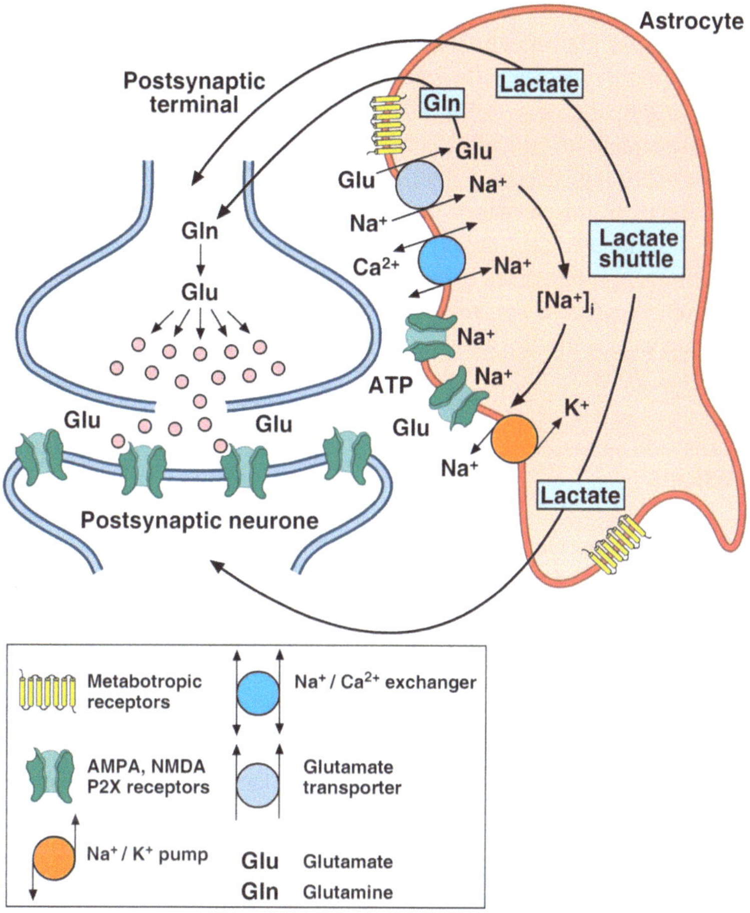

Fig. 25.2 *Possible physiological roles for $[Na^+]_i$ signalling. Local signalling mediated by ionotropic receptors and transporters in astroglial perisynaptic processes.* Synaptic release of neurotransmitters (glutamate and/or ATP) activates ionotropic receptors and glutamate transporters, which generate Na^+ influx and $[Na^+]_i$ elevation. Increases in $[Na^+]_i$ can assume a signalling role through modulating neurotransmitter transporters, switching the reverse mode of NCX and stimulating Na^+/K^+ pumps. This in turn can affect synaptic transmission and plasticity by modulating the time kinetic of glutamate removal from the cleft, through stimulating local metabolic support via lactate shuttle and through increase in extracellular K^+ buffering by the Na^+/K^+ pump

turn trigger exocytotic release of neurotransmitters from astroglia as have been demonstrated in several experimental paradigms (Benz et al. 2004; Paluzzi et al. 2007; Reyes et al. 2011).

Intracellular Na^+ is coupled to several other homeostatic systems. In particular, changes in $[Na^+]_i$ directly modulate $H^+/OH^-/HCO_3^-$ transport systems, which are fundamental for pH homeostasis, both intra- and extracellular (Deitmer and Rose 2010). Further, $[Na^+]_i$ directly controls the uptake of glutamate and GABA. Increases in $[Na^+]_i$ can significantly slow down or even reverse glutamate and/or GABA transporters. In fact, GABA transporter, because of its stoichiometry, is the most susceptible to regulation by $[Na^+]_i$, and even moderate rises in cytosolic Na^+ concentration can trigger GABA release through the reversed transporter (Heja et al. 2009; Wu et al. 2007). In addition, changes in $[Na^+]_i$ affect the activity of glutamine synthetase that further influences glutamate homeostasis (Benjamin 1987).

The second important target of $[Na^+]_i$ is the Na^+/K^+-ATPase, which has been found to co-localise with NCX in cortical astrocytes at plasma membrane-endoplasmic reticulum junctions in the perisynaptic processes (Blaustein et al. 2002; Juhaszova and Blaustein 1997). The Na^+/K^+-ATPase in turn plays a critical role in regulation of neuronal-glial lactate shuttle (Magistretti 2006, 2009). In this scenario, local $[Na^+]_i$ increases will stimulate local supply of active synapses with energy substrate. In addition, $[Na^+]_i$ rise stimulates glucose uptake at the endfeet level thus supporting the lactate shuttle (Voutsinos-Porche et al. 2003).

25.8 Conclusions

Rapid fluctuations in cytosolic Na^+ in the astroglial processes, controlled by concerted activity of plasmalemmal Na^+ permeable receptors and channels together with Na^+ transporters and exchangers, may represent an additional layer of complexity in intracellular signalling, complementing more studied Ca^{2+} signalling and contributing to local bidirectional communication between a single synapse and perisynaptic glial processes.

References

C. Agulhon, J. Petravicz, A.B. McMullen, E.J. Sweger, S.K. Minton, S.R. Taves, K.B. Casper, T.A. Fiacco, K.D. McCarthy, What is the role of astrocyte calcium in neurophysiology? Neuron **59**, 932–946 (2008)

F. Ashour, J. Deuchars, Electron microscopic localisation of P2X4 receptor subunit immunoreactivity to pre- and post-synaptic neuronal elements and glial processes in the dorsal vagal complex of the rat. Brain Res. **1026**, 44–55 (2004)

A.M. Benjamin, Influence of Na^+, K^+, and Ca^{2+} on glutamine synthesis and distribution in rat brain cortex slices: a possible linkage of glutamine synthetase with cerebral transport processes and energetics in the astrocytes. J. Neurochem. **48**, 1157–1164 (1987)

M. Bennay, J. Langer, S.D. Meier, K.W. Kafitz, C.R. Rose, Sodium signals in cerebellar Purkinje neurons and Bergmann glial cells evoked by glutamatergic synaptic transmission. Glia **56**, 1138–1149 (2008)

B. Benz, G. Grima, K.Q. Do, Glutamate-induced homocysteic acid release from astrocytes: possible implication in glia-neuron signaling. Neuroscience **124**, 377–386 (2004)

Y. Bernardinelli, P.J. Magistretti, J.Y. Chatton, Astrocytes generate Na^+-mediated metabolic waves. Proc. Natl. Acad. Sci. U. S. A. **101**, 14937–14942 (2004)

M.P. Blaustein, M. Juhaszova, V.A. Golovina, P.J. Church, E.F. Stanley, Na/Ca exchanger and PMCA localization in neurons and astrocytes: functional implications. Ann. N. Y. Acad. Sci. **976**, 356–366 (2002)

N. Burnashev, A. Villarroel, B. Sakmann, Dimensions and ion selectivity of recombinant AMPA and kainate receptor channels and their dependence on Q/R site residues. J. Physiol. (Lond.) **496**, 165–173 (1996)

J.Y. Chatton, L. Pellerin, P.J. Magistretti, GABA uptake into astrocytes is not associated with significant metabolic cost: implications for brain imaging of inhibitory transmission. Proc. Natl. Acad. Sci. U. S. A. **100**, 12456–12461 (2003)

B.A. Clark, B. Barbour, Currents evoked in Bergmann glial cells by parallel fibre stimulation in rat cerebellar slices. J. Physiol. (Lond.) **502**(Pt 2), 335–350 (1997)

D.F. Condorelli, F. Conti, V. Gallo, F. Kirchhoff, G. Seifert, C. Steinhauser, A. Verkhratsky, X. Yuan, Expression and functional analysis of glutamate receptors in glial cells. Adv. Exp. Med. Biol. **468**, 49–67 (1999)

F. Conti, A. Minelli, N.C. Brecha, Cellular localization and laminar distribution of AMPA glutamate receptor subunits mRNAs and proteins in the rat cerebral cortex. J. Comp. Neurol. **350**, 241–259 (1994)

F. Conti, S. DeBiasi, A. Minelli, M. Melone, Expression of NR1 and NR2A/B subunits of the NMDA receptor in cortical astrocytes. Glia **17**, 254–258 (1996)

N.C. Danbolt, Glutamate uptake. Prog. Neurobiol. **65**, 1–105 (2001)

J.W. Deitmer, C.R. Rose, Ion changes and signalling in perisynaptic glia. Brain Res. Rev. **63**, 113–129 (2010)

R. DiPolo, L. Beauge, The calcium pump and sodium-calcium exchange in squid axons. Annu. Rev. Physiol. **45**, 313–324 (1983)

C.L. Floyd, F.A. Gorin, B.G. Lyeth, Mechanical strain injury increases intracellular sodium and reverses Na^+/Ca^{2+} exchange in cortical astrocytes. Glia **51**, 35–46 (2005)

H. Franke, J. Grosche, H. Schadlich, U. Krugel, C. Allgaier, P. Illes, P2X receptor expression on astrocytes in the nucleus accumbens of rats. Neuroscience **108**, 421–429 (2001)

H. Franke, A. Gunther, J. Grosche, R. Schmidt, S. Rossner, R. Reinhardt, H. Faber-Zuschratter, D. Schneider, P. Illes, P2X7 receptor expression after ischemia in the cerebral cortex of rats. J. Neuropathol. Exp. Neurol. **63**, 686–699 (2004)

M. Fumagalli, R. Brambilla, N. D'Ambrosi, C. Volonte, M. Matteoli, C. Verderio, M.P. Abbracchio, Nucleotide-mediated calcium signaling in rat cortical astrocytes: role of P2X and P2Y receptors. Glia **43**, 218–230 (2003)

A. Gadea, A.M. Lopez-Colome, Glial transporters for glutamate, glycine and GABA I. Glutamate transporters. J. Neurosci. Res. **63**, 453–460 (2001)

V. Gallo, C.A. Ghiani, Glutamate receptors in glia: new cells, new inputs and new functions. Trends Pharmacol. Sci. **21**, 252–258 (2000)

J.R. Geiger, T. Melcher, D.S. Koh, B. Sakmann, P.H. Seeburg, P. Jonas, H. Monyer, Relative abundance of subunit mRNAs determines gating and Ca^{2+} permeability of AMPA receptors in principal neurons and interneurons in rat CNS. Neuron **15**, 193–204 (1995)

W.F. Goldman, P.J. Yarowsky, M. Juhaszova, B.K. Krueger, M.P. Blaustein, Sodium/calcium exchange in rat cortical astrocytes. J. Neurosci. **14**, 5834–5843 (1994)

V.A. Golovina, Visualization of localized store-operated calcium entry in mouse astrocytes. Close proximity to the endoplasmic reticulum. J. Physiol. (Lond.) **564**, 737–749 (2005)

M. Grimaldi, M. Maratos, A. Verma, Transient receptor potential channel activation causes a novel form of [Ca $^{2+}$]I oscillations and is not involved in capacitative Ca $^{2+}$ entry in glial cells. J. Neurosci. **23**, 4737–4745 (2003)

N. Hamilton, S. Vayro, F. Kirchhoff, A. Verkhratsky, J. Robbins, D.C. Gorecki, A.M. Butt, Mechanisms of ATP- and glutamate-mediated calcium signaling in white matter astrocytes. Glia **56**, 734–749 (2008)

L. Heja, P. Barabas, G. Nyitrai, K.A. Kekesi, B. Lasztoczi, O. Toke, G. Tarkanyi, K. Madsen, A. Schousboe, A. Dobolyi, M. Palkovits, J. Kardos, Glutamate uptake triggers transporter-mediated GABA release from astrocytes. PLoS One **4**, e7153 (2009)

M.T. Heneka, J.J. Rodriguez, A. Verkhratsky, Neuroglia in neurodegeneration. Brain Res. Rev. **63**, 189–211 (2010)

L. Hertz, H.R. Zielke, Astrocytic control of glutamatergic activity: astrocytes as stars of the show. Trends Neurosci. **27**, 735–743 (2004)

P. Illes, A. Verkhratsky, G. Burnstock, H. Franke, P2X receptors and their roles in astroglia in the central and peripheral nervous system. Neuroscientist (2011 in press; doi 1073858411418524)

T. Isa, S. Itazawa, M. Iino, K. Tsuzuki, S. Ozawa, Distribution of neurones expressing inwardly rectifying and Ca^{2+}-permeable AMPA receptors in rat hippocampal slices. J. Physiol. (Lond.) **491**, 719–733 (1996)

S.I. Itazawa, T. Isa, S. Ozawa, Inwardly rectifying and Ca^{2+}-permeable AMPA-type glutamate receptor channels in rat neocortical neurons. J. Neurophysiol. **78**, 2592–2601 (1997)

R. Jabs, E. Guenther, K. Marquordt, T.H. Wheeler-Schilling, Evidence for P2X3, P2X4, P2X5 but not for P2X7 containing purinergic receptors in Muller cells of the rat retina. Brain Res. Mol. Brain Res. **76**, 205–210 (2000)

G. James, A.M. Butt, P2X and P2Y purinoreceptors mediate ATP-evoked calcium signalling in optic nerve glia in situ. Cell Calcium **30**, 251–259 (2001)

M. Juhaszova, M.P. Blaustein, Na^{+} pump low and high ouabain affinity alpha subunit isoforms are differently distributed in cells. Proc. Natl. Acad. Sci. U. S. A. **94**, 1800–1805 (1997)

R. Kanjhan, G.D. Housley, P.R. Thorne, D.L. Christie, D.J. Palmer, L. Luo, A.F. Ryan, Localization of ATP-gated ion channels in cerebellum using P2x2R subunit-specific antisera. Neuroreport **7**, 2665–2669 (1996)

R. Karadottir, P. Cavelier, L.H. Bergersen, D. Attwell, NMDA receptors are expressed in oligodendrocytes and activated in ischaemia. Nature **438**, 1162–1166 (2005)

H. Kettenmann, B.R. Ransom (eds.), *Neuroglia* (OUP, Oxford, 2005)

L. Kiedrowski, J.T. Wroblewski, E. Costa, Intracellular sodium concentration in cultured cerebellar granule cells challenged with glutamate. Mol. Pharmacol. **45**, 1050–1054 (1994)

H.K. Kimelberg, M. Nedergaard, Functions of astrocytes and their potential as therapeutic targets. Neurotherapeutics **7**, 338–353 (2010)

H.K. Kimelberg, S. Pang, D.H. Treble, Excitatory amino acid-stimulated uptake of $22Na^{+}$ in primary astrocyte cultures. J. Neurosci. **9**, 1141–1149 (1989)

S. Kirischuk, H. Kettenmann, A. Verkhratsky, Na^{+}/Ca^{2+} exchanger modulates kainate-triggered Ca^{2+} signaling in Bergmann glial cells in situ. FASEB J. **11**, 566–572 (1997)

S. Kirischuk, H. Kettenmann, A. Verkhratsky, Membrane currents and cytoplasmic sodium transients generated by glutamate transport in Bergmann glial cells. Pflugers Arch. **454**, 245–252 (2007)

T. Knopfel, E. Guatteo, G. Bernardi, N.B. Mercuri, Hyperpolarization induces a rise in intracellular sodium concentration in dopamine cells of the substantia nigra pars compacta. Eur. J. Neurosci. **10**, 1926–1929 (1998)

T. Kondoh, T. Nishizaki, H. Aihara, N. Tamaki, NMDA-responsible, APV-insensitive receptor in cultured human astrocytes. Life Sci. **68**, 1761–1767 (2001)

M. Kukley, J.A. Barden, C. Steinhauser, R. Jabs, Distribution of P2X receptors on astrocytes in juvenile rat hippocampus. Glia **36**, 11–21 (2001)

U. Lalo, Y. Pankratov, F. Kirchhoff, R.A. North, A. Verkhratsky, NMDA receptors mediate neuron-to-glia signaling in mouse cortical astrocytes. J. Neurosci. **26**, 2673–2683 (2006)

U. Lalo, Y. Pankratov, S.P. Wichert, M.J. Rossner, R.A. North, F. Kirchhoff, A. Verkhratsky, P2X1 and P2X5 subunits form the functional P2X receptor in mouse cortical astrocytes. J. Neurosci. **28**, 5473–5480 (2008)

U. Lalo, O. Palygin, R.A. North, A. Verkhratsky, Y. Pankratov, Age-dependent remodelling of ionotropic signalling in cortical astroglia. Aging Cell **10**, 392–402 (2011a)

U. Lalo, Y. Pankratov, V. Parpura, A. Verkhratsky, Ionotropic receptors in neuronal-astroglial signalling: what is the role of "excitable" molecules in non-excitable cells. Biochim. Biophys. Acta **1813**, 992–1002 (2011b)

U. Lalo, A. Verkhratsky, Y. Pankratov, Ionotropic ATP receptors in neuronal-glial communication. Semin. Cell Dev. Biol. **22**, 220–228 (2011c)

A. Loesch, G. Burnstock, Electron-immunocytochemical localization of P2X1 receptors in the rat cerebellum. Cell Tissue Res. **294**, 253–260 (1998)

T. Lopez, A.M. Lopez-Colome, A. Ortega, NMDA receptors in cultured radial glia. FEBS Lett. **405**, 245–248 (1997)

P.J. Magistretti, Neuron-glia metabolic coupling and plasticity. J. Exp. Biol. **209**, 2304–2311 (2006)

P.J. Magistretti, Role of glutamate in neuron-glia metabolic coupling. Am. J. Clin. Nutr. **90**, 875S–880S (2009)

E.B. Malarkey, Y. Ni, V. Parpura, Ca^{2+} entry through TRPC1 channels contributes to intracellular Ca^{2+} dynamics and consequent glutamate release from rat astrocytes. Glia **56**, 821–835 (2008)

T. Matsuda, K. Takuma, E. Nishiguchi, H. Hashimoto, J. Azuma, A. Baba, Involvement of Na^{+}-Ca^{2+} exchanger in reperfusion-induced delayed cell death of cultured rat astrocytes. Eur. J. Neurosci. **8**, 951–958 (1996)

V. Matyash, H. Kettenmann, Heterogeneity in astrocyte morphology and physiology. Brain Res. Rev. **63**, 2–10 (2010)

I. Micu, Q. Jiang, E. Coderre, A. Ridsdale, L. Zhang, J. Woulfe, X. Yin, B.D. Trapp, J.E. McRory, R. Rehak, G.W. Zamponi, W. Wang, P.K. Stys, NMDA receptors mediate calcium accumulation in myelin during chemical ischaemia. Nature **439**, 988–992 (2006)

A. Minelli, P. Castaldo, P. Gobbi, S. Salucci, S. Magi, S. Amoroso, Cellular and subcellular localization of Na^{+}-Ca^{2+} exchanger protein isoforms, NCX1, NCX2, and NCX3 in cerebral cortex and hippocampus of adult rat. Cell Calcium **41**, 221–234 (2007)

T. Muller, T. Moller, T. Berger, J. Schnitzer, H. Kettenmann, Calcium entry through kainate receptors and resulting potassium-channel blockade in Bergmann glial cells. Science **256**, 1563–1566 (1992)

M. Nedergaard, J.J. Rodriguez, A. Verkhratsky, Glial calcium and diseases of the nervous system. Cell Calcium **47**, 140–149 (2010)

T. Nishizaki, T. Matsuoka, T. Nomura, T. Kondoh, N. Tamaki, Y. Okada, Store Ca^{2+} depletion enhances NMDA responses in cultured human astrocytes. Biochem. Biophys. Res. Commun. **259**, 661–664 (1999)

N.A. Oberheim, X. Wang, S. Goldman, M. Nedergaard, Astrocytic complexity distinguishes the human brain. Trends Neurosci. **29**, 547–553 (2006)

J.F. Oliveira, T. Riedel, A. Leichsenring, C. Heine, H. Franke, U. Krugel, W. Norenberg, P. Illes, Rodent cortical astroglia express in situ functional P2X7 receptors sensing pathologically high ATP concentrations. Cereb. Cortex **21**, 806–820 (2011)

S.G. Owe, P. Marcaggi, D. Attwell, The ionic stoichiometry of the GLAST glutamate transporter in salamander retinal glia. J. Physiol. (Lond.) **577**, 591–599 (2006)

S. Paluzzi, S. Alloisio, S. Zappettini, M. Milanese, L. Raiteri, M. Nobile, G. Bonanno, Adult astroglia is competent for Na^{+}/Ca^{2+} exchanger-operated exocytotic glutamate release triggered by mild depolarization. J. Neurochem. **103**, 1196–1207 (2007)

O. Palygin, U. Lalo, A. Verkhratsky, Y. Pankratov, Ionotropic NMDA and P2X1/5 receptors mediate synaptically induced Ca^{2+} signalling in cortical astrocytes. Cell Calcium **48**, 225–231 (2010)

O. Palygin, U. Lalo, Y. Pankratov, Distinct pharmacological and functional properties of NMDA receptors in mouse cortical astrocytes. Br. J. Pharmacol. **163**, 1755–1766 (2011)

Y. Pankratov, U. Lalo, O.A. Krishtal, A. Verkhratsky, P2X receptors and synaptic plasticity. Neuroscience **158**, 137–148 (2009)

V. Parpura, V. Grubisic, A. Verkhratsky, Ca^{2+} sources for the exocytotic release of glutamate from astrocytes. Biochim. Biophys. Acta **1813**, 984–991 (2011)

O. Peters, S.L. Palay, H. deF Webster, *The Fine Structure of the Nervous System* (Oxford University Press, Oxford, 1991)

A. Pisani, P. Calabresi, A. Tozzi, G. Bernardi, T. Knopfel, Early sodium elevations induced by combined oxygen and glucose deprivation in pyramidal cortical neurons. Eur. J. Neurosci. **10**, 3572–3574 (1998)

P. Pizzo, A. Burgo, T. Pozzan, C. Fasolato, Role of capacitative calcium entry on glutamate-induced calcium influx in type-I rat cortical astrocytes. J. Neurochem. **79**, 98–109 (2001)

D.G. Puro, J.P. Yuan, N.J. Sucher, Activation of NMDA receptor-channels in human retinal Muller glial cells inhibits inward-rectifying potassium currents. Vis. Neurosci. **13**, 319–326 (1996)

R.C. Reyes, A. Verkhratsky, V. Parpura, Plasmalemmal Na^{+}/Ca^{2+} exchanger modulates Ca^{2+}-dependent exocytotic release of glutamate from rat cortical astrocytes ASN Neuro 4(1). pii: e00075. doi: 10.1042/AN20110059 (2012)

J.J. Rodriguez, M. Olabarria, A. Chvatal, A. Verkhratsky, Astroglia in dementia and Alzheimer's disease. Cell Death Differ. **16**, 378–385 (2009)

H. Rojas, C. Colina, M. Ramos, G. Benaim, E.H. Jaffe, C. Caputo, R. DiPolo, Na^{+} entry via glutamate transporter activates the reverse Na^{+}/Ca^{2+} exchange and triggers Cai^{2+}-induced Ca^{2+} release in rat cerebellar Type-1 astrocytes. J. Neurochem. **100**, 1188–1202 (2007)

C.R. Rose, B.R. Ransom, Intracellular sodium homeostasis in rat hippocampal astrocytes. J. Physiol. (Lond.) **491**(Pt 2), 291–305 (1996a)

C.R. Rose, B.R. Ransom, Mechanisms of H^{+} and Na^{+} changes induced by glutamate, kainate, and D-aspartate in rat hippocampal astrocytes. J. Neurosci. **16**, 5393–5404 (1996b)

C.R. Rose, B.R. Ransom, Gap junctions equalize intracellular Na^{+} concentration in astrocytes. Glia **20**, 299–307 (1997)

M.G. Salter, R. Fern, NMDA receptors are expressed in developing oligodendrocyte processes and mediate injury. Nature **438**, 1167–1171 (2005)

C.G. Schipke, C. Ohlemeyer, M. Matyash, C. Nolte, H. Kettenmann, F. Kirchhoff, Astrocytes of the mouse neocortex express functional N-methyl-D-aspartate receptors. FASEB J. **15**, 1270–1272 (2001)

G. Seifert, C. Steinhauser, Glial cells in the mouse hippocampus express AMPA receptors with an intermediate Ca^{2+} permeability. Eur. J. Neurosci. **7**, 1872–1881 (1995)

G. Seifert, C. Steinhauser, Ionotropic glutamate receptors in astrocytes. Prog. Brain Res. **132**, 287–299 (2001)

C. Steinhäuser, V. Gallo, News on glutamate receptors in glial cells. Trends Neurosci. **19**, 339–345 (1996)

C. Strubing, G. Krapivinsky, L. Krapivinsky, D.E. Clapham, TRPC1 and TRPC5 form a novel cation channel in mammalian brain. Neuron **29**, 645–655 (2001)

C. Strubing, G. Krapivinsky, L. Krapivinsky, D.E. Clapham, Formation of novel TRPC channels by complex subunit interactions in embryonic brain. J. Biol. Chem. **278**, 39014–39019 (2003)

K. Takuma, T. Matsuda, H. Hashimoto, S. Asano, A. Baba, Cultured rat astrocytes possess Na^+-Ca^{2+} exchanger. Glia **12**, 336–342 (1994)

A. Verkhratsky, Calcium ions and integration in neural circuits. Acta Physiol (Oxf.) **187**, 357–369 (2006)

A. Verkhratsky, Neuronismo y reticulismo: neuronal-glial circuits unify the reticular and neuronal theories of brain organization. Acta Physiol (Oxf.) **195**, 111–122 (2009)

A. Verkhratsky, Physiology of neuronal-glial networking. Neurochem. Int. **57**, 332–343 (2011)

A. Verkhratsky, A. Butt, *Glial Neurobiology. A textbook* (Wiley, Chichester, 2007)

A. Verkhratsky, F. Kirchhoff, Glutamate-mediated neuronal-glial transmission. J. Anat. **210**, 651–660 (2007a)

A. Verkhratsky, F. Kirchhoff, NMDA receptors in Glia. Neuroscientist **13**, 28–37 (2007b)

A. Verkhratsky, C. Steinhauser, Ion channels in glial cells. Brain Res. Brain Res. Rev. **32**, 380–412 (2000)

A. Verkhratsky, R.K. Orkand, H. Kettenmann, Glial calcium: homeostasis and signaling function. Physiol. Rev. **78**, 99–141 (1998)

A. Verkhratsky, O.A. Krishtal, G. Burnstock, Purinoceptors on neuroglia. Mol. Neurobiol. **39**, 190–208 (2009)

A. Verkhratsky, V. Parpura, J.J. Rodriguez, Where the thoughts dwell: the physiology of neuronal-glial "diffuse neural net". Brain Res. Rev. **66**, 133–151 (2011)

B. Voutsinos-Porche, G. Bonvento, K. Tanaka, P. Steiner, E. Welker, J.Y. Chatton, P.J. Magistretti, L. Pellerin, Glial glutamate transporters mediate a functional metabolic crosstalk between neurons and astrocytes in the mouse developing cortex. Neuron **37**, 275–286 (2003)

Y. Wu, W. Wang, A. Diez-Sampedro, G.B. Richerson, Nonvesicular inhibitory neurotransmission via reversal of the GABA transporter GAT-1. Neuron **56**, 851–865 (2007)

N. Zerangue, M.P. Kavanaugh, Flux coupling in a neuronal glutamate transporter. Nature **383**, 634–637 (1996)

D. Ziak, A. Chvatal, E. Sykova, Glutamate-, kainate- and NMDA-evoked membrane currents in identified glial cells in rat spinal cord slice. Physiol. Res. **47**, 365–375 (1998)

New Roles of NCX in Glial Cells: Activation of Microglia in Ischemia and Differentiation of Oligodendrocytes

26

Francesca Boscia, Carla D'Avanzo, Anna Pannaccione, Agnese Secondo, Antonella Casamassa, Luigi Formisano, Natascia Guida, Antonella Scorziello, Gianfranco Di Renzo, and Lucio Annunziato

Abstract

The initiation of microglial responses to the ischemic injury involves modifications of calcium homeostasis. Changes in $[Ca^{2+}]_i$ levels have also been shown to influence the developmental processes that accompany the transition of human oligodendrocyte precursor cells (OPCs) into mature myelinating oligodendrocytes and are required for the initiation of myelination and remyelination processes.

We investigated the regional and temporal changes of NCX1 protein in microglial cells of the peri-infarct and core regions after permanent middle cerebral artery occlusion (pMCAO). Interestingly, 3 and 7 days after pMCAO, NCX1 signal strongly increased in the round-shaped microglia invading the infarct core. Cultured microglial cells from the core displayed increased NCX1 expression as compared with contralateral cells and showed enhanced NCX activity in the reverse mode of operation. Similarly, NCX activity and NCX1 protein expression were significantly enhanced in BV2 microglia exposed to oxygen and glucose deprivation, whereas

F. Boscia • C. D'Avanzo • A. Pannaccione • A. Secondo
A. Casamassa • L. Formisano • N. Guida • A. Scorziello
G. Di Renzo
Division of Pharmacology, Department of Neuroscience, School of Medicine, 'Federico II' University of Naples, Via Pansini 5, Naples 80131, Italy

L. Annunziato, M.D. (✉)
Division of Pharmacology, Department of Neuroscience, School of Medicine, "Federico II" University of Naples
Via S. Pansini, 5, Naples 80131, Italy
e-mail: lannunzi@unina.it

L. Annunziato (ed.), *Sodium Calcium Exchange: A Growing Spectrum of Pathophysiological Implications*, Advances in Experimental Medicine and Biology 961, DOI 10.1007/978-1-4614-4756-6_26,

NCX2 and NCX3 were downregulated. Interestingly, in NCX1-silenced cells, $[Ca^{2+}]_i$ increase induced by hypoxia was completely prevented. The upregulation of NCX1 expression and activity observed in microglia after pMCAO suggests a relevant role of NCX1 in modulating microglia functions in the postischemic brain.

Next, we explored whether calcium signals mediated by NCX1, NCX2, or NCX3 play a role in oligodendrocyte maturation. Functional studies, as well as mRNA and protein expression analyses, revealed that NCX1 and NCX3, but not NCX2, were divergently modulated during OPC differentiation into oligodendrocyte. In fact, while NCX1 was downregulated, NCX3 was strongly upregulated during the oligodendrocyte development. Whereas the knocking down of the NCX3 isoform in OPCs prevented the upregulation of the myelin protein markers CNPase and MBP, its overexpression induced their upregulation. Furthermore, NCX3 knockout mice exhibited not only a reduced size of spinal cord but also a marked hypomyelination, as revealed by the decrease in MBP expression and by the accompanying increase in OPCs number. Our findings indicate that calcium signaling mediated by NCX3 plays a crucial role in oligodendrocyte maturation and myelin formation.

Keywords

Na^+/Ca^{2+} exchanger • NCX1 • NCX3 • Microglia • Oligodendrocyte precursor cells (OPCs) • Oligodendrocyte • Cerebral ischemia • MCAO • Myelin

26.1 Introduction

Glial cells account for about 90 % of all cells in the human brain. During recent years, a large number of studies have dramatically changed the image of glia as silent and passive supportive brain cells. In fact, there is now growing recognition that glia cells are as important as neurons for brain function. Glial cells are involved in almost every type of brain pathology. Indeed, alteration of intracellular calcium levels in response to neurodegenerative insults or to ischemic injury plays a crucial role in inducing and maintaining glial responses in the injured brain (Lipton 1999). Reactions of glial cells to the insults are of critical importance for the progress of neural pathology and include the reaction of astroglia (reactive astrogliosis), microglia (activation of microglia), and oligodendroglia (Wallerian degeneration, demyelination, and remyelination).

The Na^+/Ca^{2+} exchanger (NCX), a transmembrane domain protein, which, by operating in a bidirectional way, couples the efflux of Ca^{2+} to the influx of Na^+ into the cell or, vice versa, the influx of Ca^{2+} to the efflux of Na^+, is involved in the regulation of diverse neuronal and glial cell functions (Annunziato et al. 2004). NCX is involved in regulating intracellular Ca^{2+} concentration under pathological conditions and has been proposed as a potential therapeutic target in different disorders of the nervous system, including ischemia–reperfusion injury, (Pignataro et al. 2004; Molinaro et al. 2008), demyelinating conditions such as multiple sclerosis (MS) (Craner et al. 2004a, b), and spinal cord injury (Li et al. 2000; Tomes and Agrawal 2002).

Although all three NCX isoforms have been described in neurons, astrocytes, oligodendrocytes, and microglia (Quednau et al. 1997; Nagano et al. 2004), little information about their role in the different glial cell populations under physiological or pathological conditions has been reported.

26.2 NCX1 Expression and Functional Activity Is Upregulated in Phagocytic Microglia in the Postischemic Brain

The importance of NCX function in the ischemic brain has been already highlighted by recent findings showing that the specific knocking down of the ncx1 gene as well as ncx2 and ncx3 dramatically increases the extent of the ischemic lesion in rats and mice (Pignataro et al. 2004; Molinaro et al. 2008). Recently, our research group showed that NCX1 transcripts display a different expression pattern in the ischemic core and in the peri-infarct regions after permanent middle cerebral artery occlusion (pMCAO) in rats (Boscia et al. 2006); whether these modifications also occur in glial cells in the postischemic brain still remained unknown.

Previous in vitro studies provided evidence that among the different NCX genes, NCX1 is the most highly expressed in BV2 microglia (Quednau et al. 1997; Newell et al. 2007). Interestingly, the direct exposure of cultured microglia to interferon or nitric oxide enhanced NCX1 transcripts levels (Nagano et al. 2004; Matsuda et al. 2006). More recently, it has been demonstrated that NCX1 activity is relevant for microglial motility (Ifuku et al. 2007) and that *ncx1−/−* embryos have no detectable microglia in the brain (Ginhoux et al. 2010). These important findings together with the relevance of NCX1 function in ischemic conditions led us to investigate NCX1 expression and activity in microglia in response to the ischemic injury. By means of immunohistochemistry, double immunofluorescence, and confocal microscopy, we investigated the regional and temporal changes of NCX1 protein in microglia of the core and peri-ischemic areas 1, 3, and 7 days after pMCAO as compared with the contralateral undamaged area. The results of this study demonstrated that 1 day after pMCAO, NCX1 protein expression was detected only in some microglial cells located adjacent to the soma of neurons in the infarct core. More interestingly, 3 and 7 days after pMCAO, NCX1 signal progressively increased in microglia invading the infarct core. In these cells, NCX1 expression was limited to the round phagocytic phenotype (Boscia et al. 2009), which represents the final stage of microglia activation (Hanisch and Kettenmann 2007; Ito et al. 2001).

To further explore NCX1 expression and NCX activity in microglial cells of the infarct core, we used primary glial cultures dissociated *ex vivo* from the cortical core of ischemic rat brain and the corresponding contralateral region 7 days after pMCAO (Fig. 26.1). Double-labeling experiments of NCX1 protein with the microglial marker IB4 revealed that NCX1 immunoreactivity in microglia isolated from the contralateral region was barely detectable. By contrast, in microglia isolated from the core, NCX1 signal was intensely expressed on the plasma membrane (Fig. 26.1a–c). Then we assessed NCX activity in microglial cells of the ischemic core and the corresponding contralateral region by using Na^+-dependent $[Ca^{2+}]_i$ increase monitored by FURA-2 microfluorimetry (Fig. 26.1d). The perfusion of Na^+-free solution was followed by a fast linear rise in $[Ca^{2+}]_i$ that was largely greater in IB4+ microglia obtained from the core if compared to both IB4(−) cells of the same ischemic region and IB4+ cells obtained from the corresponding contralateral area (Fig. 26.1d, e). Finally, to dissect the direct effects of hypoxic conditions on NCX1 expression and activity, we exposed BV2 microglial cell line to 1-h OGD followed by 24–48 h of reoxygenation. In these experimental conditions, a significant increase in NCX1 protein expression occurred 24 h after reoxygenation, whereas a significant decrease in both NCX2 and NCX3 proteins occurred both at earlier and later times of reoxygenation (Fig. 26.2a–c). In microglia exposed to OGD I_{NCX}, recorded both in the forward and reverse modes of operation, were significantly higher than those detected under normoxia (Fig. 26.3a, b). To examine whether the altered Ca^{2+} handling was specifically related to

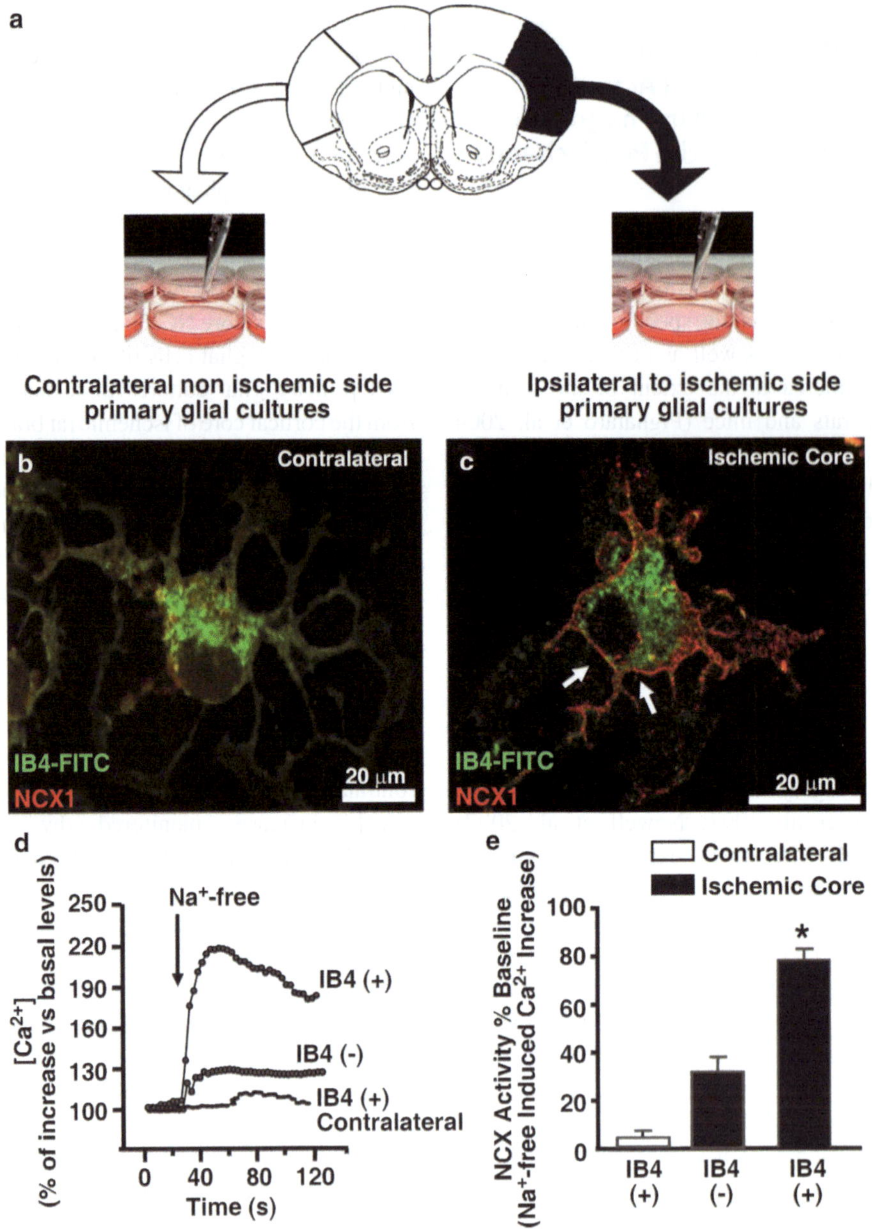

Fig. 26.1 (**a**) Schematic diagram of primary glial cell cultures obtained from the ischemic brain. (**b, c**) Colocalization of NCX1 with IB4-FITC in microglia obtained from the contralateral or core region. Scale bars: 20 μm; (**d**) single-cell traces representative of the Na^+ free effect on $[Ca^{2+}]_i$ in IB4(−) and $IB4^+$ cells from the core. (**e**) Quantification of Na^+-free induced $[Ca^{2+}]_i$ increase measured as Δ% of peak versus basal values in $IB4^+$ and IB4(−) cells isolated from the core and in $IB4^+$ cells from the contralateral cortex ($n = 20$ per group). $^*p < 0.05$ versus all groups

NCX1 expression and activity increase, we knocked down NCX1 with short-interfering RNAs in BV2 microglia. When NCX1-silenced BV2 cells were exposed to OGD plus 6 h of reoxygenation, the increase in $[Ca^{2+}]_i$ was completely prevented (Fig. 26.3c–e). This finding, together with the results showing that the protein expression NCX2 and NCX3 significantly decreased

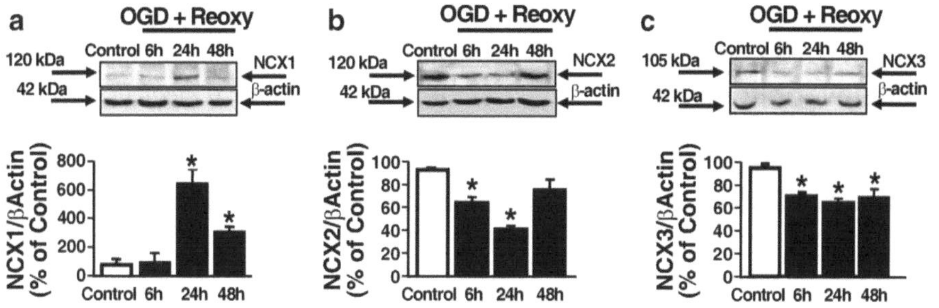

Fig. 26.2 (**a**-**c**) NCX1–3 protein levels and their densitometric analysis following OGD/reoxygenation

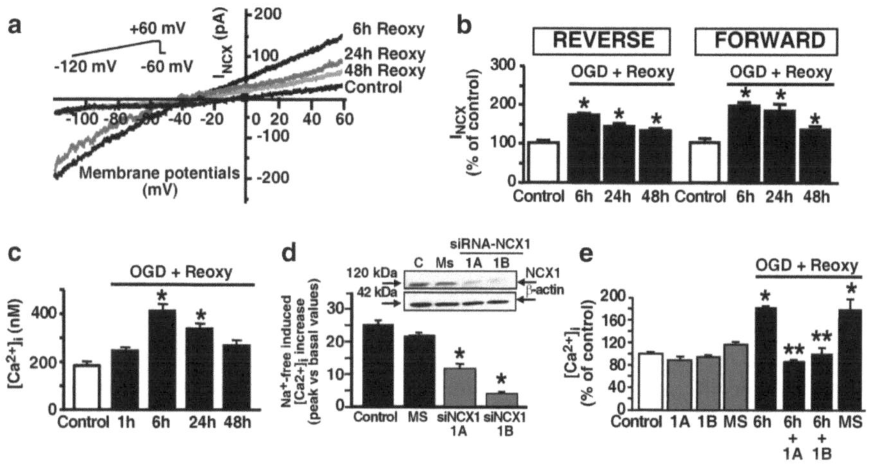

Fig. 26.3 (**a**) I_{NCX} traces recorded from BV2 microglia under normoxia and OGD/reoxygenation. (**b**) I_{NCX} quantification after OGD/reoxygenation is expressed as percentage of the current densities recorded under control conditions (n = 20 per group). (**c**) Quantification of $[Ca^{2+}]_i$ in microglia under normoxia and OGD/reoxygenation. (**d**) NCX1 protein expression (*upper* panel) and activity (*lower* panel) in NCX1-silenced BV2 microglia with siRNAs 1a and 1b. (**e**) Quantification of $[Ca^{2+}]_i$ in NCX1-silenced microglial cells after OGD plus 6 h of reoxygenation (n = 35–60 per group). *p < 0.05 versus control, **p < 0.05 versus OGD/Reoxy

after OGD, demonstrates the relevant role played by the NCX1 isoform in round phagocytic microglia during hypoxic conditions. In fact, microglia serve as scavenger cells in the injured brain. Indeed, phagocytic clearance of structures that have lost their function such as necrotic debris of cells, dendrites, and myelin is considered beneficial and represents a prerequisite for repair attempts in cerebral ischemia (Hanisch and Kettenmann 2007). Whether the increase in NCX1 expression and activity could contribute to the beneficial actions exerted by microglial activity during the different phases of microglia activation after ischemia cannot be determined at the moment. However, the importance of NCX1 function in the brain under cerebral ischemia suggests that the increased activity of NCX1 in microglia of the postischemic brain might exert a protective role.

26.3 NCX3 Exchanger Plays a Relevant Role in the Progression of Oligodendrocyte Precursor Cells (OPCs) into Oligodendrocytes

The main function of oligodendrocyte cells is the production of myelin which insulates axons in the CNS and facilitates the fast saltatory conduction of action potential. Demyelination describes a pathological process targeted at the oligodendrocytes (as it occur in multiple sclerosis (MS), ischemia, or other demyelinating diseases), in which myelin sheaths are lost around an axon with consequent alteration of conduction and impairment in sensation, movement, cognition, or other functions depending on which nerves are involved (Franklin and French-Constant 2008). Oligodendrocytes develop from oligodendrocyte precursor cells (OPCs) through distinct phenotypic stages that can be identified by the sequential expression of specific markers characteristic of progenitors (A2B5, NG2; PDGF alpha receptor) or mature and myelinating oligodendrocytes (CNPase, myelin-associated glycoprotein, MAG; myelin basic protein, MBP). A considerable number of OPCs do persist in the adult brain and may respond to demyelination by proliferation, migration, differentiation, and formation of new myelin sheaths or remyelination. In many cases, remyelination fails, and demyelinated axons permanently die. This appears to be due to the failure of OPC to proliferate and differentiate in brain lesions (Chong and Chan 2010). Thus, understanding how oligodendrocyte development proceeds and what factors govern the differentiation fate of OPCs is crucial to discover new effective therapeutic targets for demyelinating diseases, such as MS.

It is now becoming increasingly clear that changes in $[Ca^{2+}]_i$ levels not only influence the developmental processes that accompany the transition of OPCs into mature myelinating oligodendrocytes but also intervene in the initiation of myelination and remyelination processes (Barres et al. 1990). It has been reported that all three NCX isoform transcripts are expressed in oligodendrocytes (Quednau et al. 1997). Whether calcium signals mediated by NCX1, NCX2, and NCX3 might be involved in OPC maturation and myelin formation has never been investigated.

To address these questions, we first examined their expression and, then, by using patch clamp in whole-cell configuration and single-cell FURA-2 microfluorimetry, we studied their functional activity during oligodendrocyte development either in the human oligodendrocyte progenitor MO3.13 cell line or in primary rat OPC cultures (Boscia et al. 2012). Functional studies revealed an upregulation of NCX function when MO3.13 or primary OPCs were differentiated into oligodendrocyte phenotype with phorbol myristate acetate (PMA) or with thyroid hormones T3 and T4, respectively. To examine the contribution of each NCX isoform to the changes in NCX activity observed during oligodendrocyte differentiation, we performed quantitative RT-PCR, Western blotting, and immunocytochemical studies with selective NCX primers or antibodies. Expression analyses revealed that NCX1 and NCX3, but not NCX2, were divergently modulated during OPC differentiation into oligodendrocyte phenotype. In fact, whereas NCX1 was downregulated, NCX3 was strongly upregulated during oligodendrocyte development (Fig. 26.4a, b). The importance of calcium signaling mediated by NCX3 during oligodendrocyte development and myelin formation was further supported by our findings showing that the knocking down of NCX3 expression and activity by siRNA strategy in OPC cultures prevented the upregulation of the myelin proteins CNPase and MBP, whereas its overexpression induced the upregulation of these two myelin markers. Finally, to assess whether NCX3 might have a role in CNS myelination, we analyzed the expression of myelin markers in oligodendrocytes of mice lacking NCX3. Interestingly, we found that NCX3 knockout mice exhibit hypomyelination that is accompanied by a reduction

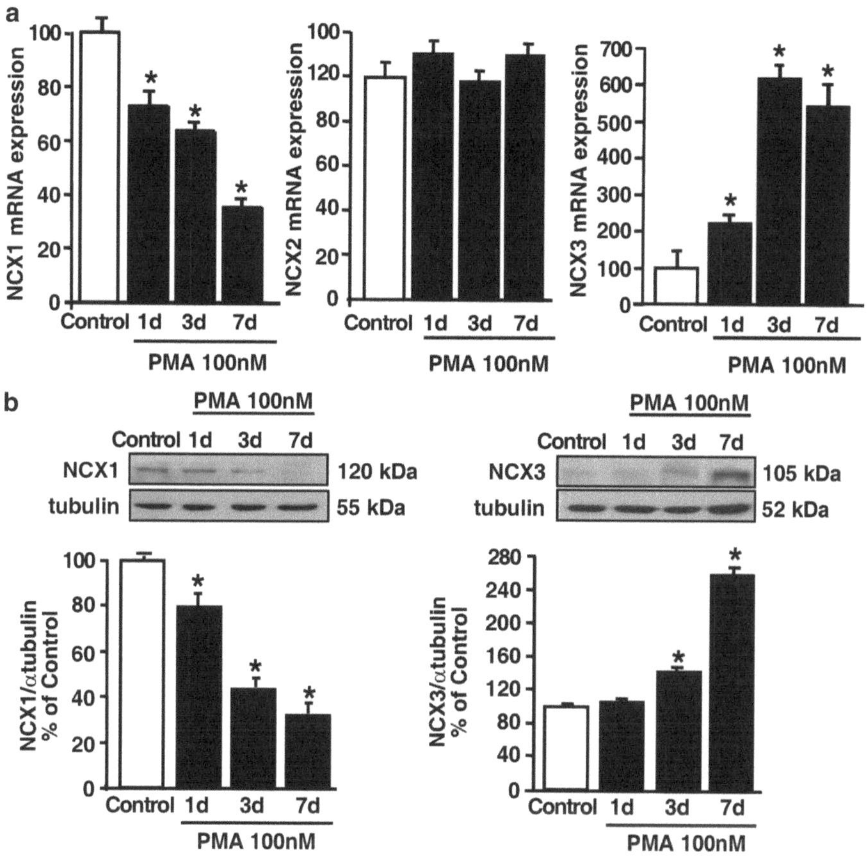

Fig. 26.4 (**a**) Real-time PCR of NCX1 (*left*), NCX2 (*middle*), and NCX3 (*right*) mRNA expression in MO3.13 cells under control conditions and following 100-nM PMA exposure for 1, 3, and 7 days. Data were normalized on the basis of the ribosomal L-19 levels and expressed as percentage of controls. Values represent means±SEM (n=4). *p<0.05 versus controls. (**b**) Western blot and densitometric analysis of NCX1 (*left*) and NCX3 (*right*) protein levels in MO3.13 cells under control conditions and following 100-nM PMA exposure for 1, 3, and 7 days. Data were normalized on the basis of α-tubulin levels and expressed as percentage of controls. Values represent means±S.E.M. (n=3–4). *$p < 0.05$ versus controls

of spinal cord size (Fig. 26.5a). Immunoblotting and quantitative immunofluorescence analyses for the myelin marker MBP and the axonal neurofilament marker NF200 revealed that both markers were reduced in *ncx3*–/– mice when compared to congenic *ncx*3+/+ mice. By contrast, the oligodendrocyte progenitor cell marker NG2 was upregulated in *ncx3*–/– mice (Fig. 26.5b). As far as concern the phenotype of *ncx3*–/– mice, it has been demonstrated that NCX3-deficient mice display skeletal muscle fiber necrosis and impaired neuromuscular transmission that is clinically associated with reduced motor activity, weakness of forelimb muscles, and fatigability (Sokolow et al. 2004). According to our results, the symptoms observed in *ncx3*–/– mice may not be exclusively attributed to alterations at the neuromuscular junction or the skeletal fibers but may also result from defects in CNS myelination.

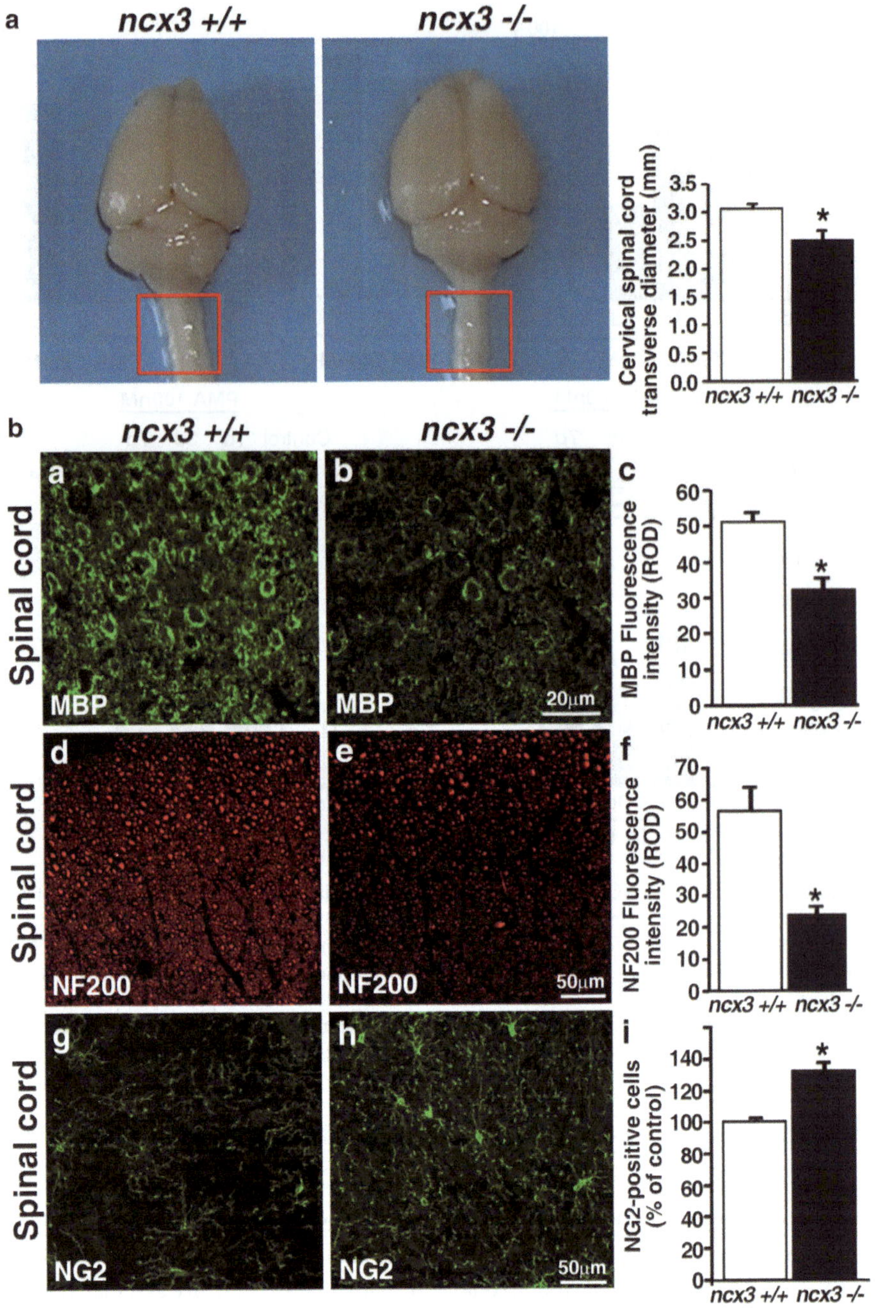

Fig. 26.5 (**a**) Representative images of 4 months *ncx3*+/+ (*left*) and *ncx3*–/– (*right*) whole brain and spinal cord samples. *Red* boxes indicate the reduction in spinal cord size between *ncx3*+/+ and *ncx3*–/– mice. (**b**) (*a-b*) Distribution of MBP immunoreactivity, (*c*) quantification of its fluorescence intensity in the *white* matter of the spinal cord from *ncx3*+/+ and *ncx3*–/– mice, (*d, e*) distribution of NF-200 immunoreactivity, (*f*) quantification of its fluorescence intensity in the *white* matter of the spinal cord from *ncx3*+/+ and *ncx3*–/– mice, (*g, h*) distribution of NG2 immunoreactivity, and (*i*) quantification of NG2-positive cells in the *white* matter of the spinal cord from ncx3+/+ and ncx3–/– mice. For cells counting 6–10 microscope fields were counted for each section, and six sections for each animal were analyzed. Values represent means ± S.E.M. (n = 3–4). *$p < 0.05$ versus ncx3+/+. Scale bars in **b**: (*a, b*) 20 µm; (*d, e*) 50 µm; (*g, h*) 50 µm

26.4 Conclusions

Collectively, our studies suggest important roles of NCX1 and NCX3 isoforms in microglia and oligodendrocytes, respectively. In particular, the upregulation of NCX1 expression and activity observed in microglial cells after pMCAO suggests the relevant role played by NCX1 in modulating microglial functions in the postischemic brain. A thorough understanding of NCX1 function in microglia may help to provide novel disease-modifying approaches in cerebral ischemia.

Furthermore, our findings, by providing new insights into the molecular and cellular mechanisms involved in OPC development, demonstrate for the first time that calcium signaling mediated by NCX3 isoform is crucially involved in oligodendrocyte maturation and myelin formation. Further studies are required to reveal whether alterations in NCX3 activity might contribute to OPC dysfunction in demyelinating diseases and whether its modulation might be therapeutically relevant.

Acknowledgments This work was supported by COFIN2008, Ricerca-Sanitaria RF-FSL352059 Ricerca finalizzata 2006, Ricerca-Oncologica 2006, Progetto-Strategico 2007, Progetto Ordinario 2007, Ricerca finalizzata 2009, and Ricerca-Sanitaria progetto Ordinario by Ministero della Salute 2008 all to LA.

References

L. Annunziato, G. Pignataro, G.F. Di Renzo, Pharmacology of brain Na^+/Ca^{2+} exchanger: from molecular biology to therapeutic perspectives. Pharmacol. Rev. **56**, 633–654 (2004)

B.A. Barres, W.J. Koroshetz, K.J. Swartz, L.L. Chun, D.P. Corey, Ion channel expression by white matter glia: the O-2A glial progenitor cell. Neuron **4**, 507–524 (1990)

F. Boscia, R. Gala, G. Pignataro, A. De Bartolomeis, M. Cicale, A. Ambesi Impiombato, G. Di Renzo, L. Annunziato, Permanent focal brain ischemia induces isoform-dependent changes in the pattern of Na^+/Ca^{2+} exchanger gene expression in the ischemic core, peri-infarct area, and intact brain regions. J. Cereb. Blood Flow Metab. **26**, 502–517 (2006)

F. Boscia, R. Gala, A. Pannaccione, A. Secondo, A. Scorziello, G.F. Di Renzo, L. Annunziato, NCX1 expression and functional activity increase in microglia invading the infarct core. Stroke **40**, 3608–3617 (2009)

F. Boscia, C. D'Avanzo, A. Pannaccione, A. Secondo, A. Casamassa, L. Formisano, N. Guida, L. Annunziato, Silencing or knocking out the Na^+/Ca^{2+} exchanger-3 (NCX3) impairs oligodendrocyte differentiation. Cell Death Differ. **19**, 562–572 (2012)

S.Y. Chong, J.R. Chan, Tapping into the glial reservoir: cells committed to remaining uncommitted. J. Cell Biol. **188**, 305–312 (2010)

M.J. Craner, B.C. Hains, A.C. Lo, J.A. Black, S.G. Waxman, Co-localization of sodium channel Nav1.6 and the sodium–calcium exchanger at sites of axonal injury in the spinal cord in EAE. Brain **127**, 294–303 (2004a)

M.J. Craner, J. Newcombe, J.A. Black, C. Hartle, M.L. Cuzner, S.G. Waxman, Molecular changes in neurons in multiple sclerosis: altered axonal expression of Nav1.2 and Nav1.6 sodium channels and Na^+/Ca^{2+} exchanger. Proc. Natl. Acad. Sci. U. S. A. **101**, 8168–8173 (2004b)

R.J.M. Franklin, C. French-Constant, Remyelination in the CNS: from biology to therapy. Nat. Rev. Neurosci. **9**, 839–855 (2008)

F. Ginhoux, M. Greter, M. Leboeuf, S. Nandi, P. See, S. Gokhan, M.F. Mehler, S.J. Conway, L.G. Ng, E.R. Stanley, I.M. Samokhvalov, M. Merad, Fate mapping analysis reveals that adult microglia derive from primitive macrophages. Science **330**, 841–845 (2010)

U.K. Hanisch, H. Kettenmann, Microglia: active sensor and versatile effector cells in the normal and pathologic brain. Nat. Neurosci. **10**, 1387–1394 (2007)

M. Ifuku, K. Färber, Y. Okuno, Y. Yamakawa, T. Miyamoto, C. Nolte, V.F. Merrino, S. Kita, T. Iwamoto, I. Komuro, B. Wang, G. Cheung, E. Ishikawa, H. Ooboshi, M. Bader, K. Wada, H. Kettenmann, M. Noda, Bradykinininduced microglial migration mediated by B1-bradykinin receptors depends on Ca^{2+} influx via reverse-mode activity of the Na^+/Ca^{2+} exchanger. J. Neurosci. **27**, 13065–13073 (2007)

D. Ito, K. Tanaka, S. Suzuki, T. Dembo, Y. Fukuuchi, Enhanced expression of Iba1, ionized calcium-binding adapter molecule 1, after transient focal cerebral ischemia in rat brain. Stroke **32**, 1208–1215 (2001)

S. Li, Q. Jiang, P.K. Stys, Important role of reverse Na^+–Ca^{2+} exchange in spinal cord white matter injury at physiological temperature. J. Neurophysiol. **84**, 1116–1119 (2000)

P. Lipton, Ischemic cell death in brain neurons. Physiol. Rev. **79**, 1431–1568 (1999)

T. Matsuda, T. Nagano, M. Takemura, A. Baba, Topics on the Na^+/Ca^{2+} exchanger: responses of Na^+/Ca^{2+} exchanger to interferon-gamma and nitric oxide in cultured microglia. J. Pharmacol. Sci. **102**, 22–26 (2006)

P. Molinaro, O. Cuomo, G. Pignataro, F. Boscia, R. Sirabella, R. Gala, S. Sokolow, A. Herchuelz, S. Schurmans, G. Di Renzo, L. Annunziato, Targeted disruption of NCX3 gene leads to a worsening of ischemic brain damage. J. Neurosci. **28**, 1179–1184 (2008)

T. Nagano, Y. Kawasaki, A. Baba, M. Takemura, T. Matsuda, Up-regulation of Na^+-Ca^{2+} exchange activity by interferon-gamma in cultured rat microglia. J. Neurochem. **90**, 784–791 (2004)

E.W. Newell, E.F. Stanley, L.C. Schlichter, Reversed Na^+/Ca^{2+} exchange contributes to Ca^{2+} influx and respiratory burst in microglia. Channels (Austin) **1**, 366–376 (2007)

G. Pignataro, R. Gala, O. Cuomo, A. Tortiglione, L. Giaccio, P. Castaldo, R. Sirabella, C. Matrone, A. Canitano, S. Amoroso, G.F. Di Renzo, L. Annunziato, Two sodium/calcium exchanger gene products, NCX1 and NCX3, play a major role in the development of permanent focal cerebral ischemia. Stroke **35**, 2566–2570 (2004)

B.D. Quednau, D.A. Nicoll, K.D. Philipson, Tissue specificity and alternative splicing of the Na^+/Ca^+2 exchanger isoforms NCX1, NCX2, and NCX3 in rat. Am. J. Physiol. **272**, C1250–C1261 (1997)

S. Sokolow, M. Manto, P. Gailly, J. Molgó, C. Vandebrouck, J.M. Vanderwinden, A. Herchuelz, S. Schurmans, Impaired neuromuscular transmission and skeletal muscle fiber necrosis in mice lacking Na/Ca exchanger 3. J. Clin. Invest. **113**, 265–273 (2004)

D.J. Tomes, S.K. Agrawal, Role of Na^+–Ca^{2+} exchanger after traumatic or hypoxic/ischemic injury to spinal cord white matter. Spine J. **2**, 35–40 (2002)

Human Macrophages and Monocytes Express Functional Na^+/Ca^{2+} Exchangers 1 and 3

27

Rosaria I. Staiano, Francescopaolo Granata, Agnese Secondo, Angelica Petraroli, Stefania Loffredo, Lucio Annunziato, Massimo Triggiani, and Gianni Marone

Abstract

The Na^+/Ca^{2+} exchanger (NCX) is a plasma membrane protein that can switch Na^+ and Ca^{2+} in either direction to maintain the homeostasis of intracellular Ca^{2+}. A family of three genes (*NCX1, NCX2,* and *NCX3*) has been identified in neurons and muscle cells. NCX activity has also been reported in certain immune cells (e.g., mast cells). We have examined the expression and function of these NCX isoforms in the human monocytes and lung macrophages. Monocytes were purified from peripheral blood of healthy donors. Macrophages (HLM) were isolated and purified from the lung parenchyma of patients undergoing thoracic surgery. NCX1 and NCX3, but not NCX2, were expressed in HLM and monocytes at both mRNA and protein level. Exposure to Na^+-free medium induced a significant increase in intracellular calcium concentration ($[Ca^{2+}]_i$) in both cell types, suggesting that NCX isoforms expressed on these cells were functionally active. This response was completely abolished by the NCX inhibitor 5-(N-4-chlorobenzyl)-20,40-dimethylbenzamil (CB-DMB). In addition, incubation of macrophages with Na^+-free medium induced a marked release of TNF-α. Preincubation of HLM with CB-DMB and RNAi-mediated knockdown of NCX1 blocked TNF-α release. Our results demonstrate that human macrophages and monocytes express NCX1 and

R.I. Staiano • F. Granata • A. Petraroli • S. Loffredo
Division of Clinical Immunology and Allergy,
School of Medicine, University of Naples Federico II,
Via S. Pansini 5, Naples, 80131, Italy

A. Secondo • L. Annunziato
Division of Pharmacology, Department of Neuroscience,
School of Medicine, University of Naples Federico II,
Naples, Italy

M. Triggiani • G. Marone (✉)
Division of Clinical Immunology and Allergy,
School of Medicine, University of Naples Federico II,
Via S. Pansini 5, Naples, 80131, Italy

Center for Basic and Clinical Immunology
Research (CISI), Naples, Italy
e-mail: marone@unina.it

L. Annunziato (ed.), *Sodium Calcium Exchange: A Growing Spectrum of Pathophysiological Implications*, Advances in Experimental Medicine and Biology 961, DOI 10.1007/978-1-4614-4756-6_27,

NCX3 that operate in a bidirectional manner to restore $[Ca^{2+}]_i$ to generate Ca^{2+} signals and to induce TNF-α production. We suggest that NCX may modulate Ca^{2+} homeostasis and proinflammatory functions in human macrophages and monocytes.

Keywords
Monocytes • Macrophages • Na^+/Ca^{2+} exchanger • Ca^{2+}-signaling • Cytokines

27.1 Introduction

27.1.1 The Mononuclear Phagocyte System

The mononuclear phagocyte system (MPS) is defined on the basis of ontogeny and phagocytic activity and includes blood monocytes and the diverse network of tissue macrophages and dendritic cells (DC) (Biswas and Mantovani 2010). MPS has a primary role in innate immunity, recruitment of immune cells, tissue inflammation, and remodeling (1–3) (Biswas and Mantovani 2010; Lawrence and Natoli 2011; Mantovani et al. 2004). MPS cells derive from a common $CD34^+$ bone marrow progenitor, differentiating into monocytes which are released in peripheral blood (Fogg et al. 2006), where they circulate for several days before entering tissues and constituting the tissue macrophage populations (Gordon and Taylor 2005).

Mature monocytes in the peripheral circulation are heterogeneous; they vary in size and have different degrees of granularity and varied nuclear morphology (Biswas and Mantovani 2010). A wide spectrum of proinflammatory, metabolic, and immune stimuli elicits increased recruitment of monocytes to peripheral sites, where differentiation into macrophages and DCs occurs, contributing to host defense and tissue remodeling and repair (Gordon and Taylor 2005). The functional and phenotypic diversity of tissue macrophages reflects a complex interplay between intrinsic differentiation pathways and environmental inputs received from neighboring cells (Biswas and Mantovani 2010). In fact, macrophages are found in nearly every tissue in the body. The resident cells in different tissues show distinct phenotypes, including microglial cells of the brain, Langerhans cells in the skin, bone-resorbing osteoclasts in the skeletal system, alveolar macrophages in the lung, and Kupffer cells in the liver (Gordon and Taylor 2005).

These cells behave as the sentinels of the innate immune system and monitor for early signs of infection or tissue damage (Biswas and Mantovani 2010). There is compelling evidence that, depending on the microenvironment, macrophages can acquire at least two distinct functional phenotypes, which derive from two mutually exclusive activation programs: M1 macrophages are the classically activated macrophages, whereas M2 are alternatively activated macrophages (Biswas and Mantovani 2010; Martinez et al. 2009). These activation programs were initially defined by their antimicrobial activities: classical activation occurs in response to products derived from or associated with bacterial infections, such as lipopolysaccharide (LPS) and interferon-γ (IFN-γ), and results in highly inflammatory macrophages with high phagocytic and bactericidal potential. M1 macrophages are characterized by the ability to release large amounts of proinflammatory cytokines, such as IL-12, IL-23, and TNF-α; reactive oxygen and nitrogen intermediates; higher expression of major histocompatibility complex class II and costimulatory molecules; efficient antigen presentation; and microbicidal or tumoricidal activity. M1 cells mediate resistance to intracellular pathogens and tumors and elicit tissue disruptive reactions (Biswas and Mantovani 2010; Martinez et al. 2009). In contrast, alternative activation occurs in response to products derived from or associated with parasitic infections and IL-4 and IL-13 and promotes antiparasitic functions as well as those involved in tissue repair and remodeling. In general, M2 cells participate in parasite clearance, dampening of inflammation, promotion of tissue remodeling, angiogenesis, tumor progression, and immunoregulatory functions (Biswas and Mantovani 2010; Martinez et al. 2009; Mosser and Edwards 2008). Several cytokines can govern M2 polarization. IL-4 and IL-13 direct M2 polarization of macrophages during helminth infection and allergy (Raes et al. 2005). In addition, IL-33, a cytokine of the IL-1 family associated with M2 polarization, amplifies IL-13-induced polarization of alveolar macrophages to an M2 phenotype (Kurowska-Stolarska et al. 2009). However, typical M1 and M2 phenotypes are extremes of a spectrum in a galaxy of functional states (Mosser and Edwards 2008).

Another polarized macrophage phenotype is defined as tumor-associated macrophages (TAMs). These cells infiltrate tumor tissues and play an important role in subversion of adaptive immunity and in inflammatory circuits that promote tumor growth and progression (Allavena and Mantovani 2012; Biswas and Mantovani 2010). TAM show a transcriptional profile quite distinct from those M1 and M2 macrophages (Martinez et al. 2006). Such macrophages generally have an $IL\text{-}12^{lo}IL\text{-}10^{hi}$ phenotype and show impaired expression of reactive nitrogen intermediates, less antigen presentation and tumoricidal capacity, and high expression of angiogenic factors (VEGF, EGF, and semaphorin 4D) and metalloproteases (Biswas et al. 2006; Sierra et al. 2008; Torroella-Kouri et al. 2009).

Due to the ability of macrophages to reprogram their functions, it should be considered that M1 and M2 as well as TAM phenotypes might not be stably differentiated subsets and each subpopulation remains responsive to the microenvironment stimuli. This classification of macrophage subsets provides a useful working scheme; however, it unlikely fully represents the complexity of the transitional states of macrophage activation, which is often fine-tuned in response to different microenvironments (Biswas and Mantovani 2010).

27.1.2 Sodium-Calcium Exchanger in Immune Cells

Calcium signals are essential in the maturation, differentiation, and activation of cells of the immune response. Monocytes and macrophages rely on Ca^{2+} signaling to start their activation programs, leading to the production and release of proinflammatory mediators (Shumilina et al. 2011; Triggiani et al. 2001). In these cells, cytoplasmic free Ca^{2+} has vital second-messenger effects in the regulation of other fundamental functions, such as differentiation, migration, and apoptosis (Shumilina et al. 2011; Triggiani et al. 2001).

The Na^+/Ca^{2+} exchanger (NCX) is a plasma membrane protein involved in the homeostasis of intracellular Ca^{2+} (Philipson and Nicoll 2000). NCX is a bidirectional ion transporter that catalyzes the exchange of Na^+ and Ca^{2+} (transport ratio $\approx$ 3:1) depending on their electrochemical gradients (Annunziato et al. 2004). Under physiological conditions, NCX's primary role is to extrude Ca^{2+} from cells using the Na^+ gradient across the cell membrane (forward mode of operation) (Blaustein and Lederer 1999). However, in some cases it can contribute to the Ca^{2+} influx into the cells by working in reverse mode (i.e., coupling Ca^{2+} influx with Na^+ efflux) (Bouwman et al. 2006; Ifuku et al. 2007). Three isoforms of NCX (NCX1, NCX2, and NCX3) have been cloned so far in mammals, outlining a multigene family whose members share a similar molecular structure (Lytton 2007). NCX1 is prevalent in mammalian tissues such as heart, brain, and kidney, whereas the expression of NCX2 and NCX3 appears limited to the brain and skeletal muscle (Lytton 2007). Non-excitable cells such as renal epithelial cells (Schmitt et al. 1999), exocrine acinar cells (Petersen 1992), and platelets (Roberts et al. 2004) may also have NCX activity.

There is some evidence that immune cells may express NCX to maintain intracellular Ca^{2+} homeostasis as well as to act on Ca^{2+} signaling. A study suggesting an NCX activity in immune cells was conducted by Simchowitz and Cragoe (Simchowitz and Cragoe 1988). These authors identified a Na^+/Ca^{2+} exchange in human neutrophils which had a $3Na^+:1Ca^{2+}$ stoichiometry. They further described that this exchange could enhance calcium influx after the stimulation of neutrophils with a chemotactic factor (N-formyl-methionyl-leucyl-phenylalanine: fMLP) (Simchowitz and Cragoe 1988). Subsequently, the involvement of NCX in the uptake of extracellular Ca^{2+} after stimulation of neutrophils with fMLP has been confirmed (Tintinger and Anderson 2004). NCX activity has been identified also in human T cells. It has been reported that the pharmacological inhibition of NCX impaired mitogen-induced proliferation of these cells (Wacholtz et al. 1992). Additional evidence of the functional presence of NCX in T lymphocytes was provided (Balasubramanyam et al. 1994). The functional presence of NCX has been shown also in murine macrophages (Donnadieu and Trautmann 1993).

The study of the physiologic role of NCX has been impaired in the past by the lack of either specific inhibitors or antibodies. Therefore, the previous studies provided only indirect evidence for the presence of NCX in certain immune cells. More recently, it has been demonstrated the expression of NCX3 mRNA in a mast cell line (RBL) (Aneiros et al. 2005). The latter paper extended previous observations suggesting NCX activity in rat peritoneal mast cells (Alfonso et al. 1999; Praetorius et al. 1998) and in RBL cells (Rumpel et al. 2000).

Given our interest in the modulation of human monocytes and macrophages (Granata et al. 2010; Triggiani et al. 2007), we become interested in the possibility that the cells express NCXs. In this scenario, we demonstrated for the first time the expression of NCX1 and NCX3 in human primary monocytes and macrophages. Moreover, we characterized the possible role of this protein either in participation of immune response of macrophages either in restoring Ca^{2+} homeostasis after cells activation (Staiano et al. 2009).

27.2 Human Macrophages and Monocytes Express Functional Na^+/Ca^{2+} Exchangers 1 and 3

We have investigated the expression and function of NCXs in primary human macrophages purified from the lung parenchyma of patients undergoing thoracic surgery (Granata et al. 2010) (HLM) and monocytes isolated from peripheral blood of healthy donors (Triggiani et al. 2007). A first series of experiments were conducted to evaluate the expression of the exchanger both at mRNA and protein level. RT-PCR experiments done on different preparations of HLM and monocytes using target-specific primers for the three NCX isoforms showed the presence of mRNA for *NCX1* and *NCX3* but not for *NCX2* in both cell populations (Fig. 27.1a).

Subsequently, it was examined whether HLM and monocytes also synthesized NCX1 and NCX3 at protein level, using Western blot (WB) analysis. To this aim, it was necessary to perform a series of preliminary experiments to set the optimal condition for the NCX protein analysis. In fact, these proteins show a different electrophoretic mobility and immunoreactivity when cell extracts were exposed to reducing or nonreducing conditions (Santacruz-Toloza et al. 2000). WB experiments confirmed that both macrophages and monocytes express one major immunoreactive band of approximately 110 kDa which corresponds to the full-length NCX1 (Fig. 27.1b). When the same cell lysates were immunoblotted with anti-NCX3 antibody, HLM and monocytes showed a major immunoreactive band of approximately 100 kDa, the expected size for NCX3 protein (Fig. 27.1c). On equal protein loading, densitometric analysis of the full-length NCX1 (~110 kDa band) and NCX3 (~100 kDa band) in the three different samples indicated that the HLM signals were significantly more intense than monocytes. These results indicate that human lung macrophages contain more NCX1 and NCX3 proteins than monocytes. It might be possible that the expression of NCX could be modulated during monocyte differentiation in macrophages.

In a second series of experiments, we have characterized NCX activity in human monocytes and macrophages. Because NCX can operate in two directions depending on the transmembrane gradient of Na^+ and Ca^{2+} ions (Annunziato et al. 2004), we evaluated the reverse mode of operation by exposing the cells to a single pulse of Na^+-free medium (Secondo et al. 2007). In particular, NCX activity was assessed by a single-cell Fura-2 AM microfluorimetry as the capacity

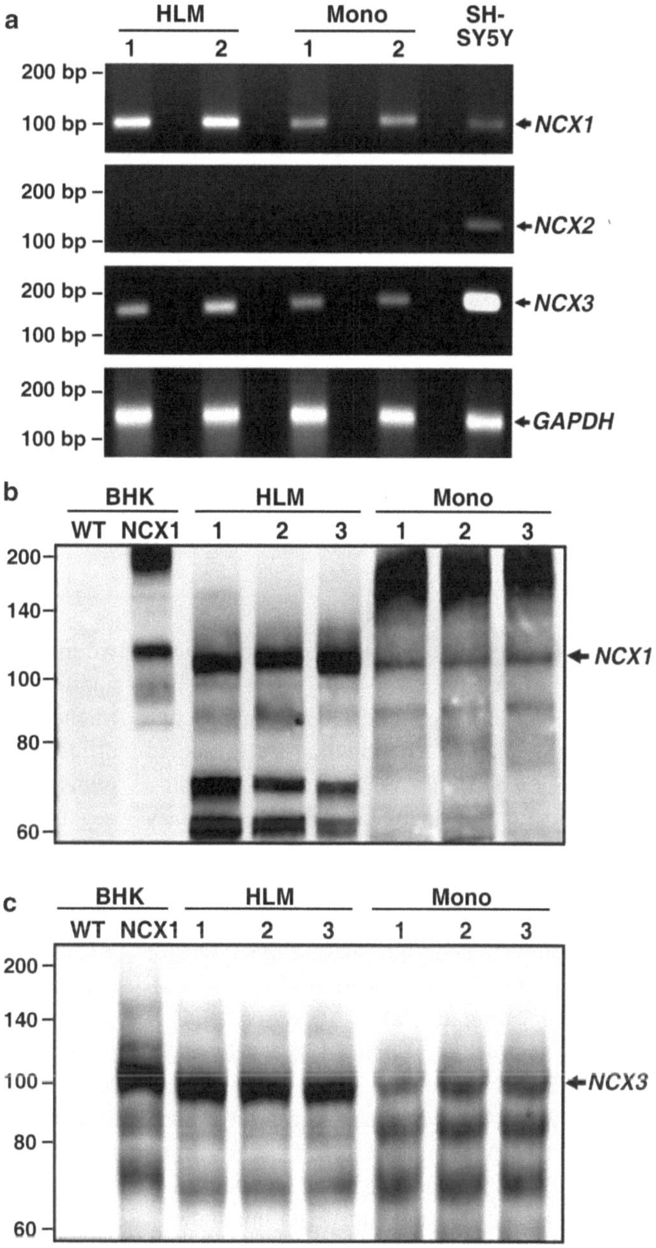

Fig. 27.1 *Expression of NCX1 and NCX3 in human lung macrophages (HLM) and monocytes.* (**a**) Expression of mRNA for NCX1, NCX2, and NCX3. NCX1 (*first row*), NCX2 (*second row*), NCX3 (*third row*), and GAPDH (*last row*) RT-PCR specific amplification products from two preparations of HLM and blood monocytes (Mono). SH-SY5Y cells were used as positive control. (**b, c**) Detection of NCX1 and NCX3 in HLM and monocytes. Protein extracts (40 μg per sample) from three preparations of HLM and Mono were separated by electrophoresis under reducing, non-denaturing conditions and transferred to nitrocellulose. Membranes were blotted with anti-NCX1 (*panel b*) or anti-NCX3 (*panel c*). Protein extracts (10 μg) of BHK cells, either stably transfected with NCX1 or NCX3 or mock-transfected (WT), were used as positive and negative controls, respectively. The *arrow* indicates the band corresponding to the full-length NCX1 (*panel b*) and NCX3 (*panel c*) (Staiano et al. 2009, Copyright Wiley-VCH Verlag GmbH & Co. KGaA. Reproduced with permission)

to generate $[Ca^{2+}]_i$ increase in macrophages and monocytes. Na^+-free medium induced a rapid, linear increase in $[Ca^{2+}]_i$ in both HLM (Fig. 27.2a) and monocytes (data not shown). To verify that the Ca^{2+} signal induced by Na^+-free medium in HLM and monocytes was due to NCX activity, cells were preincubated (1 h at 37 °C) with the amiloride derivative CB-DMB (10 μM)—a pan inhibitor of all three NCX isoforms (Rogister et al. 2001; Secondo et al. 2007)—before exposure to Na^+-free medium. CB-DMB completely blocked the Ca^{2+} influx elicited by Na^+-free perfusion in HLM (Fig. 27.2b) and monocytes (data not shown). These results indicate that NCX expressed by human macrophages and monocytes was functionally active and that $[Ca^{2+}]_i$

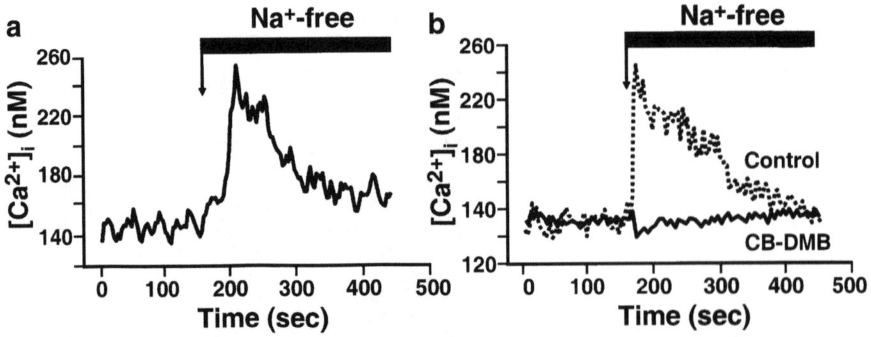

Fig. 27.2 *NCX activity in HLM.* (**a**) Effect of incubation in Na^+-free solution on the intracellular calcium concentration ($[Ca^{2+}]_i$) in HLM. Single-cell trace representative of the effect of Na^+-free solution on $[Ca^{2+}]_i$ in HLM. The trace is representative of 25–40 cells studied in three different experiments. (**b**) Effect of CB-DMB on the increase in $[Ca^{2+}]_i$ induced by Na^+-free solution in HLM and monocytes. Superimposed single-cell traces for $[Ca^{2+}]_i$ in HLM perfused with Na^+-free solution in the absence (Control) or presence of CB-DMB (10 μM, 60 min preincubation). Each trace is representative of 25–40 cells studied in three different experiments (Staiano et al. 2009, Copyright Wiley-VCH Verlag GmbH & Co. KGaA. Reproduced with permission)

increase induced by a Na^+ gradient is due to the activation of NCX.

Macrophages and monocytes rely on Ca^{2+} signaling to activate cell functions relevant to inflammation and immune regulation, including the production of cytokines and chemokines (Johansson et al. 2005; Mayne et al. 2000; Triggiani et al. 2001). We have examined the effect of Na^+-free medium on the release of two major proinflammatory products of macrophages and monocytes: TNF-α and IL-8 (Granata et al. 2010; Triggiani et al. 2001). Activation of NCX by Na^+-free medium induced the release of TNF-α from macrophages comparable to that caused by LPS (1 μg/ml) used as a positive control (Fig. 27.3a) (Sweet et al. 2001), whereas the release of IL-8 was not increased by incubation of HLM in Na^+-free medium (data not shown). The release of TNF-α is anticipated by the increase of mRNA, as indicated by real-time PCR experiments (data not shown). Moreover, preincubation (60 min at 37 °C) of HLM with CB-DMB before the exposure to Na^+-free conditions completely blocked TNF-α release (Fig. 27.3a). Thus, activation of NCX in reverse mode is associated with increased mRNA expression and release of TNF-α, but not of IL-8.

To further explore the role of NCX in the production of TNF-α induced by Na^+-free medium, we used an RNA interference (RNAi)-mediated approach to silence NCX1 expression in HLM. Silencing of NCX1 was performed by using three different silencing RNA (siRNA) oligonucleotides and a validated irrelevant sequence as a negative control. The successful of silencing was evaluated by western blot analysis. All three siRNA oligonucleotides (siRNA-S7, siRNA-S8, and siRNA-S9) clearly reduced NCX1 protein in HLM as compared to the cells treated with irrelevant oligonucleotides (data not shown). In parallel experiments, we evaluated Na^+-free-induced production of TNF-α in transfected HLM. Silencing of NCX1 with the three siRNA constructs (S7, S8, S9) significantly reduced the release of TNF-α induced by Na^+-free medium (Fig. 27.3b). Transfection of HLM with the irrelevant oligonucleotides (Sham) did not significantly modify the release of TNF-α induced by Na^+-free exposure as compared to non-transfected HLM (Wild Type). Inhibition of TNF-α production in NCX1-silenced HLM was specific for Na^+-free conditions since LPS-induced release of TNF-α was not influenced by transfection with all siRNA oligonucleotides (data not shown). These data demonstrate that NCX1 is the isoform primarily involved in Na^+-free-dependent production of TNF-α from human macrophages.

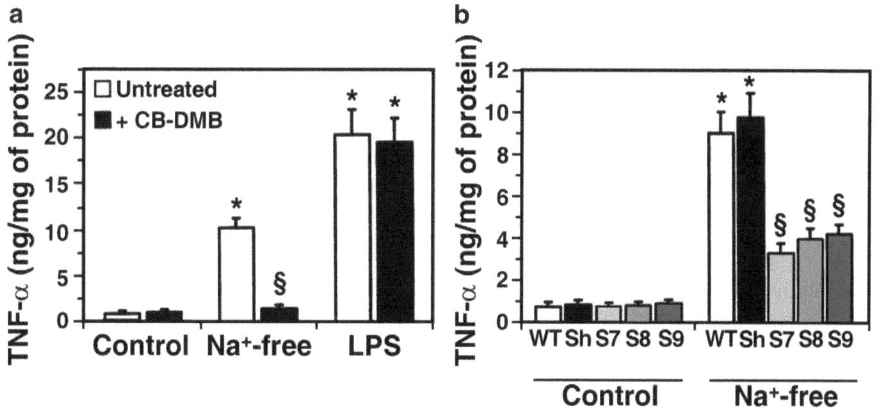

Fig. 27.3 *Role of NCX activation in HLM immune responses.* (**a**) Effect of Na^+-free solution on TNF-α production in HLM. Cells were preincubated (37 °C, 1 h) with or without CB-DMB (10 μM) and then incubated (37 °C, 4 h) with medium alone (Krebs-Ringer saline solution), Na^+-free solution, or LPS (1 μg/ml). At the end of incubation, the supernatant was collected and centrifuged (1,000×g, 4 °C, 5 min). TNF-α was determined by ELISA. The values are expressed as ng of TNF-α per mg of protein. The data are the mean ± SEM of three experiments. *$p<0.05$ *vs.* control and **$p<0.05$ *vs.* untreated as determined by Student's paired *t-test.* (**b**) Role of NCX1 in the production of TNF-α induced by Na^+-free medium in HLM. Silencing of NCX1 was performed by using HiPerfect Transfection® Kit (Qiagen). HLM either non-transfected (wild type) or transfected with siRNA oligonucleotides (S7, S8, and S9) and the irrelevant oligonucleotides (Sham) were incubated (37 °C, 4 h) with medium alone (Krebs-Ringer saline solution) or Na^+-free medium. At the end of incubation, the supernatant was collected and centrifuged (1000×g, 4 °C, 5 min). TNF-α was determined by ELISA. The values are expressed as ng of TNF-α per mg of protein. The data are the mean ± SEM of three experiments. *$p<0.05$ *vs.* the respective control and **$p<0.05$ *vs.* sham as determined by Student's paired *t-test* (Staiano et al. 2009, Copyright Wiley-VCH Verlag GmbH & Co. KGaA. Reproduced with permission)

27.3 Conclusions

The Na^+/Ca^{2+} exchanger is a key element regulating intracellular Ca^{2+} concentrations in various excitable cells where it is primarily involved in the maintenance of intracellular Ca^{2+} homeostasis (Blaustein and Lederer 1999; Philipson and Nicoll 2000). Our study demonstrates that two human inflammatory cells, lung macrophages and blood monocytes, constitutively express NCX1 and NCX3, but not NCX2. NCX can operate in bidirectional manner in these cells because Na^+-gradient-induced activation of the exchanger is coupled to an increase in $[Ca^{2+}]_i$ and its inhibition during Ca^{2+}-spike events is associated with a delay in the restoration of $[Ca^{2+}]_i$. Importantly, activation of NCX promotes TNF-α production in human macrophages by operating in reverse mode.

A major strength of this study was that NCX expression in primary macrophages and their precursor monocytes was evaluated at different levels (gene transcription, protein synthesis, and functional activity). In these cells, NCX1 and NCX3 were expressed at both mRNA and protein levels and efficiently exchanged Na^+ with Ca^{2+} to maintain $[Ca^{2+}]_i$ homeostasis. Therefore, they constitutively synthesize mRNA for NCX1 and NCX3 and translate them into proteins that are functionally active.

Blood monocytes differentiate into macrophages when they migrate into different tissues, so these two cells represent different stages of maturation of the same lineage (Biswas and Mantovani 2010; Fogg et al. 2006; Gordon and Taylor 2005; Lawrence and Natoli 2011; Mantovani et al. 2004). The fact that NCX1 and NCX3 are present in both monocytes and macrophages indicates that these molecules are expressed early in the development of MPS cells and lasts up to the final steps of cell maturation. Thus, NCX may have an important role in regulating functions in circulating monocytes and differentiated tissue macrophages.

Although detectable in both cells, mRNA and protein of the full-length NCX1 and NCX3 were expressed at a higher level in macrophages than monocytes, suggesting that the expression of the full-length NCX1 and NCX3 increases when monocytes differentiate into macrophages. It is known that induction or upregulation of gene transcription occurs when monocytes differentiate into macrophages (Biswas and Mantovani 2010; Lawrence and Natoli 2011; Mantovani et al. 2004). Thus, it is conceivable that differentiation of tissue macrophages is associated with upregulation of NCX1 and NCX3. Additional studies appear necessary to explore whether specific NCX isoforms play a role in the differentiation program of human macrophages.

Expression of NCX1 and NCX3 enables the macrophages and monocytes to respond in the classical model of activation, i.e., exposure to Na^+-free solution. The Ca^{2+} signal elicited in these cells by Na^+-free perfusion was comparable in terms of magnitude and duration. The increase was blocked by the amiloride derivative CB-DMB (Rogister et al. 2001; Secondo et al. 2007). Thus, the overall functional and pharmacological properties of the NCX activity characterized in human macrophages and monocytes appear to be similar to those in other cells and indicate that the exchanger can be involved in the generation of Ca^{2+} signals by operating in reverse mode.

A novel finding of our study was that Na^+-free-induced activation of NCX promoted expression and release of TNF-α, which is one of the main proinflammatory cytokines produced by macrophages. To our knowledge, this is the first evidence that this ion exchanger is involved in this key function of immune cells. Interestingly, this effect of NCX activation appears selective since incubation of HLM in Na^+-free conditions does not influence IL-8 production. We also demonstrate that silencing NCX1 in HLM significantly reduces the release of TNF-α induced by Na^+-free medium. These data indicate that NCX1 is the isoform primarily involved in Na^+-free-dependent production of TNF-α from macrophages even though a role for NCX3 cannot be excluded. Rapid activation of NCX activity seems sufficient to induce this response since brief exposure of macrophages to Na^+-free medium (10 min) was just as effective as prolonged incubation (4 h) in inducing TNF-α release. Ca^{2+} is involved in the production of TNF-α and other cytokines in macrophages (Johansson et al. 2005; Mayne et al. 2000; Triggiani et al. 2001), and many stimuli that activate cytokine production also generate Ca^{2+} signals (Liu et al. 2007; Zhou et al. 2006). However, these stimuli raise cytoplasmic Ca^{2+} concentrations by energy-dependent mechanisms related to intracellular intermediates. Again, the fact that NCX can operate without cellular intermediates or energy consumption may enable the macrophages to produce TNF-α even when other biochemical machineries are impaired by the reduced availability of ATP. This concept could be extended to several other functions of monocytes and macrophages that rely on Ca^{2+} signaling for their activation programs. The fact that the Ca^{2+} signals mediated by NCX induce the expression and release of TNF-α, but not IL-8 production, suggests that NCX activation is preferentially involved in selective functions of macrophages.

The co-expression of two exchangers (NCX1 and NCX3) in macrophages and monocytes may appear redundant. Current knowledge indicates that the overall functional and pharmacological properties of the three exchangers are similar (Linck et al. 1998; Lytton 2007). Therefore, under certain circumstances, the loss of function of one exchanger may be compensated by the other. For example, NCX3 might compensate for NCX1 when the intracellular ATP levels are greatly reduced (e.g., during tissue ischemia or anoxia) and the function of NCX1 is impaired (Secondo et al. 2007). Another possibility is that the two NCX isoforms expressed in mononuclear phagocytes are coupled to distinct cellular functions. This hypothesis is supported by a study in murine osteoblasts which also selectively express NCX1 and NCX3 (Sosnoski and Gay 2008). Our results demonstrate that silencing NCX1 in human macrophages significantly reduces the release of TNF-α induced by Na^+-free medium, thereby suggesting that this phenomenon is primarily dependent on NCX1 activity. Further studies will explore whether NCX1 and NCX3 expressed in mononuclear phagocytes are selectively involved in other cell functions.

In conclusion, we have provided the first evidence that human monocytes and macrophages constitutively express two functionally active forms of the Na^+/Ca^{2+} exchanger. Understanding the role of these exchangers in regulating monocyte/macrophage proinflammatory functions may reveal novel pharmacological targets to influence inflammatory and immune responses.

References

A. Alfonso, J. Lago, M.A. Botana, M.R. Vieytes, L.M. Botana, Characterization of the Na^+/Ca^{2+} exchanger on rat mast cells. Evidence for a functional role on the regulation of the cellular response. Cell. Physiol. Biochem. **9**, 53–71 (1999)

P. Allavena, A. Mantovani, Immunology in the clinic review series; focus on cancer: tumour-associated macrophages: undisputed stars of the inflammatory tumour microenvironment. Clin. Exp. Immunol. **167**, 195–205 (2012)

E. Aneiros, S. Philipp, A. Lis, M. Freichel, A. Cavalie, Modulation of Ca^{2+} signaling by Na^+/Ca^{2+} exchangers in mast cells. J. Immunol. **174**, 119–130 (2005)

L. Annunziato, G. Pignataro, G.F. Di Renzo, Pharmacology of brain Na^+/Ca^{2+} exchanger: from molecular biology to therapeutic perspectives. Pharmacol. Rev. **56**, 633–654 (2004)

M. Balasubramanyam, C. Rohowsky-Kochan, J.P. Reeves, J.P. Gardner, Na^+/Ca^{2+} exchange-mediated calcium entry in human lymphocytes. J. Clin. Invest. **94**, 2002–2008 (1994)

S.K. Biswas, L. Gangi, S. Paul, T. Schioppa, A. Saccani, M. Sironi, B. Bottazzi, A. Doni, B. Vincenzo, F. Pasqualini, L. Vago, M. Nebuloni, A. Mantovani, A. Sica, A distinct and unique transcriptional program expressed by tumor-associated macrophages (defective NF-kappaB and enhanced IRF-3/STAT1 activation). Blood **107**, 2112–2122 (2006)

S.K. Biswas, A. Mantovani, Macrophage plasticity and interaction with lymphocyte subsets: cancer as a paradigm. Nat. Immunol. **11**, 889–896 (2010)

M.P. Blaustein, W.J. Lederer, Sodium/calcium exchange: its physiological implications. Physiol. Rev. **79**, 763–854 (1999)

R.A. Bouwman, K. Salic, F.G. Padding, E.C. Eringa, B.J. van Beek-Harmsen, T. Matsuda, A. Baba, R.J. Musters, W.J. Paulus, J.J. de Lange, C. Boer, Cardioprotection via activation of protein kinase C-delta depends on modulation of the reverse mode of the Na^+/Ca^{2+} exchanger. Circulation **114**, I226–I232 (2006)

E. Donnadieu, A. Trautmann, Is there a Na^+/Ca^{2+} exchanger in macrophages and in lymphocytes? Pflugers Arch. **424**, 448–455 (1993)

D.K. Fogg, C. Sibon, C. Miled, S. Jung, P. Aucouturier, D.R. Littman, A. Cumano, F. Geissmann, A clonogenic bone marrow progenitor specific for macrophages and dendritic cells. Science **311**, 83–87 (2006)

S. Gordon, P.R. Taylor, Monocyte and macrophage heterogeneity. Nat. Rev. Immunol. **5**, 953–964 (2005)

F. Granata, A. Frattini, S. Loffredo, R.I. Staiano, A. Petraroli, D. Ribatti, R. Oslund, M.H. Gelb, G. Lambeau, G. Marone, M. Triggiani, Production of vascular endothelial growth factors from human lung macrophages induced by group IIA and group X secreted phospholipases A2. J. Immunol. **184**, 5232–5241 (2010)

M. Ifuku, K. Farber, Y. Okuno, Y. Yamakawa, T. Miyamoto, C. Nolte, V.F. Merrino, S. Kita, T. Iwamoto, I. Komuro, B. Wang, G. Cheung, E. Ishikawa, H. Ooboshi, M. Bader, K. Wada, H. Kettenmann, M. Noda, Bradykinin-induced microglial migration mediated by B1-bradykinin receptors depends on Ca^{2+} influx via reverse-mode activity of the Na^+/Ca^{2+} exchanger. J. Neurosci. **27**, 13065–13073 (2007)

U. Johansson, C. Lawson, M. Dabare, D. Syndercombe-Court, A.C. Newland, G.L. Howells, M.G. Macey, Human peripheral blood monocytes express protease receptor-2 and respond to receptor activation by production of IL-6, IL-8, and IL-1{beta}. J. Leukoc. Biol. **78**, 967–975 (2005)

M. Kurowska-Stolarska, B. Stolarski, P. Kewin, G. Murphy, C.J. Corrigan, S. Ying, N. Pitman, A. Mirchandani, B. Rana, N. van Rooijen, M. Shepherd, C. McSharry, I.B. McInnes, D. Xu, F.Y. Liew, IL-33 amplifies the polarization of alternatively activated macrophages that contribute to airway inflammation. J. Immunol. **183**, 6469–6477 (2009)

T. Lawrence, G. Natoli, Transcriptional regulation of macrophage polarization: enabling diversity with identity. Nat. Rev. Immunol. **11**, 750–761 (2011)

B. Linck, Z. Qiu, Z. He, Q. Tong, D.W. Hilgemann, K.D. Philipson, Functional comparison of the three isoforms of the Na^+/Ca^{2+} exchanger (NCX1, NCX2, NCX3). Am. J. Physiol. **274**, C415–C423 (1998)

H. Liu, H. Zhang, H.J. Forman, Silica induces macrophage cytokines through phosphatidylcholine-specific phospholipase C with hydrogen peroxide. Am. J. Respir. Cell Mol. Biol. **36**, 594–599 (2007)

J. Lytton, Na^+/Ca^{2+} exchangers: three mammalian gene families control Ca^{2+} transport. Biochem. J. **406**, 365–382 (2007)

A. Mantovani, A. Sica, S. Sozzani, P. Allavena, A. Vecchi, M. Locati, The chemokine system in diverse forms of macrophage activation and polarization. Trends Immunol. **25**, 677–686 (2004)

F.O. Martinez, S. Gordon, M. Locati, A. Mantovani, Transcriptional profiling of the human monocyte-to-macrophage differentiation and polarization: new

molecules and patterns of gene expression. J. Immunol. **177**, 7303–7311 (2006)
F.O. Martinez, L. Helming, S. Gordon, Alternative activation of macrophages: an immunologic functional perspective. Annu. Rev. Immunol. **27**, 451–483 (2009)
M. Mayne, C.P. Holden, A. Nath, J.D. Geiger, Release of calcium from inositol 1,4,5-trisphosphate receptor-regulated stores by HIV-1 Tat regulates TNF-alpha production in human macrophages. J. Immunol. **164**, 6538–6542 (2000)
D.M. Mosser, J.P. Edwards, Exploring the full spectrum of macrophage activation. Nat. Rev. Immunol. **8**, 958–969 (2008)
O.H. Petersen, Stimulus-secretion coupling: cytoplasmic calcium signals and the control of ion channels in exocrine acinar cells. J. Physiol. **448**, 1–51 (1992)
K.D. Philipson, D.A. Nicoll, Sodium-calcium exchange: a molecular perspective. Annu. Rev. Physiol. **62**, 111–133 (2000)
H.A. Praetorius, U.G. Friis, J. Praetorius, T. Johansen, Evidence for a Na^+/Ca^{2+} exchange mechanism in rat peritoneal mast cells. Pflugers Arch. **437**, 86–93 (1998)
G. Raes, L. Brys, B.K. Dahal, J. Brandt, J. Grooten, F. Brombacher, G. Vanham, W. Noel, P. Bogaert, T. Boonefaes, A. Kindt, R. Van den Bergh, P.J. Leenen, P. De Baetselier, G.H. Ghassabeh, Macrophage galactose-type C-type lectins as novel markers for alternatively activated macrophages elicited by parasitic infections and allergic airway inflammation. J. Leukoc. Biol. **77**, 321–327 (2005)
D.E. Roberts, A. McNicol, R. Bose, Mechanism of collagen activation in human platelets. J. Biol. Chem. **279**, 19421–19430 (2004)
F. Rogister, D. Laeckmann, P. Plasman, F. Van Eylen, M. Ghyoot, C. Maggetto, J. Liegeois, J. Geczy, A. Herchuelz, J. Delarge, B. Masereel, Novel inhibitors of the sodium-calcium exchanger: benzene ring analogues of N-guanidino substituted amiloride derivatives. Eur. J. Med. Chem. **36**, 597–614 (2001)
E. Rumpel, U. Pilatus, A. Mayer, I. Pecht, Na^+-dependent Ca^{2+} transport modulates the secretory response to the Fcepsilon receptor stimulus of mast cells. Biophys. J. **79**, 2975–2986 (2000)
L. Santacruz-Toloza, M. Ottolia, D.A. Nicoll, K.D. Philipson, Functional analysis of a disulfide bond in the cardiac Na^+-Ca^{2+} exchanger. J. Biol. Chem. **275**, 182–188 (2000)
R. Schmitt, D.H. Ellison, N. Farman, B.C. Rossier, R.F. Reilly, W.B. Reeves, I. Oberbaumer, R. Tapp, S. Bachmann, Developmental expression of sodium entry pathways in rat nephron. Am. J. Physiol. **276**, F367–F381 (1999)
A. Secondo, R.I. Staiano, A. Scorziello, R. Sirabella, F. Boscia, A. Adornetto, V. Valsecchi, P. Molinaro, L.M. Canzoniero, G. Di Renzo, L. Annunziato, BHK cells transfected with NCX3 are more resistant to hypoxia followed by reoxygenation than those transfected with NCX1 and NCX2: possible relationship with mitochondrial membrane potential. Cell Calcium **42**, 521–535 (2007)
E. Shumilina, S.M. Huber, F. Lang, Ca^{2+} signaling in the regulation of dendritic cell functions. Am. J. Physiol. Cell Physiol. **300**, C1205–C1214 (2011)
J.R. Sierra, S. Corso, L. Caione, V. Cepero, P. Conrotto, A. Cignetti, W. Piacibello, A. Kumanogoh, H. Kikutani, P.M. Comoglio, L. Tamagnone, S. Giordano, Tumor angiogenesis and progression are enhanced by Sema4D produced by tumor-associated macrophages. J. Exp. Med. **205**, 1673–1685 (2008)
L. Simchowitz, E.J. Cragoe Jr., Na^+-Ca^{2+} exchange in human neutrophils. Am. J. Physiol. **254**, C150–C164 (1988)
D.M. Sosnoski, C.V. Gay, NCX3 is a major functional isoform of the sodium-calcium exchanger in osteoblasts. J. Cell. Biochem. **103**, 1101–1110 (2008)
R.I. Staiano, F. Granata, A. Secondo, A. Petraroli, S. Loffredo, A. Frattini, L. Annunziato, G. Marone, M. Triggiani, Expression and function of Na^+/Ca^{2+} exchangers 1 and 3 in human macrophages and monocytes. Eur. J. Immunol. **39**, 1405–1418 (2009)
M.J. Sweet, B.P. Leung, D. Kang, M. Sogaard, K. Schulz, V. Trajkovic, C.C. Campbell, D. Xu, F.Y. Liew, A novel pathway regulating lipopolysaccharide-induced shock by ST2/T1 via inhibition of Toll-like receptor 4 expression. J. Immunol. **166**, 6633–6639 (2001)
G.R. Tintinger, R. Anderson, Counteracting effects of NADPH oxidase and the Na^+/Ca^{2+} exchanger on membrane repolarisation and store-operated uptake of Ca^{2+} by chemoattractant-activated human neutrophils. Biochem. Pharmacol. **67**, 2263–2271 (2004)
M. Torroella-Kouri, R. Silvera, D. Rodriguez, R. Caso, A. Shatry, S. Opiela, D. Ilkovitch, R.A. Schwendener, V. Iragavarapu-Charyulu, Y. Cardentey, N. Strbo, D.M. Lopez, Identification of a subpopulation of macrophages in mammary tumor-bearing mice that are neither M1 nor M2 and are less differentiated. Cancer Res. **69**, 4800–4809 (2009)
M. Triggiani, M. Gentile, A. Secondo, F. Granata, A. Oriente, M. Taglialatela, L. Annunziato, G. Marone, Histamine induces exocytosis and IL-6 production from human lung macrophages through interaction with H1 receptors. J. Immunol. **166**, 4083–4091 (2001)
M. Triggiani, A. Petraroli, S. Loffredo, A. Frattini, F. Granata, P. Morabito, R.I. Staiano, A. Secondo, L. Annunziato, G. Marone, Differentiation of monocytes into macrophages induces the upregulation of histamine H1 receptor. J. Allergy Clin. Immunol. **119**, 472–481 (2007)
M.C. Wacholtz, E.J. Cragoe Jr., P.E. Lipsky, A Na^+-dependent Ca^{2+} exchanger generates the sustained increase in intracellular Ca^{2+} required for T cell activation. J. Immunol. **149**, 1912–1920 (1992)
X. Zhou, W. Yang, J. Li, Ca^{2+}- and protein kinase C-dependent signaling pathway for nuclear factor-kappaB activation, inducible nitric-oxide synthase expression, and tumor necrosis factor-alpha production in lipopolysaccharide-stimulated rat peritoneal macrophages. J. Biol. Chem. **281**, 31337–31347 (2006)

Part IX

NCX in the Heart and Vascular Smooth Muscle

New Insights into the Contribution of Arterial NCX to the Regulation of Myogenic Tone and Blood Pressure

28

Jin Zhang

Abstract

Plasma membrane protein Na^+/Ca^{2+} exchanger (NCX) in vascular smooth muscle (VSM) cells plays an important role in intracellular Ca^{2+} homeostasis, Ca^{2+} signaling, and arterial contractility. Recent evidence in intact animals reveals that VSM NCX type 1 (NCX1) is importantly involved in the control of arterial blood pressure (BP) in the normal state and in hypertension. Increased expression of vascular NCX1 has been implicated in human primary pulmonary hypertension and several salt-dependent hypertensive animal models. Our aim is to determine the molecular and physiological mechanisms by which vascular NCX influences vasoconstriction and BP normally and in salt-dependent hypertension. Here, we describe the relative contribution of VSM NCX1 to Ca^{2+} signaling and arterial contraction, including recent data from transgenic mice ($NCX1^{smTg/Tg}$, overexpressors; $NCX1^{sm-/-}$, knockouts) that has begun to elucidate the specific contributions of NCX to BP regulation. Arterial contraction and BP correlate with the level of NCX1 expression in smooth muscle: $NCX1^{sm-/-}$ mice have decreased arterial myogenic tone (MT), vasoconstriction, and low BP. $NCX1^{smTg/Tg}$ mice have high BP and are more sensitive to salt; their arteries exhibit upregulated transient receptor potential canonical channel 6 (TRPC6) protein, increased MT, and vasoconstriction. These observations suggest that NCX is a key component of certain distinct signaling pathways that activate VSM contraction in response to stretch (i.e., myogenic response) and to activation of certain G-protein-coupled receptors. Arterial NCX expression and mechanisms that control the local (sub-plasma membrane) Na^+ gradient, including cation-selective receptor-operated channels containing TRPC6, regulate arterial Ca^{2+} and constriction, and thus BP.

J. Zhang (✉)
Department of Physiology, University of Maryland
School of Medicine, 655 W. Baltimore Street,
Baltimore, MD 21201, USA
e-mail: jzhan002@umaryland.edu

L. Annunziato (ed.), *Sodium Calcium Exchange: A Growing Spectrum of Pathophysiological Implications*, Advances in Experimental Medicine and Biology 961, DOI 10.1007/978-1-4614-4756-6_28,

Keywords
Vascular smooth muscle • NCX1 knockdown • NCX1 overexpression • Calcium • Myogenic tone • Vasoconstriction • Blood pressure • Membrane potential • TRPC • SEA0400

The role that the plasma membrane (PM) Na^+/Ca^{2+} exchanger, (NCX), might play in contraction of vascular smooth muscle (VSM) cells (VSMCs) has long been controversial. In spite of the broad agreement with the view that the exchanger mediates the extrusion of Ca^{2+} in many types of cells (e.g., heart and neurons), knowledge about NCX function in VSMCs under physiological conditions has been incomplete (Horiguchi et al. 2001; Kim et al. 1999; Rebolledo et al. 2006; Zhang et al. 2005a, b, 2010b; Zheng and Wang 2007). Recent evidence now shows that NCX-mediated Ca^{2+} flux in arterial smooth muscle cells is physiologically relevant to many signaling pathways, such as those involved in the maintenance of basal vascular tone and agonist-induced vasoconstriction.

Among the three distinct gene-coded isoforms of the mammalian NCX family (NCX1, NCX2, NCX3) (Li et al. 1994; Nicoll et al. 1990, 1996) and the many tissue-specific splice variants that have been cloned, NCX1 is the dominant isoform (Iwamoto et al. 2004b; Lee et al. 1994; Nakasaki et al. 1993; Quednau et al. 1997; Raina et al. 2008) and NCX1.3 and NCX1.7 are the predominant splice variants in VSMCs (Quednau et al. 1997), although NCX2, NCX3 (Kashihara et al. 2009; Raina et al. 2008), and the K^+-dependent Na^+/ca^{2+} exchanger (NCKX) (Dong et al. 2006) are also reported in smooth muscle tissues. Given the relative predominant expression and function of NCX1 (Kashihara et al. 2009; Raina et al. 2008; Zhang et al. 2010b), here, we will use "NCX" to refer to "NCX1" in VSMCs.

The primary physiological role of NCX in normal healthy cells has been widely regarded as that of a Ca^{2+} extrusion system, with the exchanger operating in the "Ca^{2+} efflux" or so-called forward mode. Typical examples are cardiac and neuronal cells, in which NCX plays a dominant role in extruding Ca^{2+} after an action potential (Blaustein and Lederer 1999; Nicoll et al. 1990). Indeed, in view of its high exchange capacity and relatively low affinity for Ca^{2+} (Blaustein et al. 2002), NCX is well suited for rapid recovery from a high intracellular Ca^{2+} concentration ($[Ca^{2+}]_i$). Hence, NCX is pivotal in maintaining Ca^{2+} homeostasis and protecting cells from Ca^{2+} overload and subsequent injury.

Unlike cardiac myocytes, which have rhythmic action potentials and contraction, VSMCs in intact small arteries normally produce sustained or "tonic" contraction, even in the basal ("resting") state. This contraction is an intrinsic property of VSMCs in response to stretch/pressure that actually determines the background tone of an artery. Hence, the arterial background tone is called *myogenic* tone (MT), which is a major contributor to total peripheral vascular resistance (TPR). Furthermore, under in vivo conditions, VSMCs also receive neural, hormonal, circulating, and local (paracrine) chemical signals. In response to these stimuli, VSMCs may produce transient (phasic) or sustained (tonic) contractions, or they may relax. The signal for contraction is elevated $[Ca^{2+}]_i$, resulting from either Ca^{2+} influx from the extracellular fluid (ECF) via "Ca^{2+} entry" systems or Ca^{2+} release from intracellular stores. Ca^{2+} influx from ECF usually causes tonic vasoconstriction, and Ca^{2+} release from intracellular stores results in phasic vasoconstriction. Both tonic and phasic vasoconstrictions depend, primarily, but not completely, upon elevation of $[Ca^{2+}]_i$ (Somlyo and Somlyo 2003). Therefore, the regulation of $[Ca^{2+}]_i$ in VSMCs, which includes maintaining appropriate levels of $[Ca^{2+}]_i$ for MT, elevating $[Ca^{2+}]_i$ for phasic or further tonic contraction, and restoring $[Ca^{2+}]_i$ to lower levels for relaxation (Floyd and Wray 2007), is tightly controlled by Ca^{2+} channels and transporters.

The fact that NCX can mediate both Ca^{2+} extrusion and Ca^{2+} entry makes it possible for NCX to be involved in all the processes that regulate $[Ca^{2+}]_i$ and consequently, vasoconstriction, TPR, and BP. As a bidirectional exchanger, NCX can transfer Ca^{2+} either into or out of VSMCs in exchange for Na^+ in a coupling ratio of $3Na^+:1Ca^{2+}$, and the direction of transfer is determined by both membrane potential (E_m) and the transmembrane gradients of Na^+ and Ca^{2+} (Blaustein and Lederer 1999). Nevertheless, the factors that determine in which mode NCX operates in vivo, E_m, local $[Na^+]$ and local $[Ca^{2+}]$, are unknown. Furthermore, regulation of NCX1 (Reeves et al. 2007) by Ca^{2+} and other intracellular molecules may, at different times, differentially affect the magnitude of Ca^{2+} influx or efflux and could therefore also affect whether net Ca^{2+} influx or Ca^{2+} efflux occurs over the long term in the living animal. At any given time, however, the direction of the flux through NCX is determined only by the electrochemical driving force.

Recently, genetically engineered mice with smooth-muscle-specific increased or reduced expression of NCX1 ($NCX1^{smTg/Tg}$ and $NCX1^{sm-/-}$, respectively) (Blaustein et al. 2009; Iwamoto et al. 2004b; Zhang et al. 2010b; Zhao et al. 2011) and a specific NCX inhibitor, SEA0400 (Iwamoto et al. 2004a, b; Matsuda et al. 2001), have proved useful in studying the physiological role of smooth muscle NCX. These studies suggest that VSM NCX1 mediates a net Ca^{2+} influx that is required for the maintenance of basal arterial tone and agonist-induced vasoconstriction in small resistance arteries and consequently BP, under basal physiological conditions. As reviewed here, it seems likely that in small arteries in living animals, NCX can operate in either Ca^{2+} influx or efflux mode, depending on the prevailing conditions. The evidence, however, supports the idea that in the basal state of the animal, the predominant mode is Ca^{2+} influx. We will discuss how this can arise in small arteries that have developed MT, the condition of arterial vasoconstriction that contributes substantially to TPR, and thus to BP. Possible mechanisms by which arterial NCX is involved in salt-dependent hypertension will also be discussed.

28.1 Specific Localization of NCX in VSMCs

NCX in VSMCs exhibits a specific localization (Fig. 28.1) that is closely related to its function. As revealed by immunocytochemical studies from both freshly isolated and cultured VSMCs, NCX is confined to plasmamembrane (PM) microdomains that are adjacent to underlying sarcoplasmic reticulum (SR) (Juhaszova et al. 1994; Moore et al. 1993). This is different from the uniform distribution of plasmamembrane Ca^{2+}-ATPase (PMCA). Interestingly, some Na^+-coupled transporting systems, such as transient receptor potential canonical channels (TRPCs) and/or Na^+ pumps with the $\alpha 2$ subunits are also localized to the PM of this restricted "buffer barrier" fuzzy space (Lederer et al. 1991; van Breemen et al. 1995). Notably, in the past decade, evidence has accumulated for the colocalization of NCX and TRPC6, and $\alpha 2$ Na^+ pump, and their possible functional interactions in VSMCs (Arnon et al. 2000; Baryshnikov et al. 2009; Gonzales et al. 2010; Lee et al. 2006; Lynch et al. 2008; Maruyama et al. 2006; Poburko et al. 2007; Pritchard et al. 2010).

The colocalization and coupled function of NCX with $\alpha 2$ Na^+ pumps, and TRPCs appearing to operate as a functional unit that regulates local $[Na^+]$ and $[Ca^{2+}]$ within PM-SR junctions ($[Na^+]_{Sub\text{-}PM}$ and $[Ca^{2+}]_{Sub\text{-}PM}$, respectively), might indirectly affect global $[Na^+]_i$ or $[Ca^{2+}]_i$. In this case, a putative elevation of $[Na^+]_{Sub\text{-}PM}$ (Poburko et al. 2007; Fameli et al. 2009), resulting from either $\alpha 2$ Na^+ pump inhibition (e.g., by endogenous ouabain in salt-dependent hypertension) or TRPC activation (e.g., upon G protein-coupled receptor (GPCR) activation or sympathetic nerve activity (SNA)), facilitates "Ca^{2+} influx" mode and/or inhibits "Ca^{2+} efflux" mode NCX. The increased net Ca^{2+} influx, therefore, allows more Ca^{2+} to be sequestered into SR Ca^{2+} stores by sarco-/endoplasmic reticulum Ca^{2+}-ATPase (SERCA), resulting in enhanced SR Ca^{2+} loading. Furthermore, following agonist-induced store depletion, the functional coupling between NCX and TRPCs also provides Ca^{2+} required to generate Ca^{2+} waves (Dai et al. 2010; Lee et al.

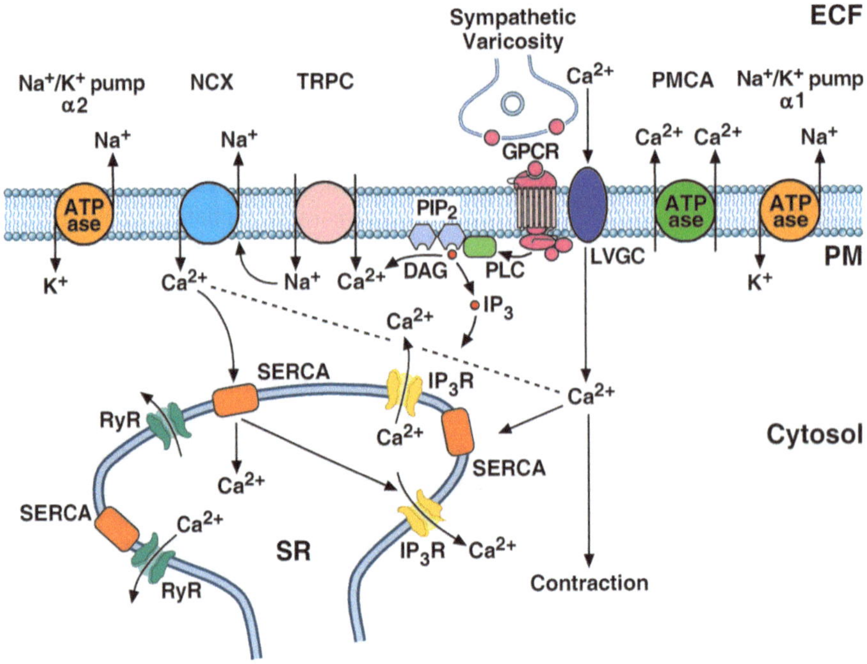

Fig. 28.1 *Diagram of the plasma membrane (PM)-sarcoplasmic reticulum (SR) junction showing the colocalization of NCX with TRPC and Na⁺ pumps with α2 subunits and their functional interaction in modulating local [Na⁺] and [Ca²⁺], SR Ca²⁺ stores, and Ca²⁺ signaling in vascular smooth muscle cells (VSMCs).* Activation of TRPC6 or inhibition of α2 Na⁺ pump lead to accumulation of [Na⁺] in PM-SR junctions ($[Na^+]_{Sub\text{-}PM}$), thereby increasing Ca^{2+} entry and reducing Ca^{2+} extrusion by NCX in the junctional space, resulting in elevated $[Ca^{2+}]$ in PM-SR junctions ($[Ca^{2+}]_{Sub\text{-}PM}$). This enhances Ca^{2+} signaling, myogenic tone, and GPCR-induced vasoconstriction in blood vessels. *DAG* diacylglycerol; *ECF* extracellular fluid; *GPCR* G-protein-coupled receptor; *IP_3* inositol 1,4,5-trisphosphate; *IP_3R* IP_3-sensitive receptor; *LVGC* L-type voltage-gated Ca^{2+} channel; *NCX* Na^+/ Ca^{2+} exchanger; *PIP_2* phosphatidylinositol 4,5-bisphosphate; *PLC* phospholipase C; *PMCA* plasma membrane Ca^{2+}-ATPase; *RyR* ryanodine-sensitive receptor; *SERCA* sarco-/endoplasmic reticulum Ca^{2+}-ATPase; *TRPC* transient receptor potential canonical channel

2001; Poburko et al. 2007; Zang et al. 2001). As a consequence, vasoconstrictions induced by agonist activation via inositol 1,4,5-trisphosphate (IP_3)-sensitive store release are enhanced. Therefore, NCX, coupling with other Na^+ and Ca^{2+} transporters, serves as a determinant of $[Ca^{2+}]$ in the SR ($[Ca^{2+}]_{SR}$) and regulates Ca^{2+} signaling in arterial contraction.

The specific localization of NCX may reflect its functional properties: Although NCX has a low affinity for Ca^{2+}, it is localized to the PM-SR junctions that presumably have a higher $[Ca^{2+}]$ (Fameli et al. 2009) than the bulk cytoplasm. Meanwhile, with its high turnover rate and exchange capacity, NCX is capable of binding and transporting Ca^{2+} rapidly upon a Ca^{2+} surge to help maintain Ca^{2+} homeostasis in the myocytes. In this sense, the function of VSM NCX is similar to that of cardiac NCX.

28.2 Regulation of NCX Activity

It is well known that NCX activity is regulated by several physiological factors including intracellular Ca^{2+}, Na^+, and phosphatidylinositol 4,5-bisphosphate (PIP2) (see review by Reeves et al. 2007). Briefly, increases in $[Na^+]_i$ decrease NCX activity (Na^+-dependent inactivation), and increases in $[Ca^{2+}]_i$ increase activity through allosteric effects (allosteric Ca^{2+} activation). PIP_2 modulates these effects. Furthermore, Na^+-dependent inactivation increases the $[Ca^{2+}]_i$ required for allosteric activation, while Ca^{2+} inhibits the ability of Na^+ to cause inactivation. Possible protein kinase A-mediated phosphorylation of NCX remains controversial (Morad et al. 2011). In cardiac myocytes, Reeves and his colleagues (2007) have proposed that the net effect of these phenomena is that NCX activity is regulated not on a beat-to-beat basis but rather by the time-integral of Ca^{2+} transients occurring over multiple beats.

The extent to which these concepts might apply to VSMCs is presently unknown. Interestingly, the $[Ca^{2+}]_i$ required for allosteric half-activation is ~150–400 nM (Reeves et al. 2007). This is precisely the range of $[Ca^{2+}]_i$ believed to exist in the cytoplasm in arterial smooth muscle cells that have developed MT (Knot and Nelson 1998; Matchkov et al. 2002; Zhang et al. 2005b). Nevertheless, given the specific localization of VSM NCX, where $[Ca^{2+}]_{Sub\text{-}PM}$ is presumably higher (~2 μM) as discussed below, allosteric regulation of NCX activity by Ca^{2+} is still uncertain in VSMCs. It is noteworthy that up- or downregulation of NCX will not change the direction of NCX operation, but it will influence the magnitude of the exchange current, either inward or outward. Therefore, physiological regulation of NCX is potentially important in determining the net effect of NCX in regulating $[Ca^{2+}]_i$, in a situation in which NCX operates in both Ca^{2+} influx and efflux modes at different times during normal physiological function, and when $[Na^+]_{Sub\text{-}PM}$ and $[Ca^{2+}]_{Sub\text{-}PM}$ are varying over time.

28.3 VSM NCX Contributes to Arterial Myogenic Tone (MT) via Net Ca^{2+} Influx

MT, the basal level of arterial tone, on top of which arteries respond to neurohumoral stimuli, is important in virtually all aspects of vascular function, both physiological and pathological (Davis and Hill 1999; Hill et al. 2006). In hypertension, MT is a likely component of the initial autoregulatory response to tissue overperfusion when BP rises as a consequence of volume expansion. The exact signaling transduction pathway underlying MT, especially the mechanisms by which VSMCs transduce mechanical stimuli (strech/pressure) to membrane depolarization and subsequent opening of LVGCs, is unresolved. Knot and Nelson (1998) revealed that an increase in "arterial wall $[Ca^{2+}]$" is the key process when intraluminal (transmural) pressure was increased. Studies published recently, however, clearly suggest that other regulatory mechanisms such as increased Ca^{2+} sensitivity is also involved (D'Angelo et al. 1997; Johnson et al. 2009; Matchkov et al. 2002). In addition, vascular remodeling (structural) may also allow wall stress to be borne by mechanisms other than active contraction mediated by activation of myosin light-chain kinase (MLCK) and/or inhibition of myosin light-chain phosphatase (MLCP) (see Review by Martinez-Lemus et al. 2009). Cytoplasmic Ca^{2+} influences artery diameter through a myriad of mechanisms, most of which converge on MLCK and MLCP (for short-term changes), and gene regulation (for longer term changes, such as molecular remodeling, i.e., changes in protein expression). Regulation of $[Ca^{2+}]_i$, the factor that ultimately influences artery diameter, is therefore pivotal. In VSMCs, NCX is one of only a small number of molecules (NCX, PMCA, L-type voltage-gated channels (LVGCs), TRPCs, and several other types of PM Ca^{2+} channels) that determine $[Ca^{2+}]_i$ in the steady state.

Although there have been extensive studies on the role of VSM NCX in regulating (agonist-induced) contractions, isolated cells or arteries on a wire myograph have been used in most of the experiments. As mentioned above, however, VSMCs in living animals are controlled by numerous factors that can influence the mode of operation of NCX. When VSMCs are isolated, most of the in vivo factors (such as neural innervation, hormonal stimulation, transmural pressure regulation, and intercellular connections) are lost. Arteries on a wire myograph do not develop true MT, even when stretched. Consequently, they are relatively hyperpolarized (e.g., E_m in rat small mesenteric arteries is ~−59 mV, Mulvany et al. 1982), compared to pressurized arteries (see below) or arteries in vivo, and, as discussed further below, their permeability to Na^+ may be lower than arteries that have developed MT. These experimental conditions (i.e., arteries on a wire myograph) lead to a prediction that favors the Ca^{2+} efflux mode of NCX. This prediction, however, may be misleading because in arteries that have developed MT or in vivo, the driving force for Ca^{2+} flux through NCX may be greatly altered, as discussed next.

Isolated pressurized arteries are in a presumably more physiological state, although they also lack tonic autonomic nerve activity, the normal influences of endothelium (there usually is no flow) and circulating hormones. Nevertheless, isolated pressurized (e.g., pressurized to 70 mm Hg at 37 °C) arteries (1) exhibit elevated $[Ca^{2+}]_i$ (e.g., ~200 nM in rat cerebral arteries, Knot and Nelson 1998; ~220 nM in rat femoral small arteries, Matchkov et al. 2002; ~185 nM in mouse mesenteric small arteries, Zhang et al. 2005b) compared to that in arteries on a wire myograph, for example, ~120 nM in rat mesenteric resistance arteries (Jensen et al. 1993); (2) are relatively depolarized (e.g., ~−40 mV in rat cerebral arteries, Gonzales et al. 2010; ~−43 mV, calculated from Knot and Nelson 1998; Kotecha and Hill 2005); and (3) have higher calcium sensitivity as a result of certain signaling pathways, particularly Rho-Rho kinase activation (Johnson et al. 2009; Matchkov et al. 2002; Somlyo and Somlyo 2003), which enhances sensitization of VSMCs to some minimal change in $[Ca^{2+}]_i$, and is important in several other signaling pathways.

How might NCX be involved in maintaining the appropriate $[Ca^{2+}]_i$ required for MT? Specifically, in which mode, Ca^{2+} efflux or influx, does NCX operate in VSMCs under the physiological basal condition, in which arteries have developed MT? Recently, by combined usage of transgenic mice ($NCX1^{smTg/Tg}$ and $NCX1^{sm-/-}$) and pharmacological interventions, we have demonstrated the contribution of NCX to MT. We found that isolated, pressurized mesenteric small arteries from $NCX1^{sm-/-}$ mice exhibit greatly reduced MT and myogenic reactivity (MR), the ability of the artery to constrict or dilate in response to a rapid increase or decrease in intraluminal pressure, (Hill et al. 2006; Zhang et al. 2010b). Vasoconstriction induced by lowering $[Na^+]_o$, resulting from decreased Ca^{2+} extrusion and/or increased Ca^{2+} entry via NCX, is remarkably reduced in $NCX1^{sm-/-}$ arteries. Furthermore, acute inhibition of NCX by SEA0400 reduces MT in wild type (WT) (Zhang et al. 2005b, 2010b), but not in $NCX1^{sm-/-}$ mouse arteries (Zhang et al. 2010b). These impaired functions reflect the reduced NCX protein expression in mesenteric arteries (Zhang et al. 2010b). Consistent with this observation, Raina et al. (2008) showed that MR is impaired by antisense oligonuleotide knockdown of NCX in isolated rat cremaster muscle arterioles. Also, Kashihara et al. (2009) reported that NCX-mediated Ca^{2+} entry is involved in the development, but not the steady-state maintenance, of myogenic constriction in rat posterior cerebral arteries. Our preliminary study (J. Wang and J. Zhang, unpublished) shows that, conversely, in $NCX1^{smTg/Tg}$ mouse arteries, MT and MR are significantly enhanced. These results indicate that NCX contributes to arterial tone by mediating Ca^{2+} entry.

As mentioned above, thermodynamically, the direction of NCX operation is determined by E_m and the chemical activities of intracellular and extracellular Ca^{2+} and Na^+ in the local (sub-PM) microdomains. In the steady state, the direction of operation is determined by the sign of the driving force, $E_m - E_{NCX}$, with a positive value corresponding to outward current and Ca^{2+} influx. E_{NCX} is the reversal potential of NCX, which is defined as $3E_{Na} - 2E_{Ca}$. Then, the operating mode of NCX can be predicted by a simulated electrochemical potential calculation (Fig. 28.2). The E_{NCX} is ~23 mV at 37 °C with typical values of ion concentration in VSMCs in PM-SR junctions: $[Na^+]_o$: 140 mM; $[Na^+]_{Sub\text{-}PM}$: 10 mM; $[Ca^{2+}]_o$: 2 mM; $[Ca^{2+}]_{Sub\text{-}PM}$: 2 μM (Fameli et al. 2009; Török 2007). E_m in isolated cells is ~−20 mV (Fameli et al. 2009), then the driving force is negative (~−43 mV), which favors an inward current and Ca^{2+} efflux via NCX. Arteries stretched on a wire myograph are depolarized to ~−59 mV (Mulvany et al. 1982), and the driving force is ~−82 mV, which still favors Ca^{2+} efflux. In cannulated arteries with MT, for example, rat cerebral arteries pressurized to 70 mm Hg, E_m is ~−40 mV

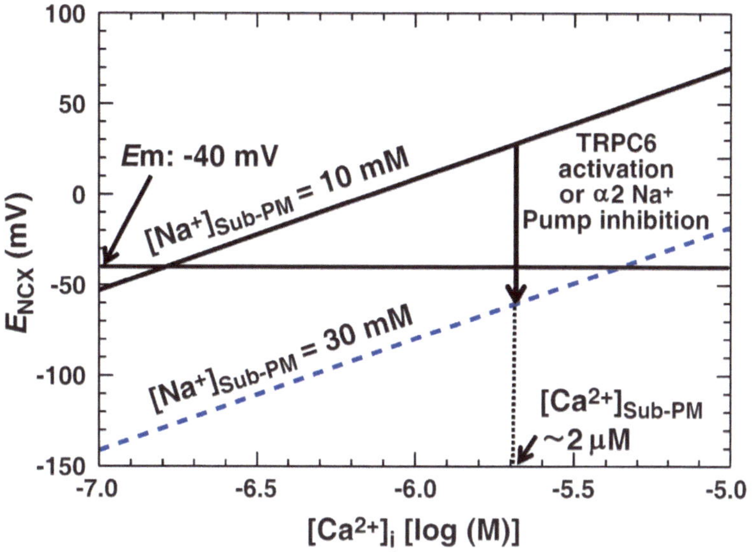

Fig. 28.2 *Simulated effect of $[Na^+]_{Sub\text{-}PM}$ and $[Ca^{2+}]_{Sub\text{-}PM}$ on the reversal potential of NCX (E_{NCX}) in VSMCs.* $E_{NCX} = 3E_{Na} - 2E_{Ca}$, where E_{Na} is the electrochemical potential of Na^+ and E_{Ca} that of Ca^{2+}. The E_{NCX} is ~23 mV at 37 °C with typical values of ion concentration in VSMCs: $[Na^+]_o$: 140 mM; $[Na^+]_{Sub\text{-}PM}$: 10 mM; $[Ca^{2+}]_o$: 2 mM; $[Ca^{2+}]_{sub\text{-}PM}$: 2 μM (Fameli et al. 2009; Török 2007). E_{NCX} shifts from ~23 to ~−62 mV when $[Na^+]_{Sub\text{-}PM}$ is increased from 10 to 30 mM, as the result of either TRPC6 activation or α2 Na^+ pump inhibition (Fameli et al. 2009). This shifting of E_{NCX} reverses NCX from "efflux" to "influx" mode in pressurized arteries that have depolarized membrane potential (E_m, e.g., the E_m is ~−40 mV in rat cerebral arteries pressurized to 70 mmHg (blue line) (Gonzales et al. 2010)), as the direction of NCX is determined by $E_m - E_{NCX}$. *$[Ca^{2+}]_i$* intracellular Ca^{2+} concentration; *$[Ca^{2+}]_o$* extracellular Ca^{2+} concentration; *$[Na^+]_o$* extracellular Na^+ concentration

(Gonzales et al. 2010), the driving force is ~−63 mV and would also favor Ca^{2+} efflux of the exchanger. When $[Na^+]_{Sub\text{-}PM}$ is assumed to be elevated, however, for example, as a result of TRPC6 activation, from 10 to 30 mM (Fameli et al. 2009), the E_{NCX} will be ~−62 mV (Fig. 28.2). At the same E_m (~−40 mV), the driving force will already be positive (~+22 mV), which will favor Ca^{2+} *entry* of the exchanger. Thus, a relatively small increase in $[Na^+]_{Sub\text{-}PM}$ can markedly change the driving force of Ca^{2+} flux through NCX. This is significant because blocking TRPC6 or TRPM4 activity hyperpolarizes E_m and attenuates MR/MT (Earley et al. 2004; Welsh et al. 2002; Gonzales et al. 2010), which suggests that TRPCs are physiologically active in arteries with tone. Hence, in pressurized arteries with MT and arteries in vivo, NCX is likely to operate in the "Ca^{2+} influx" mode. It is important to note that the putative existence of localized regions of elevated $[Na^+]_{Sub\text{-}PM}$ requires theoretical, strategically placed physical barriers to Na^+ diffusion in the sub-PM regions (Fameli et al. 2009).

With respect to the role of NCX in MT, however, there may seem to be a paradox. Earlier we emphasized that NCX does not affect global $[Ca^{2+}]_i$ but affects $[Ca^{2+}]_{sub\text{-}PM}$, and consequently, $[Ca^{2+}]_{SR}$. Thus, a question is: how does $[Ca^{2+}]_{SR}$ affect MT? In VSM SR, there are two types of Ca^{2+} release channels: ryanodine- and IP_3-sensitive receptors (RyRs and IP_3Rs, respectively) (Golovina and Blaustein 1997; Lamont and Wier 2004; Westcott and Jackson 2011). Activation of RyRs induces Ca^{2+} sparks and vasodilation (Nelson et al. 1995), whereas IP_3R activation mediates Ca^{2+} waves and vasoconstriction (Lamont and Wier 2004). Westcott and Jackson (2011) reported that inhibition of IP_3Rs decreases global $[Ca^{2+}]_i$ and dilates hamster muscle feed arteries and arterioles. Also, Mufti et al. (2010) reported that Ca^{2+} waves facilitate tone development by providing a portion of the Ca^{2+} required to signal for inhibition of MLCP and activation of MLCK. In contrast, our work in murine mesenteric small arteries (Zacharia et al. 2007) showed that Ca^{2+} waves decrease in frequency during MT development, suggesting only a minor role for SR Ca^{2+} waves in MT. Finally, ryanodine, which abolishes Ca^{2+} waves by depletion of stored Ca^{2+}, increases tone in some studies of cerebral arteries (Knot and Nelson 1998) but has no effect in others (cremaster arterioles; Westcott and Jackson 2011) or decreases MT in other studies of cerebral arteries (Mufti et al. 2010). Thus, although the degree to which Ca^{2+} waves might affect MT is not certain, the available evidence suggests a role for stored Ca^{2+} in MT in some types of arteries, with possible differences between species.

Indeed, the net Ca^{2+} influx mode of NCX under resting physiological condition is further supported by in vivo studies: intra-femoral artery injection of SEA0400 caused a direct vasodilatation and an increase in blood flow (Iwamoto et al. 2004b). Also, VSM-specific overexpression of NCX1.3 resulted in significantly elevated $[Ca^{2+}]_i$ and enhanced agonist-induced vasoconstriction, both of which are antagonized by SEA0400 (J. Zhang and W.G. Wier, unpublished). Thus, we conclude that under resting conditions in arteries with MT, NCX mediates a small, net Ca^{2+} entry. At any given instant in time, however, in the living animal, it seems possible that NCX may operate in either influx or efflux mode, in different VSMCs and in different arteries, depending on the activity of that particular artery, as controlled by the autonomic nervous system, endothelium, and all the other regulators. Nevertheless, it appears that over a short period of time the sum of all NCX activity in all the arteries that determine BP (i.e., the net flux) is Ca^{2+} influx mode.

To support the net "Ca^{2+} influx" mode of VSM NCX in arteries in vivo, a direct measurement of VSM $[Ca^{2+}]_i$ within intact arteries showing higher $[Ca^{2+}]_{SR}$ and $[Ca^{2+}]_i$ in NCX1$^{smTg/Tg}$ and lower $[Ca^{2+}]_{SR}$ and $[Ca^{2+}]_i$ in NCX1$^{sm-/-}$ mouse arteries is critical. New advances in intravital microscopy, such as multispectral intravital fluorescent resonance energy transfer (FRET) imaging (Zhang et al. 2010a), or two-photon microscopy, have recently been used to visualize $[Ca^{2+}]_i$ and diameter changes of true small resistance arteries under resting condition and in response to intravenously or locally administered agonists within intact living animals. Further experiments for the comparison of arterial $[Ca^{2+}]_i$ between WT and NCX mutants could be achieved by using NCX mutant mice that also express the exogenous MLCK FRET "biosensor," a "ratiometric" indicator (Zhang et al. 2010a).

It is worth noting that even a small change in arterial diameter, as occurs in the smooth-muscle-specific NCX mutants, has a large effect on BP. According to Poiseuille's law, resistance to flow, *R*, is inversely proportional to the fourth power of the internal radius, *r* (i.e., R□ 1/r4). The larger diameter (reduced tone) in arteries with reduced NCX activity should have profound physiological implications. For example, reduced NCX activity decreases MT by ~15 % (Zhang et al. 2005b, 2010b), that is, an increase of internal diameter from 85 to 95 μm; this would decrease *R* by 70 % and markedly decrease BP.

28.4 Altered VSM NCX Activity Correlates with BP Change

The roles of VSM NCX in the regulation of arterial MT and certain GPCR-mediated vasoconstrictions as observed in vitro seem also to be active in vivo and ultimately control BP. The crucial evidence for such activation is that BP correlates with the level of NCX1 expression in VSMCs: knockout of smooth-muscle-specific NCX1 decreases BP (Zhang et al. 2010b; Zhao et al. 2011), and overexpression of smooth-muscle-specific NCX1 increases BP (Blaustein et al. 2009; Iwamoto et al. 2004b). The clear phenotype change in both NCX1$^{sm-/-}$ and NCX1$^{smTg/Tg}$ mice is quite interesting because cardiac-specific NCX1 knockout mice have a relatively normal cardiac phenotype (Chen et al. 2010; Henderson et al. 2004) despite the well-known importance of NCX function in the heart. The loss of NCX1 in cardiac myocytes is compensated by a reduced LVGC current and a shortened action potential resulting from accelerated Ca^{2+}-dependent LVGC inactivation and augmented transient outward K^+ current (Henderson et al. 2004; Pott et al. 2007a, b). In contrast, although LVGC current is also markedly reduced in VSMCs in our NCX1$^{sm-/-}$ mice (Ren et al. 2010; Zhang et al. 2010b), the mechanism of the current decline is different. The reduced Ca^{2+} entry via NCX lowers $[Ca^{2+}]_i$, thereby reducing protein kinase C (PKC) activation that lowers LVGC activation. Also, the lower LVGC activation is not a direct consequence of reduced NCX1 expression, and it does not directly cause the reduced MT and BP in these animals (Zhang et al. 2010b). Furthermore, BP elevation in NCX1$^{smTg/Tg}$ mice seems not to be the direct result of overexpressed NCX in renal vasculature, as suppressed NCX activity in the afferent arterioles and unchanged NCX activity in efferent arterioles have been observed in Dahl salt-sensitive hypertensive rats (Nelson et al. 1999), suggesting that the activity of renal vasculature NCX is not directly related to BP change. This assumption is also supported by the normal renal function in NCX1$^{sm-/-}$ mice (Zhao et al. 2011), indicating that BP reduction is not the direct effect of renal dysfunction, although renal vascular responses to angiotensin II are attenuated (Zhao et al. 2011).

Emerging evidence has shown that NCX is also importantly involved in the control of BP in salt-dependent hypertension (Ashida et al. 1997; Blaustein et al. 2009; Iwamoto et al. 2004b; Kiraku et al. 2000; Pulina et al. 2010; Zulian et al. 2010). Overexpression of smooth-muscle-specific NCX1 increases BP and facilitates hypertension following high salt diet treatment in mice (Blaustein et al. 2009; Iwamoto et al. 2004b), whereas global NCX1 heterozygous mice (NCX1$^{+/-}$) are resistant to deoxycorticosterone acetate (DOCA)-salt hypertension (Iwamoto et al. 2004b). Furthermore, increased expression (Taniguchi et al. 2004) and activity (Chen and Roufogalis 1994; Taniguchi et al., 2004) of VSM NCX contribute to contractile dysfunction in spontaneously hypertensive rats. NCX-mediated Ca^{2+} entry contributes to elevated arterial smooth muscle $[Ca^{2+}]_i$ in both human pulmonary hypertension patients (Zhang et al. 2007) and mice overexpressing Na^+/H^+ exchanger under high salt treatment (Kiraku et al. 2000). NCX is also reported to be critically involved in the forms of hypertension (Adducin mutation, ACTH-induced, renal insufficiency models) that involve increased circulating endogenous Na^+ pump inhibitors, such as endogenous ouabain (Blaustein et al. 2011; Hamlyn et al. 1982; Hasegawa et al. 1987). When ouabain inhibits the high affinity α2 Na^+ pumps in VSMCs, $[Na^+]_{sub\text{-}PM}$ rises; this facilitates Ca^{2+} entry through NCX, resulting in vasoconstriction (Blaustein et al. 2011; Iwamoto et al. 2004b; Zhang et al. 2005b, 2010b). The detailed postulated endogenous ouabain-α2 Na^+ pump-NCX pathway has been well documented in our recent review (Blaustein et al. 2011). These results indicate that VSM NCX acts primarily as a Ca^{2+} entry pathway for regulating arterial tone and BP, especially under sodium-retaining conditions. Thus, to elucidate the mechanism by which NCX regulates BP is important in hypertension study.

Most forms of hypertension, including human essential hypertension and salt-induced animal hypertension, are often accompanied by an increase in SNA in small arteries (Grassi 2010). The overstimulation of the sympathetic nervous system may involve not only increased sympathetic outflow mediated centrally but changes in neuromuscular transmission at sympathetic neuroeffector junctions as well. In vivo, VSMCs receive neural innervation primarily from the sympathetic nervous system through adrenergic, purinergic, or neuropeptide Y (NPY) receptors, depending on different types of arteries. Thus, three co-transmitters, namely, norepinephrine (NE), ATP, and NPY, are involved in SNA (Donoso et al. 1997). In DOCA-salt hypertensive rats, for example, there is increased release of NE from the sympathetic nerve terminals in mesenteric arteries and an increase in the amplitude of experimental neurogenic contractions (Luo et al. 2003; Tsuda et al. 1989). It is also reported that sympathetic nerves from spontaneous hypertensive rat mesenteric arteries contain elevated levels of NE (Hirst and Edwards 1989). In humans, essential hypertension is often neurogenic with high rates of NE spillover (Esler et al. 2001).

Released neurotransmitters activate GPCR signaling pathways resulting in depolarization and Ca^{2+} entry. Van Breemen and colleagues showed that "Ca^{2+} entry" mode of NCX in VSMCs is activated by adrenergic receptor activation (Dai et al. 2010) and purinergic receptor activation (ATP) (Poburko et al. 2007; Syyong et al. 2007). In arteries, activation of purinergic $P2X_1$ receptors produces local Ca^{2+} transients (Lamont and Wier 2002), and neural activation of adrenoceptors produces Ca^{2+} waves (Lamont et al. 2003). Thus, we expect that Ca^{2+} influx via NCX is also involved in sympathetic neurovascular transmission since both types of receptors are activated during SNA. Also, increased NCX activity, as in NCX1$^{smTg/Tg}$ mice, would be expected to enhance vascular sensitivity to sympathetic neurotransmitters (mainly, NE).

How NCX is involved in vasoconstrictions induced by SNA/GPCR activation is also explained by the specific colocalization of NCX with TRPC6 (Fig. 28.1). The GPCRs within arterial smooth muscle cells are linked to $G_{q/11}$-phospholipase C (PLC) pathway (e.g., NE, ATP receptors) and/or to the $G_{12/13}$-Rho/Rho kinase pathway (e.g., Ang II receptor) (Maguire and Davenport 2005; Wirth et al. 2008). All three neurotransmitters activate the dual signaling $G_{q/11}$-PLC-PIP_2-IP_3/ diacylglycerol (DAG) pathway. The typical vasoconstriction induced by activation of $G_{q/11}$-PLC-linked GPCR is initiated by IP_3-sensitive SR Ca^{2+} release and maintained by Ca^{2+} entry from the ECF (Poburko et al. 2007; Syyong et al. 2007; Wirth et al. 2008). Specifically, GPCR stimulation leads to activation of PLC, PLC then cleaves PIP_2 to form IP_3 and DAG. IP_3 triggers SR Ca^{2+} release and the initial "phasic" vasoconstriction, which increases SR-dependent Ca^{2+} waves that activate MLCK. DAG activates the Na^+ permeable TRPC6, leading to local Na^+ accumulation and Ca^{2+} influx via NCX (Poburko et al. 2007; Syyong et al. 2007). Hence, the putative cellular mechanisms by which NCX is involved in salt-dependent SNA/GPCR-pathway include modifying both IP_3-sensitive SR Ca^{2+} release and DAG-TRPC6 signaling pathways. In addition, Ca^{2+} increased in VSMCs in response to stimuli has to be rapidly removed or buffered such that $[Ca^{2+}]_i$ can decrease to the resting level during muscle relaxation. The decrease in $[Ca^{2+}]_i$ is usually accomplished by two types of transport systems in almost all cell types: Ca^{2+} extrusion to the ECF by PMCA and NCX and Ca^{2+} re-sequestration by SERCA and mitochondria. In cardiac myocytes, NCX is the dominant mechanism for Ca^{2+} removal. In VSMCs, during repolarization (after periods of activity, when $[Ca^{2+}]_i$ is elevated, particularly after Ca^{2+} spikes), NCX also primarily mediates Ca^{2+} efflux to clear Ca^{2+} during relaxation, helps reset the $[Ca^{2+}]_i$ to the normal level, and prevents Ca^{2+} overload. For example, reduced NCX1 activity prolongs agonist responses by delaying the return of $[Ca^{2+}]_i$ to the resting level in cultured VSMCs (Slodzinski et al. 1995; Slodzinski and Blaustein 1998) and intact arteries (Tsang et al. 2003; Yamanaka et al. 2003).

It is noteworthy that in seeking the mechanisms of salt-induced hypertension, the upregulation of vascular NCX1 and/or TRPCs has been implicated in human primary pulmonary hypertension (Yu et al. 2004; Zhang et al. 2007) and several salt-dependent hypertensive animal models, including DOCA-salt hypertension (Bae et al. 2007), Milan hypertensive rats (Zulian et al. 2010), spontaneous hypertensive rats (Liu et al. 2009; Taniguchi et al. 2004), ouabain-induced hypertension (Pulina et al. 2010), and our NCX1$^{smTg/Tg}$ mice (M. Li, M.P. Blaustein and J. Zhang, unpublished). Thus, elucidating the functional role of SNA/GPCR-TRPC6-NCX pathway in VSMCs helps understand the specific mechanisms by which salt increases arterial constriction and TPR, and thus BP elevation in hypertension.

28.5 Conclusion

We have suggested, in this chapter and other related reports (Blaustein et al. 2009; Iwamoto et al. 2004b; Zhang et al. 2005b, 2010b; Zhao et al. 2011), that in VSMCs the normal working mode of the exchanger is "Ca^{2+} entry" rather than "Ca^{2+} extrusion" under basal physiological conditions. Thus, upregulation of the exchanger results in overloaded SR Ca^{2+} stores, elevated $[Ca^{2+}]_i$, augmented MT, and vasoconstrictions to stimuli that cause increased TPR and BP. The exact adaptation mechanisms and "go-between" molecules that link the increased VSM NCX protein expression and hypertension, however, await clarification.

Therefore, in a "paradigm shift," NCX is no longer conceived to be simply a "homeostatic housekeeper" molecule that keeps $[Ca^{2+}]_i$ low by extruding Ca^{2+}. Rather, it is recognized as a *key component of certain distinct signaling pathways* that activate smooth muscle contraction in response to stretch and to activation of certain GPCRs (Coleman and Khalil 2002; Fellner and Arendshorst 2008; Hashimoto et al. 2006; Lagaud et al. 1999; Lee et al. 2001; Poburko et al. 2007; Syyong et al. 2007; Zhang et al. 2010b). Ion transport through NCX appears to change as a consequence of specific physiological signals and mechanisms in vivo that ultimately control BP. To elucidate the functional role of the NCX signaling pathway in BP regulation and in hypertension could lead to health benefits for humans with hypertension and is also of experimental significance. A selective NCX inhibitor could be useful therapy for hypertension.

References

A. Arnon, J.M. Hamlyn, M.P. Blaustein, Na^+ entry via store-operated channels modulates Ca^{2+} signaling in arterial myocytes. Am. J. Physiol. Cell Physiol. **278**, C163–C173 (2000)

T. Ashida, H. Yoshimi, Y. Kawano, H. Matsuoka, T. Omae, Effect of cilazapril and salt on Ca^{2+} extrusion in arterial smooth muscle of Dahl rats. Am. J. Hypertens. **10**, 107S–111S (1997)

Y.M. Bae, A. Kim, Y.J. Lee, W. Lim, Y.H. Noh, E.J. Kim, J. Kim, T.K. Kim, S.W. Park, B. Kim, S.I. Cho, D.K. Kim, W.K. Ho, Enhancement of receptor-operated cation current and TRPC6 expression in arterial smooth muscle cells of deoxycorticosterone acetate-salt hypertensive rats. J. Hypertens. **25**, 809–817 (2007)

S.G. Baryshnikov, M.V. Pulina, A. Zulian, C.I. Linde, V.A. Golovina, Orai1, a critical component of store-operated Ca^{2+} entry, is functionally associated with Na^+/Ca^{2+} exchanger and plasma membrane Ca^{2+} pump in proliferating human arterial myocytes. Am. J. Physiol. Cell Physiol. **297**, C1103–C1112 (2009)

M.P. Blaustein, W.J. Lederer, Sodium/calcium exchange: its physiological implications. Physiol. Rev. **79**, 763–854 (1999)

M.P. Blaustein, M. Juhaszova, V.A. Golovina, P.J. Church, E.F. Stanley, Na/Ca exchanger and PMCA localization in neurons and astrocytes: functional implications. Ann. N. Y. Acad. Sci. **976**, 356–366 (2002)

M.P. Blaustein, J. Zhang, L. Chen, H. Song, H. Raina, S.P. Kinsey, M. Izuka, T. Iwamoto, M.I. Kotlikoff, J.B. Lingrel, K.D. Philipson, W.G. Wier, J.M. Hamlyn, The pump, the exchanger, and endogenous ouabain: signaling mechanisms that link salt retention to hypertension. Hypertension **53**, 291–298 (2009)

M.P. Blaustein, F.H. Leenen, L. Chen, V.A. Golovina, J.M. Hamlyn, T.L. Pallone, J.W. Van Huysse, J. Zhang, W.G. Wier, How NaCl raises blood pressure: a new paradigm for the pathogenesis of salt-dependent hypertension. Am. J. Physiol. Heart Circ. Physiol. **302**, H1031–H1049 (2011)

S. Chen, B.D. Roufogalis, Enhanced $45Ca^{2+}$ efflux in cultured vascular smooth muscle cells from spontaneously hypertensive rats. Am. J. Hypertens. **7**, 597–602 (1994)

L. Chen, J. Zhang, X. Hu, K.D. Philipson, S.M. Scharf, The Na^+/Ca^{2+} exchanger-1 mediates left ventricular dysfunction in mice with chronic intermittent hypoxia. J. Appl. Physiol. **109**, 1675–1685 (2010)

D.A. Coleman, R.A. Khalil, Physiologic increases in extracellular sodium salt enhance coronary vasoconstriction and Ca^{2+} entry. J. Cardiovasc. Pharmacol. **40**, 58–66 (2002)

G. D'Angelo, M.J. Davis, G.A. Meininger, Calcium and mechanotransduction of the myogenic response. Am. J. Physiol. Heart Circ. Physiol. **273**, H175–H182 (1997)

J.M. Dai, H. Syyong, J. Navarro-Dorado, S. Redondo, M. Alonso, C. van Breemen, T. Tejerina, A comparative study of alpha-adrenergic receptor mediated Ca^{2+} signals and contraction in intact human and mouse vascular smooth muscle. Eur. J. Pharmacol. **629**, 82–88 (2010)

M.J. Davis, M.A. Hill, Signaling mechanisms underlying the vascular myogenic response. Physiol. Rev. **79**, 387–423 (1999)

H. Dong, Y. Jiang, C.R. Triggle, X. Li, J. Lytton, Novel role for K^+-dependent Na^+/Ca^{2+} exchangers in regulation of cytoplasmic free Ca^{2+} and contractility in arterial smooth muscle. Am. J. Physiol. Heart Circ. Physiol. **291**, H1226–H1235 (2006)

M.V. Donoso, M. Steiner, J.P. Huidobro-Toro, BIBP 3226, suramin and prazosin identify neuropeptide Y, adenosine 5'-triphosphate and noradrenaline as sympathetic cotransmitters in the rat arterial mesenteric bed. J. Pharmacol. Exp. Ther. **282**, 691–698 (1997)

S. Earley, B.J. Waldron, J.E. Brayden, Critical role for transient receptor potential channel TRPM4 in myogenic constriction of cerebral arteries. Circ. Res. **95**, 922–929 (2004)

M. Esler, M. Rumantir, D. Kaye, G. Jennings, J. Hastings, F. Socratous, G. Lambert, Sympathetic nerve biology in essential hypertension. Clin. Exp. Pharmacol. Physiol. **28**, 986–989 (2001)

N. Fameli, K.H. Kuo, C. van Breemen, A model for the generation of localized transient $[Na^+]$ elevations in vascular smooth muscle. Biochem. Biophys. Res. Commun. **389**, 461–465 (2009)

S.K. Fellner, W.J. Arendshorst, Angiotensin II-stimulated Ca^{2+} entry mechanisms in afferent arterioles: role of transient receptor potential canonical channels and reverse Na^{+}/Ca^{2+} exchange. Am. J. Physiol. Renal Physiol. **294**, F212–F219 (2008)
R. Floyd, S. Wray, Calcium transporters and signaling in smooth muscles. Cell Calcium **42**, 467–476 (2007)
V.A. Golovina, M.P. Blaustein, Spatially and functionally distinct Ca^{2+} stores in sarcoplasmic and endoplasmic reticulum. Science **275**, 1643–1648 (1997)
A.L. Gonzales, Z.I. Garcia, G.C. Amberg, S. Earley, Pharmacological inhibition of TRPM4 hyperpolarizes vascular smooth muscle. Am. J. Physiol. Cell Physiol. **299**, C1195–C1202 (2010)
G. Grassi, Sympathetic neural activity in hypertension and related diseases. Am. J. Hypertens. **23**, 1052–1060 (2010)
J.M. Hamlyn, R. Ringel, J. Schaeffer, P.D. Levinson, B.P. Hamilton, A.A. Kowarski, M.P. Blaustein, A circulating inhibitor of (Na^{+} + K^{+})ATPase associated with essential hypertension. Nature **300**, 650–652 (1982)
T. Hasegawa, F. Masugi, T. Ogihara, Y. Kumahara, Increase in plasma ouabain-like inhibitor of Na^{+}, K^{+}-ATPase with high sodium intake in patients with essential hypertension. J. Clin. Hypertens. **3**, 419–429 (1987)
T. Hashimoto, M. Kihara, J. Ishida, N. Imai, S. Yoshida, Y. Toya, A. Fukamizu, H. Kitamura, S. Umemura, Apelin stimulates myosin light chain phosphorylation in vascular smooth muscle cells. Arterioscler. Thromb. Vasc. Biol. **26**, 1267–1272 (2006)
S.A. Henderson, J.I. Goldhaber, J.M. So, T. Han, C. Motter, A. Ngo, C. Chantawansri, M.R. Ritter, M. Friedlander, D.A. Nicoll, J.S. Frank, M.C. Jordan, K.P. Roos, R.S. Ross, K.D. Philipson, Functional adult myocardium in the absence of Na^{+}-Ca^{2+} exchange: cardiac-specific knockout of NCX1. Circ. Res. **95**, 604–611 (2004)
M.A. Hill, M.J. Davis, G.A. Meininger, S.J. Potocnik, T.V. Murphy, Arteriolar myogenic signalling mechanisms: Implications for local vascular function. Clin. Hemorheol. Microcirc. **34**, 67–79 (2006)
G.D. Hirst, F.R. Edwards, Sympathetic neuroeffector transmission in arteries and arterioles. Physiol. Rev. **69**, 546–604 (1989)
S. Horiguchi, J. Watanabe, H. Kato, S. Baba, T. Shinozaki, M. Miura, M. Fukuchi, Y. Kagaya, K. Shirato, Contribution of Na^{+}/Ca^{2+} exchanger to the regulation of myogenic tone in isolated rat small arteries. Acta Physiol. Scand. **173**, 167–173 (2001)
T. Iwamoto, S. Kita, A. Uehara, I. Imanaga, T. Matsuda, A. Baba, T. Katsuragi, Molecular determinants of Na^{+}/Ca^{2+} exchange (NCX1) inhibition by SEA0400. J. Biol. Chem. **279**, 7544–7553 (2004a)
T. Iwamoto, S. Kita, J. Zhang, M.P. Blaustein, Y. Arai, S. Yoshida, K. Wakimoto, I. Komuro, T. Katsuragi, Salt-sensitive hypertension is triggered by Ca^{2+} entry via Na^{+}/Ca^{2+} exchanger type-1 in vascular smooth muscle. Nat. Med. **10**, 1193–1199 (2004b)
P.E. Jensen, M.J. Mulvany, C. Aalkjaer, H. Nilsson, H. Yamaguchi, Free cytosolic Ca^{2+} measured with Ca^{2+}-selective electrodes and fura 2 in rat mesenteric resistance arteries. Am. J. Physiol. **265**, H741–H746 (1993)
R.P. Johnson, A.F. El-Yazbi, K. Takeya, E.J. Walsh, M.P. Walsh, W.C. Cole, Ca^{2+} sensitization via phosphorylation of myosin phosphatase targeting subunit at threonine-855 by Rho kinase contributes to the arterial myogenic response. J. Physiol. **587**, 2537–2553 (2009)
M. Juhaszova, A. Ambesi, G.E. Lindenmayer, R.J. Bloch, M.P. Blaustein, Na^{+}-Ca^{2+} exchanger in arteries: identification by immunoblotting and immunofluorescence microscopy. Am. J. Physiol. Cell Physiol. **266**, C234–C242 (1994)
T. Kashihara, K. Nakayama, T. Matsuda, A. Baba, T. Ishikawa, Role of Na^{+}/Ca^{2+} exchanger-mediated Ca^{2+} entry in pressure-induced myogenic constriction in rat posterior cerebral arteries. J. Pharmacol. Sci. **110**, 218–222 (2009)
J.K. Kim, S.J. Kim, Y.C. Kim, I. So, K.W. Kim, Influence of extracellular Na^{+} removal on cytosolic Ca^{2+} concentration in smooth muscle cells of rabbit cerebral artery. J. Smooth Muscle Res. **35**, 135–145 (1999)
J. Kiraku, T. Sugiyama, T. Ashida, N. Takahashi, J. Fujii, M. Kuro-o, R. Nagai, Increases in intracellular calcium of arterial smooth muscle cells in transgenic mice overexpressing Na^{+}/H^{+} exchanger. J. Cardiovasc. Pharmacol. **35**, 511–513 (2000)
H.J. Knot, M.T. Nelson, Regulation of arterial diameter and wall [Ca^{2+}] in cerebral arteries of rat by membrane potential and intravascular pressure. J. Physiol. **508**, 199–209 (1998)
N. Kotecha, M.A. Hill, Myogenic contraction in rat skeletal muscle arterioles: smooth muscle membrane potential and Ca^{2+} signaling. Am. J. Physiol. Heart Circ. Physiol. **289**, H1326–H1334 (2005)
G.J. Lagaud, V. Randriamboavonjy, G. Roul, J.C. Stoclet, R. Andriantsitohaina, Mechanism of Ca^{2+} release and entry during contraction elicited by norepinephrine in rat resistance arteries. Am. J. Physiol. Heart Circ. Physiol. **276**, H300–H308 (1999)
C. Lamont, W.G. Wier, Evoked and spontaneous purinergic junctional Ca^{2+} transients (jCaTs) in rat small arteries. Circ. Res. **91**, 454–456 (2002)
C. Lamont, W.G. Wier, Different roles of ryanodine receptors and inositol (1,4,5)-trisphosphate receptors in adrenergically stimulated contractions of small arteries. Am. J. Physiol. Heart Circ. Physiol. **287**, H617–H625 (2004)
C. Lamont, E. Vainorius, W.G. Wier, Purinergic and adrenergic Ca^{2+} transients during neurogenic contractions of rat mesenteric small arteries. J. Physiol. **549**, 801–808 (2003)
W.J. Lederer, E. Niggli, R.W. Hadley, Sodium-calcium exchange in excitable cells: fuzzy space. Science **248**, 283 (1991)
S.L. Lee, A.S. Yu, J. Lytton, Tissue-specific expression of Na^{+}-Ca^{2+} exchanger isoforms. J. Biol. Chem. **269**, 14849–14852 (1994)

C.H. Lee, D. Poburko, P. Sahota, J. Sandhu, D.O. Ruehlmann, C. van Breemen, The mechanism of phenylephrine-mediated [Ca^{2+}]i oscillations underlying tonic contraction in the rabbit inferior vena cava. J. Physiol. **534**, 641–650 (2001)

M.Y. Lee, H. Song, J. Nakai, M. Ohkura, M.I. Kotlikoff, S.P. Kinsey, V.A. Golovina, M.P. Blaustein, Local subplasma membrane Ca^{2+} signals detected by a tethered Ca^{2+} sensor. Proc. Natl. Acad. Sci. U. S. A. **103**, 13232–13237 (2006)

Z. Li, S. Matsuoka, L.V. Hryshko, D.A. Nicoll, M.M. Bersohn, E.P. Burke, R.P. Lifton, K.D. Philipson, Cloning of the NCX2 isoform of the plasma membrane Na^+-Ca^{2+} exchanger. J. Biol. Chem. **269**, 17434–17439 (1994)

D. Liu, D. Yang, H. He, X. Chen, T. Cao, X. Feng, L. Ma, Z. Luo, L. Wang, Z. Yan, Z. Zhu, M. Tepel, Increased transient receptor potential canonical type 3 channels in vasculature from hypertensive rats. Hypertension **53**, 70–76 (2009)

M. Luo, M.C. Hess, G.D. Fink, L.K. Olson, J. Rogers, D.L. Kreulen, X. Dai, J.J. Galligan, Differential alterations in sympathetic neurotransmission in mesenteric arteries and veins in DOCA-salt hypertensive rats. Auton. Neurosci. **104**, 47–57 (2003)

R.M. Lynch, C.S. Weber, K.D. Nullmeyer, E.D. Moore, R.J. Paul, Clearance of store-released Ca^{2+} by the Na^+-Ca^{2+} exchanger is diminished in aortic smooth muscle from Na^+-K^+-ATPase alpha 2-isoform gene-ablated mice. Am. J. Physiol. Heart Circ. Physiol. **294**, H1407–H1416 (2008)

J.J. Maguire, A.P. Davenport, Regulation of vascular reactivity by established and emerging GPCRs. Trends Pharmacol. Sci. **26**, 448–454 (2005)

L. Martinez-Lemus, M. Hill, G. Meininger, The plastic nature of the vascular wall: a continuum of remodeling events contributing to control of arteriolar diameter and structure. Physiology (Bethesda) **24**, 45–57 (2009)

Y. Maruyama, Y. Nakanishi, E.J. Walsh, D.P. Wilson, D.G. Welsh, W.C. Cole, Heteromultimeric TRPC6-TRPC7 channels contribute to arginine vasopressin-induced cation current of A7r5 vascular smooth muscle cells. Circ. Res. **98**, 1520–1527 (2006)

V.V. Matchkov, O.S. Tarasova, M.J. Mulvany, H. Nilsson, Myogenic response of rat femoral small arteries in relation to wall structure and [Ca^{2+}]i. Am. J. Physiol. Heart Circ. Physiol. **283**, H118–H125 (2002)

T. Matsuda, N. Arakawa, K. Takuma, Y. Kishida, Y. Kawasaki, M. Sakaue, K. Takahashi, T. Takahashi, T. Suzuki, T. Ota, A. Hamano-Takahashi, M. Onishi, Y. Tanaka, K. Kameo, A. Baba, SEA0400, a novel and selective inhibitor of the Na^+-Ca^{2+} exchanger, attenuates reperfusion injury in the in vitro and in vivo cerebral ischemic models. J. Pharmacol. Exp. Ther. **298**, 249–256 (2001)

E.D. Moore, E.F. Etter, K.D. Philipson, W.A. Carrington, K.E. Fogarty, L.M. Lifshitz, F.S. Fay, Coupling of the Na^+/Ca^{2+} exchanger, Na^+/K^+ pump and sarcoplasmic reticulum in smooth muscle. Nature **365**, 657–660 (1993)

M. Morad, L. Cleemann, D.R. Menick, NCX1 phosphorylation dilemma: a little closer to resolution. Focus on "Full-length cardiac Na^+/Ca^{2+} exchanger 1 protein is not phosphorylated by protein kinase A". Am. J. Physiol. Cell Physiol. **300**, C970–C973 (2011)

R.E. Mufti, S.E. Brett, C.H. Tran, R. Abd El-Rahman, Y. Anfinogenova, A. El-Yazbi, W.C. Cole, P.P. Jones, S.R. Chen, D.G. Welsh, Intravascular pressure augments cerebral arterial constriction by inducing voltage-insensitive Ca^{2+} waves. J. Physiol. **588**, 3983–4005 (2010)

M.J. Mulvany, H. Nilsson, J.A. Flatman, Role of membrane potential in the response of rat small mesenteric arteries to exogenous noradrenaline stimulation. J. Physiol. **332**, 363–373 (1982)

Y. Nakasaki, T. Iwamoto, H. Hanada, T. Imagawa, M. Shigekawa, Cloning of the rat aortic smooth muscle Na^+/Ca^{2+} exchanger and tissue-specific expression of isoforms. J. Biochem. (Tokyo) **114**, 528–534 (1993)

M.T. Nelson, H. Cheng, M. Rubart, L.F. Santana, A.D. Bonev, H.J. Knot, W.J. Lederer, Relaxation of arterial smooth muscle by calcium sparks. Science **270**, 633–637 (1995)

L.D. Nelson, M.T. Unlap, J.L. Lewis, P.D. Bell, Renal arteriolar Na^+/Ca^{2+} exchange in salt-sensitive hypertension. Am. J. Physiol. Renal Physiol. **276**, F567–F573 (1999)

D.A. Nicoll, S. Longoni, K.D. Philipson, Molecular cloning and functional expression of the cardiac sarcolemmal Na^+-Ca^{2+} exchanger. Science **250**, 562–565 (1990)

D.A. Nicoll, B.D. Quednau, Z. Qui, Y.R. Xia, A.J. Lusis, K.D. Philipson, Cloning of a third mammalian Na^+-Ca^{2+} exchanger, NCX3. J. Biol. Chem. **271**, 24914–24921 (1996)

D. Poburko, C.H. Liao, V.S. Lemos, E. Lin, Y. Maruyama, W.C. Cole, C. van Breemen, Transient receptor potential channel 6-mediated, localized cytosolic [Na^+] transients drive Na^+/Ca^{2+} exchanger-mediated Ca^{2+} entry in purinergically stimulated aorta smooth muscle cells. Circ. Res. **101**, 1030–1038 (2007)

C. Pott, X. Ren, D.X. Tran, M.J. Yang, S. Henderson, M.C. Jordan, K.P. Roos, A. Garfinkel, K.D. Philipson, J.I. Goldhaber, Mechanism of shortened action potential duration in Na^+-Ca^{2+} exchanger knockout mice. Am. J. Physiol. Cell Physiol. **292**, C968–C973 (2007a)

C. Pott, M. Yip, J.I. Goldhaber, K.D. Philipson, Regulation of cardiac L-type Ca^{2+} current in Na^+-Ca^{2+} exchanger knockout mice: functional coupling of the Ca^{2+} channel and the Na^+-Ca^{2+} exchanger. Biophys. J. **92**, 1431–1437 (2007b)

T.J. Pritchard, P.S. Bowman, A. Jefferson, M. Tosun, R.M. Lynch, R.J. Paul, Na^+-K^+-ATPase and Ca^{2+} clearance proteins in smooth muscle: a functional unit. Am. J. Physiol. Heart Circ. Physiol. **299**, H548–H556 (2010)

M.V. Pulina, A. Zulian, R. Berra-Romani, O. Beskina, A. Mazzocco-Spezzia, S.G. Baryshnikov, I. Papparella, J.M. Hamlyn, M.P. Blaustein, V.A. Golovina, Upregulation of Na^+ and Ca^{2+} transporters in arterial smooth muscle from ouabain-induced hypertensive rats. Am. J. Physiol. Heart Circ. Physiol. **298**, H263–H274 (2010)

B.D. Quednau, D.A. Nicoll, K.D. Philipson, Tissue specificity and alternative splicing of the Na^+/Ca^{2+} exchanger isoforms NCX1, NCX2, and NCX3 in rat. Am. J. Physiol. Cell Physiol. **272**, C1250–C1261 (1997)

H. Raina, S.R. Ella, M.A. Hill, Decreased activity of the smooth muscle Na^+/Ca^{2+} exchanger impairs arteriolar myogenic reactivity. J. Physiol. **586**, 1669–1681 (2008)

A. Rebolledo, F. Speroni, J. Raingo, S.V. Salemme, F. Tanzi, V. Munin, M.C. Añón, V. Milesi, The Na^+/Ca^{2+} exchanger is active and working in the reverse mode in human umbilical artery smooth muscle cells. Biochem. Biophys. Res. Commun. **339**, 840–845 (2006)

J.P. Reeves, M. Condrescu, J. Urbanczyk, O. Chernysh, New modes of exchanger regulation: physiological implications. Ann. N. Y. Acad. Sci. **1099**, 64–77 (2007)

C. Ren, J. Zhang, K.D. Philipson, M.I. Kotlikoff, M.P. Blaustein, D.R. Matteson, Activation of L-type Ca^{2+} channels by protein kinase C is reduced in smooth muscle-specific Na^+/Ca^{2+} exchanger knockout mice. Am. J. Physiol. Heart Circ. Physiol. **298**, H1484–H1491 (2010)

M.K. Slodzinski, M.P. Blaustein, Physiological effects of Na^+/Ca^{2+} exchanger knockdown by antisense oligodeoxynucleotides in arterial myocytes. Am. J. Physiol. Cell Physiol. **275**, C251–C259 (1998)

M.K. Slodzinski, M. Juhaszova, M.P. Blaustein, Antisense inhibition of Na^+/Ca^{2+} exchange in primary cultured arterial myocytes. Am. J. Physiol. **269**, C1340–C1345 (1995)

A.P. Somlyo, A.V. Somlyo, Ca^{2+} sensitivity of smooth muscle and nonmuscle myosin II: modulated by G proteins, kinases, and myosin phosphatase. Physiol. Rev. **83**, 1325–1358 (2003)

H.T. Syyong, D. Poburko, N. Fameli, C. van Breemen, ATP promotes NCX-reversal in aortic smooth muscle cells by DAG-activated Na^+ entry. Biochem. Biophys. Res. Commun. **357**, 1177–1182 (2007)

S. Taniguchi, K. Furukawa, S. Sasamura, Y. Ohizumi, K. Seya, S. Motomura, Gene expression and functional activity of sodium/calcium exchanger enhanced in vascular smooth muscle cells of spontaneously hypertensive rats. J. Cardiovasc. Pharmacol. **43**, 629–637 (2004)

T.L. Török, Electrogenic Na^+/Ca^{2+}-exchange of nerve and muscle cells. Prog. Neurobiol. **82**, 287–347 (2007)

S.Y. Tsang, X. Yao, C.M. Wong, C.L. Au, Z.Y. Chen, Y. Huang, Contribution of Na^+-Ca^{2+} exchanger to pinacidil-induced relaxation in the rat mesenteric artery. Br. J. Pharmacol. **138**, 453–460 (2003)

K. Tsuda, S. Tsuda, I. Nishio, Y. Masuyama, Inhibition of norepinephrine release by presynaptic alpha 2-adrenoceptors in mesenteric vasculature preparations from chronic DOCA-salt hypertensive rats. Jpn. Heart J. **30**, 231–239 (1989)

C. van Breemen, Q. Chen, I. Laher, Superficial buffer barrier function of smooth muscle sarcoplasmic reticulum. Trends Pharmacol. Sci. **16**, 98–105 (1995)

D.G. Welsh, A.D. Morielli, M.T. Nelson, J.E. Brayden, Transient receptor potential channels regulate myogenic tone of resistance arteries. Circ. Res. **90**, 248–250 (2002)

E.B. Westcott, W.F. Jackson, Heterogeneous function of ryanodine receptors, but not IP3 receptors, in hamster cremaster muscle feed arteries and arterioles. Am. J. Physiol. Heart Circ. Physiol. **300**, H1616–H1630 (2011)

A. Wirth, Z. Benyó, M. Lukasova, B. Leutgeb, N. Wettschureck, S. Gorbey, P. Orsy, B. Horváth, C. Maser-Gluth, E. Greiner, B. Lemmer, G. Schütz, J.S. Gutkind, S. Offermanns, G12-G13-LARG-mediated signaling in vascular smooth muscle is required for salt-induced hypertension. Nat. Med. **14**, 64–68 (2008)

J. Yamanaka, J. Nishimura, K. Hirano, H. Kanaide, An important role for the Na^+-Ca^{2+} exchanger in the decrease in cytosolic Ca^{2+} concentration induced by isoprenaline in the porcine coronary artery. J. Physiol. **549**, 553–562 (2003)

Y. Yu, I. Fantozzi, C.V. Remillard, J.W. Landsberg, N. Kunichika, O. Platoshyn, D.D. Tigno, P.A. Thistlethwaite, L.J. Rubin, J.X. Yuan, Enhanced expression of transient receptor potential channels in idiopathic pulmonary arterial hypertension. Proc. Natl. Acad. Sci. U. S. A. **101**, 13861–13866 (2004)

J. Zacharia, J. Zhang, W.G. Wier, Ca^{2+} signaling in mouse mesenteric small arteries: myogenic tone and adrenergic vasoconstriction. Am. J. Physiol. Heart Circ. Physiol. **292**, H1523–H1532 (2007)

W.J. Zang, C.W. Balke, W.G. Wier, Graded alpha1-adrenoceptor activation of arteries involves recruitment of smooth muscle cells to produce 'all or none' Ca^{2+} signals. Cell Calcium **29**, 327–334 (2001)

S. Zhang, J.X. Yuan, K.E. Barrett, H. Dong, Role of Na^+/Ca^{2+} exchange in regulating cytosolic Ca^{2+} in cultured human pulmonary artery smooth muscle cells. Am. J. Physiol. Cell Physiol. **288**, C245–C252 (2005a)

J. Zhang, M.Y. Lee, M. Cavalli, L. Chen, R. Berra-Romani, C.W. Balke, G. Bianchi, P. Ferrari, J.M. Hamlyn, T. Iwamoto, J.B. Lingrel, D.R. Matteson, W.G. Wier, M.P. Blaustein, Sodium pump alpha2 subunits control myogenic tone and blood pressure in mice. J. Physiol. **569**, 243–256 (2005b)

S. Zhang, H. Dong, L.J. Rubin, J.X. Yuan, Upregulation of Na^+/Ca^{2+} exchanger contributes to the enhanced Ca^{2+} entry in pulmonary artery smooth muscle cells from patients with idiopathic pulmonary arterial hypertension. Am. J. Physiol. Cell Physiol. **292**, C2297–C2305 (2007)

J. Zhang, L. Chen, H. Raina, M.P. Blaustein, W.G. Wier, In vivo assessment of artery smooth muscle $[Ca^{2+}]i$ and MLCK activation in FRET-based biosensor mice. Am. J. Physiol. Heart Circ. Physiol. **299**, H946–H956 (2010a)

J. Zhang, C. Ren, L. Chen, M.F. Navedo, L.K. Antos, S.P. Kinsey, T. Iwamoto, K.D. Philipson, M.I. Kotlikoff, L.F. Santana, W.G. Wier, D.R. Matteson, M.P. Blaustein, Knockout of Na^+/Ca^{2+} exchanger in smooth muscle attenuates vasoconstriction and L-type Ca^{2+} channel current and lowers blood pressure. Am. J. Physiol. Heart Circ. Physiol. **298**, H1472–H1483 (2010b)

D. Zhao, J. Zhang, M.P. Blaustein, L.G. Navar, Attenuated renal vascular responses to acute angiotensin II infusion in smooth muscle-specific Na^+/Ca^{2+} exchanger knockout mice. Am. J. Physiol. Renal Physiol. **301**, F574–F579 (2011)

Y.M. Zheng, Y.X. Wang, Sodium-calcium exchanger in pulmonary artery smooth muscle cells. Ann. N. Y. Acad. Sci. **1099**, 427–435 (2007)

A. Zulian, S.G. Baryshnikov, C.I. Linde, J.M. Hamlyn, P. Ferrari, V.A. Golovina, Upregulation of Na^+/Ca^{2+} exchanger and TRPC6 contributes to abnormal Ca^{2+} homeostasis in arterial smooth muscle cells from Milan hypertensive rats. Am. J. Physiol. Heart Circ. Physiol. **299**, H624–H633 (2010)

Toward an Understanding of the Complete NCX1 Lifetime in the Cardiac Sarcolemma

29

Donald W. Hilgemann, Mei-Jung Lin, Michael Fine, Gary Frazier, and Hao-Ran Wang

Abstract

The density of Na/Ca exchangers (NCX1) in the cardiac sarcolemma, like all plasma membrane proteins, will be influenced by (and ultimately determined by) the function of membrane insertion and retrieval processes (i.e., exo- and endocytic mechanisms). Progress in understanding these processes in cardiac muscle faces many biological and methodological complexities and hurdles. As described here, we are attempting to overcome these hurdles to study more adequately the assembly and disassembly of the cardiac sarcolemma, in general, and the control of NCX1 by membrane trafficking processes in particular. First, we have developed improved noninvasive methods to monitor the cellular capacitance of cardiac tissue (NIC) over periods of hours. Thus, we can study long-term changes of total membrane area. Second, we have developed mice that express fusion proteins of NCX1 with the pHluorin green protein. Thus, we can determine the membrane disposition of NCX1, and changes thereof, *on-line* in intact cardiac muscle.

Keywords

Cardiac sarcolemma • NCX1 • Electrophysiology • Endocytosis • Exocytosis

29.1 Introduction

The cardiac sarcolemma, like the plasma membrane of all eukaryotic cells, is constantly being expanded and diminished by discrete exo- and endocytic events that alter its content of integral membrane proteins. An improved understanding of these processes, and more generally the function of the cardiac secretory pathway, is essential for a deeper understanding of cardiac physiology

D.W. Hilgemann (✉) • M.-J. Lin • M. Fine • G. Frazier
H.-R. Wang
Department of Physiology, University of Texas Southwestern Medical Center,
5323 Harry Hines Blvd., Dallas, TX 75235, USA
e-mail: donald.hilgemann@utsouthwestern.edu

L. Annunziato (ed.), *Sodium Calcium Exchange: A Growing Spectrum of Pathophysiological Implications*, Advances in Experimental Medicine and Biology 961, DOI 10.1007/978-1-4614-4756-6_29,

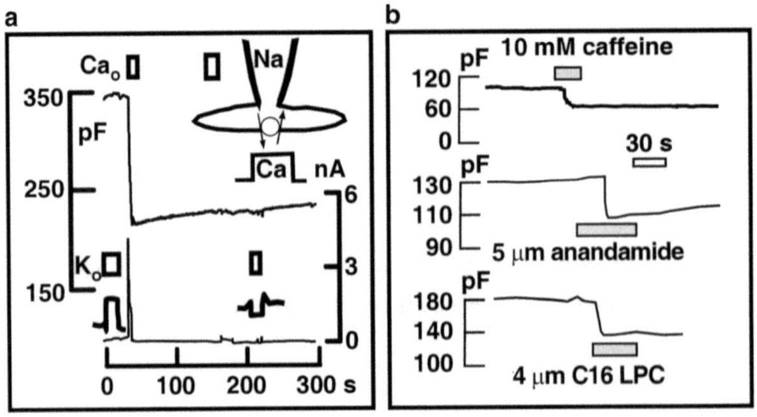

Fig. 29.1 Ca^{2+}- and amphipath-induced MEND in cardiac myocytes. (**a**) MEND induced by 3 s activation of reverse exchange current in NCX-PHL murine myocyte. 50 % MEND response with complete loss of Na/K pump current. Expanded current traces are given for application and removal of 5 mM extracellular K^+ (40 mM Na^+ and 8 mM ATP in cytoplasmic solution). (**b**) MEND induced by caffeine application for 5 s in absence of extracellular Na^+ and MEND induced by anandamide (5 μM) and lysophosphatidylcholine (4 μM) in the presence of 30 μM EGTA and no Ca^{2+}

and pathology (Best et al. 2011; Chen et al. 2009; Cusdin et al. 2008; Hayashi et al. 2010; Pongs 2009; Zorzano et al. 1997). These pathways are with certainty highly regulated and relevant to cardiac disease, but our understanding of them remains disappointingly limited. With regard to membrane transporter regulation, progress has been most impressive for glucose transporters (Fazakerley et al. 2009; Zorzano et al. 1997) that cycle large distances between the sarcolemma and perinuclear membrane depots. Connexins, such as Connexin-43, are thought to turn over entirely within 2 h (Saffitz et al. 2000), suggesting that insertion and retrieval processes must be especially powerful in the intercalated disc regions of cardiac myocytes, areas that are also rich in NCX1. For Na/H exchangers, trafficking to superficial membrane depots, which are difficult to separate from the sarcolemma optically, has been proposed recently to be a major regulatory mechanism in cardiac myocytes (Lawrence et al. 2010). In spite of the central roles that NCX1 plays in cardiac physiology and pathology (Hilgemann et al. 2006), it remains unknown how the delivery and retrieval of NCX1 are regulated and to what extent those processes are selective (i.e., cargo-specific). NCX1 and Na/K pumps sometimes appear to traffic together (Hund and Mohler 2008).

Our initial focus has been endocytosis (Fine et al. 2011; Hilgemann and Fine 2011; Lariccia et al. 2011). While many different steps leading up to protein insertion may be rate-limiting in membrane protein turnover, endocytosis will *ultimately* influence the plasmalemmal content of all membrane proteins. Our unique perspective (Fine et al. 2011; Hilgemann and Fine 2011; Lariccia et al. 2011) is that lipids themselves can play more important roles in endocytosis than heretofore suspected. In brief, we discovered that large fractions of the cell surface of many cells can be internalized in seconds in response to large Ca^{2+} transients or certain amphipathic compounds, including physiological amphipaths such as lysophosphatidylcholines (LPCs) and cannabinoids (ibid; and Fig. 29.1). Classical endocytic proteins are not involved, and ordered lipid (*Lo*) domains of the plasmalemma (>50 % of the cell surface) are preferentially internalized (Fine et al. 2011). We coined the term massive endocytosis (MEND) to describe these processes. As described in Results, constitutive endocytic processes in intact cardiac muscle appear to be mechanistically related to MEND in isolated myocytes. However, the time scales involved in intact heart are radically different from our initial results for isolated myocytes.

29.2 Results

Figure 29.1 illustrates Ca-activated MEND in mouse myocytes, using whole-cell patch clamp to monitor capacitance (i.e., membrane surface area). As shown in Fig. 29.1a, nearly 50 % of the myocyte capacitance can be lost during activation of Ca^{2+} influx via reverse NCX1 current in myocytes over-expressing an NCX1-pHluorin fusion protein (NCX1PH). Na/K pump currents are thereafter ablated (see inserts). As shown in Fig. 29.1b, release of Ca^{2+} by caffeine can cause rapid MEND of ~25 % of the SL when Ca^{2+} extrusion is inhibited by Li^+ replacement of Na^+. Further illustrated in Fig. 29.1b, the physiological amphipaths, anandamide (a cannabinoid) and LPC, can cause large MEND responses in the complete absence of Ca^{2+} (30 mM $EGTA_i$). These MEND responses occur without conductance changes, and several lines of evidence suggest that lateral and transmembrane lipid inhomogeneities play a major role in driving these responses (Fine et al. 2011; Hilgemann and Fine 2011; Lariccia et al. 2011).

That said, the plasticity of endocytosis is hardly a new discovery (Doherty and McMahon 2009; Donaldson et al. 2009; Ivanov 2008; Mayor and Pagano 2007). Neurons, like yeast (Geli and Riezman 1998), can carry out essential membrane cycling after classical players of endocytosis are deleted. Clathrin deletion (Sato et al. 2009), adapter protein 2 (AP2) deletion (Kim and Ryan 2009), and dynamin deletions (Ferguson et al. 2007) all follow a similar pattern. These recent outcomes all suggest that lipid ordering might become important in classical endocytosis, as well as MEND. Independent of lipid ordering, the results underscore a great need for new methodological approaches to analyze cardiac sarcolemma turnover. Work described next provides some new hypothesis-independent information about sarcolemma turnover in intact heart tissue.

Figure 29.2 illustrates our use of mice over-expressing NCX1PH (under MHC promoter), the pHluorin fusion being at the 1st NCX1 glycosylation site. Figure 29.2a illustrates how extracellular solutions can be manipulated to determine the disposition of NCX1 in different compartments. pHluorin (Miesenbock et al. 1998) has a pKa of about 7, and fluorescence of NCX1PH at the cell surface can be rapidly extinguished by shifting extracellular pH from 7.4 (or 7.7) to 6. NCX1PH in intracellular vesicles may or may not be fluorescent, depending on the ambient pH of the vesicles. Cytoplasmic acidification with

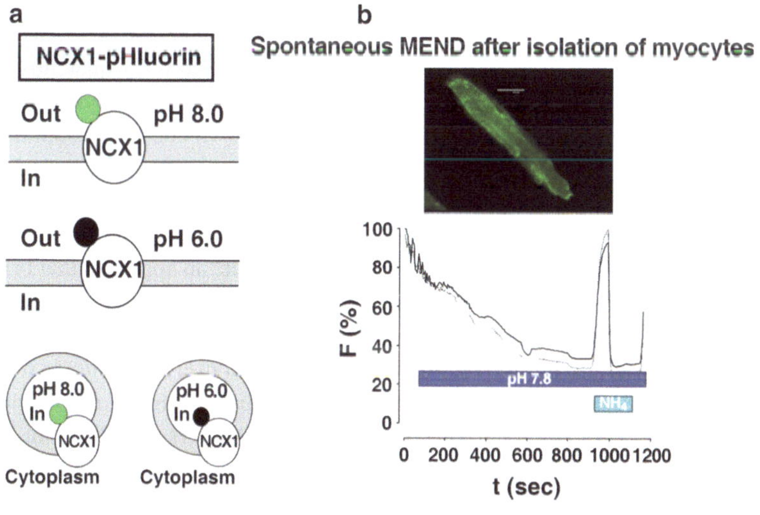

Fig. 29.2 Use of NCX1-pHluorin fusion (*NCX1PHL*) to monitor NCX1 internalization. (**a**) Four different NCX1PHL states in cells. (**b**) Spontaneous internalization of NCX1 after isolating myocytes

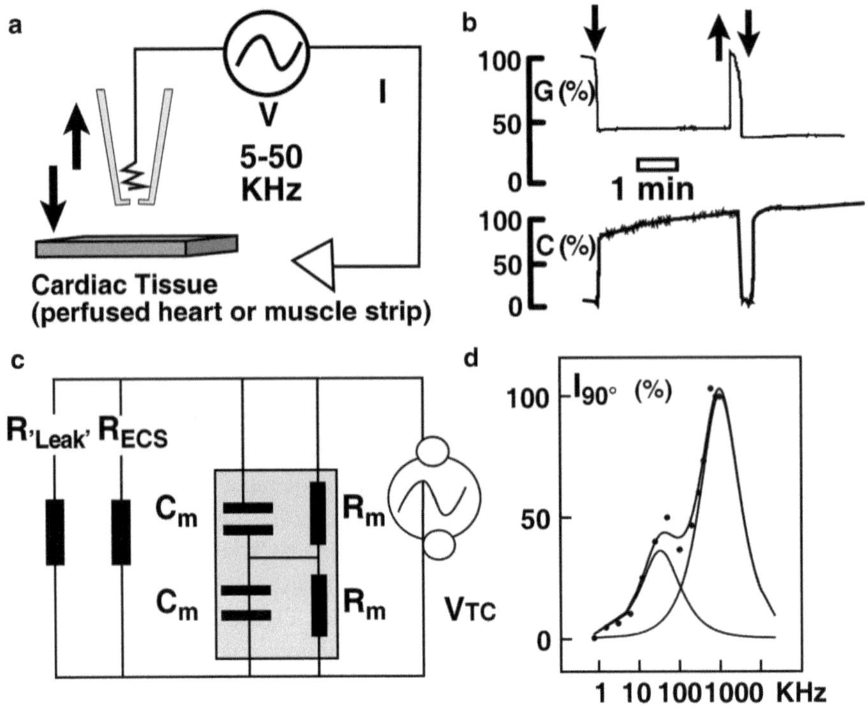

Fig. 29.3 Principles of noninvasive capacitance (NIC) signals in intact cardiac tissue. *Shaded area* represents one cell. The out-of-phase current signal at 0°, defined as NIC, amounts to 8–12 % of "in-phase" signal at 20 KHz. No NIC is encountered by recording at silicon, glass, or other non-capacitive surfaces. NIC recording is remarkably stable in many cardiac muscle preparations over periods of hours. It is routinely verified that NIC signal changes are similar in different regions of muscle preparations. See text for more explanations

25-mM extracellular Na acetate reliably extinguishes all NCXPH fluorescence not extinguished by extracellular acidification (data not shown). Alkalination of the cytoplasm with 22-mM NH_4Cl can rapidly increase fluorescence (e.g., Fig. 29.3e), thereby revealing internalized pHluorins. *To avoid extracellular pH changes, we employ impermeable buffers at very high concentrations (55-mM HEPES for pH>7).*

Figure 29.2b presents a fundamental observation made with NCX1PH in myocytes. NCX1PH fluorescence often decreases dramatically within 1 h after isolation of myocytes, especially when younger mice (2–3 months) are employed. Here, fluorescence decreases by >60 % in the nominal absence of extracellular Ca^{2+}; fluorescence can be recovered fully within seconds upon applying 25-mM NH_4Cl at constant extracellular pH. Thus, it is established that NCX1PH can be rapidly internalized after myocyte isolation. This corresponds well with our routine finding that Na/Ca exchange and Na/K pump currents decrease rapidly after isolation of myocytes from young rodents, typically in the time course of 1–2 h. Presumably, cell stress of isolation is adequate to activate MEND-related processes in myocytes, thereby underscoring long-known concerns about the physiological state of isolated myocytes.

Figure 29.3 illustrates our new electrical method to monitor cellular capacitance in intact cardiac tissue, an approach that brings substantial advantages over previous approaches in terms of utility (i.e., ability to use different cardiac tissues), ease of use, and noninvasive/benign nature. In brief, thick-walled glass pipettes are prepared with tip diameters of 0.3–0.7 m. A ball of finely coiled Ag/AgCl wire is inserted into the orifice so as to eliminate series resistance. Other types of electrodes can also be employed, but the thick glass wall of these electrodes contacts the tissue in an advantageous fashion. Using a modified patch clamp circuit that can deliver >300 μA at 2 MHz, voltage (5–10 mV) is oscillated in the tip at frequencies of 10–1,000 KHz, and signals are analyzed with standard lock-in amplifiers. The electrical field can be shown to fall off in front of the pipette tip with a space constant roughly corresponding to the tip radius. When the tip of the electrode is touched gently to the surface of cardiac tissue, as illustrated in Fig. 29.3a, b, just firm enough to reduce the conductance of the tip by about 50 %, transcellular voltage gradients are generated across cells within the tissue in front of the tip. At the surface of cardiac muscle, out-of-phase currents develop with magnitudes in the range of 5–12 % of in-phase currents. These currents reflect charging of cellular capacitances, which in turn reflect the capacitances of cell membranes arranged in series (as in epithelial cell recording; see Fig. 29.3c). In contrast to epithelia, however, which have a well-defined resistance at tight junctions, cardiac cells in the tissue experience a voltage gradient (~2 mV/100 μm) along their surfaces. Charging of cellular capacitances occurs through the extracellular tissue conductance (not the cytoplasm). This amounts to about 20 % of the conductance of physiological saline, so that charging time constants are in the microsecond domain, at least 100 times smaller than the electrical time constants of myocytes in patch clamp. Thus, NIC signals increase with increasing oscillation frequencies in the range of 5–30 kHz, a frequency range over which transmembrane voltages remain negligible (see Fig. 29.2c). When the electrode tip is brought to the surface of silicon rubber, glass, or other non-capacitive materials, the in-phase currents decrease without development of out-of-phase (capacitive) currents. Importantly, capacitances of internal membranes cannot contribute to the signals recorded, since the cytoplasm is never polarized. Figure 29.3d illustrates the frequency spectrum of a right ventricular strip from a young mouse under our standard conditions (superfused at flows >1 cm/s, 34 °C, and showing stable contractile function in response to electrical stimulation for hours). Working in the frequency range of 5–30 kHz, many results support the idea that capacitances recorded originate mostly from myocyte sarcolemma: (1) NIC signals are strongly decreased by extracting lipids from cells with β-methylcyclodextrans or detergents (at high concentrations). (2) NIC signals are increased by agents known to increase plasma membrane capacitance (e.g., some detergents and benzyl alcohol). (3) Morphometric analyses support our assumption that myocyte SL constitutes the large majority of surface membrane in intact myocyte. (4) Large NIC signal changes described next depend on electrical excitation of the tissue; this would not be the case if signals originated from non-myocytes.

Using the NIC method, we proceeded to examine whether large Ca^{2+} transients facilitated by stimulus pattern might cause endocytic responses (i.e., a decrease of capacitance). For the example shown in Fig. 29.4, we employed stimulation patterns that cause a large facilitation of contraction in isolated left atria of rabbits, known commonly as post-extrasystolic potentiation (Hilgemann and Langer 1984). In brief, isolated cardiac muscle strips are stimulated 5–20 times at a high frequency (2–4 Hz). This rapid stimulation results in a strong loading of the sarcoplasmic reticulum with Ca^{2+} because the sarcoplasmic release mechanism is rather inactive at high frequencies, while Ca^{2+} influx continues unabated. Then, after a brief period of rest, the next excitation causes a massive contraction as the excess sarcoplasmic Ca^{2+} is released. Figure 29.4 shows the conductance and capacitance records from NIC recordings with this protocol. The complete records are shown in the left panel, and the amplified periods of recording marked with ovals are shown on the right. From these results, it is clearly possible to massively potentiate contractions, and therefore Ca^{2+} transients, without not causing endocytic responses in intact cardiac tissue with a resolution of about 0.5 % of total sarcolemma area. These results would appear to negate the possibility that cardiac excitation-contraction coupling is regulated by Ca^{2+}-dependent exo- and/or endocytic events.

Figure 29.5 illustrates our ability to combine optical and electrical recordings in intact cardiac tissue, specifically using NCX1PHL-expressing

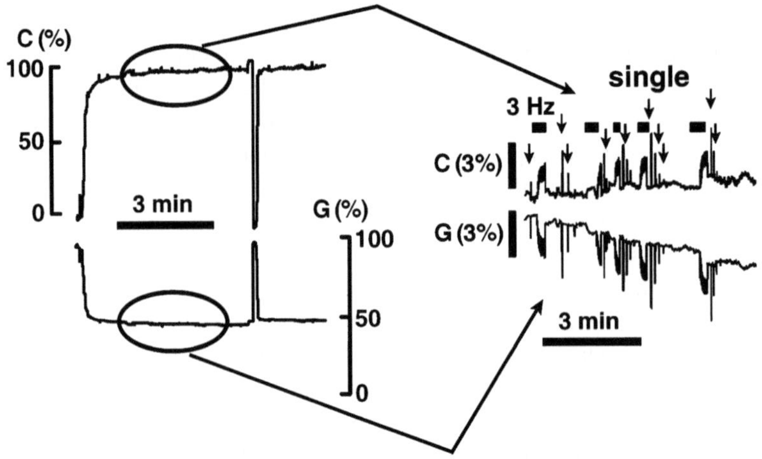

Fig. 29.4 NIC recording in an isolated left atrial strip from a rabbit. The *left panel* shows the capacitance and conductance records occurring during the induction of post-extrasystolic potentiation and its decay. The *right panel* shows amplified portions of the records. All signal changes appear to be due to contraction artifacts that follow well the contractile function of the muscle. During rapid stimulation, contractions are small. After a stimulation pause, the first contraction is greatly potentiated. The records provide no evidence for the presence of Ca^{2+}-activated endo- or exocytic events in this muscle to a resolution of about 1 % of sarcolemmal area

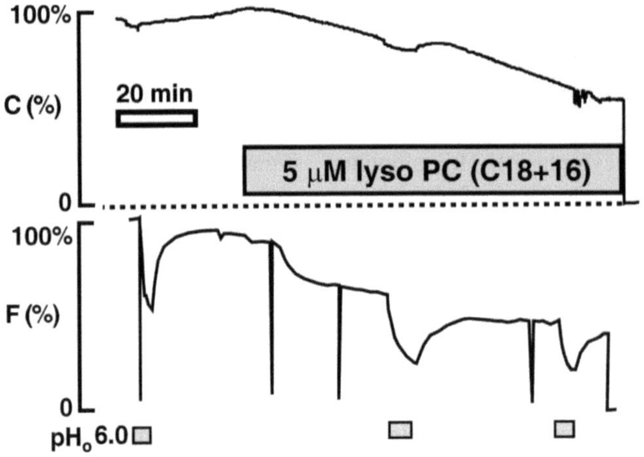

Fig. 29.5 Simultaneous use of optical NCX1-pHluorin fusion (*NCX1PHL*) recording and electrical NIC recording to monitor NCX1 internalization. Slow internalization of 50 % of NCX1 and SL is induced by the physiological amphipath, LPC, employing a 1:1 mix of C16 and C18 LPC at a total concentration of 5 μM. Note that brief periods of extracellular acidification were introduced to demonstrate that most fluorescence arises from exchangers with pHluorin facing the extracellular space

right ventricular strips (RVSs) of young adult mice. After testing several ventricular muscle preparations, we find that rapidly superfused RVSs (0.6–0.8 mm thick), employed at 32–34 °C, are advantageous. Arterial perfusion of whole hearts (or partial hearts) is in our experience not reliable for periods of hours. Comparing RV strips from young mice with papillary muscles, RVS contractile activity is more easily monitored, electrical activity is controlled just as reliably by electrical stimulation, and the function of RVSs is much more reproducible than that of murine papillary muscles. Notwithstanding these advantages, it should be born in mind that RVS are likely to have an anoxic core that may affect the interpretation of some results. In the RVS shown in Fig. 29.5, NIC is recorded simultaneously with NCX1PH fluorescence using a conventional inverted microscope with a 20× lens. Image quality is poor (not shown). However, this poor resolution brings a major advantage that fluorescence signals are nearly impervious to small changes of "focal plane" and muscle contraction. *The downward spikes in the record reflect closure of a shutter to check baseline light levels.*

The experiment shown in Fig. 29.5 demonstrates the induction of a large endocytic response to LPC (5 μM), whereby equal amounts of C18- and C16-LPC were employed together. A large loss of capacitance and reduction of NCX1PHL signals occur in parallel (>50 %). However, the changes occur over the course of nearly an hour (typical for five recordings), not within seconds as occurs in isolated myocytes. These results suggest therefore that cardiac sarcolemma can indeed turn over substantially, possibly under control of biochemical pathways leading to the generation of lipid mediators such as LPC, but the turnover is taking place over very prolonged periods of time.

29.3 Discussion

We discovered recently that large fractions of the sarcolemma can be internalized rapidly in isolated murine and rat myocytes (Lariccia et al. 2011). Either large Ca^{2+} transients or certain amphipathic compounds, including naturally occurring amphipaths (Fig. 29.1), can serve as triggers. How and if these results are relevant to physiology of the intact heart is still not certain. We have described here some initial efforts to study turnover of the cardiac sarcolemma and NCX1 in intact heart via new approaches that include both electrical and optical methods. It is clear now that myocytes in intact cardiac tissue behave drastically differently from isolated myocyte. Myocyte isolation clearly enables massive endocytic responses that are in some way inhibited from occurring in the intact myocardium, and it remains to be established under what physiological and pathological circumstances these processes may become activated in the intact heart.

The NIC method described here can be developed in a wide variety to ways that may improve its reliability and the range of applications for which it can be used. These include development of a primary current oscillator with monitoring of induced potential oscillations at different distances from the source, inclusion of a "guard" electrode around the area monitored so that the tissue monitored is perturbed more uniformly, and the use of two active electrodes to allow analysis of interacting fields in adjacent tissue areas. Three areas are of most interest to us at this time: (1) It may be possible to develop a contactless NIC method for cardiac muscle strips and tissues. (2) The development of NIC methods for cell cultures, whereby similarly large electrodes brought to a short distance from the surface of a cell culture can allow monitoring of cell culture capacitance. (3) Analysis of NIC signals in hearts of mice developing hypertrophy and eventually cardiac failure will be of much pathological interest, since T-tubule remodeling is increasingly implicated to play an important role in the etiology of heart failure (for refs., see Lariccia et al. 2011). An involvement of MEND-like processes in the pathology of heart failure seems entirely realistic. Both prolonged Ca^{2+} transients and lysolipids can cause MEND, and both are hallmarks of the myocyte signaling environment in cardiac failure. In support of this suggestion, it has been demonstrated here that LPC can indeed cause massive internalization of NCX1 and cardiac sarcolemma over a period of about 1 h.

References

J.M. Best, J.D. Foell, C.R. Buss, B.P. Delisle, R.C. Balijepalli, C.T. January, T.J. Kamp, The small GTPase Rab11b regulates degradation of surface membrane L-type Cav1.2 channels. Am. J. Physiol. Cell Physiol. **300**, C1023–C1033 (2011)

J. Chen, J. Sroubek, Y. Krishnan, Y. Li, J. Bian, T.V. McDonald, PKA phosphorylation of HERG protein regulates the rate of channel synthesis. Am. J. Physiol. Heart Circ. Physiol. **296**, H1244–H1254 (2009)

F.S. Cusdin, J.J. Clare, A.P. Jackson, Trafficking and cellular distribution of voltage-gated sodium channels. Traffic **9**, 17–26 (2008)

G.J. Doherty, H.T. McMahon, Mechanisms of endocytosis. Annu. Rev. Biochem. **78**, 857–902 (2009)

J.G. Donaldson, N. Porat-Shliom, L.A. Cohen, Clathrin-independent endocytosis: a unique platform for cell signaling and PM remodeling. Cell. Signal. **21**, 1–6 (2009)

D.J. Fazakerley, S.P. Lawrence, V.A. Lizunov, S.W. Cushman, G.D. Holman, A common trafficking route for GLUT4 in cardiomyocytes in response to insulin, contraction and energy-status signalling. J. Cell Sci. **122**, 727–734 (2009)

S.M. Ferguson, G. Brasnjo, M. Hayashi, M. Wolfel, C. Collesi, S. Giovedi, A. Raimondi, L.W. Gong, P. Ariel, S. Paradise, E. O'Toole, R. Flavell, O. Cremona, G. Miesenbock, T.A. Ryan, P. De Camilli, A selective activity-dependent requirement for dynamin 1 in synaptic vesicle endocytosis. Science **316**, 570–574 (2007)

M. Fine, M.C. Llaguno, V. Lariccia, M.J. Lin, A. Yaradanakul, D.W. Hilgemann, Massive endocytosis driven by lipidic forces originating in the outer plasmalemmal monolayer: a new approach to membrane recycling and lipid domains. J. Gen. Physiol. **137**, 137–154 (2011)

M.I. Geli, H. Riezman, Endocytic internalization in yeast and animal cells: similar and different. J. Cell Sci. **111**(Pt 8), 1031–1037 (1998)

K. Hayashi, W. Shuai, Y. Sakamoto, H. Higashida, M. Yamagishi, S. Kupershmidt, Trafficking-competent KCNQ1 variably influences the function of HERG long QT alleles. Heart Rhythm **7**, 973–980 (2010)

D.W. Hilgemann, M. Fine, Mechanistic analysis of massive endocytosis in relation to functionally defined surface membrane domains. J. Gen. Physiol. **137**, 155–172 (2011)

D.W. Hilgemann, G.A. Langer, Transsarcolemmal calcium movements in arterially perfused rabbit right ventricle measured with extracellular calcium-sensitive dyes. Circ. Res. **54**, 461–467 (1984)

D.W. Hilgemann, A. Yaradanakul, Y. Wang, D. Fuster, Molecular control of cardiac sodium homeostasis in health and disease. J. Cardiovasc. Electrophysiol. **17**(Suppl 1), S47–S56 (2006)

T.J. Hund, P.J. Mohler, Ankyrin-based targeting pathway regulates human sinoatrial node automaticity. Channels (Austin) **2**, 404–406 (2008)

A.I. Ivanov, Pharmacological inhibition of endocytic pathways: is it specific enough to be useful? Methods Mol. Biol. **440**, 15–33 (2008)

S.H. Kim, T.A. Ryan, Synaptic vesicle recycling at CNS synapses without AP-2. J. Neurosci. **29**, 3865–3874 (2009)

V. Lariccia, M. Fine, S. Magi, M.J. Lin, A. Yaradanakul, M.C. Llaguno, D.W. Hilgemann, Massive calcium-activated endocytosis without involvement of classical endocytic proteins. J. Gen. Physiol. **137**, 111–132 (2011)

S.P. Lawrence, G.D. Holman, F. Koumanor, Translocation of the Na/H exchanger (NHE1) in cardiomyocyte responses to insulin and energy status signaling. Biochem. J. **432**, 515–523 (2010)

S. Mayor, R.E. Pagano, Pathways of clathrin-independent endocytosis. Nat. Rev. Mol. Cell Biol. **8**, 603–612 (2007)

G. Miesenbock, D.A. De Angelis, J.E. Rothman, Visualizing secretion and synaptic transmission with pH-sensitive green fluorescent proteins. Nature **394**, 192–195 (1998)

O. Pongs, Ins and outs of cardiac voltage-gated potassium channels. Curr. Opin. Pharmacol. **9**, 311–315 (2009)

J.E. Saffitz, J.G. Laing, K.A. Yamada, Connexin expression and turnover: implications for cardiac excitability. Circ. Res. **86**, 723–728 (2000)

K. Sato, G.G. Ernstrom, S. Watanabe, R.M. Weimer, C.H. Chen, M. Sato, A. Siddiqui, E.M. Jorgensen, B.D. Grant, Differential requirements for clathrin in receptor-mediated endocytosis and maintenance of synaptic vesicle pools. Proc. Natl. Acad. Sci. U. S. A. **106**, 1139–1144 (2009)

A. Zorzano, L. Sevilla, M. Camps, C. Becker, J. Meyer, H. Kammermeier, P. Munoz, A. Guma, X. Testar, M. Palacin, J. Blasi, Y. Fischer, Regulation of glucose transport, and glucose transporters expression and trafficking in the heart: studies in cardiac myocytes. Am. J. Cardiol. **80**, 65A–76A (1997)

Part X

NCX Role in Hypertension, Heart Failure, Ischemia-Reperfusion, Arrhythmias and Diabetes

Cardiac Sodium-Calcium Exchange and Efficient Excitation-Contraction Coupling: Implications for Heart Disease

30

Joshua I. Goldhaber and Kenneth D. Philipson

Abstract

Cardiovascular disease is a leading cause of death worldwide, with ischemic heart disease alone accounting for >12% of all deaths, more than HIV/AIDS, tuberculosis, lung, and breast cancer combined. Heart disease has been the leading cause of death in the United States for the past 85 years and is a major cause of disability and health-care expenditures. The cardiac conditions most likely to result in death include heart failure and arrhythmias, both a consequence of ischemic coronary disease and myocardial infarction, though chronic hypertension and valvular diseases are also important causes of heart failure. Sodium-calcium exchange (NCX) is the dominant calcium (Ca^{2+}) efflux mechanism in cardiac cells. Using ventricular-specific NCX knockout mice, we have found that NCX is also an essential regulator of cardiac contractility independent of sarcoplasmic reticulum Ca^{2+} load. During the upstroke of the action potential, sodium (Na^+) ions enter the diadic cleft space between the sarcolemma and the sarcoplasmic reticulum. The rise in cleft Na^+, in conjunction with depolarization, causes NCX to transiently reverse. Ca^{2+} entry by this mechanism then "primes" the diadic cleft so that subsequent Ca^{2+} entry through Ca^{2+} channels can more efficiently trigger Ca^{2+} release from the sarcoplasmic reticulum. In NCX knockout mice, this mechanism is inoperative (Na^+ current has no effect on the Ca^{2+} transient), and excitation-contraction coupling relies upon the elevated diadic cleft Ca^{2+} that arises from the slow extrusion of cytoplasmic Ca^{2+} by the ATP-dependent sarcolemmal Ca^{2+} pump. Thus, our data support the conclusion that NCX is an important regulator of cardiac contractility. These findings suggest that manipulation of NCX may be beneficial in the treatment of heart failure.

J.I. Goldhaber (✉) • K.D. Philipson
Cedars-Sinai Heart Institute, 8700 Beverly Boulevard,
Davis Research Building 2065, Los Angeles,
CA 90048, USA
e-mail: goldhaberj@cshs.org

L. Annunziato (ed.), *Sodium Calcium Exchange: A Growing Spectrum of Pathophysiological Implications*, Advances in Experimental Medicine and Biology 961, DOI 10.1007/978-1-4614-4756-6_30,

Keywords
Sodium-calcium exchange • Excitation-contraction coupling • Heart failure • Calcium channels • Sodium current • Contractility

30.1 Introduction

Heart disease, including heart failure (HF), myocardial infarction (MI), and their complications, is a global problem accounting for more than 12% of all deaths worldwide in 2011 according to the World Health Organization (2011). In the United States, 5.8 million people carry a diagnosis of heart failure; 1.1 million are hospitalized with HF each year as a primary diagnosis, and 3.39 million patients visit an outpatient clinic annually because of HF (Roger et al. 2011). The CDC estimates the US cost of HF in 2010 to be $39.2 billion (2011). This is an enormous financial expenditure as well as disease burden. There are also one million myocardial infarctions annually in the United States. Fifty percent of patients with MI will die of arrhythmia before hospitalization. Another 5% develop cardiogenic shock, and half of these patients die as well (Roger et al. 2011). Thus, the severity and the prevalence of heart disease in the world are astounding. In this chapter, we will briefly review the pathogenesis of HF and then discuss how new insights into the role of NCX in excitation-contraction (EC) coupling may offer opportunities to improve the treatment of this debilitating disease.

30.2 Pathogenesis of Heart Failure

The pathogenesis of HF has been an intense area of investigation. Although several lines of evidence suggest that NCX activity is increased in HF and contributes to contractile dysfunction by depleting sarcoplasmic reticulum (SR) Ca^{2+} content (Studer et al. 1994; Flesch et al. 1996; Hobai and O'Rourke 2000; Hasenfuss and Pieske 2002; Armoundas et al. 2007), recent clinical advances have ignored the exchanger and instead target abnormal activation of neuroendocrine signals. Neuroendocrine activation has multiple deleterious effects but with respect to EC coupling, it is thought to lead to hyperphosphorylation of ryanodine receptors (RyRs) by kinases (PKA and/or CaMKII), leading to SR Ca^{2+} leak (Marks 2000). Neuroendocrine activation also promotes beta-adrenergic receptor downregulation and associated abnormal G protein signaling, which likewise blunts the response of LCCs and SR Ca^{2+} loading to adrenergic signals (Koch et al. 2000). Other factors contribute to contractile dysfunction: these include defective SR Ca^{2+}-ATPase activity, leading to reduced SR Ca^{2+} content (Schmidt et al. 1998); myofilament dysfunction, which decreases the contractile response to released Ca^{2+} (Hajjar and Gwathmey 1990); mitochondrial dysfunction, which leads to energy starvation; and fibrosis, which replaces myocytes with non-contracting cells (Ingwall and Weiss 2004).

30.3 Manipulating Contractility in Heart Failure

Although targeting the neuroendocrine system through the use of beta blockers, angiotensin-converting enzyme inhibitors, and aldosterone antagonists has been a relatively effective strategy to manage HF (Fonarow et al. 2011), patients still complain of fatigue, shortness of breath, and limited exercise tolerance. Ultimately, their disease progresses and hospitalizations for decompensation become more frequent as resting blood flow to vital organs decreases. Thus, another approach is necessary. One such approach is to employ inotropic agents to directly stimulate contractile function. These agents most commonly operate by further stimulating beta-adrenergic receptors, which in turn trigger a signaling cascade that results in (1) increased Ca^{2+} influx via Ca^{2+} current (I_{Ca}), (2) increased SR Ca^{2+} uptake rate (via phospholamban (PLB) phosphorylation), and (3) increased myofilament Ca^{2+} responsiveness. However, several seminal studies have established that inotropes increase mortality and morbidity in the HF population (Felker et al. 2003) despite improved pump function. For example, the ADHERE registry of >10,000 patients showed significantly higher in-hospital mortality (adjusted by propensity score) for HF patients treated with the beta agonist dobutamine or the phosphodiesterase inhibitor milrinone instead of vasodilators (Abraham et al. 2005). The ESCAPE trial of severe HF patients undergoing evaluation for heart transplantation found that those who were "electively" treated with inotropes had a 1.8-fold increase in 6-month mortality (Elkayam et al. 2007). Thus inotropes, while sometimes unavoidable in the short run, are dangerous in the long run. The problem appears to be the very thing that improves contractility: increased cellular Ca^{2+} load leading to SR Ca^{2+} overload, which has a variety of deleterious consequences including arrhythmia and cell death.

30.4 A Modern View of Excitation-Contraction Coupling in Health and Disease

Recent developments in understanding of the role of NCX in EC coupling may help reveal new and safer strategies to improve contractility than the current generation of inotropes. We have long known that a Ca^{2+}-induced Ca^{2+} release (CICR) mechanism controls EC coupling in cardiac cells (Fabiato 1983). Ca^{2+} entering through sarcolemmal L-type Ca^{2+} channels (LCCs) triggers release of Ca^{2+} by RyRs on the SR surface (London and Krueger 1986). This reaction occurs throughout the ventricular cell within functional units known as couplons (Stern et al. 1997; Franzini-Armstrong et al. 1999). These units, which are located primarily along transverse (t) tubules, permit sarcolemmal LCCs to admit Ca^{2+} into a restricted junctional region (the diadic cleft), leading to a significant rise in Ca^{2+} concentration. This Ca^{2+} gates a cluster of RyRs on the apposing membrane of the junctional SR, allowing Ca^{2+} release from the SR to generate a Ca^{2+} spark (Cheng et al. 1993). The spatial separation between couplons is sufficient to permit their local control (Stern 1992), which explains the voltage dependence of Ca^{2+} transients. However, we now know that action potentials in healthy cells trigger each couplon simultaneously in a coordinated and synchronous manner (Inoue and Bridge 2003). This synchronous activity appears to be critical for optimum contractility.

Failing cardiac muscle is characterized by the loss of synchronized Ca^{2+} release upon depolarization, as exemplified by postinfarct remodeling in the rabbit (Litwin et al. 2000). We have found similar loss of synchronization of Ca^{2+} release in rabbit cells exposed to metabolic inhibitors (Fig. 30.1), an experimental condition that recapitulates the metabolic stress of HF (Chantawansri et al. 2008). The loss of synchronization can in large part be explained by changes in the single-channel characteristics of LCCs. For example, the Ca^{2+} spark probability and distribution of spark latencies are predicted by LCC latency, open time, and opening probability (P_o). Primary changes in RyR behavior (Meissner 1994) and cellular structure (Gomez et al. 2001) may also contribute to loss of synchronization.

30.5 Excitation-Contraction Coupling in NCX KO Mice

We wondered whether NCX might alter EC coupling independent of changes in SR Ca^{2+} stores and Ca^{2+} channel activity. To explore this possibility, we took advantage of our ventricular-specific NCX knockout mice. These mice live into adulthood with normal cardiac function. Isolated cells from these mice exhibit normal resting Ca^{2+}, preserved SR Ca^{2+} stores, and normal Ca^{2+} transients in response to electrical stimulation (Henderson et al. 2004). Because NCX is absent and no other Ca^{2+} efflux mechanism increases to compensate, Ca^{2+} removal in response to caffeine-induced SR Ca^{2+} release is dramatically reduced. A major adaptation in this model appears to be a reduction in Ca^{2+} influx through LCCs and an associated increase in EC coupling gain (Pott et al. 2005). The reduced Ca^{2+} current

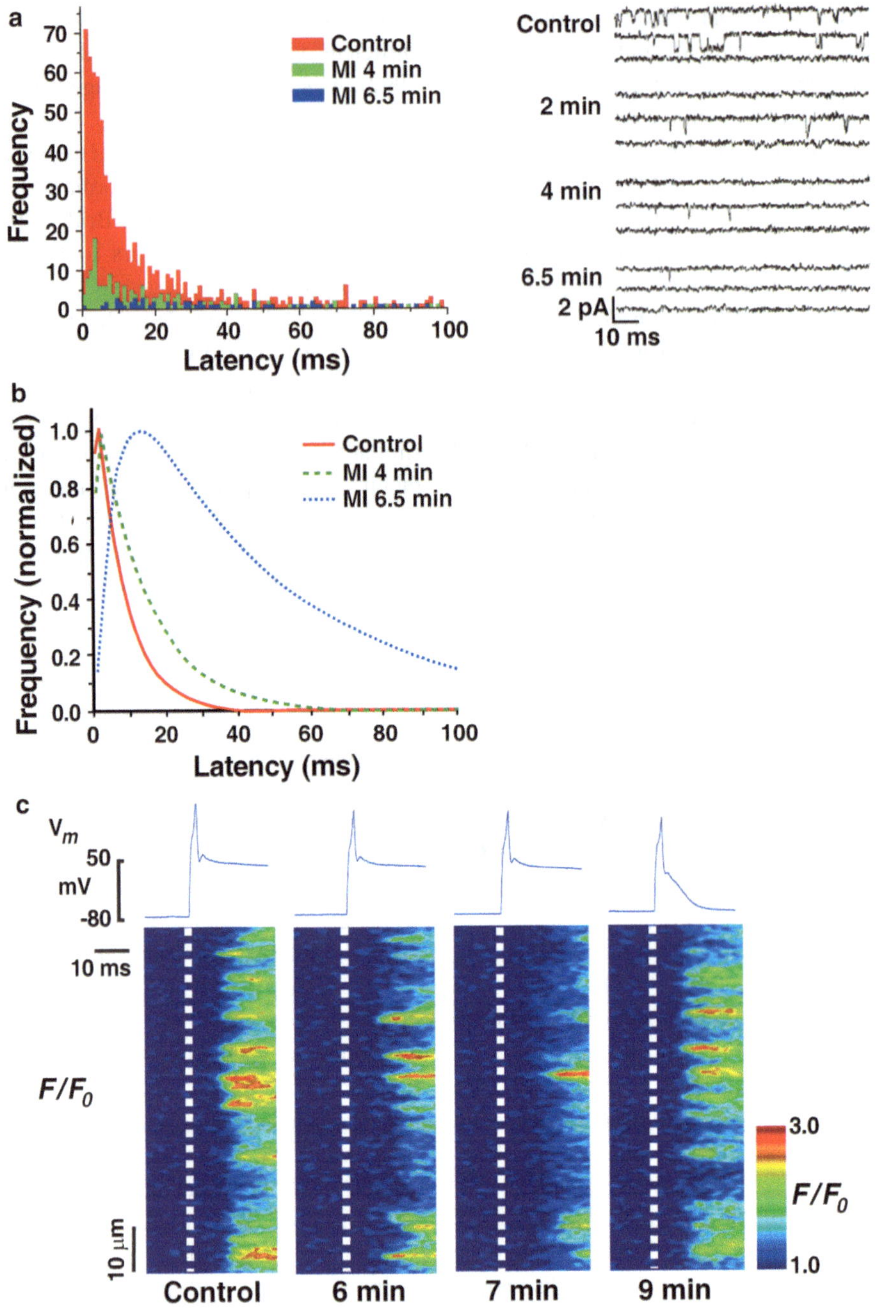

Fig. 30.1 Effect of metabolic inhibition on LCCs and triggered Ca^{2+} sparks in isolated adult rabbit ventricular myocytes. (**a**) Frequency distribution of single L-type Ca^{2+} channel latency under control conditions (*red*), and after 4 (*green*) and 6.5 (*blue*) min of oxidative and glycolytic metabolic inhibition (*MI*) with FCCP (50 nM) and 2-deoxyglucose (10 mM). Normalized frequency distribution is shown in *Panel b*. Note the increase in the proportion of delayed latencies during MI. (**c**) Shows high-speed line-scan images of Ca^{2+} sparks in response to depolarization by action potentials (shown above each image) under control conditions and during MI. Spark latency increases during MI while spark probability declines, in parallel with the increased latency and reduced open probability of LCCs as shown in (**a**) and in the inset of original single Ca^{2+} channel records. In this example, spark probability and latency returned to normal as the action potential duration became very short, the result of increased Ca^{2+} influx caused by rapid early repolarization. The ability to trigger Ca^{2+} release with a short action potential indicates that RyRs can still respond to Ca^{2+} even during advanced MI and that changes in Ca^{2+} channel behavior are the primary reason for reduced spark probability and increased latency (From Chantawansri et al. 2008 with permission)

(I_{Ca}) is caused by an increase in subsarcolemmal/diadic cleft Ca^{2+} concentration and the resulting Ca^{2+}-dependent inactivation (Pott et al. 2007a). Action potential shortening caused by upregulation of the transient outward current (I_{TO}) also limits Ca^{2+} entry during depolarization (Pott et al. 2007b). Resting Ca^{2+} sparks, the elementary events of EC coupling that reflect CICR activity at the single-couplon level, are reduced in frequency compared to wild-type cells. However, the sparks that do occur are larger and last longer (Neco et al. 2010). The frequency reduction is consistent with reduced diastolic triggering of sparks by the smaller KO I_{Ca}, and the difference in spark size is caused by the lack of NCX-mediated Ca^{2+} removal from the diadic cleft in KO cells. Spark activity and size equalize when cells from WT and KO mice are permeabilized to eliminate the influence of NCX, I_{Ca}, and differences in cleft Ca^{2+} (Neco et al. 2010). This indicates that RyR function is not responsible for differences in spark frequency and directly implicates I_{Ca} and NCX as the responsible elements.

30.6 Reverse NCX and SR Ca Release Triggering

How then does NCX affect cleft Ca^{2+} and microscopic EC coupling during depolarization? In the cardiac-specific NCX knockout (KO) mouse, effective EC coupling is dependent upon elevated diadic cleft Ca^{2+} throughout the cardiac cycle. This is made clear by experiments buffering Ca^{2+} in the cytoplasm using EGTA. Under strong Ca^{2+} buffering conditions, KO mice exhibit reduced coupling efficiency (exemplified by decreased spark number and increased spark latency), whereas wild-type (WT) mice display normal coupling (Fig. 30.2, from Neco et al. 2010). Keep in mind that under these highly buffered conditions, we expect I_{Ca} to be as large in the KO as it is in the WT (Pott et al. 2007b). The best explanation for preserved EC coupling in buffered WT cells is that NCX helps maintain coupling during depolarization. We have hypothesized that reverse NCX primes the diadic cleft with a subthreshold amount of Ca^{2+} during the initial upstroke of the action potential in response to Na^+ entry via I_{Na} into the subsarcolemmal space. Only a small amount of additional Ca^{2+} brought in by LCCs is needed to trigger release in all couplons. A similar argument was proposed by LeBlanc and Hume in 1990 when they showed that blocking I_{Na} reduced Ca^{2+} release (LeBlanc and Hume 1990). However, these authors argued that reverse NCX was a *direct* trigger. Although subsequent reports from several other groups supported LeBlanc and Hume's findings (Haworth and Goknur 1991; Nuss and Houser 1992; Kohmoto et al. 1994; Wasserstrom and Vites 1996; Lines et al. 2006), others refuted NCX ability to trigger SR Ca^{2+} release in any fashion that was remotely close to what could be triggered by I_{Ca} (Bers et al. 1990; Sham et al. 1992; Lipp and Niggli 1994; Lopez-Lopez et al. 1995; Sipido et al. 1995, 1997). Furthermore, many of the experiments supportive of Leblanc and Hume were criticized on technical grounds: poor voltage control, inadvertent activation or inactivation of Ca^{2+} channels by voltage protocols, instability in SR Ca^{2+} content, incomplete blockade of I_{Ca} by voltage-dependent blockers, and nonphysiologic intracellular Na^+ concentrations.

To address these criticisms, we once again took advantage of the NCX KO mouse and also carefully constructed voltage clamp protocols and waveforms in the shape of an action potential so as to minimize voltage errors and inactivation of I_{Ca} that might confound interpretation. In order to trigger Ca^{2+} release in the absence of I_{Na}, the action potential clamp was preceded by a linear ramp depolarization from −70 to −40 mV over a period of 1.3 s. This prepulse strategy was designed to inactivate I_{Na} without first generating the large Na^+ influx that typifies square-wave prepulses. It also prevented unwanted activation of LCCs by voltage errors produced by saturating Na^+ currents (I_{Na}) activated during the prepulse. This was verified in control experiments. Thus, we were able to expeditiously eliminate I_{Na} without the use of tetrodotoxin (TTX) and without introducing voltage errors or unplanned changes in Ca^{2+} channel activity. Using this protocol, we found that eliminating I_{Na} selectively decreases (but does not eliminate) Ca^{2+} release in WT but has no effect in NCX KO (Fig. 30.3). The absence of an effect of inactivating I_{Na} in NCX KO confirms that reverse NCX in response to rapid influx of Na^+ via I_{Na}

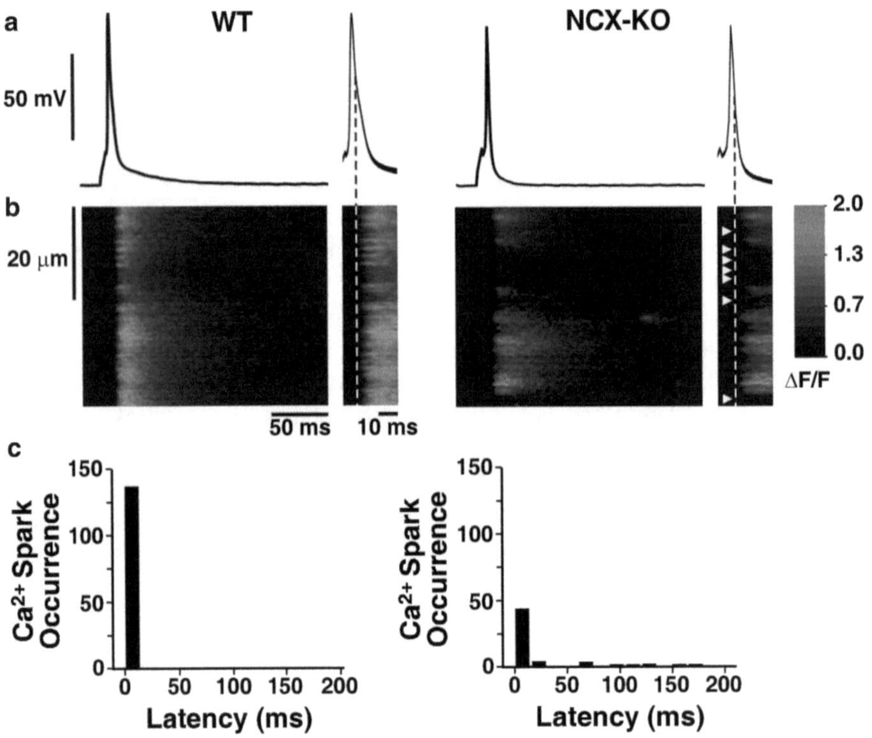

Fig. 30.2 Buffering Ca^{2+} in the diadic cleft reduces spark probability in NCX knockouts, but not in wild type. (**a**) Representative action potentials stimulated by current commands, and (**b**) corresponding high-speed (0.24 ms/line) line-scan images recorded simultaneously in representative WT and NCX KO myocytes. Cells were loaded with 1 mM fluo-3 and 3 mM EGTA via the patch pipette to buffer Ca^{2+} in the diadic cleft. Action potentials and images are also shown on a higher-resolution temporal scale (*scale bar*, 10 ms). *Dashed line* indicates the time when the earliest Ca spark was activated. *Arrowheads* mark the positions where couplons failed to activate. Fluorescence intensities are reported in self-ratioed □F/F magnitude as indicated in the adjoining palette. (**c**) Ca^{2+} spark latency histograms (15-ms bins) constructed from line-scan images recorded in WT (*left*, $n=8$ cells from four mice) and NCX KO (*right*, $n=8$ cells from four mice) myocytes. Note the increased spark latency in the KO compared to WT, a consequence of buffering the diadic cleft with EGTA. These data show that KO mice require elevated cleft Ca for efficient EC coupling, whereas WT mice are able to prime the diadic cleft with Ca^{2+} via reverse NCX (From Neco et al. 2010, with permission).

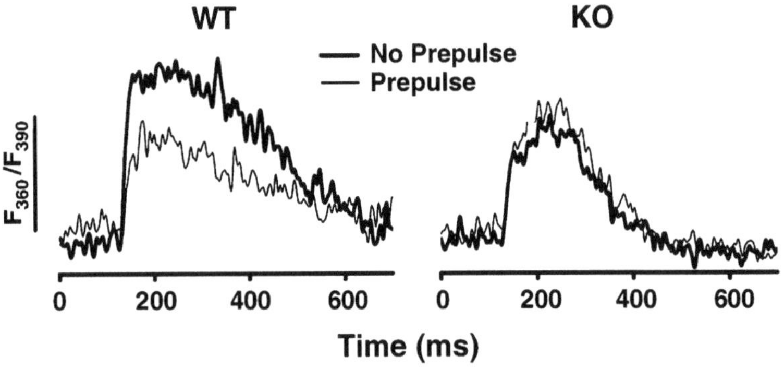

Fig. 30.3 Reverse NCX is an essential component of the Ca^{2+}-induced Ca^{2+} release mechanism of cardiac EC coupling. We used a ramp prepulse to inactivate I_{Na} immediately prior to application of an action potential voltage clamp waveform in WT (*left*) and NCX KO (*right*) myocytes. In this representative example of Ca transients recorded with Fura-2 in patch clamped myocytes, we found that in WT the prepulse markedly reduced the Ca^{2+} transient, whereas in KO there was no effect. This shows that reverse NCX driven by Na^+ entry during the upstroke of the action potential increases coupling fidelity, i.e., the probability of triggering Ca^{2+} release from the SR (From Larbig et al. 2010, with permission)

makes an important contribution to the triggering process. To confirm this finding using a different approach, we applied the Na^+ channel blocker TTX (5 μM) using a rapid solution exchange device 1 s prior to depolarization by the action potential voltage clamp. TTX rapidly and reversibly reduced the Ca^{2+} transient without reducing SR Ca^{2+} load, confirming the effect of Na-induced reverse NCX on CICR.

30.7 Importance of Na Channel Isoforms Concentrated in Transverse Tubules

Ventricular myocytes contain numerous isoforms of Na^+ channels in addition to the cardiac isoform Na_v 1.5. One group of isoforms (Na_V 1.1, 1.2, 1.3, and 1.6), often referred to collectively as "neuronal Na^+ channels," appears to be concentrated in transverse tubules (t-tubules) (Gershome et al. 2011). Blocking these channels in rats apparently has no effect on EC coupling (Brette and Orchard 2006). However, some other groups have suggested that these channels do have an effect on contractility (Maier et al. 2002). We reasoned that since the process of EC coupling in ventricular myocytes is mainly concentrated in couplons located in t-tubules, then selective inhibition of "neuronal" Na^+ channels should be sufficient to eliminate the contribution of reverse NCX to the trigger for SR Ca^{2+} release. We tested this hypothesis in rabbit, a species which is more dependent on Ca^{2+} influx from LCCs for triggering than mouse (i.e., less EC coupling gain). When we exposed rabbit cells to 100-nM TTX, a low concentration that specifically inhibits "neuronal" Na^+ channels (Goldin 2001; Catterall et al. 2005), we found reduced SR Ca^{2+} release similar to the reduction in Ca^{2+} release observed during a slow ramp prepulse and similar to that described above for mouse (Torres et al. 2010).

30.8 Essential Role of NCX in Priming the Diadic Cleft

Our results suggest that NCX plays an essential role in the process of Ca^{2+}-induced Ca^{2+} release, not simply by direct triggering of RyRs (which seems unlikely based on the relative inefficiency of NCX as demonstrated by Sham et al. (1992) and Sipido et al. (1997)), but through the following sequence of events: in response to t-tubular "neuronal" Na^+ channel activation upon depolarization, the rise in junctional Na^+ concentration activates reverse NCX which *primes* the diadic cleft with Ca^{2+}. We know that the relationship between RyR P_o and activating Ca^{2+} is sigmoid (Copello et al. 1997). The NCX-mediated priming of cleft Ca^{2+} moves Ca^{2+} concentration along the flat part of this sigmoid curve without increasing RyR P_o appreciably. However, the Ca^{2+} concentration reaches all the way to the inflection point for the steep portion of the sigmoid curve. We propose that this priming takes place during the 4 ms of the action potential that precedes activation of I_{Ca}. Subsequent Ca^{2+} entry upon activation of I_{Ca} will further raise Ca^{2+} in a concentration range where it is related steeply to RyR P_o, so that the NCX and I_{Ca} effectively sum their activities in a nonlinear fashion (Torres et al. 2010). Without this priming effect, the entry of Ca^{2+} via I_{Ca} may still be sufficient to trigger but with less efficiency than when the system is first primed by NCX. Thus, it seems that NCX is necessary to increase the coupling efficiency (Polakova et al. 2008) of CICR. In NCX KO myocytes, the cleft Ca^{2+} is elevated throughout the cardiac cycle, so further priming by I_{Na} and NCX is not required (Larbig et al. 2010).

30.9 Conclusion

These findings raise the intriguing possibility of manipulating NCX as a therapeutic tool in HF, not simply to alter Ca^{2+} efflux and SR Ca^{2+} load like a cardiac glycoside (e.g., digitalis), but rather as a way to prime the diadic cleft and maximize coupling efficiency. The goal is to provide maximum inotropic support without provoking SR Ca^{2+} overload and the consequent arrhythmias and cellular damage. The increase in Ca^{2+} entry via reverse NCX required to accomplish this increase in coupling efficiency is unknown but should be minimal (Torres et al. 2010). On the other hand, we have shown evidence that ablation of NCX substantially reduces ischemia/reperfusion injury (Imahashi et al. 2005) and may also reduce triggered arrhythmias (Nagy et al. 2004). Thus, we are faced with two opposing strategies for involving NCX in the protection and improvement of cardiac function: enhancing reverse NCX to optimize CICR and blocking NCX during acute ischemia/reperfusion to prevent Ca^{2+} overload. Unfortunately, pharmacological agonists and antagonists of the exchanger lack the specificity for these purposes and will require further development. Hopefully, new work involving structure/function of NCX (John et al. 2011) will soon lead to a new family of pharmacological agents.

In summary, we have found that knocking out NCX in the ventricle reduces LCC activity through Ca^{2+}-dependent inactivation, independent of SR Ca^{2+} load and global cytoplasmic Ca^{2+} levels, which are unchanged. The reduction in LCC activity also reduces the frequency of resting Ca^{2+} sparks. Nevertheless, the size of Ca^{2+} sparks is increased, supporting the concept that NCX resides within or at least very near couplons and thereby locally regulates the removal of diadic cleft Ca^{2+}. Conversely, we have found that effective EC coupling in mouse and rabbit requires activation of TTX-sensitive Na^+ channels in order to promote reverse NCX, which primes the diadic cleft with Ca^{2+} and increase coupling fidelity. We conclude that cardiac NCX is a key transporter responsible for normal contractility in addition to its classic function as a regulator of cellular Ca^{2+} by facilitating Ca^{2+} efflux. NCX is therefore a potentially major therapeutic target with a higher safety margin than current agents.

References

W.T. Abraham, K.F. Adams, G.C. Fonarow, M.R. Costanzo, R.L. Berkowitz, T.H. LeJemtel, M.L. Cheng, J. Wynne, In-hospital mortality in patients with acute decompensated heart failure requiring intravenous vasoactive medications: an analysis from the Acute Decompensated Heart Failure National Registry (ADHERE). J. Am. Coll. Cardiol. **46**, 57–64 (2005)

A.A. Armoundas, J. Rose, R. Aggarwal, B.D. Stuyvers, B. O'Rourke, D.A. Kass, E. Marban, S.R. Shorofsky, G.F. Tomaselli, C. William Balke, Cellular and molecular determinants of altered Ca^{2+} handling in the failing rabbit heart: primary defects in SR Ca^{2+} uptake and release mechanisms. Am. J. Physiol. Heart Circ. Physiol. **292**, H1607–H1618 (2007)

D.M. Bers, W.J. Lederer, J.R. Berlin, Intracellular Ca transients in rat cardiac myocytes: role of Na-Ca exchange in excitation-contraction coupling. Am. J. Physiol. **258**, C944–C954 (1990)

F. Brette, C.H. Orchard, No apparent requirement for neuronal sodium channels in excitation-contraction coupling in rat ventricular myocytes. Circ. Res. **98**, 667–674 (2006)

W.A. Catterall, A.L. Goldin, S.G. Waxman, International Union of Pharmacology. XLVII. Nomenclature and structure-function relationships of voltage-gated sodium channels. Pharmacol. Rev. **57**, 397–409 (2005)

Center for Disease Control and Prevention, Heart failure fact sheet (2011), http://www.cdc.gov/DHDSP/data_statistics/fact_sheets/fs_heart_failure.htm. Accessed 3 Dec 2011

C. Chantawansri, N. Huynh, J. Yamanaka, A. Garfinkel, S.T. Lamp, M. Inoue, J.H. Bridge, J.I. Goldhaber, Effect of metabolic inhibition on couplon behavior in rabbit ventricular myocytes. Biophys. J. **94**, 1656–1666 (2008)

H. Cheng, W.J. Lederer, M.B. Cannell, Calcium sparks - elementary events underlying excitation-contraction coupling in heart muscle. Science **262**, 740–744 (1993)

J.A. Copello, S. Barg, H. Onoue, S. Fleischer, Heterogeneity of Ca^{2+} gating of skeletal muscle and cardiac ryanodine receptors. Biophys. J. **73**, 141–156 (1997)

U. Elkayam, G. Tasissa, C. Binanay, L. Stevenson, M. Gheorghiade, J. Warnica, J. Young, B. Rayburn, J. Rogers, T. Demarco, Use and impact of inotropes and vasodilator therapy in hospitalized patients with severe heart failure. Am. Heart J. **153**, 98–104 (2007)

A. Fabiato, Calcium-induced release of calcium from the cardiac sarcoplasmic reticulum. Am. J. Physiol. **245**, C1–C14 (1983)

G.M. Felker, R.L. Benza, A.B. Chandler, J.D. Leimberger, M.S. Cuffe, R.M. Califf, M. Gheorghiade, C.M. O'Connor, Heart failure etiology and response to milrinone in decompensated heart failure: results from the OPTIME-CHF study. J. Am. Coll. Cardiol. **41**, 997–1003 (2003)

M. Flesch, R.H. Schwinger, F. Schiffer, K. Frank, M. Südkamp, F. Kuhn-Regnier, G. Arnold, M. Böhm, Evidence for functional relevance of an enhanced expression of the Na^+-Ca^{2+} exchanger in failing human myocardium. Circulation **94**, 992–1002 (1996)

G.C. Fonarow, C.W. Yancy, A.F. Hernandez, E.D. Peterson, J.A. Spertus, P.A. Heidenreich, Potential impact of optimal implementation of evidence-based heart failure therapies on mortality. Am. Heart J. **161**, 1024–1030 (2011). e1023

C. Franzini-Armstrong, F. Protasi, V. Ramesh, Shape, size, and distribution of Ca^{2+} release units and couplons in skeletal and cardiac muscles. Biophys. J. **77**, 1528–1539 (1999)

C. Gershome, E. Lin, H. Kashihara, L. Hove-Madsen, G.F. Tibbits, Colocalization of voltage-gated Na^+ channels with the Na^+/Ca^{2+} exchanger in rabbit cardiomyocytes during development. Am. J. Physiol. Heart Circ. Physiol. **300**, H300–H311 (2011)

A.L. Goldin, Resurgence of sodium channel research. Annu. Rev. Physiol. **63**, 871–894 (2001)

A.M. Gomez, S. Guatimosim, K.W. Dilly, G. Vassort, W.J. Lederer, Heart failure after myocardial infarction - Altered excitation-contraction coupling. Circulation **104**, 688–693 (2001)

R.J. Hajjar, J.K. Gwathmey, Direct evidence of changes in myofilament responsiveness to Ca^{2+} during hypoxia and reoxygenation in myocardium. Am. J. Physiol. **259**, H784–H795 (1990)

G. Hasenfuss, B. Pieske, Calcium cycling in congestive heart failure. J. Mol. Cell. Cardiol. **34**, 951–969 (2002)

R.A. Haworth, A.B. Goknur, Control of the Na-Ca exchanger in isolated heart cells. II. Beta-dependent activation in normal cells by intracellular calcium. Circ. Res. **69**, 1514–1524 (1991)

S.A. Henderson, J.I. Goldhaber, J.M. So, T. Han, C. Motter, A. Ngo, C. Chantawansri, M.R. Ritter, M. Friedlander, D.A. Nicoll, J.S. Frank, M.C. Jordan, K.P. Roos, R.S. Ross, K.D. Philipson, Functional adult myocardium in the absence of Na^+-Ca^{2+} exchange: cardiac-specific knockout of NCX1. Circ. Res. **95**, 604–611 (2004)

I.A. Hobai, B. O'Rourke, Enhanced Ca^{2+}-activated Na^+-Ca^{2+} exchange activity in canine pacing-induced heart failure. Circ. Res. **87**, 690–698 (2000)

K. Imahashi, C. Pott, J.I. Goldhaber, C. Steenbergen, K.D. Philipson, E. Murphy, Cardiac-specific ablation of the Na^+/Ca^{2+} exchanger confers protection against ischemia/reperfusion injury. Circ. Res. **97**, 916–921 (2005)

J.S. Ingwall, R.G. Weiss, Is the failing heart energy starved?: on using chemical energy to support cardiac function. Circ. Res. **95**, 135–145 (2004)

M. Inoue, J.H. Bridge, Ca^{2+} sparks in rabbit ventricular myocytes evoked by action potentials: involvement of clusters of L-type Ca^{2+} channels. Circ. Res. **92**, 532–538 (2003)

S.A. John, B. Ribalet, J.N. Weiss, K.D. Philipson, M. Ottolia, Ca^{2+}-dependent structural rearrangements within Na^+-Ca^{2+} exchanger dimers. Proc. Natl. Acad. Sci. U. S. A. **108**, 1699–1704 (2011)

W.J. Koch, R.J. Lefkowitz, H.A. Rockman, Functional consequences of altering myocardial adrenergic receptor signaling. Annu. Rev. Physiol. **62**, 237–260 (2000)

O. Kohmoto, A.J. Levi, J.H.B. Bridge, Relation between reverse sodium-calcium exchange and sarcoplasmic reticulum calcium release in guinea pig ventricular cells. Circ. Res. **74**, 550–554 (1994)

R. Larbig, N. Torres, J.H. Bridge, J.I. Goldhaber, K.D. Philipson, Activation of reverse Na^+-Ca^{2+} exchange by the Na^+ current augments the cardiac Ca^{2+} transient: evidence from NCX knockout mice. J. Physiol. **588**, 3267–3276 (2010)

N. Leblanc, J.R. Hume, Sodium current-induced release of calcium from cardiac sarcoplasmic reticulum. Science **248**, 372–376 (1990)

G.T. Lines, J.B. Sande, W.E. Louch, H.K. Mørk, P. Grøttum, O.M. Sejersted, Contribution of the Na^+/Ca^{2+} Exchanger to Rapid Ca^{2+} Release in Cardiomyocytes. Biophys. J. **91**, 779–792 (2006)

P. Lipp, E. Niggli, Sodium current-induced calcium signals in isolated guinea-pig ventricular myocytes. J. Physiol. **474**, 439–446 (1994)

S.E. Litwin, D. Zhang, J.H. Bridge, Dyssynchronous Ca^{2+} sparks in myocytes from infarcted hearts. Circ. Res. **87**, 1040–1047 (2000)

B. London, J.W. Krueger, Contraction in voltage-clamped, internally perfused single heart cells. J. Gen. Physiol. **88**, 475–505 (1986)

J.R. Lopez-Lopez, P.S. Shacklock, C.W. Balke, W.G. Wier, Local calcium transients triggered by single L-type calcium channel currents in cardiac cells. Science **268**, 1042–1045 (1995)

S.K. Maier, R.E. Westenbroek, K.A. Schenkman, E.O. Feigl, T. Scheuer, W.A. Catterall, An unexpected role for brain-type sodium channels in coupling of cell surface depolarization to contraction in the heart. Proc. Natl. Acad. Sci. U. S. A. **99**, 4073–4078 (2002)

A.R. Marks, Cardiac intracellular calcium release channels: role in heart failure. Circ. Res. **87**, 8–11 (2000)

G. Meissner, Ryanodine receptor/Ca^{2+} release channels and their regulation by endogenous effectors. Annu. Rev. Physiol. **56**, 485–508 (1994)

Z.A. Nagy, L. Virag, A. Toth, P. Biliczki, K. Acsai, T. Banyasz, P. Nanasi, J.G. Papp, A. Varro, Selective inhibition of sodium-calcium exchanger by SEA-0400 decreases early and delayed after depolarization in canine heart. Br. J. Pharmacol. **143**, 827–831 (2004)

P. Neco, B. Rose, N. Huynh, R. Zhang, J.H. Bridge, K.D. Philipson, J.I. Goldhaber, Sodium-calcium exchange is essential for effective triggering of calcium release in mouse heart. Biophys. J. **99**, 755–764 (2010)

H.B. Nuss, S.R. Houser, Sodium-calcium exchange-mediated contractions in feline ventricular myocytes. Am. J. Physiol. **263**, H1161–H1169 (1992)

E. Polakova, A. Zahradnikova Jr., J. Pavelkova, I. Zahradnik, A. Zahradnikova, Local calcium release activation by DHPR calcium channel openings in rat cardiac myocytes. J. Physiol. (London) **586**, 3839–3854 (2008)

C. Pott, K.D. Philipson, J.I. Goldhaber, Excitation-contraction coupling in Na^+-Ca^{2+} exchanger knockout mice: reduced transsarcolemmal Ca^{2+} flux. Circ. Res. **97**, 1288–1295 (2005)

C. Pott, M. Yip, J.I. Goldhaber, K.D. Philipson, Regulation of cardiac L-type Ca^{2+} current in Na^+-Ca^{2+} exchanger knockout mice: functional coupling of the Ca^{2+} channel and the Na^+-Ca^{2+} exchanger. Biophys. J. **92**, 1431–1437 (2007a)

C. Pott, X. Ren, D.X. Tran, M.J. Yang, S. Henderson, M.C. Jordan, K.P. Roos, A. Garfinkel, K.D. Philipson, J.I. Goldhaber, Mechanism of shortened action potential duration in Na^+-Ca^{2+} exchanger knockout mice. Am. J. Physiol. Cell Physiol. **292**, C968–C973 (2007b)

V.L. Roger, A.S. Go, D.M. Lloyd-Jones, R.J. Adams, J.D. Berry, T.M. Brown, M.R. Carnethon, S. Dai, G. de Simone, E.S. Ford, C.S. Fox, H.J. Fullerton, C.

Gillespie, K.J. Greenlund, S.M. Hailpern, J.A. Heit, P.M. Ho, V.J. Howard, B.M. Kissela, S.J. Kittner, D.T. Lackland, J.H. Lichtman, L.D. Lisabeth, D.M. Makuc, G.M. Marcus, A. Marelli, D.B. Matchar, M.M. McDermott, J.B. Meigs, C.S. Moy, D. Mozaffarian, M.E. Mussolino, G. Nichol, N.P. Paynter, W.D. Rosamond, P.D. Sorlie, R.S. Stafford, T.N. Turan, M.B. Turner, N.D. Wong, J. Wylie-Rosett, Heart disease and stroke statistics–2011 update: a report from the American Heart Association. Circulation **124**, e18–e209 (2011)

U. Schmidt, R.J. Hajjar, P.A. Helm, C.S. Kim, A.A. Doye, J.K. Gwathmey, Contribution of abnormal sarcoplasmic reticulum ATPase activity to systolic and diastolic dysfunction in human heart failure. J. Mol. Cell. Cardiol. **30**, 1929–1937 (1998)

J.S. Sham, L. Cleemann, M. Morad, Gating of the cardiac Ca^{2+} release channel: the role of Na^+ current and Na^+-Ca^{2+} exchange. Science **255**, 850–853 (1992)

K.R. Sipido, E. Carmeliet, A. Pappano, Na^+ current and Ca^{2+} release from the sarcoplasmic reticulum during action potentials in guinea-pig ventricular myocytes. J. Physiol. (London) **489**, 1–17 (1995)

K.R. Sipido, M. Maes, F. Van de Werf, Low efficiency of Ca^{2+} entry through the Na^+-Ca^{2+} exchanger as trigger for Ca^{2+} release from the sarcoplasmic reticulum. A comparison between L-type Ca^{2+} current and reverse-mode Na^+-Ca^{2+} exchange. Circ. Res. **81**, 1034–1044 (1997)

M.D. Stern, Theory of excitation-contraction coupling in cardiac muscle. Biophys. J. **63**, 497–517 (1992)

M.D. Stern, G. Pizarro, E. Rios, Local control model of excitation-contraction coupling in skeletal muscle. J. Gen. Physiol. **110**, 415–440 (1997)

R. Studer, H. Reinecke, J. Bilger, T. Eschenhagen, M. Bohm, G. Hasenfuss, H. Just, J. Holtz, H. Drexler, Gene expression of the cardiac Na^+-Ca^{2+} exchanger in end-stage human heart failure. Circ. Res. **75**, 443–453 (1994)

N.S. Torres, R. Larbig, A.N. Rock, J.I. Goldhaber, J.H. Bridge, Na^+ currents are required for efficient excitation-contraction coupling in rabbit ventricular myocytes: a possible contribution of neuronal Na^+ channel to triggering Ca^{2+} release from the sarcoplasmic reticulum. J. Physiol. **588**, 4249–4260 (2010)

J.A. Wasserstrom, A.M. Vites, The role of Na^+-Ca^{2+} exchange in activation of excitation-contraction coupling in rat ventricular myocytes. J. Physiol. **493**, 529–542 (1996)

World Health Organization, The top 10 causes of death (2011), http://www.who.int/mediacentre/factsheets/fs310/en/. Accessed 3 Dec 2011

Cross Talk Between Plasma Membrane Na^+/Ca^{2+} Exchanger-1 and TRPC/Orai-Containing Channels: Key Players in Arterial Hypertension

31

Maria V. Pulina, A. Zulian, Sergey G. Baryshnikov, Cristina I. Linde, Eiji Karashima, John M. Hamlyn, Patrizia Ferrari, Mordecai P. Blaustein, and Vera A. Golovina

Abstract

Arterial smooth muscle (ASM) Na^+/Ca^{2+} exchanger type 1 (NCX1) and TRPC/Orai-containing receptor/store-operated cation channels (ROC/SOC) are clustered with α2 Na^+ pumps in plasma membrane microdomains adjacent to the underlying junctional sarcoplasmic reticulum. This arrangement enables these transport proteins to function as integrated units to help regulate local Na^+ metabolism, Ca^{2+} signaling, and arterial tone. They thus influence vascular resistance and blood pressure (BP). For instance, upregulation of NCX1 and TRPC6 has been implicated in the pathogenesis of high BP in several models of essential hypertension. The models include ouabain-induced hypertensive rats, Milan hypertensive rats, and Dahl salt-sensitive hypertensive rats, all of which exhibit elevated plasma ouabain levels. We suggest that these molecular mechanisms are key contributors to the increased vascular resistance ("whole body autoregulation") that elevates BP in essential hypertension. Enhanced expression and function of ASM NCX1 and TRPC/Orai1-containing channels in hypertension implies that these proteins are potential targets for pharmacological intervention.

M.V. Pulina • A. Zulian • S.G. Baryshnikov • C.I. Linde • E. Karashima • J.M. Hamlyn • M.P. Blaustein • V.A. Golovina (✉)
Department of Physiology, University of Maryland School of Medicine, 685 W Baltimore St. HSF 1, room 565, Baltimore, MD 21201, USA
e-mail: vgolovin@umaryland.edu

P. Ferrari
Prassis-sigma tau Research Institute, Settimo Milanese, Milan, Italy

L. Annunziato (ed.), *Sodium Calcium Exchange: A Growing Spectrum of Pathophysiological Implications*, Advances in Experimental Medicine and Biology 961, DOI 10.1007/978-1-4614-4756-6_31,

Keywords
Arterial smooth muscle cells • Ca^{2+} signaling • Store-operated channels • Receptor-operated channels • TRPC6 • Ouabain-induced hypertensive rats • Milan hypertensive strain rats • Dahl salt-sensitive hypertensive rats

31.1 Regulation of Ca^{2+} Signaling in Vascular Smooth Muscle: Important Role of Na^+/Ca^{2+} Exchanger and TRPC/Orai-Containing Channels

The contractility of vascular smooth muscle is reliant upon increases in the intracellular Ca^{2+} concentration (Lagaud et al. 1999; Zou et al. 2000; McDaniel et al. 2001). Intracellular [Ca^{2+}] is regulated by various Ca^{2+} entry, exit, and storage mechanisms in a variety of cell types including vascular myocytes (Sanders 2001). The role of voltage-gated Ca^{2+} channels (VGCCs) in Ca^{2+} entry in vascular smooth muscle cells (SMCs) has long been recognized (Nelson et al. 1990). Accumulating evidence indicates, however, that nonselective cation channels, such as store-operated, receptor-operated, and stretch-activated channels (SOCs, ROCs, and SACs, respectively) also play essential role in regulation of vascular tone by mediating the entry of Ca^{2+} and Na^+ (Welsh et al. 2002; Brayden et al. 2008). Importantly, Ca^{2+} can also either exit or enter vascular SMCs through the Na^+/Ca^{2+} exchanger type 1 (NCX1) which is driven by the Na^+ electrochemical gradient across the plasma membrane (PM) under the control of the Na^+ pump (Blaustein and Lederer 1999) and the nonselective cation channels. Na^+ entry through these channels and consequent depolarization should promote Ca^{2+} entry not only via L-type Ca^{2+} channels but also via NCX1 and, thereby, contribute to myogenic tone, which is a key in the setting of vascular resistance in hypertension (Davis and Hill 1999). Indeed, in contrast to the heart, where NCX1 primarily mediates Ca^{2+} extrusion (Dibb et al. 2007), in SMCs, which are partially depolarized in arteries with tone, NCX1 primarily mediates Ca^{2+} entry (Iwamoto et al. 2004; Kashihara et al. 2009). The implication is that multiple mechanisms, including not only VGCCs but also ROCs, SOCs, and NCX1, contribute to the maintenance of Ca^{2+} entry, elevated free cytosolic Ca^{2+} concentration ($[Ca^{2+}]_{cyt}$), and arterial tone (Welsh et al. 2002; Iwamoto et al. 2004; Kashihara et al. 2009); the latter influences total peripheral vascular resistance (TPR) and blood pressure (BP) (Blaustein and Hamlyn 2010). Numerous reports indicate that mammalian SOCs and ROCs are homo- and heterotetramers formed by members of "classical" or "canonical" family of seven proteins (TRPC1–TRPC7) that are homologous to the *Drosophila* transient receptor potential (TRP) channel (Beech et al. 2004). The diversity of heteromeric assemblies of TRPC isoforms may partly contribute to functional heterogeneity in the vasculature. Two recently discovered families of transmembrane proteins, Orai1 [also known as Ca^{2+} release-activated Ca^{2+} (CRAC) channel modulator] and STIM1 (stromal interacting molecule 1), may also contribute to SOC-mediated Ca^{2+} entry (Zhang et al. 2005b; Yeromin et al. 2006; Cahalan 2009). Orai1 may form the Ca^{2+} selectivity filter of the CRAC channel (Yeromin et al. 2006), which may be another type of SOC. The role of Orai1 in store-operated Ca^{2+} entry (SOCE) was confirmed in arterial SMCs (Baryshnikov et al. 2009; Potier et al. 2009; Ng et al. 2010). STIM1, the putative Ca^{2+} sensor in the SR, regulates SOCs and CRAC channels (Zhang et al. 2005b; Cahalan 2009). Evidence indicates that depletion of SR Ca^{2+} stores triggers STIM1 to translocate into defined SR-PM "junctional" areas in which coupling to Orai proteins can occur (Wu et al. 2006). STIM1 associates not only with Orai1 but also with TRPC proteins (Li et al. 2008; Liao et al. 2008; Yuan et al. 2009), suggesting that in some tissues, SOCs and CRAC channels are regulated by similar molecular components.

31.2 Functional Interaction of NCX1 with TRPC/Orai Proteins Within the Plasma Membrane-Junctional SR Regions

The structural integrity of PM-junctional SR units and the spatial relationship between SOCs/ROCs and the SR Ca^{2+} stores may play an important role in regulating Ca^{2+} influx and vascular tone. Indeed, in many cell types, the peripheral SR lies within 8–20 nm of the PM, with which it appears to form "junctions" (Somlyo and Franzini-Armstrong 1985). Van Breemen and colleagues (1995) postulated that in smooth muscle, Ca^{2+} entering the cells is directly accumulated by the peripheral SR ("superficial buffer barrier") before it reaches the myofilaments. Thus, diffusion of ions from the PM-SR junctions to "bulk" cytosol must be markedly restricted. This may explain why Ca^{2+} influx fails to induce contraction in some smooth muscle when SR Ca^{2+} stores are depleted by SERCA inhibitors (Flemming et al. 2002). We previously demonstrated that SOCE occurs at PM-SR junctions (Golovina 2005; Lee et al. 2006).

Ca^{2+} homeostasis in vascular SMCs is influenced not only by direct Ca^{2+} entry through TRPC channels but also by Na^+ entry through these nonselective cation channels (Arnon et al. 2000; Eder et al. 2005; Poburko et al. 2007; Fellner and Arendshorst 2008). The Ca^{2+}/Na^+ permeability ratio varies between different members of the TRPC family from 1 to 9 (Nilius et al. 2007). The entering Na^+ apparently accumulating in a restricted space between the PM and adjacent SR then also promotes Ca^{2+} entry through nearby NCX1 (Arnon et al. 2000; Eder et al. 2005; Poburko et al. 2007). We did not observe, however, significant changes in SOCE in human aortic SMCs when experiments were performed at low (5 mM) $[Na^+]_o$ (Baryshnikov et al. 2009). This may be explained by relative differences in the contribution of NCX1 to SOCE in different cell types. Silencer RNA data demonstrated a functional interaction between NCX1 and TRPC6 in rat mesenteric arterial SMCs; knockdown of NCX1 markedly (by ~25 %) reduced TRPC6 expression (Pulina et al. 2010). NCX1 is also functionally associated with Orai1 in human aortic SMCs (Baryshnikov et al. 2009). Knockdown of the Orai1 gene downregulated expression of NCX1 and PM Ca^{2+} pumps and significantly attenuated not only SOCE but also Ca^{2+} extrusion from primary cultured human aortic SMCs (Baryshnikov et al. 2009). These results imply a mutual interaction between the regulation of Ca^{2+} homeostasis by NCX1 and regulation of Ca^{2+} (and Na^+) entry through SOCs and ROCs.

The functional interaction of NCX1 with TRPCs and with Orai1 raises the possibility that they colocalize in small membrane clusters. In fact, immunocytochemical observations indicate that NCX1, TRPCs, and Orai1 are all confined to the PM microdomains that overlie the closely apposed junctional SR (Juhaszova et al. 1996; Golovina 2005; Baryshnikov et al. 2009; Zulian et al. 2010) where STIM1 accumulates after store depletion (Wu et al. 2006). These PM microdomains also contain the high ouabain affinity α2 Na^+ pumps (Juhaszova and Blaustein 1997; Lee et al. 2006; Song et al. 2006) which are involved in control of agonist-mediated vasoconstriction (Shelly et al. 2004), myogenic tone, and BP (Zhang et al. 2005a). High-power images of a portion of hASMC show that the NCX1 labeling pattern (Fig. 31.1a, *inset*) is remarkably similar to the pattern observed with antibodies directed against the Orai1 (Fig. 31.1b, *inset*). Indeed, when Fig. 31.1*ca* (NCX1, green) is overlaid on Fig. 31.1*cb* (Orai1, *red*), extensive overlap of the labels is observed (Fig. 31.1*cc*), as indicated by the large amount of yellow in the image. Notably, reactivity was not detected in the PM in the absence of the primary anti-NCX1 or anti-Orai1 antibodies (not shown). Also, both NCX1 and TRPC6 labels are distributed in a distinct reticular pattern that parallels the organization of the underlying ER-Tracker-stained SR (Zulian et al. 2010). Moreover, co-immunoprecipitation experiments provide evidence for the association of NCX1 and TRPC3 in protein complexes (Rosker et al. 2004). These findings indicate that PM microdomains that include TRPC/Orai1-containing channels, NCX1, and α2 Na^+ pumps function as integrated units that help to

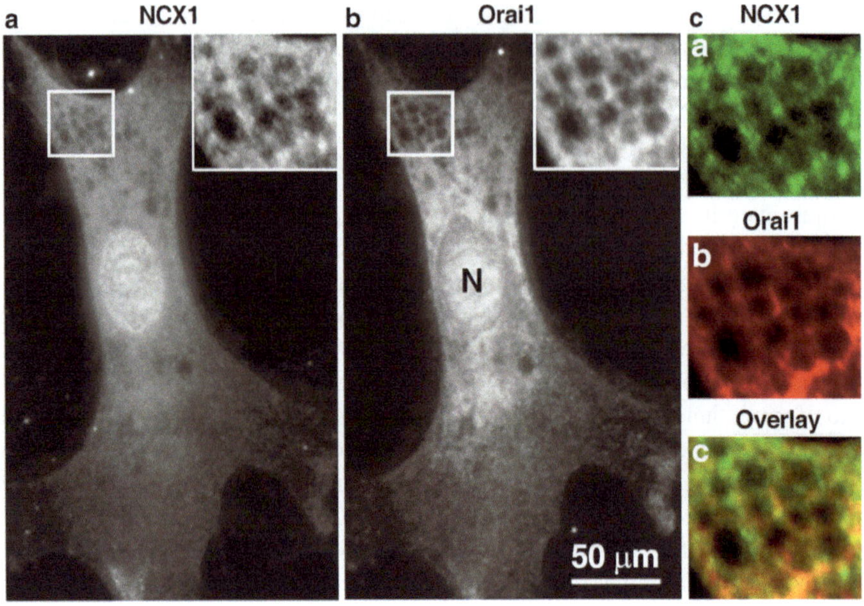

Fig. 31.1 *Immunofluorescent localization of NCX1 and Orai1 in hASMC.* (**a**, **b**) Images of cell double labeled with anti-NCX1 antibody (**a**) and anti-Orai1 antibody (**b**). *Insets* in (**a**) and (**b**) (enlargements of the *boxed areas*) indicate that NCX1 and Orai1 labels show similar distributions. *Panels* **c***a* and **c**b are pseudocolor images (*green* – anti-NCX1; *red* – anti-Orai1) of enlarged boxes from (**a**) and (**b**), respectively. *Panel* **c***c*: colocalization of NCX1 (**c***a*) and Orai1 staining (**c***b*); the *yellow, orange, and yellow/green areas* in the overlay indicate regions of overlap between the two epitopes (Reprinted with permission from Baryshnikov et al. (2009))

regulate Ca^{2+} signaling and vascular tone and are, thus, critical for the determination of vascular resistance in hypertension.

31.3 Dysregulation of NCX1 and TRPC/Orai-Containing Channels Is Implicated in the Pathogenesis of High Blood Pressure

Arterial smooth muscle NCX1 and TRPC channel proteins apparently play a key role in salt-dependent hypertension (Iwamoto et al. 2004; Bae et al. 2007). A common feature of many animal models of salt-dependent hypertension is an elevated plasma level of endogenous ouabain (Hamlyn et al. 1991; Krep et al. 1995; Kaide et al. 1999), an adrenocortical hormone (Boulanger et al. 1993; Shah et al. 1998). Plasma endogenous ouabain also is significantly elevated in ~45 % of patients with essential hypertension (Rossi et al. 1995). Moreover, in rodents, the prolonged administration of low doses of ouabain induces sustained BP elevation, termed "ouabain-induced hypertension" (Manunta et al. 1994; Pulina et al. 2010). Binding of ouabain to its receptors, the high ouabain affinity α2 Na^+ pumps, modulates Ca^{2+} signaling pathways in arterial SMCs and endothelial cells leading to increase in arterial tone, peripheral vascular resistance, and, thereby, elevated BP (Blaustein et al. 2009). In contrast to α2 Na^+ pumps, rodent α1 Na^+ pumps have a very low affinity for ouabain (O'Brien et al. 1994) and may be absent from the PM-SR junctions (Lee et al. 2006). Furthermore, neither ouabain nor adrenocorticotropic hormone (ACTH) can induce hypertension when the α2 ouabain-binding site is mutated to an ouabain-resistant form (Dostanic-Larson et al. 2005; Lorenz et al. 2008). The key molecular links that lead from ouabain's interaction with Na^+ pumps to the sustained elevation of BP are, however, not completely understood.

Table 31.1 NCX1 and SOC/ROC proteins are upregulated in many types of hypertension (HTN)

		Artery smooth muscle							
			SOCs				ROCs		
	Hypertension	NCX1	TRPC1	TRPC5	Orai1	STIM1	TRPC3	TRPC6	References
1	Ouabain[a] (vs. vehicle)	↑	↑	↔	N/R	N/R	↔	↑	Pulina et al. (2010)
2	Milan HTN[a] (vs. Milan NT)	↑	↔	↔	N/R	N/R	↔	↑	Zulian et al. (2010)
3	Dahl/SS high vs. low salt[a,b]	↑	N/R	N/R	N/R	N/R	N/R	↑	Figure 31.2 (This chapter)
4	DOCA-salt[a,b]	N/R	↔	N/R	N/R	N/R	↔	↑	Bae et al. (2007)
5	SHR (vs. WKY)	↑	↑	↑	N/R	N/R	↑	↔	Taniguchi et al. (2004), Liu et al. (2009), Chen et al. (2010)
6	Stroke-prone SHR (vs. WKY)	N/R	N/R	N/R	↑	↑	N/R	N/R	Guachini et al. (2009)
7	Human primary pulmonary HTN	↑	↔	N/R	N/R	N/R	↑	↑	Yu et al. (2004), Zhang et al. (2007a, b)

N/R no reports
[a]Hypertension associated with elevated plasma ouabain levels
[b]Salt-sensitive hypertension models

Our recent studies demonstrate that freshly dissociated mesenteric artery myocytes from ouabain hypertensive rats exhibit Ca^{2+} dysregulation: elevated resting $[Ca^{2+}]_{cyt}$ and augmented phenylephrine-induced (1 μM) Ca^{2+} signals (Pulina et al. 2010). The altered Ca^{2+} homeostasis is associated with upregulated expression of key PM transporters: α2 Na^+ pumps, NCX1, and TRPC1 and TRPC6 (Pulina et al. 2010). Inhibition of the α2 Na^+ pumps would be expected to increase $[Na^+]_{cyt}$ in the local spaces between the PM and junctional SR and reduce the Na^+ electrochemical gradient, thereby driving Ca^{2+} into the cells via adjacent NCX1. Upregulation of NCX1 (by five- to sixfold) in arterial SMCs from ouabain hypertensive rats should tend to accelerate, further, Ca^{2+} entry and promote net Ca^{2+} gain (Pulina et al. 2010). Pharmacological and genetic engineering studies reveal that NCX1 contributes to the ouabain-induced increases in arterial SMC Ca^{2+} signaling, myogenic tone, and BP (Iwamoto et al. 2004; Zhang et al. 2010). The vasoconstrictor action of ouabain (Zhang et al. 2005a), as well as ACTH, deoxycorticosterone acetate (DOCA)-salt, and Dahl-salt hypertension, is blocked by NCX inhibitors or by reduced NCX1 expression (Iwamoto et al. 2004; Zhang et al. 2010).

The arterial myocytes from ouabain hypertensive rats also exhibit increased Ca^{2+} entry through SOCs and ROCs. This effect is a consequence of upregulated expression of TRPC1 and TRPC6 proteins (~threefold and >sixfold, respectively) (Pulina et al. 2010). Importantly, the actions of in vivo ouabain on arterial Ca^{2+} signaling can be mimicked in vitro in primary cultured arterial SMCs. Like chronic in vivo ouabain infusion, prolonged (72 h) treatment of cultured arterial SMCs with nanomolar ouabain upregulates expression of NCX1 and TRPC6 and increases Ca^{2+} entry through SOCs and ROCs (Pulina et al. 2010). Thus, the aforementioned effects of in vivo ouabain administration appear to be triggered by ouabain itself and not by the increase in BP or by other circulating factors.

It is noteworthy that Milan hypertensive, (DOCA)-salt, and Dahl salt-sensitive (Dahl/SS) hypertensive rat models have elevated plasma endogenous ouabain levels (Hamlyn et al. 1991; Leenen et al. 1994; Ferrandi et al. 1997). Therefore, it is not surprising that high blood pressure in these models is also associated with upregulation of NCX1 and TRPC6 (Table 31.1) which might be triggered by the high plasma endogenous ouabain. Figure 31.2a shows that a

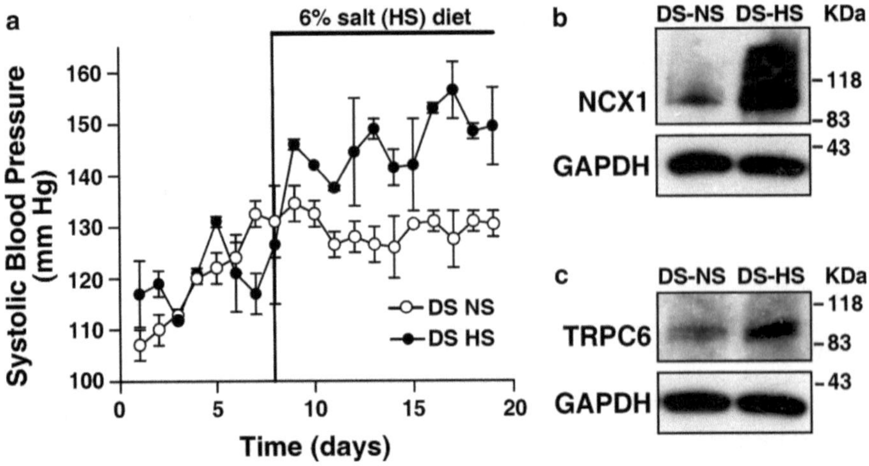

Fig. 31.2 *Upregulated NCX1 and TRPC6 in de-endothelialized mesenteric arteries from hypertensive Dahl salt-sensitive (DS) rats on high-salt (HS) diet.* (**a**) Development of hypertension in DS rats on HS diet. Normotensive male DS rats were split into two groups; one was fed normal salt (NS) and the other, HS (6 %) diet. An HS diet for 12 days markedly increases BP in DS rats (student's *t*-test, $* = P < 0.05$). (**b, c**) Western blot analysis of NCX1 (**b**) and TRPC6 (**c**) expression in mesenteric artery from DS rats on NS and HS diets. Representative Western blots are shown (30 µg/lane); comparable results were obtained in four immunoblots

high-salt diet for 12 days markedly increases blood pressure in Dahl/SS rats. NCX1 and TRPC6 are greatly upregulated in mesenteric arteries from these rats compared to Dahl/SS rats on normal salt diet (Fig. 31.2b, c). Expression of NCX1 is also increased (~13-fold) in mesenteric artery SMCs from Milan hypertensive strain (MHS) rats (Fig. 31.3a, b) (Zulian et al. 2010). MHS rats are a genetic model of hypertension with an adducin gene polymorphism linked to enhanced renal tubular Na^+ reabsorption (Ferrandi et al. 1996, 1999). Figure 31.3c, d shows that removal of extracellular Na^+ [conditions that favor Na^+ extrusion and Ca^{2+} entry via NCX1 (Blaustein and Lederer 1999)] induced a rapid increase in $[Ca^{2+}]_{cyt}$. The increase in $[Ca^{2+}]_{cyt}$ in arterial SMCs in response to removal of extracellular Na^+ ("Na^+-free"), a measure of Na^+/Ca^{2+} exchange activity, was significantly greater in arterial SMCs from MHS than from Milan normotensive strain (MNS) rats. Nevertheless, this ~30 % increment (Fig. 31.3d) was *far* smaller than the ~13-fold increase in NCX1 expression (Fig. 31.3b) (Zulian et al. 2010). This difference may be explained, in part, by buffering of NCX1-mediated Ca^{2+} entry in the PM-junctional SR regions by the SR and mitochondria that limit its diffusion into the cytosol. Indeed, the estimated transient increase in the sub-PM Ca^{2+} concentration upon substitution of extracellular Na^+ by NMDG in rat ASMCs is >13-fold greater than the observed increase in $[Ca^{2+}]_{cyt}$ (Poburko et al. 2006). Furthermore, SERCA2 expression is 2.5-fold higher in arterial myocytes from MHS than from MNS rats (Zulian et al. 2010).

Augmented expression of NCX1 and TRPC proteins has also been observed in arterial SMCs from spontaneously hypertensive rats (SHR) (Taniguchi et al. 2004; Liu et al. 2009; Chen et al. 2010) and in cultured pulmonary artery myocytes from humans with primary pulmonary arterial hypertension (PAH) (Yu et al. 2004; Zhang et al. 2007a, b) (Table 31.1). Furthermore, Giachini and colleagues (2009) demonstrated that expressions of Orai1 and STIM1, as well as SOC-dependent contraction of endothelium-denuded aortic rings, are significantly greater in stroke-prone SHR compared with WKY rats. Upregulation of TRPC6 may underlie the abnormally enhanced proliferation of pulmonary artery SMCs from PAH patients (Yu et al. 2004). A recent study identified a single-nucleotide polymorphism in the TRPC6 gene promoter that is associated with idiopathic PAH and that apparently

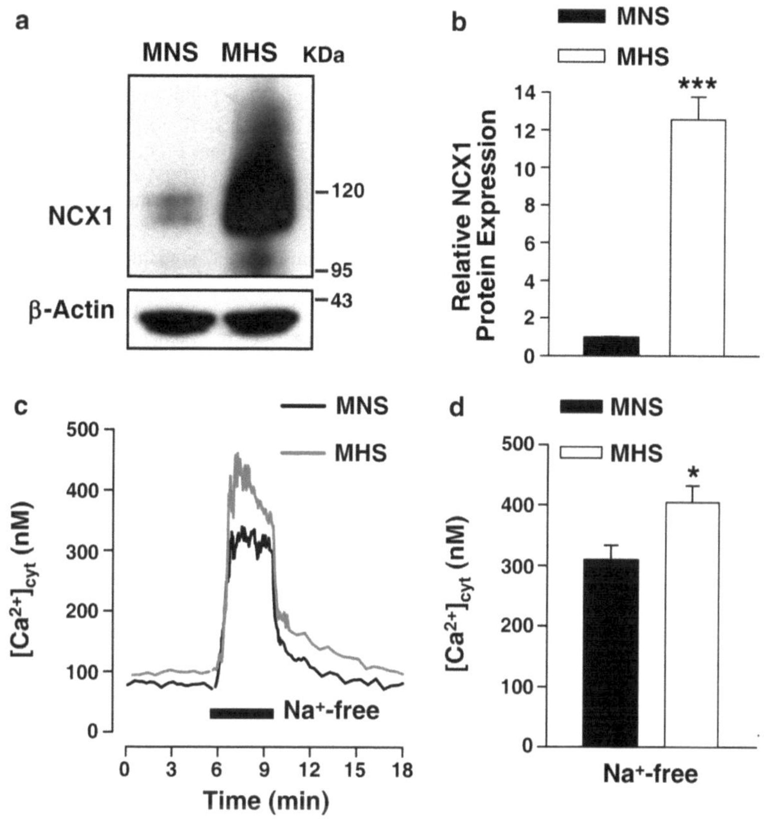

Fig. 31.3 Augmented NCX1 expression and enhanced Ca^{2+} entry via the reverse mode of NCX1 in freshly dissociated mesenteric artery myocytes from Milan hypertensive strain (MHS) rats. (**a**, **b**) Western blot analysis of NCX1 protein expression in smooth muscle cell membranes from mesenteric arteries of MNS and MHS rats. (**a**) Representative immunoblot (30 μg/lane). (**b**) Summary data are normalized to the amount of β-actin and are expressed as mean ± S.E.M. from nine immunoblots (total of 22 rats). ***P <0.001 versus MNS arterial SMCs. (**c**, **d**) Activation of the reverse mode of NCX1 in arterial ASMCs from MNS (*black*) and MHS (*gray*) rats. (**c**) Representative time course records showing changes in $[Ca^{2+}]_{cyt}$ in single arterial SMCs; time of treatment with Na^+-free solution is indicated. Nifedipine (10 μM) was added 10 min before the records shown and was maintained throughout the experiment. (**d**) Summarized data show the NCX1-mediated Ca^{2+} entry in 33 MNS and 23 MHS mesenteric artery SMCs. *P <0.05 versus MNS arterial myocytes (Reprinted with permission from Zulian et al. (2010))

influences TRPC6 activity in pulmonary artery SMCs (Yu et al. 2009). Regardless of the mechanism(s) involved in upregulation of these transport systems (NCX1, TRPC/Orai1-containing channels), they may play an important role in the development and/or maintenance of many forms of hypertension (Table 31.1).

A unique organizational arrangement of α2 Na^+ pumps, NCX1, and TRPC/Orai1 proteins at PM-SR junctions enables these transport systems to function cooperatively to help regulate Ca^{2+} signaling. As a consequence, these proteins work together to modulate arterial myogenic tone. Therefore, they likely make a key contribution to the elevated vascular resistance, a hallmark of sustained hypertension (Cowley 1992). The upregulation of these proteins appears to be some of the critical molecular mechanisms involved in the long-term, "whole body autoregulation" of vascular resistance. Enhanced expression and function of arterial smooth muscle NCX1 and TRPC/Orai1-containing channels in experimental and clinical hypertension implies that these proteins are potential targets for pharmacological intervention.

References

A. Arnon, J.M. Hamlyn, M.P. Blaustein, Na^+ entry via store-operated channels modulates Ca^{2+} signaling in arterial myocytes. Am. J. Physiol. Cell Physiol. **278**, C163–C173 (2000)

Y.M. Bae, A. Kim, Y.J. Lee, W. Lim, Y.H. Noh, E.J. Kim, J. Kim, T.K. Kim, S.W. Park, B. Kim, S.I. Cho, D.K. Kim, W.K. Ho, Enhancement of receptor-operated cation current and TRPC6 expression in arterial smooth muscle cells of deoxycorticosterone acetate-salt hypertensive rats. J. Hypertens. **25**, 809–817 (2007)

S.G. Baryshnikov, M.V. Pulina, A. Zulian, C.I. Linde, V.A. Golovina, Orai1, a critical component of store-operated Ca^{2+} entry, is functionally associated with Na^+/Ca^{2+} exchanger in proliferating human arterial myocytes. Am. J. Physiol. Cell Physiol. **297**, C1103–C1112 (2009)

D.J. Beech, K. Muraki, R. Flemming, Non-selective cationic channels of smooth muscle and the mammalian homologues of Drosophila TRP. J. Physiol. **559**, 685–706 (2004)

M.P. Blaustein, J.M. Hamlyn, Signaling mechanisms that link salt retention to hypertension: endogenous ouabain, the Na^+ pump, the Na^+/Ca^{2+} exchanger and TRPC proteins. Biochim. Biophys. Acta **1802**, 1219–1229 (2010)

M.P. Blaustein, W.J. Lederer, Sodium/calcium exchange: its physiological implications. Physiol. Rev. **79**, 763–854 (1999)

M.P. Blaustein, J. Zhang, L. Chen, H. Song, H. Raina, S.P. Kinsey, M. Izuka, T. Iwamoto, M.I. Kotlikoff, J.B. Lingrel, K.D. Philipson, W.G. Wier, J.M. Hamlyn, The pump, the exchanger, and endogenous ouabain: signaling mechanisms that link salt retention to hypertension. Hypertension **53**, 291–298 (2009)

B.R. Boulanger, M.P. Lilly, J.M. Hamlyn, J. Laredo, D. Shurtleff, D.S. Gann, Ouabain is secreted by the adrenal gland in awake dogs. Am. J. Physiol. **264**, E413–E419 (1993)

J.E. Brayden, S. Earley, M.T. Nelson, S. Reading, Transient receptor potential (TRP) channels, vascular tone and autoregulation of cerebral blood flow. Clin. Exp. Pharmacol. Physiol. **35**, 1116–1120 (2008)

M.D. Cahalan, STIMulating store-operated Ca^{2+} entry. Nat. Cell Biol. **11**, 69–77 (2009)

X. Chen, D. Yang, S. Ma, H. He, Z. Luo, X. Feng, T. Cao, L. Ma, Z. Yan, D. Liu, M. Tepel, Z. Zhu, Increased rhythmicity in hypertensive arterial smooth muscle is linked to transient receptor potential canonical channels. J. Cell. Mol. Med. **14**, 2483–2494 (2010)

A.W. Cowley Jr., Long-term control of arterial blood pressure. Physiol. Rev. **72**, 231–300 (1992)

M.J. Davis, M.A. Hill, Signaling mechanisms underlying the vascular myogenic response. Physiol. Rev. **79**, 387–423 (1999)

K.M. Dibb, H.K. Graham, L.A. Venetucci, D.A. Eisner, A.W. Trafford, Analysis of cellular calcium fluxes in cardiac muscle to understand calcium homeostasis in the heart. Cell Calcium **42**, 503–512 (2007)

I. Dostanic-Larson, J.W. Van Huysse, J.N. Lorenz, J.B. Lingrel, The highly conserved cardiac glycoside binding site of Na^+, K^+- ATPase plays a role in blood pressure regulation. Proc. Natl. Acad. Sci. U. S. A. **102**, 15845–15850 (2005)

P. Eder, M. Poteser, C. Romanin, K. Groschner, Na^+ entry and modulation of Na^+/Ca^{2+} exchange as a key mechanism of TRPC signaling. Pflugers Arch. **451**, 99–104 (2005)

S.K. Fellner, W.J. Arendshorst, Angiotensin II-stimulated Ca^{2+} entry mechanisms in afferent arterioles: role of transient receptor potential canonical channels and reverse Na^+/Ca^{2+} exchange. Am. J. Physiol. Renal Physiol. **294**, F212–F219 (2008)

M. Ferrandi, G. Tripodi, S. Salardi, M. Florio, R. Modica, P. Barassi, P. Parenti, A. Shainskaya, S. Karlish, G. Bianchi, P. Ferrari, Renal Na^+, K^+-ATPase in genetic hypertension. Hypertension **28**, 1018–1025 (1996)

M. Ferrandi, P. Manunta, S. Balzan, J.M. Hamlyn, G. Bianchi, P. Ferrari, Ouabain-like factor quantification in mammalian tissues and plasma: comparison of two independent assays. Hypertension **30**, 886–896 (1997)

M. Ferrandi, S. Salardi, G. Tripodi, P. Barassi, R. Rivera, P. Manunta, R. Goldshleger, P. Ferrari, G. Bianchi, S.J. Karlish, Evidence for an interaction between adducin and Na^+-K^+-ATPase: relation to genetic hypertension. Am. J. Physiol. **277**, H1338–H1349 (1999)

R. Flemming, A. Cheong, A.M. Dedman, D.J. Beech, Discrete store-operated calcium influx into an intracellular compartment in rabbit arteriolar smooth muscle. J. Physiol. **543**, 455–464 (2002)

F.R. Giachini, C.W. Chiao, F.S. Carneiro, V.V. Lima, Z.N. Carneiro, A.M. Dorrance, R.C. Tostes, R.C. Webb, Increased activation of stromal interaction molecule-1/Orai-1 in aorta from hypertensive rats: a novel insight into vascular dysfunction. Hypertension **53**, 409–416 (2009)

V.A. Golovina, Visualization of localized store-operated calcium entry in mouse astrocytes. Close proximity to the endoplasmic reticulum. J. Physiol. **564**, 737–749 (2005)

J.M. Hamlyn, M.P. Blaustein, S. Bova, D.W. DuCharme, D.W. Harris, F. Mandel, W.R. Mathews, J.H. Ludens, Identification and characterization of a ouabain-like compound from human plasma. Proc. Natl. Acad. Sci. U. S. A. **88**, 6259–6263 (1991)

T. Iwamoto, S. Kita, J. Zhang, M.P. Blaustein, Y. Arai, S. Yoshida, K. Wakimoto, I. Komuro, T. Katsuragi, Salt-sensitive hypertension is triggered by Ca^{2+} entry via Na^+/Ca^{2+} exchanger type-1 in vascular smooth muscle. Nat. Med. **10**, 1193–1199 (2004)

M. Juhaszova, M.P. Blaustein, Na^+ pump low and high ouabain affinity alpha subunit isoforms are differently distributed in cells. Proc. Natl. Acad. Sci. U. S. A. **94**, 1800–1805 (1997)

M. Juhaszova, H. Shimizu, M.L. Borin, R.K. Yip, E.M. Santiago, G.E. Lindenmayer, M.P. Blaustein,

Localization of the Na^+/Ca^{2+} exchanger in vascular smooth muscle, and in neurons and astrocytes. Ann. N. Y. Acad. Sci. **779**, 318–335 (1996)

J. Kaide, N. Ura, T. Torii, M. Nakagawa, T. Takada, K. Shimamoto, Effects of digoxin-specific antibody Fab fragment (Digibind) on blood pressure and renal water-sodium metabolism in 5/6 reduced renal mass hypertensive rats. Am. J. Hypertens. **12**, 611–619 (1999)

T. Kashihara, K. Nakayama, T. Matsuda, A. Baba, T. Ishikawa, Role of Na^+/Ca^{2+} exchanger-mediated Ca^{2+} entry in pressure-induced myogenic constriction in rat posterior cerebral arteries. J. Pharmacol. Sci. **110**, 218–222 (2009)

H. Krep, D.A. Price, P. Soszynski, Q.F. Tao, S.W. Graves, N.K. Hollenberg, Volume sensitive hypertension and the digoxin-like factor. Reversal by a Fab directed against digoxin in DOCA-salt hypertensive rats. Am. J. Hypertens. **8**, 921–927 (1995)

G.J. Lagaud, V. Randriamboavonjy, G. Roulm, J.C. Stoclet, R. Andriantsitohaina, Mechanism of Ca^{2+} release and entry during contraction elicited by nor-epinephrine in rat resistance arteries. Am. J. Physiol. **276**, H300–H308 (1999)

M.Y. Lee, H. Song, J. Nakai, M. Ohkura, M.I. Kotlikoff, S.P. Kinsey, V.A. Golovina, M.P. Blaustein, Local sub-plasma membrane Ca^{2+} signals detected by a tethered Ca^{2+} sensor. Proc. Natl. Acad. Sci. U. S. A. **103**, 13232–13237 (2006)

F.H. Leenen, E. Harmsen, H. Yu, Dietary sodium and central vs peripheral ouabain-like activity in Dahl salt-sensitive vs salt-resistant rats. Am. J. Physiol. **267**, H1916–H1920 (1994)

J. Li, P. Sukumar, C.J. Milligan, B. Kumar, Z.Y. Ma, C.M. Munsch, L.H. Jiang, K.E. Porter, D.J. Beech, Interactions, functions, and independence of plasma membrane STIM1 and TRPC1 in vascular smooth muscle cells. Circ. Res. **103**, e97–e104 (2008)

Y. Liao, C. Erxleben, J. Abramowitz, V. Flockerzi, M.X. Zhu, D.L. Armstrong, L. Birnbaumer, Functional interactions among Orai1, TRPCs, and STIM1 suggest a STIM-regulated heteromeric Orai/TRPC model for SOCE/Icrac channels. Proc. Natl. Acad. Sci. U. S. A. **105**, 2895–2900 (2008)

D. Liu, D. Yang, H. He, X. Chen, T. Cao, X. Feng, L. Ma, Z. Luo, L. Wang, Z. Yan, Z. Zhu, M. Tepel, Increased transient receptor potential canonical type 3 channels in vasculature from hypertensive rats. Hypertension **53**, 70–76 (2009)

J.N. Lorenz, E.L. Loreaux, I. Dostanic-Larson, V. Lasko, J.R. Schnetzer, R.J. Paul, J.B. Lingrel, ACTH-induced hypertension is dependent on the ouabain-binding site of the alpha2-Na^+, K^+-ATPase subunit. Am. J. Physiol. Heart Circ. Physiol. **295**, H273–H280 (2008)

P. Manunta, A.C. Rogowski, B.P. Hamilton, J.M. Hamlyn, Ouabain-induced hypertension in the rat: relationships among plasma and tissue ouabain and blood pressure. J. Hypertens. **12**, 549–560 (1994)

S.S. McDaniel, O. Platoshyn, J. Wang, Y. Yu, M. Sweeney, S. Krick, L.J. Rubin, J.X. Yuan, Capacitative Ca^{2+} entry in agonist- induced pulmonary vasoconstriction. Am. J. Physiol. Lung Cell. Mol. Physiol. **280**, L870–L880 (2001)

M.T. Nelson, J.B. Patlak, J.F. Worley, N.B. Standen, Calcium channels, potassium channels, and voltage dependence of arterial smooth muscle tone. Am. J. Physiol. **259**, C3–C18 (1990)

L.C. Ng, D. Ramduny, J.A. Airey, C.A. Singer, P.S. Keller, X.M. Shen, H. Tian, M. Valencik, J.R. Hume, Orai1 interacts with STIM1 and mediates capacitative Ca^{2+} entry in mouse pulmonary arterial smooth muscle cells. Am. J. Physiol. Cell Physiol. **299**, C1079–C1090 (2010)

B. Nilius, G. Owsianik, T. Voets, J.A. Peters, Transient receptor potential cation channels in disease. Physiol. Rev. **87**, 165–217 (2007)

W.J. O'Brien, J.B. Lingrel, E.T. Wallick, Ouabain binding kinetics of the rat alpha two and alpha three isoforms of the sodium-potassium adenosine triphosphate. Arch. Biochem. Biophys. **310**, 32–39 (1994)

D. Poburko, K. Potter, E. van Breemen, N. Fameli, C.H. Liao, O. Basset, U.T. Ruegg, C. van Breemen, Mitochondria buffer NCX-mediated Ca^{2+}-entry and limit its diffusion into vascular smooth muscle cells. Cell Calcium **40**, 359–371 (2006)

D. Poburko, C.H. Liao, V.S. Lemos, E. Lin, Y. Maruyama, W.C. Cole, C. van Breemen, Transient receptor potential channel 6- mediated, localized cytosolic $[Na^+]$ transients drive Na^+/Ca^{2+} exchanger-mediated Ca^{2+} entry in purinergically stimulated aorta smooth muscle cells. Circ. Res. **101**, 1030–1038 (2007)

M. Potier, J.C. Gonzalez, R.K. Motiani, I.F. Abdullaev, J.M. Bisaillon, H.A. Singer, M. Trebak, Evidence for STIM1- and Orai1- dependent store-operated calcium influx through ICRAC in vascular smooth muscle cells: role in proliferation and migration. FASEB J. **23**, 2425–2437 (2009)

M.V. Pulina, A. Zulian, R. Berra-Romani, O. Beskina, A. Mazzocco-Spezzia, S.G. Baryshnikov, I. Papparella, J.M. Hamlyn, M.P. Blaustein, V.A. Golovina, Up-regulation of Na^+ and Ca^{2+} transporters in arterial smooth muscle from ouabain hypertensive rats. Am. J. Physiol. Heart Circ. Physiol. **298**, H263–H274 (2010)

C. Rosker, A. Graziani, M. Lukas, P. Eder, M.X. Zhu, C. Romanin, K. Groschner, Ca^{2+} signaling by TRPC3 involves Na^+ entry and local coupling to the Na^+/Ca^{2+} exchanger. J. Biol. Chem. **279**, 13696–13704 (2004)

G. Rossi, P. Manunta, J.M. Hamlyn, E. Pavan, R. De Toni, A. Semplicini, A.C. Pessina, Immunoreactive endogenous ouabain in primary aldosteronism and essential hypertension: relationship with plasma renin, aldosterone and blood pressure levels. J. Hypertens. **13**, 1181–1191 (1995)

K.M. Sanders, Invited review: mechanisms of calcium handling in smooth muscles. J. Appl. Physiol. **91**, 1438–1449 (2001)

J.R. Shah, J. Laredo, B.P. Hamilton, J.M. Hamlyn, Different signaling pathways mediate stimulated secretions of endogenous ouabain and aldosterone from bovine adrenocortical cells. Hypertension **31**, 463–468 (1998)

D.A. Shelly, S. He, A. Moseley, C. Weber, M. Stegemeyer, R.M. Lynch, J. Lingrel, R.J. Paul, Na^+ pump alpha 2-isoform specifically couples to contractility in vascular smooth muscle: evidence from gene-targeted neonatal mice. Am. J. Physiol. Cell Physiol. **286**, C813–C820 (2004)

A.V. Somlyo, C. Franzini-Armstrong, New views of smooth muscle structure using freezing, deep-etching and rotary shadowing. Experientia **41**, 841–856 (1985)

H. Song, M.Y. Lee, S.P. Kinsey, D.J. Weber, M.P. Blaustein, An N-terminal sequence targets and tethers Na^+ pump alpha2 subunits to specialized plasma membrane microdomains. J. Biol. Chem. **281**, 12929–12940 (2006)

S. Taniguchi, K. Furukawa, S. Sasamura, Y. Ohizumi, K. Seya, S. Motomura, Gene expression and functional activity of sodium/calcium exchanger enhanced in vascular smooth muscle cells of spontaneously hypertensive rats. J. Cardiovasc. Pharmacol. **43**, 629–637 (2004)

C. van Breemen, Q. Chen, I. Laher, Superficial buffer barrier function of smooth muscle sarcoplasmic reticulum. Trends Pharmacol. Sci. **16**, 98–105 (1995)

D.G. Welsh, A.D. Morielli, M.T. Nelson, J.E. Brayden, Transient receptor potential channels regulate myogenic tone of resistance arteries. Circ. Res. **90**, 248–250 (2002)

M.M. Wu, J. Buchanan, R.M. Luik, R.S. Lewis, Ca^{2+} store depletion causes STIM1 to accumulate in ER regions closely associated with the plasma membrane. J. Cell Biol. **174**, 803–813 (2006)

A.V. Yeromin, S.L. Zhang, W. Jiang, Y. Yu, O. Safrina, M.D. Cahalan, Molecular identification of the CRAC channel by altered ion selectivity in a mutant of Orai. Nature **443**, 226–229 (2006)

Y. Yu, I. Fantozzi, C.V. Remillard, J.W. Landsberg, N. Kunichika, O. Platoshyn, D.D. Tigno, P.A. Thistlethwaite, L.J. Rubin, J.X. Yuan, Enhanced expression of transient receptor potential channels in idiopathic pulmonary arterial hypertension. Proc. Natl. Acad. Sci. U. S. A. **101**, 13861–13866 (2004)

Y. Yu, S.H. Keller, C.V. Remillard, O. Safrina, A. Nicholson, S.L. Zhang, W. Jiang, N. Vangala, J.W. Landsberg, J.Y. Wang, P.A. Thistlethwaite, R.N. Channick, I.M. Robbins, J.E. Loyd, H.A. Ghofrani, F. Grimminger, R.T. Schermuly, M.D. Cahalan, L.J. Rubin, J.X. Yuan, A functional single-nucleotide polymorphism in the TRPC6 gene promoter associated with idiopathic pulmonary arterial hypertension. Circulation **119**, 2313–2322 (2009)

J.P. Yuan, M.S. Kim, W. Zeng, D.M. Shin, G. Huang, P.F. Worley, S. Muallem, TRPC channels as STIM1-regulated SOCs. Channels (Austin) **3**, 221–225 (2009)

J. Zhang, M.Y. Lee, M. Cavalli, L. Chen, R. Berra-Romani, C.W. Balke, G. Bianchi, P. Ferrari, J.M. Hamlyn, T. Iwamoto, J.B. Lingrel, D.R. Matteson, W.G. Wier, M.P. Blaustein, Sodium pump alpha2 subunits control myogenic tone and blood pressure in mice. J. Physiol. **569**, 243–256 (2005a)

S.L. Zhang, Y. Yu, J. Roos, J.A. Kozak, T.J. Deerinck, M.H. Ellisman, K.A. Stauderman, M.D. Cahalan, STIM1 is a Ca^{2+} sensor that activates CRAC channels and migrates from the Ca^{2+} store to the plasma membrane. Nature **437**, 902–905 (2005b)

S. Zhang, H. Dong, L.J. Rubin, J.X. Yuan, Upregulation of Na^+/Ca^{2+} exchanger contributes to the enhanced Ca^{2+} entry in pulmonary artery smooth muscle cells from patients with idiopathic pulmonary arterial hypertension. Am. J. Physiol. Cell Physiol. **292**, C2297–C2305 (2007a)

S. Zhang, H.H. Patel, F. Murray, C.V. Remillard, C. Schach, P.A. Thistlethwaite, P.A. Insel, J.X. Yuan, Pulmonary artery smooth muscle cells from normal subjects and IPAH patients show divergent cAMP-mediated effects on TRPC expression and capacitative Ca^{2+} entry. Am. J. Physiol. Lung Cell. Mol. Physiol. **292**, L1202–L1210 (2007b)

J. Zhang, C. Ren, L. Chen, M.F. Navedo, L.K. Antos, S.P. Kinsey, T. Iwamoto, K.D. Philipson, M.I. Kotlikoff, L.F. Santana, W.G. Wier, D.R. Matteson, M.P. Blaustein, Knockout of Na^+/Ca^{2+} exchanger in smooth muscle attenuates vasoconstriction and L-type Ca^{2+} channel current and lowers blood pressure. Am. J. Physiol. Heart Circ. Physiol. **298**, H1472–H1483 (2010)

H. Zou, P.H. Ratz, M.A. Hill, Temporal aspects of Ca^{2+} and myosin phosphorylation during myogenic and norepinephrine-induced arteriolar constriction. J. Vasc. Res. **37**, 556–567 (2000)

A. Zulian, S.G. Baryshnikov, C.I. Linde, J.M. Hamlyn, P. Ferrari, V.A. Golovina, Upregulation of Na^+/Ca^{2+} exchanger and TRPC6 contributes to abnormal Ca^{2+} homeostasis in arterial smooth muscle cells from Milan hypertensive rats. Am. J. Physiol. Heart Circ. Physiol. **299**, H624–H633 (2010)

T-Tubule Remodelling and Ryanodine Receptor Organization Modulate Sodium-Calcium Exchange

32

Karin R. Sipido, Károly Acsai, Gudrun Antoons, Virginie Bito, and Niall Macquaide

Abstract

The Na^+/Ca^{2+} exchanger (NCX) is a key regulator of intracellular Ca^{2+} in cardiac myocytes, predominantly contributing to Ca^{2+} removal during the diastolic relaxation process but also modulating excitation-contraction coupling. NCX is preferentially located in the T-tubules and can be close to or within the dyad, where L-type Ca^{2+} channels face ryanodine receptors (RyRs), the Ca^{2+} release channels of the sarcoplasmic reticulum. However, especially in larger animals, not all RyRs are in dyads or adjacent to T-tubules, and a substantial fraction of Ca^{2+} release from the sarcoplasmic reticulum thus occurs at distance from NCX. This chapter deals with the functional consequences of NCX location and how NCX can modulate diastolic and systolic Ca^{2+} events. The loss of T-tubules and the effects on RyR function and NCX modulation are explored, as well as quantitative measurement of local Ca^{2+} gradients at the level of the dyadic space.

Keywords

Ryanodine receptor • Na^+/Ca^{2+} exchange • T-tubules • Hypertrophy • Heart • Myocardial infarction

K.R. Sipido, M.D., Ph.D. (✉) • V. Bito • N. Macquaide
Laboratory of Experimental Cardiology, Department of Cardiovascular Diseases, University of Leuven, Campus Gasthuisberg O/N 7th floor, Herestraat 49, B-3000 Leuven, Belgium
e-mail: Karin.Sipido@med.kuleuven.be

K. Acsai
Division of Cardiovascular Pharmacology, Hungarian Academy of Sciences, Szeged, Hungary

G. Antoons
Laboratory of Experimental Cardiology, Department of Cardiovascular Diseases, University of Leuven, Campus Gasthuisberg O/N 7th floor, Herestraat 49, B-3000 Leuven, Belgium

Department of Cardiology, Medical University of Graz, Graz, Austria

L. Annunziato (ed.), *Sodium Calcium Exchange: A Growing Spectrum of Pathophysiological Implications*, Advances in Experimental Medicine and Biology 961, DOI 10.1007/978-1-4614-4756-6_32,

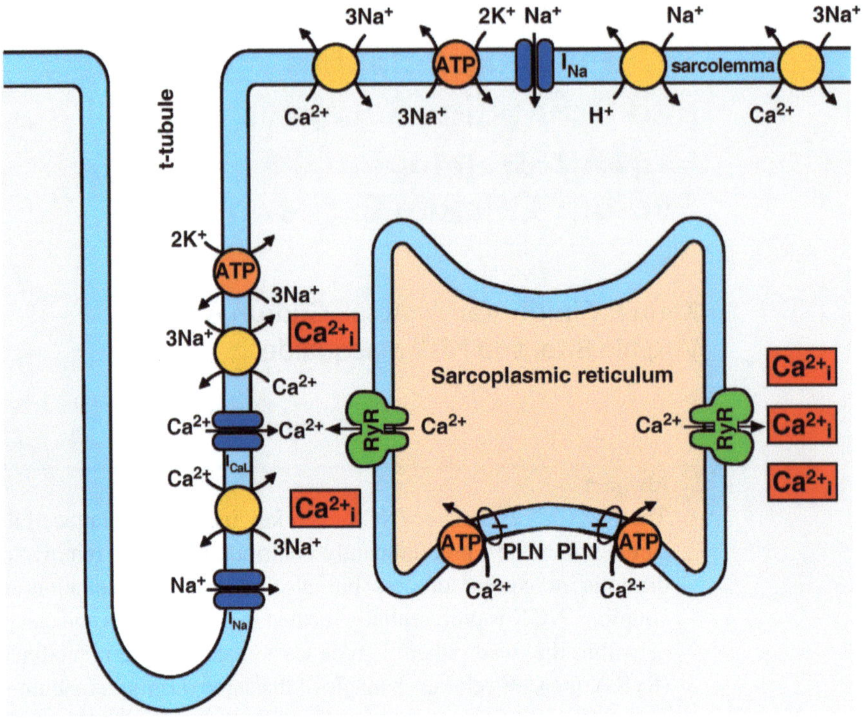

Fig. 32.1 *Schematic diagram of the Ca^{2+} cycle in a cardiac ventricular myocyte*. In a couplon, microdomains of Ca^{2+} are generated by the activation of L-type Ca^{2+} channels (LTCC) and subsequent release of Ca^{2+} from the sarcoplasmic reticulum through ryanodine receptors (RyRs). Ca^{2+} is removed from the cytosol by reuptake into the sarcoplasmic reticulum by the ATP-driven Ca^{2+} pump (SERCA), which is modulated by phospholamban, (PLN), and by efflux through NCX. NCX is located mainly in the T-tubules but not exclusively. As shown in the diagram, not all RyRs face T-tubules and LTCC; these RyRs will be activated and release Ca^{2+} from the SR through a process of diffusion and Ca^{2+}-induced Ca^{2+} release. Removal of Ca^{2+} from these sites through NCX can be delayed because of the larger distance to the membrane

32.1 Introduction

In ventricular myocytes, T-tubules provide an extensive network connected to the external sarcolemma. In rat and mouse myocytes with a high density of T-tubules, this structural organization of the surface membrane ensures fast spreading of the action potential throughout the myocytes and synchronizes the initiation of excitation-contraction coupling. Figure 32.1 represents a general scheme with the different actors involved in excitation-contraction coupling in a cardiac myocyte. Along T-tubules, couplons are areas where L-type Ca^{2+} channels (LTCC) are face ryanodine receptors (RyR) in the underlying sarcoplasmic reticulum (SR). In these specialized regions, microdomains of Ca^{2+} are generated during the opening of LTCC and the subsequent activation of RyR. It has been a matter of debate how closely Na^+/Ca^{2+} exchange (NCX) molecules are associated with this microdomain. Co-localization studies suggest it is less than 15% that is tightly associates with RyR and LTCC (Schulson et al. 2011; Scriven et al. 2002). Nevertheless, NCX is more dense in T-tubules, in the vicinity of the dyad (Despa et al. 2003), and during Ca^{2+} release from the SR, the current appears to be activated by a submembrane compartment different from bulk Ca^{2+} (Trafford et al.

1995). Independent of the finer detail, NCX in the T-tubules is essential for efficient Ca^{2+} removal during the heartbeat. Remodelling of T-tubules will affect this process, and this can be further modulated by changes in expression and Na^+ concentration. In the first part of this chapter, examples of such different scenarios are presented. Subsequently, a closer look is taken from the viewpoint of the RyR clusters close or more distant to NCX. Lastly, the issue of the distribution of NCX in and outside couplons is examined from the point of view of NCX as a reporter on Ca^{2+} within the dyadic cleft of couplons.

32.2 Loss and Reorganization of T-Tubules During Remodelling Modulates Ca^{2+} Removal and Relaxation

Several studies have reported changes in NCX expression or function with cardiac hypertrophy and heart failure, many reporting an increase, others no change or a decrease (Sipido et al. 2002a). One of the most common used assays for the Ca^{2+} removal capacity of NCX is the quantification of the rate of decline of a caffeine-induced SR Ca^{2+} release. If measured in the continuous presence of caffeine, this process relies primarily on NCX, with slower processes such as plasma membrane Ca^{2+} ATPase and mitochondria contributing to a lesser extent (Bassani et al. 1994; Sipido and Wier 1991; Varro et al. 1993). Gomez et al. were among the first to appreciate the confounding factor of changing geometry in hypertrophy (Gomez et al. 2002). Indeed, a larger surface to volume ratio (S/V) will lead to an apparent reduced NCX removal rate in the presence of a normal level of expression and function per unit of surface area. This was also recognized in the study by Quinn et al. who studied remodelling after myocardial infarction (MI) in the rabbit (Quinn et al. 2003). Despite increased protein expression levels, the removal of Ca^{2+} was reduced. This was ascribed to the loss of T-tubules with a reduction of surface area. In addition, changes in Na^+ can further modulate the removal capacity (Bers et al. 2006; Verdonck et al. 2003). In the dog with chronic atrioventricular block (cAVB), myocytes are hypertrophied. NCX protein expression is unchanged but higher subsarcolemmal Na^+, related a change in Na^+/K^+ pump activity, induces a shift in the reversal potential, with increased Ca^{2+} influx (Sipido et al. 2000). Current density during Ca^{2+} removal is also increased possibly reflecting Na^+-dependent or allosteric activation. The altered NCX function further contributes to a higher SR Ca^{2+} content in this model of compensated hypertrophy. It may also contribute to the increased susceptibility for arrhythmias (Sipido et al. 2002b).

Loss of T-tubules will by itself reduce the S/V in the absence of hypertrophy. T-tubules loss can be induced with formamide and has demonstrated that, at least in rat myocytes, NCX is more densely expressed in T-tubules (Yang et al. 2002). Consequently to both the reduced S/V and relative decrease in NCX density, Ca^{2+} removal from the cells was slowed down. Likewise loss of T-tubules can be readily seen in cultured myocytes where cell size also remains essentially unchanged. Removal of Ca^{2+} by NCX then also appears to be reduced (Louch et al. 2004).

Several reports have recently described reorganization of T-tubules as part of the remodelling process after MI or pressure overload, also in humans (Crossman et al. 2011; Louch et al. 2006; Song et al. 2006; Wei et al. 2010). NCX function was however not systematically studied. In our study of remodelling in chronic ischemia with coronary stenosis and adjacent to MI, we found that the loss of T-tubules was associated with reduced Ca^{2+} removal rate during caffeine-induced Ca^{2+} transients (Heinzel et al. 2008). However, NCX expression was unchanged and the reduced removal was ascribed to the reduced S/V. In contrast, remodelling of atrial myocytes in a sheep model of persistent atrial fibrillation (AF) leads to increased expression of NCX (Lenaerts et al. 2009). Despite a substantial reduction of the T-tubules in these myocytes

Table 32.1 Possible configurations of myocyte size, T-tubule and NCX expression have different functional consequences. The table is a non-exhaustive list

	Cell size	T-tubules	Surface: volume	NCX expression	Ca removal	Reference
Detubulation	=	↓	↓	=	↓	Yang et al. (2002)
Cell culture	=	↓	↓	=	↓	Louch et al. (2004)
Chronic ischemia	↑	↓	↓	=	↓	Heinzel et al. (2008)
Atrial fibrillation	↑	↓	↓	↑	↑	Lenaerts et al. (2009)

and cellular hypertrophy, the rate of Ca^{2+} removal by NCX during a caffeine transient is increased. In this case this may contribute to depletion of the SR.

Table 32.1 summarizes the different types of modulation of NCX as consequence of changes in sarcolemmal organization.

32.3 Organization of RyR and Distance to T-Tubular Membrane NCX Affect Spark Behaviour

One of the consequences of reorganization and/or loss of T-tubules is that a number of RyRs appear to be less directly coupled to the sarcolemma, a phenomenon coined as 'orphaned' receptors (Song et al. 2006). Following up on our study in chronic ischemia, we examined whether this would change the intrinsic behaviour of RyR by characterizing sparks in subpopulations of RyR (Biesmans et al. 2011). In the earlier study, we had demonstrated that during depolarization, areas that had a delayed rise of Ca^{2+} were areas that appeared to be devoid of T-tubules. As illustrated in Fig. 32.2, we then examined spark properties in these delayed areas and compared them to the sparks from early areas.

This analysis showed that sparks originating in early areas, near T-tubules, have a higher frequency and shorter duration. This shorter duration could be related Ca^{2+} removal by NCX since block of NCX with $NiCl_2$ eliminated differences in duration (Biesmans et al. 2011).

This observation is connected to earlier observations of a biphasic time course of the NCX current during sarcoplasmic reticulum Ca^{2+} release (Trafford et al. 1995). In this chapter, it was proposed that the early rapid decline resulted from the submembrane increase in Ca^{2+} while the second phase was the result of Ca^{2+} removal from the global rise in cytosolic Ca^{2+}. Since normal healthy pig myocytes have a number of RyRs that are far away from the membrane, we expected that such a bi-exponential time course would likely be present, as illustrated in Fig. 32.3a. Though not seen in all, the majority of cells indeed had a bi-exponential decline (Fig. 32.3, panel b CTRL).

After MI, in myocytes from the adjacent area where chronic ischemia remained because of a severe coronary stenosis, loss of T-tubules increased the fraction of non-coupled RyR (Heinzel et al. 2008). Analysis of sparks showed that the duration of sparks increased specifically in the near-membrane RyR, suggesting that there was a further loss of NCX in the membrane, despite absence of protein downregulation (Biesmans et al. 2011). The combination of a larger fraction of non-coupled RyR with reduced NCX leads to a slowing of the time course of the NCX current with fewer cells having a bi-exponential time course (Fig. 32.3b).

These data indicate that the subcellular distribution of RyR influences the time course of NCX. Less synchronized Ca^{2+} release from all RyR leads to a protracted time course of NCX with consequently a less pronounced peak. In theory, this would result in a reduced capacity for membrane depolarization though a potential link to arrhythmias remains to be explored.

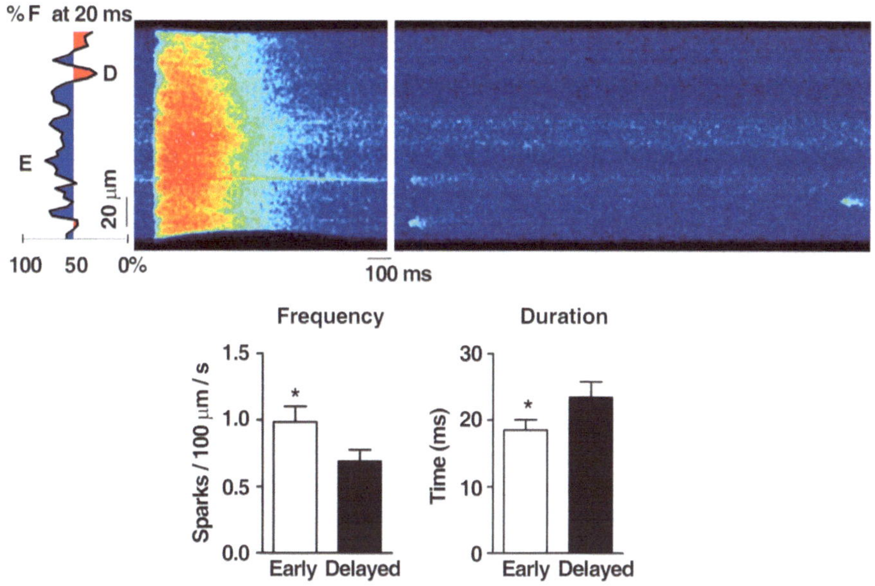

Fig. 32.2 *Different spark properties according to their proximity to T-tubules. Upper panel:* Typical example of a line scan image during and after 1-Hz stimulation. After loading the SR with conditioning pulses from −70 to +10 mV at 1 Hz, stimulation was stopped and 15 s of diastole was recorded for Ca^{2+} sparks. Sparks were assigned to early (*blue*) and delayed (*red*) release areas corresponding to their position on the scan line. *Lower panel:* Spark frequency and duration in early (presence of T-tubules) versus delayed release areas (absence of T-tubules) in normal pigs (n_{cells}=41). * denotes *P* <0.05 (Modified after Biesmans et al. 2011)

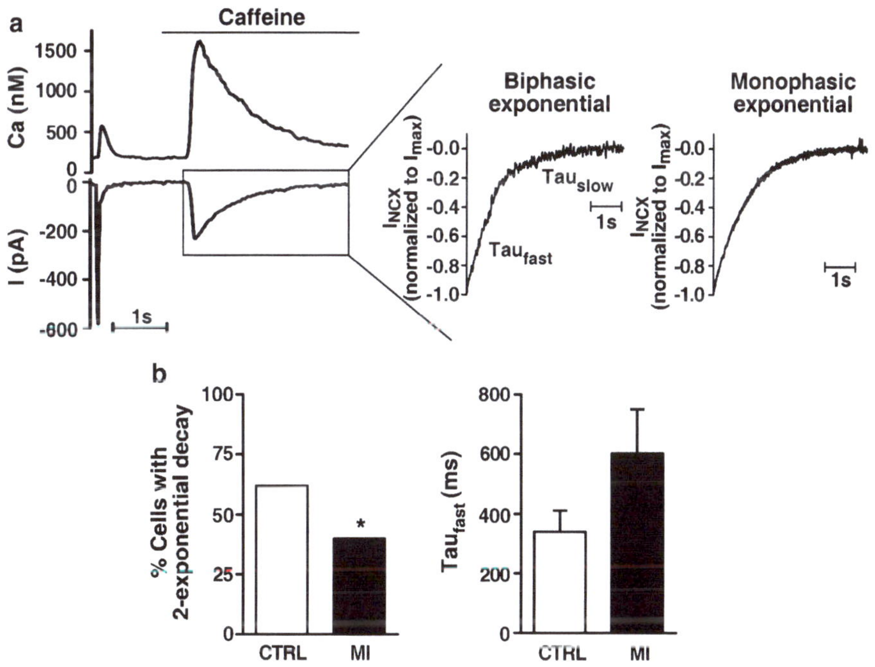

Fig. 32.3 *Slowing NCX time course with spark redistribution after MI.* (**a**) Example of current and $[Ca^{2+}]_i$ transient recording obtained during the last conditioning pulse from −70 to +10 mV and fast 10-mM caffeine application (*left*). The decay of the current was fit by a 1- or 2-exponential according to the goodness of fit (R^2 >0.95). The *right panels* are typical examples of a 2-exponential and a monophasic decay in separate myocytes from normal pigs. (**b**) *Left panel*: The percentage of cells better fit by a biphasic exponential was significantly higher in normal than in pigs with ischemic cardiomyopathy (MI) (normal pigs n_{cells} = 17; MI, n_{cells} = 12, *P* <0.05). *Right panel:* Tau of fast component of I_{NCX} decay tended to be faster in normal than in MI myocytes (normal n_{cells} = 8; MI, n_{cells} = 4). * denotes *P* <0.05 (Modified after Biesmans et al. 2011)

32.4 NCX Association with Couplons: A Sensor for Ca^{2+}

As pointed out above, not all NCX resides in T-tubules, where most couplons are located. In addition, even in T-tubules, not all NCX may reside in couplons.

The exact location of NCX is a key element in the debated issue of whether reverse mode NCX can activate the RyR. Currently, there is good evidence in favour of a modulatory effect of NCX in the activation of RyR during excitation-contraction coupling, with Ca^{2+} influx through LTCC the main source of trigger Ca^{2+} (Larbig et al. 2010). The quantitative aspects remain however under debate and are addressed elsewhere.

The discrepancy between the NCX current and the global Ca^{2+} during the rise time of the $[Ca^{2+}]_i$ transient was a first evidence that NCX, or at least part of it, was located in a submembrane compartment (reviewed in Sipido et al. 2006). This microdomain experiences a high surge of Ca^{2+} and such local gradients are expected between the dyadic cleft and cytosol. However, how much NCX exactly is in which location remains under study. A more precise location of NCX, beyond its presence in T-tubules, has been explored by Scriven and Moore (2000) in rat ventricular myocytes. Using immunostaining and advanced image analysis, they quantified co-localization with RyR as well as with LTCC. Their conclusion was that NCX was largely located outside the dyads, though it was also proposed that the functional Ca^{2+} microdomain could extend beyond the dyad (Schulson et al. 2011; Scriven et al. 2000).

Functionally, the higher NCX current has been used to extrapolate submembrane $[Ca^{2+}]_i$, based on an independent calibration of NCX in steady state conditions (Weber et al. 2002). The obtained values showed the earlier increase in submembrane Ca^{2+}, but the values were still below the values that had been proposed for dyadic cleft in computational approaches (Shannon and Bers 2004).

We recently further extended this approach by dissecting the NCX current in two components. With the assumption of negligible Ca^{2+} gradients between cytosol and surface membrane that is not in the T-tubular compartment close to release sites, a first component was calculated as the expected NCX current for the measured cytosolic Ca^{2+} transient. The remaining NCX current component should then reflect the Ca^{2+} values near release sites not reported by the cytosolic dye. The approach is illustrated in Fig. 32.4 (see Acsai et al. (2011) for more detailed description). This approach yields higher values and sharper time course when compared to the global cytosolic $[Ca^{2+}]_i$ transient but also when compared to a derivation of the total NCX current with calibration (Fig. 32.5a). The obtained peak value of around 15 μmol/L approaches some of the computational predicted values. In the same study, we also derived Ca^{2+} concentration near release sites from the Ca^{2+}-induced inactivation of the LTCC (Fig. 32.5b). Since LTCCs are located primarily within the couplons, this latter estimation should reflect the dyadic $[Ca^{2+}]$. Although this method tended to result in somewhat smaller $[Ca^{2+}]$ values (possibly due to saturation of LTCC inactivation), the similarity between values and especially the time courses is remarkable.

Does this imply that NCX is co-localizing with LTCC in the dyadic space? The derivations in this study are purely functional, hence the definition 'near release sites' rather than 'dyadic' Ca^{2+} concentrations. The data strongly suggest that this functional space is accessible to both proteins as reporters for the prevailing Ca^{2+}. However, it should be noted that the calculations put a fraction of around 15% of NCX in this space near release sites, so a quite low number compared to the total capacity. The data do not allow concluding on the active contribution of NCX in raising Ca^{2+} in this space during excitation-contraction coupling nor on the extent to which NCX would lower the Ca^{2+} in this space.

32.5 Perspectives for Future Work

Recent developments in high-resolution microscopy may shed further light on the exact location and co-localization of NCX with the dyad

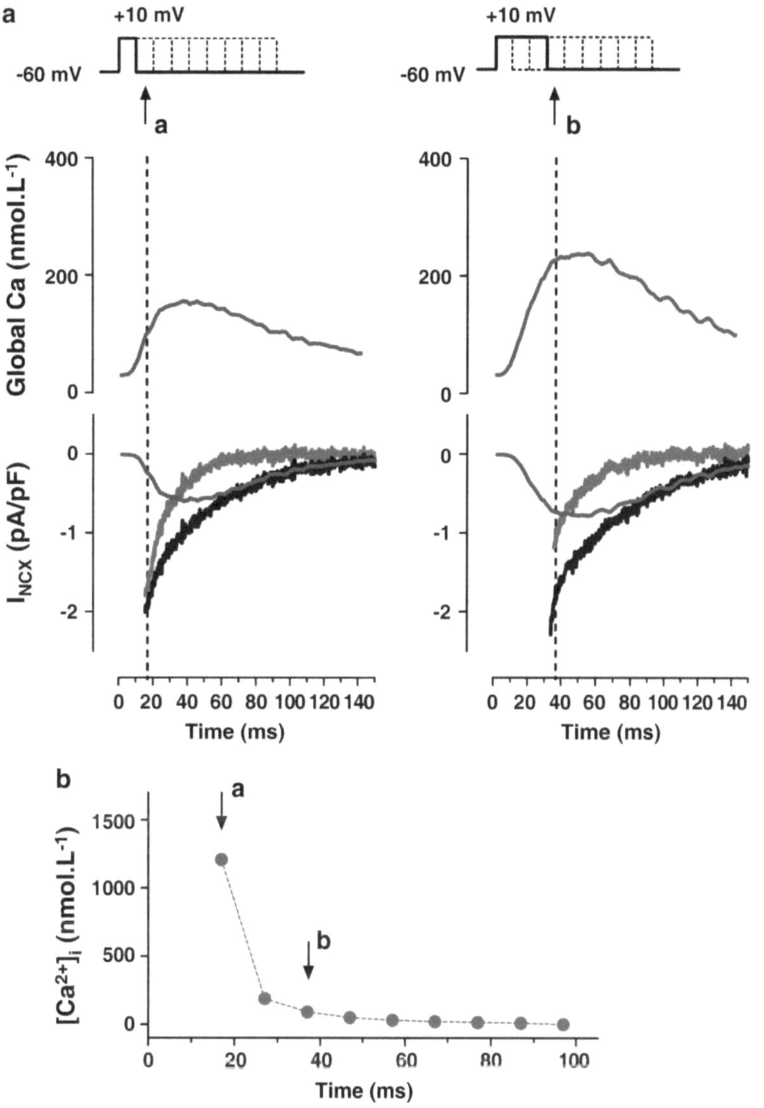

Fig. 32.4 (**a**) Two recordings at 10 ms and at 30 ms into the depolarizing step are shown with the global $[Ca^{2+}]_i$ transient and the recorded NCX tail current on repolarization (*black current trace*). Predicted values for NCX activated by the cytosolic Ca^{2+} transient are calculated based on a model equation relating NCX to Ca^{2+} (*gray trace*). The difference current between the recorded (*black trace*) and calculated (*gray trace*) then reflects NCX activated by Ca^{2+} near release sites and not reported by the cytosolic dye. This measurement is repeated every 10 ms. (**b**) $[Ca^{2+}]_{nrs}$ for during the example depolarizing step to +10 mV as calculated from the values of local NCX current at the corresponding time points (Reproduced from Acsai et al. 2011)

(Baddeley et al. 2011). The heterogeneity in RyR function within couplons and in non-coupled sites needs to be further explored in particular with regard to the feedback on NCX currents and pro-arrhythmia. Computational modelling may help to make predictions regarding arrhythmogenesis be tested in relevant animal models.

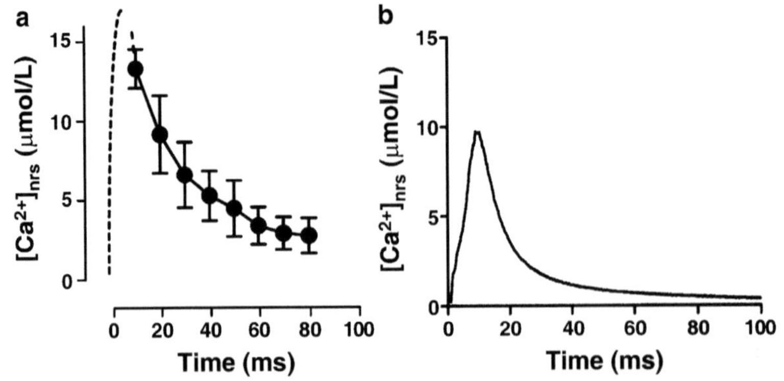

Fig. 32.5 *Both NCX and LTCC report high- and short-lived [Ca^{2+}] transients during early depolarization.* (**a**) Calculated Ca^{2+} near release sites using the method shown in Fig. 32.4 during a depolarizing step to +10 mV (mean value of four cells). (**b**) Ca^{2+} near release sites derived from Ca^{2+}-dependent inactivation of LTCCs during a similar voltage step (mean value from 11 cells) (Modified after Acsai et al. 2011)

Acknowledgements The authors acknowledge support from EC FP7/2007–2013 under grant agreement no. HEALTH-F2-2009-241526, EUTrigTreat (to K.R.S.).

References

K. Acsai, G. Antoons, L. Livshitz, Y. Rudy, K.R. Sipido, Microdomain [Ca^{2+}] near ryanodine receptors as reported by L-type Ca^{2+} and Na^+/Ca^{2+} exchange currents. J. Physiol. **589**, 2569–2583 (2011)

D. Baddeley, D. Crossman, S. Rossberger, J.E. Cheyne, J.M. Montgomery, I.D. Jayasinghe, C. Cremer, M.B. Cannell, C. Soeller, 4D super-resolution microscopy with conventional fluorophores and single wavelength excitation in optically thick cells and tissues. PLoS One **6**, e20645 (2011)

J.W. Bassani, R.A. Bassani, D.M. Bers, Relaxation in rabbit and rat cardiac cells: species-dependent differences in cellular mechanisms. J. Physiol. **476**, 279–293 (1994)

D.M. Bers, S. Despa, J. Bossuyt, Regulation of Ca^{2+} and Na^+ in normal and failing cardiac myocytes. Ann. N. Y. Acad. Sci. **1080**, 165–177 (2006)

L. Biesmans, N. Macquaide, F.R. Heinzel, V. Bito, G.L. Smith, K.R. Sipido, Subcellular heterogeneity of ryanodine receptor properties in ventricular myocytes with low T-tubule density. PLoS One **6**, e25100 (2011)

D.J. Crossman, P.N. Ruygrok, C. Soeller, M.B. Cannell, Changes in the organization of excitation-contraction coupling structures in failing human heart. PLoS One **6**, e17901 (2011)

S. Despa, F. Brette, C.H. Orchard, D.M. Bers, Na/Ca exchange and Na/K-ATPase function are equally concentrated in transverse tubules of rat ventricular myocytes. Biophys. J. **85**, 3388–3396 (2003)

A.M. Gomez, B. Schwaller, H. Porzig, G. Vassort, E. Niggli, M. Egger, Increased exchange current but normal Ca^{2+} transport via Na^+-Ca^{2+} exchange during cardiac hypertrophy after myocardial infarction. Circ. Res. **91**, 323–330 (2002)

F.R. Heinzel, V. Bito, L. Biesmans, M. Wu, E. Detre, F. von Wegner, P. Claus, S. Dymarkowski, F. Maes, J. Bogaert, F. Rademakers, J. D'Hooge, K. Sipido, Remodeling of T-tubules and reduced synchrony of Ca^{2+} release in myocytes from chronically ischemic myocardium. Circ. Res. **102**, 338–346 (2008)

R. Larbig, N. Torres, J.H. Bridge, J.I. Goldhaber, K.D. Philipson, Activation of reverse Na^+-Ca^{2+} exchange by the Na^+ current augments the cardiac Ca^{2+} transient: evidence from NCX knockout mice. J. Physiol. **588**, 3267–3276 (2010)

I. Lenaerts, V. Bito, F.R. Heinzel, R.B. Driesen, P. Holemans, J. D'Hooge, H. Heidbuchel, K.R. Sipido, R. Willems, Ultrastructural and functional remodeling of the coupling between Ca^{2+} influx and sarcoplasmic reticulum Ca^{2+} release in right atrial myocytes from experimental persistent atrial fibrillation. Circ. Res. **105**, 876–885 (2009)

W.E. Louch, V. Bito, F.R. Heinzel, R. Macianskiene, J. Vanhaecke, W. Flameng, K. Mubagwa, K.R. Sipido, Reduced synchrony of Ca^{2+} release with loss of T-tubules-a comparison to Ca^{2+} release in human failing cardiomyocytes. Cardiovasc. Res. **62**, 63–73 (2004)

W.E. Louch, H.K. Mork, J. Sexton, T.A. Stromme, P. Laake, I. Sjaastad, O.M. Sejersted, T-tubule disorganization and reduced synchrony of Ca^{2+} release in murine cardiomyocytes following myocardial infarction. J. Physiol. **574**, 519–533 (2006)

F.R. Quinn, S. Currie, A.M. Duncan, S. Miller, R. Sayeed, S.M. Cobbe, G.L. Smith, Myocardial infarction causes increased expression but decreased activity of the myocardial Na^{+}-Ca^{2+} exchanger in the rabbit. J. Physiol. **553**, 229–242 (2003)

M.N. Schulson, D.R. Scriven, P. Fletcher, E.D. Moore, Couplons in rat atria form distinct subgroups defined by their molecular partners. J. Cell Sci. **124**, 1167–1174 (2011)

D.R. Scriven, P. Dan, E.D. Moore, Distribution of proteins implicated in excitation-contraction coupling in rat ventricular myocytes. Biophys. J. **79**, 2682–2691 (2000)

D.R. Scriven, A. Klimek, K.L. Lee, E.D. Moore, The molecular architecture of calcium microdomains in rat cardiomyocytes. Ann. N. Y. Acad. Sci. **976**, 488–499 (2002)

T.R. Shannon, D.M. Bers, Integrated Ca^{2+} management in cardiac myocytes. Ann. N. Y. Acad. Sci. **1015**, 28–38 (2004)

K.R. Sipido, W.G. Wier, Flux of Ca^{2+} across the sarcoplasmic reticulum of guinea-pig cardiac cells during excitation-contraction coupling. J. Physiol. **435**, 605–630 (1991)

K.R. Sipido, P.G. Volders, S.H. de Groot, F. Verdonck, F. Van de Werf, H.J. Wellens, M.A. Vos, Enhanced Ca^{2+} release and Na/Ca exchange activity in hypertrophied canine ventricular myocytes: potential link between contractile adaptation and arrhythmogenesis. Circulation **102**, 2137–2144 (2000)

K.R. Sipido, P.G. Volders, M.A. Vos, F. Verdonck, Altered Na/Ca exchange activity in cardiac hypertrophy and heart failure: a new target for therapy? Cardiovasc. Res. **53**, 782–805 (2002a)

K.R. Sipido, P.G. Volders, M. Schoenmakers, S.H. De Groot, F. Verdonck, M.A. Vos, Role of the Na/Ca exchanger in arrhythmias in compensated hypertrophy. Ann. N. Y. Acad. Sci. **976**, 438–445 (2002b)

K.R. Sipido, A. Varro, D. Eisner, Sodium calcium exchange as a target for antiarrhythmic therapy. Handb. Exp. Pharmacol. **171**, 159–199 (2006)

L.S. Song, E.A. Sobie, S. McCulle, W.J. Lederer, C.W. Balke, H. Cheng, Orphaned ryanodine receptors in the failing heart. Proc. Natl. Acad. Sci. U. S. A. **103**, 4305–4310 (2006)

A.W. Trafford, M.E. Diaz, S.C. O'Neill, D.A. Eisner, Comparison of subsarcolemmal and bulk calcium concentration during spontaneous calcium release in rat ventricular myocytes. J. Physiol. **488**, 577–586 (1995)

A. Varro, N. Negretti, S.B. Hester, D.A. Eisner, An estimate of the calcium content of the sarcoplasmic reticulum in rat ventricular myocytes. Pflugers Arch. **423**, 158–160 (1993)

F. Verdonck, P.G. Volders, M.A. Vos, K.R. Sipido, Intracellular Na^{+} and altered Na^{+} transport mechanisms in cardiac hypertrophy and failure. J. Mol. Cell. Cardiol. **35**, 5–25 (2003)

C.R. Weber, V. Piacentino 3rd, K.S. Ginsburg, S.R. Houser, D.M. Bers, Na^{+}-Ca^{2+} exchange current and submembrane [Ca^{2+}] during the cardiac action potential. Circ. Res. **90**, 182–189 (2002)

S. Wei, A. Guo, B. Chen, W. Kutschke, Y.P. Xie, K. Zimmerman, R.M. Weiss, M.E. Anderson, H. Cheng, L.S. Song, T-tubule remodeling during transition from hypertrophy to heart failure. Circ. Res. **107**, 520–531 (2010)

Z. Yang, C. Pascarel, D.S. Steele, K. Komukai, F. Brette, C.H. Orchard, Na^{+}-Ca^{2+} exchange activity is localized in the T-tubules of rat ventricular myocytes. Circ. Res. **91**, 315–322 (2002)

33 Na^+/Ca^{2+} Exchange and the Plasma Membrane Ca^{2+}-ATPase in β-Cell Function and Diabetes

André Herchuelz, Evrard Nguidjoe, Lin Jiang, and Nathalie Pachera

Abstract

The rat pancreatic β-cell expresses two splice variants of the Na^+/Ca^{2+} exchanger 1 (NCX1) and six splice variants of the plasma membrane Ca^{2+}-ATPase (PMCA). In the β-cell, Na^+/Ca^{2+} exchange displays a high capacity, contributes to both Ca^{2+} outflow and influx and participates to the control of insulin release. Gain of function studies show that overexpression of NCX1 or PMCA2 leads to endoplasmic reticulum (ER) Ca^{2+} depletion with subsequent ER stress, decrease in β-cell proliferation and β-cell death by apoptosis. Interestingly, chronic exposure to cytokines or high free fatty acids concentration also induces ER Ca^{2+} depletion and β-cell death in diabetes. Loss of function studies shows, on the contrary, that heterozygous inactivation of NCX1 ($Ncx1^{+/-}$) leads to an increase in β-cell function (insulin production and release) and a fivefold increase in both β-cell mass and proliferation. The mutation also increases β-cell resistance to hypoxia, and $Ncx1^{+/-}$ islets show a four to seven times higher rate of diabetes cure than $Ncx1^{+/+}$ islets when transplanted in diabetic animals. Thus, downregulation of the Na^+/Ca^{2+} exchanger leads to various changes in β-cell function that are opposite to the major abnormalities seen in diabetes. In addition, the β-cell, which is an excitable cell, includes the mutually exclusive exon B in the alternative splicing region of NCX1, which confers a high sensitivity of its NCX splice variants (NCX1.3 & 1.7) to the inhibitory action of compounds like KB-R7943. This provides a unique model for the prevention and treatment of β-cell dysfunction in diabetes and following islet transplantation.

A. Herchuelz (✉) • E. Nguidjoe • L. Jiang • N. Pachera
Faculté de Médicine, Laboratoire de Pharmacodynamie et de Thérapeutique, Université Libre de Bruxelles (ULB), Route de Lennik, 808-Bâtiment GE, Brussels, B-1070, Belgium
e-mail: herchu@ulb.ac.be

L. Annunziato (ed.), *Sodium Calcium Exchange: A Growing Spectrum of Pathophysiological Implications*, Advances in Experimental Medicine and Biology 961, DOI 10.1007/978-1-4614-4756-6_33,

Keywords
Sodium-calcium exchange • Plasma membrane Ca^{2+}-ATPase • Calcium • β-cell • Diabetes

33.1 Introduction

Calcium (Ca^{2+}) plays an important role in the process of glucose-induced insulin release from the pancreatic β-cell. When stimulated by glucose, the β-cell displays a complex series of events that leads to a rise in cytosolic free-Ca^{2+} concentration ($[Ca^{2+}]_i$), that triggers insulin release.

The β-cell is equipped with a double system responsible for Ca^{2+} extrusion: the Na^+/Ca^{2+} exchanger (NCX) and the plasma membrane Ca^{2+}-ATPase (PMCA) (Carafoli 1988; Blaustein and Lederer 1999). The Na^+/Ca^{2+} exchanger is an electrogenic transporter located at the plasma membrane that couples the exchange of 3 Na^+ for 1 Ca^{2+}. The Na^+/Ca^{2+} exchanger has been cloned in 1990 (Nicoll et al. 1990), and four isoforms have been identified: NCX1, NCX2, NCX3 and NCX4 (Hryshko and Philipson 1997; On et al. 2008). The rat β-cell expresses two splice variants of the isoform NCX1, namely, NCX1.3 and NCX1.7; the mouse β-cell expresses also NCX1.2 (Van Eylen et al. 1997). PCR amplification did not yield any DNA fragment for NCX2, and NCX3 was not looked for (Van Eylen et al. 2001). In the rat pancreatic β-cell, Na^+/Ca^{2+} exchange displays a quite high capacity and participates in the control of $[Ca^{2+}]_i$ and of insulin release (Herchuelz and Plasman 1991).

The PMCA belongs to the P-type family of transport ATPases which form a phosphorylated intermediate during the reaction cycle (Carafoli 1994). The β-cell expresses the four main isoforms of the PMCA, namely, PMCA1, PMCA2, PMCA3 and PMCA4 (Kamagate et al. 2000). Six alternative splice mRNA variants, characterized at splice sites A and C, were detected in the β-cell (rPMCA1xb, 2yb, 2wb, 3za, 3zc, 4xb), plus one additional variant in pancreatic islet cells (PMCA4za). At the mRNA and protein level, five variants predominated (1xb, 2wb, 3za, 3zc, 4xb), whilst one additional isoform (4za) predominated only at the protein level. This provides evidence for the presence of PMCA2 and PMCA3 isoforms at the protein level in non-neuronal tissue. Hence, the pancreatic β-cell is equipped with multiple PMCA isoforms with possible differential regulation, providing a full range of PMCAs for $[Ca^{2+}]_i$ regulation (Kamagate et al. 2000).

In previous work, by performing loss and gain of function studies (use of antisense oligonucleotides targeting NCX1 or overexpression of NCX1), we observed that the Na^+/Ca^{2+} exchanger contributed significantly to Ca^{2+} outflow from the cell (70%) but also to Ca^{2+} entry in the β-cell. Indeed, during the upstroke of the action potentials, the Na^+/Ca^{2+} exchanger may reverse and contribute to Ca^{2+} entry (about 25 % of the initial peak) (Van Eylen et al. 1998, 2002).

By performing PMCA overexpression studies, we could demonstrate that the PMCA also contributes to Ca^{2+} outflow out of the cell (Kamagate et al. 2002). Surprisingly, in the clone showing the highest level of overexpression, the rise in $[Ca^{2+}]_i$ induced by membrane depolarization (K^+: 50 mM) was almost completely abolished. This is striking because the PMCA is considered as a high-affinity, low-capacity system at variance with the Na^+/Ca^{2+} exchanger which is a low-affinity, high-capacity system. Hence, the overexpression of a low-capacity system was not expected to reduce the rise in $[Ca^{2+}]_i$ to such an extent.

Different lines of evidence suggest that glucose, the main physiological stimulus of insulin release, stimulates β-cell Na^+/Ca^{2+} exchange activity (Plasman et al. 1990; Van Eylen et al. 2002). Previous work on the PMCA shows, on the contrary, that glucose inhibits PMCA activity (Gagliardino and Rossi 1994). To understand the respective role of these two mechanisms, we studied the effect of glucose on PMCA and NCX transcription, expression and activity in rat pancreatic islet cells (Ximenes et al. 2003). Glucose (11.1 and 22.2 mM) induced a parallel decrease in PMCA transcription, expression and activity. In contrast, the sugar induced a parallel increase in NCX transcription, expression and activity. The effects of the sugar were mimicked by the metabolizable insulin secretagogue α-ketoisocaproate and persisted in the presence of the Ca^{2+} channel blocker nifedipine. The above results are compatible with the view that when stimulated by glucose, the β-cell switches from a low-efficiency Ca^{2+} extruding mechanism, the PMCA, to a high-capacity system, the Na^{+}/Ca^{2+} exchanger, in order to better face the increase in Ca^{2+} inflow. These effects of glucose do not result from a direct effect of the sugar itself and are not mediated by the increase in intracellular free-Ca^{2+} concentration induced by the sugar (Ximenes et al. 2003).

33.2 NCX and PMCA in Diabetes

Evidences suggest that programmed cell death (apoptosis) represents the main mechanism of β-cell death in animal models of type 1 diabetes mellitus (T1DM) and possibly also in human with T1DM (Eizirik and Mandrup-Poulsen 2001). On the other hand, type 2 diabetes mellitus (T2DM) is a complex disease characterized by both insulin resistance and β-cell dysfunction. One of the earliest abnormalities occurring in this disease is the alteration in pulsatile insulin release with the suppression of the first phase of insulin response to glucose, both defects being present well before the development of overt hyperglycaemia and clinical diabetes (Utzschneider and Kahn 2004). The second phase of insulin release is also diminished, and a number of abnormalities of continuous insulin release have been observed (Porte 1991; Kahn et al. 2009). In addition to a defect in β-cell function, a reduction in islet and β-cell mass has been observed (Maclean and Ogilvie 1955; Rahier et al. 2008). This reduction could be related to increased programmed cell death (apoptosis), to a decrease in β-cell replication or both (Cnop et al. 2005).

Ca^{2+} is not solely of importance in cell signalling but may also trigger programmed cell death (apoptosis) and regulates death-specific enzymes. Therefore, the development of strategies to control Ca^{2+} homeostasis may represent a potential approach to prevent or enhance cell apoptosis. To test this hypothesis, the Na^{+}/Ca^{2+} exchanger (NCX1.7 isoform) was stably overexpressed in insulin-secreting tumoural cells (Diaz-Horta et al. 2002). NCX overexpression increased apoptosis induced by sarco-endoplasmic reticulum (ER) Ca^{2+}-ATPase (SERCA) inhibitors but not by agents increasing $[Ca^{2+}]_i$ through the opening of plasma membrane Ca^{2+} channels. NCX overexpression reduced the rise in $[Ca^{2+}]_i$ induced by all agents, depleted ER Ca^{2+} stores, sensitized the cells to Ca^{2+}-independent proapoptotic signalling pathways and reduced cell proliferation by about 40%. ER Ca^{2+} store depletion was accompanied by the activation of the ER-specific caspase (caspase 12), the activation being enhanced by SERCA inhibitors. Hence, Na^{+}/Ca^{2+} exchanger overexpression, by depleting ER Ca^{2+} stores, triggers the activation of caspase 12 and increases apoptotic cell death (Diaz-Horta et al. 2002). By increasing apoptosis and decreasing cell proliferation, overexpression of Na^{+}/Ca^{2+} exchanger may represent a new potential approach in cancer gene therapy. On the other hand, the latter results open the way to the development of new strategies to control cellular Ca^{2+} homeostasis that could on the contrary prevent the process of apoptosis that mediates, in part, β-cell destruction in T1 and T2DM (Diaz-Horta et al. 2002).

Type 1 cytokines, such as interleulin 1β (IL-1β), tumour necrosis factor-α (TNF-α) and interferon-γ (IFN-γ), are early mediators of β-cell death in T1DM (Eizirik and Mandrup-Poulsen 2001), and a combination of IL-1β + IFN-γ has been shown to decrease the expression of the ER Ca^{2+} pump SERCA2b in the β-cell (Cardozo et al. 2001a, b; Kutlu et al. 2003). Therefore, we wondered whether cytokines could not induce β-cell death by depleting ER Ca^{2+} stores, namely, by a mechanism similar to that induced by Na^{+}/Ca^{2+} overexpression. By using fura-2 and furaptra to monitor cytosolic and ER free-Ca^{2+} concentration, we could show that cytokines, like NCX overexpression, induce a severe depletion of ER Ca^{2+} stores with activation of ER stress (Cardozo et al. 2005).

Chronic exposure to high free fatty acids (FFA) concentrations causes β-cell apoptosis (Cnop et al. 2001, 2005; Maedler et al. 2001) and may contribute to the increased β-cell apoptosis rates in T2DM (Butler et al. 2003), a phenomenon called lipotoxicity (El-Assaad et al. 2003). Interestingly, we could recently show that saturated FFA also induce ER stress via ER Ca^{2+} depletion with resulting β-cell apoptosis (Cunha et al. 2008).

In summary, ER Ca^{2+} depletion with resulting ER stress appears as a mechanism common to various conditions leading to β-cell death. In order to further evaluate such a view, we examined whether PMCA overexpression may lead to a similar picture and explored in further details the pathways triggered by ER stress. On the other hand, our data also suggest that the opposite effect, namely, a reduction in Na^+/Ca^{2+} exchange or PMCA activity, may lead to an increase in ER Ca^{2+} stores and perhaps reduce β-cell apoptosis. These hypotheses have been tested and are presented in the following paragraphs (Jiang et al. 2010; Nguidjoe et al. 2011).

33.3 Effect of PMCA Overexpression on β-Cell Death

In this study, clonal β-cells (BRIN-BD11) were examined for the effect of PMCA overexpression on cytosolic, ER and mitochondrial $[Ca^{2+}]$ using a combination of aequorins with different Ca^{2+} affinities and on the ER and mitochondrial pathways of apoptosis (Jiang et al. 2010). Overexpression of PMCA decreased $[Ca^{2+}]$ in the cytosol, the ER and the mitochondria and induced apoptosis. Figure 33.1 shows that PMCA2-overexpressing clones displayed an increased rate of basal apoptosis (Fig. 33.1a). The sarco-endoplasmic reticulum Ca^{2+}-ATPase (SERCA) inhibitors cyclopiazonic acid (CPA, 50 μM) and thapsigargin (500 nM) increased apoptosis levels in all clones, but this was more marked in PMCA2-overexpressing cells (Fig. 33.1a). The rate of apoptosis induced by CPA was significantly higher in clone 2 than in clone 5, namely, in the clone showing the highest level of PMCA2 overexpression (Fig. 33.1a), indicating that Ca^{2+} depletion in itself is instrumental. PMCA2-overexpressing cells also displayed increased levels of caspase 3 cleavage compared to control cells, in the absence or presence of CPA (Fig. 33.1b). Under these conditions, the rate of caspase 3 cleavage tended to be higher in clone 2 than in clone 5.

PMCA overexpression activated the IRE1α-XBP1s but inhibited the PERK-eiF2α-CHOP and the ATF6-BiP pathways of the ER unfolded protein response. Increased Bax/Bcl-2 expression ratio (proapoptotic/antiapoptotic Bcl-2 family members) was observed in PMCA-overexpressing β-cells. This was followed by Bax translocation to the mitochondria with subsequent cytochrome c release, opening the permeability transition pore, loss of mitochondrial membrane potential and apoptosis. Interestingly, $[Ca^{2+}]$ was not solely decreased in the cytoplasm and the ER of PMCA2-overexpressing cells, but also in the mitochondria. Indeed, in various models of cell death due to ER Ca^{2+} release and ER Ca^{2+} depletion, cell death is attributed to the uptake of Ca^{2+} by the mitochondria with resulting opening of the MPTP and loss of the mitochondrial electrochemical gradient (Orrenius et al. 2003). Although the two latter changes were observed in our cells, they cannot be attributed to mitochondrial Ca^{2+} overload (Jiang et al. 2010).

Thus, PMCA2 overexpression like Ncx1 overexpression induces β-cell Ca^{2+} depletion and β-cell death, but the mechanisms leading to β cell death could be different.

33.4 Effect of Ncx1 Heterozygous Inactivation on Pancreatic β-Cell Function

If it is possible to increase apoptosis and to decrease β-cell proliferation by increasing the activity of the Na^+/Ca^{2+} exchanger, it may be possible to obtain the opposite effects by downregulating such a mechanism. In order to test this

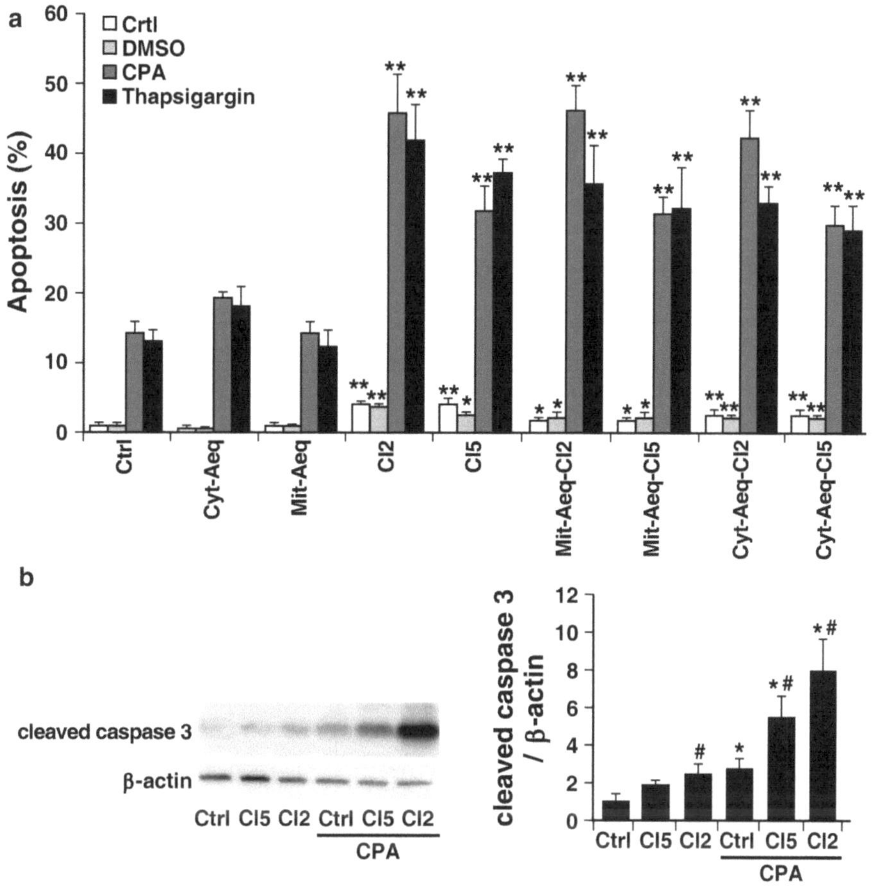

Fig. 33.1 *Effect of PMCA2 overexpression on cells viability and caspase 3 cleavage.* (**a**) Cell viability. Non-transfected BRIN-BD11 cells (Ctrl) or different clones of BRIN cells transfected with PMCA2, clone 2 (Cl 2) or 5 (Cl 5) and/or aequorin targeted to the cytosol (Cyt-Aeq) or the mitochondria (Mit-Aeq) were untreated (*white bars*) or treated for 24 h with CPA (*dark grey bars*), thapsigargin (*black bars*) or the solvent DMSO (*light grey bars*). Apoptosis levels were evaluated by observation under a microscope after HO-PI staining. The data are expressed as the percentage of apoptotic cells over the total number of cells counted ± SEM. Results are means of three to five independent experiments. **P* <0.05; ***P* <0.01; versus respective non-transfected control. (**b**) Caspase 3 cleavage. Western blot analyses of non-transfected (Ctrl) and PMCA2-transfected cells, clone 2 (Cl 2) or clone 5 (Cl 5) using an antibody directed against the cleaved caspase 3 fragment. *Left panel*: representative blot of caspase 3 and β-actin expression. *Right panel*: quantitative assessment of cleaved caspase 3 levels normalized to the β-actin levels. Results are means of five independent experiments. **P* <0.05 versus respective non-treated control. #*P* <0.05 versus respective non-transfected condition. This research was originally published in J. Biol. Chem. Jiang, L., Allagnat, F., Nguidjoe, E., Kamagate, A., Pachera, A., Vanderwinden, J.-M., Brini, M., Carafoli, E., Eizirik D. L., Cardozo, A. K. and Herchuelz, A., (2010). Plasma membrane Ca^{2+}-ATPase overexpression depletes both mitochondrial and endoplasmic reticulum Ca^{2+} stores and triggers apoptosis in insulin-secreting BRIN-BD11 cells (Reproduced from Jiang et al. (2010))

hypothesis, we generated Ncx1 heterozygous deficient mice (Ncx1$^{+/-}$) (Nguidjoe et al. 2011).

Evidence was obtained that the expression of the exchanger in the β-cell was reduced not only at the mRNA but also at the functional level. For instance, the uptake of ^{45}Ca induced by the removal of extracellular Na^+ (reverse Na^+/Ca^{2+} exchange) was reduced by half (Fig. 33.2a), while the increase in cytosolic free-Ca^{2+} concentration ($[Ca^{2+}]_i$) induced by the same manoeuvre

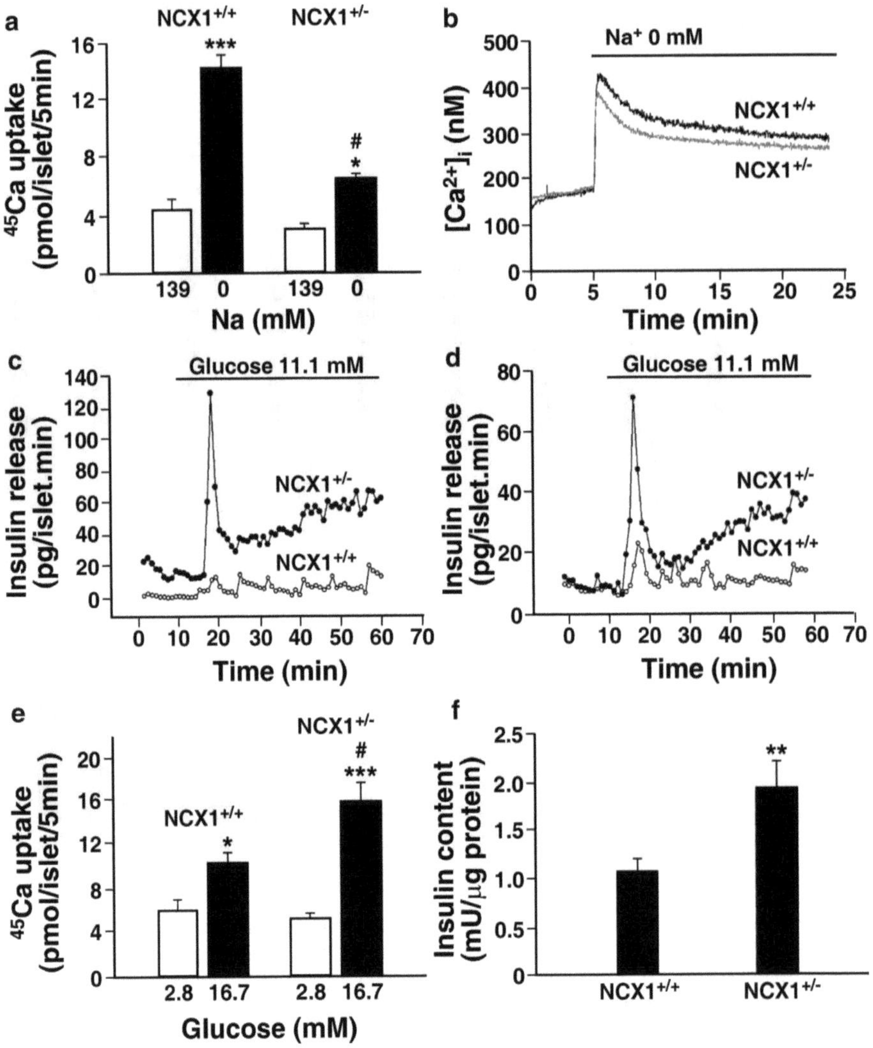

Fig. 33.2 *Effect of Ncx1 heterozygous inactivation on Na^+/Ca^{2+} exchange activity and on islet function.* (**a**) ^{45}Ca uptake in the presence and the absence of extracellular Na^+. Means ± SEM of four experiments, comprising four to five replicates each. **P* <0.05, ****P* <0.001 versus Na^+ 139 mM; #*P* <0.001 versus $Ncx1^{+/+}$. (**b**) Effect of extracellular Na^+ removal on $[Ca^{2+}]_i$ in $Ncx1^{+/+}$ and $Ncx1^{+/-}$ islets. The period of exposure to Na^+-free medium is indicated by a bar above the curves. The curves shown are the mean of seven traces in each case. (**c–d**) Effect of 11.1 mM glucose on insulin release from groups of 20 islets, (**c**) representative experiment, (**d**) mean of 4 and 6 ($Ncx1^{+/-}$) experiments. The amount of insulin released in response to glucose is about 2.5 times higher in $Ncx1^{+/-}$ than in $Ncx1^{+/+}$ islets (*P* <0.05). (**e**) Effect of glucose on ^{45}Ca uptake in $Ncx1^{+/+}$ and $Ncx1^{+/-}$ islets (*n* = 4 to 6 experiments, **P* <0.05, ****P* <0.001 versus 2.8 mM; #P <0.01 versus $Ncx1^{+/+}$ islets at 16.7 mM glucose). (**f**) Insulin content of batches of ten islets; *n* = 10 experiments ***P* <0.01 versus $Ncx1^{+/+}$ islets (Reproduced from Nguidjoe et al. (2011))

(Fig. 33.2b) was reduced by 24% (*P* <0.05) in $Ncx1^{+/-}$ compared to $Ncx1^{+/-}$ islets.

We then measured the effect of glucose-induced insulin release from pancreatic islets. Figure 33.2c, d shows the effect of an increase in glucose concentration from 2.8 to 11.1 mM on insulin release from perfused islets (c: representative experiment, d: mean of 4–6 experiments). In $Ncx1^{+/+}$ islets, glucose induced an oscillatory increase in insulin release, while in $Ncx1^{+/-}$ islets, the sugar induced a marked first phase followed by a progressive increase in insulin release with

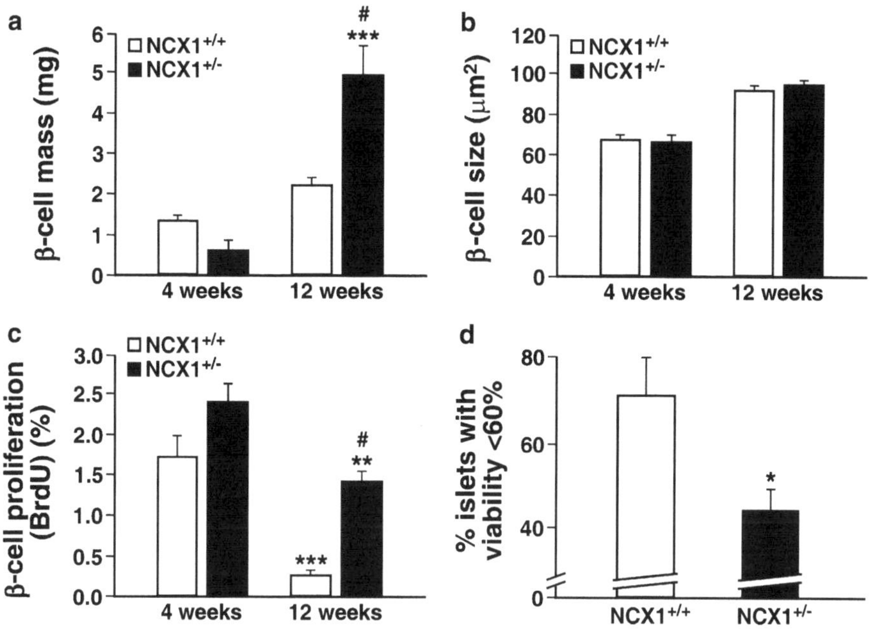

Fig. 33.3 *Effect of Ncx1 heterozygous inactivation on β-cell mass, size, proliferation and resistance to hypoxia.* Changes in β-cell mass (**a**), size (**b**), proliferation rate (**c**) between weeks 4 and 12 in Ncx1$^{+/+}$ (*open bars*) and Ncx1$^{+/-}$ mice (*closed bars*). Mean ± SEM values from 5 to 6 pancreases, respectively. (**a**) ***P <0.001 versus 4 weeks values; #P <0.001 versus Ncx1$^{+/+}$ β-cells at 12 weeks. (**c**) **P <0.01, ***P <0.001 versus respective value at 4 weeks; #P <0.01 versus Ncx1$^{+/+}$ islets at 12 weeks. (**d**) Cell viability measured in intact islets using Ho342 and PI staining after 6 h exposure to hypoxia. Mean ± SEM values from four individual experiments. *P <0.05 versus Ncx1$^{+/+}$ islets (Reproduced from Nguidjoe et al. (2011))

less clear oscillations. The amount of insulin released during the initial phase (16–20 min) and the whole period of stimulation (16–60 min) was about 2.6 and 2.4 times higher, respectively, in Ncx1$^{+/-}$ than in Ncx1$^{+/+}$ islets (P <0.02). The major increase in insulin release was attended, by an increase in glucose-induced ^{45}Ca uptake (Fig. 33.2e), in islet insulin content (Fig. 33.2f) and in proinsulin immunostaining. The increase in ^{45}Ca uptake induced by 16.7 mM glucose and insulin content was twice as high in Ncx1$^{+/-}$ than in Ncx1$^{+/+}$ islets. Taken as a whole, the data so far presented show that Ncx1 heterozygous inactivation strongly increases β-cell function including glucose-induced insulin production and release.

We then measured β-cell mass, size and proliferation. The latter parameters were measured at 4 and 12 weeks, namely, in the young and adult age (Fig. 33.3a–c). As expected, β-cell mass was increased at 12 compared to 4 weeks in both types of islets though the increase was of much larger magnitude in Ncx1$^{+/-}$ than Ncx1^{++} islets (8.8 vs. 1.6-fold increase, respectively, P <0.001, Fig. 33.3a). This increase was not due to β-cell or islet hypertrophy since no change in β-cell and islet size was observed between Ncx1$^{+/+}$ and Ncx1$^{+/-}$ mice (Fig. 33.3b). It was rather due to an increase in β-cell proliferation rate. Again, as expected, β-cell proliferation was decreased at 12 compared to 4 weeks (Fig. 33.3c), though the decrease was of lower magnitude in Ncx1$^{+/-}$ than Ncx1$^{+/+}$ islets (−40% vs. −85%, P <0.01). As a result, a 5.25 times higher proliferation rate was observed at 12 weeks in Ncx1$^{+/-}$ compared to Ncx1$^{+/+}$ mice (Fig. 33.3c).

β-cells apoptosis was also measured, but no difference could be found between Ncx1$^{+/-}$ and Ncx1$^{+/+}$ islets, whether under basal or stimulated conditions (e.g. in the presence of SERCA inhibitors or cytokines) and using different methods (TUNEL method or Ho342 and PI staining). The sole condition under which a difference could be found was when the islets were exposed to hypoxia (6 h). Thus, in $Ncx1^{+/+}$ islets, 71% of the islets showed a decrease in viability below 60% compared with 45% in $Ncx1^{+/-}$ islets when exposed to hypoxia (Fig. 33.3d).

Islet transplantation represents a valuable approach in the treatment of diabetes. However, its applicability is limited by the need to transplant a high number of islets (from two or more donors). In clinical islet transplantation, it has been estimated that up to 70% of the transplanted β-cell mass is destroyed in the early post-transplant period due to non-immune-mediated physiological stress, namely, prolonged hypoxia during the revascularization process (Emamaullee and Shapiro 2006). Therefore, we transplanted $Ncx1^{+/-}$ islets under the kidney capsule of alloxan-diabetic mice to examine their performances compared to $Ncx1^{+/+}$ islets. The rate of success of a 200 $Ncx1^{+/+}$ islet transplantation was 2/5. In comparison, the rate of success of a 100 $Ncx1^{+/-}$ islet transplantation was 4/5, whilst the rate of success of a 50 $Ncx1^{+/-}$ islets transplantation was 2/3. This suggests that the $Ncx1^{+/-}$ islets are at least 4–7 times more efficient to cure diabetes than $Ncx1^{+/+}$ islets.

Otherwise, the phenotype of $Ncx1^{+/-}$ mice appeared normal, and their glucose metabolism (*in vivo*) was similar to that of $Ncx1^{+/+}$ mice except for an increased and earlier initial peak of insulin release during the glucose tolerance test, a finding in agreement with the major increase in glucose-induced first phase insulin release (Fig. 33.2c–d) in $Ncx1^{+/-}$ islets.

Downregulation of the β-cell Na^+/Ca^{2+} exchanger is thus a unique model providing a novel concept for the prevention and treatment of T1DM and T2DM and to improve the applicability of islet transplantation.

33.5 Effect of the NCX Inhibitor KB-R7943 on NCX1 Splice Variants and Insulin Release

In a recent study, Hamming et al. 2010, using the patch-clamp technique, examined the effect of the NCX inhibitor KB-R7943 on recombinant NCX1 isoforms activity in forward mode. They observed that at variance with NCX1.1, the isoform expressed in the heart, NCX1.3 and NCX1.7, the isoforms expressed in the β-cell, exhibit significant inactivation during forward mode operation and are 15- to 18-fold more sensitive to KB-R7943 inhibition, compared with NCX1.1 ($IC_{50}s = 2.9$ and 2.4 vs. 43 μM, respectively). Because NCX1 splice variants differ only in the exon composition of the alternative splice region (ABCDEF for NCX1.1, BD and BDF for NCX1.3 and NCX1.7, respectively), the authors replaced exon A by exon B in NCX1.1 generating NCX1.11. This conferred to NCX1.11 the same sensitivity to KB-R7943 than NCX1.3 and NCX1.7. On the contrary, replacing exon B with A in NCX1.3 (NCX1.4), dramatically decreased the activity of the drug. KB-R7943 also increased glucose-induced rise in $[Ca^{2+}]_i$ and insulin release from pancreatic islets. The latter data are in agreement with our own (Nguidjoe et al. 2011). Taken together, our works suggest that inhibitors of the Na^+/Ca^{2+} exchanger may represent new therapeutic agents in diabetes that could prevent the development of T1DM and T2DM in at-risk patients, preserve residual β-cell function and mass in recent-onset DM, activate endogenous β-cell regeneration by stimulation of β-cell proliferation in DM and improve the applicability of islet transplantation.

33.6 Conclusions

Gain of NCX1 or PMCA2 function leads to ER stress, decrease in β-cell proliferation and β-cell death by apoptosis. Loss of NCX1 function leads on the contrary to an increase in β-cell function, mass and proliferation, namely, to various changes in β-cell function that are opposite to the major abnormalities seen in diabetes. Hence, we identified novel targets to preserve and protect functional β-cell mass in the treatment of both T1DM and T2DM. Because NCX1 splice variants expressed in the β-cell are about 20 times more sensitive to NCX1 inhibitors like KB-R7943 than isoforms expressed in other tissues, future work in the field may lead to the development of new therapeutic agents that could prevent the development of diabetes in at-risk patients, preserve residual β-cell function and mass in recent-onset DM and activate endogenous β-cell regeneration by stimulation of β-cell proliferation in DM.

Acknowledgements The authors thank A. Van Praet, A. Iabkriman and M.P. Berghmans (Laboratory of Pharmacology; ULB) for excellent technical support. This work has been supported by grants from the Belgian Fund for Scientific Research (FRSM 3.4593.04, 3.4527.08) and the EFSD/Novo Nordisk Programme in Diabetes Research (2005/2006).

References

M.P. Blaustein, W.J. Lederer, Sodium/calcium exchange: its physiological implications. Physiol. Rev. **79**, 763–854 (1999)

A.E. Butler, J. Janson, S. Bonner-Weir, R. Ritzel, R.A. Rizza, P.C. Butler, Beta-cell deficit and increased beta-cell apoptosis in humans with type 2 diabetes. Diabetes **52**, 102–110 (2003)

E. Carafoli, Membrane transport of calcium: an overview. Methods Enzymol. **157**, 3–11 (1988)

E. Carafoli, Biogenesis: plasma membrane calcium ATPase: 15 years of work on the purified enzyme. FASEB J. **8**, 993–1002 (1994)

A.K. Cardozo, H. Heimberg, Y. Heremans, R. Leeman, B. Kutlu, M. Kruhoffer, T. Orntoft, D.L. Eizirik, A comprehensive analysis of cytokine-induced and nuclear factor-kappa B-dependent genes in primary rat pancreatic beta-cells. J. Biol. Chem. **276**, 48879–48886 (2001a)

A.K. Cardozo, M. Kruhoffer, R. Leeman, T. Orntoft, D.L. Eizirik, Identification of novel cytokine-induced genes in pancreatic beta-cells by high-density oligonucleotide arrays. Diabetes **50**, 909–920 (2001b)

A.K. Cardozo, F. Ortis, J. Storling, Y.M. Feng, J. Rasschaert, M. Tonnesen, F. Van Eylen, T. Mandrup-Poulsen, A. Herchuelz, D.L. Eizirik, Cytokines downregulate the sarcoendoplasmic reticulum pump Ca^{2+} ATPase 2b and deplete endoplasmic reticulum Ca^{2+}, leading to induction of endoplasmic reticulum stress in pancreatic beta-cells. Diabetes **54**, 452–461 (2005)

M. Cnop, J.C. Hannaert, A. Hoorens, D.L. Eizirik, D.G. Pipeleers, Inverse relationship between cytotoxicity of free fatty acids in pancreatic islet cells and cellular triglyceride accumulation. Diabetes **50**, 1771–1777 (2001)

M. Cnop, N. Welsh, J.C. Jonas, A. Jorns, S. Lenzen, D.L. Eizirik, Mechanisms of pancreatic beta-cell death in type 1 and type 2 diabetes: many differences, few similarities. Diabetes **54**(Suppl 2), S97–S107 (2005)

D.A. Cunha, P. Hekerman, L. Ladriere, A. Bazarra-Castro, F. Ortis, M.C. Wakeham, F. Moore, J. Rasschaert, A.K. Cardozo, E. Bellomo, L. Overbergh, C. Mathieu, R. Lupi, T. Hai, A. Herchuelz, P. Marchetti, G.A. Rutter, D.L. Eizirik, M. Cnop, Initiation and execution of lipotoxic ER stress in pancreatic beta-cells. J. Cell Sci. **121**, 2308–2318 (2008)

O. Diaz-Horta, A. Kamagate, A. Herchuelz, F. Van Eylen, Na^+/Ca^{2+} exchanger overexpression induces endoplasmic reticulum-related apoptosis and caspase-12 activation in insulin-releasing BRIN-BD11 cells. Diabetes **51**, 1815–1824 (2002)

D.L. Eizirik, T. Mandrup-Poulsen, A choice of death –the signal-transduction of immune-mediated beta-cell apoptosis. Diabetologia **44**, 2115–2133 (2001)

W. El-Assaad, J. Buteau, M.L. Peyot, C. Nolan, R. Roduit, S. Hardy, E. Joly, G. Dbaibo, L. Rosenberg, M. Prentki, Saturated fatty acids synergize with elevated glucose to cause pancreatic beta-cell death. Endocrinology **144**, 4154–4163 (2003)

J.A. Emamaullee and A.M. Shapiro, Perspectives in diabetes, interventional strategies to prevent beta-cell apoptosis in islet transplantation. Diabetes **55**, 1907–1914 (2006)

J.J. Gagliardino, J.P. Rossi, Ca^{2+}-ATPase in pancreatic islets: its possible role in the regulation of insulin secretion. Diabetes Metab. Rev. **10**, 1–17 (1994)

K.S. Hamming, D. Soliman, N.J. Webster, G.J. Searle, L.C. Matemisz, D.A. Liknes, X.Q. Dai, T. Pulinilkunnil, M.J. Riedel, J.R. Dyck, P.E. Macdonald, P.E. Light, Inhibition of beta-cell sodium-calcium exchange enhances glucose-dependent elevations in cytoplasmic calcium and insulin secretion. Diabetes **59**, 1686–1693 (2010)

A. Herchuelz, P.O. Plasman, Sodium-calcium exchange in the pancreatic B cell. Ann. N. Y. Acad. Sci. **639**, 642–656 (1991)

L.V. Hryshko, K.D. Philipson, Sodium-calcium exchange: recent advances. Basic Res. Cardiol. **92**(Suppl. 1), 45–51 Review (1997)

L. Jiang, F. Allagnat, E. Nguidjoe, A. Kamagate, A. Pachera, J.-M. Vanderwinden, M. Brini, E. Carafoli, D.L. Eizirik, A.K. Cardozo, A. Herchuelz, Plasma membrane Ca^{2+}-ATPase overexpression depletes both mitochondrial and endoplasmic reticulum Ca^{2+} stores and triggers apoptosis in insulin-secreting BRIN BD11 cells. J. Biol. Chem. **285**, 30634–30643 (2010)

S.E. Kahn, S. Zraika, K.M. Utzschneider, R.L. Hull, The beta cell lesion in type 2 diabetes: there has to be a primary functional abnormality. Diabetologia **52**, 1003–1012 (2009)

A. Kamagate, A. Herchuelz, A. Bollen, F. Van Eylen, Expression of multiple plasma membrane Ca^{2+}-ATPases in rat pancreatic islet cells. Cell Calcium **27**, 231–246 (2000)

A. Kamagate, A. Herchuelz, F. Van Eylen, Plasma membrane Ca^{2+}-ATPase overexpression reduces Ca^{2+} oscillations and increases insulin release induced by glucose in insulin-secreting BRIN-BD11 cells. Diabetes **51**, 2773–2788 (2002)

B. Kutlu, A.K. Cardozo, M.I. Darville, M. Kruhoffer, N. Magnusson, T. Orntoft, D.L. Eizirik, Discovery of gene networks regulating cytokine-induced dysfunction and apoptosis in insulin-producing INS-1 cells. Diabetes **52**, 2701–2719 (2003)

N. Maclean, R.F. Ogilvie, Quantitative estimation of the pancreatic islet tissue in diabetic subjects. Diabetes **4**, 367–376 (1955)

K. Maedler, G.A. Spinas, D. Dyntar, W. Moritz, N. Kaiser, M.Y. Donath, Distinct effects of saturated and monounsaturated fatty acids on beta-cell turnover and function. Diabetes **50**, 69–76 (2001)

E. Nguidjoe, S. Sokolow, S. Bigabwa, N. Pachera, E. D'Amico, F. Allagnat, J.M. Vanderwinden, A. Sener, M. Manto, M. Depreter, J. Mast, G. Joanny, E. Montanya, J. Rahier, A.K. Cardozo, D.L. Eizirik, S. Schurmans, A. Herchuelz, Heterozygous inactivation of the Na/Ca^{2+} exchanger increases glucose-induced insulin release, beta-cell proliferation, and mass. Diabetes **60**, 2076–2085 (2011)

D.A. Nicoll, S. Longoni, K.D. Philipson, Molecular cloning and functional expression of the cardiac sarcolemmal Na^{+}-Ca^{2+} exchanger. Science **250**, 562–565 (1990)

C. On, C.R. Marshall, N. Chen, C.D. Moyes, G.F. Tibbits, Gene structure evolution of the Na^{+}-Ca^{2+} exchanger (NCX) family. BMC Evol. Biol. **8**, 127–142 (2008)

S. Orrenius, B. Zhivotovsky, P. Nicotera, Regulation of cell death: the calcium-apoptosis link. Nat. Rev. Mol. Cell Biol. **4**, 552–565 (2003)

P.O. Plasman, P. Lebrun, A. Herchuelz, Characterization of the process of sodium-calcium exchange in pancreatic islet cells. Am. J. Physiol. **259**, E844–E850 (1990)

D. Porte Jr., Banting lecture 1990. Beta-cells in type II diabetes mellitus. Diabetes **40**, 166–180 (1991)

J. Rahier, Y. Guiot, R.M. Goebbels, C. Sempoux, J.C. Henquin, Pancreatic beta-cell mass in European subjects with type 2 diabetes. Diab. Obes. Metab. **10**(Suppl 4), 32–42 (2008)

K.M. Utzschneider, S.E. Kahn, β-cell dysfunction in type 2 diabetes, in *International Textbook of Diabetes Mellitus*, ed. by R.A. DeFronzo, E. Ferrannini, H. Keen, P. Zimmet (Wiley, Hoboken, 2004), pp. 375–388

F. Van Eylen, M. Svoboda, A. Herchuelz, Identification, expression pattern and potential activity of Na/Ca^{2+} exchanger isoforms in rat pancreatic B-cells. Cell Calcium **21**, 185–193 (1997)

F. Van Eylen, C. Lebeau, J. Albuquerque-Silva, A. Herchuelz, Contribution of Na/Ca^{2+} exchange to Ca^{2+} outflow and entry in the rat pancreatic beta-cell: studies with antisense oligonucleotides. Diabetes **47**, 1873–1880 (1998)

F. Van Eylen, A. Bollen, A. Herchuelz, NCX1 Na/Ca^{2+} exchanger splice variants in pancreatic islet cells. J. Endocrinol. **168**, 517–526 (2001)

F. Van Eylen, O.D. Horta, A. Barez, A. Kamagate, P.R. Flatt, R. Macianskiene, K. Mubagwa, A. Herchuelz, Overexpression of the Na/Ca^{2+} exchanger shapes stimulus-induced cytosolic Ca^{2+} oscillations in insulin-producing BRIN-BD11 cells. Diabetes **51**, 366–375 (2002)

H.M. Ximenes, A. Kamagate, F. Van Eylen, A. Carpinelli, A. Herchuelz, Opposite effects of glucose on plasma membrane Ca^{2+}-ATPase and Na/Ca^{2+} exchanger transcription, expression, and activity in rat pancreatic beta-cells. J. Biol. Chem. **278**, 22956–22963 (2003)

Part XI

NCX Partners in Ionic Homeostasis: ASIC, NMDA, NHE and TRPC

The Na⁺/H⁺ Exchanger NHE5 Is Sorted to Discrete Intracellular Vesicles in the Central and Peripheral Nervous Systems

34

Viktoria Lukashova, Tushare Jinadasa, Alina Ilie, David Verbich, Ellis Cooper, and John Orlowski

Abstract

The pH milieu of the central and peripheral nervous systems is an important determinant of neuronal excitability, function, and survival. In mammals, neural acid–base homeostasis is coordinately regulated by ion transporters belonging to the Na^+/H^+ exchanger (NHE) and bicarbonate transporter gene families. However, the relative contributions of individual isoforms within the respective families are not fully understood. This report focuses on the NHE family, specifically the plasma membrane-type NHE5 which is preferentially transcribed in brain, but the distribution of the native protein has not been extensively characterized. To this end, we generated a rabbit polyclonal antibody that specifically recognizes NHE5. In both central (cortex, hippocampus) and peripheral (superior cervical ganglia, SCG) nervous tissue of mice, NHE5 immunostaining was punctate and highly concentrated in the somas and to lesser amounts in the dendrites of neurons. Very little signal was detected in axons. Similarly, in primary cultures of differentiated SCG neurons, NHE5 localized predominantly to vesicles in the somatodendritic compartment, though some immunostaining was also evident in punctate vesicles along the axons. NHE5 was also detected predominantly in intracellular vesicles of cultured SCG glial cells. Dual immunolabeling of SCG neurons showed that NHE5 did not colocalize with markers for early endosomes (EEA1) or synaptic vesicles (synaptophysin), but did partially colocalize with the transferrin receptor, a marker of recycling endosomes. Collectively, these data suggest that NHE5 partitions into a unique vesicular pool in neurons that shares some

V. Lukashova • T. Jinadasa • A. Ilie • D. Verbich
• E. Cooper • J. Orlowski (✉)
Department of Physiology, McGill University,
McIntyre Medical Science Bldg, 3655 Promenade
Sir-William-Osler, Montreal, QC H3G 1Y6, Canada
e-mail: john.orlowski@mcgill.ca

L. Annunziato (ed.), *Sodium Calcium Exchange: A Growing Spectrum of Pathophysiological Implications*, Advances in Experimental Medicine and Biology 961, DOI 10.1007/978-1-4614-4756-6_34,

characteristics of recycling endosomes where it may serve as an important regulated store of functional transporters required to maintain cytoplasmic pH homeostasis.

Keywords
Brain • pH homeostasis • Sodium/proton exchangers • Subcellular distribution • Imaging

34.1 Introduction

The central nervous system (CNS) is highly active metabolically relative to most other organ systems and is especially reliant on glucose and its metabolite lactate as fuel to generate the main form of biochemical energy, adenosine 5′-triphosphate (ATP), needed to sustain membrane excitability and synaptic transmission (Chih et al. 2001; Schurr et al. 1988; Tsacopoulos and Magistretti 1996; Wyss et al. 2011). Indeed, it is estimated that the brains of most vertebrate species consume between 2% and 8% of the total body energy production, but in humans this can increase to as much as 20% (Mink et al. 1981). Such elevated rates of metabolism would be anticipated to generate considerable CO_2 and H^+ (or acid equivalents) as waste products that are in dynamic equilibrium ($CO_2 + H_2O \leftrightarrow H_2CO_3 \leftrightarrow H^+ + HCO_3^-$) and must be efficiently buffered or removed to maintain a suitable pH milieu for stable protein activity and interactions, and ultimately neural function. In this regard, it is well documented that various depolarizing stimuli (e.g., repetitive spike activity, hormones, and neurotransmitters) increase metabolic acid production which elicits transient fluctuations in intra- as well as extracellular pH of many types of neurons and glial cells that, in turn, can further modulate membrane excitability (Chen and Chesler 1992a, b; Chesler 2003; Dulla et al. 2005; Hsu et al. 2000; Luckermann et al. 1997; Paalasmaa and Kaila 1996; Ransom 2000; Rose and Ransom 1996; Trapp et al. 1996). Mammals have also developed highly specialized chemosensitive cells located centrally (Mulkey et al. 2004; Putnam 2010; Severson et al. 2003; Williams et al. 2007) as well as peripherally (DeSimone and Lyall 2006; Gonzalez et al. 1992) that are exquisitely responsive to minute changes in arterial blood/interstitial CO_2/H^+ levels and are believed to modulate adaptive behaviors such as breathing, arousal, nociception, and sour taste. The underlying molecular mechanisms are not fully elucidated, but are thought to reflect alterations in the conductances of certain pH-sensitive neurotransmitter-gated receptors (Traynelis and Cull-Candy 1991; Zhai et al. 1998), voltage- and H^+-gated cation channels (Baron et al. 2002; Buckler et al. 2000; Church et al. 1998; Kiss and Korn 1999; Tombaugh and Somjen 1996; Tombaugh and Somjen 1997; Waldmann 2001), gap junctional coupling (Rorig et al. 1996), as well as activation of unique H^+-sensing G protein-coupled receptors (Huang et al. 2007; Seuwen et al. 2006). Abnormal disruptions of acid–base homeostasis have also been implicated in the progression of certain neuropathies, including ischemic and hypoxic injuries (Bondarenko and Chesler 2001; Li and Siesjo 1997; Pulsinelli 1992), cerebral edema (Kempski 2001), and seizures (Ali et al. 2006; Gu et al. 2001; Hentschke et al. 2006; Schuchmann et al. 2006; Vilas et al. 2009). Hence, precise regulation of neural pH is an important physiological process and may serve as a physiologically relevant signal to modulate nervous system function (Brookes 1997; Deitmer and Rose 1996; Putnam 2001; Putnam et al. 2004; Takahashi and Copenhagen 1996). It is therefore important to understand the mechanisms that generate and regulate neural pH homeostasis.

The molecular machinery responsible for acid–base regulation in the nervous system is similar to that of peripheral cell types and involves the coordinated activities of several distinct ion transporters resident at the plasma membrane, principally Na^+/H^+ antiporters or exchangers (NHA/NHE/NHX) in combination with one or more HCO_3^--dependent transporters in the forms of Cl^-/HCO_3^- exchangers, Na^+-dependent Cl^-/HCO_3^- exchangers, and Na^+-HCO_3^- cotransporters (for comprehensive reviews, see Casey et al. 2010; Chesler 2003; Putnam et al. 2004; Romero et al. 2004). These pH-regulating transporters exist as multiple isoforms that are distributed throughout the nervous system in a ubiquitous or cell-specific manner, suggesting unique contributions to neural pH homeostasis (Bonnet et al. 2000; Douglas et al. 2001; Havenga et al. 1994; Kobayashi et al. 1994; Ma and Haddad 1997; Rickmann et al. 2007). However, their individual regulatory properties and functions have yet to be extensively characterized, often hindered in part by the lack of suitable isoform-specific molecular probes. For the purposes of this report, further discussions will focus on the mammalian NHEs, with a specific emphasis on the NHE5 isoform which is preferentially expressed in brain.

34.1.1 Diversity of Na^+/H^+ Exchangers

The mammalian Na^+/H^+ exchanger gene family (classified as the solute carrier SLC9 family by the HUGO Gene Nomenclature Committee) consists of at least 11 structural diverse isoforms (NHE1-9/SLC9A1-9 and NHA1-2/SLC9B1-2) that assemble as homodimers and mediate the countertransport of alkali cations (Li^+, Na^+, or K^+) for H^+ across biological membranes, though their substrate specificity can differ among isoforms (Brett et al. 2005; Casey et al. 2010). The NHEs are expressed in a ubiquitous or tissue-specific manner and are differentially sorted to the plasma membrane as well as various endomembrane compartments, reflecting their involvement in diverse physiological processes. The plasma membrane-type NHEs include NHE1 to NHE5 and are generally involved in cytoplasmic or systemic pH and fluid volume homeostasis. Kinetically, they share a common mode of operation, preferentially exchanging extracellular Na^+ (and in some cases Li^+) for cytoplasmic H^+ (i.e., Na^+-selective NHEs). By contrast, NHE6 to NHE9 are present in most tissues/cells where they accumulate predominantly in organelles along the secretory and endocytic-degradative pathways. Their roles are not well defined, but crude transport measurements indicate that may operate as nonselective monovalent cation (Li^+, Na^+, or K^+)/H^+ exchangers involved in organellar pH and/or cation homeostasis. Finally, NHA1 and NHA2 are the most divergent evolutionarily and have not been extensively characterized. Analyses of the subcellular localization of NHA2 have produced divergent results, with reports indicating that it resides at the apical surface of epithelial cells (Fuster et al. 2008; Xiang et al. 2007), the endolysosomal compartment and basolateral membrane of osteoclasts (Hofstetter et al. 2010), and the inner membrane of mitochondria (Battaglino et al. 2008; Fuster et al. 2008). The basis for these differences remains obscure.

34.1.2 NHE Expression and Function in Brain

The relative contributions of NHEs to CNS function have been a subject of considerable interest not only because of their fundamental importance to cellular pH homeostasis but also because they have been implicated in a number of neuropathies. For example, recent genetic studies in humans have linked mutations in the organellar NHE6 and NHE9 isoforms to distinct neurological diseases, including mental retardation (Garbern et al. 2010; Gilfillan et al. 2008), autism (Morrow et al. 2008), and attention deficit hyperactivity disorders (de Silva et al. 2003; Franke et al. 2009; Lasky-Su et al. 2008a; Lasky-Su et al. 2008b; Markunas et al. 2010). However, at present the mechanistic bases for the observed phenotypes are unknown.

Among the plasma membrane-type isoforms, NHE1 is distributed throughout the brain as well as peripheral tissues where it fulfills basic housekeeping functions, such as the regulation of cytoplasmic pH and maintenance of cell volume. Despite its ubiquitous expression, loss of NHE1 function produces a dominant neurological phenotype. Mice containing a null NHE 1 mutation (*Nhe1*$^{-/-}$) exhibit ataxia, epileptic-like seizures, and significant postnatal mortality (Bell et al. 1999; Cox et al. 1997). These changes are associated with selective loss of neurons in the cerebellum and brainstem (Cox et al. 1997) as well as enhanced neuronal excitability in the hippocampal and cortical regions (Gu et al. 2001; Xia et al. 2003). By contrast, abnormal hyperactivation of NHE1 is thought to contribute to the progression of cerebrovascular injuries resulting from ischemic stroke. Moderate to severe reductions in cerebral blood flow significantly decrease the supply of glucose and oxygen required to maintain the high energy demands of the brain. As a consequence, ATP stores are rapidly depleted, and lactate, pyruvate, and protons accumulate due to anaerobic metabolism of the limited stores of glucose (Pulsinelli 1992). This results in rapid increases in both intracellular and extracellular acidity which disrupts the homeostasis of other ions; most notable is the excess accumulation of Na_i^+ (via the Na^+/H^+ exchanger) and Ca_i^{2+} (via the Na^+/Ca^{2+} exchanger acting in reverse mode) (Matsuda et al. 1996; Stys et al. 1991). This, in turn, precipitates a series of other cellular changes that lead to neuronal dysfunction and ultimately tissue damage, including free radical toxicity, cellular edema, apoptosis, and necrosis (Pulsinelli 1992; White et al. 2000). A role for NHE1 is most convincingly demonstrated by studies showing that specific NHE antagonists (Andreeva et al. 1992; Horikawa et al. 2001; Kuribayashi et al. 1999; Lee et al. 2009; Luo et al. 2006; Matsumoto et al. 2004; Park et al. 2005; Suzuki et al. 2002; Vornov et al. 1996), or NHE1 gene ablation (Luo et al. 2005; Wang et al. 2008), effectively mitigate neural injuries associated with ischemia both in vitro and in vivo. Taken together, these studies implicate the functional coupling of Na^+/H^+ and Na^+/Ca^{2+} exchange as critical upstream events in the pathogenesis of cerebral ischemic and reperfusion injuries.

Other plasma membrane isoforms such as NHE2 to NHE4, though mainly found in epithelia of peripheral tissues, are also present in certain areas of the brain where they likely fulfill more specialized roles (Ma and Haddad 1997). For example, NHE3 is present in cerebellar Purkinje and glial cells (Ma and Haddad 1997) as well as chemosensitive neurons of the ventrolateral medulla oblongata that modulate the rate of respiration (Wiemann et al. 1999). Its presence in the latter neurons is particularly intriguing in light of studies showing that relatively selective antagonists of NHE3 acidify and activate these neurons in vitro (Wiemann et al. 1999; Wiemann and Bingmann 2001) and elevate the central respiratory response to hypercapnia in vivo (Abu-Shaweesh et al. 2002; Kiwull-Schone et al. 2001, 2007; Wiemann et al. 2005), implicating a regulatory role for this isoform in the control of breathing rhythm.

NHE5 is unique among the plasma membrane-type isoforms by its preferential mRNA expression in the central nervous system (Attaphitaya et al. 1999; Baird et al. 1999). In transfected non-neuronal and neuronal cells, exogenous NHE5 is present and active at the plasma membrane, but interestingly the bulk of the protein (>85%) accumulates in a transferrin receptor (Tf-R)-associated recycling endosomal compartment (Szászi et al. 2002). Subsequent studies have identified a number of factors that regulate its trafficking to and from the cell surface, including the protein kinases PI3-K (Szászi et al. 2002) and CK2 (Lukashova et al. 2011) as well as ancillary interacting proteins such as secretory carrier membrane protein SCAMP2 (Diering et al. 2009), RACK1 (Onishi et al. 2007), and the clathrin endocytic adaptors β-arrestin1 and β-arrestin2 (Szabó et al. 2005). However, it remains unclear whether the subcellular distribution and regulation of the exogenously expressed NHE5 reflects the behavior of the endogenous protein in its native environment and, if so, what is the functional significance of this distribution.

In this report, we describe the subcellular location of the native NHE5 protein in selected regions of the central (cortex, hippocampus) and peripheral (superior cervical ganglia, SCG) nervous systems as well as in primary cultures of SCG neuronal and glial cells using a newly developed

NHE5 isoform-specific rabbit polyclonal antibody. Similar to transfected cell systems, the results show that NHE5 is sorted almost exclusively into a unique vesicular pool in neurons as well as glial cells that share some characteristics of recycling endosomes and which may serve as an important reservoir of functional transporters.

34.2 Experimental Procedures

34.2.1 Materials

C57BL/6 mice were from Jackson Laboratories. Restriction enzymes were from New England Biolabs. Polyvinylidene fluoride (PVDF) membrane, Millex-HV filter, and Amicon Ultra-4 centrifugation filter units were from Millipore. Lipofectamine™ was from Invitrogen (Life Technologies Corporation). α-MEM, fetal bovine serum, and penicillin–streptomycin were from GIBCO. Goat anti-rabbit or anti-mouse HRP-conjugated antibodies were from Jackson Immuno Research Laboratories, Inc. Monoclonal anti-synaptophysin was from Synaptic Systems. Monoclonal antibody against microtubule-associated protein 2 (MAP2) was purchased from Sigma-Aldrich. Monoclonal anti-EEA1, anti-transferrin antibodies were from BD Biosciences. Monoclonal anti-SMI 31 (anti-neurofilament) and anti-HA antibodies were from Covance. Goat anti-mouse or anti-rabbit Alexa Fluor secondary antibodies were from Molecular Probes (Life Technologies Corporation). Enhanced chemiluminescence system, Protein G-Sepharose 4B, glutathione sepharose 4B, and pGEX-2T bacterial expression vector were purchased from GE Healthcare Life Sciences. All other reagents were obtained from Fisher Scientific or Sigma.

34.2.2 Mouse Brain Cryosectioning and Primary Tissue Culture

All procedures for animal handling were carried out according to the guidelines of the Canadian Council on Animal Care (CCAC). Mice were anesthetized and sacrificed by cervical dislocation, followed by dissection of the brains which were immediately removed and placed on dry ice. Twenty micron thick sections were cut using a cryostat and mounted on gelatin subbed slides. Slides were dried at room temperature for 30 min. They were then kept frozen with desiccant until staining.

Mouse superior cervical ganglia (SCG) were dissected from neonatal C57BL/6 littermates and dissociated enzymatically and mechanically as previously described (Gingras et al. 2002). Briefly, ganglia were incubated for 45 min in Hank's Balanced Salt Solution/trypsin 3× (1 mg/ml) and gently triturated with a fire-polished Pasteur pipette. Dissociated cells were washed in plating media, isolated by centrifugation (800 rpm, 4 min), and resuspended in growth media. Neurons were then plated at a density of ~35 cells/mm^2 on laminin-coated ACLAR coverslips in a modified Petri dish. The neurons were incubated in 1.5 ml of Leibovitz's L-15 medium supplemented with 5% rat serum, vitamins, cofactors, penicillin, streptomycin, sodium bicarbonate, and nerve growth factor NGF 2.5 S (25 ng/ml). Neurons were maintained in 5% CO_2 at 37 °C and treated with 1-ß-D-arabinofuranoside (Ara-C, 5 mM; Sigma-Aldrich) for the first 2–3 days in culture to eliminate non-neuronal cells.

34.2.3 Polyclonal NHE5 Antibody Production

cDNA segments encoding amino acids 689–720, 689–789, and 789–896 of the cytoplasmic C-terminus of human NHE5 were amplified by polymerase chain reaction (PCR), excised with *Bam*HI and *Eco*RI, and inserted into the pGEX-2T bacterial expression vector in frame with the amino-terminal glutathione S-transferase (GST) coding sequence. After sequence confirmation, plasmids were transformed into BL21 *E. coli* cells for protein expression. Cells were lysed in 1×phosphate-buffered saline (PBS) supplemented with 0.4 mM of lysozyme and purified by overnight incubation with glutathione-conjugated sepharose beads at 4 °C. Proteins were eluted with 10 mM glutathione in 1 × PBS. Eluted proteins were concentrated using Amicon Ultra-4 centrifugal units (Millipore) and filtered via 0.45 μm Millex-HV filter. Rabbits were injected intradermally with 300 μg of antigen in complete Freund's adjuvant using standard Canadian Council on Animal Care (CCAC)-approved protocols. Subsequent booster injections were given in incomplete Freund's adjuvant every 2–3 weeks. Sera were obtained from the animals before each immunization and then every third week starting on day 30 after initial immunization.

Sodium azide was added to the sera to a final concentration of 0.05% followed by centrifugation at 15,000 × *g* for 5 min at 4 °C to clear up the serum. Clarified serum was collected and added to a 2 ml Bio-Rad protein A chromatography column at an approximate drip rate of 2 ml per minute. The flow through was passed through the column twice and washed with TBS (50 mM Tris–HCl, pH 7.4; 150 mM NaCl, 0.05% sodium azide) at a volume 10 times greater than the loaded serum. The washes were repeated until the absorbance reached less than 0.2 units above background. The column was drained of TBS without allowing the column to stand dry. Fifteen milliliters of elution buffer pH 2.7 (50 mM Glycine-HCl, pH 2.7) was added to the column, and 1 ml fractions were collected in microcentrifuge tubes each containing 100 μl of neutralization buffer (1 M Tris–HCl, pH 8.0; 1.5 M NaCl; 1 mM EDTA; 0.5% sodium azide). Each fraction was mixed immediately and placed on ice prior to collecting the next fraction. Each fraction was measured for protein concentration, and fractions registering an absorbance greater than 0.2 units over background were combined. The combined fraction's pH was measured and adjusted to approximately pH 7.4.

34.2.4 Cell Culture

Chinese hamster ovary AP-1 cells were maintained in complete α-MEM supplemented with 10% fetal bovine serum, 100 units/mL penicillin, 100 μg/mL streptomycin, and 25 mM $NaHCO_3$, pH 7.4, and incubated in an humidified atmosphere of 95% air, 5% CO_2 at 37 °C. Cells were transiently transfected with HA epitope-tagged forms of NHE1-9 using Lipofectamine and incubated for a minimum of 48 h prior to Western blot analysis.

34.2.5 SDS-PAGE and Immunoblotting

AP-1 cells were solubilized in cell lysis solution (NaCl 150 mM, Tris 20 mM, EDTA 10 mM, 0.5% Triton X-100, pH 7.4) and added to an equal amount of sample buffer (1 M Tris-Cl, 1% SDS, 10% glycerol, 0.1% bromophenol blue) and then subjected to SDS-PAGE using 10% polyacrylamide gels. Proteins were then transferred to a PVDF membrane for immunoblotting. The PVDF membrane was incubated with skim milk–PBS (5% nonfat dry skim milk and 0.1% Tween 20 in PBS, pH 7.4) for 1 h at room temperature in order to block nonspecific binding. The PVDF membrane was subsequently incubated for 1 h with the primary antibody in skim milk–PBS–Tween 20 at the following concentrations: anti-NHE5 at 1:500 and monoclonal anti-HA at 1:5,000. The membrane was washed three times for 10 min with PBS–Tween 20, and an appropriate secondary antibody was added at a concentration of 1:5,000 for 1 h at room temperature. Goat polyclonal anti-rabbit horseradish peroxidase (HRP)-conjugated secondary antibody from Upstate Biotechnology was used for blots probed with polyclonal anti-NHE5. Goat anti-mouse HRP-conjugated secondary antibody from Jackson ImmunoResearch Laboratories was used for blots probed with monoclonal anti-HA. Membranes were again washed with PBS–Tween 20 three times for 10 min each, and the immunocomplex was visualized using an enhanced chemiluminescence system (Amersham Biosciences).

34.2.6 Confocal Microscopy

Primary cultures and cryosections were fixed in 1% paraformaldehyde, permeabilized in 0.1%

Triton in 1× PBS, blocked in 10% goat serum in 1× PBS, and incubated with the primary antibody diluted in blocking solution at the following concentrations: anti-NHE5 at 1/500–1/1,000, anti-HA at 1/2,000, anti-synaptophysin 1/500, anti-neurofilament at 1/10,000, anti-MAP2 at 1/1,000, anti-transferrin at 1/300, and anti-EEA1 at 1/1,000. Images were analyzed by confocal laser scanning microscopy using a Zeiss IM 35 microscope with a 63× objective (Neofluar, N.A. 0.75). Images were analyzed using LSM AIM and Zen Software.

34.3 Results and Discussion

As mentioned above, NHE5 mRNA is widely distributed in the brain, but precise localization of the encoded protein awaits the development of isoform-specific antibodies. We demonstrated previously that an external triple HA epitope-tagged form of NHE5 ($NHE5_{HA}$) resides in both the plasma membrane and in transferrin receptor (Tf-R)-associated recycling endosomes when ectopically expressed in Chinese hamster ovary (CHO)-derived AP-1 cells (Szászi et al. 2002). Although these cells retain many exocytic and endocytic sorting mechanisms that are analogous to the more specialized membrane trafficking pathways found in polarized cells such as epithelia and neurons (Cameron et al. 1991; Chavez et al. 1996; Coorssen et al. 1996; D'Souza et al. 1998; Morimoto et al. 1995; Wilson and Colton 1997; Yoshimori et al. 1996), sorting of NHE5 in CHO/AP-1 cells may differ from that in neuronal cells. For example, the synaptic vesicle protein synaptophysin, which colocalizes with Tf-R-associated recycling endosomes when ectopically expressed in CHO cells (Cameron et al. 1991), only partially coincides with Tf-R in the somatodendritic region of hippocampal neurons and is selectively enriched in axonal synaptic vesicles that are devoid of Tf-R (Mundigl et al. 1993). Hence, it is conceivable that NHE5 may sort to discrete vesicular compartments in neuronal cells other than Tf-R-associated endosomes, and this merits further investigation.

To generate a specific antibody against NHE5, recombinant fusion proteins comprised of bacterial glutathione S-transferase (GST) linked to isoform-specific, evolutionarily conserved, C-terminal segments (amino acids 689–720, 689–789, and 789–896) of human NHE5 were constructed and injected into rabbits using standard protocols. Of the various constructs, the serum from rabbit #3568 injected with GST-NHE5(789–896) generated the strongest immunoreactive signal (*data not shown*). This serum was then subjected to affinity purification to enrich for anti-$NHE5_{3568}$ polyclonal antibodies (simply referred to as α-$NHE5_p$). To test for NHE isoform specificity, whole cell lysates of AP-1 cells stably expressing $NHE5_{HA}$ and its various HA epitope-tagged paralogs (NHE1–4, 6–9) were fractionated by SDS-PAGE and then subjected to immunoblotting using an anti-HA monoclonal antibody (α-HA_m) or α-$NHE5_p$ (Fig. 34.1a, *upper* and *lower* panels, respectively). As expected, the α-HA_m antibody detected all of the NHE isoforms which were well expressed in the AP-1 cells with the exception of $NHE4_{HA}$ which showed low levels of expression (Fig. 34.1a, *upper panel*). By contrast, when reprobing the blot with α-$NHE5_p$, a strong immunoreactive signal was detected only from cell lysates containing $NHE5_{HA}$, verifying the selectivity of the antibody for NHE5. Additional control experiments showed no detectable signal in NHE5-expressing cells labeled only with secondary antibodies or in untransfected cells (*data not shown*) as well as tissue incubated with competing NHE5 antigen (Fig. 34.1d).

To determine the distribution of NHE5 in native tissues, immunofluorescence studies were performed using murine tissue sections from the central (neonatal cortex and hippocampus CA3 region) and peripheral (superior cervical ganglia, SCG) nervous systems. Primary cultures of differentiated SCG neurons (>2 weeks) were also examined. Cryosections of each tissue were prepared

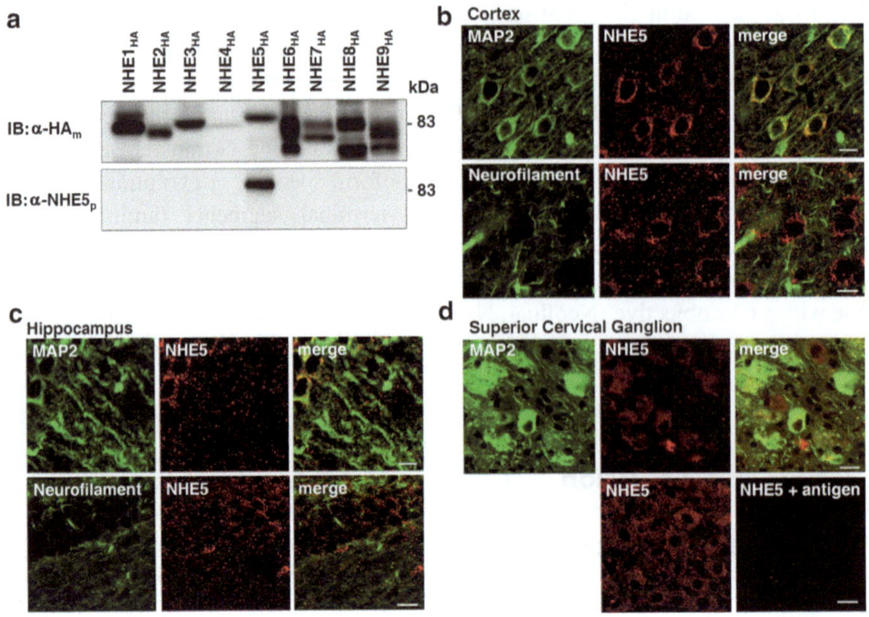

Fig. 34.1 *Immunolocalization of NHE5 in mouse nervous tissue.* (**a**) The specificity of an affinity-purified rabbit polyclonal antibody generated against amino acids 789–896 of NHE5 (α-NHE5p) was tested by probing immunoblots containing whole cell lysates of Chinese hamster ovary AP-1 cells transiently expressing HA-tagged constructs of the different NHE isoforms (NHE1-9_{HA}). The blots were probed with either a monoclonal anti-HA antibody (*upper panel*) or α-NHE5p (*lower panel*). (**b–d**) Cryosections of tissue from the cortex (**b**), hippocampus CA3 (**c**), and superior cervical ganglia (SCG) (**d**) were prepared from young mice (1 month old) and subjected to dual immunolabeling with a polyclonal antibody to native NHE5 (α-NHE5p) and either a mouse monoclonal antibody to microtubule-associated protein 2 (MAP2) to visualize cell bodies and dendritic processes or to neurofilament, a marker of axons, followed by incubation with appropriate Alexa-conjugated secondary antibodies. The SCG sections were also incubated with α-NHE5p alone, or in the presence of the NHE5-immunizing antigen for 1 h, which blocked the signal. The data reveal that NHE5 is preferentially localized in punctate vesicles in the cell bodies as well as dendrites. Scale bar: 10 μm.

from embryonic 18.5 day-old animals. For controls, some cryosections were processed with preimmune serum, without the primary antibody or with the primary antibody in the presence of the immunizing antigen. Cross-reactivity tests were also conducted with different combinations of the secondary antibodies. Sections were visualized using a Zeiss laser scanning confocal microscope.

NHE5 expression was readily detected in cortex (Fig. 34.1b), hippocampus CA3 (Fig. 34.1c), and SCG (Fig. 34.1d) and showed similar subcellular distributions, notably punctate immunostaining that was highly concentrated in the somas and dendrites, as revealed by co-staining with an antibody against microtubule-associated protein 2 (MAP2), a somatodendritic marker. Very little signal was detected in axons co-stained with an anti-neurofilament H, a recognized axonal marker. Importantly, the signal for NHE5 was blocked in the presence of excess immunizing antigen GST-NHE5(789–896), confirming the specificity of the antibody (Fig. 34.1d). Likewise, in primary cultures of differentiated SCG neurons, NHE5 localized predominantly to somatodendritic vesicles (Fig. 34.2a–d), though some signal was also detected in punctate vesicles present along the axons (Fig. 34.2e–h). Intriguingly, unlike in AP-1 cells, NHE5 showed colocalization with only a subset of vesicles containing the transferrin receptor Tf-R in the cell body and

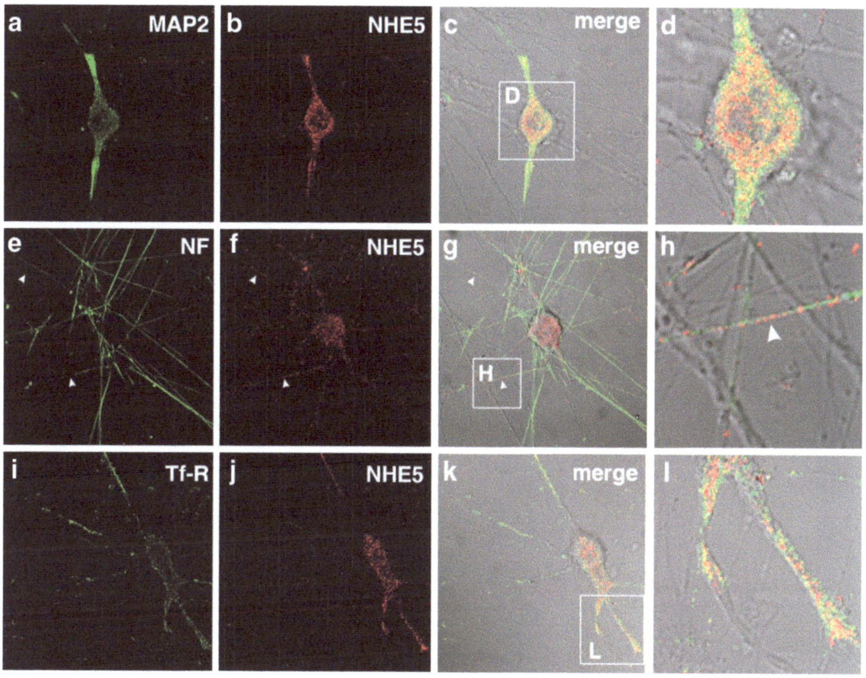

Fig. 34.2 *Localization of NHE5 in primary cultures of murine superior cervical ganglion (SCG) neurons*. Fixed and permeabilized primary cultures of SCG neurons were dual immunolabeled with α-NHE5p (**b**, **f**, **j**) and monoclonal antibodies to MAP2 (**a**) to visualize cell bodies and dendrites, neurofilament (NF) (**e**) to visualize axons, and transferrin receptor (Tf-R) (**i**) to label recycling endosomes. The merged images (**c**, **g**, **k**) were overlaid on the differential interference contrast (DIC) images of the neurons. The *boxed sections* within panels **c**, **g**, and **k** are enlarged as panels **d**, **h**, and **l**, respectively. Immunolabeling of NHE5 was detected predominantly in the soma and dendrites, and to some extent in axons (indicated by *arrowheads*)

dendrites (Fig. 34.2i–l). Moreover, NHE5 did not colocalize with the early endosome antigen 1 (EEA1) (Fig. 34.3a) nor with synaptophysin (Fig. 34.3b), a marker of synaptic vesicles. NHE5 was also detected predominantly in intracellular vesicles of cultured SCG glial cells. The presence of NHE5 at the cell surface was not readily evident in these confocal images.

During the course of the studies, Diering and colleagues (Diering et al. 2011) also reported that NHE5 is present almost exclusively in intracellular vesicles in dendrites and to a lesser extent in axons of cultured hippocampal neurons, similar to our observations. Intriguing, NHE5-containing vesicles were recruited to the cell surface of dendritic spines in response to glycine-induced activation of NMDA receptors where it modulated the pH homeostasis and activity-dependent growth of the spines, suggesting that NHE5 may play an important role in synaptic maturation and plasticity. However, the mechanistic basis for this translocation is unknown.

Collectively, these data suggest that NHE5 partitions into a discrete pool of intracellular vesicles in central and peripheral neurons as well as glial cells. With respect to neurons, these vesicles are distinct from early endosomes and synaptic vesicles and partially overlap with transferrin-containing recycling endosomes. More detailed analysis of the nature of the intracellular compartments containing NHE5 using other specific organelle markers is needed to better understand the physiological role of NHE5 in neuronal function.

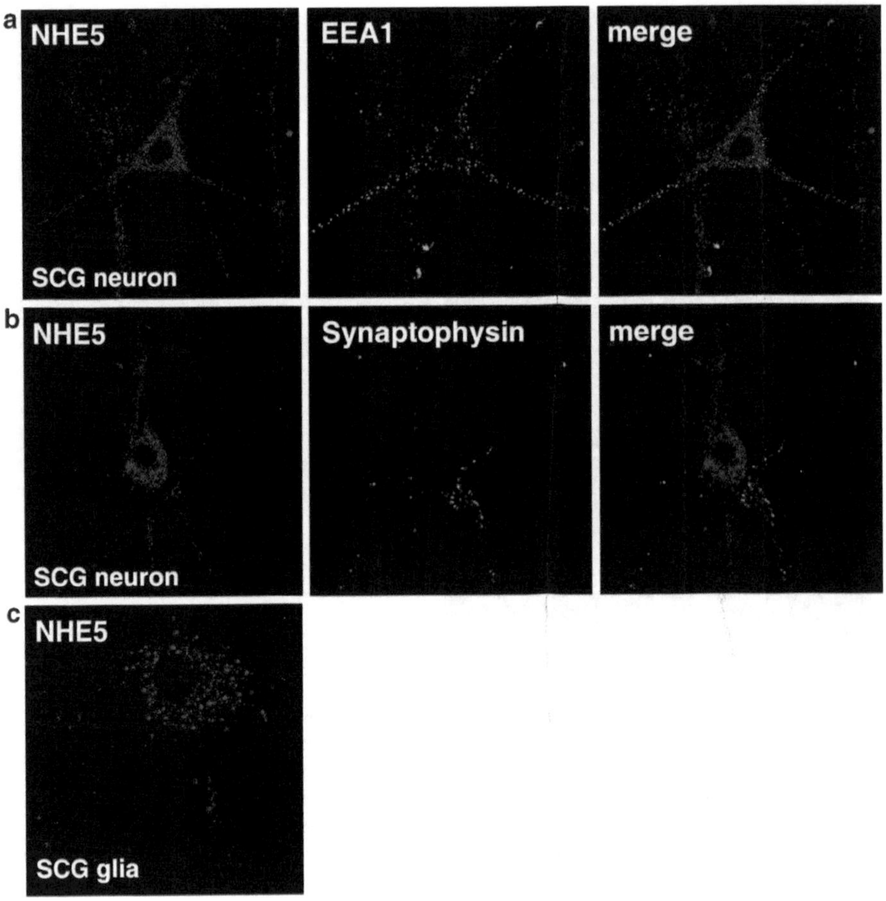

Fig. 34.3 *NHE5 is not localized to early endosomes or synaptic vesicles in primary cultures of superior cervical ganglion (SCG) neurons.* (**a**, **b**) Primary cultures SCG neurons were incubated simultaneously with α-NHE5p and monoclonal antibodies to synaptophysin and early endosomal antigen 1 (EEA1) followed by labeling with the corresponding Alexa-conjugated secondary antibodies. (**c**) Immunolabeling of NHE5 was also detected predominantly in intracellular vesicles of primary cultures of SCG glial cells

References

J.M. Abu-Shaweesh, I.A. Dreshaj, R.J. Martin, K.J. Wirth, U. Heinelt, M.A. Haxhiu, Inhibition of Na^+/H^+ exchanger type 3 reduces duration of apnea induced by laryngeal stimulation in piglets. Pediatr. Res. **52**, 459–464 (2002)

A. Ali, K.P. Pillai, F.J. Ahmad, Y. Dua, D. Vohora, Anticonvulsant effect of amiloride in pentetrazole-induced status epilepticus in mice. Pharmacol. Rep. **58**, 242–245 (2006)

N. Andreeva, B. Khodorov, E. Stelmashook, S. Sokolova, E.J. Cragoe Jr., I. Victorov, 5-(N-ethyl-N-isopropyl) amiloride and mild acidosis protect cultured cerebellar granule cells against glutamate-induced delayed neuronal death. Neuroscience **49**, 175–181 (1992)

S. Attaphitaya, K. Park, J.E. Melvin, Molecular cloning and functional expression of a rat Na^+/H^+ exchanger (NHE5) highly expressed in brain. J. Biol. Chem. **274**, 4383–4388 (1999)

N.R. Baird, J. Orlowski, E.Z. Szabó, H.C. Zaun, P.J. Schultheis, A.G. Menon, G.E. Shull, Molecular cloning, genomic organization, and functional expression of Na^+/H^+ exchanger isoform 5 (NHE5) from human brain. J. Biol. Chem. **274**, 4377–4382 (1999)

A. Baron, R. Waldmann, M. Lazdunski, ASIC-like, proton-activated currents in rat hippocampal neurons. J. Physiol. **539**, 485–494 (2002)

R.A. Battaglino, L. Pham, L.R. Morse, M. Vokes, A. Sharma, P.R. Odgren, M. Yang, H. Sasaki, P. Stashenko, NHA-oc/NHA2: a mitochondrial cation-proton antiporter selectively expressed in osteoclasts. Bone **42**, 180–192 (2008)

S.M. Bell, C.M. Schreiner, P.J. Schultheis, M.L. Miller, R.L. Evans, C.V. Vorhees, G.E. Shull, W.J. Scott, Targeted disruption of the murine *Nhe1* locus induces ataxia, growth retardation, and seizures. Am. J. Physiol. **276**, C788–C795 (1999)

A. Bondarenko, M. Chesler, Rapid astrocyte death induced by transient hypoxia, acidosis, and extracellular ion shifts. Glia **34**, 134–142 (2001)

U. Bonnet, T. Leniger, M. Wiemann, Alteration of intracellular pH and activity of CA3-pyramidal cells in guinea pig hippocampal slices by inhibition of transmembrane acid extrusion. Brain Res. **872**, 116–124 (2000)

C.L. Brett, M. Donowitz, R. Rao, Evolutionary origins of eukaryotic sodium/proton exchangers. Am. J. Physiol. Cell Physiol. **288**, C223–C239 (2005)

N. Brookes, Intracellular pH as a regulatory signal in astrocyte metabolism. Glia **21**, 64–73 (1997)

K.J. Buckler, B.A. Williams, E. Honore, An oxygen-, acid- and anaesthetic-sensitive TASK-like background potassium channel in rat arterial chemoreceptor cells. J. Physiol. **525**(Pt 1), 135–142 (2000)

P.L. Cameron, T.C. Sudhof, R. Jahn, P. De Camilli, Colocalization of synaptophysin with transferrin receptors: implications for synaptic vesicle biogenesis. J. Cell Biol. **115**, 151–164 (1991)

J.R. Casey, S. Grinstein, J. Orlowski, Sensors and regulators of intracellular pH. Nat. Rev. Mol. Cell Biol. **11**, 50–61 (2010)

R.A. Chavez, S.G. Miller, H.P. Moore, A biosynthetic regulated secretory pathway in constitutive secretory cells. J. Cell Biol. **133**, 1177–1191 (1996)

J.C. Chen, M. Chesler, Extracellular alkaline shifts in rat hippocampal slice are mediated by NMDA and non-NMDA receptors. J. Neurophysiol. **68**, 342–344 (1992a)

J.C. Chen, M. Chesler, Modulation of extracellular pH by glutamate and GABA in rat hippocampal slices. J. Neurophysiol. **67**, 29–36 (1992b)

M. Chesler, Regulation and modulation of pH in the brain. Physiol. Rev. **83**, 1183–1221 (2003)

C.P. Chih, J. He, T.S. Sly, E.L. Roberts Jr., Comparison of glucose and lactate as substrates during NMDA-induced activation of hippocampal slices. Brain Res. **893**, 143–154 (2001)

J. Church, K.A. Baxter, J.G. McLarnon, pH modulation of Ca^{2+} responses and a Ca^{2+}-dependent K^+ channel in cultured rat hippocampal neurones. J. Physiol. **511**(Pt 1), 119–132 (1998)

J.R. Coorssen, H. Schmitt, W. Almers, Ca^{2+}-triggers massive exocytosis in Chinese hamster ovary cells. EMBO J. **15**, 3787–3791 (1996)

G.A. Cox, C.M. Lutz, C.L. Yang, D. Biemesderfer, R.T. Bronson, A. Fu, P.S. Aronson, J.L. Noebels, W.N. Frankel, Sodium/hydrogen exchanger gene defect in slow-wave epilepsy mutant mice. Cell **91**, 139–148 (1997)

S. D'Souza, A. Garcia-Cabado, F. Yu, K. Teter, G.L. Lukacs, K. Skorecki, H.P. Moore, J. Orlowski, S. Grinstein, The epithelial sodium-hydrogen antiporter Na^+/H^+ exchanger 3 accumulates and is functional in recycling endosomes. J. Biol. Chem. **273**, 2035–2043 (1998)

M.G. de Silva, K. Elliott, H.H. Dahl, E. Fitzpatrick, S. Wilcox, M. Delatycki, R. Williamson, D. Efron, M. Lynch, S. Forrest, Disruption of a novel member of a sodium/hydrogen exchanger family and DOCK3 is associated with an attention deficit hyperactivity disorder-like phenotype. J. Med. Genet. **40**, 733–740 (2003)

J.W. Deitmer, C.R. Rose, pH regulation and proton signalling by glial cells. Prog. Neurobiol. **48**, 73–103 (1996)

J.A. DeSimone, V. Lyall, Salty and sour taste: sensing of sodium and protons by the tongue. Am. J. Physiol. Gastrointest. Liver Physiol. **291**, G1005–G1010 (2006)

G.H. Diering, J. Church, M. Numata, Secretory carrier membrane protein 2 regulates cell-surface targeting of brain-enriched Na^+/H^+ exchanger NHE5. J. Biol. Chem. **284**, 13892–13903 (2009)

G.H. Diering, F. Mills, S.X. Bamji, M. Numata, Regulation of dendritic spine growth through activity-dependent recruitment of brain-enriched Na^+/H^+ exchanger NHE5. Mol. Biol. Cell **22**, 2246–2257 (2011)

R.M. Douglas, B.M. Schmitt, Y. Xia, M.O. Bevensee, D. Biemesderfer, W.F. Boron, G.G. Haddad, Sodium-hydrogen exchangers and sodium-bicarbonate co-transporters: ontogeny of protein expression in the rat brain. Neuroscience **102**, 217–228 (2001)

C.G. Dulla, P. Dobelis, T. Pearson, B.G. Frenguelli, K.J. Staley, S.A. Masino, Adenosine and ATP link PCO2 to cortical excitability via pH. Neuron **48**, 1011–1023 (2005)

B. Franke, B.M. Neale, S.V. Faraone, Genome-wide association studies in ADHD. Hum. Genet. **126**, 13–50 (2009)

D.G. Fuster, J. Zhang, M. Shi, I.A. Bobulescu, S. Andersson, O.W. Moe, Characterization of the sodium/hydrogen exchanger NHA2. J. Am. Soc. Nephrol **19**, 1547–1556 (2008)

J.Y. Garbern, M. Neumann, J.Q. Trojanowski, V.M. Lee, G. Feldman, J.W. Norris, M.J. Friez, C.E. Schwartz, R. Stevenson, A.A. Sima, A mutation affecting the sodium/proton exchanger, SLC9A6, causes mental retardation with tau deposition. Brain **133**, 1391–1402 (2010)

G.D. Gilfillan, K.K. Selmer, I. Roxrud, R. Smith, M. Kyllerman, K. Eiklid, M. Kroken, M. Mattingsdal, T. Egeland, H. Stenmark, H. Sjoholm, A. Server, L. Samuelsson, A. Christianson, P. Tarpey, A. Whibley, M.R. Stratton, P.A. Futreal, J. Teague, S. Edkins, J. Gecz, G. Turner, F.L. Raymond, C. Schwartz, R.E. Stevenson, D.E. Undlien, P. Stromme, SLC9A6 mutations cause X-linked mental retardation, microcephaly, epilepsy, and ataxia, a phenotype mimicking Angelman syndrome. Am. J. Hum. Genet. **82**, 1003–1010 (2008)

J. Gingras, S. Rassadi, E. Cooper, M. Ferns, Agrin plays an organizing role in the formation of sympathetic synapses. J. Cell Biol. **158**, 1109–1118 (2002)

C. Gonzalez, L. Almaraz, A. Obeso, R. Rigual, Oxygen and acid chemoreception in the carotid body chemoreceptors. Trends Neurosci. **15**, 146–153 (1992)

X.Q. Gu, H. Yao, G.G. Haddad, Increased neuronal excitability and seizures in the Na^+/H^+ exchanger null mutant mouse. Am. J. Physiol. Cell Physiol. **281**, C496–C503 (2001)

M.J.E. Havenga, G.J. Bosman, H. Appelhans, W.J. De Grip, Expression of the anion exchanger (AE) gene family in human brain. Identification of a new AE protein: AEO. Mol. Brain Res. **25**, 97–104 (1994)

M. Hentschke, M. Wiemann, S. Hentschke, I. Kurth, I. Hermans-Borgmeyer, T. Seidenbecher, T.J. Jentsch, A. Gal, C.A. Hubner, Mice with a targeted disruption of the Cl–/HCO3– exchanger AE3 display a reduced seizure threshold. Mol. Cell. Biol. **26**, 182–191 (2006)

W. Hofstetter, M. Siegrist, A. Simonin, O. Bonny, D.G. Fuster, Sodium/hydrogen exchanger NHA2 in osteoclasts: subcellular localization and role in vitro and in vivo. Bone **47**, 331–340 (2010)

N. Horikawa, M. Nishioka, N. Itoh, Y. Kuribayashi, K. Matsui, N. Ohashi, The Na^+/H^+ exchanger SM-20220 attenuates ischemic injury in in vitro and in vivo models. Pharmacology **63**, 76–81 (2001)

K.S. Hsu, Y.C. Liang, C.C. Huang, Influence of an extracellular acidosis on excitatory synaptic transmission and long-term potentiation in the CA1 region of rat hippocampal slices. J. Neurosci. Res. **62**, 403–415 (2000)

C.W. Huang, J.N. Tzeng, Y.J. Chen, W.F. Tsai, C.C. Chen, W.H. Sun, Nociceptors of dorsal root ganglion express proton-sensing G-protein-coupled receptors. Mol. Cell. Neurosci. **36**, 195–210 (2007)

O. Kempski, Cerebral edema. Semin. Nephrol. **21**, 303–307 (2001)

L. Kiss, S.J. Korn, Modulation of N-type Ca^{2+} channels by intracellular pH in chick sensory neurons. J. Neurophysiol. **81**, 1839–1847 (1999)

H. Kiwull-Schone, M. Wiemann, S. Frede, D. Bingmann, K.J. Wirth, U. Heinelt, H.J. Lang, P. Kiwull, A novel inhibitor of the Na^+/H^+ exchanger type 3 activates the central respiratory CO2 response and lowers the apneic threshold. Am. J. Respir. Crit. Care Med. **164**, 1303–1311 (2001)

H. Kiwull-Schone, P. Kiwull, S. Frede, M. Wiemann, Role of brainstem sodium/proton exchanger 3 for breathing control during chronic acid base imbalance. Am. J. Respir. Crit. Care Med. **176**, 513–519 (2007)

S. Kobayashi, C.W. Morgans, J.R. Casey, R.R. Kopito, AE3 anion exchanger isoforms in the vertebrate retina: developmental regulation and differential expression in neurons and glia. J. Neurosci. **14**, 6266–6279 (1994)

Y. Kuribayashi, N. Itoh, M. Kitano, N. Ohashi, Cerebroprotective properties of SM-20220, a potent Na^+/H^+ exchange inhibitor, in transient cerebral ischemia in rats. Eur. J. Pharmacol. **383**, 163–168 (1999)

J. Lasky-Su, R.J. Anney, B.M. Neale, B. Franke, K. Zhou, J.B. Maller, A.A. Vasquez, W. Chen, P. Asherson, J. Buitelaar, T. Banaschewski, R. Ebstein, M. Gill, A. Miranda, F. Mulas, R.D. Oades, H. Roeyers, A. Rothenberger, J. Sergeant, E. Sonuga-Barke, H.C. Steinhausen, E. Taylor, M. Daly, N. Laird, C. Lange, S.V. Faraone, Genome-wide association scan of the time to onset of attention deficit hyperactivity disorder. Am. J. Med. Genet. B Neuropsychiatr. Genet. **147B**, 1355–1358 (2008a)

J. Lasky-Su, B.M. Neale, B. Franke, R.J. Anney, K. Zhou, J.B. Maller, A.A. Vasquez, W. Chen, P. Asherson, J. Buitelaar, T. Banaschewski, R. Ebstein, M. Gill, A. Miranda, F. Mulas, R.D. Oades, H. Roeyers, A. Rothenberger, J. Sergeant, E. Sonuga-Barke, H.C. Steinhausen, E. Taylor, M. Daly, N. Laird, C. Lange, S.V. Faraone, Genome-wide association scan of quantitative traits for attention deficit hyperactivity disorder identifies novel associations and confirms candidate gene associations. Am. J. Med. Genet. B Neuropsychiatr. Genet. **147B**, 1345–1354 (2008b)

B.K. Lee, D.H. Lee, S. Park, S.L. Park, J.S. Yoon, M.G. Lee, S. Lee, K.Y. Yi, S.E. Yoo, K.H. Lee, Y.S. Kim, S.H. Lee, E.J. Baik, C.H. Moon, Y.S. Jung, Effects of KR-33028, a novel Na^+/H^+ exchanger-1 inhibitor, on glutamate-induced neuronal cell death and ischemia-induced cerebral infarct. Brain Res. **1248**, 22–30 (2009)

P.A. Li, B.K. Siesjo, Role of hyperglycaemia-related acidosis in ischaemic brain damage. Acta Physiol. Scand. **161**, 567–580 (1997)

M. Luckermann, S. Trapp, K. Ballanyi, GABA- and glycine-mediated fall of intracellular pH in rat medullary neurons in situ. J. Neurophysiol. **77**, 1844–1852 (1997)

V. Lukashova, E.Z. Szabo, T. Jinadasa, A. Mokhov, D.W. Litchfield, J. Orlowski, CK2 phosphorylation of an acidic Ser/Thr di-isoleucine motif in the Na^+/H^+ exchanger NHE5 isoform promotes association with b-arrestin2 and endocytosis. J. Biol. Chem. **286**, 11456–11468 (2011)

J. Luo, H. Chen, D.B. Kintner, G.E. Shull, D. Sun, Decreased neuronal death in Na^+/H^+ exchanger isoform 1-null mice after in vitro and in vivo ischemia. J. Neurosci. **25**, 11256–11268 (2005)

J. Luo, H. Chen, D.B. Kintner, G.E. Shull, D. Sun, Inhibition of Na^+/H^+ exchanger isoform 1 attenuates mitochondrial cytochrome C release in cortical neurons following in vitro ischemia. Acta Neurochir. Suppl. **96**, 244–248 (2006)

E. Ma, G.G. Haddad, Expression and localization of Na^+/H^+ exchangers in rat central nervous system. Neuroscience **79**, 591–603 (1997)

C.A. Markunas, K.S. Quinn, A.L. Collins, M.E. Garrett, A.M. Lachiewicz, J.L. Sommer, E. Morrissey-Kane, S.H. Kollins, A.D. Anastopoulos, A.E. Ashley-Koch, Genetic variants in SLC9A9 are associated with measures of attention-deficit/hyperactivity disorder symptoms in families. Psychiatr. Genet. **20**, 73–81 (2010)

T. Matsuda, K. Takuma, E. Nishiguchi, S. Asano, H. Hashimoto, J. Azuma, A. Baba, Involvement of

Na^+-Ca^{2+} exchanger in reperfusion-induced delayed cell death of cultured rat astrocytes. Eur. J. Neurosci. **8**, 951–958 (1996)

Y. Matsumoto, S. Yamamoto, Y. Suzuki, T. Tsuboi, S. Terakawa, N. Ohashi, K. Umemura, Na^+/H^+ exchanger inhibitor, SM-20220, is protective against excitotoxicity in cultured cortical neurons. Stroke **35**, 185–190 (2004)

J.W. Mink, R.J. Blumenschine, D.B. Adams, Ratio of central nervous system to body metabolism in vertebrates: its constancy and functional basis. Am. J. Physiol. **241**, R203–R212 (1981)

T. Morimoto, S. Popov, K.M. Buckley, M.M. Poo, Calcium-dependent transmitter secretion from fibroblasts: modulation by synaptotagmin I. Neuron **15**, 689–696 (1995)

E.M. Morrow, S.Y. Yoo, S.W. Flavell, T.K. Kim, Y. Lin, R.S. Hill, N.M. Mukaddes, S. Balkhy, G. Gascon, A. Hashmi, S. Al-Saad, J. Ware, R.M. Joseph, R. Greenblatt, D. Gleason, J.A. Ertelt, K.A. Apse, A. Bodell, J.N. Partlow, B. Barry, H. Yao, K. Markianos, R.J. Ferland, M.E. Greenberg, C.A. Walsh, Identifying autism loci and genes by tracing recent shared ancestry. Science **321**, 218–223 (2008)

D.K. Mulkey, R.L. Stornetta, M.C. Weston, J.R. Simmons, A. Parker, D.A. Bayliss, P.G. Guyenet, Respiratory control by ventral surface chemoreceptor neurons in rats. Nat. Neurosci. **7**, 1360–1369 (2004)

O. Mundigl, M. Matteoli, L. Daniell, A. Thomas-Reetz, A. Metcalf, R. Jahn, P. De Camilli, Synaptic vesicle proteins and early endosomes in cultured hippocampal neurons: differential effects of Brefeldin A in axon and dendrites. J. Cell Biol. **122**, 1207–1221 (1993)

I. Onishi, P.J. Lin, G.H. Diering, W.P. Williams, M. Numata, RACK1 associates with NHE5 in focal adhesions and positively regulates the transporter activity. Cell. Signal. **19**, 194–203 (2007)

P. Paalasmaa, K. Kaila, Role of voltage-gated calcium channels in the generation of activity-induced extracellular pH transients in the rat hippocampal slice. J. Neurophysiol. **75**, 2354–2360 (1996)

H.S. Park, B.K. Lee, S. Park, S.U. Kim, S.H. Lee, E.J. Baik, S. Lee, K.Y. Yi, S.E. Yoo, C.H. Moon, Y.S. Jung, Effects of sabiporide, a specific Na^+/H^+ exchanger inhibitor, on neuronal cell death and brain ischemia. Brain Res. **1061**, 67–71 (2005)

W. Pulsinelli, Pathophysiology of acute ischaemic stroke. Lancet **339**, 533–536 (1992)

R.W. Putnam, Intracellular pH regulation of neurons in chemosensitive and nonchemosensitive areas of brain slices. Respir. Physiol. **129**, 37–56 (2001)

R.W. Putnam, CO_2 chemoreception in cardiorespiratory control. J. Appl. Physiol. **108**, 1796–1802 (2010)

R.W. Putnam, J.A. Filosa, N.A. Ritucci, Cellular mechanisms involved in CO_2 and acid signaling in chemosensitive neurons. Am. J. Physiol. Cell Physiol. **287**, C1493–C1526 (2004)

B.R. Ransom, Glial modulation of neural excitability mediated by extracellular pH: a hypothesis revisited. Prog. Brain Res. **125**, 217–228 (2000)

M. Rickmann, B. Orlowski, K. Heupel, E. Roussa, Distinct expression and subcellular localization patterns of $Na^+/HCO3-$ cotransporter (SLC 4A4) variants NBCe1-A and NBCe1-B in mouse brain. Neuroscience **146**, 1220–1231 (2007)

M.F. Romero, C.M. Fulton, W.F. Boron, The SLC4 family of HCO3– transporters. Pflugers Arch. **447**, 495–509 (2004)

B. Rorig, G. Klausa, B. Sutor, Intracellular acidification reduced gap junction coupling between immature rat neocortical pyramidal neurones. J. Physiol. (Lond.) **490**, 31–49 (1996)

C.R. Rose, B.R. Ransom, Mechanisms of H^+ and Na^+ changes induced by glutamate, kainate, and D-aspartate in rat hippocampal astrocytes. J. Neurosci. **16**, 5393–5404 (1996)

S. Schuchmann, D. Schmitz, C. Rivera, S. Vanhatalo, B. Salmen, K. Mackie, S.T. Sipila, J. Voipio, K. Kaila, Experimental febrile seizures are precipitated by a hyperthermia-induced respiratory alkalosis. Nat. Med. **12**, 817–823 (2006)

A. Schurr, C.A. West, B.M. Rigor, Lactate-supported synaptic function in the rat hippocampal slice preparation. Science **240**, 1326–1328 (1988)

K. Seuwen, M.G. Ludwig, R.M. Wolf, Receptors for protons or lipid messengers or both? J. Recept. Signal Transduct. Res. **26**, 599–610 (2006)

C.A. Severson, W. Wang, V.A. Pieribone, C.I. Dohle, G.B. Richerson, Midbrain serotonergic neurons are central pH chemoreceptors. Nat. Neurosci. **6**, 1139–1140 (2003)

P.K. Stys, S.G. Waxman, B.R. Ransom, Reverse operation of the Na^+-Ca^{2+} exchanger mediates Ca^{2+} influx during anoxia in mammalian CNS white matter. Ann. N. Y. Acad. Sci. **639**, 328–332 (1991)

Y. Suzuki, Y. Matsumoto, Y. Ikeda, K. Kondo, N. Ohashi, K. Umemura, SM-20220, a Na^+/H^+ exchanger inhibitor: effects on ischemic brain damage through edema and neutrophil accumulation in a rat middle cerebral artery occlusion model. Brain Res. **945**, 242–248 (2002)

E.Z. Szabó, M. Numata, V. Lukashova, P. Iannuzzi, J. Orlowski, b-Arrestins bind and decrease cell-surface abundance of the Na^+/H^+ exchanger NHE5 isoform. Proc. Natl. Acad. Sci. U. S. A. **102**, 2790–2795 (2005)

K. Szászi, A. Paulsen, E.Z. Szabó, M. Numata, S. Grinstein, J. Orlowski, Clathrin-mediated endocytosis and recycling of the neuron-specific Na^+/H^+ exchanger NHE5 isoform: regulation by phosphatidylinositol 3′-kinase and the actin cytoskeleton. J. Biol. Chem. **277**, 42623–42632 (2002)

K.I. Takahashi, D.R. Copenhagen, Modulation of neuronal function by intracellular pH. Neurosci. Res. **24**, 109–116 (1996)

G.C. Tombaugh, G.G. Somjen, Effects of extracellular pH on voltage-gated Na^+, K^+ and Ca^{2+} currents in isolated rat CA1 neurons. J. Physiol. **493**(Pt 3), 719–732 (1996)

G.C. Tombaugh, G.G. Somjen, Differential sensitivity to intracellular pH among high- and low-threshold Ca^{2+}

currents in isolated rat CA1 neurons. J. Neurophysiol. **77**, 639–653 (1997)

S. Trapp, M. Luckermann, P.A. Brooks, K. Ballanyi, Acidosis of rat dorsal vagal neurons in situ during spontaneous and evoked activity. J. Physiol. (Lond.) **496**, 695–710 (1996)

S.F. Traynelis, S.G. Cull-Candy, Pharmacological properties and H^+ sensitivity of excitatory amino acid receptor channels in rat cerebellar granule neurones. J. Physiol. **433**, 727–763 (1991)

M. Tsacopoulos, P.J. Magistretti, Metabolic coupling between glia and neurons. J. Neurosci. **16**, 877–885 (1996)

G.L. Vilas, D.E. Johnson, P. Freund, J.R. Casey, Characterization of an epilepsy-associated variant of the human Cl–/HCO3– exchanger AE3. Am. J. Physiol. Cell Physiol. **297**, C526–C536 (2009)

J.J. Vornov, A.G. Thomas, D. Jo, Protective effects of extracellular acidosis and blockade of sodium/hydrogen ion exchange during recovery from metabolic inhibition in neuronal tissue culture. J. Neurochem. **67**, 2379–2389 (1996)

R. Waldmann, Proton-gated cation channels: neuronal acid sensors in the central and peripheral nervous system. Adv. Exp. Med. Biol. **502**, 293–304 (2001)

Y. Wang, J. Luo, X. Chen, H. Chen, S.W. Cramer, D. Sun, Gene inactivation of Na^+/H^+ exchanger isoform 1 attenuates apoptosis and mitochondrial damage following transient focal cerebral ischemia. Eur. J. Neurosci. **28**, 51–61 (2008)

B.C. White, J.M. Sullivan, D.J. DeGracia, B.J. O'Neil, R.W. Neumar, L.I. Grossman, J.A. Rafols, G.S. Krause, Brain ischemia and reperfusion: molecular mechanisms of neuronal injury. J. Neurol. Sci. **179**, 1–33 (2000)

M. Wiemann, D. Bingmann, Ventrolateral neurons of medullary organotypic cultures: intracellular pH regulation and bioelectric activity. Respir. Physiol. **129**, 57–70 (2001)

M. Wiemann, J.R. Schwark, U. Bonnet, H.W. Jansen, S. Grinstein, R.E. Baker, H.J. Lang, K. Wirth, D. Bingmann, Selective inhibition of the Na^+/H^+ exchanger type 3 activates CO2/H^+-sensitive medullary neurones. Pflugers Arch. **438**, 255–262 (1999)

M. Wiemann, S. Frede, D. Bingmann, P. Kiwull, H. Kiwull-Schone, Sodium/proton exchanger 3 in the medulla oblongata and set point of breathing control. Am. J. Respir. Crit. Care Med. **172**, 244–249 (2005)

R.H. Williams, L.T. Jensen, A. Verkhratsky, L. Fugger, D. Burdakov, Control of hypothalamic orexin neurons by acid and CO2. Proc. Natl. Acad. Sci. U. S. A. **104**, 10685–10690 (2007)

J.M. Wilson, T.L. Colton, Targeting of an intestinal apical endosomal protein to endosomes in nonpolarized cells. J. Cell Biol. **136**, 319–330 (1997)

M.T. Wyss, R. Jolivet, A. Buck, P.J. Magistretti, B. Weber, In vivo evidence for lactate as a neuronal energy source. J. Neurosci. **31**, 7477–7485 (2011)

Y. Xia, P. Zhao, J. Xue, X.Q. Gu, X. Sun, H. Yao, G.G. Haddad, Na^+ channel expression and neuronal function in the Na^+/H^+ exchanger 1 null mutant mouse. J. Neurophysiol. **89**, 229–236 (2003)

M. Xiang, M. Feng, S. Muend, R. Rao, A human Na^+/H^+ antiporter sharing evolutionary origins with bacterial NhaA may be a candidate gene for essential hypertension. Proc. Natl. Acad. Sci. U. S. A. **104**, 18677–18681 (2007)

T. Yoshimori, P. Keller, M.G. Roth, K. Simons, Different biosynthetic transport routes to the plasma membrane in BHK and CHO cells. J. Cell Biol. **133**, 247–256 (1996)

J. Zhai, R.W. Peoples, C. Li, Proton inhibition of GABA-activated current in rat primary sensory neurons. Pflugers Arch. **435**, 539–545 (1998)

The Role of Na^+/H^+ Exchanger Isoform 1 in Inflammatory Responses: Maintaining H^+ Homeostasis of Immune Cells

35

Yejie Shi, Dong Kim, Marie Caldwell, and Dandan Sun

Abstract

The Na^+/H^+ exchangers (NHEs) are a family of membrane transporter proteins which catalyze the electroneutral exchange of one Na^+ for one H^+ and thus regulate intracellular pH (pH_i) and cell volume. It is shown that Na^+/H^+ exchanger isoform 1 (NHE-1), but not other isoforms, is the important mechanism in extruding H^+ and regulating pH_i in the immune system. Immune cells, such as monocytes and neutrophils, generate reactive oxygen species (ROS) and cytokines in response to various stimuli and provide the first line of defense in the immune system. NHE-1 is activated during this respiratory burst and required to maintain an optimal pH_i for the immune cells. In the central nervous system, NHE-1 is important for microglial (macrophage) activation and participates in the inflammatory response under pathological conditions including cerebral ischemia and traumatic brain injury. NHE-1 also affects Ca^{2+} homeostasis in microglia and contributes to the increase of $[Ca^{2+}]_i$ by coupling to the Na^+/Ca^{2+} exchanger (NCX) stimulation, thus regulating immune cell function and participating in ischemic cell death. A better understanding of the function of NHE-1 in inflammatory responses will provide insight into its role in brain damage under disease conditions.

Keywords

Na^+/H^+ exchange • Na^+/Ca^{2+} exchange • Inflammation • H^+ homeostasis

35.1 Introduction

The Na^+/H^+ exchangers (NHEs) are a family of membrane transport proteins which catalyze the secondary active electroneutral exchange of one Na^+ for one H^+. So far, nine NHE isoforms have been cloned in mammalian tissues. The Na^+/H^+ exchanger isoform 1 (NHE-1) was identified first

Y. Shi • D. Kim • M. Caldwell • D. Sun, M.D., Ph.D. (✉)
Department of Neurology, University of Pittsburgh Medical School, S-598 South Biomedical Science Tower (BST), 3500 Terrace Street, Pittsburgh, PA 15213, USA
e-mail: sund@upmc.edu

L. Annunziato (ed.), *Sodium Calcium Exchange: A Growing Spectrum of Pathophysiological Implications*, Advances in Experimental Medicine and Biology 961, DOI 10.1007/978-1-4614-4756-6_35,

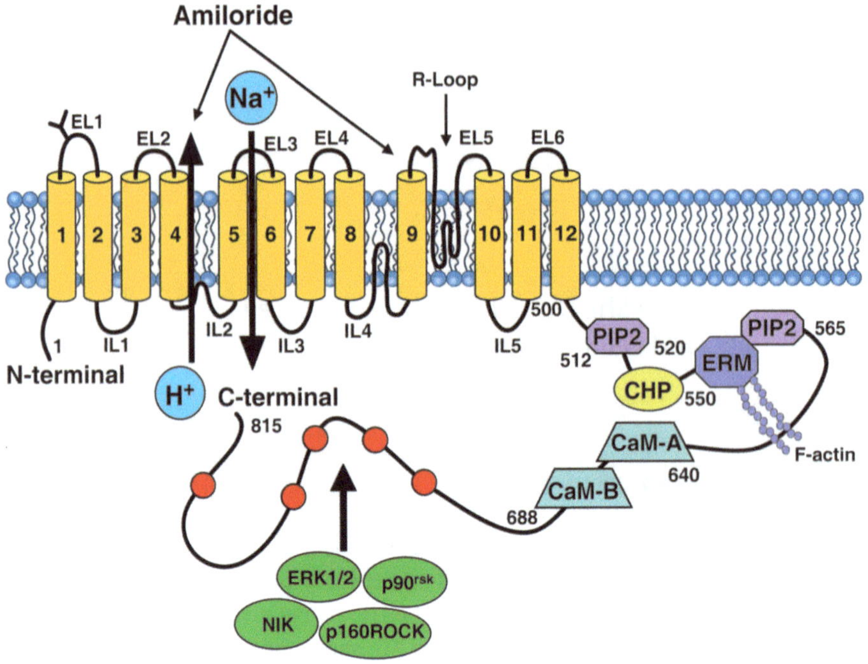

Fig. 35.1 *Basic structure and regulatory elements of NHE-1 protein*. The topology of the protein illustrates regions of the cytoplasmic domain which are important in regulation or protein–protein interaction, the positions of reentrant loops, and the membrane-associated segments. The associated factors, phosphatidylinositol 4,5-bisphosphate (PIP_2), ERM protein family (ezrin, radixin, and moesin), calmodulin (CaM), and calcineurin homologous protein (CHP), are also shown in their approximate known binding sites. "P" indicates the approximate site for phosphorylation of the cytosolic tail of the protein mediated by extracellular signal-related kinase (ERK1/2), p90 ribosomal S kinase ($p90^{RSK}$), Nck-interacting kinase (NIK), or p160 Rho-associated kinase (p160ROCK) (Luo and Sun (2007))

and is the most abundant isoform in the brain (Orlowski and Grinstein 2004). All characterized NHE isoforms consist of about 600–900 amino acids with approximately 40 % amino acid homology. NHE-1 is made of 815 amino acids with both N- and C-terminal domains (Fig. 35.1). The N-terminal domain forms 12 transmembrane (TM) domains which are highly conserved in most NHE isoforms and responsible for cation translocation (Putney et al. 2002). The less-conserved C-terminal domain is the regulatory site for modifying the exchanger activity. It comprises of distinct subdomains that can be phosphorylated by several protein kinases, including extracellular signal-related kinase (ERK1/2), p90 ribosomal S kinase ($p90^{RSK}$) (Moor et al. 2001), and p38 mitogen-activated protein kinase (MAPK) (Khaled et al. 2001).

The NHE-1 isoform is found in the plasma membrane of most mammalian cells and normally serves a "housekeeping" function (Fliegel 2005). It resides on the cell surface but can be present in discrete microdomains of the plasma membrane, such as the lamellipodia in fibroblasts (Grinstein et al. 1993), the basolateral membrane of epithelia (Biemesderfer et al. 1992), and the intercalated disks and t-tubules of cardiac myocytes (Petrecca et al. 1999). NHE-1 is involved in a number of important physiological functions. These include serving as the principal alkalinizing mechanism in many cell types in response to intracellular acidification. Together with bicarbonate transporting systems, NHE-1 plays a crucial role in maintaining cytoplasmic acid–base balance. NHE-1 also functions in restoring cell volume following osmotic stress-induced cell shrinkage by providing a major resource for Na^+ influx, which is coupled to Cl^- and H_2O uptake for cell volume recovery (Rotin and Grinstein 1989).

Cells such as macrophages and neutrophils provide the first line of defense by phagocytosis, release of cytokines and chemokines, as well as generation of reactive oxygen species (ROS). Many of these functions are coupled with changes of pH_i (De Vito 2006). NHE-1 plays an important role in immune cell function by extruding H^+ and regulating intracellular pH (pH_i) (De Vito 2006). In this chapter, we reviewed recent evidence about the expression of NHE-1 in the immune system and how NHE-1 participates in the inflammatory responses under various pathological conditions in peripheral and central nervous systems. A better understanding of the function of NHE-1 in inflammatory responses will provide insight into its role in brain damage under several disease conditions.

35.2 NHE-1 in Immune Cells

35.2.1 NHE-1 Activity in Monocytes and Dendritic Cells

Monocytes are integral cells in the immune system which migrate to infected tissues of where they differentiate into dendritic cells to elicit an immune response (Ziegler-Heitbrock et al. 2010). Monocytes and antigen-presenting dendritic cells restrict the spread of pathogens through production of superoxide via NADPH oxidase (NOX) (Soeiro-Pereira et al. 2011). NHE-1 is essential in maintaining pH_i, cell volume, and cell migration of dendritic cells and is critical for normal dendritic cell functions in response to infections (De Vito 2006). Exposure of dendritic cells to bacterial lipopolysaccharide (LPS) triggers both dendritic cell production of ROS and the activation of NHE-1. Rotte et al. showed that the result of LPS stimulation on dendritic cells includes cytosolic acidification, cell swelling, enhanced ROS production, and induction of migration (Rotte et al. 2010a). The pathways of protein kinase Akt isoform (Akt2) are responsible for LPS-mediated dendritic cell activation (Bhandaru et al. 2012). The phosphorylation of Akt2 activates the formation of the enzyme phosphoinositide 3-kinase (PI_3), an important regulator of NHE-1. Exposing dendritic cells to the PI_3-kinase inhibitor Wortmannin (LY294002) halted the effects of LPS on NHE activity, cell volume, ROS production, and migration (Rotte et al. 2010b). Taken together, these studies suggest that NHE-1 activity is required for function of dendritic cells in response to infection.

35.2.2 NHE-1 Activity in Neutrophil Cells

Neutrophils are inflammatory leukocyte cells, which are the first of the circulating leukocytes to arrive at the site of injury or infection (Bernardes-Silva et al. 2001). Neutrophils kill microorganisms by ingesting them during phagocytosis. Neutrophils rapidly produce superoxide and other ROS, known as the respiratory burst, which helps to kill microbes (Martins et al. 2003; Robinson 2008). Interestingly, NHE-1 activity is also stimulated during the respiratory burst in response to an invading substance or inflammation (Robinson 2008). NHE-1 regulates pH_i of neutrophils during the respiratory bursts by extrusion of H^+ following activation of NOX (Fukushima et al. 1996). NHE-1 is the isoform present in neutrophil cells but not NHE-2, NHE-3, or NHE-4. Exposing neutrophil cells to opsonized zymosan, a phagocytic stimulant, provoked transient cytosolic acidification and prolonged alkalinization and a result of NHE-1 stimulation, as the alkalinization was absent in Na^+-free medium and was inhibited by the NHE inhibitor methyl-methyl-propenyl-amiloride (MMPA) (Fukushima et al. 1996). Therefore, NHE-1-medaited H^+ extrusion plays a role in maintaining H^+ homeostasis in neutrophils (Fukushima et al. 1996).

In addition, NHE-1 activity is involved in neutrophil movement to locations of infections and lesions (Hayashi et al. 2008). The exact mechanisms of NHE-1 in migration of neutrophil cells are unknown. One theory is that NHE-1 controls the electroneutral exchange of sodium for hydrogen, and the change of pH_i and the increase of intracellular cations affect the cell movement (Hayashi et al. 2008). Removal of extracellular sodium impairs neutrophil movement, suggesting the important role of NHE-1 in ion translocation and cell movement function (Hayashi et al. 2008). Moreover, migration of neutrophil cells was significantly reduced in cells that lacked the NHE-1 function (Hayashi et al. 2008). On the other hand, NHE-1 causes an increase of osmolytes and water that result in cell swelling and extension of pseudopods (Hayashi et al. 2008). Taken together, NHE-1 plays an important role in neutrophil function, which is involved in H^+ extrusion during the respiratory burst and cell migration.

35.2.3 NHE-1 Activity in Microglia

Microglia are resident macrophages ubiquitously distributed throughout the central nervous system (CNS). They serve as neurological sensors and can be rapidly activated under many pathological conditions, including neurodegenerative disease, brain tumor, trauma, infection, and stroke (Graeber and Streit 2010; Jin et al. 2010; Tambuyzer et al. 2009). Activated microglia undergo a series of transformations including morphological change, proliferation, migration, and upregulation of surface markers such as CD45, CD4, and MHC Class I molecules (Tambuyzer et al. 2009). They also release biological substances upon activation, including ROS, nitrogen species, cytokines, and growth factors (Harrigan et al. 2008; Yenari et al. 2010). One important component for microglial activation and function is the NOX. NOX catalyzes the reduction reaction of molecular oxygen to superoxide anion using NADPH as an electron donor and is the major source of ROS production in microglia (Bedard and Krause 2007; Harrigan et al. 2008). While ROS production by microglia is beneficial in clearing invading pathogens from the brain, overproduction may also induce neuronal damage (Hu et al. 2011). Therefore, ROS liberated by microglial respiratory burst may play an important role in numerous neurodegenerative processes.

In the process of NOX-dependent ROS generation, H^+ accumulates inside microglia, causing depolarization and cytoplasmic acidification (De Vito 2006). NOX activity is markedly sensitive to pH_i with an optimal pH_i of 7.2 (Henderson et al. 1988). Therefore, cytoplasmic acidification may slow superoxide anion production as a result of inhibition of NOX (Swallow et al. 1993), and restoring physiological pH_i is critical for maintaining respiratory burst activity in microglia. Recent studies confirmed the expression of NHE-1 in cultured microglia and showed that NHE-1 is required to maintain an optimal pH_i and sustain microglial respiratory burst (Liu et al. 2010). In cultured microglia, activation of microglia by several stimuli depends on NHE-1-mediated H^+ homeostasis (Liu et al. 2010). Inhibition of NHE-1 with its potent inhibitor HOE 642 impaired pH_i regulation in microglia under basal conditions. HOE 642 also reduced the production of superoxide anion as well as proinflammatory cytokines IL-6, IL-1β, and TNF-α induced by LPS or in vitro ischemia (Liu et al. 2010).

Besides the direct regulatory role of NHE-1 on pH_i and its association with NOX, NHE-1 may also contribute to microglial activation via affecting intracellular Ca^{2+} signaling. Elevation of $[Na^+]_i$ and $[Ca^{2+}]_i$ was detected in activated microglia (Liu et al. 2010). Stimulation of NHE-1 activity increases Na^+ accumulation inside the cell. Increased intracellular Na^+ concentration in turn stimulates the reversal mode of the Na^+/Ca^{2+} exchange (NCX_{rev}), resulting in increased $[Ca^{2+}]_i$ which will lead to mitochondrial damage and cell death (Dong et al. 2006). It has been shown that NCX is present in cultured microglia, and its activity is stimulated by IFN-γ and NO (Nagano et al. 2004). IFN-γ causes both transient and delayed increases in NCX activity in cultured microglia, peaking at 1 h and 24 h after treatment, respectively (Nagano et al. 2004). Immunoblotting analysis revealed an increase in NCX protein expression in microglia treated with IFN-γ for 24 h (Matsuda et al. 2006). The delayed increase in NCX activity was blocked by the protein synthesis inhibitors cycloheximide and actinomycin D, while the transient increase was not (Nagano et al. 2004). These observations suggest that the transient increase of NCX activity is due to the activation of NCX function and that the delayed increase of NCX activity is due to the elevation of NCX protein expression (Matsuda et al. 2006). Pharmacological studies suggest that protein kinase C or the tyrosine kinase pathway is responsible for IFN-γ-induced transient and delayed increases in NCX activity, and the ERK signaling pathway is involved in IFN-γ-induced delayed increase in NCX activity in microglia. NCX activity is also elevated in cultured microglia after treatment with LPS for 24 h, as reflected by the Na^+-dependent Ca^{2+} uptake (Nagano et al. 2004). Thus, it is likely that NCX is involved in microglial activation, but the precise underlying mechanisms require further investigation.

35.3 NHE-1 in Inflammatory Responses

35.3.1 NHE-1 in Dendritic Cells in Response to Infection

Traditionally, T-cells induced by dendritic cells were thought to provide the main source of immunity during infection (Bennett and Chakraverty 2012). However, recent evidence suggests that dendritic cells migrate directly to the sites of neuroinflammation following cerebral ischemia independent of T-cells. The presence of dendritic cells was detected in photochemically induced focal ischemia of the cerebral cortex using a mouse photothrombosis model, ex vivo (Reichmann et al. 2002). The high levels of CD45, CD11b, and CD11c (dendritic cell markers) detected in the cells of rat cerebral ischemia implied that the ischemic cells expressed high levels of the MHC II^+ phenotype. This expression indicated that migrating dendritic cells were immature and thus unable to direct T-cell production. The presence of dendritic cells following ischemia was identified in a healthy transgenic brain expressing enhanced yellow fluorescent protein (EYFP) under the dendritic cell marker, CD11c (Felger et al. 2010). The results suggest that dendritic-like cells are associated with focal ischemia independent of T-cells (Felger et al. 2010). These studies suggest that NHE-1, which is important in the functioning of dendritic cells, might also be crucial in the aftermath of ischemia.

35.3.2 NHE-1 and Neutrophils in Brain Injury

Numerous investigations have shown a link between inflammation and an acute brain injury of a stroke or traumatic brain injury (Amor et al. 2010). Following brain ischemia, there is an almost immediate activation of resident cells, and then of circulating neutrophil cells (Bernardes-Silva et al. 2001). Minutes to hours after an ischemic stroke, ROS and proinflammatory mediators are released from the injured tissue and trigger the adhesion and transendothelial migration of circulating leukocytes (Jin et al. 2010). In hours to days after the stroke, the subacute phase, leukocytes give off cytokines and chemokines, excessive production of ROS, and matrix metalloproteinases (MMPs). They increase the inflammatory response by causing an activation of resident cells and more leukocytes to infiltrate the injury site. This causes disruption of the blood–brain barrier (BBB), brain edema, and other deleterious effects (Jin et al. 2010). Neutrophil cells accumulate in the blood vessels, which causes reduced blood flow and blocked microvessels. The lysosomes, enzymes, and free radicals released by the neutrophils increase vascular permeability, which can lead to edema. It has also been shown that neutrophil depletion decreases cerebral tissue injury (Bernardes-Silva et al. 2001). Inhibition of leukocyte infiltration into the ischemic brain via antiadhesion molecules has been shown to reduce infarct size, edema, and neurological deficits in rodent stroke models (Zhang et al. 2003). Clinically, several drugs that target neutrophil recruitment have been developed as potential therapies for ischemic stroke, but the treatment has yet to be successful (Jin et al. 2010). Clinical studies have shown that the concentration of the neutrophil cells in the regions of human cerebral infarction is directly related to the severity of brain tissue damage and poor neurological outcome after the ischemic stroke (Jin et al. 2010).

35.3.3 NHE-1 and Microglia in Ischemic Brain Injury

Ischemic brain injury is a complex process including proinflammatory responses such as microglia/macrophage activation, astrogliosis, neutrophil infiltration, and cytokine/chemokine release (Wang et al. 2007). Inflammatory responses in ischemic brains occur at days after stroke onset and last for weeks (Hsu et al. 2000). Although some of these responses are found to be helpful for tissue repair processes, they can also be cytotoxic and contribute to cell death. Since these inflammatory responses last for a longer time, they may provide potential therapeutic targets in a prolonged treatment time window for stroke.

Activated microglial cells were found in ischemic core area immediately following ischemia/reperfusion. Besides the early activation in the ischemic core area, reperfusion triggers a delayed activation of microglia in the peri-infarct area starting from 3 days post ischemia and further maintains to 7 days post ischemia. A specific inhibitor of NHE-1, HOE 642, was shown to reduce the number of activated microglia in the peri-infarct area (Shi et al. 2011). Similar results were also observed when NHE-1 gene was transgenically knockdown. Moreover, blockade of NHE-1 either by HOE 642 or genetic knockdown reduced NOX activation in the peri-infarct area (Shi et al. 2011). Proinflammatory cytokine formation after ischemia was reduced with NHE-1 blockade. These results suggest that NHE-1 plays an important role in microglial activation following cerebral ischemia through its effect in maintaining NOX activity.

NHE-1 activity may also be linked to microglial activation via stimulation of NCX_{rev} following cerebral ischemia. Boscia et al. reported NCX protein expression was increased in invading microglia of the infarct core after focal cerebral ischemia (Boscia et al. 2009). Cultured microglial cells obtained from ischemic core area displayed increased NCX-1 expression and showed enhanced NCX activity. NCX activity and NCX-1 protein expression were significantly enhanced in BV2 microglia exposed to oxygen and glucose deprivation (Boscia et al. 2009). Taken together, these data suggest that cerebral ischemia stimulates activation of both NHE-1 and NCX in microglia, and these transporters may contribute to proinflammatory responses and ischemic brain injury.

35.4 Conclusions

In this chapter, we reviewed recent experimental findings on NHE-1 in the immune system and its role in inflammatory responses in brains. NHE-1 protein is expressed in different immune cells including monocytes, neutrophils, and macrophage/microglia. NHE-1 activity is important for immune cell function by extruding H^+ and regulation of pH_i during the respiratory burst (Fig. 35.2). Thus, NHE-1 is important in the production of ROS and cytokines and is involved in inflammatory responses under many pathological conditions including infection, trauma, and ischemia. Since excessive proinflammatory response contributes to brain injury in several CNS diseases, NHE-1 may present a new therapeutic target for reducing proinflammatory responses under these pathophysiological conditions.

Acknowledgments This work was supported by NIH grants R01NS 48216 and R01NS 38118 (D. Sun). We would like to thank Douglas B. Kintner for his critical comments of the manuscript.

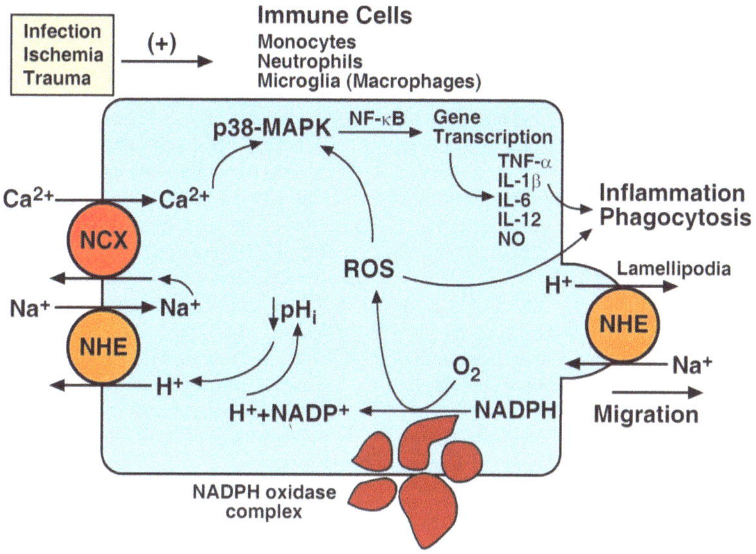

Fig. 35.2 *Role of NHE-1 in immune cells*. Immune cells act as the first line of defense in response to various stimuli by production of inflammatory substances, including ROS and cytokines. NHE-1 extrudes excessive H^+ during the respiratory burst in these cells; thus, it is essential for immune cell activation and inflammatory responses. NHE-1 may also participate in the extending of lamellipodia and immune cell movement and phagocytosis

References

S. Amor, F. Puentes, D. Baker, P. van der Valk, Inflammation in neurodegenerative diseases. Immunology **129**, 154–169 (2010)

K. Bedard, K.H. Krause, The NOX family of ROS-generating NADPH oxidases: physiology and pathophysiology. Physiol. Rev. **87**, 245–313 (2007)

C.L. Bennett, R. Chakraverty, Dendritic cells in tissues: in situ stimulation of immunity and immunopathology. Trends Immunol. **33**, 8–13 (2012)

M. Bernardes-Silva, D.C. Anthony, A.C. Issekutz, V.H. Perry, Recruitment of neutrophils across the blood-brain barrier: the role of E- and P-selectins. J. Cereb. Blood Flow Metab. **21**, 1115–1124 (2001)

M. Bhandaru, W. Yang, A. Rotte, V. Pasham, F. Lang, Regulation of Na^+/H^+ exchanger in dendritic cells by Akt2. Pflugers Arch. **463**, 355–363 (2012)

D. Biemesderfer, R.F. Reilly, M. Exner, P. Igarashi, P.S. Aronson, Immunocytochemical characterization of Na^+/H^+ exchanger isoform NHE-1 in rabbit kidney. Am. J. Physiol. **263**, F833–F840 (1992)

F. Boscia, R. Gala, A. Pannaccione, A. Secondo, A. Scorziello, G. Di Renzo, L. Annunziato, NCX1 expression and functional activity increase in microglia invading the infarct core. Stroke **40**, 3608–3617 (2009)

P. De Vito, The sodium/hydrogen exchanger: a possible mediator of immunity. Cell. Immunol. **240**, 69–85 (2006)

Z. Dong, P. Saikumar, J.M. Weinberg, M.A. Venkatachalam, Calcium in cell injury and death. Annu. Rev. Pathol. **1**, 405–434 (2006)

J.C. Felger, T. Abe, U.W. Kaunzner, A. Gottfried-Blackmore, J. Gal-Toth, B.S. McEwen, C. Iadecola, K. Bulloch, Brain dendritic cells in ischemic stroke: time course, activation state, and origin. Brain Behav. Immun. **24**, 724–737 (2010)

L. Fliegel, The Na^+/H^+ exchanger isoform 1. Int. J. Biochem. Cell Biol. **37**, 33–37 (2005)

T. Fukushima, T.K. Waddell, S. Grinstein, G.G. Goss, J. Orlowski, G.P. Downey, Na^+/H^+ exchange activity during phagocytosis in human neutrophils: role of Fcgamma receptors and tyrosine kinases. J. Cell Biol. **132**, 1037–1052 (1996)

M.B. Graeber, W.J. Streit, Microglia: biology and pathology. Acta Neuropathol. **119**, 89–105 (2010)

S. Grinstein, M. Woodside, T.K. Waddell, G.P. Downey, J. Orlowski, J. Pouyssegur, D.C. Wong, J.K. Foskett, Focal localization of the NHE-1 isoform of the Na^+/H^+ antiport: assessment of effects on intracellular pH. EMBO J. **12**, 5209–5218 (1993)

T.J. Harrigan, I.F. Abdullaev, D. Jourd'heuil, A.A. Mongin, Activation of microglia with zymosan promotes excitatory amino acid release via volume-regulated anion channels: the role of NADPH oxidases. J. Neurochem. **106**, 2449–2462 (2008)

H. Hayashi, O. Aharonovitz, R.T. Alexander, N. Touret, W. Furuya, J. Orlowski, S. Grinstein, Na^+/H^+ exchange and pH regulation in the control of neutrophil chemokinesis and chemotaxis. Am. J. Physiol. Cell Physiol. **294**, C526–C534 (2008)

L.M. Henderson, J.B. Chappell, O.T. Jones, Internal pH changes associated with the activity of NADPH oxidase of human neutrophils. Further evidence for the presence of an H^+ conducting channel. Biochem. J. **251**, 563–567 (1988)

C.Y. Hsu, S.H. Ahmed, K.R. Lees, The therapeutic time window: theoretical and practical considerations. J. Stroke Cerebrovasc. Dis. **9**(6 Pt 2), 24–31 (2000). 9, 24–31

S. Hu, W.S. Sheng, S.J. Schachtele, J.R. Lokensgard, Reactive oxygen species drive herpes simplex virus (HSV)-1-induced proinflammatory cytokine production by murine microglia. J. Neuroinflammation **8**, 123 (2011)

R. Jin, G. Yang, G. Li, Inflammatory mechanisms in ischemic stroke: role of inflammatory cells. J. Leukoc. Biol. **87**, 779–789 (2010)

A.R. Khaled, A.N. Moor, A. Li, K. Kim, D.K. Ferris, K. Muegge, R.J. Fisher, L. Fliegel, S.K. Durum, Trophic factor withdrawal: p38 mitogen-activated protein kinase activates NHE1, which induces intracellular alkalinization. Mol. Cell. Biol. **21**, 7545–7557 (2001)

Y. Liu, D.B. Kintner, V. Chanana, J. Algharabli, X. Chen, Y. Gao, J. Chen, P. Ferrazzano, J.K. Olson, D. Sun, Activation of microglia depends on Na^+/H^+ exchange-mediated H^+ homeostasis. J. Neurosci. **30**, 15210–15220 (2010)

J. Luo, D. Sun, Physiology and pathophysiology of Na^+/H^+ exchange isoform 1 in the central nervous system. Curr. Neurovasc. Res. **4**, 205–215 (2007)

P.S. Martins, E.G. Kallas, M.C. Neto, M.A. Dalboni, S. Blecher, R. Salomao, Upregulation of reactive oxygen species generation and phagocytosis, and increased apoptosis in human neutrophils during severe sepsis and septic shock. Shock **20**, 208–212 (2003)

T. Matsuda, T. Nagano, M. Takemura, A. Baba, Topics on the Na^+/Ca^{2+} exchanger: responses of Na^+/Ca^{2+} exchanger to interferon-gamma and nitric oxide in cultured microglia. J. Pharmacol. Sci. **102**, 22–26 (2006)

A.N. Moor, X.T. Gan, M. Karmazyn, L. Fliegel, Activation of Na^+/H^+ exchanger-directed protein kinases in the ischemic and ischemic-reperfused rat myocardium. J. Biol. Chem. **276**, 16113–16122 (2001)

T. Nagano, Y. Kawasaki, A. Baba, M. Takemura, T. Matsuda, Up-regulation of Na^+/Ca^{2+} exchange activity by interferon-gamma in cultured rat microglia. J. Neurochem. **90**, 784–791 (2004)

J. Orlowski, S. Grinstein, Diversity of the mammalian sodium/proton exchanger SLC9 gene family. Pflugers Arch. **447**, 549–565 (2004)

K. Petrecca, R. Atanasiu, S. Grinstein, J. Orlowski, A. Shrier, Subcellular localization of the Na^+/H^+ exchanger NHE1 in rat myocardium. Am. J. Physiol. **276**, H709–H717 (1999)

L.K. Putney, S.P. Denker, D.L. Barber, The changing face of the Na^+/H^+ exchanger, NHE1: structure, regulation, and cellular actions. Annu. Rev. Pharmacol. Toxicol. **42**, 527–552 (2002)

G. Reichmann, M. Schroeter, S. Jander, H.G. Fischer, Dendritic cells and dendritic-like microglia in focal cortical ischemia of the mouse brain. J. Neuroimmunol. **129**, 125–132 (2002)

J.M. Robinson, Reactive oxygen species in phagocytic leukocytes. Histochem. Cell Biol. **130**, 281–297 (2008)

D. Rotin, S. Grinstein, Impaired cell volume regulation in Na^+/H^+ exchange-deficient mutants. Am. J. Physiol. **257**, C1158–C1165 (1989)

A. Rotte, V. Pasham, M. Eichenmuller, H. Mahmud, N.T. Xuan, E. Shumilina, F. Gotz, F. Lang, Effect of bacterial lipopolysaccharide on Na^+/H^+ exchanger activity in dendritic cells. Cell. Physiol. Biochem. **26**, 553–562 (2010a)

A. Rotte, V. Pasham, W. Yang, M. Eichenmuller, M. Bhandaru, E. Shumilina, F. Lang, Phosphoinositide 3-kinase-dependent regulation of Na^+/H^+ exchanger in dendritic cells. Pflugers Arch. **460**, 1087–1096 (2010b)

Y. Shi, V. Chanana, J.J. Watters, P. Ferrazzano, D. Sun, Role of sodium/hydrogen exchanger isoform 1 in microglial activation and proinflammatory responses in ischemic brains. J. Neurochem. **119**, 124–135 (2011)

P. Soeiro-Pereira, A. Falcai, C. Kubo, E. Oliveira-Junior, O. Marques, E. Antunes, A. Condino-Neto, BAY 41-2272, a soluble guanylate cyclase agonist, activates human mononuclear phagocytes. Br. J. Pharmacol. (2011). doi:10.1111/j.1476-5381.2011.01764.x. Epub ahead of print

C.J. Swallow, S. Grinstein, R.A. Sudsbury, O.D. Rotstein, Relative roles of Na^+/H^+ exchange and vacuolar-type H^+ ATPases in regulating cytoplasmic pH and function in murine peritoneal macrophages. J. Cell. Physiol. **157**, 453–460 (1993)

B.R. Tambuyzer, P. Ponsaerts, E.J. Nouwen, Microglia: gatekeepers of central nervous system immunology. J. Leukoc. Biol. **85**, 352–370 (2009)

Q. Wang, X.N. Tang, M.A. Yenari, The inflammatory response in stroke. J. Neuroimmunol. **184**, 53–68 (2007)

M.A. Yenari, T.M. Kauppinen, R.A. Swanson, Microglial activation in stroke: therapeutic targets. Neurotherapeutics **7**, 378–391 (2010)

L. Zhang, Z.G. Zhang, R.L. Zhang, M. Lu, M. Krams, M. Chopp, Effects of a selective CD11b/CD18 antagonist and recombinant human tissue plasminogen activator treatment alone and in combination in a rat embolic model of stroke. Stroke **34**, 1790–1795 (2003)

L. Ziegler-Heitbrock, P. Ancuta, S. Crowe, M. Dalod, V. Grau, D.N. Hart, P.J. Leenen, Y.J. Liu, G. MacPherson, G.J. Randolph, J. Scherberich, J. Schmitz, K. Shortman, S. Sozzani, H. Strobl, M. Zembala, J.M. Austyn, M.B. Lutz, Nomenclature of monocytes and dendritic cells in blood. Blood **116**, e74–e80 (2010)

Acid-Sensing Ion Channels in Pathological Conditions

36

Xiang-Ping Chu and Zhi-Gang Xiong

Abstract

Acid-sensing ion channels (ASICs), a novel family of proton-gated amiloride-sensitive cation channels, are expressed primarily in neurons of peripheral sensory and central nervous systems. Recent studies have shown that activation of ASICs, particularly the ASIC1a channels, plays a critical role in neuronal injury associated with neurological disorders such as brain ischemia, multiple sclerosis, and spinal cord injury. In normal conditions in vitro, ASIC1a channels desensitize rapidly in the presence of a continuous acidosis or following a preexposure to minor pH drop, raising doubt for their contributions to the acidosis-mediated neuronal injury. It is now known that the properties of ASICs can be dramatically modulated by signaling molecules or biochemical changes associated with pathological conditions. Modulation of ASICs by these molecules can lead to dramatically enhanced and/or prolonged activities of these channels, thus promoting their pathological functions. Understanding of how ASICs behave in pathological conditions may help define new strategies for the treatment and/or prevention of neuronal injury associated with various neurological disorders.

Keywords

Acid-sensing ion channel • Acidosis • Ischemia • Neuron • Modulation

36.1 Introduction

In physiological conditions, extracellular pH (pH_o) is maintained at ~7.4 through various H^+ transporting mechanisms (Chesler 1990). In pathological conditions such as tissue inflammation, ischemic stroke, traumatic brain injury, and epileptic seizure, a marked reduction of pH_o, a condition termed acidosis, takes place

X.-P. Chu (✉)
Department of Basic Medical Science, School of Medicine, University of Missouri-Kansas City, 2411 Holmes Street, Kansas City, MO, USA
e-mail: chux@umkc.edu

Z.-G. Xiong (✉)
Department of Neurobiology, Morehouse School of Medicine, 720 Westview Drive SW, Atlanta, GA 30310, USA
e-mail: zxiong@msm.edu

L. Annunziato (ed.), *Sodium Calcium Exchange: A Growing Spectrum of Pathophysiological Implications*, Advances in Experimental Medicine and Biology 961, DOI 10.1007/978-1-4614-4756-6_36,

(Ljunggren et al. 1974; Nedergaard et al. 1991; Rehncrona 1985; Revici et al. 1949; Siesjo 1988; Siesjo et al. 1996; Sutherland et al. 2000; Tombaugh and Sapolsky 1993). In brain ischemia, for example, the shortage of oxygen, due to the lack of blood supply, results in increased anaerobic glycolysis which leads to lactic acid accumulation (Siesjo et al. 1996; Tombaugh and Sapolsky 1993). Accumulation of lactic acid, along with H^+ release from ATP hydrolysis, results in a dramatic decrease in tissue pH. In addition, cessation of local circulation results in carbon dioxide accumulation and carbonic acid buildup, which also contribute to the decrease of brain pH (Siesjo 1988).

For many years, acidosis has been known to be associated with neuronal injury (Siesjo 1988; Siesjo et al. 1996; Tombaugh and Sapolsky 1993). However, the underlying mechanism was not entirely clear. Low tissue pH may cause cell injury through nonselective denaturation of proteins and nucleic acids (Kalimo et al. 1981), hinderance of postischemic metabolic recovery by inhibiting mitochondrial energy metabolism, impairment of postischemic blood flow via vascular edema (Hillered et al. 1985), stimulation of pathologic free radical formation (Rehncrona et al. 1989), and inhibition of astrocytic glutamate uptake which may contribute to excitatory neuronal injury (Swanson et al. 1995). Recent finding that acidosis, at the level commonly seen in neurological disorders, can induce neuronal injury by activating a distinct family of ligand-gated cation channels, the acid-sensing ion channels (ASICs), has dramatically changed the view of acid signaling and provided a novel therapeutic target for neuroprotection (Benveniste and Dingledine 2005; Huang and McNamara 2004; Voilley 2004; Wemmie et al. 2006; Xiong et al. 2004, 2007; Yermolaieva et al. 2004).

Following the cloning and characterization of the first ASIC subunit (ASIC1a) in 1997 (Waldmann et al. 1997), six additional subunits of ASICs (1b1, 1b2, 2a, 2b, 3, and 4) have been identified (Wemmie et al. 2006). ASICs are voltage-independent, amiloride-sensitive, and Na^+-selective cation channels belonging to the degenerin/epithelial Na^+ channel (DEG/ENaC) superfamily (Alvarez et al. 2000). Expression of 1a, 2a, 2b, and 4 has been demonstrated in CNS neurons, while all other ASICs, except ASIC4, are expressed in peripheral sensory neurons. Nonneuronal tissues such as vascular smooth muscle cells (Grifoni et al. 2008), some glial cells (Feldman et al. 2008; Huang et al. 2010), and bone (Jahr et al. 2005) have been shown to express ASICs.

Each ASIC subunit consists of two transmembrane domains (TM1 and TM2) and a large cysteine-rich extracellular loop, with the pre-TM2 region essential for ion permeability and the gating of these channels (Bassler et al. 2001; Krishtal 2003; Waldmann et al. 1997) (Fig. 36.1). Although a tetrameric assembly was initially proposed for functional ASICs, the recent success of crystal structure of chicken ASIC1a channels has revealed a trimeric assembly (Jasti et al. 2007). Among all ASICs, the homomeric ASIC1a channel is of particular interest because of its wide distribution, high sensitivity to acid, and permeability to Ca^{2+} (Waldmann et al. 1997; Xiong et al. 2004; Yermolaieva et al. 2004).

In peripheral sensory neurons, ASICs are implicated in nociception (Benson et al. 1999; Bevan and Yeats 1991; Bohlen et al. 2011; Chen et al. 2002; Deval et al. 2008; Krishtal and Pidoplichko 1981; Mazzuca et al. 2007; Sluka et al. 2003; Ugawa et al. 2002; Wu et al. 2004), mechanosensation (Page et al. 2005; Price et al. 2000, 2001), and taste transduction (Lin et al. 2002; Ugawa et al. 2003; Ugawa 2003). In CNS, ASIC1a is involved in synaptic plasticity, learning/memory, fear conditioning (Wemmie et al. 2002, 2003), and retinal physiology (Ettaiche et al. 2006). A number of studies have demonstrated that activation or sensitization of Ca^{2+}-permeable ASIC1a channels is involved in acidosis-mediated neuronal injury (Friese et al. 2007; Gao et al. 2005; Li et al. 2011; Pignataro et al. 2011; Xiong et al. 2004), disclosing a novel neuroprotective target (Xiong et al. 2008).

Under experimental conditions (e.g., in patch-clamp recordings), ASICs are activated only by rapid pH drops, and the currents of most ASIC subtypes, particularly the homomeric ASIC1a channels, desensitize rapidly in the continuous presence of acidic pH (Fig. 36.2a). Preexposure of ASIC1a channels to very small

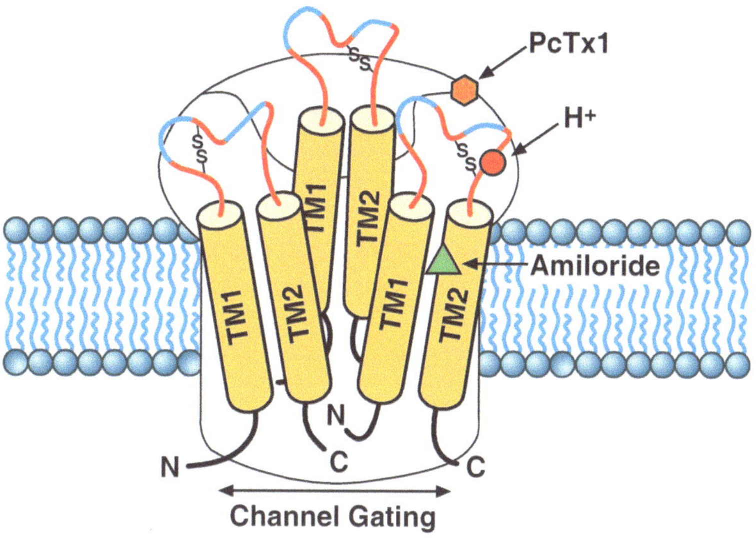

Fig. 36.1 Schematic representation of ASIC structure. Each ASIC subunit has two transmembrane domains (TM1 and TM2), short intracellular N and C-terminals, and a large cysteine-rich extracellular loop, with the pre-TM2 region critical for ion permeability and the gating of the channel. Amiloride, a nonselective ASIC blocker, binds to channel pore; while PcTx1, a selective ASIC1a blocker, binds to extracellular site of the channel

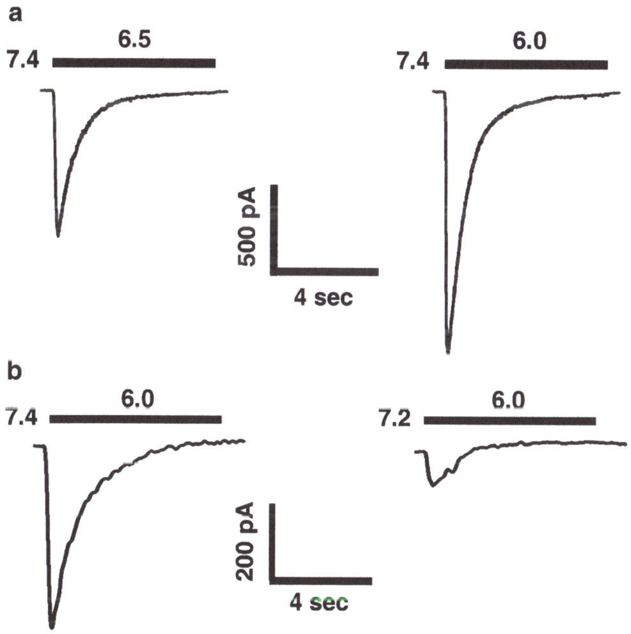

Fig. 36.2 Desensitization of ASIC current in cultured mouse cortical neurons. (**a**) ASIC current was activated by lowering the extracellular pH from 7.4 to 6.5 or 6.0. The current decays rapidly in the continuous presence of reduced pH. (**b**) ASIC current was activated by lowering the extracellular pH to 6.0 from a preset pH of 7.4 or 7.2. Preincubation of cells at pH 7.2 dramatically inhibited the ASIC current activated at pH 6.0

pH decrease (e.g., from 7.4 to 7.2) that does not activate the channel itself also suppresses the channel activity in response to subsequent large pH drops (Fig. 36.2b). Thus, whether a significant amount of ASIC current can be activated in the pathological conditions and whether the effects of ASIC activation could be long lasting are important questions crucial in determining the functional significance of these channels. In this regard, recent findings showing that endogenous signaling molecules and biochemical changes associated with various pathological conditions can dramatically alter the properties of ASICs have provided desirable answer to the questions. Understanding the modulation of ASICs by molecules or biochemical changes associated with pathological conditions will help in understanding the pathological functions of these channels. It will also aid in the development of new and effective strategies for the treatment and/or prevention of neuronal injury associated with various neurological disorders.

36.2 Lactate

In the normal brain, the concentration of lactate is at ~1 mM. Following ischemia, anaerobic metabolism of glucose leads to increased production of lactate. A concentration of lactate between 12 and 20 mM has been reported (Schurr 2002; Schurr and Rigor 1998). Increased concentration of lactate correlates well with increased degree of brain injury (Siesjo et al. 1996), though the detailed mechanism is unclear. The finding that lactate enhances the ASIC current provided a new interpretation (Immke and McCleskey 2001).

In sensory neurons that innervate the heart, Immke and McCleskey first demonstrated that the addition of 15 mM lactate dramatically increased the amplitude of ASIC current activated by a moderate pH drop to ~7.0 (Immke and McCleskey 2001). Applications of the same concentration of lactate at pH values that do not activate ASICs (e.g., pH 8.0 or pH 7.4) caused no response. Thus, lactate acts by potentiating but not activating the ASICs. In COS-7 transfected with different subunits of ASICs, both ASIC3 and ASIC1a currents were potentiated by lactate (Immke and McCleskey 2001). The effect of lactate on ASIC current persists in excised cell-free membrane patches suggesting that no second messenger or signaling cascade is involved. Since lactate has the ability to chelate the divalent cations including Ca^{2+} and Mg^{2+}, which have a modulatory role on various membrane receptors and ion channels (Hess et al. 1986; Xiong and MacDonald 1999; Zhou and Jones 1995), the authors hypothesized that potentiation of the ASIC currents may be due to the chelation of divalent cations by lactate. Indeed, adjusting the concentrations of Ca^{2+} and Mg^{2+} eliminated the effect of lactate, whereas reducing the divalent concentrations mimicked the effect of lactate (Immke and McCleskey 2001). Other monocarboxylic acids which have divalent cation chelation ability also potentiated the ASIC current. Similar to the cardiac sensory neurons, potentiation of the ASIC current by lactate has been reported in other neurons such as cerebellar Purkinje neurons (Allen and Attwell 2002).

36.3 Ca^{2+}

Ca^{2+} is one of the most important modulators of various voltage-gated and ligand-gated ion channels including ASICs. Coapplication of Ca^{2+} with acidic solution reduces ASIC currents (Gao et al. 2004; Immke and McCleskey 2003; Paukert et al. 2004; Waldmann et al. 1997). Similarly, pretreatment followed by continuous presence of the extracellular Ca^{2+} also inhibits the ASIC currents (de Weille and Bassilana 2001; Wang et al. 2006). Investigation into the mechanisms underlying Ca^{2+} modulation of ASICs has led to the finding that Ca^{2+} decreases the affinity of ASICs for H^+ (Babini et al. 2002; Immke and McCleskey 2003). It has been proposed that, at a pH of 7.4, ASICs (e.g., ASIC3 channels) are closed because of the Ca^{2+} blockade. As the pH_o is decreased, binding of H^+ to the channel displaces Ca^{2+} from its binding site, leading to opening of the channel (Immke and McCleskey 2003). For ASIC1a channels, different models have been proposed (Paukert et al. 2004; Zhang et al. 2006). Paukert and colleagues showed that two negatively charged residues near the entrance of the channel pore, E425 and D432, are crucial for Ca^{2+} blockade of the ASIC1a channel (Paukert et al. 2004). They proposed that there are more than one Ca^{2+} binding sites, one that mediates blocking and one mediates modulation, that exist on the ASIC1a channel. In pathological conditions such as brain ischemia, significant decrease of extracellular Ca^{2+} concentration ($[Ca^{2+}]_e$) takes place (Ekholm et al. 1995;

Hansen and Zeuthen 1981). Reduction in $[Ca^{2+}]_e$ is expected to relieve the inhibition of ASICs, thus increasing the activity and pathological function of these channels.

36.4 Arachidonic Acid

Arachidonic acid (AA) is a polyunsaturated fatty acid present in the phospholipids of all cell membranes. It is one of the most abundant fatty acids in the brain. In addition to being involved in cellular signaling as a lipid second messenger involved in the regulation of signaling enzymes such as PLC and PKC (Farooqui and Horrocks 2006; Sang and Chen 2006), AA plays important roles in various pathological conditions. For example, it is a key intermediate in inflammatory responses. In addition, the role of AA in the pathophysiology of neurological disorders including ischemic brain injury has been well documented (Farooqui et al. 2006; Farooqui and Horrocks 2006; Muralikrishna and Hatcher 2006; Siesjo and Katsura 1992). Following brain ischemia, the rise of intracellular Ca^{2+} concentration leads to the activation of phospholipase A2, resulting in increased production of lipid mediators including AA (Farooqui and Horrocks 2006; Muralikrishna and Hatcher 2006; Rehncrona et al. 1982). High concentrations of lipid mediators cause neurotoxicity (Farooqui and Horrocks 2006).

Earlier studies have shown that AA has effects on a variety of voltage-gated and ligand-gated ion channels (Angelova and Muller 2006; Hu et al. 2006; Keros and McBain 1997; Mignen et al. 2005; Miller et al. 1992; Nagano et al. 1995; Paoletti and Ascher 1994). Particularly, it potentiates the opening of NMDA-gated channels (Casado and Ascher 1998; Miller et al. 1992; Paoletti and Ascher 1994), which might contribute to its neurotoxicity.

Allen and Attwell were the first to show that AA also enhances the activity of ASICs (Allen and Attwell 2002). In rat cerebellar Purkinje cells, bath perfusion of solution with 5 or 10 μM AA produced a large increase in the amplitude of ASIC current. In addition to the peak current, application of AA enhanced or induced a sustained component of the ASIC current (Allen and Attwell 2002). The potentiation of the ASIC current by AA appears to be independent of its derivatives, since an agent known to block the breakdown of AA did not affect its capacity to potentiate the ASIC current (Allen and Attwell 2002). The molecular mechanism for AA potentiation of ASICs is not clear. One possibility is that insertion of AA into the membrane induces membrane stretch and that the ASICs are stretch sensitive, as are NMDA channels (Casado and Ascher 1998). This explanation is supported by the finding that perfusion of neurons with hypotonic saline, which causes cell swelling and membrane stretch, mimicked the potentiation of ASIC currents by AA (Allen and Attwell 2002). Subsequent studies by Smith et al. also suggested that AA potentiates ASIC activation by a direct mechanism (Smith et al. 2007). They confirmed that inhibition of AA metabolism had no effect on the potentiation of ASIC1a. However, the potentiation of single ASIC2a channels by AA could also be observed in cell-free patches, suggesting that membrane stretch is not required.

36.5 Spermine

Spermine is a polyvalent cation whose extracellular concentration fluctuates significantly within the nervous system. Although spermine is involved in various physiological processes, high concentration of spermine can induce neuronal depolarization and cytoplasmic Ca^{2+} overload, which might lead to neuronal damage (Toninello et al. 2004). Following ischemia, the activity of ornithine decarboxylase (ODC), a rate-limiting enzyme responsible for polyamine synthesis, is enhanced, leading to increased concentration of spermine and neuronal injury (Kindy et al. 1994).

The mechanism underlying increased neuronal injury by spermine has not been entirely clarified. Spermine can modulate NMDA receptor function (Benveniste and Mayer 1993; Rock and Macdonald 1995). Thus, several studies have attempted to link its neurotoxicity to increased activation of NMDA receptors. However, these studies have yielded inconsistent results (Johnson 1998; Li et al. 2007). Additional mechanisms are likely involved in spermine-mediated neurotoxicity.

Babini and colleagues were the first to demonstrate that spermine potentiates the activities of ASICs (Babini et al. 2002). Subsequent studies by Duan et al. showed that extracellular spermine exacerbated ischemic neuronal injury through sensitization of ASIC1a channels to extracellular acidosis (Duan et al. 2011). Pharmacological blockade of ASIC1a or deletion of the ASIC1 gene reduced the enhancing effect of spermine on ischemic neuronal damage, both in cultured neurons and in a mouse model of focal ischemia. Spermine also reduced desensitization of ASIC1a in the open state and accelerated recovery from desensitization in response to repeated acid stimulation. Enhanced channel activity was accompanied by increased acid-induced neuronal depolarization and cytoplasmic Ca^{2+} overload, which may explain the exacerbation of neuronal damage caused by spermine. Thus, extracellular spermine contributes to ischemic neuronal injury, at least in part, by enhancing the activity of ASIC1a channels (Duan et al. 2011).

36.6 Proteases

Brain injury is accompanied by increased protease activity (Gingrich and Traynelis 2000). Blood-derived proteases such as thrombin, tissue plasminogen activator, and plasmin can gain access to CNS interstitial spaces due to a compromised blood–brain barrier (Gingrich and Traynelis 2000; Vivien and Buisson 2000). Previous studies have demonstrated that proteases modulate the activities of various ion channels including ENaC, which belongs to the same superfamily as ASICs (Chraibi et al. 1998; Holt et al. 2001; Nicole et al. 2001), suggesting that ASICs could be a target. Indeed, subsequent studies by Poirot and colleagues showed that the activity of ASIC1a is modulated by serine proteases (Poirot et al. 2004). Exposure of CHO cells stably expressing ASIC1a channels to trypsin or other serine proteases (e.g., proteinase K and chymotrypsin) shifted the pH dependence of activation and steady-state inactivation of the ASIC1a channels to more acidic pH values. As a consequence, protease exposure leads to a decrease in ASIC1a activity when currents are activated by a pH drop from 7.4. Interestingly, if the channel is activated from a basal pH of 7, a condition relevant to ischemia, protease exposure increases rather than decreases the ASIC1a activity. In addition, protease treatment dramatically accelerates the recovery rate of ASIC1a channels from desensitization (Poirot et al. 2004). The effects of proteases on ASICs involve proteolysis of the channel protein, as the capacity of trypsin to modulate the ASIC1a channels was eliminated with soybean trypsin inhibitor or modification of trypsin's catalytic site with TLCK, a reagent that irreversibly modifies a histidine residue in the catalytic site of trypsin. Cleavage of the channel protein was confirmed by Western blot showing reduction of a 64-kDa ASIC protein band to a lower-molecular-weight band of 49 kDa (Poirot et al. 2004). Further studies demonstrated that trypsin cleaves ASIC1a subunits at Arg-145 in the N-terminal part of the extracellular loop. The cleavage site is between a highly conserved sequence and a sequence that is critical for ASIC1a inhibition by PcTx1 (Vukicevic et al. 2006). Since activation of ASIC1a is involved in acidosis-mediated ischemic brain injury, modulation of ASIC1a by proteases could be relevant to its pathological function in brain ischemia.

36.7 CaMKII

Ca^{2+}/calmodulin (CaM)-dependent protein kinase II (CaMKII) is the most abundant kinase isoform in brain, particularly enriched in neurons. It is a multifunctional protein kinase that regulates normal neuronal function. It is regulated by multisite phosphorylation, which can alter the enzyme activity and targeting to cellular microdomains through interactions with binding proteins. CaMKII is a major mediator of the function of excitatory glutamate receptors. Activation of glutamate receptors triggers an increase of intracellular Ca^{2+} and an autophosphorylation of CaMKII at T286. This process makes the kinase active independent from Ca^{2+} stimulation, a transition required for synaptic plasticity. Increased CaMKII activity was also implicated in the regulation of neuronal death, though the detailed mechanism was unclear.

Recent studies by Gao et al. suggest that during ischemia, coupling between the NMDARs/CaMKII cascade and the ASIC1a channel contributes significantly to acidotoxic neuronal death (Gao et al. 2005). They demonstrated that global brain ischemia in rats results in an increased phosphorylation of ASIC1a by CaMKII at Ser478 and Ser479. This phosphorylation sensitizes the channel to low pH, exacerbating cell death by allowing increased calcium conductance. The phosphorylation is a result of the activation of NR2B-containing NMDA receptors and an increase of intracellular Ca^{2+} during ischemia.

Consistent with the previous study (Xiong et al. 2004), they observed an enhancement of ASIC currents by oxygen–glucose deprivation (OGD). Inhibition of CaMKII with KN-93 or CaMKIINtide abolished the enhancement of ASIC currents, indicating an involvement of CaMKII. Consistent with the involvement of CaMKII, they showed an increased phosphorylation of ASIC1a after transient global ischemia which can be blocked by intracerebroventricular administration of CaMKII inhibitor KN-93 or CaMKIINtide. In addition, there was an increased coimmunoprecipitation of ASIC1a with CaMKIIα after global ischemia, which can be inhibited by KN-93 or CaMKIINtide. Pharmacological inhibition of CaMKII phosphorylation of ASIC1a with KN-93 or mutation of ASIC1a at Ser478 and Ser479 produced neuroprotection. Thus, phosphorylation of ASIC1a by CaMKII at the site of Ser478 and Ser479 plays an essential role in ischemia-induced cell death.

36.8 Nitric Oxide

Nitric oxide (NO) is an important reactive oxygen/nitrogen species which has a variety of physiological and pathological functions (Star 1993). Endogenous synthesis of NO is catalyzed by nitric oxide synthase, which is Ca^{2+} dependent. During ischemia, overactivation of the glutamate receptors and subsequent increases in intracellular Ca^{2+} lead to activation of the Ca^{2+}-dependent neuronal form of nitric oxide synthase (nNOS), resulting in an increased production of NO (Bolanos and Almeida 1999; Schulz et al. 1997). NO can also be released by activated microglia (Boje and Arora 1992). NO regulates the protein function by two main pathways: an indirect mechanism that involves the production of cGMP and the activation of protein kinase G, and a direct mechanism that involves modification of the tertiary structure of proteins by *S*-nitrosylating the thiol side chains of cysteine residues (Jaffrey et al. 2001). Increased NO production is known to increase neuronal injury (Boje and Arora 1992). Although the formation of a strong oxidant of peroxynitrite is likely involved in cell injury, other mechanisms cannot be excluded. Cadiou and colleagues were the first to report that ASICs are a target of NO. NO donor S-nitroso-N-acetylpenicillamine (SNAP) potentiates proton-gated currents in DRG neurons and in CHO cells expressing each of the ASIC subunits. Modulators of the cGMP/PKG pathway had no effect on the potentiation, but in excised patches from CHO cells expressing ASIC2a, the potentiation could be reversed by externally applying reducing agents. NO therefore has a direct external effect on ASICs, probably through oxidization of cysteine residues (Cadiou et al. 2007).

Consistent with the potentiation of ASIC currents by NO, subsequent studies by Jett et al. demonstrated that acid-induced cell injury is potentiated by NO donor (Jetti et al. 2010). They showed that, at a pH of 6.1, death rates of Neuro2A cells expressing ASIC1 channels were significantly higher than the cells that do not express ASICs. Amiloride, a blocker of ASICs, protected the cell from acid injury, suggesting that acid injury is mediated by activating ASICs. Sodium nitroprusside, a potent NO donor, not only increased the ASIC-mediated currents but also increased acid-induced cell death.

36.9 Dynorphins

Dynorphins are endogenous opioid neuropeptides abundantly expressed in the CNS. They are involved in a variety of physiologic functions including antinociception and neuroendocrine signaling and may be protective to neurons and oligodendroglia via their opioid receptor-mediated effects. However, under pathophysiological conditions where dynorphin levels are substantially elevated, these peptides are excitotoxic, partially through actions at glutamate receptors (Hauser et al. 1999). The excitotoxic actions of dynorphins require supraphysiological concentrations or prolonged tissue exposure. Thus, dynorphins can have either a protective or destructive action in neurons and glia, and the net effect may depend upon the distribution of receptors in a particular region and the amount of dynorphins released.

Recently, Sherwood and Askwith reported that at high concentrations, dynorphin A and big dynorphin potentiate acid-activated currents in cortical neurons and in CHO cells expressing homomeric ASIC1a subunits (Sherwood and Askwith 2009). The potentiation of the ASIC1a activity was mediated through a limitation of steady-state desensitization of the channel. The potentiation of ASIC1a activity by dynorphin was not mediated by opioid receptor activation but through a direct interaction with ASIC1a. Alteration of steady-state desensitization by dynorphins enhanced ASIC1a-triggered neuronal injury during prolonged acidosis. Thus, ASIC1a is a new nonopioid receptor target for dynorphins, and dynorphins can enhance ischemic brain injury by preventing steady-state desensitization of ASIC1a channels.

36.10 FMRFamide

FMRFamide and structurally related peptides are abundant in invertebrate nervous systems where they function as neurotransmitters and neuromodulators. Although FMRFamide itself has not been isolated in mammals, several FMRFamide-related peptides exist in the mammalian nervous system. FMRFamide and related peptides are generally thought to exert their physiological roles through G-protein-coupled receptors (Lingueglia et al. 2006). However, two ionotropic receptors involved in the function of these peptides have recently been identified. FMRFamide-gated Na^+ channel (FaNaC), which is a neuronal Na^+ channel in invertebrates, is directly activated by micromolar concentrations of FMRFamide and RFamide-related peptides (RFRPs) (Lingueglia et al. 2006). In addition, ASICs, which share significant structure and sequence homology with FaNaC in the mammalian nervous system, can be modulated by FMRFamide and RFRPs. FMRFamide and RFRPs such as neuropeptide FF are incapable of generating any ASIC currents on their own but significantly potentiate ASIC currents in sensory neurons and in heterologous expression systems (Askwith et al. 2000; Catarsi et al. 2001; Xie et al. 2003). In addition to their effects on the amplitude of ASIC currents, FMRFamide and RFRPs also reduce the rate of current desensitization (Askwith et al. 2000; Catarsi et al. 2001; Xie et al. 2003).

36.11 Insulin

Insulin and insulin receptors are expressed at high levels in discrete regions within the CNS, and insulin is released by depolarization in cultured CNS neurons (Wozniak et al. 1993). In addition to its conventional role in glucose uptake, insulin has been shown to act as a neuromodulator of many brain functions, such as food intake, neuronal growth, and maturation. It has also been shown that insulin administration protects neurologic function in cerebral ischemia in nondiabetic rats (LeMay et al. 1988). The exact mechanism was not clear. Almost 15 years ago, Wan and colleagues first demonstrated that insulin promotes surface expression of inhibitory $GABA_A$ receptors, thereby potentiating $GABA_A$-receptor-mediated synaptic inhibition. This effect may partially explain the neuroprotective effect of insulin in nondiabetic animals.

Very recently, Chai and colleagues demonstrated that insulin plays an important role in regulating the level of surface expression of ASIC1a channels (Chai et al. 2010). This finding provided an alternative explanation for the protective effect of insulin. In CHO cells expressing ASIC1a subunit, serum depletion, a condition mimicking ischemia, induced a significant increase in the level of ASIC1a surface expression. As a result, the ASIC1a current was dramatically enhanced. Among the components of serum, insulin was identified as the key factor that maintains a low level of ASIC1a expression on the plasma membrane. Removing insulin from culture medium increased the trafficking of the ASIC1a channels to the surface membrane. Similarly, neurons subjected to insulin depletion increased the surface expression of ASIC1a subunit with resultant potentiation of ASIC1a currents. Thus, in the normal condition where insulin is present, ASIC1a is predominantly localized to the endoplasmic reticulum. Under conditions of substrate depletion (e.g., following brain ischemia), the lack of insulin may stimulate translocation of the ASIC1a channels to the cell surface, thus increasing the function of these channels (Chai et al. 2010). Since activation of ASIC1a channels plays a critical role in acidosis-mediated ischemic brain injury, inhibition of ASIC1a expression by insulin may account partially for its neuroprotective effect.

36.12 Conclusion

ASICs represent new biological components in neurons. Increasing evidence indicates the involvement of these channels in both physiological and pathological processes such as nociception, mechanosensation, taste transduction, synaptic plasticity, learning/memory, and acidosis-mediated neurodegeneration. Although in most electrophysiological recordings ASIC responses appear to be transient in nature, various biochemical changes, largely occur in pathological conditions, dramatically enhance the amplitude and reduce the desensitization of the ASIC current (Fig. 36.3) or increase the recovery of ASICs from desensitization. Knowing the changes of ASIC property and the mechanisms of modulation in pathological conditions is, with no doubt, critical for understanding the pathological functions of these channels and for establishing effective therapeutic interventions.

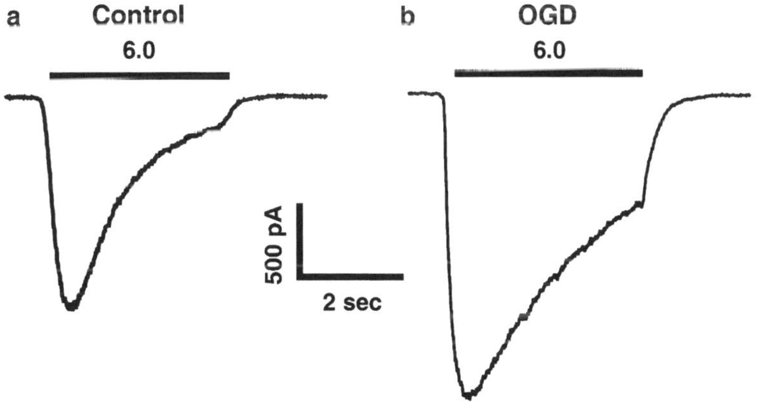

Fig. 36.3 Oxygen–glucose deprivation (OGD), an in vitro model of ischemia, increases the amplitude but reduces the desensitization of ASIC current in mouse cortical neurons

Acknowledgments XPC is supported by American Heart Association Scientist Development Grant 0735092 N, University of Missouri Research Board, and University of Missouri-Kansas City School of Medicine start-up funding. The work in ZGX's lab is supported in part by NIH R01NS047506, R01NS066027, UL1 RR025008, U54 RR026137, AHA 0840132 N, and ALZ IIRG-10-173350.

References

N.J. Allen, D. Attwell, Modulation of ASIC channels in rat cerebellar Purkinje neurons by ischemia-related signals. J. Physiol. **543**, 521–529 (2002)

D.R. Alvarez, C.M. Canessa, G.K. Fyfe, P. Zhang, Structure and regulation of amiloride-sensitive sodium channels. Annu. Rev. Physiol. **62**, 573–594 (2000)

P. Angelova, W. Muller, Oxidative modulation of the transient potassium current IA by intracellular arachidonic acid in rat CA1 pyramidal neurons. Eur. J. Neurosci. **23**, 2375–2384 (2006)

C.C. Askwith, C. Cheng, M. Ikuma, C. Benson, M.P. Price, M.J. Welsh, Neuropeptide FF and FMRFamide potentiate acid-evoked currents from sensory neurons and proton-gated DEG/ENaC channels. Neuron **26**, 133–141 (2000)

E. Babini, M. Paukert, H.S. Geisler, S. Grunder, Alternative splicing and interaction with di- and polyvalent cations control the dynamic range of acid-sensing ion channel 1 (ASIC1). J. Biol. Chem. **277**, 41597–41603 (2002)

E.L. Bassler, T.J. Ngo-Anh, H.S. Geisler, J.P. Ruppersberg, S. Grunder, Molecular and functional characterization of acid-sensing ion channel (ASIC) 1b. J. Biol. Chem. **276**, 33782–33787 (2001)

C.J. Benson, S.P. Eckert, E.W. McCleskey, Acid-evoked currents in cardiac sensory neurons: a possible mediator of myocardial ischemic sensation. Circ. Res. **84**, 921–928 (1999)

M. Benveniste, R. Dingledine, Limiting stroke-induced damage by targeting an acid channel. N. Engl. J. Med. **352**, 85–86 (2005)

M. Benveniste, M.L. Mayer, Multiple effects of spermine on N-methyl-D-aspartic acid receptor responses of rat cultured hippocampal neurones. J. Physiol. **464**, 131–163 (1993)

S. Bevan, J. Yeats, Protons activate a cation conductance in a sub-population of rat dorsal root ganglion neurones. J. Physiol. **433**, 145–161 (1991)

C.J. Bohlen, A.T. Chesler, R. Sharif-Naeini, K.F. Medzihradszky, S. Zhou, D. King, E.E. Sanchez, A.L. Burlingame, A.I. Basbaum, D. Julius, A heteromeric Texas coral snake toxin targets acid-sensing ion channels to produce pain. Nature **479**, 410–414 (2011)

K.M. Boje, P.K. Arora, Microglial-produced nitric oxide and reactive nitrogen oxides mediate neuronal cell death. Brain Res. **587**, 250–256 (1992)

J.P. Bolanos, A. Almeida, Roles of nitric oxide in brain hypoxia-ischemia. Biochim. Biophys. Acta **1411**, 415–436 (1999)

H. Cadiou, M. Studer, N.G. Jones, E.S. Smith, A. Ballard, S.B. McMahon, P.A. McNaughton, Modulation of acid-sensing ion channel activity by nitric oxide. J. Neurosci. **27**, 13251–13260 (2007)

M. Casado, P. Ascher, Opposite modulation of NMDA receptors by lysophospholipids and arachidonic acid: common features with mechanosensitivity [in process citation]. J. Physiol. **513**, 317–330 (1998)

S. Catarsi, K. Babinski, P. Seguela, Selective modulation of heteromeric ASIC proton-gated channels by neuropeptide FF. Neuropharmacology **41**, 592–600 (2001)

S. Chai, M. Li, D. Branigan, Z.G. Xiong, R.P. Simon, Activation of acid-sensing ion channel 1a (ASIC1a) by surface trafficking. J. Biol. Chem. **285**, 13002–13011 (2010)

C.C. Chen, A. Zimmer, W.H. Sun, J. Hall, M.J. Brownstein, A. Zimmer, A role for ASIC3 in the modulation of high-intensity pain stimuli. Proc. Natl. Acad. Sci. U. S. A. **99**, 8992–8997 (2002)

M. Chesler, The regulation and modulation of pH in the nervous system. Prog. Neurobiol. **34**, 401–427 (1990)

A. Chraibi, V. Vallet, D. Firsov, S.K. Hess, J.D. Horisberger, Protease modulation of the activity of the epithelial sodium channel expressed in Xenopus oocytes. J. Gen. Physiol. **111**, 127–138 (1998)

J. de Weille, F. Bassilana, Dependence of the acid-sensitive ion channel, ASIC1a, on extracellular Ca^{2+} ions. Brain Res. **900**, 277–281 (2001)

E. Deval, J. Noel, N. Lay, A. Alloui, S. Diochot, V. Friend, M. Jodar, M. Lazdunski, E. Lingueglia, ASIC3, a sensor of acidic and primary inflammatory pain. EMBO J. **27**, 3047–3055 (2008)

B. Duan, Y.Z. Wang, T. Yang, X.P. Chu, Y. Yu, Y. Huang, H. Cao, J. Hansen, R.P. Simon, M.X. Zhu, Z.G. Xiong, T.L. Xu, Extracellular spermine exacerbates ischemic neuronal injury through sensitization of ASIC1a channels to extracellular acidosis. J. Neurosci. **31**, 2101–2112 (2011)

A. Ekholm, T. Kristian, B.K. Siesjo, Influence of hyperglycemia and of hypercapnia on cellular calcium transients during reversible brain ischemia. Exp. Brain Res. **104**, 462–466 (1995)

M. Ettaiche, E. Deval, M. Cougnon, M. Lazdunski, N. Voilley, Silencing acid-sensing ion channel 1a alters cone-mediated retinal function. J. Neurosci. **26**, 5800–5809 (2006)

A.A. Farooqui, L.A. Horrocks, Phospholipase A2-generated lipid mediators in the brain: the good, the bad, and the ugly. Neuroscientist **12**, 245–260 (2006)

A.A. Farooqui, W.Y. Ong, L.A. Horrocks, Inhibitors of brain phospholipase A2 activity: their neuropharmacological effects and therapeutic importance for the treatment of neurologic disorders. Pharmacol. Rev. **58**, 591–620 (2006)

D.H. Feldman, M. Horiuchi, K. Keachie, E. McCauley, P. Bannerman, A. Itoh, T. Itoh, D. Pleasure, Characterization of acid-sensing ion channel expression in oligodendrocyte-lineage cells. Glia **56**, 1238–1249 (2008)

M.A. Friese, M.J. Craner, R. Etzensperger, S. Vergo, J.A. Wemmie, M.J. Welsh, A. Vincent, L. Fugger, Acid-sensing ion channel-1 contributes to axonal degeneration in autoimmune inflammation of the central nervous system. Nat. Med. **13**, 1483–1489 (2007)

J. Gao, L.J. Wu, L. Xu, T.L. Xu, Properties of the proton-evoked currents and their modulation by Ca^{2+} and Zn^{2+} in the acutely dissociated hippocampus CA1 neurons. Brain Res. **1017**, 197–207 (2004)

J. Gao, B. Duan, D.G. Wang, X.H. Deng, G.Y. Zhang, L. Xu, T.L. Xu, Coupling between NMDA receptor and acid-sensing ion channel contributes to ischemic neuronal death. Neuron **48**, 635–646 (2005)

M.B. Gingrich, S.F. Traynelis, Serine proteases and brain damage – is there a link? Trends Neurosci. **23**, 399–407 (2000)

S.C. Grifoni, N.L. Jernigan, G. Hamilton, H.A. Drummond, ASIC proteins regulate smooth muscle cell migration. Microvasc. Res. **75**, 202–210 (2008)

A.J. Hansen, T. Zeuthen, Extracellular ion concentrations during spreading depression and ischemia in the rat brain cortex. Acta Physiol. Scand. **113**, 437–445 (1981)

K.F. Hauser, J.K. Foldes, C.S. Turbek, Dynorphin A (1–13) neurotoxicity in vitro: opioid and non-opioid mechanisms in mouse spinal cord neurons. Exp. Neurol. **160**, 361–375 (1999)

P. Hess, J.B. Lansman, R.W. Tsien, Calcium channel selectivity for divalent and monovalent cations. Voltage and concentration dependence of single channel current in ventricular heart cells. J. Gen. Physiol. **88**, 293–319 (1986)

L. Hillered, M.L. Smith, B.K. Siesjo, Lactic acidosis and recovery of mitochondrial function following forebrain ischemia in the rat. J. Cereb. Blood Flow Metab. **5**, 259–266 (1985)

J.C. Holt, M. Lioudyno, G. Athas, M.M. Garcia, P. Perin, P.S. Guth, The effect of proteolytic enzymes on the alpha9-nicotinic receptor-mediated response in isolated frog vestibular hair cells. Hear. Res. **152**, 25–42 (2001)

H.Z. Hu, R. Xiao, C. Wang, N. Gao, C.K. Colton, J.D. Wood, M.X. Zhu, Potentiation of TRPV3 channel function by unsaturated fatty acids. J. Cell. Physiol. **208**, 201–212 (2006)

Y. Huang, J.O. McNamara, Ischemic stroke: "acidotoxicity" is a perpetrator. Cell **118**, 665–666 (2004)

C. Huang, Z.L. Hu, W.N. Wu, D.F. Yu, Q.J. Xiong, J.R. Song, Q. Shu, H. Fu, F. Wang, J.G. Chen, Existence and distinction of acid-evoked currents in rat astrocytes. Glia **58**, 1415–1424 (2010)

D.C. Immke, E.W. McCleskey, Lactate enhances the acid-sensing Na^{+} channel on ischemia-sensing neurons. Nat. Neurosci. **4**, 869–870 (2001)

D.C. Immke, E.W. McCleskey, Protons open acid-sensing ion channels by catalyzing relief of Ca^{2+} blockade. Neuron **37**, 75–84 (2003)

S.R. Jaffrey, H. Erdjument-Bromage, C.D. Ferris, P. Tempst, S.H. Snyder, Protein S-nitrosylation: a physiological signal for neuronal nitric oxide. Nat. Cell Biol. **3**, 193–197 (2001)

H. Jahr, M. van Driel, G.J. van Osch, H. Weinans, J.P. van Leeuwen, Identification of acid-sensing ion channels in bone. Biochem. Biophys. Res. Commun. **337**, 349–354 (2005)

J. Jasti, H. Furukawa, E.B. Gonzales, E. Gouaux, Structure of acid-sensing ion channel 1 at 1.9 A resolution and low pH. Nature **449**, 316–323 (2007)

S.K. Jetti, S.M. Swain, S. Majumder, S. Chatterjee, V. Poornima, A.K. Bera, Evaluation of the role of nitric oxide in acid sensing ion channel mediated cell death. Nitric Oxide **22**, 213–219 (2010)

T.D. Johnson, Polyamines and cerebral ischemia. Prog. Drug Res. **50**, 193–258 (1998)

H. Kalimo, S. Rehncrona, B. Soderfeldt, Y. Olsson, B.K. Siesjo, Brain lactic acidosis and ischemic cell damage: 2. Histopathology. J. Cereb. Blood Flow Metab. **1**, 313–327 (1981)

S. Keros, C.J. McBain, Arachidonic acid inhibits transient potassium currents and broadens action potentials during electrographic seizures in hippocampal pyramidal and inhibitory interneurons. J. Neurosci. **17**, 3476–3487 (1997)

M.S. Kindy, Y. Hu, R.J. Dempsey, Blockade of ornithine decarboxylase enzyme protects against ischemic brain damage. J. Cereb. Blood Flow Metab. **14**, 1040–1045 (1994)

O. Krishtal, The ASICs: signaling molecules? Modulators? Trends Neurosci. **26**, 477–483 (2003)

O.A. Krishtal, V.I. Pidoplichko, A receptor for protons in the membrane of sensory neurons may participate in nociception. Neuroscience **6**, 2599–2601 (1981)

D.R. LeMay, L. Gehua, G.B. Zelenock, L.G. D'Alecy, Insulin administration protects neurologic function in cerebral ischemia in rats. Stroke **19**, 1411–1419 (1988)

J. Li, K.M. Doyle, T. Tatlisumak, Polyamines in the brain: distribution, biological interactions, and their potential therapeutic role in brain ischaemia. Curr. Med. Chem. **14**, 1807–1813 (2007)

M.H. Li, K. Inoue, H.F. Si, Z.G. Xiong, Calcium-permeable ion channels involved in glutamate receptor-independent ischemic brain injury. Acta Pharmacol. Sin. **32**, 734–740 (2011)

W. Lin, T. Ogura, S.C. Kinnamon, Acid-activated cation currents in rat vallate taste receptor cells. J. Neurophysiol. **88**, 133–141 (2002)

E. Lingueglia, E. Deval, M. Lazdunski, FMRFamide-gated sodium channel and ASIC channels: a new class of ionotropic receptors for FMRFamide and related peptides. Peptides **27**, 1138–1152 (2006)

B. Ljunggren, K. Norberg, B.K. Siesjo, Influence of tissue acidosis upon restitution of brain energy metabolism following total ischemia. Brain Res. **77**, 173–186 (1974)

M. Mazzuca, C. Heurteaux, A. Alloui, S. Diochot, A. Baron, N. Voilley, N. Blondeau, P. Escoubas, A. Gelot, A. Cupo, A. Zimmer, A.M. Zimmer, A. Eschalier, M. Lazdunski, A tarantula peptide against pain via ASIC1a channels and opioid mechanisms. Nat. Neurosci. **10**, 943–945 (2007)

O. Mignen, J.L. Thompson, T.J. Shuttleworth, Arachidonate-regulated Ca^{2+}-selective (ARC) channel activity is modulated by phosphorylation and involves an A-kinase anchoring protein. J. Physiol. **567**, 787–798 (2005)

B. Miller, M. Sarantis, S.F. Traynelis, D. Attwell, Potentiation of NMDA receptor currents by arachidonic acid. Nature **355**, 722–725 (1992)

A.R. Muralikrishna, J.F. Hatcher, Phospholipase A2, reactive oxygen species, and lipid peroxidation in cerebral ischemia. Free Radic. Biol. Med. **40**, 376–387 (2006)

N. Nagano, Y. Imaizumi, M. Watanabe, Modulation of calcium channel currents by arachidonic acid in single smooth muscle cells from vas deferens of the guinea-pig. Br. J. Pharmacol. **116**, 1887–1893 (1995)

M. Nedergaard, R.P. Kraig, J. Tanabe, W.A. Pulsinelli, Dynamics of interstitial and intracellular pH in evolving brain infarct. Am. J. Physiol. **260**, R581–R588 (1991)

O. Nicole, F. Docagne, C. Ali, I. Margaill, P. Carmeliet, E.T. MacKenzie, D. Vivien, A. Buisson, The proteolytic activity of tissue-plasminogen activator enhances NMDA receptor-mediated signaling. Nat. Med. **7**, 59–64 (2001)

A.J. Page, S.M. Brierley, C.M. Martin, M.P. Price, E. Symonds, R. Butler, J.A. Wemmie, L.A. Blackshaw, Different contributions of ASIC channels 1a, 2, and 3 in gastrointestinal mechanosensory function. Gut **54**, 1408–1415 (2005)

P. Paoletti, P. Ascher, Mechanosensitivity of NMDA receptors in cultured mouse central neurons. Neuron **13**, 645–655 (1994)

M. Paukert, E. Babini, M. Pusch, S. Grunder, Identification of the Ca^{2+} blocking site of acid-sensing ion channel (ASIC) 1: implications for channel gating. J. Gen. Physiol. **124**, 383–394 (2004)

G. Pignataro, O. Cuomo, E. Esposito, R. Sirabella, G. Di Renzo, L. Annunziato, ASIC1a contributes to neuroprotection elicited by ischemic preconditioning and postconditioning. Int. J. Physiol. Pathophysiol. Pharmacol. **3**, 1–8 (2011)

O. Poirot, M. Vukicevic, A. Boesch, S. Kellenberger, Selective regulation of acid-sensing ion channel 1 by serine proteases. J. Biol. Chem. **279**, 38448–38457 (2004)

M.P. Price, G.R. Lewin, S.L. McIlwrath, C. Cheng, J. Xie, P.A. Heppenstall, C.L. Stucky, A.G. Mannsfeldt, T.J. Brennan, H.A. Drummond, J. Qiao, C.J. Benson, D.E. Tarr, R.F. Hrstka, B. Yang, R.A. Williamson, M.J. Welsh, The mammalian sodium channel BNC1 is required for normal touch sensation. Nature **407**, 1007–1011 (2000)

M.P. Price, S.L. McIlwrath, J. Xie, C. Cheng, J. Qiao, D.E. Tarr, K.A. Sluka, T.J. Brennan, G.R. Lewin, M.J. Welsh, The DRASIC cation channel contributes to the detection of cutaneous touch and acid stimuli in mice. Neuron **32**, 1071–1083 (2001)

S. Rehncrona, Brain acidosis. Ann. Emerg. Med. **14**, 770–776 (1985)

S. Rehncrona, E. Westerberg, B. Akesson, B.K. Siesjo, Brain cortical fatty acids and phospholipids during and following complete and severe incomplete ischemia. J. Neurochem. **38**, 84–93 (1982)

S. Rehncrona, H.N. Hauge, B.K. Siesjo, Enhancement of iron-catalyzed free radical formation by acidosis in brain homogenates: differences in effect by lactic acid and CO2. J. Cereb. Blood Flow Metab. **9**, 65–70 (1989)

E. Revici, E. Stoopen, E. Frenk, R.A. Ravich, The painful focus. II. The relation of pain to local physico-chemical changes. Bull. Inst. Appl. Biol. **1**, 21 (1949)

D.M. Rock, R.L. Macdonald, Polyamine regulation of *N*-methyl-D-aspartate receptor channels. Annu. Rev. Pharmacol. Toxicol. **35**, 463–482 (1995)

N. Sang, C. Chen, Lipid signaling and synaptic plasticity. Neuroscientist **12**, 425–434 (2006)

J.B. Schulz, R.T. Matthews, T. Klockgether, J. Dichgans, M.F. Beal, The role of mitochondrial dysfunction and neuronal nitric oxide in animal models of neurodegenerative diseases. Mol. Cell. Biochem. **174**, 193–197 (1997)

A. Schurr, Lactate, glucose and energy metabolism in the ischemic brain. Int. J. Mol. Med. **10**, 131–136 (2002)

A. Schurr, B.M. Rigor, Brain anaerobic lactate production: a suicide note or a survival kit? Dev. Neurosci. **20**, 348–357 (1998)

T.W. Sherwood, C.C. Askwith, Dynorphin opioid peptides enhance acid-sensing ion channel 1a activity and acidosis-induced neuronal death. J. Neurosci. **29**, 14371–14380 (2009)

B.K. Siesjo, Acidosis and ischemic brain damage. Neurochem. Pathol. **9**, 31–88 (1988)

B.K. Siesjo, K. Katsura, Ischemic brain damage: focus on lipids and lipid mediators. Adv. Exp. Med. Biol. **318**, 41–56 (1992)

B.K. Siesjo, K. Katsura, T. Kristian, Acidosis-related damage. Adv. Neurol. **71**, 209–233 (1996)

K.A. Sluka, M.P. Price, N.M. Breese, C.L. Stucky, J.A. Wemmie, M.J. Welsh, Chronic hyperalgesia induced by repeated acid injections in muscle is abolished by the loss of ASIC3, but not ASIC1. Pain **106**, 229–239 (2003)

E.S. Smith, H. Cadiou, P.A. McNaughton, Arachidonic acid potentiates acid-sensing ion channels in rat sensory neurons by a direct action. Neuroscience **145**, 686–698 (2007)

R.A. Star, Nitric oxide. Am. J. Med. Sci. **306**, 348–358 (1993)

S.P. Sutherland, S.P. Cook, E.W. McCleskey, Chemical mediators of pain due to tissue damage and ischemia. Prog. Brain Res. **129**, 21–38 (2000)

R.A. Swanson, K. Farrell, R.P. Simon, Acidosis causes failure of astrocyte glutamate uptake during hypoxia. J. Cereb. Blood Flow Metab. **15**, 417–424 (1995)

G.C. Tombaugh, R.M. Sapolsky, Evolving concepts about the role of acidosis in ischemic neuropathology. J. Neurochem. **61**, 793–803 (1993)

A. Toninello, M. Salvi, B. Mondovi, Interaction of biologically active amines with mitochondria and their role in the mitochondrial-mediated pathway of apoptosis. Curr. Med. Chem. **11**, 2349–2374 (2004)

S. Ugawa, Identification of sour-taste receptor genes. Anat. Sci. Int. **78**, 205–210 (2003)

S. Ugawa, T. Ueda, Y. Ishida, M. Nishigaki, Y. Shibata, S. Shimada, Amiloride-blockable acid-sensing ion channels are leading acid sensors expressed in human nociceptors. J. Clin. Invest. **110**, 1185–1190 (2002)

S. Ugawa, T. Yamamoto, T. Ueda, Y. Ishida, A. Inagaki, M. Nishigaki, S. Shimada, Amiloride-insensitive currents of the acid-sensing ion channel-2a (ASIC2a)/ ASIC2b heteromeric sour-taste receptor channel. J. Neurosci. **23**, 3616–3622 (2003)

D. Vivien, A. Buisson, Serine protease inhibitors: novel therapeutic targets for stroke? J. Cereb. Blood Flow Metab. **20**, 755–764 (2000)

N. Voilley, Acid-sensing ion channels (ASICs): new targets for the analgesic effects of non-steroid anti-Inflammatory drugs (NSAIDs). Curr. Drug Targets Inflamm. Allergy **3**, 71–79 (2004)

M. Vukicevic, G. Weder, A. Boillat, A. Boesch, S. Kellenberger, Trypsin cleaves acid-sensing ion channel 1a in a domain that is critical for channel gating. J. Biol. Chem. **281**, 714–722 (2006)

R. Waldmann, G. Champigny, F. Bassilana, C. Heurteaux, M. Lazdunski, A proton-gated cation channel involved in acid-sensing. Nature **386**, 173–177 (1997)

W. Wang, B. Duan, H. Xu, L. Xu, T.L. Xu, Calcium-permeable acid-sensing ion channel is a molecular target of the neurotoxic metal ion lead. J. Biol. Chem. **281**, 2497–2505 (2006)

J.A. Wemmie, J. Chen, C.C. Askwith, A.M. Hruska-Hageman, M.P. Price, B.C. Nolan, P.G. Yoder, E. Lamani, T. Hoshi, J.H. Freeman, M.J. Welsh, The acid-activated ion channel ASIC contributes to synaptic plasticity, learning, and memory. Neuron **34**, 463–477 (2002)

J.A. Wemmie, C.C. Askwith, E. Lamani, M.D. Cassell, J.H. Freeman Jr., M.J. Welsh, Acid-sensing ion channel 1 is localized in brain regions with high synaptic density and contributes to fear conditioning. J. Neurosci. **23**, 5496–5502 (2003)

J.A. Wemmie, M.P. Price, M.J. Welsh, Acid-sensing ion channels: advances, questions and therapeutic opportunities. Trends Neurosci. **29**, 578–586 (2006)

M. Wozniak, B. Rydzewski, S.P. Baker, M.K. Raizada, The cellular and physiological actions of insulin in the central nervous system. Neurochem. Int. **22**, 1–10 (1993)

L.J. Wu, B. Duan, Y.D. Mei, J. Gao, J.G. Chen, M. Zhuo, L. Xu, M. Wu, T.L. Xu, Characterization of acid-sensing ion channels in dorsal horn neurons of rat spinal cord. J. Biol. Chem. **279**, 43716–43724 (2004)

J. Xie, M.P. Price, J.A. Wemmie, C.C. Askwith, M.J. Welsh, ASIC3 and ASIC1 mediate FMRFamide-related peptide enhancement of H^+-gated currents in cultured dorsal root ganglion neurons. J. Neurophysiol. **89**, 2459–2465 (2003)

Z.G. Xiong, J.F. MacDonald, Sensing of extracellular calcium by neurones. Can. J. Physiol. Pharmacol. **77**, 715–721 (1999)

Z.G. Xiong, X.M. Zhu, X.P. Chu, M. Minami, J. Hey, W.L. Wei, J.F. MacDonald, J.A. Wemmie, M.P. Price, M.J. Welsh, R.P. Simon, Neuroprotection in ischemia: blocking calcium-permeable acid-sensing ion channels. Cell **118**, 687–698 (2004)

Z.G. Xiong, X.P. Chu, R.P. Simon, Acid sensing ion channels–novel therapeutic targets for ischemic brain injury. Front. Biosci. **12**, 1376–1386 (2007)

Z.G. Xiong, G. Pignataro, M. Li, S.Y. Chang, R.P. Simon, Acid-sensing ion channels (ASICs) as pharmacological targets for neurodegenerative diseases. Curr. Opin. Pharmacol. **8**, 25–32 (2008)

O. Yermolaieva, A.S. Leonard, M.K. Schnizler, F.M. Abboud, M.J. Welsh, Extracellular acidosis increases neuronal cell calcium by activating acid-sensing ion channel 1a. Proc. Natl. Acad. Sci. U. S. A. **101**, 6752–6757 (2004)

P. Zhang, F.J. Sigworth, C.M. Canessa, Gating of acid-sensitive ion channel-1: release of Ca^{2+} block vs. allosteric mechanism. J. Gen. Physiol. **127**, 109–117 (2006)

W. Zhou, S.W. Jones, Surface charge and calcium channel saturation in bullfrog sympathetic neurons. J. Gen. Physiol. **105**, 441–462 (1995)

Nonselective Cation Channels and Links to Hippocampal Ischemia, Aging, and Dementia

37

John F. MacDonald, Jillian C. Belrose, Yu-Feng Xie, and Michael F. Jackson

Abstract

Stroke is a very strong risk factor for dementia. Furthermore, ischemic stroke and Alzheimer's disease (AD) share a number of overlapping mechanisms of neuron loss and dysfunction, including those induced by the inappropriate activation of N-methyl-D-aspartate receptors (NMDARs). These receptors form a major subtype of excitatory glutamate receptor. They are nonselective cation channels with appreciable Ca^{2+} permeability, and their overactivation leads to neurotoxicity in the cortex and hippocampus. NMDARs have therefore been therapeutic targets in both conditions, but they have failed in the treatment of stroke, and there is limited rationale for using them in treating AD. In this chapter, we discuss current understanding of subtypes of NMDARs and their potential roles in ischemic stroke and AD. We also discuss the properties of several other nonselective cation channels, transient receptor potential melastatin 2 and 7 channels, and their implications in linking these conditions.

J.F. MacDonald (✉)
Department of Anatomy and Cell Biology, University of Western Ontario, London, ON N6A 5K8, Canada

Robarts Research Institute, Molecular Brain Research Group, University of Western Ontario, 100 Perth Drive, London, ON N6A 5K8, Canada

Department of Physiology and Pharmacology, University of Western Ontario, London, ON N6A 5K8, Canada
e-mail: jfmacdonald@robarts.ca

J.C. Belrose
Department of Anatomy and Cell Biology, University of Western Ontario, London, ON N6A 5K8, Canada

Robarts Research Institute, Molecular Brain Research Group, University of Western Ontario, 100 Perth Drive, London, ON N6A 5K8, Canada

Y.-F. Xie
Robarts Research Institute, Molecular Brain Research Group, University of Western Ontario, 100 Perth Drive, London, ON N6A 5K8, Canada

M.F. Jackson
Robarts Research Institute, Molecular Brain Research Group, University of Western Ontario, 100 Perth Drive, London, ON N6A 5K8, Canada

Department of Physiology and Pharmacology, University of Western Ontario, London, ON N6A 5K8, Canada

L. Annunziato (ed.), *Sodium Calcium Exchange: A Growing Spectrum of Pathophysiological Implications*, Advances in Experimental Medicine and Biology 961, DOI 10.1007/978-1-4614-4756-6_37,

Keywords
N-methyl-D-aspartate receptors (NMDA receptors) • Transient receptor potential melastatin channels (TRPM7, TRPM2) • Src family kinases • Stroke • Alzheimer's disease

37.1 Introduction

High blood pressure, diabetes, and obesity are vascular risk factors that greatly exacerbate the occurrence of major strokes (Sahathevan et al. 2012). Less well recognized is their role in causing "covert" strokes, which, in the absence of any catastrophic vascular event, are characterized by deficits in cognition and memory or the early onset of dementia. The incidence of "covert" to "evident" strokes increases substantially in the aging population. Indeed, covert strokes afflict as many as 25 % of the population aged 70–75 as compared to 7 % of evident strokes, and, in the case of Canada, as many as 450,000 individuals may suffer covert strokes as a consequence of minor blockages of small cerebral vessels leading to vascular dementia (Hakim 2007). There is also evidence that stroke and Alzheimer's disease (AD) interact and share the same risk factors (Greenberg and Jin 2006; Sun et al. 2006). In the case of "evident" strokes, the primary standard treatment is the use of thrombolytic drugs such as tPA (tissue plasminogen activator), which must be administered within 3 h of stroke onset. Paradoxically, the reperfusion achieved by these drugs administered at later times may initiate further cell damage (reperfusion injury) (Ikeda et al. 2003; de la Perez and Davalos 2007). Additionally, their use is inappropriate for vascular strokes given that they promote hemorrhaging in the brain. Clearly there is a major need to identify alternative therapeutic approaches. Given the aging population, the developed world is facing a major combined crisis of increased frequency of strokes followed by increasing prevalence of survivors with dementia. In addition to preventative strategies to prevent their occurrences, two major clinical objectives should be met: (1) to minimize the immediate damage in strokes and (2) to prevent the subsequent potential for later cell loss, cognitive deficits, and the early onset of dementia.

37.1.1 Acute Responses of Central Neurons to Ischemia

An ischemic episode initiates a variety of immediate stresses upon the metabolism and the ionic homeostasis of central neurons. These stresses include the activation of glutamate receptors (Choi and Rothman 1990); the loading of cells with Na^+ and water (osmotic stress) (Rothman 1984; Rothman 1983, 1985, 1986); the loading of neurons with Ca^{2+}, due to the activation of *N*-methyl-D-aspartate receptors (NMDARs) (MacDonald et al. 2006c; Morris and Trippenbach 1993; Pumain and Heinemann 1985); and the reversal of sodium–calcium exchangers (NCX) (Annunziato et al. 2007; Bano et al. 2005) (these exchangers form a major topic of the present symposium). The metabolic stresses accompanying limited perfusion and poor oxygenation result in the production of reactive oxygen species (ROS) and nitrosylative species (RNS, e.g., peroxynitrite), as well as extracellular acidification with activation of acid-sensing ion channels (ASICs) which also contribute to a delayed loss of neurons in the penumbra region (Annunziato et al. 2007; Chu et al. 2011; Li et al. 2010; MacDonald et al. 2006c; Pignataro et al. 2011; Xiong et al. 2004). The penumbra region is defined as the regions of neuronal tissue that have been compromised by poor perfusion but nevertheless survive the initial stroke event. It is recognized that neurons in these regions will go on to die at extended periods following recovery of vascular perfusion (delayed cell death). Ischemia is associated with the activation of various other nonselective cation channels such as transient receptor potential melastatin 7 and 2 channels (TRPM7, TRPM2) (Tymianski 2011) and pannexin 1 channels (Thompson et al. 2006), which can also contribute to the subsequent loss of neurons.

37.1.2 Failure of Glutamate Channel Blockers in Ischemic Stroke

Ischemia causes a massive release of glutamate coupled to an inappropriate activation of postsynaptic glutamate channels (MacDonald et al. 2006c). Many glutamate receptors are nonselective cation-permeable ion channels (permeable to Na^+, K^+, and sometimes Ca^{2+}), and many cells in the penumbral region receive enough blood that they do not die during ischemia but go on to perish long after exposure to glutamate. This "delayed cell death" occurs hours or days later, and it is believed to be dependent upon the activity of Ca^{2+}-permeable NMDARs (Choi 1985). Human trials of NMDAR antagonists have proven unsuccessful in spite of promising work using these blockers to reduce cell damage in animal models of stroke (Hoyte et al. 2004; Muir and Lees 2003). A variety of factors likely contribute to this failure including a lack of an appreciation that NMDARs play an almost universal role in excitatory synaptic transmission. For example, blocking these receptors has powerful psychomimetic effects and it induces a "psychotic-like" state. Indeed, antagonists of NMDARs mimic the positive (e.g., psychosis) as well as negative symptoms (e.g., learning and cognitive deficits) of schizophrenia in healthy subjects. They also exacerbate these symptoms in patients with the disease. The trade-off in potentially protecting neurons in stroke versus inducing "psychotic-like" symptoms, as well as major cognitive deficits, likely doomed to failure the approach of blocking NMDARs. Nevertheless, as our understanding of NMDARs has progressed, it has become clear that there are alternative ways to target the neurotoxic effects of these receptors in stroke and in AD without resorting to receptor blockade (Yang et al. 2011a).

37.1.3 Why Study CA1 Pyramidal Neurons in Stroke and Dementia?

The hippocampus is a well-characterized structure that has been repeatedly implicated in learning and memory. This structure is of particular interest in studies of stroke due to the high sensitivity of CA1 pyramidal neurons to damage and loss in global ischemia (Mattson and Kater 1989; Wang and Michaelis 2010). Furthermore, alterations in hippocampal excitatory synaptic function and disruption of neurogenesis are correlated with much of the early and mild cognitive deficits of AD (Götz et al. 2011a; Ittner and Götz 2011; Mu and Gage 2011). CA1 neurons and their afferent inputs (e.g., Schaffer collaterals of CA3 neurons) are perhaps, with respect to excitatory synaptic transmission and plasticity (e.g., long-term potentiation, LTP, and long-term depression, LTD), the best characterized of central connections. Therefore, they serve as a popular model of learning and memory. Advantages of studying CA1 neurons include the ease of experimentation in situ in hippocampal slices (electrophysiology, immunohistochemistry, biochemistry, etc.). Also CA1 neurons can be acutely isolated from these slices and electrophysiological studies performed under rigorous pharmacological conditions (Lu et al. 1999). CA1 neurons can also be grown and studied in primary tissue cultures where in vitro experiments can be designed that are technically not feasible in slices or in vivo (Lu et al. 2001).

37.2 Glutamate Receptors as Targets in Stroke and Dementia

37.2.1 NMDARs and CA1 Hippocampal Synapses

NMDARs are tetrameric nonselective channels composed of dimers of NR1 and NR2 subunits (GluN1, GluN1, GluN2x, GluN2x) (Inanobe et al. 2005) (and in some cases GluN3A,B). (*Subunits are currently signified using IUPHAR nomenclature where NR is replaced with GluN.*)

Glutamate binds directly to GluN2 subunits and the co-agonist glycine to the GluN1 subunits (Lynch and Guttmann 2001), but binding of both is required for gating of the channels. CA1 pyramidal neurons predominantly express GluN1,1,2A,2A and GluN1,1,2B,2B receptors (referred to as GluN2ARs and GluN2BRs, respectively) (Thompson et al. 2000, 2002), and they are the predominant subtypes found at Schaffer collateral (CA1) synapses (Cull-Candy and Leszkiewicz 2004). There is also accumulating evidence for functional triheteromeric receptors at CA1 synapses (Gladding and Raymond 2011; Gray et al. 2011; Rauner and Kohr 2011), although in the CA1 region the biochemical content of these receptors appears to be much less than that of dimeric GluN2ARs and GluN2BRs (Al-Hallaq et al. 2007).

GluN2B is dominantly expressed in the early postnatal brain (Brigman et al. 2010) and plays an essential role in the development and maturation of hippocampal synapses (Hestrin 1992). In embryonic cultures, GluN2ARs and GluN2BRs may be more strongly expressed at synaptic and extrasynaptic regions, respectively (Li et al. 2002). Channel properties (e.g., kinetics, permeability, and pharmacology), associated scaffolding proteins (e.g., postsynaptic density protein 95, PSD-95), and trafficking mechanisms differ for each of these subunits as do their contributions to the generation of long-term potentiation (LTP) and long-term depression (LTD) (Cull-Candy and Leszkiewicz 2004; Gielen et al. 2009). These forms of excitatory synaptic plasticity are intimately involved in learning and memory at the cellular level in the hippocampus.

37.2.2 NMDAR Subtypes and Neurotoxicity

There is general agreement that an excessive influx of Ca^{2+} into central neurons can lead to "delayed cell death," but it is also recognized that not all pathways of Ca^{2+} entry are as efficacious in producing excitotoxicity. This led to the "source specificity" theory where the entry of Ca^{2+} via NMDARs is more closely coupled to the toxic effects of Ca^{2+} than is the entry via other mechanisms, such as voltage-dependent Ca^{2+} channels (Sattler et al. 1998; Tymianski et al. 1993). Over the last 10 years, this theory has required substantial modification. Various studies have shown that different subtypes of NMDARs differ in their "source specificity" to induce cell death. For example, overactivation of the GluN2B subtype of NMDA receptor leads to a PSD-95-linked stimulation of NOS and subsequent neuronal death (Sattler et al. 1999, 2000), and an interfering peptide ("GluN2B9c") that reduced the interactions of GluN2B (but not GluN2A) with a PDZ domain of the scaffolding protein PSD-95 also greatly diminished ischemic cell loss in rodent central nervous system without disrupting hippocampal synaptic function or plasticity (Aarts et al. 2002).

NMDA receptors composed of the GluN2B subunit appear to signal via discrete "cell death" signaling pathways (Liu et al. 2007). In cortical cultures from relatively young animals, activation of GluN2BRs proved neurotoxic while reduced GluN2AR stimulation paradoxically enhanced toxicity implying that the GluN2AR subtype is neuroprotective (Cui et al. 2007). A further twist on the "source specificity" theory came from observations that selective stimulation of extrasynaptic NMDARs is neurotoxic while stimulation of synaptic NMDARs is actually neuroprotective (Hardingham 2006, 2009; Hardingham and Bading 2003; Liu et al. 2007; Sattler and Tymianski 2000; Sattler et al. 2000; Vanhoutte and Bading 2003; Zhang et al. 2011). Although initially consistent with a predominant extrasynaptic location for GluN2BRs, there is now considerable controversy about whether or not there is any preference of receptor subtype location in adult CA1 neurons (Gladding and Raymond 2011). The principle that extrasynaptic receptors are more prone to block by fast NMDAR blockers has also been proposed to underline the effectiveness of memantine in the treatment of AD symptoms (Xia et al. 2010). At this time it is not clear whether or not NMDAR subtype or NMDAR differences in regional receptor geography or potentially both can account for neurotoxic versus neuroprotective effects of glutamate.

37.2.3 Src Family Kinases and NMDARs in Stroke and AD

Ischemia is associated with the over activation of GluN2BRs that are bound to a protein complex via the scaffolding protein postsynaptic density 95 (PSD-95) (Gladding and Raymond 2011; Tymianski 2011). PSD-95 interacts directly with nitric oxide synthase (NOS) permitting a Ca^{2+}-dependent activation of this enzyme that can initiate the production of toxic peroxynitrite (Sattler et al. 1999). Other players in this complex include two members of the Src family kinases (SFK), Src and Fyn. Src and Fyn are found in high concentrations in the postsynaptic density (Jiang et al. 2008), and they play key roles in regulating NMDARs. Ischemia enhances the binding of Src to PSD-95 and stimulates its activation resulting in an increased tyrosine phosphorylation of GluN2A (Liu et al. 2005; Wang et al. 2010). Fyn is also stimulated in ischemia and chronic activation of this kinase may contribute to AD (Trepanier et al. 2012; Yang et al. 2011a). Fyn is more likely to target GluN2B as a substrate at the synapse (Jiang et al. 2008; Salter and Kalia 2004). Inhibition of SFK provides neuroprotection for CA1 neurons in ischemic stroke models (Hou et al. 2007; Jiang et al. 2008), but it is unclear if this occurs via inhibition of Src or Fyn. If one accepts the premise that GluN2ARs provide neuroprotection and Src selectively enhances GluNR2A activity (Liu et al. 2007), then blocking Src should exacerbate cell damage, not provide protection. Part of the explanation for this apparent paradox may arise from a lack of selectivity of the Src inhibitors employed (Hou et al. 2007) as they equally block Src and Fyn kinases (Jiang et al. 2008).

Fyn kinase has been extensively implicated in AD and related tauopathies (Crews and Masliah 2010; Trepanier et al. 2012; Yang et al. 2011a). Ittner and others recently provided evidence for a scenario whereby pathological hyperphosphorylation of tau protein recruits Fyn kinase into a complex, which is then translocated from the cell body into the dendrites in close proximity to excitatory spines and synapses (Gotz et al. 2011a, b; Ittner et al. 2010). The increased presence of Fyn at the synapses then selectively enhances the phosphorylation of GluN2B subunits and favors neurotoxicity. This is based upon the observation that the Tat-GluN2B9c interfering peptide employed for protection in ischemic strokes (Aarts et al. 2002) inhibited the spontaneous seizures, prevented the decreases in spatial learning, and enhanced the survival of a mouse model of AD (Ittner et al. 2010). One possible interpretation is that a chronic increase in Fyn-dependent phosphorylation of GluN2BRs leads to increased Ca^{2+} loading, perhaps via extrasynaptically located receptors, that ultimately results in excitotoxic damage to the neurons. The interfering peptide would simply decrease the ability of Fyn to phosphorylate and enhance the activity of GluN2BRs (Ittner et al. 2010).

A recent hypothesis is that production of soluble oligomers of beta-amyloid (Aβ) triggers processes early in the onset of AD, which are responsible for dysfunction of excitatory synaptic transmission, particularly in regions such as the hippocampus (Baum et al. 2010; Braithwaite et al. 2006; Snyder et al. 2005; Venkitaramani et al. 2007). Indeed, acute applications of Aβ strongly inhibit the induction of both hippocampal NMDAR-dependent LTP and LTD (Selkoe 2011). The inhibition of LTP has been attributed to the relatively selective and excessive activation of GluNB2Rs (Li et al. 2011). Amyloid β oligomers may subsequently alter synaptic plasticity through alterations in the PI3K–Akt–GSK3 Aβ signaling pathway (Jo et al. 2011; Peineau et al. 2007).

Lombroso et al. (Baum et al. 2010; Braithwaite et al. 2006; Kurup et al. 2010; Snyder et al. 2005; Venkitaramani et al. 2007) have proposed an alternative NMDAR-dependent model of the effects of Aβ. In this model, Aβ interacts in concert with α7 nicotinic receptors to generate an influx of Ca^{2+} that leads to the stimulation of striatal-enriched tyrosine phosphatase (STEP). Fyn kinase is dephosphorylated and inactivated by STEP. Thus, Aβ would ultimately reduce GluN2B phosphorylation at a regulatory tyrosine (1472) site, which is responsible for controlling the trafficking of NMDARs to the cell surface (Snyder et al. 2005). The ultimate effect of Aβ is to decrease surface expression of GluN2B by both reduced exocytosis and enhanced endocytosis of this subunit (Braithwaite et al. 2006; Nguyen et al. 2002; Paul et al. 2003; Snyder et al. 2005). Acute applications of Aβ also decreased the amplitude of NMDAR-mediated currents in cultured neurons (Snyder et al. 2005). Again there is a paradox presented by these results, as enhanced GluN2BR activity is most often associated with enhanced "neurotoxicity" and, simplistically, a reduction in the surface expression of these receptors should be "neuroprotective," not toxic. Of course it is possible that internalization of the receptors is responsible for the neurotoxic effects of Aβ. Clearly it is important to understand how SFK actually regulate NMDAR activity, plasticity, and cell viability.

37.2.4 SFK and Regulation of CA1 Synaptic Plasticity

37.2.4.1 SFK-Dependent Amplification and Triggering of LTP

Basal synaptic transmission at the CA1 synapse is predominantly mediated by α-amino-3-hydroxyl-5-methyl-4-isoxazole-propionate receptors (AMPARs) due to the voltage-dependent block of NMDARs by Mg^{2+}. Rapid relief of the Mg^{2+} block during repetitive high-frequency stimulation leads to a large influx of Ca^{2+}, and the induction of long-term potentiation (LTP) via CamKII activation and downstream increases in the gating and surface expression of AMPARs (Soderling and Derkach 2000; Song and Huganir 2002). However, this is not the complete story. Relief of the Mg^{2+} block alone is insufficient to stimulate enough Ca^{2+} entry to trigger LTP; the activity of NMDARs must also be amplified by SFK and specifically by Src (Bartos et al. 2010; Huang et al. 2001; Lu et al. 1998; MacDonald et al. 2006b; Salter and Kalia 2004). Moreover, the recruitment of NMDARs during basal transmission is limited not only by Mg^{2+} but also by the active suppression of Src activity through phosphorylation of Y527 by C-terminal Src kinase (Csk) (Xu et al. 2008) and/or by STEP (Pelkey et al. 2002). Indeed, relief of Src suppression by a functional inhibitory antibody against Csk or STEP is sufficient to induce LTP. Therefore, Src is both necessary and sufficient for the induction of LTP.

Induction of long-term depression (LTD) at the CA1 synapse also requires NMDARs (Yashiro and Philpot 2008) (there is also an NMDAR-independent, mGluR1 form of LTD (Luscher and Huber 2010)). During the induction of LTD by low-frequency stimulation, the resulting slower and more modest Ca^{2+} influx favors the recruitment of phosphatases, such as protein phosphatase 1 (PP1) and calcineurin, leading to downregulation of AMPAR function (Malenka and Bear 2004). Src kinases inhibit LTD in cerebellar neurons, but their role in LTD has not been extensively examined at CA1 synapses (Tsuruno et al. 2008). In summary, mechanisms that determine the activity of NMDARs during induction of plasticity can serve to govern the triggering of LTP and LTD (MacDonald et al. 2006a, b).

37.2.4.2 GluN2 Receptors and Direction of Hippocampal Synaptic Plasticity

In parallel to the differential roles of GluN2ARs and GluN2BRs in pathological conditions, two competing hypotheses have been proposed for the role of NMDAR subtypes in synaptic plasticity. The first is that the absolute contribution of GluN2ARs versus GluN2BRs determines the direction of synaptic plasticity. For example, at CA1 synapses, GluN2ARs have been associated preferentially with LTP, and GluN2BRs with LTD (Bartlett et al. 2007; Brigman et al. 2010; Fox et al. 2006; Liu et al. 2004; Massey et al. 2004; Wong et al. 2007). This has been robustly challenged by the counterhypothesis that it is the ratio of GluN2A to GluN2B (GluN2A/GluN2B) which determines the direction of plasticity (Berberich et al. 2005, 2007). The "ratio" hypothesis arose from long-term experiments of visual deprivation in the kitten cortex, which is associated with a reduction in GluN2A/GluN2B and an enhanced LTP. This change has been attributed to a reduction in the relative surface expression of the GluN2A subunit (Chen and Bear 2007; Cho et al. 2009; Philpot et al. 2007; Smith et al. 2009;

Yoon et al. 2009). Priming stimulations in hippocampal slices can also enhance LTP by decreasing GluN2A/GluN2B, while increasing this ratio was associated with increased LTD (Xu et al. 2009). Note that there is strong evidence that both GluN2ARs and GluN2BRs contribute to the induction of LTP, at least under some conditions (Yashiro and Philpot 2008). It is not that surprising that the role of these receptor subtypes in the induction of plasticity is more complex than simply one causes LTD and the other LTP, given that both generate an influx of Ca^{2+}.

In recent experiments we demonstrated that several different types of G-protein-coupled receptors (GPCRs) are capable of selectively activating Src or Fyn kinases in CA1 hippocampal neurons (Yang et al. 2011b), for example, pituitary adenylate cyclase-activating peptide acting through the PAC1 receptor signals via $G\alpha q$ subunits to activate Src and not Fyn kinase. In contrast, dopamine D1 and D5 receptor agonists stimulated Fyn and not Src kinase by stimulating $G\alpha s$ subunits. The intracellular application of recombinant Src, versus Fyn, into acutely isolated CA1 pyramidal neurons selectively potentiated GluN2ARs and GluN2BRs, respectively. The selectivity of this signaling was confirmed using peptides that directly mimic the "unique domains" of Src and Fyn, respectively. These peptides prevent the association of endogenous Src and Fyn to scaffolding proteins that permit interactions with the appropriate substrate, in this case, GluN2A or GluN2B, respectively. This presented an opportunity to use GPCR agonists to selectively enhance either GluN2ARs or GluN2BRs. Both types of GPCRs enhance NMDA-induced currents and $EPSCs_{NMDA}$ in CA1, but their effects on synaptic plasticity were quite different. We observed that enhancing GluN2ARs lowered the threshold for induction of LTP while LTD was unexpectedly favored by potentiation of GluN2BRs (Yang et al. 2011b). This was done using a protocol where the stimulus frequency was varied over a range to induce either LTD or LTP at CA1 synapses. These changes in plasticity were indeed due to the selective phosphorylation of GluN2ARs by Src and GluN2BRs by Fyn because cells and slices from knock-in mice expressing mutated forms of each subunit, in which a key tyrosine phosphorylation site was substituted with phenylalanine, eliminated the associated effects (Yang et al. 2011b). The findings that Src and Fyn kinases differentially regulate subtypes of GluN2Rs likely have some bearing on the pathological activation of these kinases both during ischemic stroke and in their potential contribution to dysfunction of excitatory synaptic transmission in dementia. If GluN2ARs are coupled to "neuroprotective" mechanism and GluN2BRs to "neurotoxicity," then targeting of Src versus Fyn signaling might provide various rationales for the design of "source-specific" therapies.

37.3 TRPM2/7 Channels as Targets in Stroke and Dementia

37.3.1 Transient Receptor Potential (TRP) Channels

In mammals multiple genes encode 25 different types of TRP subunits that fall into three major families: TRPC (canonical), TRPV (vallinoid), and TRPM (melastatin) (Clapham et al. 2001; Ramsey et al. 2006). Channels in this superfamily of primary interest to the study of stroke and dementia include TRPM7 and TRPM2. Both of these channels have appreciable Ca^{2+} permeability, both can be activated by ROS and RNS (Tymianski 2011), both are implicated in ischemia-induced cell death, and both have been associated with Parkinson's-like dementia (Hermosura et al. 2005, 2008; Hermosura and Garruto 2007).

37.3.1.1 TRPM7 and Stroke

TRPM7 (and its heteromer TRPM6) possess a functional α-phosphokinase domain in their C-termini, and they serve as major regulators of Mg^{2+} (Bates-Withers et al. 2011). We examined the role of TRPM7 channels in the cell death induced by oxygen–glucose deprivation (OGD) in cultured neurons and showed they are responsible for a component of cell death in this preparation (Aarts et al. 2003). At that time, due to the lack of specific blockers for this channel, we employed an RNA interference (RNAi) approach to knock down the expression of TRPM7 in cortical and hippocampal neurons (Aarts et al. 2003; Wei et al. 2007). More recently we have shown that reduction in the expression of TRPM7 protein in the hippocampus, through viral delivery of RNAi targeting this channel, provides substantial protection of CA1 pyramidal neurons and synapses in an in vivo model of global ischemia (Sun et al. 2009). Our work served to validate the TRPM7 channel as a possible target for therapy in stroke. The mechanism of TRPM7-induced toxicity in stroke is likely coupled to its capacity to permit the entry of not only Ca^{2+} but also Zn^{2+}, which is also responsible for neurotoxicity (Inoue et al. 2010).

37.3.1.2 TRPM7 and Dementia

The activity of TRPM7 channels is likely important for the maintenance of appropriate levels of intracellular Mg^{2+} in many different cell types. There is evidence that these channels play a role in the presynaptic release of acetylcholine (Brauchi et al. 2008), but there is little or no evidence that they participate in synaptic transmission in the hippocampus or other regions of the central nervous system. TRPM7–/– mice are nonviable, although conditional knockouts have been successfully employed (Jin et al. 2008, 2011) to study these channels in these mice without causing an ultimate disruption of Mg^{2+} homeostasis. Further, RNAi knockdowns did not affect the viability of CA1 neurons (Sun et al. 2009). Genetic analysis in mice has also implicated TRPM7 in forms of neuronal degeneration ranging from multiple sclerosis to stroke and to AD (Tseveleki et al. 2010). Even more intriguing are the recent demonstrations that abrogation of presenilins enhances Ca^{2+} and Zn^{2+} entry mediated by TRPM7 channels by means of a phosphatidylinositol 4,5-bisphosphate (PIP2)-dependent mechanism. In contrast, several presenilin mutations characteristic of familial AD reduced TRPM7-mediated influx (Oh et al. 2011).

37.3.1.3 TRPM2 Stroke and Dementia

Mori and colleagues first identified and cloned a unique nonselective cation channel (TRPM2) that acts as a mediator of cell death in direct response to changes in the redox status of HEK293 cells (Hara et al. 2002) (Moran et al. 2011; Takahashi et al. 2011). Changes in redox status were induced by application of various agents capable of generating reactive oxygen (e.g., H_2O_2) and nitrogen (e.g., NO donor SNAP) species. TRPM2 channels possess an intracellular NUDT9-H domain (Perraud et al. 2003). A NUDIX motif within this domain is critical for the binding and gating of these channels by ADP-ribose (ADPR) (Perraud et al. 2001), a previously underappreciated intracellular signaling molecule itself derived from nicotinamide adenine dinucleotide (NAD^+). NAD^+ serves as a major coenzyme involved in cellular redox reactions. Thus, the identification of ADPR as an intracellular agonist for TRPM2 linked its activity directly to the redox state of a cell. TRPM2 channels are also Ca^{2+} permeable, and intracellular Ca^{2+} sensitizes their gating by ADPR (MacDonald et al. 2006c). Furthermore, ADPR concentrations are enhanced both by oxidative stress and potentially following damage to DNA (MacDonald et al. 2006c).

TRPM2 mRNA is expressed in the human brain, and its expression is increased in a transient middle cerebral artery occlusion (tMCAO) stroke model in the rat (Fonfria et al. 2006a, b). The TRPM2 channel blocker clotrimazole is also neuroprotective (Isaev et al. 2002) although it fails to protect CA1 neurons from H_2O_2 (Bai and Lipski 2010). Furthermore, TRPM2 channels, through their presence in microglia, are believed to contribute to the death of neurons following applications of either H_2O_2 or Aβ (Fonfria et al. 2005, 2006a; McNulty and Fonfria 2005). The pathological role of TRPM2 channels in stroke (Fonfria et al. 2006a) and other degenerative diseases such as AD is intriguing but still largely unexplored (Fonfria et al. 2005; Jia et al. 2011; Yamamoto et al. 2007). The expression of ADPR-activated currents in central neurons has been controversial with one report that microglia and not cultured cerebellar granule neurons or astrocytes express the message for TRPM2 for functional TRPM2 (Kraft et al. 2004). In contrast, ADP-ribose-activated currents have been observed in cultured striatal neurons (Hill et al. 2006), and very small (10's of pA) ADP-ribose-activated currents are recorded in acutely isolated cortical neurons loaded with high concentrations (1 mM) of ADP-ribose (Kaneko et al. 2006). Identification of ADP-ribose-activated responses has been limited and based largely upon correlative comparisons with the gating and pharmacological properties of recombinant TRPM2 expressed in cell lines. A TRPM2 immunochemical signal was observed in cultured cortical neurons, and this signal was depressed by expression of an RNAi targeting this protein (Kaneko et al. 2006).

Using in situ hybridization techniques, we recently demonstrated the presence of TRPM2 message in adult CA1 pyramidal neurons as well as a strong immunocytochemical signal for the protein in CA1 pyramidal neurons of the hippocampus. We also identified and characterized TRPM2 currents in cultured hippocampal and CA1 neurons (Olah et al. 2009). This identification has been confirmed using cultured hippocampal neurons from TRPM2−/− mice (Xie et al. 2011). A recent paper has also linked juvenile myoclonic epilepsy and mutation of the EF-hand motif-containing protein to the regulation of TRPM2 currents and cell death (Katano et al. 2012).

37.3.1.4 TRPM2 Channel Currents in Cultured Hippocampal Neurons?

TRPM2 recombinant channels expressed in cell lines are gated by intracellular applications of ADP-ribose (ADPR); therefore, we began our initial studies by including it in the patch pipettes. Estimates of the intracellular physiological concentrations of ADPR vary from 4–5 μM (Heiner et al. 2006) (Yamada et al. 2006) in granulocytes and erythrocytes to 60–90 μM in Jurkat T cells (Gasser et al. 2006). Very high concentrations of ADPR (1–3 mM) activate a large inward current in hippocampal neurons (Olah et al. 2009). However, there is an important caveat to this observation in that these ADPR-activated currents are only observed with some regularity in the oldest of our primary cultures (>3–4 weeks) suggesting a relationship with the time of neurons in culture and TRPM2 function. This differs from the strong expression of glutamate and GABA currents, which are maximally expressed as early as 2 weeks prior to ADPR-activated currents. This time-dependent expression of TRPM2 currents was not related to alterations in the expression of TRPM2 mRNA but was related to an age-dependent reduction of intracellular glutathione. Reducing glutathione in young hippocampal neurons revealed TRPM2 currents, and restoring its levels in the older cultures depressed TRPM2 currents (Belrose et al. 2012).

Lower concentrations (300 μM) of ADPR fail to activate TRPM2-like currents even in older cultured neurons unless an additional stimulation protocol is applied to enhance Ca^{2+} entry. Stimulation of either voltage-dependent Ca^{2+} channels and/or NMDARs is required for the activation of TRPM2-like currents during intracellular applications of ADPR. This was not unexpected as intracellular Ca^{2+} is required for ADPR gating of TRPM2 channels (McHugh et al. 2003). Specifically, we identified two different protocols of activation including (a) repeated, brief applications of NMDA and glycine (Olah et al. 2009) or (b) repeated depolarizing voltage ramps (Olah et al. 2009; Xie et al. 2011). These TRPM2-like currents demonstrated a variety of properties that are similar to the behavior of recombinant TRPM2 including a near-linear current–voltage relationship, a lack of sensitivity to applications of extracellular divalent cations, and elimination of gating when extracellular Ca^{2+} was replaced with Ba^{2+}. They were also blocked by three known pharmacological blockers of recombinant TRPM2 channels (Olah et al. 2009), and the currents were also reduced by a partial knockdown of TRPM2 protein achieved using an RNAi approach (Olah et al. 2009). More convincingly, the TRPM2-like currents, evoked using the voltage-ramp protocol, were entirely absent in cultured hippocampal neurons from TRPM2−/− mice (Xie et al. 2011; Yamamoto et al. 2008). Some residual non-TRPM2 current was observed in TRPM2−/− neurons when the NMDAR protocol was employed, but this was likely due to the NMDA-dependent activation of pannexin 1 channels, which we previously observed using more substantial applications of NMDA (Thompson et al. 2008). Using slices from TRPM2−/− mice, we were able to show that TRPM2 channels do not appear to play a direct role in excitatory synaptic transmission at CA1 neurons. However, NMDAR-dependent LTD was absent in knock-out slices likely due to a chronic inactivation of GSK3β in TRPM2–/– slices (Xie et al. 2011).

37.4 Conclusions and the Myth of Channel Time Course

There is a common belief that nonselective cation channels such as NMDA, ASICs, pannexins, and TRPM2/7 can only play a role during the acute stage of stroke-induced damage, largely based on observations that NMDAR blockers must be given within a few hours of an ischemic stroke in order to be neuroprotective (at least in rodent stroke models). In contrast, there is considerable interest in the role of NMDARs in the synaptic dysfunction and cell loss of chronic and progressive degenerative diseases such as AD and Huntington's disease (Baum et al. 2010; Gladding and Raymond 2011; Gotz et al. 2011b). Even though ischemic stroke is characterized by subsequent loss of penumbral neurons, which may occur days after the stroke, it is also associated with a predisposition to much slower progressive cell loss leading to dementia. Neurons surviving the stroke likely continue to experience various stresses, which can lead to continued activation of nonselective cation channels long after the initial stroke. Indeed, ischemia can alter the expression of TRPM2 channels as much as 1 month following the initial insult (Fonfria et al. 2006a). Likely, the duration of influence of each type of nonselective cation channels to the pathology following stroke is not concurrent, and a temporal cascade of channel events may occur. Of course, downstream of nonselective ion channels, there are a multitude of potential intracellular events including mitochondrial dysfunction, inflammatory-induced alterations in cell function, enzyme activation (e.g., caspases), DNA damage, changes in gene and protein expression, and changes in glia/neuron/vascular interactions (Kriz 2006; Sun et al. 2006). In turn, these changes in cell metabolism and signaling alter the activity of many channels and transporters thus greatly increasing the potential complexity of the responses to ischemia and ultimately to related degenerative processes in the central nervous system.

References

M. Aarts, Y. Liu, L. Liu, S. Besshoh, M. Arundine, J.W. Gurd, Y.T. Wang, M.W. Salter, M. Tymianski, Treatment of ischemic brain damage by perturbing NMDA receptor- PSD-95 protein interactions. Science **298**, 846–850 (2002)

M. Aarts, K. Iihara, W.L. Wei, Z.G. Xiong, M. Arundine, W. Cerwinski, J.F. MacDonald, M. Tymianski, A key role for TRPM7 channels in anoxic neuronal death. Cell **115**, 863–877 (2003)

R.A. Al-Hallaq, T.P. Conrads, T.D. Veenstra, R.J. Wenthold, NMDA di-heteromeric receptor populations and associated proteins in rat hippocampus. J. Neurosci. **27**, 8334–8343 (2007)

L. Annunziato, M. Cataldi, G. Pignataro, A. Secondo, P. Molinaro, Glutamate-independent calcium toxicity: introduction. Stroke **38**, 661–664 (2007)

J.-Z. Bai, J. Lipski, Differential expression of TRPM2 and TRPV4 channels and their potential role in oxidative stress-induced cell death in organotypic hippocampal culture. Neurotoxicology **31**, 204–218 (2010)

D. Bano, K.W. Young, C.J. Guerin, R. Lefeuvre, N.J. Rothwell, L. Naldini, R. Rizzuto, E. Carafoli, P. Nicotera, Cleavage of the plasma membrane Na^+/Ca^{2+} exchanger in excitotoxicity. Cell **120**, 275–285 (2005)

T.E. Bartlett, N.J. Bannister, V.J. Collett, S.L. Dargan, P.V. Massey, Z.A. Bortolotto, S.M. Fitzjohn, Z.I. Bashir, G.L. Collingridge, D. Lodge, Differential roles of NR2A and NR2B-containing NMDA receptors in LTP and LTD in the CA1 region of two-week old rat hippocampus. Neuropharmacology **52**, 60–70 (2007)

J.A. Bartos, J.D. Ulrich, H. Li, M.A. Beazely, Y. Chen, J.F. MacDonald, J.W. Hell, Postsynaptic clustering and activation of Pyk2 by PSD-95. J. Neurosci. **30**, 449–463 (2010)

C. Bates-Withers, R. Sah, D. Clapham, TRPM7, the Mg^{2+} inhibited channel and kinase. Adv. Exp. Med. Biol. **704**, 173–256 (2011)

M. Baum, P. Kurup, J. Xu, P. Lombroso, A STEP forward in neural function and degeneration. Commun. Integr. Biol. **3**, 419–441 (2010)

J.C. Belrose, Y.F. Xie, L.J. Gierszewski, J.F. Macdonald, M.F. Jackson, Loss of glutathione homeostasis associated with neuronal senescence facilitates TRPM2 channel activation in cultured hippocampal pyramidal neurons. Mol. Brain **5**, 11 (2012)

S. Berberich, P. Punnakkal, V. Jensen, V. Pawlak, P.H. Seeburg, O. Hvalby, G. Kohr, Lack of NMDA receptor subtype selectivity for hippocampal long-term potentiation. J. Neurosci. **25**, 6907–6910 (2005)

S. Berberich, V. Jensen, O. Hvalby, P.H. Seeburg, G. Kohr, The role of NMDAR subtypes and charge transfer during hippocampal LTP induction. Neuropharmacology **52**, 77–86 (2007)

S. Braithwaite, S. Paul, A. Nairn, P. Lombroso, Synaptic plasticity: one STEP at a time. Trends Neurosci. **29**, 452–460 (2006)

S. Brauchi, G. Krapivinsky, L. Krapivinsky, D.E. Clapham, TRPM7 facilitates cholinergic vesicle fusion with the plasma membrane. Proc. Natl. Acad. Sci. U. S. A. **105**, 8304–8308 (2008)

J.L. Brigman, T. Wright, G. Talani, S. Prasad-Mulcare, S. Jinde, G.K. Seabold, P. Mathur, M.I. Davis, R. Bock, R.M. Gustin, R.J. Colbran, V.A. Alvarez, K. Nakazawa, E. Delpire, D.M. Lovinger, A. Holmes, Loss of GluN2B-containing NMDA receptors in CA1 hippocampus and cortex impairs long-term depression, reduces dendritic spine density, and disrupts learning. J. Neurosci. **30**, 4590–4600 (2010)

W.S. Chen, M.F. Bear, Activity-dependent regulation of NR2B translation contributes to metaplasticity in mouse visual cortex. Neuropharmacology **52**, 200–214 (2007)

K.K. Cho, L. Khibnik, B.D. Philpot, M.F. Bear, The ratio of NR2A/B NMDA receptor subunits determines the qualities of ocular dominance plasticity in visual cortex. Proc. Natl. Acad. Sci. U. S. A. **106**, 5377–5382 (2009)

D.W. Choi, Glutamate neurotoxicity in cortical cell culture is calcium dependent. Neurosci. Lett. **58**, 293–297 (1985)

D.W. Choi, S.M. Rothman, The role of glutamate neurotoxicity in hypoxic-ischemic neuronal death. Annu. Rev. Neurosci. **13**, 171–182 (1990)

X.P. Chu, C.J. Papasian, J.Q. Wang, Z.G. Xiong, Modulation of acid-sensing ion channels: molecular mechanisms and therapeutic potential. Int. J. Physiol. Pathophysiol. Pharmacol. **3**, 288–309 (2011)

D.E. Clapham, L.W. Runnels, C. Strubing, The TRP ion channel family. Nat. Rev. Neurosci. **2**, 387–396 (2001)

L. Crews, E. Masliah, Molecular mechanisms of neurodegeneration in Alzheimer's disease. Hum. Mol. Genet. **19**(R1), R12–R20 (2010)

H. Cui, A. Hayashi, H.S. Sun, M.P. Belmares, C. Cobey, T. Phan, J. Schweizer, M.W. Salter, Y.T. Wang, R.A. Tasker, D. Garman, J. Rabinowitz, P.S. Lu, M. Tymianski, PDZ protein interactions underlying NMDA receptor-mediated excitotoxicity and neuroprotection by PSD-95 inhibitors. J. Neurosci. **27**, 9901–9915 (2007)

S.G. Cull-Candy, D.N. Leszkiewicz, Role of distinct NMDA receptor subtypes at central synapses. Sci. STKE. **2004**, re16 (2004)

O.N. de la Perez, A. Davalos, Neuroprotection in cerebral infarction: the opportunity of new studies. Cerebrovasc. Dis. **24**(Suppl 1), 153–156 (2007)

E. Fonfria, I.C. Marshall, I. Boyfield, S.D. Skaper, J.P. Hughes, D.E. Owen, W. Zhang, B.A. Miller, C.D. Benham, S. McNulty, Amyloid beta-peptide(1–42) and hydrogen peroxide-induced toxicity are mediated by TRPM2 in rat primary striatal cultures. J. Neurochem. **95**, 715–723 (2005)

E. Fonfria, C. Mattei, K. Hill, J.T. Brown, A. Randall, C.D. Benham, S.D. Skaper, C.A. Campbell, B. Crook, P.R. Murdock, J.M. Wilson, F.P. Maurio, D.E. Owen, P.L. Tilling, S. McNulty, TRPM2 is elevated in the tMCAO stroke model, transcriptionally regulated, and functionally expressed in C13 microglia. J. Recept. Signal Transduct. Res. **26**, 179–198 (2006a)

E. Fonfria, P.R. Murdock, F.S. Cusdin, C.D. Benham, R.E. Kelsell, S. McNulty, Tissue distribution profiles of the human TRPM cation channel family. J. Recept. Signal Transduct. Res. **26**, 159–178 (2006b)

C.J. Fox, K.I. Russell, Y.T. Wang, B.R. Christie, Contribution of NR2A and NR2B NMDA subunits to bidirectional synaptic plasticity in the hippocampus in vivo. Hippocampus **16**, 907–915 (2006)

A. Gasser, G. Glassmeier, R. Fliegert, M.F. Langhorst, S. Meinke, D. Hein, S. Kruger, K. Weber, I. Heiner, N. Oppenheimer, J.R. Schwarz, A.H. Guse, Activation of T cell calcium influx by the second messenger ADP-ribose. J. Biol. Chem. **281**, 2489–2496 (2006)

M. Gielen, R.B. Siegler, L. Mony, J.W. Johnson, P. Paoletti, Mechanism of differential control of NMDA receptor activity by NR2 subunits. Nature **459**, 703–707 (2009)

C. Gladding, L. Raymond, Mechanisms underlying NMDA receptor synaptic/extrasynaptic distribution and function. Mol. Cell. Neurosci. **48**, 308–328 (2011)

J. Gotz, A. Eckert, M. Matamales, L.M. Ittner, X. Liu, Modes of Abeta toxicity in Alzheimer's disease. Cell. Mol. Life Sci. **68**, 3359–3375 (2011a)

J. Gotz, A. Ittner, L.M. Ittner, Tau-targeted treatment strategies in Alzheimer's disease. Br. J. Pharmacol. **165**, 1246–1259 (2011b)

J.A. Gray, Y. Shi, H. Usui, M.J. During, K. Sakimura, R.A. Nicoll, Distinct modes of AMPA receptor suppression at developing synapses by GluN2A and GluN2B: single-cell NMDA receptor subunit deletion in vivo. Neuron **71**, 1085–1101 (2011)

D.A. Greenberg, K. Jin, Neurodegeneration and neurogenesis: focus on Alzheimer's disease. Curr. Alzheimer Res. **3**, 25–28 (2006)

A. Hakim, Vascular disease: the tsunami of health care. Stroke **38**, 3296–3597 (2007)

Y. Hara, M. Wakamori, M. Ishii, E. Maeno, M. Nishida, T. Yoshida, H. Yamada, S. Shimizu, E. Mori, J. Kudoh, N. Shimizu, H. Kurose, Y. Okada, K. Imoto, Y. Mori, LTRPC2 Ca^{2+}-permeable channel activated by changes in redox status confers susceptibility to cell death. Mol. Cell **9**, 163–173 (2002)

G.E. Hardingham, Pro-survival signalling from the NMDA receptor. Biochem. Soc. Trans. **34**, 936–938 (2006)

G.E. Hardingham, Coupling of the NMDA receptor to neuroprotective and neurodestructive events. Biochem. Soc. Trans. **37**, 1147–1160 (2009)

G.E. Hardingham, H. Bading, The Yin and Yang of NMDA receptor signalling. Trends Neurosci. **26**, 81–89 (2003)

I. Heiner, J. Eisfeld, M. Warnstedt, N. Radukina, E. Jungling, A. Luckhoff, Endogenous ADP-ribose enables calcium-regulated cation currents through TRPM2 channels in neutrophil granulocytes. Biochem. J. **398**, 225–232 (2006)

M. Hermosura, R. Garruto, TRPM7 and TRPM2-Candidate susceptibility genes for Western Pacific ALS and PD? Biochim. Biophys. Acta **1772**, 822–857 (2007)

M. Hermosura, H. Nayakanti, M. Dorovkov, F. Calderon, A. Ryazanov, D. Haymer, R. Garruto, A TRPM7 variant shows altered sensitivity to magnesium that may contribute to the pathogenesis of two Guamanian neurodegenerative disorders. Proc. Natl. Acad. Sci. U. S. A. **102**, 11510–11515 (2005)

M. Hermosura, A. Cui, R. Go, B. Davenport, C. Shetler, J. Heizer, C. Schmitz, G. Mocz, R. Garruto, A.-L. Perraud, Altered functional properties of a TRPM2 variant in Guamanian ALS and PD. Proc. Natl. Acad. Sci. U. S. A. **105**, 18029–18063 (2008)

S. Hestrin, Developmental regulation of NMDA receptor-mediated synaptic currents at a central synapse. Nature **357**, 686–689 (1992)

K. Hill, N.J. Tigue, R.E. Kelsell, C.D. Benham, S. McNulty, M. Schaefer, A.D. Randall, Characterisation of recombinant rat TRPM2 and a TRPM2-like conductance in cultured rat striatal neurones. Neuropharmacology **50**, 89–97 (2006)

X.-Y. Hou, Y. Liu, G.-Y. Zhang, PP2, a potent inhibitor of Src family kinases, protects against hippocampal CA1 pyramidal cell death after transient global brain ischemia. Neurosci. Lett. **420**, 235–244 (2007)

L. Hoyte, P.A. Barber, A.M. Buchan, M.D. Hill, The rise and fall of NMDA antagonists for ischemic stroke. Curr. Mol. Med. **4**, 131–136 (2004)

Y.Q. Huang, W.Y. Lu, D.W. Ali, K.A. Pelkey, G.M. Pitcher, Y.M. Lu, H. Aoto, J.C. Roder, T. Sasaki, M.W. Salter, J.F. MacDonald, CAK beta/Pyk2 kinase is a signaling link for induction of long-term potentiation in CA1 hippocampus. Neuron **29**, 485–496 (2001)

K. Ikeda, H. Negishi, Y. Yamori, Antioxidant nutrients and hypoxia/ischemia brain injury in rodents. Toxicology **189**, 55–61 (2003)

A. Inanobe, H. Furukawa, E. Gouaux, Mechanism of partial agonist action at the NR1 subunit of NMDA receptors. Neuron **47**, 71–84 (2005)

K. Inoue, D. Branigan, Z.-G. Xiong, Zinc-induced neurotoxicity mediated by transient receptor potential melastatin 7 channels. J. Biol. Chem. **285**, 7430–7439 (2010)

N.K. Isaev, E.V. Stelmashook, U. Dirnagl, N.A. Andreeva, L. Manuhova, V.S. Vorobjev, I.N. Sharonova, V.G. Skrebitsky, I.V. Victorov, J. Katchanov, M. Weih, D.B. Zorov, Neuroprotective effects of the antifungal drug clotrimazole. Neuroscience **113**, 47–53 (2002)

L. Ittner, J. Götz, Amyloid-β and tau–a toxic pas de deux in Alzheimer's disease. Nat. Rev. Neurosci. **12**, 65–137 (2011)

L.M. Ittner, Y.D. Ke, F. Delerue, M. Bi, A. Gladbach, J. van Eersel, H. Wolfing, B.C. Chieng, M.J. Christie, I.A. Napier, A. Eckert, M. Staufenbiel, E. Hardeman, J. Gotz, Dendritic function of tau mediates amyloid-beta toxicity in Alzheimer's disease mouse models. Cell **142**, 387–397 (2010)

J. Jia, S. Verma, S. Nakayama, N. Quillinan, M.R. Grafe, P.D. Hurn, P.S. Herson, Sex differences in neuroprotection provided by inhibition of TRPM2 channels following experimental stroke. J. Cereb. Blood Flow Metab. **31**, 2160–2168 (2011)

X. Jiang, D. Mu, V. Biran, J. Faustino, S. Chang, C. Rincón, R. Sheldon, D. Ferriero, Activated Src kinases interact with the N-methyl-D-aspartate receptor after neonatal brain ischemia. Ann. Neurol. **63**, 632–673 (2008)

J. Jin, B.N. Desai, B. Navarro, A. Donovan, N.C. Andrews, D.E. Clapham, Deletion of Trpm7 disrupts embryonic development and thymopoiesis without altering Mg^{2+} homeostasis. Science **322**, 756–760 (2008)

J. Jin, L.J. Wu, J. Jun, X. Cheng, H. Xu, N.C. Andrews, D.E. Clapham, The channel kinase, TRPM7, is required for early embryonic development. Proc. Natl. Acad. Sci. U. S. A. **109**, E225–E233 (2011)

J. Jo, D.J. Whitcomb, K.M. Olsen, T.L. Kerrigan, S.C. Lo, G. Bru-Mercier, B. Dickinson, S. Scullion, M. Sheng, G. Collingridge, K. Cho, Abeta(1–42) inhibition of LTP is mediated by a signaling pathway involving caspase-3, Akt1 and GSK-3beta. Nat. Neurosci. **14**, 545–547 (2011)

S. Kaneko, S. Kawakami, Y. Hara, M. Wakamori, E. Itoh, T. Minami, Y. Takada, T. Kume, H. Katsuki, Y. Mori, A. Akaike, A critical role of TRPM2 in neuronal cell death by hydrogen peroxide. J. Pharmacol. Sci. **101**, 66–76 (2006)

M. Katano, T. Numata, K. Aguan, Y. Hara, S. Kiyonaka, S. Yamamoto, T. Miki, S. Sawamura, T. Suzuki, K. Yamakawa, Y. Mori, The juvenile myoclonic epilepsy-related protein EFHC1 interacts with the redox-sensitive TRPM2 channel linked to cell death. Cell Calcium **51**, 179–185 (2012)

R. Kraft, C. Grimm, K. Grosse, A. Hoffmann, S. Sauerbruch, H. Kettenmann, G. Schultz, C. Harteneck, Hydrogen peroxide and ADP-ribose induce TRPM2-mediated calcium influx and cation currents in microglia. Am. J. Physiol. Cell Physiol. **286**, C129–C137 (2004)

J. Kriz, Inflammation in ischemic brain injury: timing is important. Crit. Rev. Neurobiol. **18**, 145–157 (2006)

P. Kurup, Y. Zhang, J. Xu, D.V. Venkitaramani, V. Haroutunian, P. Greengard, A.C. Nairn, P.J. Lombroso, Abeta-mediated NMDA receptor endocytosis in Alzheimer's disease involves ubiquitination of the tyrosine phosphatase STEP61. J. Neurosci. **30**, 5948–5957 (2010)

B. Li, N. Chen, T. Luo, Y. Otsu, T.H. Murphy, L.A. Raymond, Differential regulation of synaptic and extra-synaptic NMDA receptors. Nat. Neurosci. **5**, 833–834 (2002)

M. Li, K. Inoue, D. Branigan, E. Kratzer, J.C. Hansen, J.W. Chen, R.P. Simon, Z.G. Xiong, Acid-sensing ion channels in acidosis-induced injury of human brain neurons. J. Cereb. Blood Flow Metab. **30**, 1247–1260 (2010)

H.-B. Li, M. Jackson, K. Yang, C. Trepanier, M. Salter, B. Orser, J. Macdonald, Plasticity of synaptic GluN receptors is required for the Src-dependent induction of long-term potentiation at CA3-CA1 synapses. Hippocampus **21**, 1053–1114 (2011)

L. Liu, T.P. Wong, M.F. Pozza, K. Lingenhoehl, Y. Wang, M. Sheng, Y.P. Auberson, Y.T. Wang, Role of NMDA receptor subtypes in governing the direction of hippocampal synaptic plasticity. Science **304**, 1021–1024 (2004)

Y. Liu, G.-Y. Zhang, J.-Z. Yan, T.-L. Xu, Suppression of Pyk2 attenuated the increased tyrosine phosphorylation of NMDA receptor subunit 2A after brain ischemia in rat hippocampus. Neurosci. Lett. **379**, 55–63 (2005)

Y. Liu, T.P. Wong, M. Aarts, A. Rooyakkers, L. Liu, T.W. Lai, D.C. Wu, J. Lu, M. Tymianski, A.M. Craig, Y.T. Wang, NMDA receptor subunits have differential roles in mediating excitotoxic neuronal death both in vitro and in vivo. J. Neurosci. **27**, 2846–2857 (2007)

Y.M. Lu, J.C. Roder, J. Davidow, M.W. Salter, Src activation in the induction of long-term potentiation in CA1 hippocampal neurons. Science **279**, 1363–1367 (1998)

W.Y. Lu, Z.G. Xiong, S. Lei, B.A. Orser, E. Dudek, M.D. Browning, J.F. MacDonald, G-protein-coupled receptors act via protein kinase C and Src to regulate NMDA receptors. Nat. Neurosci. **2**, 331–338 (1999)

W. Lu, H. Man, W. Ju, W.S. Trimble, J.F. MacDonald, Y.T. Wang, Activation of synaptic NMDA receptors induces membrane insertion of new AMPA receptors and LTP in cultured hippocampal neurons. Neuron **29**, 243–254 (2001)

C. Luscher, K.M. Huber, Group 1 mGluR-dependent synaptic long-term depression: mechanisms and implications for circuitry and disease. Neuron **65**, 445–459 (2010)

D.R. Lynch, R.P. Guttmann, NMDA receptor pharmacology: perspectives from molecular biology. Curr. Drug Targets **2**, 215–231 (2001)

J.F. MacDonald, M.F. Jackson, M.A. Beazely, G protein-coupled receptors control NMDARs and metaplasticity in the hippocampus. Biochim. Biophys. Acta **1768**, 941–951 (2006a)

J.F. MacDonald, M.F. Jackson, M.A. Beazely, Hippocampal long-term synaptic plasticity and signal amplification of NMDA receptors. Crit. Rev. Neurobiol. **18**, 71–84 (2006b)

J.F. MacDonald, Z.G. Xiong, M.F. Jackson, Paradox of Ca^{2+} signaling, cell death and stroke. Trends Neurosci. **29**, 75–81 (2006c)

R.C. Malenka, M.F. Bear, LTP and LTD: an embarrassment of riches. Neuron **44**, 5–21 (2004)

P.V. Massey, B.E. Johnson, P.R. Moult, Y.P. Auberson, M.W. Brown, E. Molnar, G.L. Collingridge, Z.I. Bashir, Differential roles of NR2A and NR2B-containing NMDA receptors in cortical long-term potentiation and long-term depression. J. Neurosci. **24**, 7821–7828 (2004)

M.P. Mattson, S.B. Kater, Development and selective neurodegeneration in cell cultures from different hippocampal regions. Brain Res. **490**, 110–125 (1989)

D. McHugh, R. Flemming, S.Z. Xu, A.L. Perraud, D.J. Beech, Critical intracellular Ca^{2+} dependence of transient receptor potential melastatin 2 (TRPM2) cation channel activation. J. Biol. Chem. **278**, 11002–11006 (2003)

S. McNulty, E. Fonfria, The role of TRPM channels in cell death. Pflugers Arch. **451**, 235–242 (2005)

M. Moran, M. McAlexander, T. Bíró, A. Szallasi, Transient receptor potential channels as therapeutic targets. Nat. Rev. Drug Discov. **10**, 601–621 (2011)

M.E. Morris, T. Trippenbach, Changes in extracellular $[K^+]$ and $[Ca^{2+}]$ induced by anoxia in neonatal rabbit medulla. Am. J. Physiol. **264**, R761–R769 (1993)

Y. Mu, F.H. Gage, Adult Hippocampal Neurogenesis and its Role in Alzheimer's Disease. Mol. Neurodegener. **6**, 85 (2011)

K.W. Muir, K.R. Lees, Excitatory amino acid antagonists for acute stroke. Cochrane Database Syst. Rev. **2003**, CD001244 (2003)

T.H. Nguyen, J. Liu, P.J. Lombroso, Striatal enriched phosphatase 61 dephosphorylates Fyn at phosphotyrosine 420. J. Biol. Chem. **277**, 24274–24279 (2002)

H. Oh, Y. Chun, Y. Kim, S. Youn, S. Shin, M. Park, T.-W. Kim, S. Chung, Modulation of transient receptor potential melastatin related 7 (TRPM7) channel by presenilins. Dev. Neurobiol. **18** (2011). doi:10.1002/dneu.22001

M. Olah, M. Jackson, H. Li, Y. Perez, H.-S. Sun, S. Kiyonaka, Y. Mori, M. Tymianski, J. MacDonald, Ca^{2+}-dependent induction of TRPM2 currents in hippocampal neurons. J. Physiol. **587**, 965–1044 (2009)

S. Paul, A.C. Nairn, P. Wang, P.J. Lombroso, NMDA-mediated activation of the tyrosine phosphatase STEP regulates the duration of ERK signaling. Nat. Neurosci. **6**, 34–42 (2003)

S. Peineau, C. Taghibiglou, C. Bradley, T.P. Wong, L. Liu, J. Lu, E. Lo, D. Wu, E. Saule, T. Bouschet, P. Matthews, J.T. Isaac, Z.A. Bortolotto, Y.T. Wang, G.L. Collingridge, LTP inhibits LTD in the hippocampus via regulation of GSK3beta. Neuron **53**, 703–717 (2007)

K.A. Pelkey, R. Askalan, S. Paul, L.V. Kalia, T.H. Nguyen, G.M. Pitcher, M.W. Salter, P.J. Lombroso, Tyrosine phosphatase STEP is a tonic brake on induction of long-term potentiation. Neuron **34**, 127–138 (2002)

A.L. Perraud, A. Fleig, C.A. Dunn, L.A. Bagley, P. Launay, C. Schmitz, A.J. Stokes, Q. Zhu, M.J. Bessman, R. Penner, J.P. Kinet, A.M. Scharenberg, ADP-ribose gating of the calcium-permeable LTRPC2 channel revealed by Nudix motif homology. Nature **411**, 595–599 (2001)

A.L. Perraud, C. Schmitz, A.M. Scharenberg, TRPM2 Ca^{2+} permeable cation channels: from gene to biological function. Cell Calcium **33**, 519–531 (2003)

B.D. Philpot, K.K. Cho, M.F. Bear, Obligatory role of NR2A for metaplasticity in visual cortex. Neuron **53**, 495–502 (2007)

G. Pignataro, O. Cuomo, E. Esposito, R. Sirabella, R.G. Di, L. Annunziato, ASIC1a contributes to neuroprotection elicited by ischemic preconditioning and postconditioning. Int. J. Physiol. Pathophysiol. Pharmacol. **3**, 1–8 (2011)

R. Pumain, U. Heinemann, Stimulus- and amino acid-induced calcium and potassium changes in rat neocortex. J. Neurophysiol. **53**, 1–16 (1985)

I.S. Ramsey, M. Delling, D.E. Clapham, An introduction to trp channels. Annu. Rev. Physiol. **68**, 619–647 (2006)

C. Rauner, G. Kohr, Triheteromeric NR1/NR2A/NR2B receptors constitute the major N-methyl-D-aspartate receptor population in adult hippocampal synapses. J. Biol. Chem. **286**, 7558–7566 (2011)

S.M. Rothman, Synaptic activity mediates death of hypoxic neurons. Science **220**, 536–537 (1983)

S. Rothman, Synaptic release of excitatory amino acid neurotransmitter mediates anoxic neuronal death. J. Neurosci. **4**, 1884–1891 (1984)

S.M. Rothman, The neurotoxicity of excitatory amino acids is produced by passive chloride influx. J. Neurosci. **5**, 1483–1489 (1985)

S.M. Rothman, Glutamate and anoxic neuronal death in vitro. Adv. Exp. Med. Biol. **203**, 687–695 (1986)

R. Sahathevan, A. Brodtmann, G. Donnan, Dementia, stroke, and vascular risk factors; a review. Int. J. Stroke **7**, 61–134 (2012)

M.W. Salter, L.V. Kalia, SRC kinases: a hub for NMDA receptor regulation. Nat. Rev. Neurosci. **5**, 317–328 (2004)

R. Sattler, M. Tymianski, Molecular mechanisms of calcium-dependent excitotoxicity. J. Mol. Med. **78**, 3–13 (2000)

R. Sattler, M.P. Charlton, M. Hafner, M. Tymianski, Distinct influx pathways, not calcium load, determine neuronal vulnerability to calcium neurotoxicity. J. Neurochem. **71**, 2349–2364 (1998)

R. Sattler, Z. Xiong, W.Y. Lu, M. Hafner, J.F. MacDonald, M. Tymianski, Specific coupling of NMDA receptor activation to nitric oxide neurotoxicity by PSD-95 protein. Science **284**, 1845–1848 (1999)

R. Sattler, Z. Xiong, W.Y. Lu, J.F. MacDonald, M. Tymianski, Distinct roles of synaptic and extrasynaptic NMDA receptors in excitotoxicity. J. Neurosci. **20**, 22–33 (2000)

D. Selkoe, Alzheimer's disease. Cold Spring Harb. Perspect. Biol. **3**, a004457 (2011)

G.B. Smith, A.J. Heynen, M.F. Bear, Bidirectional synaptic mechanisms of ocular dominance plasticity in visual cortex. Philos. Trans. R. Soc. Lond. B Biol. Sci. **364**, 357–367 (2009)

E.M. Snyder, Y. Nong, C.G. Almeida, S. Paul, T. Moran, E.Y. Choi, A.C. Nairn, M.W. Salter, P.J. Lombroso, G.K. Gouras, P. Greengard, Regulation of NMDA receptor trafficking by amyloid-beta. Nat. Neurosci. **8**, 1051–1058 (2005)

T.R. Soderling, V.A. Derkach, Postsynaptic protein phosphorylation and LTP. Trends Neurosci. **23**, 75–80 (2000)

I. Song, R.L. Huganir, Regulation of AMPA receptors during synaptic plasticity. Trends Neurosci. **25**, 578–588 (2002)

X. Sun, G. He, H. Qing, W. Zhou, F. Dobie, F. Cai, M. Staufenbiel, L.E. Huang, W. Song, Hypoxia facilitates Alzheimer's disease pathogenesis by up-regulating BACE1 gene expression. Proc. Natl. Acad. Sci. U. S. A. **103**, 18727–18732 (2006)

H.S. Sun, M.F. Jackson, L.J. Martin, K. Jansen, L. Teves, H. Cui, S. Kiyonaka, Y. Mori, M. Jones, J.P. Forder, T.E. Golde, B.A. Orser, J.F. MacDonald, M. Tymianski, Suppression of hippocampal TRPM7 protein prevents delayed neuronal death in brain ischemia. Nat. Neurosci. **12**, 1300–1307 (2009)

N. Takahashi, D. Kozai, R. Kobayashi, M. Ebert, Y. Mori, Roles of TRPM2 in oxidative stress. Cell Calcium **50**, 279–366 (2011)

C.L. Thompson, D.L. Drewery, H.D. Atkins, F.A. Stephenson, P.L. Chazot, Immunohistochemical localization of N-methyl-D-aspartate receptor NR1, NR2A, NR2B and NR2C/D subunits in the adult mammalian cerebellum. Neurosci. Lett. **283**, 85–88 (2000)

C.L. Thompson, D.L. Drewery, H.D. Atkins, F.A. Stephenson, P.L. Chazot, Immunohistochemical localization of N-methyl-D-aspartate receptor subunits in the adult murine hippocampal formation: evidence for a unique role of the NR2D subunit. Brain Res. Mol. Brain Res. **102**, 55–61 (2002)

R.J. Thompson, N. Zhou, B.A. MacVicar, Ischemia opens neuronal gap junction hemichannels. Science **312**, 924–927 (2006)

R. Thompson, M. Jackson, M. Olah, R. Rungta, D. Hines, M. Beazely, J. MacDonald, B. MacVicar, Activation of pannexin-1 hemichannels augments aberrant bursting in the hippocampus. Science **322**, 1555–1564 (2008)

C.H. Trepanier, M.F. Jackson, J.F. MacDonald, Regulation of NMDA receptors by the tyrosine kinase Fyn. FEBS J. **279**, 12–19 (2012)

V. Tseveleki, R. Rubio, S.-S. Vamvakas, J. White, E. Taoufik, E. Petit, J. Quackenbush, L. Probert, Comparative gene expression analysis in mouse models for multiple sclerosis, Alzheimer's disease and stroke for identifying commonly regulated and disease-specific gene changes. Genomics **96**, 82–173 (2010)

S. Tsuruno, S.Y. Kawaguchi, T. Hirano, Src-family protein tyrosine kinase negatively regulates cerebellar long-term depression. Neurosci. Res. **61**, 329–332 (2008)

M. Tymianski, Emerging mechanisms of disrupted cellular signaling in brain ischemia. Nat. Neurosci. **14**, 1369–1373 (2011)

M. Tymianski, M.P. Charlton, P.L. Carlen, C.H. Tator, Source specificity of early calcium neurotoxicity in cultured embryonic spinal neurons. J. Neurosci. **13**, 2085–2104 (1993)

P. Vanhoutte, H. Bading, Opposing roles of synaptic and extrasynaptic NMDA receptors in neuronal calcium signalling and BDNF gene regulation. Curr. Opin. Neurobiol. **13**, 366–371 (2003)

D.V. Venkitaramani, J. Chin, W.J. Netzer, G.K. Gouras, S. Lesne, R. Malinow, P.J. Lombroso, Beta-amyloid modulation of synaptic transmission and plasticity. J. Neurosci. **27**, 11832–11837 (2007)

X. Wang, E.K. Michaelis, Selective neuronal vulnerability to oxidative stress in the brain. Front. Aging Neurosci. **2**, 12 (2010)

W.-W. Wang, S.-Q. Hu, C. Li, C. Zhou, S.-H. Qi, G.-Y. Zhang, Transduced PDZ1 domain of PSD-95 decreases Src phosphorylation and increases nNOS (Ser847) phosphorylation contributing to neuroprotection after cerebral ischemia. Brain Res. **1328**, 162–232 (2010)

W.L. Wei, H.S. Sun, M.E. Olah, X. Sun, E. Czerwinska, W. Czerwinski, Y. Mori, B.A. Orser, Z.G. Xiong, M.F. Jackson, M. Tymianski, J.F. MacDonald, TRPM7 channels in hippocampal neurons detect levels of extracellular divalent cations. Proc. Natl. Acad. Sci. U. S. A. **104**, 16323–16328 (2007)

T.P. Wong, J.G. Howland, J.M. Robillard, Y. Ge, W. Yu, A.K. Titterness, K. Brebner, L. Liu, J. Weinberg, B.R. Christie, A.G. Phillips, Y.T. Wang, Hippocampal long-term depression mediates acute stress-induced spatial memory retrieval impairment. Proc. Natl. Acad. Sci. U. S. A. **104**, 11471–11476 (2007)

P. Xia, H.S. Chen, D. Zhang, S.A. Lipton, Memantine preferentially blocks extrasynaptic over synaptic NMDA receptor currents in hippocampal autapses. J. Neurosci. **30**, 11246–11250 (2010)

Y.F. Xie, J.C. Belrose, G. Lei, M. Tymianski, Y. Mori, J.F. MacDonald, M.F. Jackson, Dependence of NMDA/GSK3beta mediated metaplasticity on TRPM2 channels at hippocampal CA3-CA1 synapses. Mol. Brain **4**, 44 (2011)

Z.G. Xiong, X.M. Zhu, X.P. Chu, M. Minami, J. Hey, W.L. Wei, J.F. MacDonald, J.A. Wemmie, M.P. Price, M.J. Welsh, R.P. Simon, Neuroprotection in ischemia; blocking calcium-permeable Acid-sensing ion channels. Cell **118**, 687–698 (2004)

J. Xu, M. Weerapura, M.K. Ali, M.F. Jackson, H. Li, G. Lei, S. Xue, C.L. Kwan, M.F. Manolson, K. Yang, J.F. MacDonald, X.M. Yu, Control of excitatory synaptic transmission by C-terminal Src kinase. J. Biol. Chem. **283**, 17503–17514 (2008)

Z. Xu, R.Q. Chen, Q.H. Gu, J.Z. Yan, S.H. Wang, S.Y. Liu, W. Lu, Metaplastic regulation of long-term potentiation/long-term depression threshold by activity-dependent changes of NR2A/NR2B ratio. J. Neurosci. **29**, 8764–8773 (2009)

K. Yamada, N. Hara, T. Shibata, H. Osago, M. Tsuchiya, The simultaneous measurement of nicotinamide adenine dinucleotide and related compounds by liquid chromatography/electrospray ionization tandem mass spectrometry. Anal. Biochem. **352**, 282–285 (2006)

S. Yamamoto, T. Wajima, Y. Hara, M. Nishida, Y. Mori, Transient receptor potential channels in Alzheimer's disease. Biochim. Biophys. Acta **1772**, 958–967 (2007)

S. Yamamoto, S. Shimizu, S. Kiyonaka, N. Takahashi, T. Wajima, Y. Hara, T. Negoro, T. Hiroi, Y. Kiuchi, T. Okada, S. Kaneko, I. Lange, A. Fleig, R. Penner, M. Nishi, H. Takeshima, Y. Mori, TRPM2-mediated Ca^{2+} influx induces chemokine production in monocytes that aggravates inflammatory neutrophil infiltration. Nat. Med. **14**, 738–747 (2008)

K. Yang, J. Belrose, C.H. Trepanier, G. Lei, M.F. Jackson, J.F. MacDonald, Fyn, a potential target for Alzheimer's disease. J. Alzheimers Dis. **27**, 243–252 (2011a)

K. Yang, C. Trepanier, B. Sidhu, Y.F. Xie, H. Li, G. Lei, M.W. Salter, B.A. Orser, T. Nakazawa, T. Yamamoto, M.F. Jackson, J.F. MacDonald, Metaplasticity gated through differential regulation of GluN2A versus GluN2B receptors by Src family kinases. EMBO J. **31**, 805–816 (2011b)

K. Yashiro, B.D. Philpot, Regulation of NMDA receptor subunit expression and its implications for LTD, LTP, and metaplasticity. Neuropharmacology **55**, 1081–1094 (2008)

B.J. Yoon, G.B. Smith, A.J. Heynen, R.L. Neve, M.F. Bear, Essential role for a long-term depression mechanism in ocular dominance plasticity. Proc. Natl. Acad. Sci. U. S. A. **106**, 9860–9865 (2009)

S.J. Zhang, B. Buchthal, D. Lau, S. Hayer, O. Dick, M. Schwaninger, R. Veltkamp, M. Zou, U. Weiss, H. Bading, A signaling cascade of nuclear calcium-CREB-ATF3 activated by synaptic NMDA receptors defines a gene repression module that protects against extrasynaptic NMDA receptor-induced neuronal cell death and ischemic brain damage. J. Neurosci. **31**, 4978–4990 (2011)

ERRATUM TO

Chapter 17

New Insights in Mitochondrial Calcium Handling by Sodium/Calcium Exchanger

Antonella Scorziello, Claudia Savoia, Agnese Secondo, Francesca Boscia, Maria Jose Sisalli, Alba Esposito, Annalisa Carlucci, Pasquale Molinaro, Luca Lignitto, Gianfranco Di Renzo, Antonio Feliciello, and Lucio Annunziato

L, Annunziato (ed.) *Sodium Calcium Exchange: A Growing Spectrum of Pathophysiological Implications*, Advances in Experimental Medicine and Biology 961, DOI 10.1007/978-1-4614-4756-6_17, pp. 203–209

DOI 10.1007/978-1-4614-4756-6_38

The publisher regrets that in the print and online versions of this title, the following author name and affiliation for this chapter were not included:

Annagrazia Adornetto
Division of Pharmacology
Department of Neuroscience
School of Medicine
Federico II University of Naples
Naples, Italy

The online version of the original chapter can be found at
http://dx.doi.org/10.1007/978-1-4614-4756-6_17

List of Participants

ARGENTINA
Beaugé Louis
Berberian Graciela
Millon Maria Beatriz Bistue

AUSTRALIA
Cross Jane

BELGIUM
Herchuelz André
Sipido Karin R.

CANADA
Hryshko Larry V.
Lytton Jonathan
MacDonald John F.
Orlowski John
Roberts Diane
Schnetkamp Paul P. M.
Szerencsei Robert
Tibbits Glen
Tory Lisa C

FRANCE
Levitsky Dmitri O.

HUNGARY
Acsai Karoli
Toth Andras

ISRAEL
Boyman Liron
Gilardi Moshe
Khananshvili Daniel
Nita Iulia Ioana
Noda Mami
Rahamimoff Hannah
Sekler Israel

ITALY
Annunziato Lucio
Bernardini Renato
Boscia Francesca
Cantarella Giuseppina
Cantile Maria
Casamassa Antonella
Cataldi Mauro
Cerullo Pierpaolo
Cuomo Ornella
Di Benedetto Giulia
Di Renzo Gianfranco
Esposito Alba
Formisano Luigi
Guida Natascia
Landolina Nadine
Marone Gianni
Molinaro Pasquale
Nasti Annamaria Assunta
Omura Katia
Pannaccione Anna
Pignataro Giuseppe
Ronsisvalle Nicole
Scollo Mimmo
Scorziello Antonella
Secondo Agnese
Sirabella Rossana
Stucchi Michela
Trimarco Bruno
Valsecchi Valeria
Viggiano Davide
Vinciguerra Antonio

JAPAN
Iwamoto Takahiro
Kuroda Hidetaka
Matsuoka Satoshi
Ota Yuki
Sato Masaki
Shibukawa Yoshiyuki
Sobhan Ubaidus
Takeuchi Ayako
Takuma Kazuhiro
Tsumura Maki

LITHUANIA
Juska Alfonsas

MEXICO
Espinosa-Tanguma Ricardo

NETHERLAND
Baartscheer Antonius

NEW ZELAND
Empson Ruth M

NORWAY
Carlson Cathrine
Louch William E
Swift Fredrik
Wanichawan Pimthanya

SLOVAK REPUBLIC
Hudecova Sona
Krizanova Olga
Lehotsky Jan

SOUTH KOREA
Choi Kyung-Chul
Hyun Sang-Hwan
Jeung Eui-Bae
Kim Joon-Chul
Woo Sun-Hee

SWITZERLAND
Hilge Mark

TAIWAN
Lung-Sen Kao
Yang YA-CHI

UK
Abramov Andrey
Andrikopoulos Petros
Eisner David A.
Ginger Rebecca
Verkhratsky Alexej
Wray Susan

USA
Blaustein Mordecai P
Chen Jau-Nian
Cheung Joseph Y.
Ginsburg Kenneth
Goldhaber Joshua
Golovina Vera
Hilgemann Donald
Menick Donald R.
Ottolia Michela
Philipson Kenneth
Ren Xiaoyan
Shi Yejie
Sun Dandan
Wang JuFang
Xiong Zhigang
Zhang Jin
Zheng Lei

VENEZUELA
Dipolo Reinaldo

L. Annunziato (ed.), *Sodium Calcium Exchange: A Growing Spectrum of Pathophysiological Implications*, Advances in Experimental Medicine and Biology 961, DOI 10.1007/978-1-4614-4756-6,

Author Index

L. Annunziato (ed.), *Sodium Calcium Exchange: A Growing Spectrum of Pathophysiological Implications*, Advances in Experimental Medicine and Biology 961, DOI 10.1007/978-1-4614-4756-6,

Subject Index

L. Annunziato (ed.), *Sodium Calcium Exchange: A Growing Spectrum of Pathophysiological Implications*, Advances in Experimental Medicine and Biology 961, DOI 10.1007/978-1-4614-4756-6,

O

P

MIX
Papier aus verantwortungsvollen Quellen
Paper from responsible sources
FSC® C105338

If you have any concerns about our products,
you can contact us on
ProductSafety@springernature.com

In case Publisher is established outside the EU,
the EU authorized representative is:
Springer Nature Customer Service Center GmbH
Europaplatz 3, 69115 Heidelberg, Germany

Printed by Libri Plureos GmbH
in Hamburg, Germany